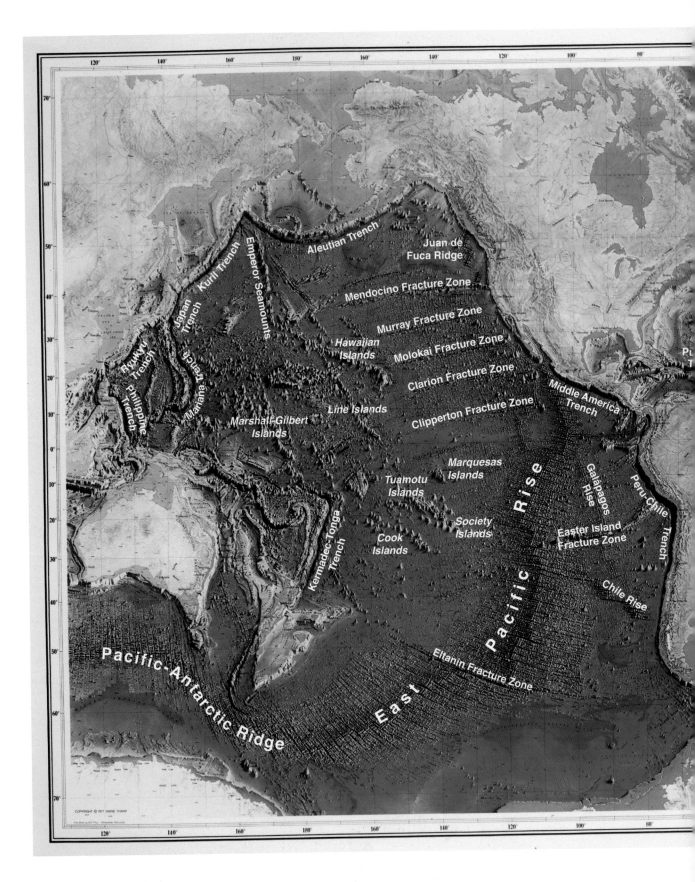

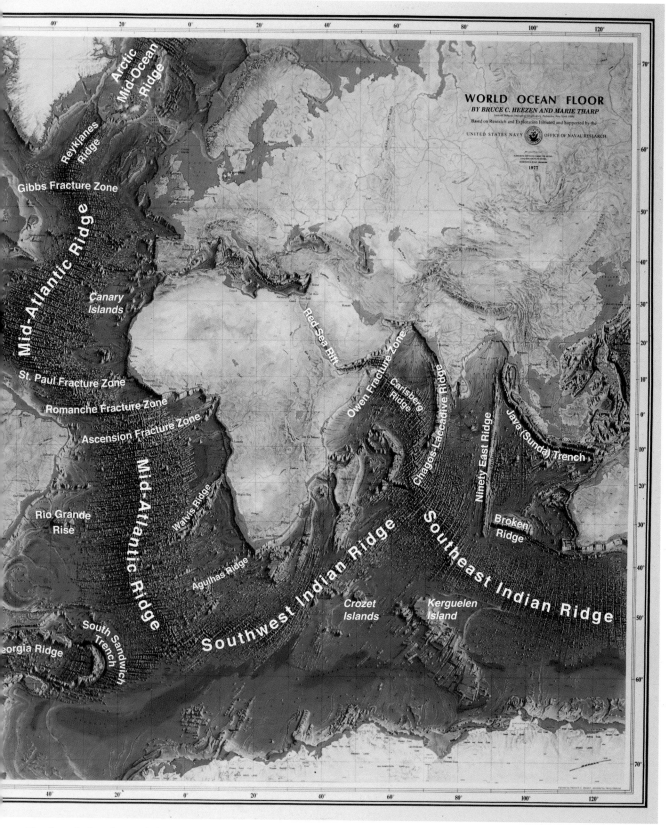

The "map" that changed the world. This visualization, which shows what Earth's surface would look like if all the water in the oceans were emptied and exaggerates sea floor features for clarity, was created by geologist Bruce Heezen and mapmaker Marie Tharp in 1977, based on ship soundings and sonar data. For the first time in the history of the world, the shape of Earth's sea floor could be observed, including its many diverse features.

Essentials of
Oceanography

ELEVENTH EDITION

Alan P. Trujillo

DISTINGUISHED TEACHING PROFESSOR
PALOMAR COLLEGE

Harold V. Thurman

PROFESSOR EMERITUS
MT. SAN ANTONIO COLLEGE

PEARSON

Boston   Columbus   Indianapolis   New York   San Franciso   Upper Saddle River
Amsterdam   Cape Town   Dubai   London   Madrid   Milan   Munich   Paris   Montréal   Toronto
Delhi   Mexico City   São Paulo   Sydney   Hong Kong   Seoul   Singapore   Taipei   Tokyo

Acquisitions Editor: Andrew Dunaway
Senior Marketing Manager: Maureen McLaughlin
Senior Project Editor: Crissy Dudonis
Assistant Editor: Sean Hale
Assistant Editor: Michelle White
Senior Marketing Assistant: Nicola Houston
Director of Development: Jennifer Hart
Art Development: Jay McElroy
Media Producer: Tod Regan
Media Editor: Miriam Adrianowicz
Managing Editor, Geosciences and Chemistry: Gina M. Cheselka
Project Manager, Production: Janice Stangel
Full Service/Composition: PreMediaGlobal
Project Manager, Full Service: PreMediaGlobal
Illustrations: Imagineering and Justin Hoffman
Cartography: International Mapping
Image Permissions Manager: Maya Melenchuk
Photo Researcher: Carly Bergey
Text Permissions Manager: Alison Bruckner
Text Permissions Project Manager: Liz Kincaid, PreMediaGlobal
Design Manager: Derek Bacchus
Cover and Interior Design: Gary Hespenheide
Operations Specialist: Jeffrey Sargent
Cover Image Credit: Splash News/Corbis

Library of Congress Cataloging-in-Publication Data

Trujillo, Alan P.
    Essentials of oceanography / Alan P. Trujillo, Harold V. Thurman.—Eleventh ed.
        p. cm.
    Includes index.
    ISBN 978-0-321-81405-0
    1. Oceanography—Textbooks. I. Thurman, Harold V. II. Title.
    GC11.2.T49 2013
    551.46—dc23
                                        2012042270

3 4 5 6 7 8 9 10–V011–16 15 14 13

www.pearsonhighered.com

ISBN-10: 0-321-81405-3; ISBN-13: 978-0-321-81405-0 (Student Edition)
ISBN-10: 0-321-82088-6; ISBN-13: 978-0-321-82088-4 (Instructor's Review Copy)

*In loving memory of Ann Trujillo, my mom
and the kindest person I've known*

—Al Trujillo

For Vanessa and Will

—Hal Thurman

This 11th edition of *Essentials of Oceanography* is dedicated to the memory of Harold V. Thurman, who passed away at the age of 78 on December 29, 2012, at The Villages Regional Hospital in Florida.

Harold graduated in 1952 from Salem Community High School in Salem, Illinois, and then attended Bacone College and Oklahoma State University, where he was a member of Sigma Phi Epsilon. On July 2, 1955, Harold married his childhood sweetheart, Iantha (Ann) Lou Britton. Two years later, Harold graduated with a Bachelor's degree in geology. He then went to work for Kerr McGee Corporation in Oklahoma, South America, and the Gulf of Mexico as a petroleum geologist. After receiving a Master of Arts degree in education, Harold became a college professor and an oceanography textbook author.

Professor Thurman began teaching at Mt. San Antonio College (Mt. SAC) in Walnut, California, in 1968 as a temporary teacher and taught Physics 1 (a surveying class) and three Physical Geology labs. In 1970, he taught his first class of General Oceanography. It was from this experience that he decided to write a textbook on oceanography and received a contract with Charles E. Merrill Publishing Company in 1973. The first edition of his book *Introductory Oceanography* was released in 1975. Harold authored or co-authored over 20 editions of textbooks that include *Introductory Oceanography, Essentials of Oceanography, Physical Geology, Marine Biology,* and *Oceanography Laboratory Manual,* many of which are still being used today throughout the world. In addition, he contributed to the *World Book Encyclopedia* on the topics of "Arctic Ocean," "Atlantic Ocean," Indian Ocean," and "Pacific Ocean."

Harold very much enjoyed teaching but he eventually retired from Mt. SAC in 1995 so that he and his wife Ann could travel more freely. In February 2002, they settled in The Villages in Florida. In 2010, he was inducted into the Salem Community High School Academic Foundation Hall of Fame for his publishing accomplishments. In 2012, his textbook *Essentials of Oceanography* received a "Texty" Textbook Excellence Award from the Text and Academic Authors Association for its excellence in the areas of content, presentation, appeal, and teachability. Hal's writing expertise, knowledge about the ocean, and easy-going demeanor will be dearly missed.

ABOUT THE AUTHORS

ALAN P. TRUJILLO Al Trujillo is a Distinguished Teaching Professor in the Earth, Space, and Aviation Sciences Department at Palomar College in San Marcos, California. He received his bachelor's degree in geology from the University of California at Davis and his master's degree in geology from Northern Arizona University, afterward working for several years in industry. Al began teaching at Palomar in 1990. In 1997, he was awarded Palomar's Distinguished Faculty Award for Excellence in Teaching, and in 2005 he received Palomar's Faculty Research Award. He has co-authored *Introductory Oceanography* with Hal Thurman and is a contributing author for the textbooks *Earth* and *Earth Science*. In addition to writing and teaching, Al works as a naturalist and lecturer aboard natural history expedition vessels in Alaska and the Sea of Cortez/Baja California. His research interests include beach processes, sea cliff erosion, and active teaching techniques. He enjoys photography, and he collects sand as a hobby. Al and his wife, Sandy, have two children, Karl and Eva.

HAROLD V. THURMAN Hal Thurman retired in May 1994, after 24 years of teaching in the Earth Sciences Department at Mt. San Antonio College in Walnut, California. Interest in geology led to a bachelor's degree from Oklahoma A&M University, followed by seven years working as a petroleum geologist, mainly in the Gulf of Mexico, where his interest in the oceans developed. He earned a master's degree from California State University at Los Angeles and then joined the Earth Sciences faculty at Mt. San Antonio College. Other books that Hal has co-authored include *Introductory Oceanography* with Alan Trujillo as well as a marine biology textbook. He also wrote articles on the Pacific, Atlantic, Indian, and Arctic Oceans for the 1994 edition of *World Book Encyclopedia* and served as a consultant on the National Geographic publication *Realms of the Sea*. He still enjoys going to sea on vacations with his wife, Iantha.

BRIEF CONTENTS

CONTENTS

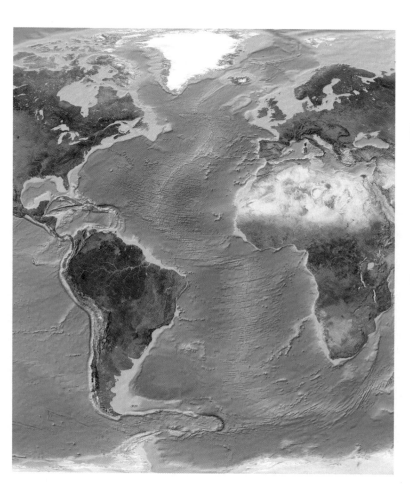

PREFACE

To the Student

Welcome! You're about to embark on a journey that is far from ordinary. Over the course of this term, you will discover the central role the oceans play in the vast global system of which you are a part.

This book's content was carefully developed to provide a foundation in science by examining the vast body of oceanic knowledge. This knowledge includes information from a variety of scientific disciplines—geology, chemistry, physics, and biology—as they relate to the oceans. However, no formal background in any of these disciplines is required to successfully master the subject matter contained within this book. Our desire is to have you take away from your oceanography course much more than just a collection of facts. Instead, we want you to develop a fundamental understanding of how the oceans work and why the oceans behave the way that they do.

This book is intended to help you in your quest to know more about the oceans. Taken as a whole, the components of the ocean—its sea floor, chemical constituents, physical components, and life-forms—comprise one of Earth's largest interacting, interrelated, and interdependent systems. Because human activities impact Earth systems, it is important to understand not only how the oceans operate but also how the oceans interact with Earth's other systems (such as its atmosphere, biosphere, and hydrosphere) as part of a larger picture. Thus, this book uses a systems approach to highlight the interdisciplinary relationships among oceanographic phenomena and how those phenomena affect other Earth systems.

To that end—and to help you make the most of your study time—we focused the presentation in this book by organizing the material around three essential components:

1. **CONCEPTS:** General ideas derived or inferred from specific instances or occurrences (for instance, the concept of density can be used to explain why the oceans are layered)

2. **PROCESSES:** Actions or occurrences that bring about a result (for instance, the process of waves breaking at an angle to the shore results in the movement of sediment along the shoreline)

3. **PRINCIPLES:** Rules or laws concerning the functioning of natural phenomena or mechanical processes (for instance, the principle of sea floor spreading suggests that the geographic positions of the continents have changed through time)

Interwoven within these concepts, processes, and principles are hundreds of photographs, illustrations, real-world examples, and applications that make the material relevant and accessible (and maybe sometimes even *entertaining*) by bringing science to life.

Ultimately, it is our hope that by understanding how the oceans work, you will develop a new awareness and appreciation of all aspects of the marine environment and its role in Earth systems. To this end, the book has been written for you, the student of the oceans. So enjoy and immerse yourself! You're in for an exciting ride.

Al Trujillo
Harold Thurman

To the Instructor

This eleventh edition of *Essentials of Oceanography* is designed to accompany an introductory college-level course in oceanography taught to students who have no formal background in mathematics or science. As in previous editions, the goal of this edition of the textbook is to clearly present the relationships of scientific principles to ocean phenomena in an engaging and meaningful way.

This edition has greatly benefited from being thoroughly reviewed by hundreds of students who made numerous suggestions for improvement. Comments by former students about the book include, *"I have really enjoyed the oceanography book we've used this semester. It had just the right mix of graphics, text, and user-friendliness that really held my interest."* and *"What I really liked about the book is that it's a welcoming textbook—open and airy. You could almost read it at bedtime like a story because of all the interesting pictures."*

This edition has been reviewed in detail by a host of instructors from leading institutions across the country. One reviewer of the tenth edition described the text as follows: *"Use this text. It is engaging and thoughtful. It is current and useful. It challenges the students to reach, yet it is thorough and clear in the execution of the text. The pictures tell the story and the diagrams are understandable. It is student-friendly and it is instructor-friendly. This text does fit the bill for teaching and learning oceanography."*

In 2012, the tenth edition of *Essentials of Oceanography* received a Textbook Excellence Award called a "Texty" from the Text and Academic Authors Association (TAA). The Texty award recognizes written works for their excellence in the areas of content, presentation, appeal, and teachability. The publisher, Pearson Education, nominated the book for the award, and the textbook was critically reviewed by a panel of expert judges.

The 16-chapter format of this textbook is designed for easy coverage of the material in a 15- or 16-week semester. For courses taught on a 10-week quarter system, instructors may need to select those chapters that cover the topics and concepts of primary relevance to their course. Chapters are self-contained and can thus be covered in any order. Following the introductory chapter (Chapter 1, which covers the general geography of the oceans; a historical perspective of oceanography; the reasoning behind the scientific method; and a discussion of the origin of Earth, the atmosphere, the oceans, and life itself), the

A USER'S GUIDE FOR STUDENTS: HOW TO READ A SCIENCE TEXTBOOK

Have you known someone who could scan a reading assignment or sleep with it under their pillow and somehow absorb all the information? Studies have shown that those people haven't really committed anything to long-term memory. For most of us, it takes a focused, concentrated effort to gain knowledge through reading. Interestingly, if you have the proper motivation and reading techniques, you can develop excellent reading comprehension. What is the best way to read a science textbook such as this one that contains many new and unfamiliar terms?

One common mistake is to approach reading a science textbook as one would read a newspaper, magazine, or novel. Instead, many reading instructors suggest using the SQ4R reading technique, which is based on research about how the brain learns. The SQ4R technique includes these steps:

1. **Survey:** Read the title, introduction, major headings, first sentences, concept statements, review questions, summary, and study aids to become familiar with the content in advance.

2. **Question:** Have questions in mind when you read. If you can't think of any good questions, use the chapter questions as a guide.

3. **Read:** Read flexibly through the chapter, using short time periods to accomplish the task one section at a time (not all in one sitting).

4. **Recite:** Answer the chapter questions. Take notes after each section and review your notes before you move on.

5. **(w)Rite:** Write summaries and/or reflections on what you've read. Write answers to the questions in Step 2.

6. **Review:** Review the text using the strategy in the survey step. Take the time to review your end-of-section notes as well as your summaries.

To help you study most effectively, this textbook includes many study aids that are designed to be used with the SQ4R technique. For example, each chapter includes a word cloud of key terms, a list of learning objectives that are tied to the Essential Concepts throughout

the chapter, review Concept Check questions embedded at the end of each section, and an Essential Concepts Review that includes a chapter summary, study resources, and critical thinking questions.

Here are some additional reading tips that may seem like common sense but are often overlooked:

- Don't attempt to do your reading when you are tired, distracted, or agitated.
- Break up your reading into manageable sections. Don't save it all until the last minute.
- Take a short break if your concentration begins to fade. Listen to music, call a friend, have a snack, or drink some water. Then return to your reading.

Remember that being a successful student is hard work; it is not something one does in his/her spare time. With a little effort in applying the SQ4R reading technique, you will begin to see a difference in what you remember from your reading.

---

four major academic disciplines of oceanography are represented in the following chapters:

- Geological oceanography (Chapters 2–4 and parts of Chapters 10 and 11)
- Chemical oceanography (Chapter 5 and part of Chapter 11)
- Physical oceanography (Chapters 6–9 and parts of Chapters 10 and 11)
- Biological oceanography (Chapters 12–15)

One of the most significant additions to the book in the previous edition was a new interdisciplinary chapter, "The Oceans and Climate Change" (Chapter 16), which focuses on the important environmental issue of human-caused global climate change and its impact on the ocean. This chapter has been updated in this edition with new information and enhanced graphics.

We strongly believe that oceanography is at its best when it links together several scientific disciplines and shows how they are interrelated in the oceans. Therefore, this interdisciplinary approach is a key element of every chapter.

What's New in This Edition?

Changes in this edition are designed to increase the readability, relevance, and appeal of this book. Major changes include the following:

- A word cloud has been added at the beginning of each chapter. Each word cloud uses different font sizes to show the most important vocabulary terms within the chapter and directs students to the glossary to discover the meaning of any terms they don't already know.
- An enhanced learning objectives section, titled Essential Concepts, has been added at the beginning of each chapter. It highlights the most important points of each chapter and establishes a discrete learning path.
- A new active learning pedagogy divides chapter material into easily digestible chunks, which makes studying easier and assists student learning. (Cognitive science research shows that the ability to "chunk" information is essential to enhancing learning and memory.)
- Review Questions now close each major section of the book; these Concept Checks allow and encourage students to pause and test their knowledge as they proceed through the chapter.

- Key Concepts are now reorganized as Essential Concepts that reinforce the learning path and tie into each section's Concept Checks.
- A new section titled Essential Concepts Review has been included at the end of each chapter. This section is tied to each chapter-opening Essential Concept; it includes a summary and new Study Resources and Critical Thinking Question sections.
- Information has been updated throughout the text to include some of the most recent developments in oceanography, such as the 2010 Gulf of Mexico oil spill, the 2011 Japanese earthquake and resulting tsunami, the hurricane called Superstorm Sandy in 2012, and updated material in Chapter 16, "The Oceans and Climate Change."
- The improved illustration package of new photos, satellite images, and figures makes oceanographic topics more accessible and interesting.
- A host of new artwork has been created by marine biologist and Digital Graphic Artist Justin Hofman.
- The labeling of all figures has been standardized to make them more appealing and consistent throughout.
- A total of 60 Web Animations have been included from Pearson's Geoscience Animations Library, featuring state-of-the-art computer animations created by Al Trujillo and a panel of geoscience educators.
- A total of 10 new Geoscience Animations have been added to help students visualize some of the most challenging oceanographic concepts, including 10 new animations that have been specifically designed for this edition.
- Web links have been included to more than 50 hand-picked Web videos that show important oceanographic processes in action.
- New QR codes have been embedded in the text to allow students to use their mobile devices to link directly to selected Web videos.
- Encounter Earth call-outs have been added that illustrate interesting oceanographic features with interactive online maps.
- This edition places greater emphasis on the ocean's role in Earth systems.
- The information in 12 Diving Deeper features (boxed material) has been moved to the text to indicate their importance as topics within oceanography.
- Other accessory Diving Deeper features have been migrated online to MasteringOceanography as Bonus Web Content.
- The remaining Diving Deeper features are organized around the following four themes:
  - **HISTORICAL FEATURES**, which focus on historical developments in oceanography that tie into chapter topics
  - **RESEARCH METHODS IN OCEANOGRAPHY**, which highlight how oceanographic knowledge is obtained
  - **OCEANS AND PEOPLE**, which illustrate the interaction of humans and the ocean environment
  - **FOCUS ON THE ENVIRONMENT**, which emphasize environmental issues that are an increasingly important component of the book
- All text in the chapters has been thoroughly reviewed and edited in a continued effort to refine the style and clarity of the writing.
- A detailed list of specific chapter-by-chapter changes is available at www2.palomar.edu/users/atrujillo.

In addition, this edition continues to offer some of the previous edition's most popular features, including the following:

- Scientific accuracy and thorough coverage of oceanography topics
- "Students Sometimes Ask …" questions, which present actual student questions along with the authors' answers
- Use of the international metric system (Système International [SI] units), with comparable English system units in parentheses
- Explanation of word etymons (*etumon* = sense of a word) as new terms are introduced, in an effort to demystify scientific terms by showing what the terms actually mean
- Use of **bold print** on key terms, which are defined when they are introduced and are described in the glossary
- A reorganized "Essential Concepts Review" summary at the end of each chapter
- All the resources from **mygeoscienceplace.com**, are now available within **MasteringOceanography**, which features chapter-specific Essential Concepts, eText, Bonus Web Content, Geoscience Animations, Web Videos, Web Destinations, and three Test Yourself quiz modules

For the Student

- **MASTERINGOCEANOGRAPHY™** delivers engaging, dynamic learning opportunities—focused on course objectives and responsive to each student's progress—that are proven to help students absorb course material and understand difficult concepts. MasteringOceanography includes:
  - **STUDY AREA**, which is designed to be a one-stop resource for students to acquire study help and serve as a launching pad for further exploration. Content for the site was written by author Al Trujillo and is tied, chapter-by-chapter, to the text. The Study Area is organized around a four-step learning pathway:
    1. *Review*, which contains **Essential Concepts** as learning objectives.
    2. *Read*, which contains the **eText** and **Bonus Web Content**.
    3. *Visualize*, which contains **Geoscience Animations**, **Web Videos**, and **Web Destinations**. **Geoscience Animations** were created by a team of geoscience educators and include a suite of 60 visualizations that help students understand complex oceanographic concepts and processes by allowing the user to control the action. For example, students can fully examine how an animation develops by replaying it, controlling its pace, and stopping and starting the animation anywhere in its sequence. In order to facilitate effective study, Al Trujillo has written an accompanying narration and assessment quiz questions including hints and specific wrong-answer feedback for each animation. **Web Videos** include more than 50 hand-selected short video clips of oceanographic processes in action. **Web Destinations** include links to some of the best oceanography sites on the Web.

4. *Test Yourself*, which contains three **Test Yourself** modules, including multiple-choice and true/false, multiple-answer, and image-labeling exercises. Answers, once submitted, are automatically graded for instant feedback.

- **RSS FEEDS**, which allow students to subscribe and stay up-to-date on oceanographic discoveries.

- **WORD STUDY TOOLS** such as flashcards and a searchable on-line glossary to help make the most of students study time.

- **THE PEARSON eTEXT** gives students complete access to a digital version of the text whenever and wherever you have access to the Internet. eText pages look exactly like the printed text, offering powerful new portability and functionality.

- Instructor Manual and TESTGEN are also available on the Instructor Resource Center (IRC) on DVD

For the Instructor

- **MASTERINGOCEANOGRAPHY** help instructors maximize class time with easy-to-assign, customizable, and automatically graded assessments that motivate students to learn outside of class and arrive prepared for lecture or lab.

- MasteringOceanography provides a rich and flexible set of course materials to get instructors started quickly, including pre-built assignments that instructors can use as is or customize to fit their needs.

- MasteringOceanography provides quick and easy access to information on student performance against learning outcomes. Instructors can quickly add their own learning outcomes, or use publisher-provided ones, to track student performance.

- The MasteringOceanography gradebook and diagnostic tools capture the step-by-step work of every student, providing unique insight into class performance.

- Assignable items in MasteringOceanography include:

  - **ENCOUNTER OCEANS ACTIVITIES**, which provide interactive explorations of oceanography concepts using Google Earth™. Students work through the activities in Google Earth and then test their knowledge by answering the assessment questions, which include hints and specific wrong-answer feedback.

  - **GEOSCIENCE ANIMATION ACTIVITIES** illuminate the most difficult-to-understand topics in oceanography. The animation activities include audio narration, a text transcript, and assignable multiple-choice questions with specific wrong-answer feedback.

  - **COACHING ACTIVITIES**, which consist of sophisticated, high-impact visuals that ask students to demonstrate their knowledge by synthesizing and analyzing core concepts using higher-order thinking skills.

  - **READING QUESTIONS**, which can be assigned to ensure that students read the textbook before coming to class. These questions help your students stay on track, become more engaged in lecture, and allows them to check their understanding of the content.

- **INSTRUCTOR MANUAL (DOWNLOAD ONLY)** contains learning objectives, chapter outlines, answers to embedded end-of-section questions, and suggested short demonstrations to spice up your lectures. The Instructor Manual is available in both Microsoft Word® and Adobe PDF formats.

- **TESTGEN® COMPUTERIZED TEST BANK (DOWNLOAD ONLY)** is a computerized test generator that lets instructors view and edit *Test Bank* questions, transfer questions to tests, and print the test in a variety of customized formats. The *Test Bank* includes nearly 1200 multiple-choice, fill-in-the-blank, and short-answer/essay questions. All questions are tied to the chapter's learning outcomes, rated based on Bloom's taxonomy of learning domains, and include the section in which each question's answer can be found. The *Test Bank* is also available in Microsoft Word® and is uploadable into Blackboard course management system. *Test Bank* questions are available in both Microsoft Word® and Adobe PDF formats.

- **INSTRUCTOR RESOURCE CENTER (IRC) ON DVD** puts all your lecture resources in one easy-to-reach place:

  - **GEOSCIENCE ANIMATIONS:** An extensive collection of 60 animations from the Pearson Geoscience Animation Library can be shown in class to help students understand some of the most difficult-to-visualize topics of oceanography.

  - **POWERPOINT® PRESENTATIONS:** The IRC on DVD includes three PowerPoint® files for each chapter so that you can cut down on your preparation time, no matter what your lecture needs:

    1. **EXCLUSIVELY ART:** This file provides all the photos, art, and tables from the text, in order, loaded into PowerPoint® slides.

    2. **LECTURE OUTLINE:** This file averages 35 slides per chapter and includes customizable lecture outlines with supporting art.

    3. **CLASSROOM RESPONSE SYSTEM (CRS) QUESTIONS:** Authored for use in conjunction with classroom response systems, this PowerPoint® allows you to electronically poll your class for responses to questions, pop quizzes, attendance, and more.

  - **TRANSPARENCY ACETATES:** Provided electronically, every table and most of the illustrations in this edition are available to be printed out as full-color, projection-enhanced transparencies.

Acknowledgments

We are indebted to many individuals for their helpful comments and suggestions during the revision of this book. Al Trujillo is indebted to his colleagues at Palomar Community College for their keen interest in the project, for allowing him to use some of their creative ideas in the book, and for continuing to provide valuable feedback. A particularly big thank you goes to Patty Deen for her continuing support and for recognizing the contribution the book makes to the Oceanography Program at Palomar College. In addition, Cari Gomes reviewed parts of the book during the last stages of the production process and provided valuable corrections when I was unavailable.

Adam Petrusek of Charles University, Prague, Czech Republic, deserves special recognition for his many suggestions for improving the text. Adam translated a previous edition of the textbook into Czech, and in the process, he thoroughly reviewed every part of it.

Many people were instrumental in helping the text evolve from its manuscript stage. My chief liaison at Pearson Education, Geoscience Editor Andrew Dunaway, suggested many of the new ideas in the book to make it more student-friendly and expertly guided the project. Copy Editor Kitty Wilson did a superb job of editing the manuscript, catching many English and other grammar errors, including obscure errors that had persisted throughout several previous editions. Geosciences Senior Project Editor Crissy Dudonis kept the book on track by making sure deadlines were met along the way and facilitated the distribution of various versions of the manuscripts. Media Producers Lee Ann Doctor and Tod Regan helped create the electronic supplements that accompany this book, including MasteringOceanography and all of its outstanding features. The animations studio Bridge360 crafted the new animations and added additional ideas, which led to great improvements. The Pearson art studios did a beautiful job of modernizing and updating all maps and many of the figures to make them more consistent throughout. Art Development Editor Jay McElroy reviewed every single piece of art throughout the text and suggested many improvements to make the figures more clear. Marine biologist and talented Digital Graphic Artist Justin Hofman supplied a host of new figures featuring realistic marine organisms that greatly improved the art program. The artful design elements of the text, including its color scheme, text wrapping, and end-of-chapter features, was developed by Layout Designer Gary Hespenheide, in conjunction with Pearson's design department. New photos were researched and secured by Research and Permissions Manager Carly Bergey of PreMediaGlobal. Last but not least, Senior Production Manager Lindsay Bethoney of PreMediaGlobal deserves special recognition for her persistence and encouragement during the many long hours of turning the manuscript into the book you see today.

Al Trujillo thanks his former student Rich Yonts for reviewing the entire manuscript in page proofs and catching many typos and other errors. Al Trujillo would also like to thank his students, whose questions provided the material for the "Students Sometimes Ask … " sections and whose continued input has proved invaluable for improving the text. Because scientists (and all good teachers) are always experimenting, thanks also for allowing yourselves to be a captive audience with which to conduct my experiments.

Al Trujillo also thanks his patient and understanding family for putting up with his absence during the long hours of preparing "The Book." Finally, appreciation is extended to the chocolate manufacturers Hershey, See's, and Ghirardelli, for providing inspiration. A heartfelt thanks to all of you!

Many other individuals (including several anonymous reviewers) have provided valuable technical reviews for this and previous works. The following reviewers are gratefully acknowledged:

Patty Anderson, *Scripps Institution of Oceanography*
William Balsam, *University of Texas at Arlington*
Tsing Bardin, *City College of San Francisco*
Steven Benham, *Pacific Lutheran University*
Lori Bettison-Varga, *College of Wooster*
Thomas Bianchi, *Tulane University*
David Black, *University of Akron*
Mark Boryta, *Consumnes River College*
Laurie Brown, *University of Massachusetts*
Kathleen Browne, *Rider University*
Nancy Bushell, *Kauai Community College*

Chatham Callan, *Hawaii Pacific University*
Mark Chiappone, *Miami-Dade College-Homestead Campus*
Chris Cirmo, *State University of New York, Cortland*
G. Kent Colbath, *Cerritos Community College*
Thomas Cramer, *Brookdale Community College*
Richard Crooker, *Kutztown University*
Cynthia Cudaback, *North Carolina State University*
Warren Currie, *Ohio University*
Hans Dam, *University of Connecticut*
Dan Deocampo, *California State University, Sacramento*
Richard Dixon, *Texas State University*
Holly Dodson, *Sierra College*
Joachim Dorsch, *St. Louis Community College*
Wallace Drexler, *Shippensburg University*
Walter Dudley, *University of Hawaii*
Iver Duedall, *Florida Institute of Technology*
Jessica Dutton, *Adelphi University*
Charles Ebert, *State University of New York, Buffalo*
Jiasong Fang, *Hawaii Pacific University*
Kenneth Finger, *Irvine Valley College*
Benjamin Giese, *Texas A&M University*
Cari Gomes, *MiraCosta College*
Dave Gosse, *University of Virginia*
Carla Grandy, *City College of San Francisco*
John Griffin, *University of Nebraska, Lincoln*
Gary Griggs, *University of California, Santa Cruz*
Joseph Holliday, *El Camino Community College*
Mary Anne Holmes, *University of Nebraska, Lincoln*
Timothy Horner, *California State University, Sacramento*
Alan Jacobs, *Youngstown State University*
Ron Johnson, *Old Dominion University*
Uwe Richard Kackstaetter, *Metropolitan State University of Denver*
Eryn Klosko, *State University of New York, Westchester Community College*
M. John Kocurko, *Midwestern State University*
Lawrence Krissek, *Ohio State University*
Paul LaRock, *Louisiana State University*
Gary Lash, *State University of New York, Fredonia*
Richard Laws, *University of North Carolina*
Richard Little, *Greenfield Community College*
Stephen Macko, *University of Virginia, Charlottesville*
Chris Marone, *Pennsylvania State University*
Matthew McMackin, *San Jose State University*
James McWhorter, *Miami-Dade Community College*
Gregory Mead, *University of Florida*
Keith Meldahl, *MiraCosta College*
Nancy Mesner, *Utah State University*
Chris Metzler, *MiraCosta College*
Johnnie Moore, *University of Montana*
P. Graham Mortyn, *California State University, Fresno*
Andrew Muller, *Millersville University*
Andrew Muller, *Utah State University*
Jay Muza, *Florida Atlantic University*
Jennifer Nelson, *Indiana University–Purdue University at Indianapolis*
Jim Noyes, *El Camino Community College*
Sarah O'Malley, *Maine Maritime Academy*
B. L. Oostdam, *Millersville University*
William Orr, *University of Oregon*
Donald Palmer, *Kent State University*
Nancy Penncavage, *Suffolk County Community College*
Curt Peterson, *Portland State University*
Edward Ponto, *Onondaga Community College*
Donald Reed, *San Jose State University*

Randal Reed, *Shasta College*
M. Hassan Rezaie Boroon, *California State University, Los Angeles*
Cathryn Rhodes, *University of California, Davis*
James Rine, *University of South Carolina*
Felix Rizk, *Manatee Community College*
Angel Rodriguez, *Broward Community College*
Beth Simmons, *Metropolitan State College of Denver*
Jill Singer, *State University of New York, Buffalo*
Arthur Snoke, *Virginia Polytechnic Institute*
Pamela Stephens, *Midwestern State University*
Dean Stockwell, *University of Alaska, Fairbanks*
Scott Stone, *Fairfax High School, Virginia*
Lenore Tedesco, *Indiana University–Purdue University at Indianapolis*
Shelly Thompson, *West High School*
Craig Tobias, *University of North Carolina, Wilmington*
M. Craig VanBoskirk, *Florida Community College at Jacksonville*
Bess Ward, *Princeton University*
Jackie Watkins, *Midwestern State University*
Arthur Wegweiser, *Edinboro University of Pennsylvania*
Diana Wenzel, *Seminole State College of Florida*
John White, *Louisiana State University*
Katryn Wiese, *City College of San Francisco*
John Wormuth, *Texas A&M University*
Memorie Yasuda, *Scripps Institution of Oceanography*

Although this book has benefited from careful review by many individuals, the accuracy of the information rests with the authors. If you find errors or have comments about the text, please contact us.

Al Trujillo
Department of Earth, Space, and Aviation Sciences
Palomar College
1140 W. Mission Rd.
San Marcos, CA 92069
atrujillo@palomar.edu
Web: www2.palomar.edu/users/atrujillo
Oceanography blog for students: www2.palomar.edu/pages/deepdeepocean4students/
Oceanography blog for instructors: www2.palomar.edu/pages/deepdeepocean4instructors/

Hal Thurman
17580 SE 88th Covington Circle
The Villages, FL 32162
ilhvt@embarqmail.com

"If there is magic on this planet, it is contained in water."
—Loren Eiseley, American educator and natural science writer (1907–1977)

About Our Sustainability Initiatives

Pearson recognizes the environmental challenges facing this planet, as well as acknowledges our responsibility in making a difference. This book is carefully crafted to minimize environmental impact. The binding, cover, and paper come from facilities that minimize waste, energy consumption, and the use of harmful chemicals. Pearson closes the loop by recycling every out-of-date text returned to our warehouse.

Along with developing and exploring digital solutions to our market's needs, Pearson has a strong commitment to achieving carbon-neutrality. As of 2009, Pearson became the first carbon- and climate-neutral publishing company. Since then, Pearson remains strongly committed to measuring, reducing, and offsetting our carbon footprint.

The future holds great promise for reducing our impact on Earth's environment, and Pearson is proud to be leading the way. We strive to publish the best books with the most up-to-date and accurate content, and to do so in ways that minimize our impact on Earth. To learn more about our initiatives, please visit **www.pearson.com/responsibility**.

PEARSON

Each chapter is organized into sections that start with learning goals and end with assessment questions tied to those learning goals. The end-of-chapter material is also organized by the chapter's sections, helping students remain focused on the essential concepts throughout the chapter.

Essential elements form a path to successful learning

◄ Each chapter opens with a list of learning objectives called Essential Concepts, which provides a roadmap to the chapter. Each chapter section focuses on one main Essential Concept, and many sections include sub Essential Concepts.

ESSENTIAL CONCEPTS

At the end of this chapter, you should be able to:

2.1 Describe the evidence that supports continental drift

2.2 Describe the evidence that supports plate tectonics

2.3 Discuss the ocean and land features that occur at plate boundaries

2.4 Summarize some of the applications of plate tectonics

2.5 Demonstrate an understanding of how Earth has changed in the past and how it will look in the future

2.6 Discuss how plate tectonics is used as a working model to help explain features and processes on Earth

CONCEPT CHECK 2.3

1 Most lithospheric plates contain both oceanic- and continental-type crust. Use plate boundaries to explain why this is true.

2 Describe the differences between oceanic ridges and oceanic rises. Include in your answer why these differences exist.

3 Convergent boundaries can be divided into three types, based on the type of crust contained on the two colliding plates. Compare and contrast the different types of convergent boundaries that result from these collisions.

4 Describe the differences in earthquake magnitudes that occur between the three types of plate boundaries and explain why these differences occur.

▲ Each chapter section ends with a Concept Check, which asks students to stop and check their understanding of the Essential Concept before moving on to the next section.

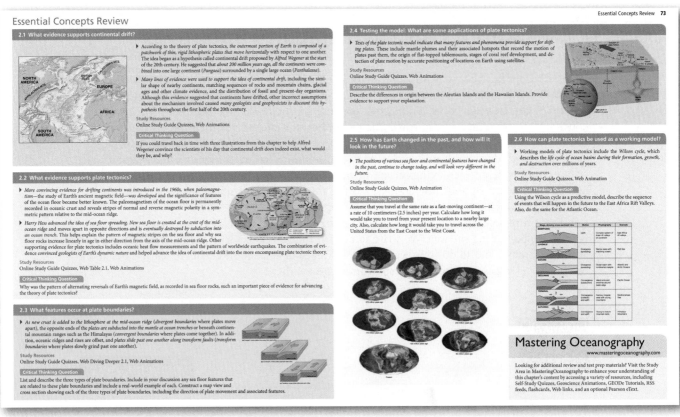

▲ Each chapter ends with the Essential Concepts Review, which simplifies the study process. Also organized by section, this review provides a visual summary of the chapter's key concepts and figures, and also includes study resources and Critical Thinking Questions.

Highly visual and interactive tools demystify oceanography by enabling students to see oceanographic processes in action.

Dynamic visuals and integrated media bring oceanography to life

A wide variety of video clips that show various oceanographic processes in action are referenced throughout the book. These clips can be accessed via students' mobile devices using Quick Response (QR) codes found within the textbook's pages.

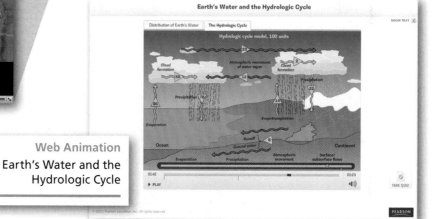

Geoscience Animation icons link the narrative to state-of-the-art animations and help students visualize challenging oceanographic concepts.

Web Animation
Earth's Water and the Hydrologic Cycle

The new edition includes over 50 new figures, designed to make oceanographic topics even more accessible and easier to visualize.

Figure 12.7 **Benthos (bottom dwellers): Representative intertidal and shallow subtidal forms.** Schematic drawing of various benthos organisms. Organisms include **(1)** sponges, **(2)** sand dollars, **(3)** crinoid, **(4)** sea anemones (open and closed), **(5)** barnacles, **(6)** mussels, **(7)** sea urchin, **(8)** sea cucumber, **(9)** sea hare, **(10)** shore crab, **(11)** sea star, **(12)** abalone, **(13)** ghost crab, **(14)** lug worm, **(15)** annelid worm, and **(16)** clam.

A focus on the environment and climate change throughout the book addresses these important issues in a variety of ways.

Timely and topical: The environment and climate change

The new edition includes a chapter entirely dedicated to the topic of climate change. Chapter 16 includes detailed coverage of greenhouse gases and the unintended and severe changes in the ocean, such as ocean warming, more intense hurricanes, increasing ocean acidity, changes in deep-water circulation, melting of polar ice, and rising sea level.

Diving deeper: Focus on the Environment discuss interesting and current environmental issues, emphasizing their relevance to the world's oceans.

Everyday topics in a real world context help students relate oceanography to their lives while engaging them in how oceanography is studied.

Turning interest into engagement

The new edition includes a variety of features, including Historical Features, Research Methods in Oceanography, and Oceans and People. These features foster multi-dimensional understanding with captivating examples and stories.

The popular "Students Sometimes Ask" feature answers often-entertaining questions posed by real students.

STUDENTS SOMETIMES ASK . . .

I've been to Hawaii and seen a black sand beach, which forms when lava flows into the ocean and is broken up by waves. Is the black sand hydrogenous sediment?

No. Many active volcanoes in the world have black sand beaches that are created when waves break apart dark-colored volcanic rock. The material that produces the black sand is derived from a continent or an island, so it is considered lithogenous sediment. Even though molten lava sometimes flows into the ocean, the resulting black sand could never be considered hydrogenous sediment because the lava was never *dissolved* in water.

MasteringOceanography

The MasteringOceanography™ platform delivers engaging, dynamic learning opportunities—focused on course objectives and responsive to each student's progress—that are proven to help students absorb course material and understand difficult concepts. With a easy-to-use interface, MasteringOceanography provides a wide variety of activities and assessments, an interactive gradebook, powerful diagnostics, and complete customization capabilities, as well as instructor and student resources.

Help students visualize oceanographic processes

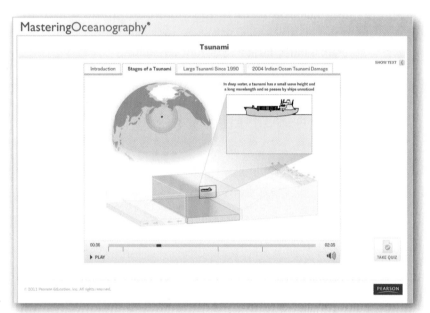

Geoscience Animations illuminate the most difficult-to-visualize topics from across the physical geosciences, including the field of oceanography, and more. The animation activities include audio narration, a text transcript, and assignable assessment questions and specific wrong-answer feedback.

Encounter Activities provide rich, interactive explorations of oceanography concepts using the dynamic features of **Google Earth**™ to visualize and explore Earth and its oceans. All explorations include corresponding Google Earth KMZ media files, and questions include hints and specific wrong-answer feedback to help coach students towards mastery of the concepts.

Engage students outside of class

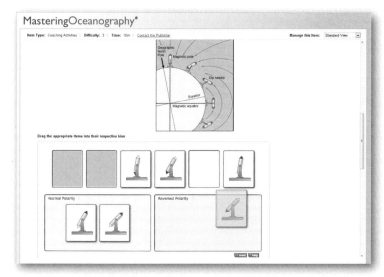

Coaching Activities challenge learners by involving them in activities that require higher-order thinking skills such as the synthesis, analysis, and application of the toughest topics in oceanography.

Oceanography videos include Earth Report Videos and Video Field Trips. Videos from Television for the Environment's Earth Report series recognize the efforts of individuals around the world to unite and protect the planet. Professors can assign questions related to each video, which include hints and specific wrong-answer feedback.

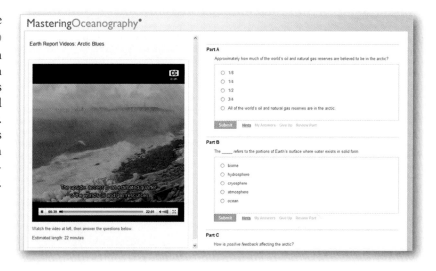

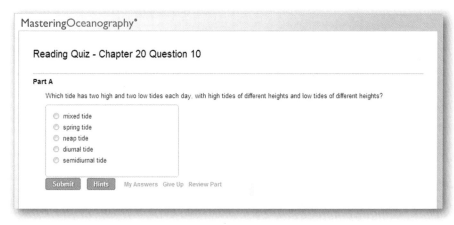

Reading Quizzes, help students keep on track, be more engaged in lecture, and check their understanding of the content. Each quiz included specific wrong-answer feedback and hints.

MasteringOceanography

Quickly monitor student results

With the Mastering gradebook and diagnostics, instructors will be better informed about students' progress than ever before. Mastering captures the step-by-step work of every student—including wrong answers submitted, hints requested, and time taken at every step of every problem—all providing unique insight into the most common misconceptions of the class.

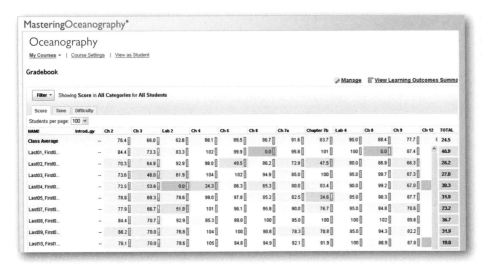

The Gradebook records all scores for automatically graded assignments. Shades of red highlight struggling students and challenging assignments.

Diagnostics provide unique insight into class and student performance. With a single click, charts summarize the most difficult problems, vulnerable students, grade distribution, and score improvement over the duration of the course.

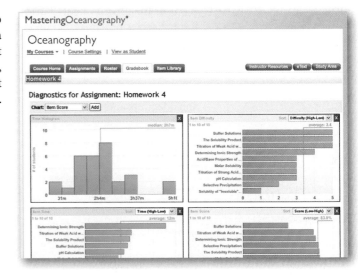

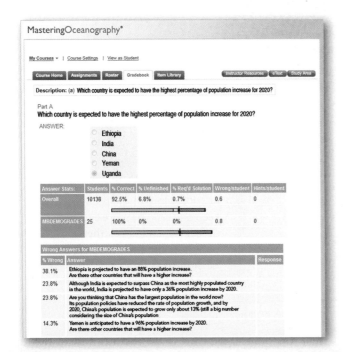

With a single click, Individual Student Performance Data provides at-a-glance statistics into each individual student's performance, including time spent on the problem, number of hints opened, and number of wrong and correct answers submitted.

Quickly measure student performance against Learning Outcomes

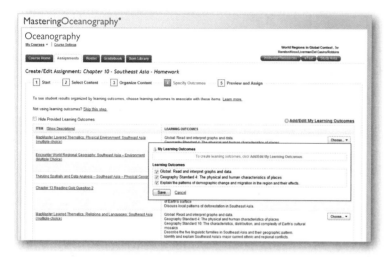

Learning Outcomes

MasteringOceanography provides quick and easy access to information on student performance against learning outcomes and makes it easy for instructors to share those results.

Instructors can:

- Quickly add learning outcomes or use publisher-provided ones to track student performance and report it to administration.

- View class and individual student performance against specific learning outcomes.

- Effortlessly export results to a spreadsheet and further customize and/or share with chairs, deans, administrators, and/or accreditation boards.

Easy to customize

Customize publisher-provided problems or quickly add instructor's own. MasteringOceanography makes it easy for instructors to edit any questions or answers, import instructor's own questions, and quickly add images, links, and files to further enhance the student experience.

Upload instructor's own video and audio files from their hard drive to share with students, as well as record video from their computer's webcam directly into MasteringOceanography—no plug-ins required. Students can download video and audio files to their local computer or launch them in Mastering to view the content.

Pearson eText gives students access to *Essentials of Oceanography,* **Eleventh Edition** whenever and wherever they can access the Internet. The eText pages look exactly like the printed text, and include powerful interactive and customization functions. Users can create notes, highlight text in different colors, create bookmarks, zoom, click hyperlinked words and phrases to view definitions, and view as a single page or as two pages. Pearson eText also links students to associated media files, enabling them to view an animation as they read the text, and offers a full-text search and the ability to save and export notes. The Pearson eText also includes embedded URLs in the chapter text with active links to the Internet.

NEW! The Pearson eText app is a great companion to Pearson's eText browser-based book reader. It allows existing subscribers who view their Pearson eText titles on a Mac or PC to additionally access their titles in a bookshelf on the iPad® either online or via download.

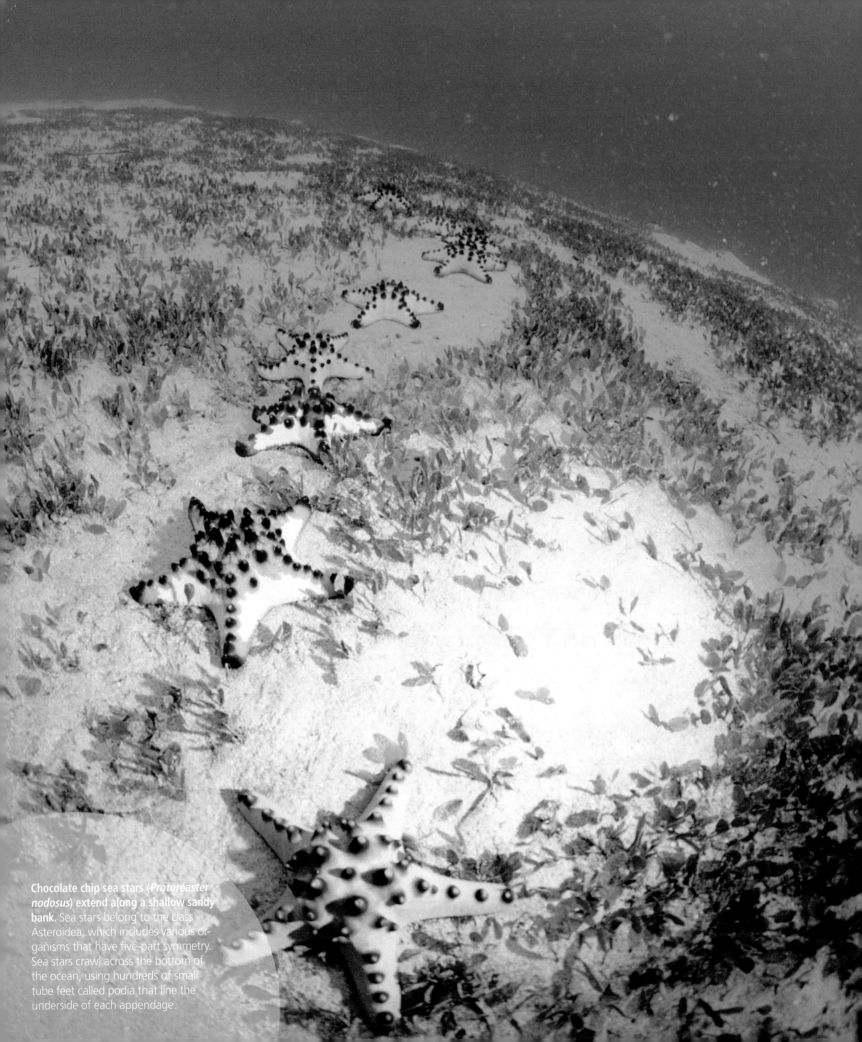

Chocolate chip sea stars (*Protoreaster nodosus*) extend along a shallow sandy bank. Sea stars belong to the class Asteroidea, which includes various organisms that have five-part symmetry. Sea stars crawl across the bottom of the ocean, using hundreds of small tube feet called podia that line the underside of each appendage.

Intro

Figure I.1 Word cloud of commonly used words in this Introduction. The most commonly used words in this Introduction are shown by larger font sizes in this word cloud, which is a visual aid for identifying important terms. For example, in one glance at the word cloud, you can see the important topics in this Introduction. Look for word clouds of important vocabulary terms on the opening page of each chapter throughout this book.

"Oceanography is not so much a science as a collection of scientists who find common cause in trying to understand the complex nature of the ocean. In the vast salty seas that encompass the earth, there is plenty of room for persons trained in physics, chemistry, biology, and engineering to practice their specialties. Thus, an oceanographer is any scientifically trained person who spends much of his [or her] career on ocean problems."

—Willard Bascom, Oceanographer and Explorer (1980)

INTRODUCTION

Welcome to a book about the oceans. As you read this book, we hope that it elicits a sense of wonder and a spirit of curiosity about our watery planet. The ocean represents many different things to different people. To some, it is a wilderness of beauty and tranquility, a refuge from hectic civilized lives. Others see it as a vast recreational area that inspires either rest or physical challenge. To others, it is a mysterious place that is full of unknown wonders. And to others, it is a place of employment unmatched by any on land. To be sure, its splendor has inspired artists, writers, and poets for centuries. Whatever your view, we hope that understanding the way the oceans work will increase your appreciation of the marine environment. Above all, take time to admire the oceans.

Essentials of Oceanography was first written to help students develop *an awareness about the marine environment*—that is, develop an appreciation for the oceans by learning about oceanic processes (how the oceans behave) and their interrelationships (how physical entities are related to one another in the oceans). In this eleventh edition, our goal is the same: to give the reader the scientific background to understand the basic principles underlying oceanic phenomena. In this way, one can then make informed decisions about the oceans in the years to come. We hope that some of you will be inspired so much by the oceans that you will continue to study them formally or informally in the future. (For those who may be considering a life-long career in oceanography, see **Appendix V**, "Careers in Oceanography," for some tips and practical advice.)

I.1 What Is Oceanography?

Oceanography[1] (*ocean* = the marine environment, *graphy* = description of) is literally the description of the marine environment. Although the term was first coined in the 1870s, at the beginning of scientific exploration of the oceans, this definition does not fully portray the extent of what oceanography encompasses: Oceanography does much more than just *describe* marine phenomena. Oceanography could be more accurately called the scientific study of all aspects of the marine environment (**Figure I.1**). Hence, the field of study called oceanography could (and maybe *should*) be called oceanology (*ocean* = the marine en-

---
[1]Note that all bolded words are key vocabulary terms that are defined in the glossary at the end of this book.

vironment, *ology* = the study of). However, the science of studying the oceans has traditionally been called *oceanography*. It is also called *marine science* and includes the study of the water of the ocean, the life within it, and the (not so) solid Earth beneath it.

Since prehistoric time, people have used the oceans as a means of transportation and as a source of food. However, the importance of ocean processes has been studied technically only since the 1930s. The impetus for these studies began with the search for offshore petroleum and expanded with the emphasis on ocean warfare during World War II. In fact, it was during World War II that the great expansion in oceanography, which continues today, began. The realization by governments of the importance of marine problems and their readiness to make money available for research, the growth in the number of ocean scientists at work, and the increasing sophistication of scientific equipment have made it feasible to study the ocean on a scale and to a degree of complexity never before attempted nor even possible. Historically, those who have made their living fishing in the ocean have gone where the physical processes of the oceans offer good fishing. But how marine life interrelates with ocean geology, chemistry, and physics to create good fishing grounds has been more or less a mystery until only recently, when scientists in these disciplines began to investigate the oceans with new technology. Along with these expanded studies came the realization of how much of an impact humans are beginning to have on the ocean. As a result, much recent research has been concerned with documenting human impacts on the ocean.

Oceanography is typically divided into different academic disciplines (or subfields) of study. The four main disciplines of oceanography that are covered in this book are as follows:

- **GEOLOGICAL OCEANOGRAPHY**, which is the study of the structure of the sea floor and how the sea floor has changed through time; the creation of sea floor features; and the history of sediments deposited on it.

  - **CHEMICAL OCEANOGRAPHY**, which is the study of the chemical composition and properties of seawater; how to extract certain chemicals from seawater; and the effects of pollutants.

  - **PHYSICAL OCEANOGRAPHY**, which is the study of waves, tides, and currents; the ocean–atmosphere relationship that influences weather and climate; and the transmission of light and sound in the oceans.

  - **BIOLOGICAL OCEANOGRAPHY**, which is the study of the various oceanic life-forms and their relationships to one another; adaptations to the marine environment; and developing sustainable methods of harvesting seafood.

Other disciplines include ocean engineering, marine archaeology, and marine policy. Because the study of oceanography often examines in detail all the different disciplines of oceanography, it is frequently described as being an *interdisciplinary* science, or one covering all the disciplines of science as they apply to the oceans (**Figure I.2**). The content of this book includes the broad range of interdisciplinary science topics that comprises the field of oceanography. In essence, this is a book about *all* aspects of the oceans.

I.2 How Are Earth's Oceans Unique?

The oceans are the largest and most prominent feature on Earth. In fact, they are the single most defining feature of our planet. As viewed from space, our planet is a beautiful blue, white, and brown globe (see

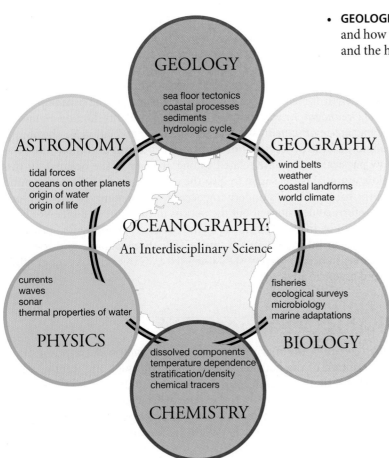

Figure I.2 A Venn diagram showing the interdisciplinary nature of oceanography. Oceanography is an interdisciplinary science that overlaps into many scientific disciplines.

the chapter-opening photo in Chapter 1). It is our oceans of liquid water that set us apart in the solar system (**Figure I.3**).

No other body in the solar system has a confirmed ocean, but recent satellite missions have revealed some tantalizing possibilities. For example, the spidery network of fluid-filled cracks on Jupiter's moon Europa (**Figure I.4**) almost certainly betrays the presence of an ocean of liquid water beneath its icy surface. Two other moons of Jupiter, Ganymede and Callisto, may also have liquid oceans of water beneath their cold, icy crust. Yet another possibility for a nearby world with an ocean beneath its icy surface is Saturn's tiny moon Enceladus, which displays geysers of water vapor and ice that have recently been analyzed and, remarkably, contain salt. And evidence continues to mount that Saturn's giant moon Titan has seas of liquid hydrocarbons, suggesting that Titan may be the only other body in the solar system besides Earth known to have stable liquid at its surface. All these moons are enticing targets for space missions to search for signs of extraterrestrial life.

Further outside the solar system, nearly 1000 exoplanets have been discovered orbiting other star systems, including a few rocky exoplanets that are Earth-sized and may be orbiting their Sun-like stars at just the right distance for water to remain liquid, potentially sustaining life. Still, the fact that our planet has so much water, *and in the liquid form*, is unique in the solar system and beyond.

The oceans determine where our continents end and have thus shaped political boundaries and human history. The oceans conceal many features; in fact, the majority of Earth's geographic features are on the ocean floor. Remarkably, there was once more known about the surface of the moon than about the floor of the oceans! Fortunately, our knowledge of both has increased dramatically over the past few decades.

The oceans influence climate and weather all over the globe—even in continental areas far from any ocean—through an intricate pattern of currents and heating/cooling mechanisms that scientists are only now beginning to understand. The oceans are also the lungs of the planet, taking carbon dioxide gas out of the atmosphere and replacing it with oxygen gas Some scientists have estimated that the oceans supply as much as 70% of the oxygen that humans breathe.

The oceans are essential to all life and are in large part responsible for the development of life on Earth, providing a stable environment in which life could evolve over millions of years. Today, the oceans contain the greatest number of living things on the planet, from microscopic bacteria and algae to the largest life-form alive today (the blue whale). Interestingly, water is the major component of nearly every life-form on Earth, and our own body fluid chemistry is remarkably similar to the chemistry of seawater.

The oceans hold many secrets waiting to be discovered, and new discoveries about the oceans are made nearly every day. The oceans are a source of food, minerals, and energy that remains largely untapped. More than half of the world population lives in coastal areas near the oceans, taking advantage of the mild climate, an inexpensive form of transportation, and vast recreational opportunities. Unfortunately, the oceans are also the dumping ground for many of society's wastes. In fact, the oceans are currently showing alarming changes caused by pollution, overfishing, invasive species, and climate change, among other things.

I.3 What Is Rational Use Of Technology?

Many stresses have been put on the oceans by our ever-increasing human population. For instance, population studies reveal that about 50% of world population—some 3.5 billion people—live along coasts, and more than 80% of all

Figure I.3 Relative sizes of the spheres of water on Earth. This image shows all of Earth's liquid water using three orbs of proportional sizes. The big sphere is all liquid water in the world, 97% of which is seawater. The next smallest sphere represents a subset of the larger sphere, showing freshwater in the ground, lakes, swamps, and rivers. The tiny speck below it represents an even smaller subset of all the water—just the freshwater in lakes and rivers.

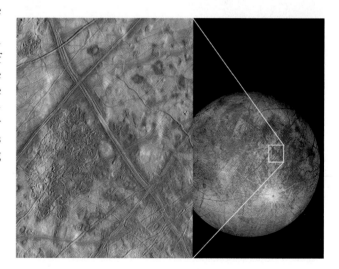

Figure I.4 Jupiter's moon Europa. Europa's network of dark fluid-filled cracks suggests the presence of an ocean beneath its icy surface.

Diving Deeper **Introduction**

OCEAN LITERACY: WHAT SHOULD PEOPLE KNOW ABOUT THE OCEAN?

The ocean is the defining feature of our planet. Accordingly, there is great interest in developing *ocean literacy*, which means understanding the ocean's influence on humans as well as humans' influence on the ocean. For example, scientists and educators agree that an ocean-literate person:

- Understands the essential principles and fundamental concepts about the functioning of the ocean.

- Can communicate about the ocean in a meaningful way.

- Is able to make informed and responsible decisions regarding the ocean and its resources.

To achieve this goal, ocean educators and experts have developed the **Seven Principles of Ocean Literacy**. The following ideas are what everyone—especially those who successfully pass a college course in oceanography or marine science—should understand about the ocean:

1. Earth has one big ocean with many features.

2. The ocean and life in the ocean shape the features of Earth.

3. The ocean is a major influence on weather and climate.

4. The ocean makes Earth habitable.

5. The ocean supports a great diversity of life and ecosystems.

6. The ocean and humans are inextricably interconnected.

7. The ocean is largely unexplored.

This book is intended to help all people achieve ocean literacy. For more information about the Seven Principles of Ocean Literacy, see http://oceanliteracy.wp2.coexploration.org/

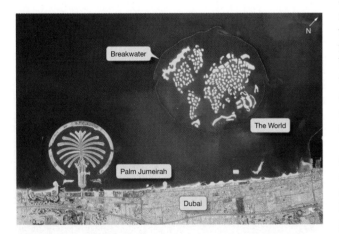

Figure I.5 Dubai's artificial islands: Palm Jumeirah and The World. These artificial islands, which jut audaciously into the Persian Gulf off Dubai, United Arab Emirates, were created specifically for coastal housing tracts and commercial buildings and resemble the shape of a palm tree and a globe of the world as seen from space. Even though the islands have doubled Dubai's shoreline, they have disrupted the region's coastal ecosystem. Palm Jumeirah is about 10 kilometers (6 miles) wide. Photo taken by an astronaut aboard the International Space Station in 2010.

Americans live within an hour's drive from an ocean or the Great Lakes. In the future, these figures are expected to increase. In the United States, for example, 8 of the 10 largest cities are in coastal environments, and about 3600 people move to the coast every day. By 2025, as much as 75% of the global population is expected to live at the coast. Clearly, coastal regions are desirable places to live (**Figure I.5**).

Human migration to the coasts will further mar fragile coastal ocean ecosystems. Specifically, the surge in coastal population has resulted in increasing amounts of industrial waste and sewage disposal at sea, large volumes of polluted runoff, and increasing use of seawater for human benefit (for example, using seawater as a coolant in coastal power plants). Coastal habitats have also come under intense pressure as more development has occurred, resulting in coastal wetlands being filled in even though they are vital to the cleansing of runoff waters, and their diversity and high productivity support coastal fisheries. In addition, all this development and its ecological impacts will pose immense challenges for coastal communities, including increased demands on energy, infrastructure, and the supply of freshwater.

Although it may seem as if humans have severely and irreversibly damaged the oceans, our impact has mostly been felt in coastal areas, where ocean depths are shallow and ocean circulation is restricted. The world's oceans are a vast habitat that has not yet been permanently damaged. Generally, humans have been able to inflict only minor damage here and there along the margins of the oceans. However, as human population increases and technology makes us more powerful, the threat of irreversible harm becomes greater. For instance, human-caused emissions are starting to affect the basic chemistry of the oceans. And in the open ocean (areas far from shore), deep-ocean mining and nuclear waste disposal have been proposed. How do we as a society deal with these human impacts on the marine environment? How do we fairly regulate the ocean's use? One of the first steps is to understand how the oceans work and, at the same time, increase our understanding of how human actions are affecting the oceans.

If used wisely, our technology can actually reduce the threat of irreversible harm. Which path will we take? We all need to carefully evaluate our own actions and the effects those actions have on the environment. In addition, we need to make conscientious decisions about those we elect to public office. Some of you may even have direct responsibility for initiating legislation that affects our environment. It is our hope that you, as a student of the marine environment, will gain enough knowledge while studying oceanography to help your community (and perhaps even your nation) make rational use of technology in the oceans.

The blue marble next generation. This composite image of satellite data shows Earth's interrelated atmosphere, oceans, and land—including human presence. Its various layers include the land surface, sea ice, ocean, cloud cover, city lights, and the hazy edge of Earth's atmosphere.

Before you begin reading this chapter, use the glossary at the end of this book to discover the meanings of any of these words you don't already know:

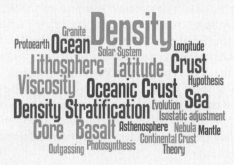

INTRODUCTION TO PLANET "EARTH"

ESSENTIAL CONCEPTS

At the end of this chapter, you should be able to:

1.1 Compare the characteristics of Earth's oceans

1.2 Discuss how early exploration of the oceans was achieved

1.3 Describe the nature of scientific inquiry

1.4 Explain how Earth and the solar system were formed

1.5 Explain how Earth's atmosphere and oceans were formed

1.6 Discuss why life is thought to have originated in the oceans

1.7 Demonstrate an understanding of how old Earth is

"When you're circling the Earth every 90 minutes, what becomes clearest is that it's mostly water; the continents look like they're floating objects."

—Loren Shriver, NASA astronaut (2008)

It seems perplexing that our planet is called "Earth" when 70.8% of its surface is covered by oceans. Many early human cultures that lived near the Mediterranean (*medi* = middle, *terra* = land) Sea envisioned the world as being composed of large landmasses surrounded by marginal bodies of water (**Figure 1.1**). How surprised they must have been when they ventured into the larger oceans of the world. Our planet is misnamed "Earth" because we live on the land portion of the planet. If we were marine animals, our planet would probably be called "Ocean," "Water," "Hydro," "Aqua," or even "Oceanus" to indicate the prominence of Earth's oceans.

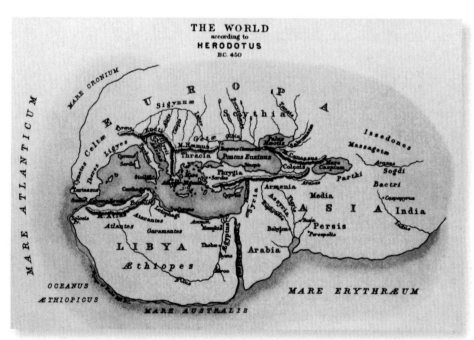

Figure 1.1 An early map of the world. The world according to the Greek Herodotus in 450 B.C., showing the prominence of the Mediterranean Sea surrounded by the continents of Europe, Libya (Africa), and Asia. Seas are labeled "mare" and are shown as bands of water encircling the land.

1.1 How Many Oceans Exist on Earth?

When one examines a world map (**Figure 1.2**), it's easy to appreciate the impressive extent of Earth's oceans. Notice that *the oceans dominate the surface area of the globe*. For those people who have traveled by boat across an ocean (or even flown across one in an airplane), the one thing that immediately strikes them is that the oceans are enormous. Notice, also, that *the oceans are interconnected* and form a single continuous body of seawater, which is why the oceans are commonly referred to as a "world ocean" (singular, not plural). For instance, a vessel at sea can travel from one ocean to another, whereas it is impossible to travel on land from one continent to most others without crossing an ocean. In addition, *the volume of the oceans is immense*. For example, the oceans comprise the planet's largest habitat and contain 97.2% of all the water on or near Earth's surface.

The Four Principal Oceans, Plus One

Our world ocean can be divided into four principal oceans plus an additional ocean, based on the shape of the ocean basins and the positions of the continents (Figure 1.2).

PACIFIC OCEAN   The **Pacific Ocean** is the world's largest ocean, covering more than half of the ocean surface area on Earth (**Figure 1.3b**). The Pacific Ocean is the single largest geographic feature on the planet, spanning more than one-third of Earth's entire surface. The Pacific Ocean is so large that *all* of the continents could fit into the space occupied by it—with room left over! Although the Pacific Ocean is also the deepest ocean in the world (**Figure 1.3c**), it contains many small tropical islands. It was named in 1520 by explorer Ferdinand Magellan's party in honor of the fine weather they encountered while crossing into the Pacific (*paci* = peace) Ocean.

ATLANTIC OCEAN   The **Atlantic Ocean** is about half the size of the Pacific Ocean and is not quite as deep (Figure 1.3). It separates the Old World (Europe, Asia, and Africa) from the New World (North and South America). The Atlantic Ocean was named after Atlas, who was one of the Titans in Greek mythology.

INDIAN OCEAN   The **Indian Ocean** is slightly smaller than the Atlantic Ocean and has about the same average depth (Figure 1.3). It is mostly in the Southern Hemisphere (south of the equator, or below 0 degrees latitude in Figure 1.2). The Indian Ocean was named for its proximity to the subcontinent of India.

ARCTIC OCEAN   The **Arctic Ocean** is about 7% the size of the Pacific Ocean and is only a little more than one-quarter as deep as the rest of the oceans (Figure 1.3). Although it has a permanent layer of sea ice at the surface, the ice is only a few meters thick. The Arctic Ocean was named after its location in the Arctic region, which exists beneath the northern constellation Ursa Major, otherwise known as the Big Dipper, or the Bear (*arktos* = bear).

SOUTHERN OCEAN, OR ANTARCTIC OCEAN   Oceanographers recognize an additional ocean near the continent of Antarctica in the Southern Hemisphere (Figure 1.2). Defined by the meeting of currents near Antarctica called the Antarctic Convergence, the **Southern Ocean**, or **Antarctic Ocean**, is really the portions of the Pacific, Atlantic, and Indian Oceans south of about 50 degrees south latitude. This ocean was named for its location in the Southern Hemisphere.

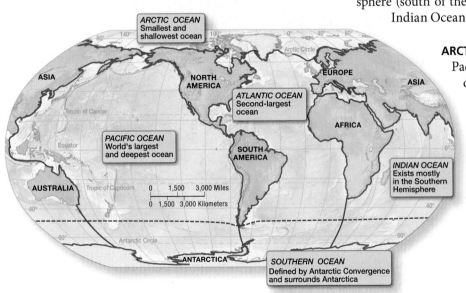

Figure 1.2 Earth's oceans. Map showing the four principal oceans (outlined in red), plus the Southern Ocean, or Antarctic Ocean.

Oceans versus Seas: What Are the Seven Seas?

What is the difference between a sea and an ocean? In common use, the terms "sea" and "ocean" are often used interchangeably. For instance, a *sea* star lives in the *ocean*; the *ocean* is full of *sea*water; *sea* ice forms in the *ocean*; and one might stroll the *sea*shore while living in *ocean*-front property. Technically, however, a *sea* is defined as follows:

- Smaller and shallower than an ocean (this is why the Arctic Ocean might be more appropriately considered a sea)
- Composed of salt water (although some inland "seas," such as the Caspian Sea in Asia, are actually large lakes with relatively high salinity)
- Somewhat enclosed by land (although some seas, such as the Sargasso Sea in the Atlantic Ocean, are defined by strong ocean currents rather than by land)
- Directly connected to the world ocean

"Sailing the seven seas" is a familiar phrase in literature and song, but are there really seven seas? To the ancients, the term "seven" often meant "many," and before the 15th century, Europeans considered the main seas of the world to be:

1. The Red Sea
2. The Mediterranean Sea
3. The Persian Gulf
4. The Black Sea
5. The Adriatic Sea
6. The Caspian Sea
7. The Indian Ocean

Today, the world ocean is generally divided into the four principal oceans plus one. If the oceans are counted as seas and the Pacific and Atlantic Oceans are arbitrarily split at the equator, a more modern count of the seven "seas" would be (1) the North Pacific, (2) the South Pacific, (3) the North Atlantic, (4) the South Atlantic, (5) the Indian, (6) the Arctic, and (7) the Southern, or Antarctic.

COMPARING THE OCEANS TO THE CONTINENTS Figure 1.3d shows that the average depth of the world's oceans is 3682 meters[1] (12,080 feet). This means that there must be some extremely deep areas in the ocean to offset the shallow areas close to shore. Figure 1.3d also shows that the deepest depth in the oceans (the Challenger Deep region of the Mariana Trench, which is near Guam) is a staggering 11,022 meters (36,161 feet) below sea level.

How do the continents compare to the oceans? Figure 1.3d shows that the average height of the continents is only 840 meters (2756 feet), illustrating that the average height of the land is not very far above sea level. The highest mountain

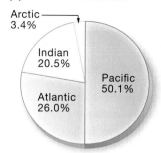
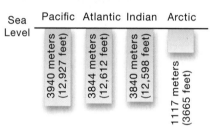
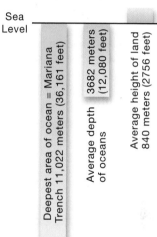

Figure 1.3 Ocean size and depth. (a) Relative proportions of land and ocean on Earth's surface. **(b)** Relative size of the four principal oceans. **(c)** Average ocean depth. **(d)** Comparing average and maximum depth of the oceans to average and maximum height of land.

Web Animation
Earth's Water and the Hydrologic Cycle

[1]Throughout this book, metric measurements are used (and the corresponding English measurements follow in parentheses). See Appendix I, "Metric and English Units Compared," for conversion factors between the two systems of units.

STUDENTS SOMETIMES ASK . . .

Have humans ever explored the deepest ocean trenches? Could anything live there?

Humans have indeed visited the deepest part of the oceans—where there is crushing high pressure, complete darkness, and near-freezing water temperatures—and they first did so over half a century ago! In January 1960, U.S. Navy Lt. Don Walsh and explorer Jacques Piccard descended to the bottom of the Challenger Deep region of the Mariana Trench in the *Trieste*, a deep-diving bathyscaphe (*bathos* = depth, *scaphe* = a small ship) (**Figure 1.4**). At 9906 meters (32,500 feet), the men heard a loud cracking sound that shook the cabin. They were unable to see that a 7.6-centimeter (3-inch) Plexiglas viewing port had cracked (miraculously, it held for the rest of the dive). More than five hours after leaving the surface, they reached the bottom, at 10,912 meters (35,800 feet)— a record depth for human descent. They did observe some small organisms that are adapted to life in the deep: a flatfish, a shrimp, and some jellies.

In 2012, film icon James Cameron made a historic solo dive to the Mariana Trench in his submersible *DEEPSEA CHALLENGER* (**Figure 1.5**). On the seven-hour round trip voyage, Cameron spent about three hours at the deepest spot on the planet to take photographs and collect samples for scientific research. Other notable voyages to the deep ocean in submersibles are discussed in **Web Diving Deeper 1.3**.

ESSENTIAL CONCEPT 1.1B

The deepest part of the ocean is the Mariana Trench in the Pacific Ocean. It is 11,022 meters (36,161 feet) deep and has been visited only twice by humans: once in 1960 and more recently in 2012.

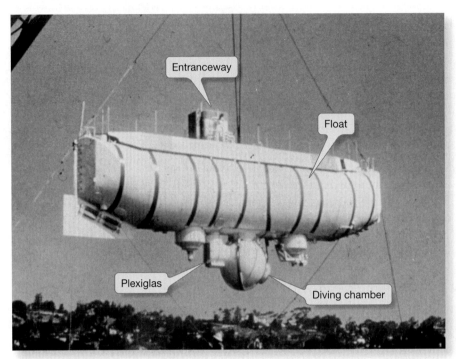

Figure 1.4 The U.S. Navy's bathyscaphe *Trieste*. The *Trieste* suspended on a crane before its record-setting deep dive in 1960. The 1.8-meter- (6-foot)-diameter diving chamber (the round ball below the float) accommodated two people and had steel walls 7.6 centimeters (3 inches) thick.

Figure 1.5 James Cameron emerges from the submersible *DEEPSEA CHALLENGER* after his solo dive to the Mariana Trench. In 2012, famous moviemaker James Cameron completed a record-breaking solo dive to the bottom of the Marina Trench, becoming only the third human to visit the deepest spot on Earth.

CONCEPT CHECK 1.1

1 How did the view of the ocean by early Mediterranean cultures influence the naming of planet Earth?

2 What is the difference between an ocean and a sea? Name the seven seas (both ancient and modern versions).

3 Compare the depth of the Mariana Trench to the height of the tallest mountains on Earth.

in the world (the mountain with the greatest height above sea level) is Mount Everest in the Himalaya Mountains of Asia, at 8850 meters (29,035 feet). Even so, Mount Everest is a full 2172 meters (7126 feet) shorter than the Mariana Trench is deep. The mountain with the *greatest total height* from base to top is Mauna Kea on the island of Hawaii in the United States. It measures 4206 meters (13,800 feet) above sea level and 5426 meters (17,800 feet) from sea level down to its base, for a total height of 9632 meters (31,601 feet). The total height of Mauna Kea is 782 meters (2566 feet) higher than Mount Everest, but it is still 1390 meters (4560 feet) shorter than the Mariana Trench is deep. Therefore, no mountain on Earth is taller than the Mariana Trench is deep.

1.2 How Was Early Exploration Of the Oceans Achieved?

The ocean's huge extent over the surface area of Earth has not prevented humans from exploring its furthest reaches. Since early times, humans have developed technology that has allowed civilizations to travel across large stretches of open ocean. Today, we can cross even the Pacific Ocean in less than a day by airplane. Even so, much of the deep ocean remains out of reach and woefully unexplored. In fact, the surface of the Moon has been mapped more accurately than most parts of the sea floor. Yet satellites at great distances above Earth are being used to gain knowledge about our watery home.

Early History

Humankind probably first viewed the oceans as a source of food. Archeological evidence suggests that when boat technology was developed about 40,000 years ago, people probably traveled the oceans. Most likely, their vessels were built to move upon the ocean's surface and transport oceangoing people to new fishing grounds. The oceans also provided an inexpensive and efficient way to move large and heavy objects, facilitating trade and interaction between cultures.

PACIFIC NAVIGATORS The peopling of the Pacific Islands (Oceania) is somewhat perplexing because there is no evidence that people actually evolved on these islands. Their presence required travel over hundreds or even thousands of kilometers of open ocean from the continents (probably in small vessels of that time—double canoes, outrigger canoes, or balsa rafts) as well as remarkable navigation skills (**Diving Deeper 1.1**). The islands in the Pacific Ocean are widely scattered, so it is likely that only a fortunate few of the voyagers made landfall and that many others perished during voyages. **Figure 1.6** shows the three major inhabited island regions in the Pacific Ocean: Micronesia (*micro* = small, *nesia* = islands), Melanesia (*mela* = black, *nesia* = islands), and Polynesia (*poly* = many, *nesia* = islands), which covers the largest area.

No written records of Pacific human history exist before the arrival of Europeans in the 16th century. Nevertheless, the movement of Asian peoples into Micronesia and Melanesia is easy to imagine because distances between islands are relatively short. In Polynesia, however, large distances separate island groups, which must have presented great challenges to ocean voyagers. Easter Island, for example, at the southeastern corner of the triangular-shaped Polynesian Islands

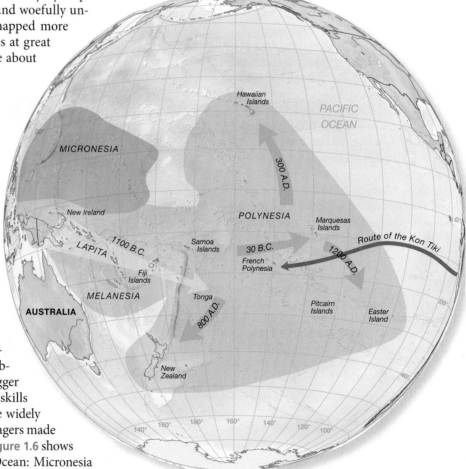

Figure 1.6 The peopling of the Pacific islands. The major island groups of the Pacific Ocean are Micronesia (*brown shading*), Melanesia (*peach shading*), and Polynesia (*green shading*). The "Lapita people" present in New Ireland 5000–4000 B.C. can be traced to Fiji, Tonga, and Samoa by 1100 B.C. (*yellow arrow*). Green arrows show the peopling of distant islands throughout Polynesia. The route of Thor Heyerdahl's balsa raft *Kon Tiki* is also shown (*red arrow*).

HOW DO SAILORS KNOW WHERE THEY ARE AT SEA?: FROM STICK CHARTS TO SATELLITES

How do you know where you are in the ocean, without roads, signposts, or any land in sight? How do you determine the distance to a destination? How do you find your way back to a good fishing spot or where you have discovered sunken treasure? Sailors have relied on a variety of navigation tools to help answer questions such as these by being able to locate where they are at sea.

Some of the first navigators were the Polynesians. Remarkably, the Polynesians were able to successfully navigate to small islands located at great distances across the Pacific Ocean. These early navigators must have been very aware of the marine environment and been able to read subtle differences in the ocean and sky. The tools they used to help them navigate between islands included the Sun and Moon, the nighttime stars, the behavior of marine organisms, various ocean properties, and an ingenious device called a *stick chart* (**Figure 1A**). Stick charts are like a map that accurately depicts the dominant pattern of ocean waves. By orienting their vessels relative to these regular ocean wave directions, sailors could successfully navigate at sea. The bent wave directions let them know when they were getting close to an island—even one that was located beyond the horizon.

Figure 1A Navigational stick chart. This bamboo stick chart of Micronesia's Marshall Islands shows islands (represented by shells at the junctions of the sticks), regular ocean wave direction (represented by the straight strips), and waves that bend around islands (represented by the curved strips). Similar stick charts were used by early Polynesian navigators.

The importance of knowing where you are at sea is illustrated by a tragic incident in 1707, when a British battle fleet was more than 160 kilometers (100 miles) off course and ran aground in the Isles of Scilly near England, with the loss of four ships and nearly 2000 men. **Latitude** (location north or south) was relatively easy to determine at sea by measuring the position of the Sun and stars using a device called a *sextant* (*sextant* = sixth, in reference to the instrument's arc, which is one-sixth of a circle) (**Figure 1B**).

The accident occurred because the ship's crew had no way of keeping track of their **longitude** (location east or west; see Appendix III, "Latitude and Longitude on Earth"). To determine longitude, which is a function of time, it was necessary to know the time difference between a reference meridian and when the Sun was directly overhead of a ship at sea (noon local time). The pendulum-driven clocks in use in the early 1700s, however, would not work for long on a rocking ship at sea. In 1714, the British Parliament offered a £20,000 prize (about $20 million today) for developing a device that would work well enough at sea to determine longitude within half a degree, or 30 nautical miles (34.5 statute miles), after a voyage to the West Indies.

A cabinetmaker in Lincolnshire, England, named **John Harrison** began working in 1728 on such a timepiece, which was dubbed the

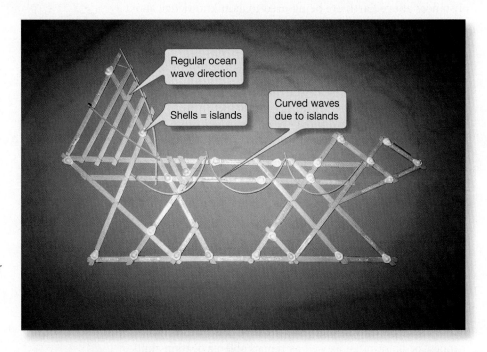

Regular ocean wave direction

Shells = islands

Curved waves due to islands

region, is more than 1600 kilometers (1000 miles) from Pitcairn Island, the next nearest island. Clearly, a voyage to the Hawaiian Islands must have been one of the most difficult because Hawaii is more than 3000 kilometers (2000 miles) from the nearest inhabited islands, the Marquesas Islands (Figure 1.6).

Archeological evidence suggests that humans from New Guinea may have occupied New Ireland as early as 4000 or 5000 B.C. However, there is little evidence of human travel farther into the Pacific Ocean before 1100 B.C. By then, the *Lapita people*[2],

---

[2]In recent years, a combination of genetic, linguistic, and archaeological evidence has suggested that the forebears of the Lapita people—and thus Polynesians—originated in Taiwan, just off the coast of China.

chronometer (*chrono* = time, *meter* = measure) Harrison's first chronometer, H-1, was successfully tested in 1736, but he received only £500 of the prize because the device was deemed too complex, costly, and fragile. Eventually, his more compact fourth version, H-4—which resembles an oversized pocket watch (**Figure 1C**)—was tested during a trans-Atlantic voyage in 1761. Upon reaching Jamaica, it was so accurate that it had lost only *five seconds* of time, a longitude error of only 0.02 degree, or 1.2 nautical miles (1.4 statute miles)! Although Harrison's chronometer greatly exceeded the requirements of the government, the committee in charge of the prize withheld payment, mostly because the astronomers on the board wanted the solution to come from measurement of the stars. Because the committee refused to award him the prize without further proof, a second sea trial was conducted in 1764,

which confirmed his success. Harrison was reluctantly granted £10,000. Only when King George III intervened in 1773 did Harrison finally receive the remaining prize money and recognition for his life work—at age 80.

Today, navigating at sea relies on the *Global Positioning System (GPS)*, which was initiated in the 1970s by the U.S. Department of Defense. Initially designed for military purposes but now available for a variety of civilian uses, GPS relies on a system of 24 satellites that send continuous radio signals to the surface. Position is determined by very accurate measurement of the time of travel of radio signals from at least four of the satellites to receivers on board a ship (or on land). Thus, a vessel

can determine its exact latitude and longitude to within a few meters—a small fraction of the length of most ships. Navigators from days gone by would be amazed at how quickly and accurately a vessel's location can be determined, but they might say that it has taken all the adventure out of navigating at sea.

Figure 1C John Harrison and his chronometer H-4. Painting (circa 1735) of John Harrison holding his chronometer H-4, which was his life's work. The timepiece H-4 proved to be a vital technological breakthrough that allowed the determination of longitude at sea and won Harrison the prize for solving the longitude problem.

Figure 1B Using a handheld sextant. This sextant is similar to the ones used by early navigators to determine latitude.

a group of early settlers who produced a distinctive type of pottery, had traveled on to Fiji, Tonga, and Samoa (Figure 1.6, *yellow arrow*). From there, Polynesians sailed on to the Marquesas (about 30 B.C.), which appear to have been the starting point for voyages to the far reaches of the island Pacific (Figure 1.6, *green arrows*), including the Hawaiian Islands (about 300 A.D.) and New Zealand (about 800 A.D.). Surprisingly, new genetic research suggests that Polynesians populated Easter Island relatively recently, about 1200 A.D.

Despite the obvious Polynesian backgrounds of the Hawaiians, the Maori of New Zealand, and the Easter Islanders, an adventurous biologist/anthropologist named **Thor Heyerdahl** proposed that voyagers from South America may

Figure 1.7 *Kon Tiki.* In 1947, Thor Heyerdahl sailed this authentic balsa raft named *Kon Tiki* from South America to Polynesia to show that ancient South American cultures may have completed similar voyages.

have reached islands of the South Pacific before the coming of the Polynesians. To prove his point, in 1947 he sailed the *Kon Tiki*—a balsa raft designed like those that were used by South American navigators at the time of European discovery (**Figure 1.7**)—from South America to the Tuamotu Islands, a journey of more than 11,300 kilometers (7000 miles) (Figure 1.6, *red arrow*). Although the remarkable voyage of the *Kon Tiki* demonstrates that early South Americans could have traveled to Polynesia just as easily as early Asian cultures, anthropologists can find no evidence of such a migration. Further, comparative DNA studies show a strong genetic relationship between the peoples of Easter Island and Polynesia but none between these groups and natives in coastal North or South America.

EUROPEAN NAVIGATORS The first Mediterranean people known to have developed the art of navigation were the **Phoenicians**, who lived at the eastern end of the Mediterranean Sea, in the present-day area of Egypt, Syria, Lebanon, and Israel. As early as 2000 B.C., they investigated the Mediterranean Sea, the Red Sea, and the Indian Ocean. The first recorded circumnavigation of Africa, in 590 B.C., was made by the Phoenicians, who had also sailed as far north as the British Isles.

The Greek astronomer-geographer **Pytheas** sailed northward in 325 B.C. using a simple yet elegant method for determining latitude (one's position north or south) in the Northern Hemisphere. His method involved measuring the angle between an observer's line of sight to the North Star and line of sight to the northern horizon.[3] Despite Pytheas's method for determining latitude, it was still impossible to accurately determine longitude (one's position east or west).

One of the key repositories of scientific knowledge at the time was the **Library of Alexandria** in Alexandria, Egypt, which was founded in the 3rd century B.C. by *Alexander the Great*. It housed an impressive collection of written knowledge that attracted scientists, poets, philosophers, artists, and writers who studied and researched there. The Library of Alexandria soon became the intellectual capital of the world, featuring history's greatest accumulation of ancient writings.

As long ago as 450 B.C., Greek scholars became convinced that Earth was round, using lines of evidence such as the way ships disappeared beyond the horizon and the shadows of Earth that appeared during eclipses of the Moon. This inspired the Greek **Eratosthenes** (pronounced "AIR-uh-TOS-thuh-neez") (276–192 B.C.), the second librarian at the Library of Alexandria, to cleverly use the shadow of a stick in a hole in the ground and elementary geometry to determine Earth's circumference. His value of 40,000 kilometers (24,840 miles) compares well with the true value of 40,032 kilometers (24,875 miles) known today.

An Egyptian-Greek named **Claudius Ptolemy** (c. 85 A.D.–c. 165 A.D.) produced a map of the world in about 150 A.D. that represented the extent of Roman knowledge at that time. The map included the continents of Europe, Asia, and Africa, as did earlier Greek maps, but it also included vertical lines of longitude and horizontal lines of latitude, which had been developed by Alexandrian scholars. Moreover, Ptolemy showed the known seas to be surrounded by land, much of which was as yet unknown and proved to be a great enticement to explorers.

Ptolemy also introduced an (erroneous) update to Eratosthenes's surprisingly accurate estimate of Earth's circumference. Ptolemy wrongly depended on flawed calculations and an overestimation of the size of Asia, and as a result, he determined Earth's circumference to be 29,000 kilometers (18,000 miles), which is about 28% too small. Remarkably, nearly 1500 years later, Ptolemy's error caused explorer Christopher Columbus to believe he had encountered parts of Asia rather than a new world.

The Middle Ages

After the destruction of the Library of Alexandria in 415 A.D. (in which all of its contents were burned) and the fall of the Roman Empire in 476 A.D., the achievements

---

[3] Pytheas's method of determining latitude is featured in Appendix III, "Latitude and Longitude on Earth."

of the Phoenicians, Greeks, and Romans were mostly lost. Some of the knowledge, however, was retained by the *Arabs*, who controlled northern Africa and Spain. The Arabs used this knowledge to become the dominant navigators in the Mediterranean Sea area and to trade extensively with East Africa, India, and southeast Asia. The Arabs were able to trade across the Indian Ocean because they had learned how to take advantage of the seasonal patterns of monsoon winds. During the summer, when monsoon winds blow from the southwest, ships laden with goods would leave the Arabian ports and sail eastward across the Indian Ocean. During the winter, when the trade winds blow from the northeast, ships would return west.[4]

Meanwhile, in the rest of southern and eastern Europe, Christianity was on the rise. Scientific inquiry counter to religious teachings was actively suppressed, and the knowledge gained by previous civilizations was either lost or ignored. As a result, the Western concept of world geography degenerated considerably during these so-called *Dark Ages*. For example, one notion envisioned the world as a disk with Jerusalem at the center.

In northern Europe, the **Vikings** of Scandinavia, who had excellent ships and good navigation skills, actively explored the Atlantic Ocean (**Figure 1.8**). Late in the 10th century, aided by a period of worldwide climatic warming, the Vikings colonized Iceland. In about 981, **Erik "the Red" Thorvaldson** sailed westward from Iceland and discovered Greenland. He may also have traveled further westward to Baffin Island. He returned to Iceland and led the first wave of Viking colonists to Greenland in 985. **Bjarni Herjólfsson** sailed from Iceland to join the colonists, but he sailed too far southwest and is thought to be the first Viking to have seen what is now called Newfoundland. Bjarni did not land but instead returned to the new colony at Greenland. **Leif Eriksson**, son of Erik the Red, became intrigued by Bjarni's stories about the new land Bjarni had seen. In 995, Leif bought Bjarni's ship and set out from Greenland for the land that Bjarni had seen to the southwest. Leif spent the winter in that portion of North America and named the land *Vinland* (now Newfoundland, Canada) after the grapes that were found there. Climatic cooling and inappropriate farming practices for the region caused these Viking colonies in Greenland and Vinland to struggle and die out by about 1450.

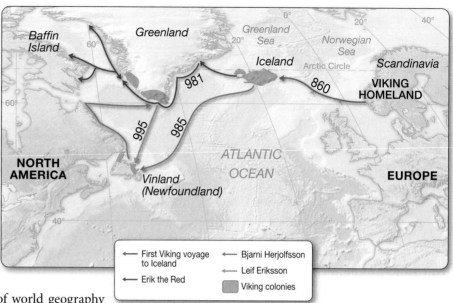

Figure 1.8 Viking colonies in the North Atlantic. Map showing the routes and dates of Viking explorations and the locations of the colonies that were established in Iceland, Greenland, and parts of North America.

The Age Of Discovery in Europe

The 30-year period from 1492 to 1522 is known as the **Age of Discovery**. During this time, Europeans explored the continents of North and South America, and the globe was circumnavigated for the first time. As a result, Europeans learned the true extent of the world's oceans and that human populations existed elsewhere on newly "discovered" continents and islands with cultures vastly different from those familiar to European voyagers.

Why was there such an increase in ocean exploration during the Age of Discovery? One reason was that Sultan Mohammed II had captured Constantinople (the capital of eastern Christendom) in 1453, a conquest that isolated Mediterranean port cities from the riches of India, Asia, and the East Indies (modern-day Indonesia). As a result, the Western world had to search for new eastern trade routes by sea.

The Portuguese, under the leadership of **Prince Henry the Navigator** (1392–1460), led a renewed effort to explore outside Europe. The prince established a marine institution at Sagres to improve Portuguese sailing skills. The treacherous journey around the tip of Africa was a great obstacle to an alternative trade route.

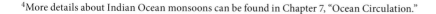

[4]More details about Indian Ocean monsoons can be found in Chapter 7, "Ocean Circulation."

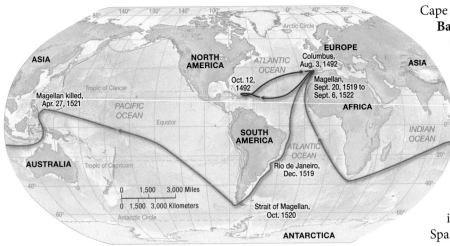

Figure 1.9 Voyages of Columbus and Magellan. Map showing the dates and routes of Columbus's first voyage and the first circumnavigation of the globe by Magellan's party.

Cape Agulhas (at the southern tip of Africa) was first rounded by **Bartholomeu Diaz** in 1486. He was followed in 1498 by **Vasco da Gama**, who continued around the tip of Africa to India, thus establishing a new eastern trade route to Asia.

Meanwhile, the Italian navigator and explorer **Christopher Columbus** was financed by Spanish monarchs to find a new route to the East Indies across the Atlantic Ocean. During Columbus's first voyage in 1492, he sailed west from Spain and made landfall after a two-month journey (**Figure 1.9**). Columbus believed that he had arrived in the East Indies somewhere near India, but Earth's circumference had been substantially underestimated, so he was unaware that he had actually arrived in uncharted territory in the Caribbean. Upon his return to Spain and the announcement of his discovery, additional voyages were planned. During the next 10 years, Columbus made three more trips across the Atlantic.

Even though Christopher Columbus is widely credited with discovering North America, he never actually set foot on the continent.[5] Still, his journeys inspired other navigators to explore the "New World." For example, in 1497, only five years after Columbus's first voyage, the Italian navigator and explorer **Giovanni Caboto**, who was also known as **John Cabot**, landed somewhere on the northeastern coast of North America. Later, Europeans first saw the Pacific Ocean in 1513, when **Vasco Núñez de Balboa** attempted a land crossing of the Isthmus of Panama and sighted a large ocean to the west from atop a mountain.

The culmination of the Age of Discovery was a remarkable circumnavigation of the globe initiated by **Ferdinand Magellan** (Figure 1.9). Magellan left Spain in September 1519, with five ships and 280 sailors. He crossed the Atlantic Ocean, sailed down the eastern coast of South America, and traveled through a passage to the Pacific Ocean at 52 degrees south latitude, now named the Strait of Magellan in his honor. After landing in the Philippines in March, 1521, Magellan was killed about a month later in a fight with the inhabitants of these islands. **Juan Sebastian del Caño** completed the circumnavigation by taking the last of the ships, the *Victoria*, across the Indian Ocean, around Africa, and back to Spain in 1522. After three years, just one ship and 18 men completed the voyage.

Following these expeditions, the Spanish initiated many other voyages to take gold from the Aztec and Inca cultures in Mexico and South America. The English and Dutch, meanwhile, used smaller, more maneuverable ships to rob the gold from bulky Spanish galleons, which resulted in many confrontations at sea. The maritime dominance of Spain ended when the English defeated the Spanish Armada in 1588. With control of the seas, the English thus became the dominant world power—a status they retained until early in the 20th century.

The Beginning Of Voyaging for Science

The English realized that increasing their scientific knowledge of the oceans would help maintain their maritime superiority. For this reason, Captain **James Cook** (1728–1779), an English navigator and prolific explorer (**Figure 1.10**), undertook three voyages of scientific discovery with the ships *Endeavour*, *Resolution*, and *Adventure* between 1768 and 1779. He searched for the continent Terra Australis ("Southern Land," or Antarctica) and concluded that it lay beneath or beyond the extensive ice fields of the southern oceans, if it existed at all. Cook also mapped many previously unknown islands, including the South Georgia, South Sandwich, and Hawaiian Islands. During his last voyage, Cook searched for the

[5]For more information about the voyages of Columbus, see Diving Deeper 6.1 in Chapter 6, "Air–Sea Interaction."

STUDENTS SOMETIMES ASK . . .

What is NOAA? What is its role in oceanographic research?

NOAA (pronounced "NO-ah") stands for National Oceanic and Atmospheric Administration and is the branch of the U.S. Department of Commerce that oversees oceanographic research. Scientists at NOAA work to ensure wise use of ocean resources through the National Ocean Service, the National Oceanographic Data Center, the National Marine Fisheries Service, and the National Sea Grant Office. Other U.S. government agencies that work with oceanographic data include the U.S. Naval Oceanographic Office, the Office of Naval Research, the U.S. Coast Guard, and the U.S. Geological Survey (coastal processes and marine geology). The NOAA Website is at **www.noaa.gov**. Recently, federal officials developed the National Ocean Policy Implementation Plan, which proposes moving NOAA to the Department of the Interior so that agencies dealing with natural resources would all be grouped within the same department.

fabled "northwest passage" from the Pacific Ocean to the Atlantic Ocean and stopped in Hawaii, where he was killed in a skirmish with native Hawaiians.

Cook's expeditions added greatly to the scientific knowledge of the oceans. He determined the outline of the Pacific Ocean and was the first person known to cross the Antarctic Circle in his search for Antarctica. Cook initiated systematic sampling of subsurface water temperatures, measuring winds and currents, taking *soundings* (which are depth measurements that, at the time, were taken by lowering a long rope with a weight on the end to the sea floor), and collecting data on coral reefs. Cook also discovered that a shipboard diet containing the German staple sauerkraut prevented his crew from contracting scurvy, a disease that incapacitated sailors. Scurvy is caused by a vitamin C deficiency, and the cabbage used to make sauerkraut contains large quantities of vitamin C. Prior to Cook's discovery about preventing scurvy, the malady claimed more lives than all other types of deaths at sea, including contagious disease, gunfire, and shipwreck. In addition, by proving the value of John Harrison's chronometer as a means of determining longitude (see Diving Deeper 1.1), Cook made possible the first accurate maps of Earth's surface, some of which are still in use today.

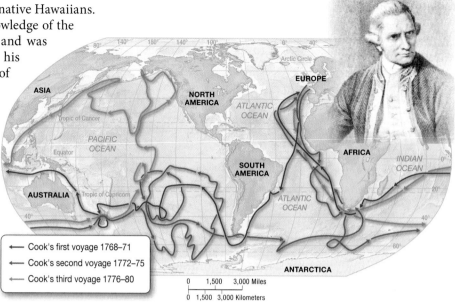

Figure 1.10 Captain James Cook (1728–1779) and his voyages of exploration. Routes taken by Captain James Cook (*inset*) on his three scientific voyages, which initiated scientific exploration of the oceans. Cook was killed in 1779 in Hawaii during his third voyage.

History Of Oceanography...To Be Continued

Much has changed since the early days of studying the oceans, when scientists used buckets, nets, and lines deployed from ships. And yet, some things remain the same. For example, going to sea aboard ships continues to be a mainstay of ocean science. Also, even though efforts to monitor the ocean are getting bigger and more sophisticated, vast swaths of the marine world remain unknown.

Today, oceanographers employ many high-technology tools, such as state-of-the-art research vessels that routinely use sonar to map the sea floor, remotely operated data collection devices, drifting buoys, robotics, sea floor observation networks, sophisticated computer models, and Earth-orbiting satellites. Many of these tools are featured throughout this book. Further, additional events in the history of oceanography can be found as Diving Deeper features in subsequent chapters. These boxed features are identified by the "Historical Feature" theme, and each introduces an important historical event that is related to the subject of that particular chapter.

ESSENTIAL CONCEPT 1.2

The ocean's large size did not prohibit early explorers from venturing into all parts of the ocean for discovery, trade, or conquest. Voyaging for science began relatively recently, and many parts of the ocean remain unknown.

CONCEPT CHECK 1.2

① While the Arabs dominated the Mediterranean region during the Middle Ages, what were the most significant ocean-related events taking place in northern Europe?

② Describe the important events in oceanography that occurred during the Age of Discovery in Europe.

③ List some of the major achievements of Captain James Cook.

1.3 What Is the Nature Of Scientific Inquiry?

In modern society, scientific studies are increasingly used to substantiate the need for action. However, there is often little understanding of how science operates. For instance, how certain are we about a particular scientific theory? How are facts different from theories?

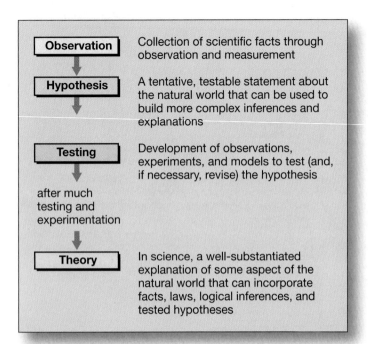

Observation	Collection of scientific facts through observation and measurement
Hypothesis	A tentative, testable statement about the natural world that can be used to build more complex inferences and explanations
Testing	Development of observations, experiments, and models to test (and, if necessary, revise) the hypothesis

after much testing and experimentation

Theory	In science, a well-substantiated explanation of some aspect of the natural world that can incorporate facts, laws, logical inferences, and tested hypotheses

Figure 1.11 The scientific method. A flow diagram showing the main steps in the scientific method, along with definitions of terms. In actuality, the process of science is often less formally conducted than is implied in this figure.

The overall goal of science is to discover underlying patterns in the natural world and then to use this knowledge to make predictions about what should or should not be expected to happen given a certain set of circumstances. Scientists develop explanations about the causes and effects of various natural phenomena (such as why Earth has seasons or what the structure of matter is). This work is based on an assumption that all natural phenomena are controlled by understandable physical processes and the same physical processes operating today have been operating throughout time. Consequently, science has demonstrated remarkable power in allowing scientists to describe the natural world accurately, to identify the underlying causes of natural phenomena, and to better predict future events that rely on natural processes.

Science supports the explanation of the natural world that best explains all available observations. Scientific inquiry is formalized into what is called the **scientific method**, which is used to formulate scientific theories (**Figure 1.11**).

Observations

The scientific method begins with *observations*, which are occurrences we can measure with our senses. They are things we can manipulate, see, touch, hear, taste, or smell, often by experimenting with them directly or by using sophisticated tools (such as a microscope or telescope) to sense them. If an observation is repeatedly confirmed—that is, made so many times that it is assumed to be completely valid—then it can be called a *scientific fact*.

Hypothesis

As observations are being made, the human mind attempts to sort out the observations in a way that reveals some underlying order or pattern in the objects or phenomena being observed. This sorting process—which involves a lot of trial and error—seems to be driven by a fundamental human urge to make sense of our world. This is how **hypotheses** (*hypo* = under, *thesis* = an arranging) are made. A hypothesis is sometimes labeled as an informed or educated guess, but it is more than that. A hypothesis is a tentative, testable statement about the general nature of the phenomena observed. In other words, a hypothesis is an initial idea of how or why things happen in nature.

Suppose we want to understand why whales *breach* (that is, why whales sometimes leap entirely out of water). After scientists observe breaching many times, they can organize their observations into a hypothesis. For instance, one hypothesis is that a breaching whale is trying to dislodge parasites from its body. Scientists often have multiple working hypotheses (for example, whales may use breaching to communicate with other whales). If a hypothesis cannot be tested, it is not scientifically useful, no matter how interesting it might seem.

Testing

Hypotheses are used to predict certain occurrences that lead to further research and the refinement of those hypotheses. For instance, the hypothesis that a breaching whale is trying to dislodge its parasites suggests that breaching whales have more parasites than whales that don't breach. Analyzing the number of parasites on breaching versus nonbreaching whales would either support that hypothesis or cause it to be recycled and modified. If observations clearly suggest that the hypothesis is incorrect (that is, the hypothesis is *falsified*), then it must be dropped, and other alternative explanations of the facts must be considered.

In science, the validity of any explanation is determined by its coherence with observations in the natural world and its ability to predict further observations. Only after much testing and experimentation—usually done by many experimenters using a wide variety of repeatable tests—does a hypothesis gain validity where it can be advanced to the next step.

Theory

If a hypothesis has been strengthened by additional observations and if it is successful in predicting additional phenomena, then it can be advanced to what is called a **theory** (*theoria* = a looking at). A theory is a well-substantiated explanation of some aspect of the natural world that can incorporate facts, laws (descriptive generalizations about the behavior of an aspect of the natural world), logical inferences, and tested hypotheses. A theory is not a guess or a hunch. Rather, it is an understanding that develops from extensive observation, experimentation, and creative reflection.

In science, theories are formalized only after many years of testing and verifying predictions. Thus, scientific theories have been rigorously scrutinized to the point where most scientists agree that they are the best explanation of certain observable facts. Examples of prominent, well-accepted theories that are held with a very high degree of confidence include biology's theory of evolution (which is discussed later in this chapter) and geology's theory of plate tectonics (which is covered in the next chapter).

Theories and the Truth

We've seen how the scientific method is used to develop theories, but does science ever arrive at the undisputed "truth"? Science never reaches an absolute truth because we can never be certain that we have all the observations, especially considering that new technology will be available in the future to examine phenomena in different ways. New observations are always possible, so the nature of scientific truth is subject to change. Therefore, it is more accurate to say that science arrives at that which is *probably* true, based on the available observations.

It is not a downfall of science that scientific ideas are modified as more observations are collected. In fact, the opposite is true. Science is a process that depends on reexamining ideas as new observations are made. Thus, science progresses when new observations yield new hypotheses and modification of theories. As a result, science is littered with hypotheses that have been abandoned in favor of later explanations that fit new observations. One of the best known is the idea that Earth was at the center of the universe, a proposal that was supported by the apparent daily motion of the Sun, Moon, and stars around Earth.

The statements of science should never be accepted as the "final truth." Over time, however, they generally form a sequence of increasingly more accurate statements. Theories are the endpoints in science and do not turn into facts through accumulation of evidence. Nevertheless, the data can become so convincing that the accuracy of a theory is no longer questioned. For instance, the *heliocentric* (*helios* = sun, *centric* = center) *theory* of our solar system states that Earth revolves around the Sun rather than vice versa. Such concepts are supported by such abundant observational and experimental evidence that they are no longer questioned in science.

Is there really such a formal method to science as the scientific method suggests? Actually, the work of scientists is much less formal and is not always done in a clearly logical and systematic manner. Like detectives analyzing a crime scene, scientists use ingenuity and serendipity, visualize models, and sometimes follow hunches in order to unravel the mysteries of nature.

Finally, a key component of verifying scientific ideas is through the peer review process. Once scientists make a discovery, their aim is to get the word out to the scientific community about their results. This is typically done via a published paper, but a draft of the manuscript is first checked by other experts to see if the work has been conducted according to scientific standards and the conclusions are valid. Normally, corrections are suggested and the paper is revised before it is published. This process is a strength of the scientific community and helps weed out inaccurate or poorly formed ideas.

STUDENTS SOMETIMES ASK . . .

How can I accept a scientific idea if it's just a theory?

When most people use the word "theory" in everyday life, it usually means an idea or a guess (such as the all-too-common "conspiracy theory"), but the word has a much different meaning in science. In science, a theory is not a guess or a hunch. It's a well-substantiated, well-supported, well-documented explanation for observations about the natural world. It's a powerful tool that ties together all the facts about something, providing an explanation that fits all the observations and can even be used to make predictions. In science, theory is the ultimate goal; it's the well-proven explanation of how things work.

There is also a misconception that in science, once a theory is proven, it becomes a *law*. That's not how it works. In science, we collect facts, or observations, we use laws to describe them, and we use a theory to explain them. For example, the *law of gravity* is a description of the force; then there is the *theory of gravitational attraction*, which explains why the force occurs. Theories don't get "promoted" to a law by an abundance of proof, and so a theory never becomes a law. This doesn't mean that scientists are unsure of theories; in fact, theories are as close to proven as anything in science can be. So, don't discount a scientific idea because it's "just a theory."

ESSENTIAL CONCEPT 1.3

Science supports the explanation of the natural world that best explains all available observations. Because new observations can modify existing theories, science is always developing.

1.4 How Were Earth and the Solar System Formed?

Earth is the third of eight major planets[6] in our **solar system** that revolve around the Sun (**Figure 1.12**). Evidence suggests that the Sun and the rest of the solar system formed about 5 billion years ago from a huge cloud of gas and space dust called a **nebula** (*nebula* = a cloud). Astronomers base this hypothesis on the orderly nature of our solar system and the consistent age of meteorites (pieces of the early solar system). Using sophisticated telescopes, astronomers have also been able to observe distant nebula in various stages of formation (**Figure 1.13**). In addition, more than 300 planets have been discovered outside our solar system—including one that is about the size of Earth—by detecting the telltale wobble of distant stars.

The Nebular Hypothesis

According to the **nebular hypothesis** (**Figure 1.14**), all bodies in the solar system formed from an enormous cloud composed mostly of hydrogen and helium, with only a small percentage of heavier elements. As this huge accumulation of gas and dust revolved around its center, the Sun began to form as magnetic fields and turbulence worked with the force of gravity to concentrate particles. In its early stages, the diameter of the Sun may have equaled or exceeded the diameter of our entire planetary system today.

(a) Features and relative sizes of the Sun and the eight major planets of the solar system.

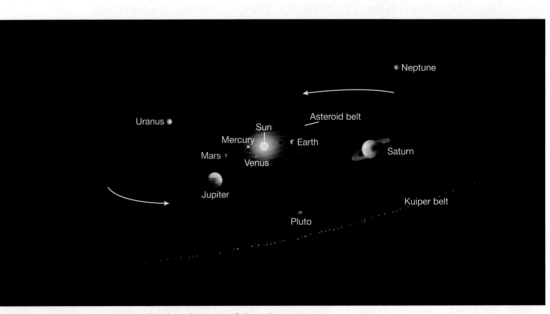

(b) Orbits and relative positions of various features of the solar system.

Figure 1.12 The solar system. Schematic views of the solar system, which includes the Sun and eight major planets.

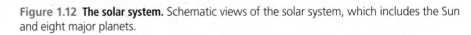

[6]Pluto, which used to be considered the ninth planet in our solar system, was reclassified by the International Astronomical Union as a "dwarf planet" in 2006, along with other similar bodies.

As the nebular matter that formed the Sun contracted, small amounts of it were left behind in eddies, which are similar to small whirlpools in a stream. The material in these eddies was the beginning of the **protoplanets** (*proto* = original, *planetes* = wanderer) and their orbiting satellites, which later consolidated into the present planets and their moons.

Protoearth

Protoearth looked very different from Earth today. Its size was larger than today's Earth, and there were neither oceans nor any life on the planet. In addition, the structure of the deep Protoearth is thought to have been *homogenous* (*homo* = alike, *genous* = producing), which means that it had a uniform composition throughout. The structure of Protoearth changed, however, when its heavier constituents migrated toward the center to form a heavy core.

During this early stage of formation, many meteorites from space bombarded Protoearth (**Figure 1.15**). In fact, a leading theory states that the Moon was born in the aftermath of a titanic collision between a Mars-size planet named *Theia* and Protoearth. While most of Theia was swallowed up and incorporated into the magma ocean it created on impact, the collision also flung a small world's worth of vaporized and molten rock into orbit. Over time, this debris coalesced into a sphere and created Earth's orbiting companion, the Moon.

During this early formation of the protoplanets and their satellites, the Sun condensed into such a hot, concentrated mass that forces within its interior began releasing energy through a process known as a **fusion** (*fusus* = melted) **reaction**. A fusion reaction occurs when temperatures reach tens of millions of degrees and hydrogen **atoms** (*a* = not, *tomos* = cut) combine to form helium atoms, releasing large amounts of energy.[7] Not only does the Sun emit light, it also emits *ionized* (electrically charged) particles that make up the *solar wind*. During the early stages

Figure 1.13 The Ghost Head Nebula. NASA's Hubble Space Telescope image of the Ghost Head Nebula (NGC 2080), which is a site of active star formation.

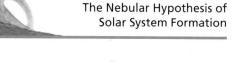
Web Animation
The Nebular Hypothesis of Solar System Formation

Figure 1.14 The nebular hypothesis of solar system formation. (a) A huge cloud of dust and gases (a nebula) contracts. **(b)** Most of the material is gravitationally swept toward the center, producing the Sun, while the remainder flattens into a disk. **(c)** Small eddies are created by the circular motion. **(d)** In time, most of the remaining debris forms the planets and their moons.

[7]Fusion in stars also combines higher elements to form even higher elements, such as carbon. As a result, all matter—even the matter that comprises our bodies—originated as stardust long ago.

Figure 1.15
Protoearth. An artist's conception of what Earth looked like early in its development.

of formation of the solar system, this solar wind blew away the nebular gas that remained from the formation of the planets and their satellites.

Meanwhile, the protoplanets closest to the Sun (including Earth) were heated so intensely by solar radiation that their initial atmospheres (mostly hydrogen and helium) boiled away. In addition, the bombardment of Earth by ionized solar radiation (which causes atoms to release particles) and the subsequent heat given off by protoplanets caused them to drastically shrink in size. As the protoplanets continued to contract, heat was produced deep within their cores from the spontaneous disintegration of atoms, called *radioactivity* (*radio* = ray, *acti* = to cause).

Density and Density Stratification

Density, which is an extremely important physical property of matter, is defined as mass per unit volume. In common terms, an easy way to think about density is that it is a measure of *how heavy something is for its size*. For instance, an object that has a low density is light for its size (like a dry sponge, foam packing, or a surfboard). Conversely, an object that has a high density is heavy for its size (like cement, most metals, or a large container full of water). Note that density has nothing to do with the *thickness* of an object; some objects (like a stack of foam packing) can be thick but have low density. In reality, density is related to molecular packing, with higher packing of molecules into a certain space resulting in higher density. Density is an extremely important concept that will be discussed in many other chapters in this book. For example, the density of Earth's layers dramatically affects their positions within Earth (Chapter 2), the density of air masses affects their properties (Chapter 6), and the density of water masses influences their position and movement (Chapter 7).

The release of internal heat was so intense that Earth's surface became molten. Once Earth became a ball of hot liquid rock, the elements were able to segregate according to their densities in a process called **density stratification** (*strati* = a layer, *fication* = making), which occurs because of *gravitational separation*. The highest-density materials (primarily iron and nickel) concentrated in the core, whereas progressively lower-density components (primarily rocky material) formed concentric spheres around the core. If you've ever noticed how oil-and-vinegar salad dressing settles out into a lower-density top layer (the oil) and a higher-density bottom layer (the vinegar), then you've seen how density stratification causes separate layers to form.

Earth's Internal Structure

As a result of density stratification, Earth became a layered sphere based on density, with the highest-density material found near the center of Earth and the lowest-density material located near the surface. Let's examine Earth's internal structure and the characteristics of its layers.

CHEMICAL COMPOSITION VERSUS PHYSICAL PROPERTIES The cross-sectional view of Earth in **Figure 1.16** shows that Earth's inner structure can be subdivided according to its chemical composition (the chemical makeup of Earth materials) or its physical properties (how the rocks respond to increased temperature and pressure at depth).

CHEMICAL COMPOSITION Based on chemical composition, Earth consists of three layers: the **crust**, the **mantle**, and the **core** (Figure 1.16). If Earth were reduced to the size of an apple, then the crust would be its thin skin. It extends from the surface to an average depth of about 30 kilometers (20 miles). The crust is composed of relatively low-density rock, consisting mostly of various *silicate minerals* (common rock-forming minerals with silicon and oxygen). There are two types of crust—oceanic and continental—which will be discussed in the next section.

Immediately below the crust is the mantle. It occupies the largest volume of the three layers and extends to a depth of about 2885 kilometers (1800 miles). The mantle is composed of relatively high-density iron and magnesium silicate rock.

Beneath the mantle is the core. It forms a large mass from 2885 kilometers (1800 miles) to the center of Earth at 6371 kilometers (3960 miles). The core is composed of even higher-density metal (mostly iron and nickel).

PHYSICAL PROPERTIES Based on physical properties, Earth is composed of five layers (Figure 1.16): the **inner core**, the **outer core**, the **mesosphere** (*mesos* = middle, *sphere* = ball), the **asthenosphere** (*asthenos* = weak, *sphere* = ball), and the **lithosphere** (*lithos* = rock, *sphere* = ball).

The lithosphere is Earth's cool, rigid, outermost layer. It extends from the surface to an average depth of about 100 kilometers (62 miles) and includes the crust plus the topmost portion of the mantle. The lithosphere is *brittle* (*brytten* = to shatter), meaning that it will fracture when force is applied to it. As will be discussed in Chapter 2, "Plate Tectonics and the Ocean Floor," the plates involved in plate tectonic motion are the plates of the lithosphere.

Beneath the lithosphere is the asthenosphere. The asthenosphere is *plastic* (*plasticus* = molded), meaning that it will flow when a gradual force is applied to it. It extends from about

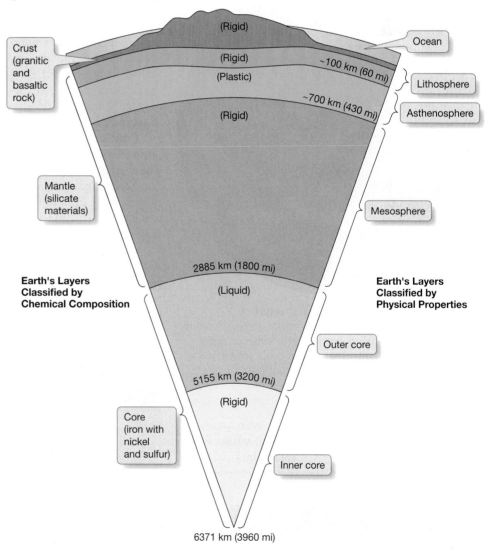

Figure 1.16 Comparison of Earth's chemical composition and physical properties. A cross-sectional view of Earth, showing Earth's layers classified by chemical composition along the left side of the diagram. For comparison, Earth's layers classified by physical properties are shown along the right side of the diagram.

STUDENTS SOMETIMES ASK . . .

How do we know about the internal structure of Earth?

You might suspect that the internal structure of Earth has been sampled directly. However, humans have never penetrated beneath the crust! The internal structure of Earth is determined by using indirect observations. Every time there is an earthquake, waves of energy (called *seismic waves*) penetrate Earth's interior. Seismic waves change their speed and are bent and reflected as they move through zones having different properties. An extensive network of monitoring stations around the world detects and records this energy. The data are analyzed and used to determine the structure of Earth's interior (**Figure 1.17**).

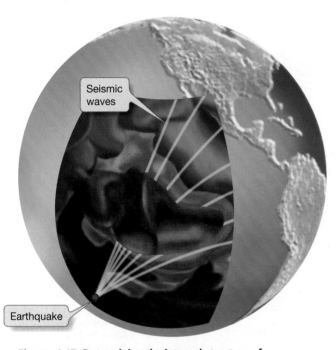

Figure 1.17 Determining the internal structure of Earth. When an earthquake occurs (*red dot*, below), it sends seismic waves (*yellow lines*) through Earth's interior, which is shown diagrammatically. Detection of these seismic waves around the globe reveals information about the structure, composition, and properties of the deep Earth.

Web Animation
How Seismic Waves Reveal
Earth's Internal Layers

ESSENTIAL CONCEPT 1.4

Earth has differences in composition and physical properties that create layers such as the brittle lithosphere and the plastic asthenosphere, which is capable of flowing slowly over time.

100 kilometers (62 miles) to 700 kilometers (430 miles) below the surface, which is the base of the upper mantle. At these depths, it is hot enough to partially melt portions of most rocks.

Beneath the asthenosphere is the mesosphere. The mesosphere extends to a depth of about 2885 kilometers (1800 miles), which corresponds to the middle and lower mantle. Although the asthenosphere deforms plastically, the mesosphere is rigid, most likely due to the increased pressure at these depths.

Beneath the mesosphere is the core. The core consists of the outer core, which is liquid and capable of flowing, and the inner core, which is rigid and does not flow. Again, the increased pressure at the center of Earth keeps the inner core from flowing.

NEAR THE SURFACE The top portion of Figure 1.18 shows an enlargement of Earth's layers closest to the surface.

Lithosphere The lithosphere is a relatively cool, rigid shell that includes all the crust and the topmost part of the mantle. In essence, the topmost part of the mantle is attached to the crust and the two act as a single unit, approximately 100 kilometers (62 miles) thick. The expanded view in Figure 1.18 shows that the crust portion of the lithosphere is further subdivided into oceanic crust and continental crust, which are compared in Table 1.1.

Oceanic versus Continental Crust **Oceanic crust** underlies the ocean basins and is composed of the igneous rock **basalt**, which is dark colored and has a relatively high density of about 3.0 grams per cubic centimeter.[8] The average thickness of the oceanic crust is only about 8 kilometers (5 miles). Basalt originates as molten magma beneath Earth's crust (typically from the mantle), some of which comes to the surface during underwater sea floor eruptions.

Continental crust is composed mostly of the lower-density and lighter-colored igneous rock **granite**.[9] It has a density of about 2.7 grams per cubic centimeter. The average thickness of the continental crust is about 35 kilometers (22 miles) but may reach a maximum of 60 kilometers (37 miles) beneath the highest mountain ranges. Most granite originates beneath the surface as molten magma that cools and hardens within Earth's crust. No matter which type of crust is at the surface, it is all part of the lithosphere.

Asthenosphere The asthenosphere is a relatively hot, plastic region beneath the lithosphere. It extends from the base of the lithosphere to a depth of about 700 kilometers (430 miles) and is entirely contained within the upper mantle. The asthenosphere can deform without fracturing if a force is applied slowly. This means that it has the ability to flow but has high **viscosity** (*viscosus* = sticky). Viscosity is a measure of a

TABLE 1.1 COMPARING OCEANIC AND CONTINENTAL CRUST

	Oceanic crust	Continental crust
Main rock type	Basalt (dark-colored igneous rock)	Granite (light-colored igneous rock)
Density (grams per cubic centimeter)	3.0	2.7
Average thickness	8 kilometers (5 miles)	35 kilometers (22 miles)

[8]Water has a density of 1.0 grams per cubic centimeter. Thus, basalt with a density of 3.0 grams per cubic centimeter is three times denser than water.

[9]At the surface, continental crust is often covered by a relatively thin layer of surface sediments. Below these, granite can be found.

substance's resistance to flow.[10] Studies indicate that the high-viscosity asthenosphere is flowing slowly through time; this has important implications for the movement of lithospheric plates.

ISOSTATIC ADJUSTMENT Isostatic (*iso* = equal, *stasis* = standing) **adjustment**—the vertical movement of crust—is the result of the buoyancy of Earth's lithosphere as it floats on the denser, plasticlike asthenosphere below. **Figure 1.19**, which shows a container ship floating in water, provides an example of isostatic adjustment. It shows that an empty ship floats high in the water. Once the ship is loaded with cargo, though, the ship undergoes isostatic adjustment and floats lower in the water (but hopefully won't sink!). When the cargo is unloaded, the ship isostatically adjusts itself and floats higher again.

Similarly, both continental and oceanic crust float on the denser mantle beneath. Oceanic crust is denser than continental crust, however, so oceanic crust floats lower in the mantle because of isostatic adjustment. Oceanic crust is also thin, which creates low areas for the oceans to occupy. Areas where the continental crust is thickest (such as large mountain ranges on the continents) float higher than continental crust of normal thickness, also because of isostatic adjustment. These mountains are similar to the top of a floating iceberg—they float high because there is a very thick mass of crustal material beneath them, plunged deeper into the asthenosphere. Thus, tall mountain ranges on Earth are composed of a great thickness of crustal material that in essence keeps them buoyed up.

Areas that are exposed to an increased or decreased load experience isostatic adjustment. For instance, during the most recent ice age (which occurred during the Pleistocene Epoch between about 1.8 million and 10,000 years ago), massive ice sheets alternately covered and exposed northern regions such as Scandinavia and northern Canada. The additional weight of ice several kilometers thick caused these areas to isostatically adjust themselves lower in the mantle. Since the end of the most recent ice age, the reduced load on these areas caused by the melting of ice caused these areas to rise and experience **isostatic rebound**, which continues today. The rate at which isostatic rebound occurs gives scientists important information about the properties of the upper mantle.

Further, isostatic adjustment provides additional evidence for the movement of Earth's tectonic plates. Because continents isostatically adjust themselves by moving *vertically*, they must not be firmly fixed in one position on Earth. As a result, the plates that contain these continents should certainly be able to move *horizontally* across Earth's surface. This remarkable idea will be explored in more detail in the next chapter.

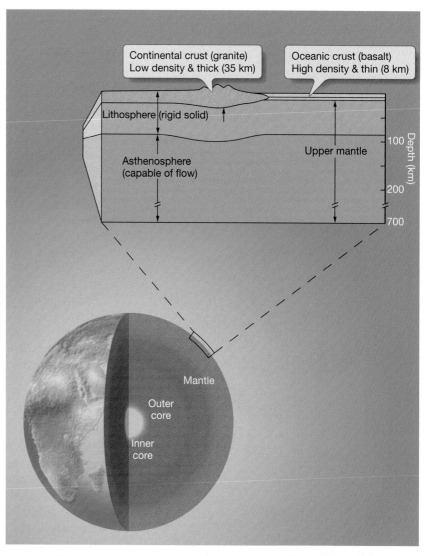

Figure **1.18** **Internal structure of Earth.** Enlargement (*top*) shows that the rigid lithosphere includes the crust (either continental or oceanic) plus the topmost part of the mantle to a depth of about 100 kilometers (60 miles). Beneath the lithosphere, the plastic asthenosphere extends to a depth of 700 kilometers (430 miles).

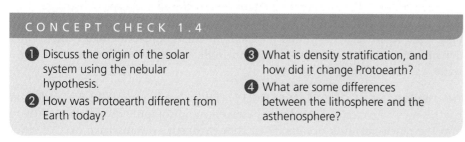

CONCEPT CHECK 1.4

1 Discuss the origin of the solar system using the nebular hypothesis.

2 How was Protoearth different from Earth today?

3 What is density stratification, and how did it change Protoearth?

4 What are some differences between the lithosphere and the asthenosphere?

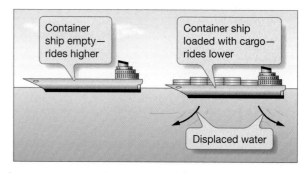

Figure **1.19** **A container ship experiences isostatic adjustment.** A ship will ride higher in water when it is empty and will ride lower in water when it is loaded with cargo, illustrating the principle of isostatic adjustment.

---

[10]Substances that have high viscosity (a high resistance to flow) include toothpaste, honey, tar, and Silly Putty; a common substance that has low viscosity is water. Note that a substance's viscosity often changes with temperature. For instance, as honey is heated, it flows more easily.

H₂O vapor and other gases

(a)

H₂O vapor and other gases

(b)

(c)

Figure 1.20 Formation of Earth's oceans. Early in Earth's history, widespread volcanic activity released large amounts of water vapor (H_2O vapor) and smaller quantities of other gases. As Earth cooled, the water vapor **(a)** condensed into clouds and **(b)** fell to Earth's surface, where it accumulated to form the oceans **(c)**.

ESSENTIAL CONCEPT 1.5

Originally, Earth had no oceans. The oceans (and atmosphere) came from inside Earth as a result of outgassing and were present by at least 4 billion years ago.

1.5 How Were Earth's Atmosphere and Oceans Formed?

The formation of Earth's atmosphere is related to the formation of the oceans; both are a direct result of density stratification.

Origin Of Earth's Atmosphere

Where did the atmosphere come from? As previously mentioned, Earth's initial atmosphere consisted of leftover gases from the nebula, but those particles were blown out to space by the Sun's solar wind. After that, a second atmosphere was most likely expelled from inside Earth by a process called **outgassing**. During the period of density stratification, the lowest-density material contained within Earth was composed of various gases. These gases rose to the surface and were expelled to form Earth's early atmosphere.

What was the composition of these atmospheric gases? They are believed to have been similar to the gases emitted from volcanoes, geysers, and hot springs today: mostly water vapor (steam), with small amounts of carbon dioxide, hydrogen, and other gases. The composition of this early atmosphere was not, however, the same composition as today's atmosphere. The composition of the atmosphere changed over time because of the influence of life (as will be discussed shortly) and possibly because of changes in the mixing of material in the mantle.

Origin Of Earth's Oceans

Where did the oceans come from? Similarly, their origin is linked directly to the origin of the atmosphere. Because outgassing releases mostly water vapor, this was the primary source of water on Earth, including supplying the oceans with water. **Figure 1.20** shows that as Earth cooled, the water vapor released to the atmosphere during outgassing condensed, fell to Earth, and accumulated in low areas. Evidence suggests that by at least 4 billion years ago, most of the water vapor from outgassing had accumulated to form the first permanent oceans on Earth.

Recent research, however, suggests that not all water came from inside Earth. Comets, being about half water, were once widely held to be the source of Earth's oceans. During Earth's early development, space debris left over from the origin of the solar system bombarded the young planet, and there could have been plenty of water supplied to Earth. However, spectral analyses of the chemical composition of three comets—Halley, Hyakutake, and Hale-Bopp—during near-Earth passes they made in 1986, 1996, and 1997, respectively, revealed a crucial chemical difference between the hydrogen in comet ice and that in Earth's water. If similar comets supplied large quantities of water to Earth, much of Earth's water would still exhibit the telltale type of hydrogen identified in these comets. Recently, however, comets that originated from the outer solar system's Kuiper belt have been analyzed, and their water in the form of ice contains nearly the correct type of hydrogen that is found in Earth's water. It now appears that these comets could have contributed water to an early Earth and points to an emerging picture of a complex and dynamic evolution of the early solar system. Still, it seems likely that most of Earth's water was derived from outgassing.

THE DEVELOPMENT OF OCEAN SALINITY The relentless rainfall that landed on Earth's rocky surface dissolved many elements and compounds and carried them into the newly forming oceans. Even though Earth's oceans have existed since early in the formation of the planet, its chemical composition must have changed. This is because the high carbon dioxide and sulfur dioxide content in the early atmosphere would have created a very acidic rain, capable of dissolving greater amounts of minerals in the crust than occurs today. In addition, volcanic gases such as chlorine became dissolved in the atmosphere. As rain fell and washed to the ocean, it carried some of these dissolved compounds, which accumulated in the newly forming oceans.[11] Eventually, a balance between inputs and outputs was reached, producing an ocean with a chemical composition similar to today's oceans. Further aspects of the oceans' salinity are explored in Chapter 5, "Water and Seawater."

CONCEPT CHECK 1.5

❶ Describe the origin of Earth's oceans.

❷ Describe the origin of Earth's atmosphere. How is its origin related to the origin of Earth's oceans?

❸ Have the oceans always been salty? Why or why not?

1.6 Did Life Begin in the Oceans?

The fundamental question of how life began on Earth has puzzled humankind since ancient times and has recently received a great amount of scientific study. The evidence required to understand our planet's prebiotic environment and the events that led to first living systems is scant and difficult to decipher. Still, the inventory of current views on life's origin reveals a broad assortment of opposing positions. One recent hypothesis is that the organic building blocks of life may have arrived embedded in meteors, comets, or cosmic dust. Alternatively, life may have originated around hydrothermal vents—hot springs—on the deep-ocean floor. Yet another idea is that life originated in certain minerals that acted as chemical catalysts within rocks deep below Earth's surface.

According to the fossil record on Earth, the earliest-known life-forms were primitive bacteria that lived in sea floor rocks about 3.5 billion years ago. Unfortunately, Earth's geologic record for these early times is so sparse and the rocks are so deformed by Earth processes that the rocks no longer reveal life's precursor molecules. In addition, there is no direct evidence of Earth's environmental conditions (such as its temperature, ocean acidity, or the exact composition of the atmosphere) at the time of life's origin. Still, it is clear that the basic building blocks for the development of life were available from materials already present on the early Earth. And the oceans were the most likely place for these materials to interact and produce life.

The Importance Of Oxygen to Life

Oxygen, which comprises almost 21% of Earth's present atmosphere, is essential to human life for two reasons. First, our bodies need oxygen to "burn" (*oxidize*) food, releasing energy to our cells. Second, oxygen in the upper atmosphere in the form of *ozone* (*ozone* = to smell[12]) protects the surface of Earth from most of the Sun's harmful ultraviolet radiation (which is why the atmospheric ozone hole over Antarctica has generated such concern).

Evidence suggests that Earth's early atmosphere (the product of outgassing) was different from Earth's initial hydrogen–helium atmosphere and different from the mostly nitrogen–oxygen atmosphere of today. The early atmosphere probably contained large percentages of water vapor and carbon dioxide and smaller percentages

---

[11]Note that some of these dissolved components were removed or modified by chemical reactions between ocean water and rocks on the sea floor.

[12]Ozone gets its name because of its pungent, irritating odor.

of hydrogen, methane, and ammonia but very little free oxygen (oxygen that is not chemically bound to other atoms). Why was there so little free oxygen in the early atmosphere? Oxygen may well have been outgassed, but oxygen and iron have a strong affinity for each other.[13] As a result, iron in Earth's early crust would have reacted with the outgassed oxygen immediately, removing it from the atmosphere.

Without oxygen in Earth's early atmosphere, moreover, there would have been no ozone layer to block most of the Sun's ultraviolet radiation. The lack of a protective ozone layer may, in fact, have played a crucial role in several of life's most important developmental milestones.

Stanley Miller's Experiment

In 1952, **Stanley Miller** (Figure 1.21b)—then a 22-year-old graduate student of chemist Harold Urey at the University of Chicago—conducted a laboratory experiment that had profound implications about the development of life on Earth. In Miller's experiment, he exposed a mixture of carbon dioxide, methane, ammonia, hydrogen, and water (the components of the early atmosphere and ocean) to ultraviolet light (from the Sun) and an electrical spark (to imitate lightning) (Figure 1.21a). By the end of the first day, the mixture turned pink and after a week it was a deep, muddy brown, indicating the formation of a large assortment of organic molecules, including amino acids—which are the basic components of life—and other biologically significant compounds.

Miller's now-famous laboratory experiment of a simulated primitive Earth in a bottle—which has been duplicated and confirmed numerous times since—demonstrated that vast amounts of organic molecules could have been produced in Earth's early oceans, often called a "prebiotic soup." This prebiotic soup, perhaps spiced by extraterrestrial molecules aboard comets, meteorites, or interplanetary dust, was fueled by raw materials from volcanoes, certain minerals in sea floor rocks, and undersea hydrothermal vents. On early Earth, the mixture was energized by lightning, cosmic rays, and the planet's own internal heat, and it is thought to have created life's precursor molecules about 4 billion years ago.

Figure 1.21 Creation of organic molecules. Schematic view of the apparatus **(a)** that Stanley Miller **(b)** constructed to simulate early Earth conditions, which created organic molecules.

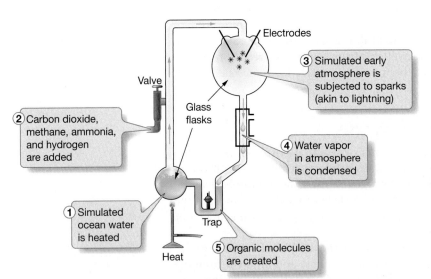

(a) Laboratory apparatus used by Stanley Miller to simulate the conditions of the early atmosphere and the oceans. The experiment produced various organic molecules and suggests that the basic components of life were created in a "prebiotic soup" in the oceans.

(b) Stanley Miller in 1999, with his famous apparatus in the foreground.

[13]As an example of the strong affinity of iron and oxygen, consider how common rust—a compound of iron and oxygen—is on Earth's surface.

Exactly how these simple organic compounds in the prebiotic soup assembled themselves into more complex molecules—such as proteins and DNA—and then into the first living entities remains one of the most tantalizing questions in science. Recent research suggests that with the vast array of organic compounds available in the prebiotic soup, several kinds of chemical reactions led to increasingly elaborate molecular structures. In fact, research suggests that small, simple molecules could have acted as templates, or "molecular midwifes," in helping the building blocks of life's genetic material form long chains and thus may have assisted in the formation of longer, more elaborate molecular complexes. Among these complexes, some began to carry out functions associated with the basic molecules of life. As the products of one generation became the building blocks for another, even more complex molecules, or polymers, emerged over many generations that could store and transfer information. Such genetic polymers ultimately became encapsulated within cell-like membranes that were also present in Earth's primitive broth. The resulting cell-like complexes thereby housed self-replicating molecules capable of multiplying—and hence evolving—genetic information. Many specialists consider this emergence of genetic replication to be the true origin of life.

> **ESSENTIAL CONCEPT 1.6A**
>
> Organic molecules were produced in a simulation of Earth's early atmosphere and ocean, suggesting that life most likely originated in the oceans.

Evolution and Natural Selection

Every living organism that inhabits Earth today is the result of **evolution** by the process of **natural selection** that has been occurring since life first existed on Earth. The theory of evolution states that groups of organisms adapt and change with the passage of time, causing descendants to differ morphologically and physiologically from their ancestors (**Diving Deeper 1.2**). Certain advantageous traits are naturally selected and passed from one generation to the next. Evolution is the process by which various **species** (*species* = a kind) have been able to inhabit increasingly numerous environments on Earth.

As we shall see, when species adapt to Earth's various environments, they can also modify the environments in which they live. This modification can be localized or nearly global in scale. For example, when plants emerged from the oceans and inhabited the land, they changed Earth from a harsh and bleak landscape as barren as that of the Moon to one that is green and lush.

Plants and Animals Evolve

The very earliest forms of life were probably **heterotrophs** (*hetero* = different, *tropho* = nourishment). Heterotrophs require an external food supply, which was abundantly available in the form of nonliving organic matter in the ocean around them. **Autotrophs** (*auto* = self, *tropho* = nourishment), which can manufacture their own food supply, evolved later. The first autotrophs were probably similar to present-day **anaerobic** (*an* = without, *aero* = air) bacteria, which live without atmospheric oxygen. They may have been able to derive energy from inorganic compounds at deep-water hydrothermal vents using a process called **chemosynthesis** (*chemo* = chemistry, *syn* = with, *thesis* = an arranging).[14] In fact, the recent detection of microbes deep within the ocean crust as well as the discovery of 3.2-billion-year-old microfossils of bacteria from deep-water marine rocks support the idea of life's origin on the deep-ocean floor in the absence of light.

PHOTOSYNTHESIS AND RESPIRATION Eventually, more complex single-celled autotrophs evolved. They developed a green pigment called **chlorophyll** (*chloro* = green, *phyll* = leaf), which captures the Sun's energy through cellular **photosynthesis** (*photo* = light, *syn* = with, *thesis* = an arranging). In photosynthesis (**Figure 1.22**, *top*), plant cells capture light energy and store it as sugars. In cellular **respiration** (*respirare* = to breathe) (Figure 1.22, *middle*), sugars are oxidized with oxygen,

---

[14]More details about chemosynthesis are discussed in Chapter 15, "Animals of the Benthic Environment."

Figure 1.22 Photosynthesis (*top*), respiration (*middle*), and representative reactions viewed chemically (*bottom*). The process of photosynthesis, which is accomplished by plants, is represented in the upper panel. The second panel shows respiration, which is done by animals. Both processes are shown chemically in the third panel.

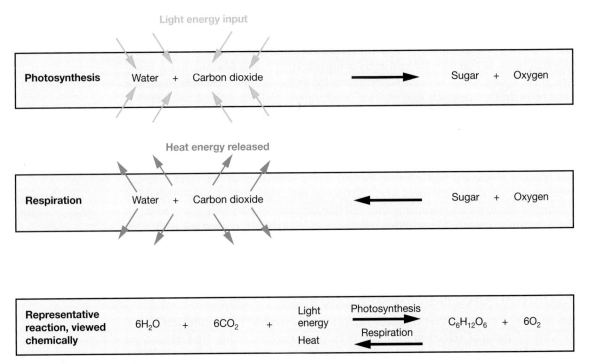

releasing stored energy that is used as a source of energy by the organism that consumes the plant to carry on various life processes.

Not only are photosynthesis and respiration chemically opposite processes, they are also complementary because the products of photosynthesis (sugars and oxygen) are used during respiration and the products of respiration (water and carbon dioxide) are used in photosynthesis (Figure 1.22, *bottom*). Thus, autotrophs (algae and plants) and heterotrophs (most bacteria and animals) have developed a mutual need for each other.

The oldest fossilized remains of organisms are primitive photosynthetic bacteria recovered from rocks formed on the sea floor about 3.5 billion years ago. However, the oldest rocks containing iron oxide (rust)—an indicator of an oxygen-rich atmosphere—did not appear until about 2.4 billion years ago. Thus, photosynthetic organisms needed about a billion years to develop and begin producing abundant free oxygen in the atmosphere. At the same time, when a large amount of oxygen-rich (ferric) iron sank to the base of the mantle, it may have been heated by the core, risen as a plume to the ocean floor, and begun releasing large amounts of oxygen through outgassing about 2.5 billion years ago.

THE GREAT OXIDATION EVENT/OXYGEN CRISIS Based on the chemical makeup of certain rocks, Earth's atmosphere became oxygen rich about 2.45 billion years ago—called the *great oxidation event*—and fundamentally changed Earth's ability to support life. Particularly for anaerobic bacteria, which had grown successfully in an oxygen-free world, all this oxygen was nothing short of a catastrophe! This is because the increased atmospheric oxygen caused the ozone concentration in the upper atmosphere to build up, thereby shielding Earth's surface from ultraviolet radiation—and effectively eliminating anaerobic bacteria's food supply of organic molecules. (Recall that Stanley Miller's experiment created organic molecules but needed ultraviolet light.) In addition, oxygen (particularly in the presence of light) is highly reactive with organic matter. When anaerobic bacteria are exposed to oxygen and light, they are killed instantaneously. By 1.8 billion years ago, the atmosphere's oxygen content had increased to such a high level that it began causing the extinction of many anaerobic organisms. Nonetheless, descendants of such bacteria survive on Earth today in isolated microenvironments that are dark and free of oxygen, such as deep in soil or rocks, in landfills, and inside other organisms.

ESSENTIAL CONCEPT 1.6B

Life on Earth has evolved over time and changed Earth's environment. For example, abundant photosynthetic organisms created today's oxygen-rich atmosphere.

Although oxygen is very reactive with organic matter and can even be toxic, it also yields nearly 20 times more energy than anaerobic respiration—a fact that some organisms exploited. For example, blue-green algae, which are also known as *cyanobacteria* (*kuanos* = dark blue), adapted to and thrived in this new oxygen-rich environment. In doing so, they altered the composition of the atmosphere.

CHANGES TO EARTH'S ATMOSPHERE Remarkably, the development and successful evolution of photosynthetic organisms are greatly responsible for the world as we know it today (**Figure 1.23**). By the trillions, these microscopic organisms transformed the planet by capturing the energy of the Sun to make food and releasing oxygen as a waste product. By this process, these organisms reduced the high amount of carbon dioxide in the early atmosphere and gradually replaced it with free oxygen. This created a third and final atmosphere on Earth: one that is oxygen rich (about 21% today). Little by little, these tiny organisms turned the atmosphere into breathable air, opening the way to the diversity of life that followed.

The graph in **Figure 1.24** shows how the concentration of atmospheric oxygen has varied during the past 600 million years. When atmospheric oxygen concentrations are high, organisms thrive, and rapid speciation occurs. At such times in the past, insects grew to gargantuan proportions, reptiles took to the air, and the forerunners of mammals developed a warm-blooded metabolism. More oxygen was dissolved in the oceans, too, and so marine biodiversity increased. At other times when atmospheric oxygen concentrations fell precipitously, biodiversity was smothered. In fact, some of the planet's worst mass extinctions are associated with sudden drops in atmospheric oxygen.

The remains of ancient plants and animals buried in oxygen-free environments have become the oil, natural gas, and coal deposits of today. These deposits, which are called *fossil fuels*, provide more than 90% of the energy humans consume to power modern society. In essence, humans depend not only on the food energy stored in today's plants but also on the energy stored in plants during the geologic past—in the form of fossil fuels.

Because of increased burning of fossil fuels for home heating, industry, power generation, and transportation during the industrial age, the atmospheric concentration of carbon dioxide and other gases that help warm the atmosphere has increased, too. Scientists predict that these human emissions will increase global warming and cause serious environmental problems in the not-too-distant future. This phenomenon is referred to as the atmosphere's *enhanced greenhouse effect* and is discussed in Chapter 16, "The Oceans and Climate Change."

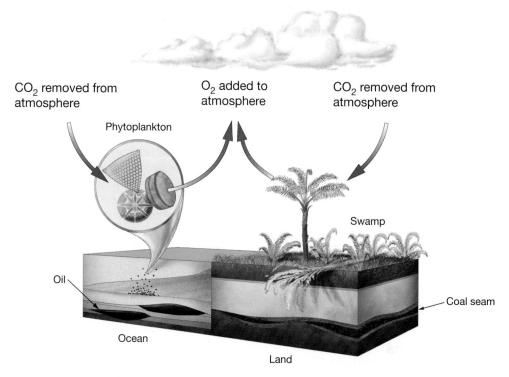

CO₂ removed from atmosphere — O₂ added to atmosphere — CO₂ removed from atmosphere

Phytoplankton

Swamp

Oil — Coal seam

Ocean — Land

Figure 1.23 The effect of plants on Earth's environment. As microscopic photosynthetic cells (*inset*) became established in the ocean, Earth's atmosphere was enriched in oxygen and depleted of carbon dioxide. As organisms died and accumulated on the ocean floor, some of their remains were converted to oil and gas. The same process occurred on land, sometimes producing coal.

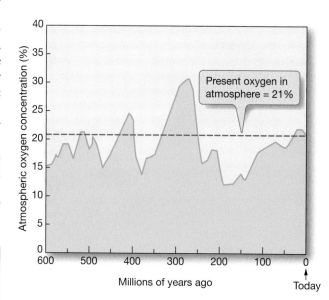

Figure 1.24 Atmospheric oxygen concentration. Graph showing how the concentration of oxygen in the atmosphere has varied during the past 600 million years. Today's oxygen level is about 21%. Low oxygen levels are closely associated with major extinction events, whereas high oxygen levels are associated with rapid speciation, including species gigantism.

CONCEPT CHECK 1.6

1 How does the presence of oxygen in our atmosphere help reduce the amount of ultraviolet radiation that reaches Earth's surface?

2 What was Stanley Miller's experiment, and what did it help demonstrate?

3 Earth has had three atmospheres (initial, early, and present). Describe the composition and origin of each one.

THE VOYAGE OF HMS *BEAGLE*: HOW IT SHAPED CHARLES DARWIN'S THINKING ABOUT THE THEORY OF EVOLUTION

"Nothing in biology makes sense except in the light of evolution."

—*Geneticist Theodosius Dobzhansky (1973)*

To help explain how biologic processes operating in nature were responsible for producing the many diverse and remarkable species on Earth, the English naturalist **Charles Darwin** (1809–1882) proposed the *theory of evolution* by natural selection, which he referred to as "common descent with modification." Many of the observations upon which he based the theory were made aboard the vessel HMS *Beagle* during its famous expedition from 1831 to 1836 that circumnavigated the globe (**Figure 1D**).

Darwin became interested in natural history during his student days at Cambridge University, where he was studying to become a minister. Because of the influence of John Henslow, a professor of botany, he was selected to serve as an unpaid naturalist on the HMS *Beagle*. The *Beagle* sailed from Devonport, England, on December 27, 1831, under the command of Captain Robert Fitzroy. The major objective of the voyage was to complete a survey of the coast of Patagonia (Argentina) and Tierra del Fuego and to make chronometric measurements. The voyage allowed the 22-year-old Darwin—who was often seasick—to disembark at various locations and study local plants and animals. What particularly influenced his thinking about evolution were the discovery of fossils in South America, the different tortoises throughout the Galápagos Islands, and the identification of 14 closely related species of Galápagos finches. These finches differ greatly in the configuration of their beaks (Figure 1D, *left inset*), which are suited to their diverse feeding habitats. After his return to England, Darwin noted the adaptations of finches and other organisms living in different habitats and concluded that all organisms change slowly over time as products of their environment.

Darwin recognized the similarities between birds and mammals and reasoned that they must have evolved from reptiles. Patiently making observations over many years, he also noted the similar skeletal framework of species such as bats, horses, giraffes, elephants, porpoises, and humans, which led him to establish relationships between various groups. Darwin suggested that the differences between species were the result of adaptation over time to different environments and modes of existence.

In 1858, Darwin hastily published a summary of his ideas about natural selection because fellow naturalist *Alfred Russel Wallace*, working half a world away cataloguing species in what is now Indonesia, had independently discovered the same idea. A year later, Darwin published his remarkable masterwork *On the Origin of Species by Means of Natural Selection* (Figure 1D, *right inset*), in which he provided extensive and compelling evidence that all living beings—including humans—have evolved from a common ancestor. At the time, Darwin's ideas were highly controversial because they stood in stark conflict with what most people believed about the origin of humans. Darwin also produced important publications on subjects as diverse as barnacle biology, carnivorous plants, and the formation of coral reefs.

Over 150 years later, Darwin's theory of evolution is so well established by evidence and reproducible experiment that it is considered a landmark influence in the scientific understanding of the underlying biologic processes operating in nature. Discoveries made since Darwin's time—including genetics and the structure of DNA—confirm how the process of evolution works. In fact, most of Darwin's ideas have been so thoroughly accepted by scientists that they are now the underpinnings of the modern study of biology. That's why the name *Darwin* is synonymous with evolution. In 2009, to commemorate Darwin's birth and his accomplishments, the Church of England even issued this formal apology to Darwin: *"The Church of England owes you an apology for misunderstanding you and, by getting our first reaction wrong, encouraging others to misunderstand you still."*

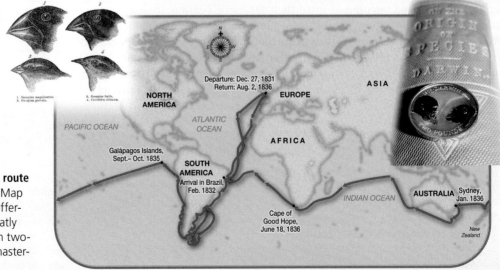

Figure 1D Charles Darwin: Galápagos finches, route of the HMS *Beagle*, and the *Origin of Species*. Map showing the route of the HMS *Beagle*, beak differences in Galápagos finches (*left inset*) that greatly influenced Charles Darwin, and the new British two-pound coin commemorating Darwin and his masterwork *On the Origin of Species* (*right inset*).

1.7 How Old Is Earth?

How can Earth scientists tell how old a rock is? It can be a difficult task to tell if a rock is thousands, millions, or even billions of years old—unless the rock contains telltale fossils. Fortunately, Earth scientists can determine how old most rocks are by using the radioactive materials contained within rocks. In essence, this technique involves reading a rock's internal "rock clock."

Radiometric Age Dating

Most rocks on Earth (as well as those from outer space) contain small amounts of radioactive materials such as uranium, thorium, and potassium. These radioactive materials spontaneously break apart or decay into atoms of other elements. Radioactive materials have a characteristic **half-life**, which is the time required for one-half of the atoms in a sample to decay to other atoms. The older the rock is, the more radioactive material will have been converted to decay product. Analytical instruments can accurately measure the amount of radioactive material and the amount of resulting decay product in rocks. By comparing these two quantities, the age of the rock can thus be determined. Such dating is referred to as **radiometric** (*radio* = radioactivity, *metri* = measure) **age dating** and is an extremely powerful tool for determining the age of rocks.

    Figure 1.25 shows an example of how radiometric age dating works. It shows how uranium 235 decays into lead 207 at a rate where one-half of the atoms turn into lead every 704 million years. By counting the number of each type of atom in a rock sample, one can tell how long it has been decaying (as long as the sample does not gain or lose atoms). Using uranium and other radioactive elements and applying this same technique, hundreds of thousands of rock samples have been age dated from around the world.

The Geologic Time Scale

The ages of rocks on Earth are shown in the **geologic time scale** (**Figure 1.26**; see **Web Diving Deeper 1.2**), which lists the names of the geologic time periods as well as important advances in the development of life-forms on Earth. Initially, the divisions between geologic periods were based on major extinction episodes as recorded in the fossil record. As radiometric age dates became available, they were also included on the geologic time scale. The oldest-known rocks on Earth, for example, are about 4.3 billion years old, and the oldest-known crystals within rocks have been dated at up to 4.4 billion years old.[15] In all, the time scale indicates that Earth is 4.6 billion years old, but few rocks survived its molten youth, a time when Earth was being bombarded by meteorites.

ESSENTIAL CONCEPT 1.7

Earth scientists can accurately determine the age of most rocks by analyzing their radioactive components, some of which indicate that Earth is 4.6 billion years old.

Web Animation
Radioactive Decay

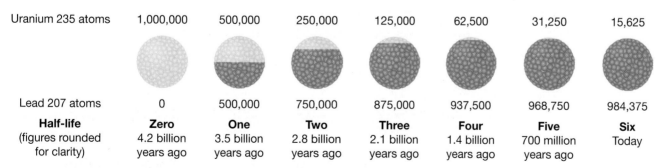

Uranium 235 atoms	1,000,000	500,000	250,000	125,000	62,500	31,250	15,625
Lead 207 atoms	0	500,000	750,000	875,000	937,500	968,750	984,375
Half-life (figures rounded for clarity)	**Zero** 4.2 billion years ago	**One** 3.5 billion years ago	**Two** 2.8 billion years ago	**Three** 2.1 billion years ago	**Four** 1.4 billion years ago	**Five** 700 million years ago	**Six** Today

Figure 1.25 Radiometric age dating. During one half-life, half of all radioactive uranium 235 atoms decay into lead 207. With each successive half-life, half of the remaining radioactive uranium atoms convert to lead. By counting the number of each type of atom in a rock sample, the rock's age can be determined.

---

[15]Recent research suggests that crystals this old imply that significant continental crust must have formed on Earth early on, perhaps by nearly 4.5 billion years ago.

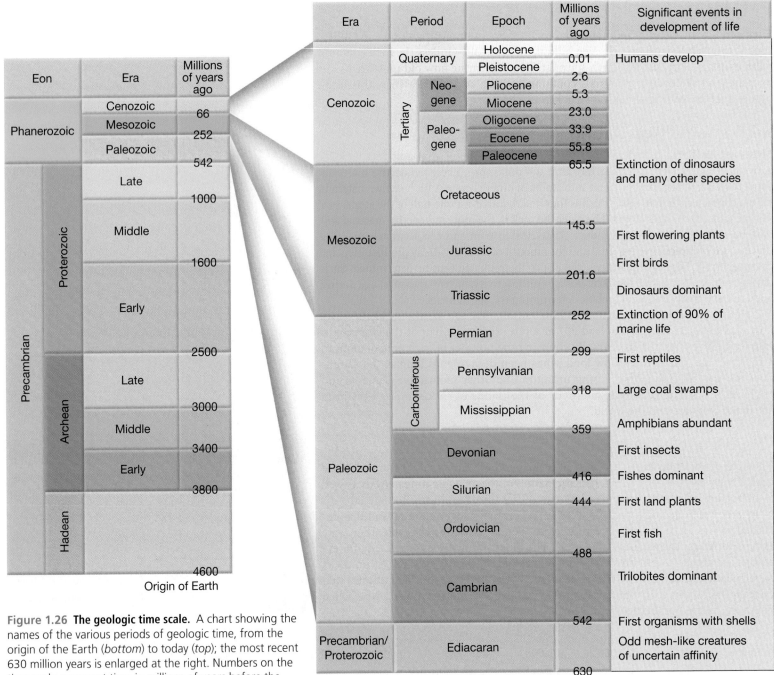

Figure 1.26 The geologic time scale. A chart showing the names of the various periods of geologic time, from the origin of the Earth (*bottom*) to today (*top*); the most recent 630 million years is enlarged at the right. Numbers on the time scale represent time in millions of years before the present; significant advances in the development of plants and animals on Earth are also shown.

Eon	Era	Millions of years ago
Phanerozoic	Cenozoic	66
	Mesozoic	252
	Paleozoic	542
Precambrian	Proterozoic — Late	1000
	Proterozoic — Middle	1600
	Proterozoic — Early	2500
	Archean — Late	3000
	Archean — Middle	3400
	Archean — Early	3800
	Hadean	4600

Origin of Earth

Era	Period	Epoch	Millions of years ago	Significant events in development of life
Cenozoic	Quaternary	Holocene	0.01	Humans develop
		Pleistocene	2.6	
	Tertiary — Neogene	Pliocene	5.3	
		Miocene	23.0	
	Tertiary — Paleogene	Oligocene	33.9	
		Eocene	55.8	
		Paleocene	65.5	Extinction of dinosaurs and many other species
Mesozoic	Cretaceous		145.5	First flowering plants
	Jurassic		201.6	First birds
	Triassic		252	Dinosaurs dominant
Paleozoic	Permian		299	Extinction of 90% of marine life
	Carboniferous — Pennsylvanian		318	First reptiles
	Carboniferous — Mississippian		359	Large coal swamps
	Devonian		416	Amphibians abundant / First insects
	Silurian		444	Fishes dominant / First land plants
	Ordovician		488	First fish
	Cambrian		542	Trilobites dominant
Precambrian/ Proterozoic	Ediacaran		630	First organisms with shells / Odd mesh-like creatures of uncertain affinity

Essential Concepts Review

1.1 How many oceans exist on Earth?

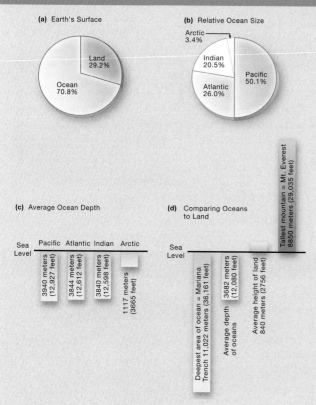

(a) Earth's Surface

Land 29.2%

Ocean 70.8%

(b) Relative Ocean Size

Arctic 3.4%

Indian 20.5%

Pacific 50.1%

Atlantic 26.0%

(c) Average Ocean Depth

Sea Level

Pacific — 3940 meters (12,927 feet)

Atlantic — 3844 meters (12,612 feet)

Indian — 3840 meters (12,598 feet)

Arctic — 1117 meters (3665 feet)

(d) Comparing Oceans to Land

Sea Level

Deepest area of ocean = Mariana Trench 11,022 meters (36,161 feet)

Average depth of oceans 3682 meters (12,080 feet)

Average height of land 840 meters (2756 feet)

Tallest mountain = Mt. Everest 8850 meters (29,035 feet)

▶ *Water covers 70.8% of Earth's surface.* The world ocean is a *single interconnected body of water,* which is large in size and volume. It can be divided into *four principal oceans* (the Pacific, Atlantic, Indian, and Arctic Oceans), plus an additional ocean (the Southern Ocean, or Antarctic Ocean). Even though there is a technical distinction between a sea and an ocean, the two terms are used interchangeably. In comparing the oceans to the continents, it is apparent that *the average land surface does not rise very far above sea level* and that *there is not a mountain on Earth that is as tall as the ocean is deep.*

Study Resources
Online Study Guide Quizzes, Web Diving Deeper 1.1 and 1.3

Critical Thinking Question

NASA has discovered a new planet that has an ocean. Using today's technology, how would you propose studying that ocean, all that's in it, and the sea floor beneath it? Assume an unlimited budget.

1.2 How was early exploration of the oceans achieved?

▶ In the Pacific, *people who populated the Pacific Islands may have been the first great navigators.* In the Western world, the *Phoenicians* were making remarkable voyages as well. Later the *Greeks, Romans,* and *Arabs* made significant contributions and advanced oceanographic knowledge. During the Middle Ages, the *Vikings* colonized Iceland and Greenland and made voyages to North America.

▶ The *Age of Discovery* in Europe renewed the Western world's interest in exploring the unknown. It began with the voyage of *Christopher Columbus* in 1492 and ended in 1522 with the first circumnavigation of Earth by a voyage initiated by *Ferdinand Magellan. Captain James Cook* was one of the first to explore the ocean for scientific purposes.

Study Resources
Online Study Guide Quizzes

Critical Thinking Question

Throughout history, what navigation techniques have enabled sailors to navigate in the open ocean far from land?

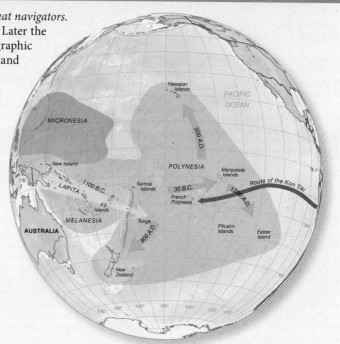

Essential Concepts Review (cont.)

1.3 What is the nature of scientific inquiry?

▶ The *scientific method* is used to understand the occurrence of physical events or phenomena and can be stated as *science supports the explanation of the natural world that best explains all available observations.* Steps in the scientific method include making *observations* and establishing *scientific facts*; forming one or more *hypotheses* (a tentative, testable statement about the general nature of the phenomena observed); extensive *testing* and *modification of hypotheses*; and, finally, developing a *theory* (a well-substantiated explanation of some aspect of the natural world that can incorporate facts, laws, logical inferences, and tested hypotheses). Science never arrives at the absolute "truth"; rather, *science arrives at what is probably true* based on the available observations and can *continually change because of new observations.*

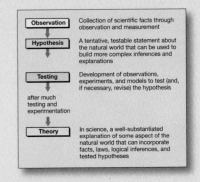

Study Resources
Online Study Guide Quizzes

Critical Thinking Question
What is the difference between a fact and a theory? Can either (or both) be revised?

1.4 How were Earth and the solar system formed?

▶ Our *solar system*, consisting of *the Sun and eight major planets*, probably formed from a *huge cloud of gas and space dust* called a *nebula*. According to the *nebular hypothesis*, the nebular matter contracted to form the Sun, and the planets were formed from eddies of material that remained. The Sun, composed of hydrogen and helium, was massive enough and concentrated enough to emit *large amounts of energy* from fusion. The Sun also emitted *ionized particles that swept away any nebular gas* that remained from the formation of the planets and their satellites.

▶ *Protoearth*, more massive and larger than Earth today, *was molten and homogenous.* The *initial atmosphere*, composed mostly of hydrogen and helium, *was later driven off into space* by intense solar radiation. Protoearth began a period of rearrangement called *density stratification* and formed a *layered internal structure based on density*, resulting in the development of the *crust, mantle,* and *core*. Studies of Earth's internal structure indicate that brittle plates of the *lithosphere* are riding on a plastic, high-viscosity *asthenosphere*. Near the surface, the lithosphere is composed of *continental* and *oceanic crust*. Continental crust consists mostly of granite and oceanic crust consists mostly of basalt. *Continental crust is lower in density, lighter in color, and thicker than oceanic crust.* Both types float isostatically on the denser mantle below.

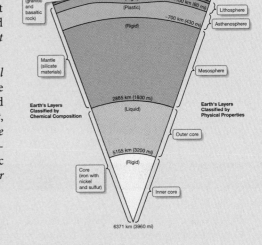

Study Resources
Online Study Guide Quizzes, Web Animations

Critical Thinking Question
Describe how the chemical composition of Earth's interior differs from its physical properties. Include specific examples.

1.5 How were Earth's atmosphere and oceans formed?

▶ *Outgassing produced an early atmosphere* rich in water vapor and carbon dioxide. Once Earth's surface cooled sufficiently, the *water vapor condensed and accumulated to give Earth its oceans. Rainfall on the surface dissolved compounds that,* when carried to the ocean, *made it salty.*

Study Resources
Online Study Guide Quizzes

Critical Thinking Question
Compare the two ways in which Earth was supplied with enough water to have an ocean. Which is likely to have contributed most of the water on Earth?

1.6 Did life begin in the oceans?

▶ *Life is thought to have begun in the oceans.* Stanley Miller's experiment showed that ultraviolet radiation from the Sun and hydrogen, carbon dioxide, methane, ammonia, and inorganic molecules from the oceans may have combined to produce *organic molecules such as amino acids.* Certain combinations of these molecules eventually produced *heterotrophic organisms* (which cannot make their own food) that were probably similar to present-day anaerobic bacteria. Eventually, *autotrophs evolved* that had the ability to make their own food through *chemosynthesis.* Later, some cells developed *chlorophyll,* which made *photosynthesis* possible and led to the *development of plants.*

▶ *Photosynthetic organisms altered the environment* by extracting carbon dioxide from the atmosphere and also by releasing free oxygen, thereby creating today's *oxygen-rich atmosphere.* Eventually, both *plants and animals evolved* into forms that could survive on land.

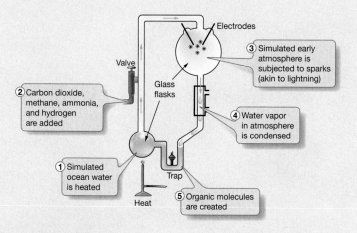

Study Resources
Online Study Guide Quizzes

Critical Thinking Question

Discuss why the great oxidation event is also called the oxygen crisis.

1.7 How old is Earth?

▶ *Radiometric age dating* is used to determine the age of most rocks. Information from extinctions of organisms and from age dating rocks comprises the *geologic time scale,* which indicates that Earth has experienced a long history of changes since *its origin 4.6 billion years ago.*

Era	Period		Epoch	Millions of years ago	Significant events in development of life
Cenozoic	Quaternary		Holocene		Humans develop
			Pleistocene	0.01	
	Tertiary	Neo-gene	Pliocene	2.6	
			Miocene	5.3	
		Paleo-gene	Oligocene	23.0	
			Eocene	33.9	
			Paleocene	55.8	
Mesozoic	Cretaceous			65.5	Extinction of dinosaurs and many other species
	Jurassic			145.5	First flowering plants
					First birds
	Triassic			201.6	Dinosaurs dominant
	Permian			252	Extinction of 90% of marine life

Study Resources
Online Study Guide Quizzes, Web Diving Deeper 1.2, Web Animation

Critical Thinking Question

Construct a representation of the geologic time scale, using an appropriate quantity of any substance (other than dollar bills or toilet paper, which are used as examples in Web Diving Deeper 1.2). Be sure to indicate some of the major changes that have occurred on Earth since its origin.

MasteringOceanography™

www.masteringoceanography.com

Looking for additional review and test prep materials? Visit the Study Area in MasteringOceanography to enhance your understanding of this chapter's content by accessing a variety of resources, including Self-Study Quizzes, Geoscience Animations, RSS feeds, flashcards, Web links, and an optional Pearson eText.

Tall mountains created by tectonic uplift. Tall coastal mountains such as these in Glacier Bay National Park in southeastern Alaska have been uplifted by plate tectonic processes, creating a large amount of relief. Some of the uplifted rocks here have come from distant areas and include parts of the sea floor.

PLATE TECTONICS AND THE OCEAN FLOOR

Before you begin reading this chapter, use the glossary at the end of this book to discover the meanings of any of these words you don't already know:

Atoll Rift valley Barrier reef Volcanic arc Continental drift Subduction Transform fault Ocean trench Convergent boundary Asthenosphere Mantle Hotspot Divergent boundary Plate Tectonics Alfred Wegener Sea floor spreading Lithosphere Nematath Transform boundary Wilson cycle Convection Pangaea Panthalassa Fringing reef Mid-ocean ridge

ESSENTIAL CONCEPTS

At the end of this chapter, you should be able to:

2.1 Describe the evidence that supports continental drift

2.2 Describe the evidence that supports plate tectonics

2.3 Discuss the ocean and land features that occur at plate boundaries

2.4 Summarize some of the applications of plate tectonics

2.5 Demonstrate an understanding of how Earth has changed in the past and how it will look in the future

2.6 Discuss how plate tectonics is used as a working model to help explain features and processes on Earth

"It is just as if we were to refit the torn pieces of a newspaper by matching their edges and then check whether the lines of print run smoothly across. If they do, there is nothing left but to conclude that the pieces were in fact joined in this way."

—Alfred Wegener, The Origins of Continents and Oceans *(1915)*

Each year at various locations around the globe, several thousand earthquakes and dozens of volcanic eruptions occur, both of which indicate how remarkably dynamic our planet is. These events have occurred throughout history, constantly changing the surface of our planet, yet only a few decades ago, most scientists believed the continents were stationary over geologic time. Since that time, a bold new theory has been advanced that helps explain surface features and phenomena on Earth, including:

- The worldwide locations of volcanoes, faults, earthquakes, and mountain building
- Why mountains on Earth haven't been eroded away
- The origin of most landforms and ocean floor features
- How the continents and ocean floor formed and why they are different
- The continuing development of Earth's surface
- The distribution of past and present life on Earth

This revolutionary new theory is called **plate tectonics** (*plate* = plates of the lithosphere; *tekton* = to build), or "the new global geology." According to the theory of plate tectonics, the outermost portion of Earth is composed of a patchwork of thin, rigid plates[1] that move horizontally with respect to one another, like icebergs floating on water. As a result, the continents are mobile and move about on Earth's surface, controlled by forces deep within Earth.

The interaction of these plates as they move builds features of Earth's crust (such as mountain belts, volcanoes, and ocean basins). For example, the tallest mountain range on Earth is the Himalaya Mountains that extend through India, Nepal, and Bhutan. This mountain range contains rocks that were deposited millions of years ago in a shallow sea, providing testimony of the power and persistence of plate tectonic activity.

In this chapter, we'll examine the early ideas about the movement of plates, the evidence for those ideas, and how they led to the theory of plate tectonics. Then we'll explore features of plate boundaries and some applications of plate tectonics, including what our planet may look like in the future.

---

[1]These thin, rigid plates are pieces of the lithosphere that comprise Earth's outermost portion and contain oceanic and/or continental crust, as described in Chapter 1.

STUDENTS SOMETIMES ASK . . .

How long has plate tectonics been operating on Earth? Will it ever stop?

It's difficult to say with certainty how long plate tectonics has been operating because our planet has been so dynamic since its early history, regularly recycling most of Earth's crust. However, recently discovered ancient volcanic rock sequences uplifted onto Greenland show telltale characteristics of tectonic activity and suggest that plate tectonics has been operating for at least the past 3.8 billion years of Earth history.

Plate motion has typically been assumed to be an active and continuous process, with new sea floor constantly being formed while old sea floor is being destroyed. Recent research, however, suggests that plates may move more actively at times, then slow down or even stop, and then start up again. The reasons for this intermittent plate motion appear to be related to plate distribution and changes in the amount of heat released from Earth.

Looking into the future, the forces that drive plates will likely decrease until plates no longer move. This is because plate tectonic processes are powered by heat released from within Earth (which is of a finite amount). The erosional work of water, however, will continue to erode Earth's features. What a different world it will be then—an Earth with no earthquakes, no volcanoes, and no mountains. Flatness will prevail!

2.1 What Evidence Supports Continental Drift?

Alfred Wegener (Figure 2.1), a German meteorologist and geophysicist, was the first to advance the idea of mobile continents in 1912. He envisioned that the continents were slowly drifting across the globe and called his idea **continental drift**. Let's examine the evidence that Wegener compiled that led him to formulate the idea of drifting continents.

Fit Of the Continents

The idea that continents—particularly South America and Africa—fit together like pieces of a jigsaw puzzle originated with the development of reasonably accurate world maps. As far back as 1620, Sir Francis Bacon wrote about how the continents appeared to fit together. However, little significance was given to this idea until 1912, when Wegener used the shapes of matching shorelines on different continents as a supporting piece of evidence for continental drift.

Wegener suggested that during the geologic past, the continents collided to form a large landmass, which he named **Pangaea** (*pan* = all, *gaea* = Earth) (Figure 2.2). Further, a huge ocean, called **Panthalassa** (*pan* = all, *thalassa* = sea) surrounded Pangaea. Panthalassa included several smaller seas, including the **Tethys** (*Tethys* = a Greek sea goddess) **Sea**. Wegener's evidence indicated that as Pangaea began to split apart, the various continental masses started to drift toward their present geographic positions.

Wegener's attempt at matching shorelines revealed considerable areas of crustal overlap and large gaps. Some of the differences could be explained by material deposited by rivers or eroded from coastlines. What Wegener didn't know at the time was that the shallow parts of the ocean floor close to shore are underlain by materials similar to those beneath continents. In the early 1960s, Sir Edward Bullard and two associates used a computer program to fit the continents together (Figure 2.3). Instead of using the shorelines of the continents as Wegener had done, Bullard achieved the best fit (for example, with minimal overlaps or gaps) by using a depth of 2000 meters (6560 feet) below sea level. This depth corresponds to halfway between the shoreline and the deep-ocean basins; as such, it represents the true edge of the continents. By using this depth, the continents fit together remarkably well.

Matching Sequences Of Rocks and Mountain Chains

If the continents were once together, as Wegener had hypothesized, then evidence should appear in rock sequences that were originally continuous but are now separated by large distances. To test the idea of drifting continents, geologists began comparing the rocks along the edges of continents with rocks found in adjacent positions on matching continents. They wanted to see if the rocks had similar types, ages, and structural styles (the type and degree of deformation). In some areas, younger rocks had been deposited during the millions of years since the continents separated, covering the rocks that held the key to the past history of the continents. In other areas, the rocks had been eroded away. Nevertheless, in many other areas, the key rocks were present.

Moreover, these studies showed that many rock sequences from one continent were identical to rock sequences on an adjacent continent—although the two

Figure 2.1 **Alfred Wegener, circa 1912–1913.** Alfred Wegener (1880–1930), shown here in his research station in Greenland, was one of the first scientists to suggest that continents are mobile.

were separated by an ocean. In addition, mountain ranges that terminated abruptly at the edge of a continent continued on another continent across an ocean basin, with identical rock sequences, ages, and structural styles. **Figure 2.4** shows, for example, how similar rocks from the Appalachian Mountains in North America match up with identical rocks from the British Isles and the Caledonian Mountains in Europe.

Wegener noted the similarities in rock sequences on both sides of the Atlantic and used the information as a supporting piece of evidence for continental drift. He suggested that mountains such as those seen on opposite sides of the Atlantic formed during the collision when Pangaea was formed. Later, when the continents split apart, once-continuous mountain ranges were separated. Confirmation of this idea exists in a similar match with mountains extending from South America through Antarctica and across Australia.

Glacial Ages and Other Climate Evidence

Wegener also noticed the occurrence of past glacial activity in areas that are now tropical and suggested that it, too, provided supporting evidence for drifting continents. Currently, the only places in the world where thick continental *ice sheets* occur are in the polar regions of Greenland and Antarctica. However, evidence of ancient glaciation is found in the lower-latitude regions of South America, Africa, India, and Australia.

These deposits, which have been dated at 300 million years old, indicate one of two possibilities: (1) There was a worldwide **ice age** at that time, and even tropical areas were covered by thick ice, or (2) some continents that are now in tropical areas were once located much closer to one of the poles. It is unlikely that the entire world was covered by ice 300 million years ago because coal deposits from the same geologic age now present in North America and Europe originated as vast semitropical swamps. Thus, a reasonable conclusion is that some of the continents must have been closer to the poles than they are today.

Another type of glacial evidence indicates that certain continents have moved from more polar regions during the past 300 million years. When glaciers flow, they move and abrade the underlying rocks, leaving grooves that indicate the direction of flow. The arrows in **Figure 2.5a** show how the glaciers would have flowed away from the South Pole on Pangaea 300 million years ago. The direction of flow is consistent with the grooves found on many continents today (**Figure 2.5b**), providing additional evidence for drifting continents.

Many examples of plant and animal fossils indicate very different climates than today. Two such examples are fossil palm trees in Arctic Spitsbergen and coal deposits in Antarctica. Earth's past environments can be interpreted from these rocks because plants and animals need specific environmental conditions in which to live. Corals, for example, generally need seawater above 18 degrees centigrade (°C) or 64 degrees Fahrenheit (°F) in order to survive. When fossil corals are found in areas that are cold today, two explanations seem most plausible: (1) Worldwide climate has changed dramatically or (2) the rocks have moved from their original location.

As explained in Chapter 16, "The Oceans and Climate Change," natural processes have caused Earth's climate to change in the geologic past. Although dramatic shifts in Earth's climate might help explain climate evidence such as fossils that seem out of place today, the distribution of these fossils could also be explained by drifting continents. Unaware of the changes in Earth's climate that are known by Earth scientists today, Wegener suggested that the out-of-place fossils as well as other climate evidence provided support for the slow movement of the continents and added another item to a growing list of evidence.

Distribution Of Organisms

To add credibility to his argument for the existence of the supercontinent of Pangaea, Wegener cited documented cases of several fossil organisms found on different landmasses that could not have crossed the vast oceans presently separating the continents.

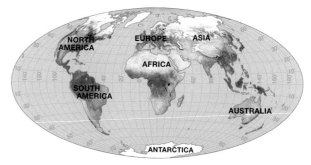

(a) The positions of the continents today.

(b) The positions of the continents about 200 million years ago, showing the supercontinent of Pangaea and the single large ocean, Panthalassa.

Figure 2.2 Reconstruction of Pangaea.

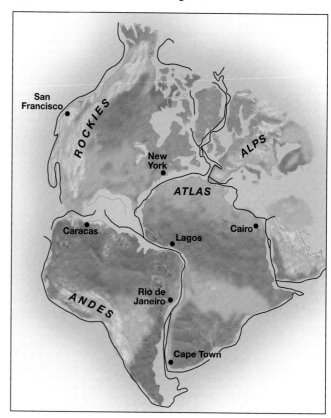

Figure 2.3 An early computer fit of the continents. Map showing the 1960s fit of the continents using a depth of 2000 meters (6560 feet) (*black lines*), which is the true edge of the ocean basin. The results indicate a remarkable match, with few overlaps and minimal gaps. Note that the present-day shorelines of the continents are shown with blue lines.

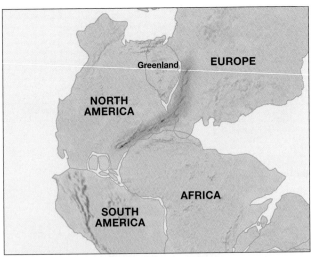

(a) Positions of the continents about 300 million years ago, showing how mountain ranges with similar age, type, and structure form one continuous belt.

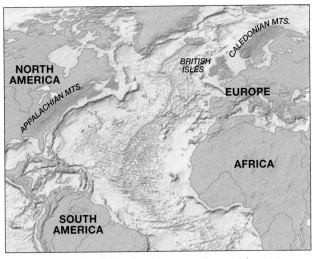

(b) Present-day positions of continents and mountain ranges.

Figure 2.4 Matching mountain ranges across the North Atlantic Ocean.

Web Animation
Breakup of Pangaea

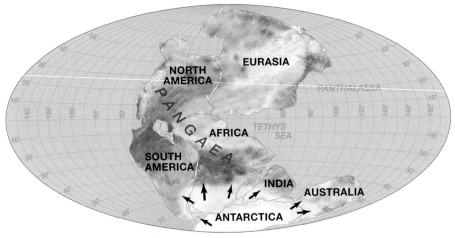

(a) Reconstruction of the supercontinent Pangaea, showing the area covered by glacial ice about 300 million years ago. Arrows indicate the direction of ice flow.

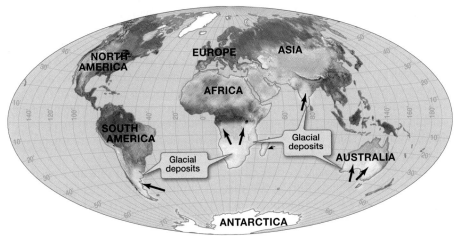

(b) The positions of the continents today, showing how the glacial deposits have moved.

Figure 2.5 Ice age on Pangaea.

For example, the fossil remains of **Mesosaurus** (*meso* = middle, *saurus* = lizard), an extinct, presumably aquatic reptile that lived about 250 million years ago, are located only in eastern South America and western Africa (**Figure 2.6**). If *Mesosaurus* had been strong enough to swim across an ocean, why aren't its remains more widely distributed?

Wegener's idea of continental drift provided an elegant solution to this problem. He suggested that the continents were closer together in the geologic past, so *Mesosaurus* didn't have to be a good swimmer to leave remains on two different continents. Later, after *Mesosaurus* became extinct, the continents moved to their present-day positions, and a large ocean now separates the once-connected landmasses. Other examples of similar fossils on different continents include those of plants, which would have had a difficult time traversing a large ocean.

Before continental drift, several ideas were proposed to help explain the curious pattern of these fossils, such as the existence of island stepping stones or a land bridge. It was even suggested that at least one pair of land-dwelling *Mesosaurus* survived the arduous journey across several thousand kilometers of open ocean by rafting on floating logs. However, there is no evidence to support the idea of island stepping stones or a land bridge, and the idea of *Mesosaurus* rafting across an ocean seems implausible.

Wegener also cited the distribution of present-day organisms as evidence to support the concept of drifting continents. For example, modern organisms with

similar ancestries clearly had to evolve in isolation during the past few million years. Most obvious of these are the Australian marsupials (such as kangaroos, koalas, and wombats), which have a distinct similarity to the marsupial opossums found in the Americas.

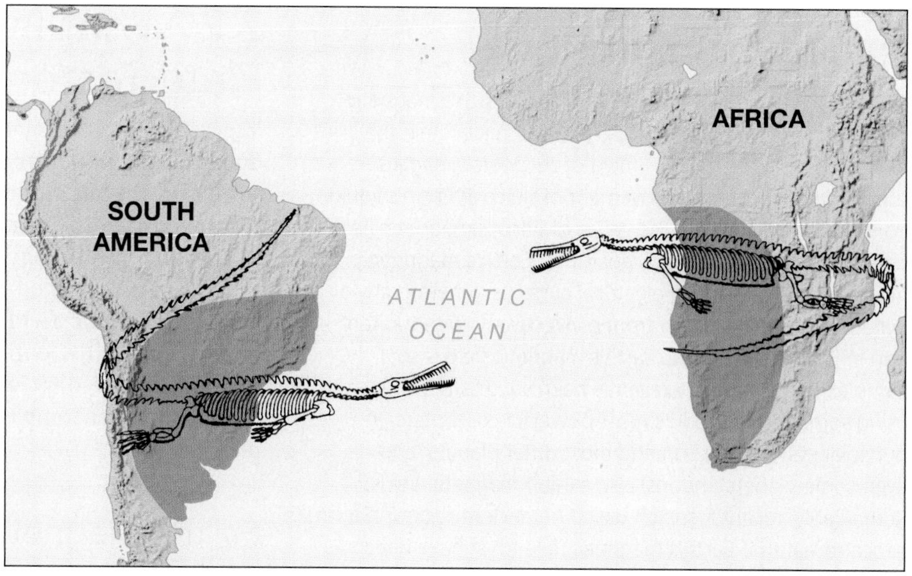

Objections to the Continental Drift Model

Wegener first published his ideas in *The Origins of Continents and Oceans* in 1915, but the book did not attract much attention until it was translated into English, French, Spanish, and Russian in 1924. From that point until his death in 1930,[2] Wegener's drift hypothesis received much hostile criticism—and sometimes open ridicule—from the scientific community because of the mechanism he proposed for the movement of the continents. Wegener suggested that the continents plowed through the ocean basins to reach their present-day positions and that the leading edges of the continents deformed into mountain ridges because of the drag imposed by ocean rocks. Further, the driving mechanism

Figure 2.6 Fossils of *Mesosaurus*. Fossils of the aquatic reptile *Mesosaurus*, which lived about 250 million years ago, are found only in South America and Africa. The limited distribution of *Mesosaurus* fossils suggest that these two continents were once joined.

he proposed was a combination of the gravitational attraction of Earth's equatorial bulge and tidal forces from the Sun and Moon.

Scientists rejected the idea as too fantastic and contrary to the laws of physics. Debate over the mechanism of drift concentrated on the long-term behavior of the substrate and the forces that could move continents laterally. Material strength calculations, for example, showed that ocean rock was too strong for continental rock to plow through it. Further, analysis of gravitational and tidal forces indicated that they were too small to move the great continental landmasses. Even without an acceptable mechanism, many geologists who studied rocks in South America and Africa accepted continental drift because it was consistent with the rock record. North American geologists—most of whom were unfamiliar with these Southern Hemisphere rock sequences—remained highly skeptical.

As compelling as his evidence may seem today, Wegener was unable to convince the scientific community as a whole of the validity of his ideas. Although his hypothesis was correct in principle, it contained several incorrect details, such as the driving mechanism for continental motion and how continents move across ocean basins. In order for any scientific viewpoint to gain wide acceptance, it must explain all available observations and have supporting evidence from a wide variety of scientific fields. This supporting evidence would not come until more details of the nature of the ocean floor were revealed, which, along with new technology that enabled scientists to determine the original positions of rocks on Earth, provided additional observations in support of drifting continents.

ESSENTIAL CONCEPT 2.1

Alfred Wegener used a variety of interdisciplinary information from land to support continental drift. However, he did not have a suitable mechanism or any information about the sea floor.

CONCEPT CHECK 2.1

1 When did the supercontinent of Pangaea exist? What was the ocean that surrounded the supercontinent called?

2 Regarding glacial ages, why is it unlikely that the entire world was covered by ice 300 million years ago?

3 Cite the lines of evidence Alfred Wegener used to support his idea of continental drift. Why did scientists of the time doubt that continents had drifted?

---

[2]Wegener perished in 1930 while trying to establish a year-round meteorological station atop the Greenland ice sheet.

STUDENTS SOMETIMES ASK . . .

What causes Earth's magnetic field?

Studies of Earth's magnetic field and research in the field of *magnetodynamics* suggest that convective movement of fluids in Earth's liquid iron–nickel outer core is the cause of Earth's magnetic field. The most widely accepted view is that Earth's magnetic field is created by strong electrical currents generated by a dynamo process resulting from convective flow of molten iron in Earth's outer core. Earth's magnetic field is so complex that it has only recently been successfully modeled using some of the world's most powerful computers. In our solar system, the Sun and most other planets (and even some planets' moons) also exhibit magnetic fields. Remarkably, recent research based on ancient rocks in South Africa reveal that Earth's magnetic field must have been present by 3.45 billion years ago.

2.2 What Evidence Supports Plate Tectonics?

Very little new information about Wegener's continental drift hypothesis was introduced between the time of Wegener's death in 1930 and the early 1950s. However, studies of the sea floor using sonar that were initiated during World War II and continued after the war provided critical evidence in support of drifting continents. In addition, technology unavailable in Wegener's time enabled scientists to analyze the way rocks retained the signature of Earth's **magnetic field**. These developments caused scientists to reexamine continental drift and advance it into the more encompassing theory of plate tectonics.

Earth's Magnetic Field and Paleomagnetism

Earth's magnetic field is shown in **Figure 2.7**. The invisible lines of magnetic force that originate within Earth and travel out into space resemble the magnetic field produced by a large bar magnet.[3] Similar to Earth's magnetic field, the ends of a

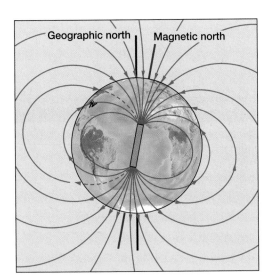

(a) Earth's magnetic field generates invisible lines of magnetic force similar to a large bar magnet. Note that magnetic north and true north are not in exactly the same location.

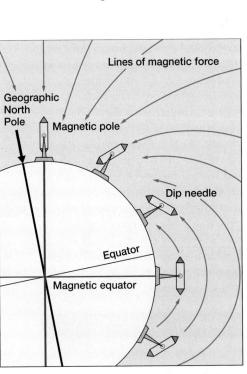

(b) Earth's magnetic field causes a dip needle to align parallel to the lines of magnetic force and change orientation with increasing latitude. Consequently, an approximation of latitude can be determined based on the dip angle.

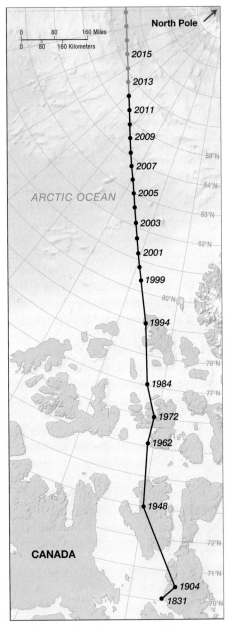

(c) Map showing the location of Earth's north magnetic pole since 1831 (*black*) and its projected location in the future (*green*).

Figure 2.7 Earth's magnetic field.

---

[3]The properties of a magnetic field can be explored easily enough with a bar magnet and some iron particles. Place the iron particles on a table and place a bar magnet nearby. Depending on the strength of the magnet, you should get a pattern resembling that in Figure 2.7a.

bar magnet have opposite polarities (labeled either + and – or N for north and S for south) that cause magnetic objects to align parallel to its magnetic field. In addition, notice in Figures 2.7b and 2.7c that Earth's geographic North Pole (the rotational axis) and Earth's magnetic north pole (magnetic north) do not coincide.

ROCKS AFFECTED BY EARTH'S MAGNETIC FIELD   Igneous (*igne* = fire, *ous* = full of) **rocks** solidify from molten **magma** (*magma* = a mass) either underground or after volcanic eruptions at the surface that produce **lava** (*lavare* = to wash). Nearly all igneous rocks contain **magnetite**, a naturally magnetic iron mineral. Particles of magnetite in magma align themselves with Earth's magnetic field because magma and lava are fluid. Once molten material is cooled to a certain temperature,[4] internal magnetite particles are frozen into position, thereby recording the angle of Earth's magnetic field at that place and time. In essence, grains of magnetite serve as tiny compass needles that record the strength and orientation of Earth's magnetic field. Unless the rock is heated to the temperature where magnetite grains are again mobile, these magnetite grains contain information about the magnetic field where the rock originated, regardless of where the rock subsequently moves.

Magnetite is also deposited in sediments. As long as the sediment is surrounded by water, the magnetite particles can align themselves with Earth's magnetic field. After sediment is buried and solidifies into **sedimentary** (*sedimentum* = settling) **rock**, the particles are no longer able to realign themselves if they are subsequently moved. Thus, magnetite grains in sedimentary rocks also contain information about the magnetic field where the rock originated. Although other rock types have been used successfully to reveal information about Earth's ancient magnetic field, the most reliable ones are igneous rocks that have high concentrations of magnetite such as **basalt**, which is the rock type that comprises oceanic crust.

PALEOMAGNETISM   The study of Earth's ancient magnetic field is called **paleomagnetism** (*paleo* = ancient). Scientists who study paleomagnetism analyze magnetite particles in rocks to determine not only their north–south direction but also their angle relative to Earth's surface. The degree to which a magnetite particle points into Earth is called its **magnetic dip**, or *magnetic inclination*.

Magnetic dip is directly related to latitude. Figure 2.7b shows that a dip needle does not dip at all at Earth's magnetic equator. Instead, the needle lies horizontal to Earth's surface. At Earth's magnetic north pole, however, a dip needle points straight into the surface. A dip needle at Earth's magnetic south pole is also vertical to the surface, but it points out instead of in. Thus, magnetic dip increases with increasing latitude, from 0 degrees at the magnetic equator to 90 degrees at the magnetic poles. Because magnetic dip is retained in magnetically oriented rocks, measuring the dip angle reveals the latitude at which the rock initially formed. Done with care, paleomagnetism is an extremely powerful tool for interpreting where rocks first formed. Based on paleomagnetic studies, convincing arguments could finally be made that the continents had drifted relative to one another.

APPARENT POLAR WANDERING   When magnetic dip data for rocks on the continents were used to determine the apparent position of the magnetic north pole over time, it appeared that the magnetic pole was wandering, or moving, through time. For example, Figure 2.8a shows

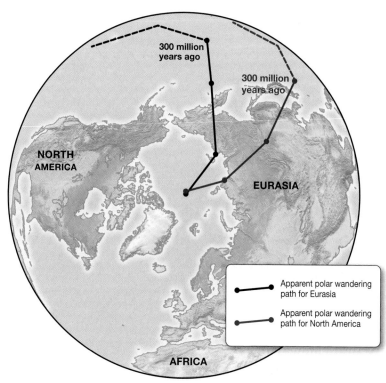

(a) The apparent magnetic polar wandering paths for North America and Eurasia (*red* and *black lines*, respectively) resulted in a dilemma because they were not in alignment.

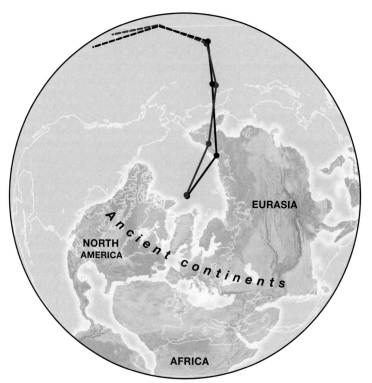

(b) The positions of the magnetic polar wandering paths more closely coincide when the landmasses are assembled.

Figure 2.8 Apparent polar wandering paths.

[4]This temperature is called the *Curie point* and is named after French physicist Pierre Curie. For typical rocky Earth materials, it is about 550°C or 1022°F.

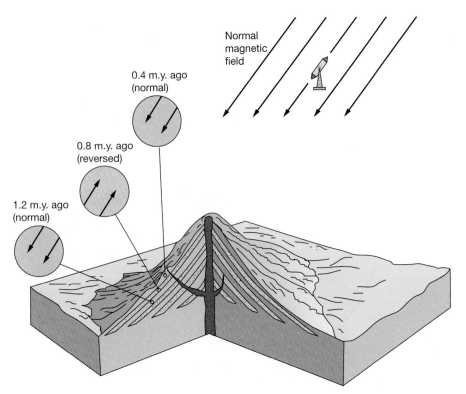

Figure 2.9 Paleomagnetism preserved in rocks. The switching of Earth's magnetic polarity through time is preserved in rocks such as these lava flows.

Web Animation
Flipping of Earth's Magnetic Field

STUDENTS SOMETIMES ASK . . .

What changes to Earth's environment would occur when the magnetic poles reverse?

During a reversal, compasses would likely show incorrect directions, and people could have difficulty navigating. The same goes for some fish, birds, and mammals that sense the magnetic field during migrations (see **Diving Deeper 2.1**). The decrease in strength of the magnetic field also reduces the protection that the field provides for life-forms against cosmic rays and particles coming from the Sun, and this could disrupt low-Earth-orbiting satellites as well as some communication and power grid systems. Also, Earth's *aurora borealis* (the Northern Lights) and its counterpart *aurora australis* (the Southern Lights), which are natural light displays in the sky, might be visible at much lower latitudes. On the bright side, we know that life on Earth has successfully survived previous magnetic reversals, so reversals might not be as dangerous as they are sometimes portrayed (such as in the 2003 science fiction film *The Core*, which is full of scientific inaccuracies).

the magnetic **polar wandering curves** for North America and Eurasia. Both curves have a similar shape but, for all rocks older than about 70 million years, the pole determined from North American rocks lies to the west of that determined from Eurasian rocks. There can be only one north magnetic pole at any given time, however, and it is unlikely that its position has changed very much through time because it always nearly coincides with Earth's rotational axis. This discrepancy implies that the magnetic pole remained stationary while North America and Eurasia moved relative to the pole and relative to each other. **Figure 2.8b** shows that when the continents are moved into the positions they occupied when they were part of Pangaea, the two wandering curves match up, providing strong evidence that the continents have moved throughout geologic time.

MAGNETIC POLARITY REVERSALS Magnetic compasses on Earth today follow lines of magnetic force and point toward magnetic north. It turns out, however, that the **polarity** (the directional orientation of the magnetic field) has reversed itself periodically throughout geologic time. In essence, the north and south magnetic poles switch. **Figure 2.9** shows how ancient rocks have recorded the switching of earth's magnetic polarity through time.

Paleomagnetic studies reveal that about 170 major reversals have occurred in the past 76 million years. The pattern of switching of Earth's magnetic field is highly irregular but occurs about every 250,000 years or so. On average, it takes several hundred to several thousand years for a change in polarity to occur; such a change is identified in rock sequences by a gradual decrease in the intensity of the magnetic field of one polarity, followed by a gradual increase in the intensity of the magnetic field of opposite polarity.

Earth's magnetic north pole—which does not coincide with the geographic North Pole—was first located near Boothia Peninsula in the Canadian Arctic in 1831; since that time, it's been migrating northwest by about 50 kilometers (30 miles) each year (Figure 2.7c). If this rate continues, Earth's magnetic pole will pass within 400 kilometers (250 miles) of the geographic North Pole in 2018 and will be in Siberia by 2050. In addition, evidence suggests that Earth's magnetic field has also been weakening during the past 2000 years, which may be an indication that Earth's current "normal" polarity may reverse itself. In fact, the last major reversal of Earth's magnetic poles occurred 780,000 years ago, which suggests that the next one is overdue.

PALEOMAGNETISM AND THE OCEAN FLOOR Paleomagnetism had certainly proved its usefulness on land, but, up until the mid-1950s, it had only been conducted on continental rocks. Would the ocean floor also show variations in magnetic polarity? To test this idea, the U.S. Coast and Geodetic Survey, in conjunction with scientists from Scripps Institution of Oceanography, undertook an extensive deep-water mapping program off Oregon and Washington in 1955. Using a sensitive instrument called a **magnetometer** (*magneto* = magnetism, *meter* = measure), which is towed behind a research vessel, the scientists spent several weeks at sea, moving back and forth in a regularly spaced pattern, measuring Earth's magnetic field and how it was affected by the magnetic properties of rocks on the ocean floor.

When the scientists analyzed their data, they found that the entire surveyed area had a pattern of north–south stripes in a surprisingly regular and alternating pattern of above-average and below-average magnetism. What was even more surprising was that the pattern appeared to be symmetrical with

DO SEA TURTLES (AND OTHER ANIMALS) USE EARTH'S MAGNETIC FIELD FOR NAVIGATION?

Sea turtles travel great distances across the open ocean so they can lay their eggs on the island where they themselves were hatched. Interestingly, this behavior of migrating to remote, isolated islands thousands of kilometers from the closest continent may have gradually evolved as ocean basins widened due to plate movement, separating the turtles' feeding and breeding grounds. How do the turtles know where an island is located, and how do they navigate at sea during their long voyage?

Radio tagging of green sea turtles (*Chelonia mydas*; **Figure 2A**) indicates that during their migration, they often travel in an essentially straight-line path to reach their destination. One hypothesis suggests that, like the Polynesian navigators (see Diving Deeper 1.1), sea turtles use wave direction to help them steer. However, studies of their migration route reveal that green sea turtles continue along a straight-line path that is independent of wave direction.

Research in *magnetoreception*, which is the study of an animal's ability to sense magnetic fields, suggests that sea turtles may use Earth's magnetic field for navigation. For instance, turtles can distinguish between different magnetic inclination angles, which in effect would allow them to sense latitude. Sea turtles can also distinguish magnetic field intensity, which gives a rough indication of longitude. By sensing these two magnetic field properties, a sea turtle could determine its position at sea and navigate to a tiny island thousands of kilometers away. Like any other good navigator, sea turtles may also use other tools, such as olfactory (scent) clues, Sun angles, local landmarks, and oceanographic phenomena.

A diverse assortment of other animals may also use magnetic properties to navigate. For example, some whales and dolphins may detect and follow the magnetic stripes on the sea floor during their movements, which may help to explain why whales sometimes beach themselves. In addition, certain bacteria use the magnetic mineral magnetite to align themselves parallel to Earth's magnetic field. Subsequently, magnetite has been linked to the "homing" ability of dozens of species, including fish, butterflies, honeybees, birds, turtles, lobsters, cows, and even humans. Even though scientists don't know what the magnetic field looks like to other animals, a variety of studies have investigated how these animals may sense it. For example, studies have suggested that sensing Earth's magnetic field relies on magnetically receptive fibers of nerves that send impulses to the brain, a GPS-like signal that comes from the inner ear, or even the detection of subatomic properties of light that are affected by the magnetic field as they enter the eye. Some animals may also use multiple magnetic sensory organs.

Do humans have an innate ability to use Earth's magnetic field for navigation? Studies conducted on humans indicate that the majority of people can identify north after being blindfolded and disoriented. Interestingly, many people point *south* instead of north, but this direction is along the lines of magnetic force. Similarly, migratory animals that rely on magnetism for navigation will not be confused by a reversal in Earth's magnetic field and will still be able to get to where they need to go. Exactly how organisms navigate with their magnetic sense seems likely to remain one of the most puzzling questions in sensory biology.

Figure 2A Green sea turtle. Green sea turtles (*Chelonia mydas*) get their name from the green-colored fat tissue in their bodies. They are listed as a threatened species and are protected by international law.

respect to a long mountain range that was fortuitously in the middle of their survey area.

Detailed paleomagnetism studies of this and other areas of the sea floor confirmed a similar pattern of alternating stripes of above-average and below-average magnetism. These stripes are called **magnetic anomalies** (*a* = without, *nomo* = law; an anomaly is a departure from normal conditions). The ocean floor had embedded in it a regular pattern of alternating magnetic stripes unlike anywhere on land.

Figure 2.10 shows that the mantle is moving in large circles. Is the mantle molten?

No. Because the mantle is often depicted as flowing in convective motion, a common misconception is that the mantle is molten. Seismic studies reveal that the mantle is unambiguously greater than 99% solid, although it does have the ability to flow S-L-O-W-L-Y over time (hence the arrows in the figure). The only places where the mantle is partially molten are (1) underneath the mid-ocean ridge, where release of pressure causes melt to form; (2) in the mantle wedge above a subducting plate, where dehydration reactions in the descending plate cause melting; and (3) in isolated mantle plumes, which are discussed later in this chapter. Make no mistake about it: The vast majority of the mantle is composed of hot, solid rock.

Researchers had a difficult time explaining why the ocean floor had such a regular pattern of magnetic anomalies. Nor could they explain how the sequence on one side of the underwater mountain range matched the sequence on the opposite side—in essence, they were a mirror image of each other. To understand how this pattern could have formed, more information was needed about ocean floor features and their origin.

Sea Floor Spreading and Features Of the Ocean Basins

Geologist **Harry Hess** (1906–1969), when he was a U.S. Navy captain in World War II, developed the habit of leaving his depth recorder on at all times while his ship was traveling at sea. After the war, compilation of these and many other depth records showed extensive mountain ridges near the centers of ocean basins and extremely deep, narrow trenches at the edges of ocean basins. In 1962, Hess published *History of Ocean Basins*, which contained the idea of **sea floor spreading** and the associated circular movement of rock material in the mantle—**convection** (*con* = with, *vect* = carried) **cells**—as the driving mechanism (**Figure 2.10**). He suggested that new ocean crust was created at the ridges, split apart, moved away from the ridges, and later disappeared back into the deep Earth at trenches. Mindful of the resistance of North American scientists to the idea of continental drift, Hess referred to his own work as "geopoetry."

As it turns out, Hess's initial ideas about sea floor spreading have been confirmed. The **mid-ocean ridge** (Figure 2.10) is a continuous underwater mountain range that winds through every ocean basin in the world and resembles the seam on a baseball. It is entirely volcanic in origin, wraps one-and-a-half times around the globe, and rises more than 2.5 kilometers (1.5 miles) above the surrounding deep-ocean floor. It even rises above sea level in places such as Iceland. New ocean floor forms at the crest, or axis, of the mid-ocean ridge. By the process of sea floor spreading, new ocean floor is split in two and carried away from the axis, replaced by the upwelling of volcanic material that fills the void with new strips of sea floor. Sea floor spreading occurs along the axis of the

Web Animation
Sea Floor Spreading and Plate Boundaries

Figure 2.10 Processes of plate tectonics. Hot molten rock comes to the surface at the mid-ocean ridge and moves outward by the process of sea floor spreading. Eventually, sea floor is destroyed at ocean trenches, where the process of subduction occurs. Slow circular movement of hot rock in the mantle produces convection cells.

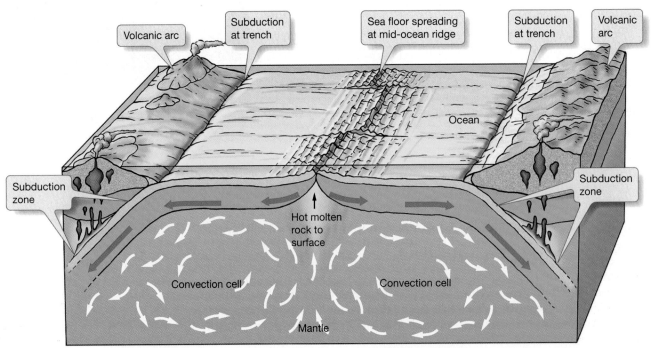

mid-ocean ridge, which is referred to as a **spreading center**. One way to think of the mid-ocean ridge is as a zipper that is being pulled apart. Thus, Earth's zipper (the mid-ocean ridge) is becoming unzipped!

At the same time, ocean floor is being destroyed at deep **ocean trenches**. Trenches are the deepest parts of the ocean floor and resemble a narrow crease or trough (Figure 2.10). Some of the largest earthquakes in the world occur near these trenches; they are caused by a plate bending downward and slowly plunging back into Earth's interior. This process is called **subduction** (*sub* = under, *duct* = lead), and the sloping area from the trench along the downward-moving plate is called a **subduction zone**.

In 1963, geologists **Frederick Vine** and **Drummond Matthews** of Cambridge University combined the seemingly unrelated pattern of magnetic sea floor stripes with the process of sea floor spreading to explain the perplexing pattern of alternating and symmetric magnetic stripes on the sea floor (**Figure 2.11**). Vine and Matthews interpreted the pattern of above-average and below-average magnetic polarity episodes embedded in sea floor rocks to be caused by Earth's magnetic field alternating between "normal" polarity (similar to today's magnetic pole position in the north) and "reversed" polarity (with the magnetic pole to the south). They proposed that the pattern could be created when newly formed rocks at the mid-ocean ridge are magnetized with whichever polarity exists on Earth during their formation. As those rocks are slowly moved away from the crest of the mid-ocean

Web Animation
Sea floor Spreading and
Rock Magnetism

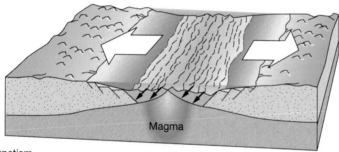
(a) Period of normal magnetism

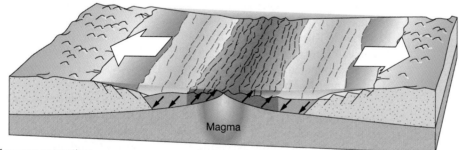
(b) Period of reverse magnetism

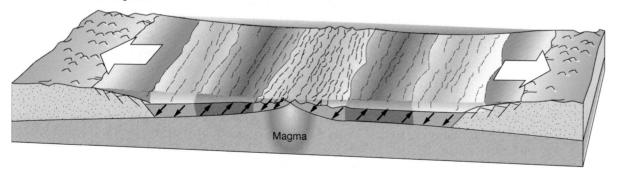
(c) Period of normal magnetism

Figure 2.11 Magnetic evidence of sea floor spreading. As new basalt is added to the ocean floor at mid-ocean ridges, it is magnetized according to Earth's existing magnetic field. This produces a pattern of normal and reversed magnetic polarity "stripes" that are identical on either side of the mid-ocean ridge.

ridge, they maintain their original polarity, and subsequent rocks record the periodic switches of Earth's magnetic polarity. The result is an alternating pattern of magnetic polarity stripes that are symmetric with respect to the mid-ocean ridge.

The pattern of alternating reversals of Earth's magnetic field as recorded in the sea floor was the most convincing piece of evidence set forth to support the concept of sea floor spreading—and, as a result, continental drift. However, the continents weren't plowing through the ocean basins as Wegener had envisioned. Instead, the ocean floor was a conveyer belt that was being continuously formed at the mid-ocean ridge and destroyed at the trenches, with the continents just passively riding along on the conveyer. By the late 1960s, most geologists had changed their stand on continental drift in light of this new evidence.

Other Evidence from the Ocean Basins

Even though the tide of scientific opinion had indeed switched to favor a mobile Earth, additional evidence from the ocean floor would further support the ideas of continental drift and sea floor spreading.

AGE OF THE OCEAN FLOOR   In the late 1960s, an ambitious deep-sea drilling program was initiated to test the existence of sea floor spreading. One of the program's primary missions was to drill into and collect ocean floor rocks for radiometric age dating. If sea floor spreading does indeed occur, then the youngest sea floor rocks would be atop the mid-ocean ridge, and the ages of rocks would increase on either side of the ridge in a symmetric pattern.

The map in **Figure 2.12**, showing the age of the ocean floor beneath deep-sea deposits, is based on the pattern of magnetic stripes verified with thousands of radiometrically age-dated samples. It shows that the ocean floor is youngest along the mid-ocean ridge, where new ocean floor is created, and the age of rocks increases with increasing distance in either direction away from the axis of the ridge. The symmetric pattern of ocean floor ages confirms that the process of sea floor spreading must indeed be occurring.

The Atlantic Ocean has the simplest and most symmetric pattern of age distribution in Figure 2.12. The pattern results from the newly formed Mid-Atlantic Ridge that rifted Pangaea apart. The Pacific Ocean has the least symmetric pattern because many subduction zones surround it. For example, ocean floor east of the East Pacific Rise that is older than 40 million years has already been subducted. The ocean floor in the northwestern Pacific, about 180 million years old, has not yet been subducted. A portion of the East Pacific Rise has even disappeared under North America. The age bands in the Pacific Ocean are wider than those in the Atlantic and Indian Oceans, which suggests that the rate of sea floor spreading is greatest in the Pacific Ocean.

Recall from Chapter 1 that the ocean is at least 4 billion years old. However, the oldest ocean floor is only 180 million years old (or 0.18 billion years old), and the majority of the ocean floor is not even half that old (see Figure 2.12). How could the ocean floor be so incredibly young, while the oceans themselves are so phenomenally old? According to plate tectonic theory, new ocean floor is created at the mid-ocean ridge by sea floor spreading and moves off the ridge to eventually be subducted and remelted in the mantle. In this way, the ocean floor keeps regenerating itself. The floor beneath the oceans today is not the same one that existed beneath the oceans 4 billion years ago.

If the rocks that comprise the ocean floor are so young, why are continental rocks so old? Using radiometric age dating, scientists have determined that the oldest rocks on land are about 4 billion years old. Many other continental rocks approach this age, implying that the same processes that constantly

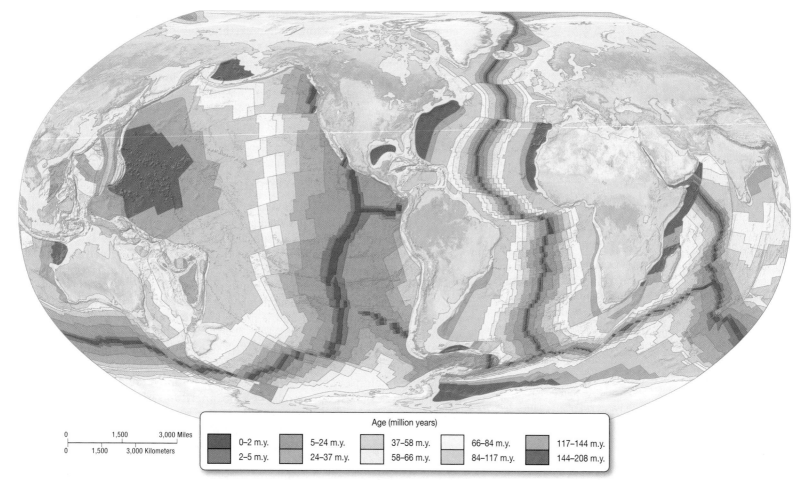

Figure 2.12 Age of the ocean crust beneath deep-sea deposits. The youngest rocks (*bright red areas*) are found along the mid-ocean ridge. Farther away from the mid-ocean ridge, the rocks increase linearly in age in either direction. Ages shown are in millions of years before present.

renew the sea floor do not operate on land. Rather, evidence suggests that continental rocks, because of their low density, do not get recycled by the process of sea floor spreading, and thus they remain at Earth's surface for long periods of time.

HEAT FLOW The heat from Earth's interior is released to the surface as **heat flow.** Current models indicate that this heat moves to the surface with magma in convective motion. Most of the heat is carried to regions of the mid-ocean ridge spreading centers (see Figure 2.10). Cooler portions of the mantle descend along subduction zones to complete each circular-moving convection cell.

Heat flow measurements show that the amount of heat flowing to the surface along the mid-ocean ridge can be up to eight times greater than the average amount flowing to other parts of Earth's crust. Additionally, heat flow at deep-sea trenches, where ocean floor is subducted, can be as little as one-tenth the average. Increased heat flow at the mid-ocean ridge and decreased heat flow at subduction zones is what would be expected based on thin crust at the mid-ocean ridge and a double thickness of crust at the trenches (see Figure 2.10).

WORLDWIDE EARTHQUAKES *Earthquakes* are sudden releases of energy caused by fault movement or volcanic eruptions. The map in **Figure 2.13a** shows

Currently, plates move an average of 2 to 12 centimeters (1 to 5 inches) per year, which is about as fast as a person's fingernails grow. A person's fingernail growth is dependent on many factors, including heredity, gender, diet, and amount of exercise but averages about 8 centimeters (3 inches) per year. This may not sound very fast, but the plates have been moving for millions of years. Over a very long time, even an object moving slowly will eventually travel a great distance. For instance, fingernails growing at a rate of 8 centimeters (3 inches) per year for 1 million years would be 80 kilometers (50 miles) long!

Evidence shows that the plates were moving faster millions of years ago than they are moving today. Geologists can determine the rate of plate motion in the past by analyzing the width of new oceanic crust produced by sea floor spreading, since fast spreading produces more sea floor rock. (Using this relationship and by examining Figure 2.12, you should be able to determine whether the Pacific Ocean or the Atlantic Ocean had a faster spreading rate.) Recent studies using this same technique indicate that about 50 million years ago, India attained a speed of 19 centimeters (7.5 inches) per year. Other research indicates that about 530 million years ago, plate motions may have been as high as 30 centimeters (1 foot) per year! What caused these rapid bursts of plate motion? Geologists are not sure why plates moved more rapidly in the past, but increased heat release from Earth's interior is a likely mechanism.

that most large earthquakes occur along ocean trenches, reflecting the energy released during subduction. Other earthquakes occur along the mid-ocean ridge, reflecting the energy released during sea floor spreading. Still others occur along major faults in the sea floor and on land, reflecting the energy released when moving plates contact other plates along their edges. When you examine the two maps in Figure 2.13, notice how closely the pattern of major earthquakes matches the locations of plate boundaries. This is because most earthquakes worldwide are created by plates interacting with each other at their margins.

The Acceptance Of a Theory

The accumulation of lines of evidence such as those mentioned in this section, along with many other lines of evidence in support of moving continents, has convinced scientists of the validity of continental drift. Since the late 1960s, the concepts of continental drift and sea floor spreading have been united into a much more encompassing theory known as plate tectonics, which describes the movement of the outermost portion of Earth and the resulting creation of continental and sea floor features. These tectonic plates are pieces of the **lithosphere** (*lithos* = rock, *sphere* = ball) that float on the more fluid **asthenosphere** (*asthenos* = weak, *sphere* = ball) below.[5]

What forces drive plate motion? Although several mechanisms have been proposed for the force (or forces) responsible for driving this motion, none of them are able to explain all aspects of plate movement. However, recent research based on a simple model of lithosphere and mantle interactions suggests that two major tectonic forces may act in unison on the leading edges of subducting plates (slabs) to generate the observed plate movements: (1) *slab pull*, which is generated by a subducting plate that pulls the rest of the plate behind it, and (2) *slab suction*, which is created as a subducting plate drags against the viscous mantle and causes the mantle to flow in toward the subduction zone, thereby sucking in nearby plates much in the same way pulling a plug from a bathtub draws floating objects toward it. Other models that include upper mantle viscosity variations suggest that mantle flow contributes to either driving or resisting plate motions. Moreover, the unequal distribution of heat within Earth—which was thought to be the underlying driving force for plate motions—may not be an adequate mechanism to explain plate movement. Although modelers continue their research into the forces that drive plate motions, studies are hampered by the inaccessibility of Earth's mantle.

Since the acceptance of the theory of plate tectonics, much research has focused on understanding various features associated with plate boundaries.

[5]See Chapter 1 for a discussion of properties of the lithosphere and asthenosphere.

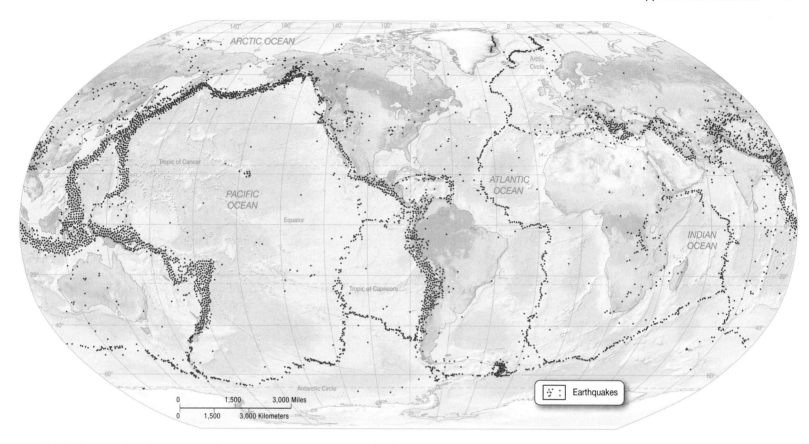

(a)   Distribution of earthquakes with magnitude equal to or greater than M_W = 5.0 for the period 1980-1990.

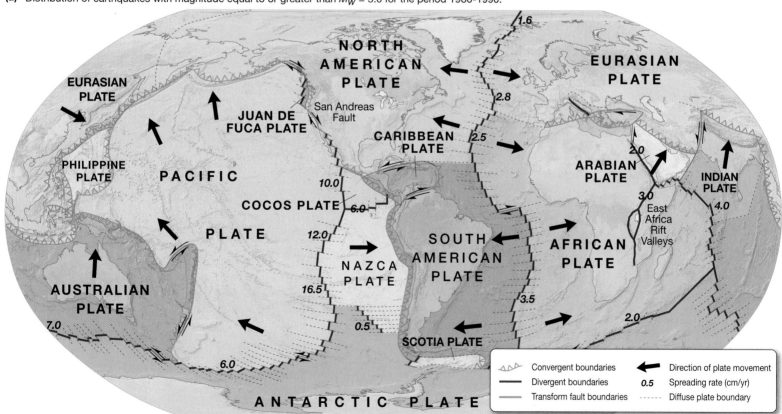

(b)   Plate boundaries define the major tectonic plates (shaded), with arrows indicating the direction of motion

Figure 2.13 **Earthquakes and tectonic plate boundaries.**   Matching world maps of **(a)** earthquakes and **(b)** tectonic plates show that most earthquakes occur along plate boundaries.

Web Animation
Relationship Between Plate
Boundaries and Features

2.3 What Features Occur at Plate Boundaries?

Plate boundaries—where plates interact with each other—are associated with a great deal of tectonic activity, such as mountain building, volcanic activity, and earthquakes. In fact, the first clues to the locations of plate boundaries were the dramatic tectonic events that occur there. For example, Figure 2.13 shows the close correspondence between worldwide earthquakes and plate boundaries. Further, **Figure 2.13b** shows that Earth's surface is composed of seven major plates along with many smaller ones. Close examination of Figure 2.13b shows that the boundaries of plates do not always follow coastlines and, as a consequence, nearly all plates contain both oceanic and continental crust.[6] Notice also that about 90% of plate boundaries occur on the sea floor.

There are three types of plate boundaries, as shown in **Figure 2.14**. **Divergent** (*di* = apart, *vergere* = to incline) **boundaries** are found along oceanic ridges where new lithosphere is being added. **Convergent** (*con* = together, *vergere* = to incline) **boundaries** are found where plates are moving together and one plate subducts beneath the other. **Transform** (*trans* = across, *form* = shape) **boundaries** are found where lithospheric plates slowly grind past one another. **Table 2.1** summarizes characteristics, tectonic processes, features, and examples of these plate boundaries.

Web Animation
Motion at Plate Boundaries

Figure 2.14 The three types of plate boundaries.

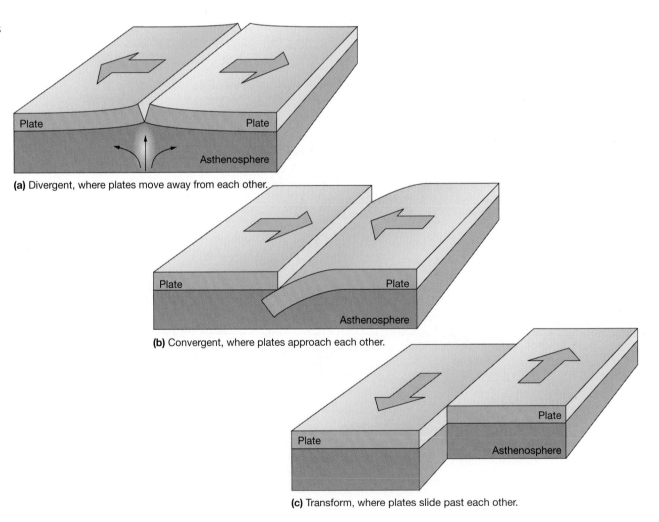

(a) Divergent, where plates move away from each other.

(b) Convergent, where plates approach each other.

(c) Transform, where plates slide past each other.

---

[6]For a review of the differences between (basaltic) oceanic and (granitic) continental crust, see Chapter 1.

TABLE 2.1   CHARACTERISTICS, TECTONIC PROCESSES, FEATURES, AND EXAMPLES OF PLATE BOUNDARIES

Plate boundary	Plate movement	Crust types	Sea floor created or destroyed?	Tectonic process	Sea floor feature(s)	Geographic examples
Divergent plate boundaries	Apart ← →	Oceanic-oceanic	New sea floor is created	Sea floor spreading	Mid-ocean ridge; volcanoes; young lava flows	Mid-Atlantic Ridge, East Pacific Rise
		Continental-continental	As a continent splits apart, new sea floor is created	Continental rifting	Rift valley; volcanoes; young lava flows	East Africa Rift Valleys, Red Sea, Gulf of California
Convergent plate boundaries	Together → ←	Oceanic-continental	Old sea floor is destroyed	Subduction	Trench; volcanic arc on land	Peru–Chile Trench, Andes Mountains
		Oceanic-oceanic	Old sea floor is destroyed	Subduction	Trench; volcanic arc as islands	Mariana Trench, Aleutian Islands
		Continental-continental	N/A	Collision	Tall mountains	Himalaya Mountains, Alps
Transform plate boundaries	Past each other → ←	Oceanic	N/A	Transform faulting	Fault	Mendocino Fault, Eltanin Fault (between mid-ocean ridges)
		Continental	N/A	Transform Faulting	Fault	San Andreas Fault, Alpine Fault (New Zealand)

Divergent Boundary Features

Divergent plate boundaries occur where two plates move apart, such as along the crest of the mid-ocean ridge, where sea floor spreading creates new oceanic lithosphere (**Figure 2.15**). A common feature along the crest of the mid-ocean ridge is a **rift valley**, which is a central downdropped linear depression (**Figure 2.16**). Pull-apart faults located along the central rift valley show that the plates are *continuously being pulled apart* rather than being pushed apart by the upwelling of material beneath the mid-ocean ridge. Upwelling of magma beneath the mid-ocean ridge is simply filling in the void left by the separating plates of lithosphere. In the process, sea floor spreading produces about 20 cubic kilometers (4.8 cubic miles) of new ocean crust worldwide each year.

Figure 2.17 shows how the development of a mid-ocean ridge creates an ocean basin. Initially, molten material rises to the surface, causing upwarping and thinning of the crust. Volcanic activity produces vast quantities of high-density basaltic rock. As the plates begin to move apart, a linear rift valley is formed, and volcanism continues. Further splitting apart of the land—a process called **rifting**—and more spreading cause the area to drop below sea level. When this occurs, the rift valley eventually floods with seawater, and a young linear sea is formed. After millions of years of sea floor spreading, a full-fledged ocean basin is created, with a mid-ocean ridge in the middle of the two landmasses.

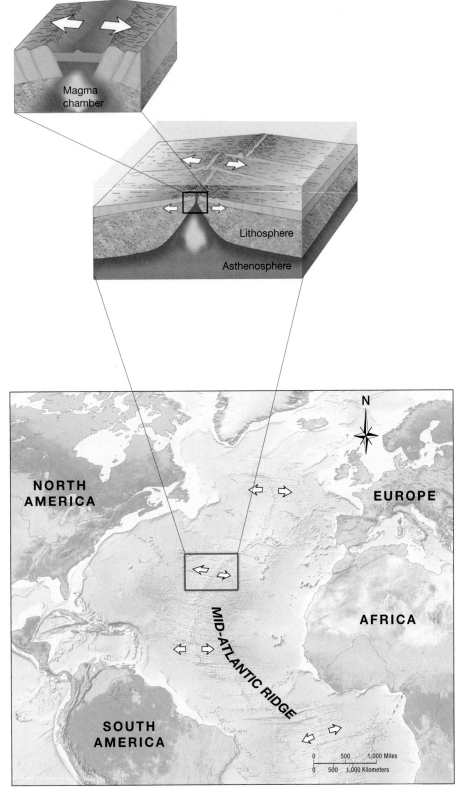

Figure 2.15 Divergent boundary at the Mid-Atlantic Ridge. Most divergent plate boundaries occur along the crest of the mid-ocean ridge, where sea floor spreading creates new oceanic lithosphere.

Figure 2.16 Rift valley in Iceland. View along a rift valley looking south from Laki volcano in Iceland, which sits atop the Mid-Atlantic Ridge. The rift valley is marked by the linear row of volcanoes extending from the bottom of the photo to the horizon that are split in half. Note the bus (*red circle*) for scale.

Two different stages of ocean basin development are shown in the map of East Africa in **Figure 2.18**. First, the rift valleys are actively pulled apart and are at the rift valley stage of formation. Second, the Red Sea is at the linear sea stage. It has rifted apart so far that the land has dropped below sea level. The Gulf of California in Mexico is another linear sea. The Gulf of California and the Red Sea are two of the youngest seas in the world, having been created only a few million years ago. If plate motions continue rifting the plates apart in these areas, they will eventually become large oceans.

OCEANIC RISES VERSUS OCEANIC RIDGES The rate at which the sea floor spreads apart varies along the mid-ocean ridge and dramatically affects its appearance. Faster spreading, for instance, produces broader and less rugged segments of the global mid-ocean ridge system. This is because fast-spreading segments of the mid-ocean ridge produce vast amounts of rock, which move away from the spreading

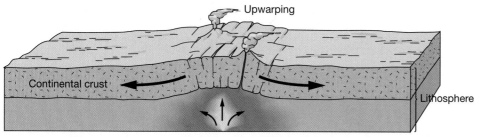

Upwarping

Continental crust   Lithosphere

(a) A shallow heat source develops under a continent, causing initial upwarping and volcanic activity.

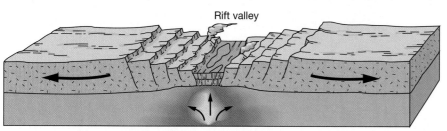

Rift valley

(b) Movement apart creates a rift valley.

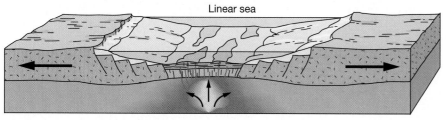

Linear sea

(c) With increased spreading, a linear sea is formed.

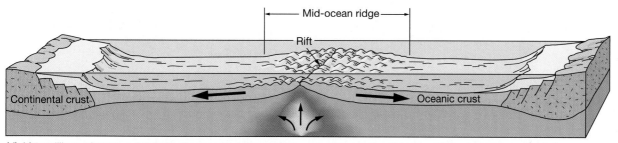

Mid-ocean ridge

Rift

Continental crust   Oceanic crust

(d) After millions of years, a full-fledged ocean basin is created, separating continental pieces that were once connected.

Figure 2.17 **Formation of an ocean basin by sea floor spreading.** Sequence of events in the formation of an ocean basin.

Web Animation
Formation of an Ocean Basin
by Sea Floor Spreading

center at a rapid rate and consequently undergo less thermal contraction and subsidence than do slower-spreading segments. In addition, central rift valleys on slow-spreading segments tend to be larger and better developed.

The gently sloping and fast-spreading parts of the mid-ocean ridge are called **oceanic rises**. For example, the **East Pacific Rise** (Figure 2.19, *bottom*) between the Pacific and Nazca Plates is a broad, low, gentle swelling of the sea floor with a small, indistinct central rift valley and has a spreading rate as high as 16.5 centimeters (6.5 inches) per year.[7] Conversely, steeper-sloping and slower-spreading areas of the mid-ocean ridge are called **oceanic ridges**. For instance, the **Mid-Atlantic Ridge** (Figure 2.19, *top*) between the South American and African Plates is a tall, steep, rugged oceanic ridge that has an average spreading rate of 2.5 centimeters (1 inch) per year and stands as much as 3000 meters (10,000 feet) above the surrounding sea floor. Its prominent central rift valley is as much as 32 kilometers (20 miles) wide and averages 2 kilometers (1.2 miles) deep.

---

[7]The spreading rate is the total widening rate of an ocean basin resulting from motion of both plates away from a spreading center.

Figure 2.18 East Africa Rift Valleys and associated features.

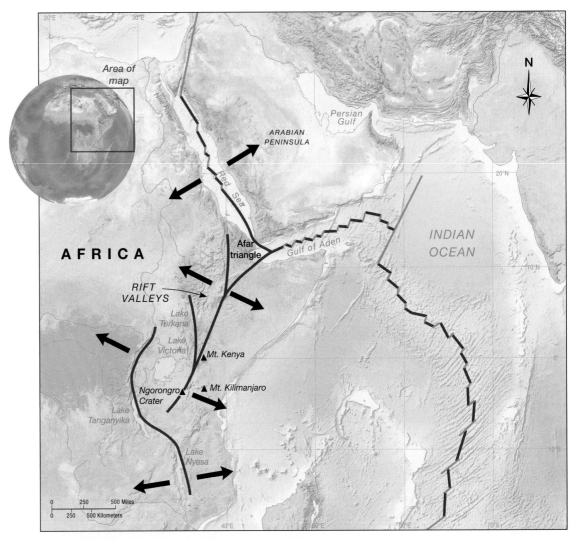

(a) Parts of east Africa are splitting apart (*arrows*), creating a series of linear downdropped rift valleys (*red lines*) along with prominent volcanoes (*triangles*). Similarly, the Red Sea and Gulf of Aden have split apart so far that they are now below sea level. The mid-ocean ridge in the Indian Ocean has experienced similar stages of development.

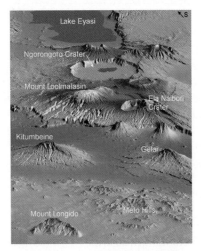

(b) Land elevation perspective view, looking southwest along part of the East Africa Rift in Tanzania, showing the downdropped Lake Eyasi and numerous volcanic peaks and craters of the Crater Highlands. Color indicates elevation, where green is lower elevation and brown/white is higher.

(c) Photo of a rift that formed in 2005 after seismic activity and a volcanic eruption of Mount Dabbahu in Ethiopia's Afar triangle, Africa; note people at left for scale.

Recently, a new class of spreading centers called *ultra-slow spreading centers* has also been recognized. These spreading centers, which were discovered along the Southwest Indian and Arctic segments of the mid-ocean ridge, are characterized by spreading rates less than 2 centimeters (0.8 inch) per year, a deep rift valley, and volcanoes that occur only at widely spaced intervals. The ultra-slow ridges are spreading so slowly, in fact, that Earth's mantle itself is exposed on the ocean floor in great slabs of rock between these volcanoes.

EARTHQUAKES ASSOCIATED WITH DIVERGENT BOUNDARIES   The amount of energy released by earthquakes along divergent plate boundaries is closely related to the spreading rate. The faster the sea floor spreads, the less energy is released in each earthquake. Earthquake intensity is usually measured on a scale called the **seismic moment magnitude** (M_w), which reflects the energy released to create very long-period seismic waves. Because it more adequately represents larger-magnitude earthquakes, the moment magnitude scale is now the most commonly used magnitude scale for earthquakes, replacing the well-known *Richter scale*. Earthquakes in the rift valley of the slow-spreading Mid-Atlantic Ridge reach a maximum magnitude of about $M_w = 6.0$, whereas those occurring along the axis of the fast-spreading East Pacific Rise seldom exceed $M_w = 4.5$.[8]

Convergent Boundary Features

Convergent boundaries—where two plates move together and collide—result in the destruction of ocean crust as one plate plunges below the other and is remelted in the mantle. One feature associated with a convergent plate boundary is a deep-ocean trench, which is a deep and narrow depression on the sea floor that marks the beginning of the subduction zone. Another feature is an arc-shaped row of highly active and explosively erupting volcanoes called a **volcanic arc** that parallels the trench and occurs above the subduction zone. Volcanic arcs are formed by the downgoing plate in the subduction zone heating up and releasing superheated gases—mostly water—that cause the overlying mantle wedge above the subducting plate to partially melt. This molten rock, which is more buoyant than the rock around it, slowly rises up to the surface and feeds the active volcanoes.

Figure 2.20 shows the three subtypes of convergent boundaries that result from interactions between the two different types of crust (oceanic and continental).

OCEANIC–CONTINENTAL CONVERGENCE   When an oceanic plate and a continental plate converge, the denser oceanic plate is subducted (Figure 2.20a). The oceanic plate becomes heated as it is subducted into the asthenosphere and releases superheated gases that partially melt the overlying mantle, which rises to the surface through the overriding continental plate. The rising basalt-rich magma mixes with the *granite* of the continental crust, producing lava in volcanic eruptions at the

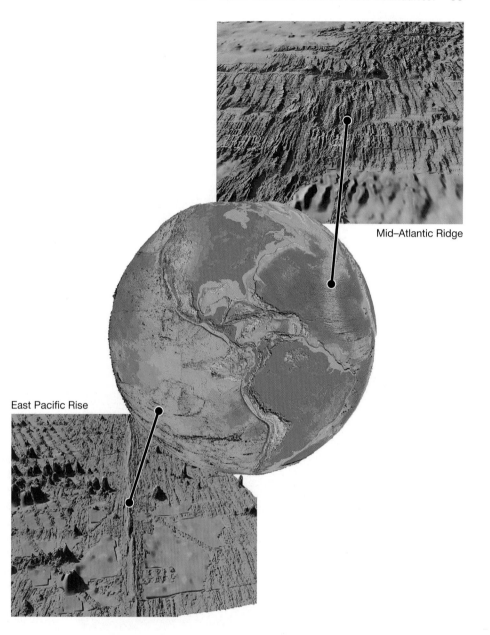

Mid–Atlantic Ridge

East Pacific Rise

Figure 2.19 Comparing oceanic rises and ridges. Perspective views of the ocean floor based on satellite bathymetry showing differences between oceanic rises and ridges. The fast-spreading East Pacific Rise (*left*) is a broad, low, gentle swelling of the mid-ocean ridge that lacks a prominent rift valley. The slow-spreading Mid-Atlantic Ridge (*above*) is a tall, steep, rugged portion of the mid-ocean ridge with a prominent central rift valley.

---

[8]Note that each one-unit increase of earthquake magnitude represents an increase of energy release of about 30 times.

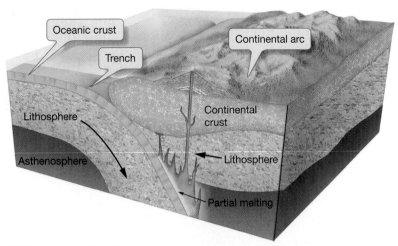

(a) Oceanic–continental convergence, where denser oceanic crust subducts and a continental arc is created.

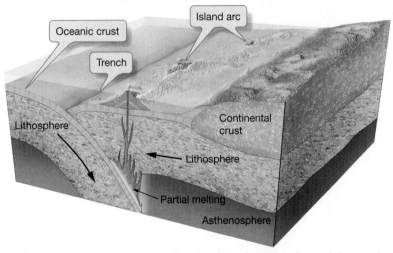

(b) Oceanic–oceanic convergence, where the older, denser sea floor subducts and an oceanic island arc is formed.

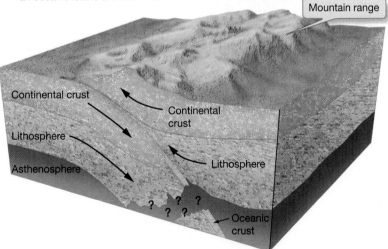

(c) Continental–continental convergence, where continental crust is too low in density to subduct. Instead, a tall uplifted mountain range is created.

Figure 2.20 The three subtypes of convergent plate boundaries and their associated features.

surface that is intermediate in composition between basalt and granite. One type of volcanic rock with this composition is called *andesite*, named after the Andes Mountains of South America because it is so common there. Because andesite magma is more viscous than basalt magma and contains such high gas content, andesitic volcanic eruptions are usually quite explosive and have historically been very destructive. The result of this volcanic activity on the continent above the subduction zone produces a type of volcanic arc called a **continental arc**. Continental arcs are created by andesitic volcanic eruptions and by the folding and uplifting associated with plate collision.

If the spreading center producing the subducting plate is far enough from the subduction zone, an oceanic trench becomes well developed along the margin of the continent. The Peru–Chile Trench is an example, and the Andes Mountains are the associated continental arc produced by partially melting the mantle above the subducting plate. If the spreading center producing the subducting plate is close to the subduction zone, however, the trench is not nearly as well developed. This is the case where the Juan de Fuca Plate subducts beneath the North American Plate off the coasts of Washington and Oregon to produce the Cascade Mountains continental arc (**Figure 2.21**). Here, the Juan de Fuca Ridge is so close to the North American Plate that the subducting lithosphere is less than 10 million years old and has not cooled enough to become very deep. In addition, the large amount of sediment carried to the ocean by the Columbia River has filled most of the trench with sediment. Many of the Cascade volcanoes of this continental arc have been active within the past 100 years. Most recently, Mount St. Helens erupted in May 1980, killing 62 people.

OCEANIC–OCEANIC CONVERGENCE When two oceanic plates converge, the denser oceanic plate is subducted (Figure 2.20b). Typically, the older oceanic plate is denser because it has had more time to cool and contract. This type of convergence produces the deepest trenches in the world, such as the Mariana Trench in the western Pacific Ocean. Similar to oceanic–continental convergence, the subducting oceanic plate becomes heated, releases superheated gases, and partially melts the overlying mantle. This buoyant molten material rises to the surface and fuels the active volcanoes, which occur as an arc-shaped row of volcanic islands that is a type of volcanic arc called an **island arc**. The molten material is mostly basaltic because there is no mixing with granitic rocks from the continents, and the eruptions are not nearly as destructive. Examples of island arc/trench systems are the West Indies' Leeward and Windward Islands/Puerto Rico Trench in the Caribbean Sea and the Aleutian Islands/Aleutian Trench in the North Pacific Ocean.

CONTINENTAL–CONTINENTAL CONVERGENCE When two continental plates converge, which one is subducted? You might expect that the older of the two (which is most likely the denser one) will be subducted. Continental lithosphere forms differently than oceanic lithosphere, however, and old continental lithosphere is no denser than young continental lithosphere. It turns out that *neither* subducts because they are both too low in density to be pulled very far down into the mantle. Instead, a tall uplifted mountain range is created by the collision of the two plates (Figure 2.20c). These mountains are composed of folded and deformed sedimentary rocks originally deposited on the sea floor that previously separated the two continental plates. The intervening oceanic

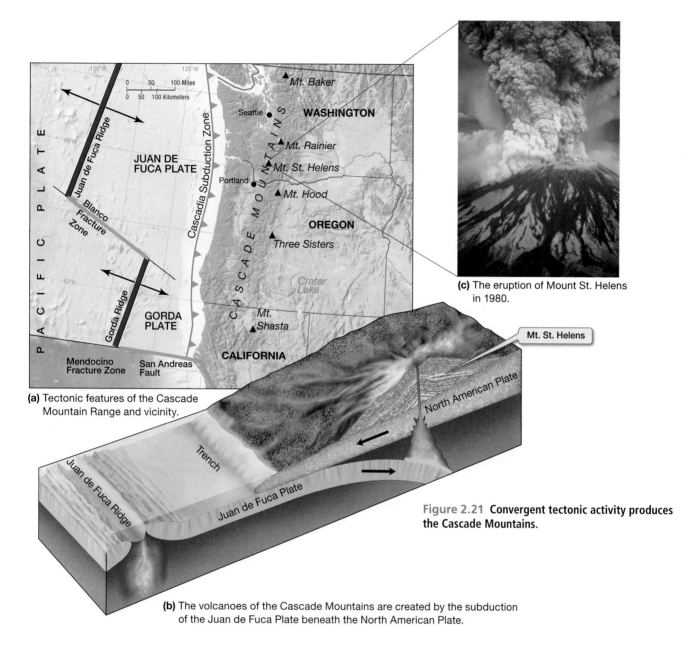

(c) The eruption of Mount St. Helens in 1980.

Mt. St. Helens

(a) Tectonic features of the Cascade Mountain Range and vicinity.

Figure 2.21 **Convergent tectonic activity produces the Cascade Mountains.**

(b) The volcanoes of the Cascade Mountains are created by the subduction of the Juan de Fuca Plate beneath the North American Plate.

crust between the two plates is subducted beneath such mountains as the plates collide. A prime example of continental–continental convergence is the collision of India with Asia (Figure 2.22). It began 45 million years ago and has created the Himalaya Mountains, presently the tallest mountains on Earth.

EARTHQUAKES ASSOCIATED WITH CONVERGENT BOUNDARIES Both spreading centers and trench systems are characterized by earthquakes, but in different ways. Spreading centers have shallow earthquakes, usually less than 10 kilometers (6 miles) deep. Earthquakes in the trenches, on the other hand, vary from near the surface down to 670 kilometers (415 miles) deep, which are the deepest earthquakes in the world. These earthquakes are clustered in a band about 20 kilometers (12.5 miles) thick that closely corresponds to the location of the subduction zone. In fact, the subducting plate in a convergent plate boundary can be traced below the surface by examining the pattern of successively deeper earthquakes extending from the trench.

Many factors combine to produce large earthquakes at convergent boundaries. The forces involved in convergent-plate-boundary collisions are enormous. Huge lithospheric slabs of rock are relentlessly pushing against each other, and the

Web Animation
Collapse of Mount St. Helens

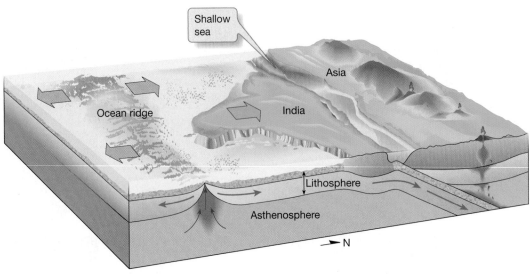

(a) Sea floor spreading along the mid-ocean ridge south of India caused the collision of India with Asia, which began about 45 million years ago.

Web Animation
Convergent Margins: India-Asia Collision

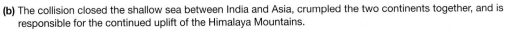

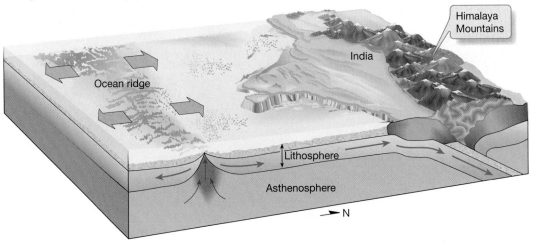

(b) The collision closed the shallow sea between India and Asia, crumpled the two continents together, and is responsible for the continued uplift of the Himalaya Mountains.

(c) View of Ladakh, India, with the snow-capped Himalaya Mountains in the background.

Figure 2.22 **The collision of India with Asia.**

subducting plate must actually bend as it dives below the surface. In addition, thick crust associated with convergent boundaries tends to store more energy than the thinner crust at divergent boundaries. Also, mineral structure changes occur at the higher pressures encountered deep below the surface, which are thought to produce changes in volume that lead to some of the most powerful earthquakes in the world. In fact, the largest earthquake ever recorded was the 1960 Chilean earthquake near the Peru–Chile Trench, which had a magnitude of $M_w = 9.5$!

Transform Boundary Features

A global sea floor map (such as the one inside the front cover of this book) shows that the mid-ocean ridge is offset by many large features oriented perpendicular (at right angles) to the crest of the ridge. What causes these offsets? They are formed because spreading at a mid-ocean ridge only occurs perpendicular to the axis of a ridge, and all parts of a plate must move together. As a result, offsets are oriented perpendicular to the ridge and parallel to each other to accommodate spreading of a linear ridge system on a spherical Earth. In addition, the offsets allow different segments of the mid-ocean ridge to spread apart at different rates. These offsets—called **transform faults**—give the mid-ocean ridge a zigzag appearance. Thousands of these transform faults, some large and some small, dissect the global mid-ocean ridge.

OCEANIC VERSUS CONTINENTAL TRANSFORM FAULTS   There are two types of transform faults. The first and most common type occurs wholly on the ocean floor and is called an **oceanic transform fault**. The second type cuts across a continent and is called a **continental transform fault**. Regardless of type, though, transform faults *always* occur between two segments of a mid-ocean ridge, as shown in **Figure 2.23**.

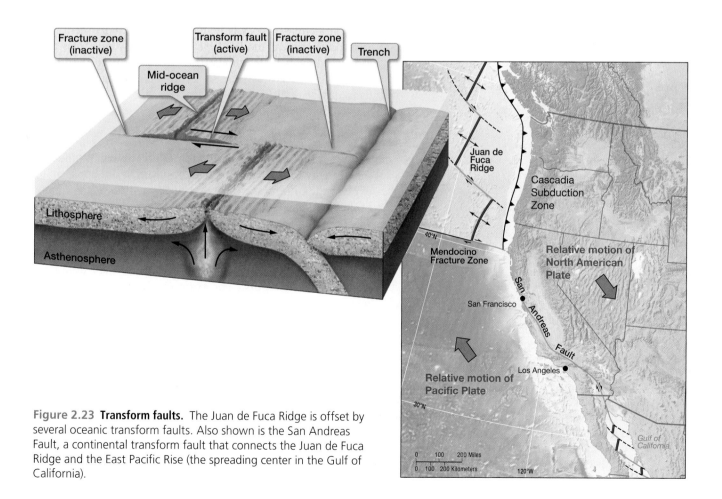

Figure 2.23 Transform faults. The Juan de Fuca Ridge is offset by several oceanic transform faults. Also shown is the San Andreas Fault, a continental transform fault that connects the Juan de Fuca Ridge and the East Pacific Rise (the spreading center in the Gulf of California).

ESSENTIAL CONCEPT 2.3

The three main types of plate boundaries are divergent (plates moving apart, such as at the mid-ocean ridge), convergent (plates moving together, such as at an ocean trench), and transform (plates sliding past each other, such as at a transform fault).

EARTHQUAKES ASSOCIATED WITH TRANSFORM BOUNDARIES   The movement of one plate past another—a process called **transform faulting**—produces shallow but often strong earthquakes in the lithosphere. Magnitudes of $M_w = 7.0$ have been recorded along some oceanic transform faults. One of the best-studied faults in the world is California's **San Andreas Fault**, a continental transform fault that runs from the Gulf of California past San Francisco and beyond into northern California. Because the San Andreas Fault cuts through continental crust, which is much thicker than oceanic crust, earthquakes are considerably larger than those produced by oceanic transform faults, sometimes up to $M_w = 8.5$.

CONCEPT CHECK 2.3

1. Most lithospheric plates contain both oceanic- and continental-type crust. Use plate boundaries to explain why this is true.
2. Describe the differences between oceanic ridges and oceanic rises. Include in your answer why these differences exist.
3. Convergent boundaries can be divided into three types, based on the type of crust contained on the two colliding plates. Compare and contrast the different types of convergent boundaries that result from these collisions.
4. Describe the differences in earthquake magnitudes that occur between the three types of plate boundaries and explain why these differences occur.

2.4 Testing the Model: What Are Some Applications Of Plate Tectonics?

One of the strengths of plate tectonic theory is how it unifies so many seemingly separate events into a single consistent model. Let's look at a few examples that illustrate how plate tectonic processes can be used to explain the origin of features that, up until the acceptance of plate tectonics, were difficult to explain.

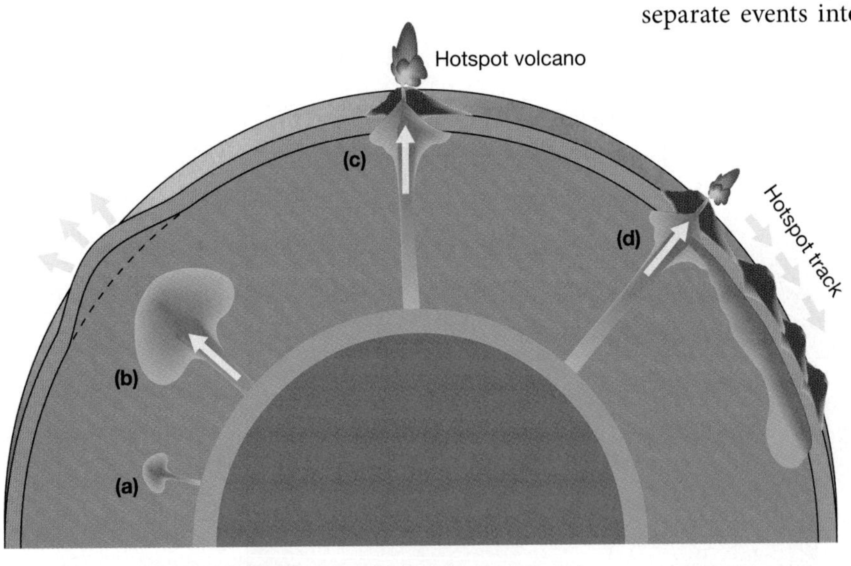

Hotspots and Mantle Plumes

Although the theory of plate tectonics helped explain the origin of many features near plate boundaries, it did not seem to explain the origin of **intraplate** (*intra* = within, *plate* = plate of the lithosphere) **features** that are far from any plate boundary. For instance, how can plate tectonics explain volcanic islands near the middle of a plate? Areas of intense volcanic activity that remain in more or less the same location over long periods of geologic time and are unrelated to plate boundaries are called **hotspots**.[9] For example, the continuing volcanism in Yellowstone National Park and Hawaii are caused by hotspots.

Why is there so much volcanic activity at hotspots? The plate tectonic model infers that hotspot volcanism is caused by the presence of **mantle plumes** (*pluma* = a soft feather), which are columnar areas of hot molten rock that arise from deep within the mantle (**Figure 2.24**). Mantle plumes can be identified by researchers who measure how fast seismic waves from earthquakes travel below ground; the underlying principle is that seismic waves move more slowly through

Figure 2.24 Origin and development of mantle plumes and hotspots. According to the plume hypothesis, **(a)** a plume of hot buoyant material detaches from the deep mantle or the core–mantle boundary; **(b)** the plume rises more rapidly in its conduit than the plume head can push through viscous mantle, which inflates the head and elevates Earth's surface; **(c)** decompression near the surface partially melts the plume head, which comes to the surface and creates a hotspot volcano; and **(d)** the volcano is carried away by plate motion as the plume tail continues to feed subsequent volcanoes, creating a hotspot track (*nematath*). Diagram not to scale.

[9]Note that a hotspot is different from either a volcanic arc or a mid-ocean ridge (both of which are related to plate boundaries), even though all are marked by a high degree of volcanic activity.

hot rock than cold. Seismic studies suggest that several types of mantle plumes exist: Some come from the core–mantle boundary, while others have a shallower source. Recent research reveals that the core–mantle boundary is not a simple smooth dividing zone but has many regional variations, which has implications for the development of mantle plumes. Because mantle plumes themselves cannot be directly sampled and the thin plume conduits are difficult to resolve using seismic data, their existence has been difficult to confirm. As a result, there is currently vigorous scientific debate regarding mantle plumes and volcanism at hotspots.[10] In fact, new studies suggest that some mantle plumes are neither deep phenomena nor fixed in position over geologic time, as assumed in the standard plume model.

Worldwide, more than 100 hotspots have been active within the past 10 million years. **Figure 2.25** shows the global distribution of prominent hotspots today. In general, hotspots do not coincide with plate boundaries. Notable exceptions are those that are near divergent boundaries where the lithosphere is thin, such as at the Galápagos Islands and Iceland. In fact, Iceland straddles the Mid-Atlantic Ridge (a divergent plate boundary). It is also directly over a 150-kilometer- (93-mile)-wide mantle plume, which accounts for its remarkable amount of volcanic activity—so much that it has caused Iceland to be one of the few areas of the global mid-ocean ridge that rise high above sea level.

Throughout the Pacific Plate, many island chains are oriented in a northwestward–southeastward direction. The most intensely studied of these is the **Hawaiian**

Web Animation
Tectonic Settings of Volcanic
Activity

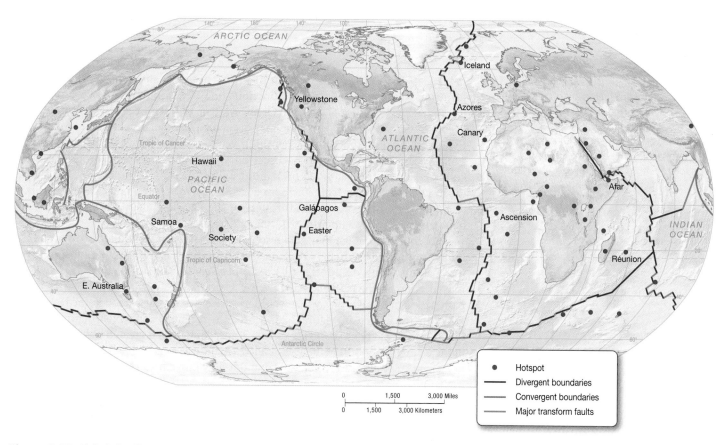

Figure 2.25 Global distribution of prominent hotspots. Prominent hotspots are shown by red dots; the locations of plate boundaries are also shown. The majority of the world's hotspots are not associated with plate boundaries; those that are tend to occur along divergent plate boundaries, where the lithosphere is thin.

---

[10]For more information about this debate, see www.MantlePlumes.org.

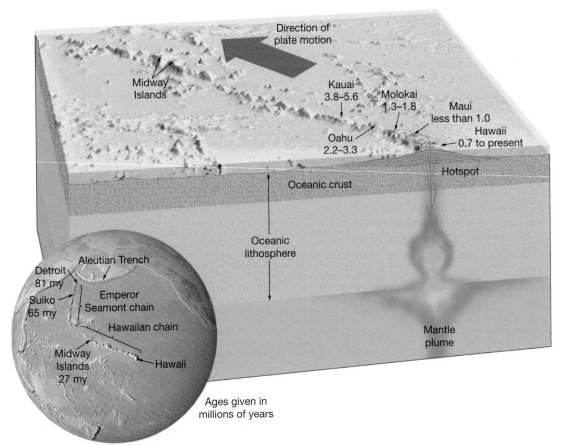

Figure 2.26 Hawaiian Islands–Emperor Seamount Chain. The chain of volcanoes that extends from Hawaii to the Aleutian Trench was created by the movement of the Pacific Plate over the Hawaiian hotspot. The sharp bend in the Hawaiian–Emperor chain (*globe*) was created by a combination of the changing motion of the Pacific Plate and the slow movement of the Hawaiian hotspot itself. Numbers represent radiometric age dates in millions of years before present.

Islands–Emperor Seamount chain in the northern Pacific Ocean (**Figure 2.26**). What created this chain of more than 100 intraplate volcanoes that stretch over 5800 kilometers (3000 miles)? Further, what caused the prominent bend in the overall direction that occurs in the middle of the chain?

To help answer these questions, let's examine the ages of the volcanoes in the chain. Every volcano in the chain has long since become extinct, except the volcano Kilauea on the island of Hawaii, which is the southeasternmost island of the chain. The age of volcanoes progressively increases northwestward from Hawaii (Figure 2.26). To the northwest, the volcanoes increase in age past Suiko Seamount (65 million years old) to Detroit Seamount (81 million years old), near the Aleutian Trench.

These age relationships suggest that the Pacific Plate has steadily moved northwestward, while the underlying mantle plume has remained relatively stationary. The resulting Hawaiian hotspot created each of the volcanoes in the chain. As the plate moved, it carried the active volcano off the hotspot, and a new volcano began forming, younger in age than the previous one. A chain of extinct volcanoes that is progressively older as one travels away from a hotspot is called a **nematath** (*nema* = thread, *tath* = dung or manure), or a *hotspot track* (see Figure 2.24). Evidence suggests that about 47 million years ago, the Pacific Plate shifted from a northerly to a northwesterly direction. This change in plate motion can account for the bend (large elbow) about halfway through the chain , separating the Hawaiian Islands from the Emperor Seamounts (see Figure 2.26). If this is true, then other hotspot tracks throughout the Pacific Plate should show a similar bend at roughly the same time, but most do not.

Recent research that may help resolve this disparity indicates that hotspots do not remain completely stationary. In fact, several studies have shown that most hotspots move at less than 1 centimeter (0.4 inch) per year, but some, like Hawaii, may have moved faster in the geologic past. Even if Hawaii's hotspot had moved faster in the past, it did not do so in a way that would have created the sharp bend in the Hawaiian–Emperor track seen in Figure 2.26. Moreover, recent plate reconstructions suggest that the observed bend in the Hawaiian–Emperor chain was created by a combination of the changing motion of the Pacific Plate (mainly as a result of changes in plate motions near Australia and Antarctica) and the slow movement of the Hawaiian hotspot itself. Other hotspot tracks may have been at least partially created by motion of their hotspots as well. Remarkably, hotspots appear to be moving in exactly the opposite direction of plates, so hotspots may still be useful for tracking plate motions.

In the future, what will become of Hawaii—the island that currently resides on the hotspot? Based on the hotspot model, the island will be carried to the northwest, off the hotspot, become inactive, and eventually be subducted into the Aleutian Trench, like all the rest of the volcanoes in the chain to the north of it. In turn, other volcanoes will build up over the hotspot. In fact, a 3500-meter (11,500-foot) volcano named **Loihi** already exists 32 kilometers (20 miles) southeast of Hawaii. Still 1 kilometer (0.6 mile) below sea level, Loihi is volcanically active and, based on its current rate of activity, it should reach the surface sometime between 30,000 and 100,000 years from now. As it builds above sea level, it will become the newest island in the long chain of volcanoes created by the Hawaiian hotspot.

Seamounts and Tablemounts

Many areas of the ocean floor (most notably on the Pacific Plate) contain tall volcanic peaks that resemble some volcanoes on land. These large volcanoes are called **seamounts** if they are cone shaped on top, like an upside-down ice cream cone. Some volcanoes are flat on top—unlike anything on land—and are called **tablemounts**, or **guyots**, after Princeton University's first geology professor, Arnold Guyot.[11] Until the theory of plate tectonics, it was unclear how the differences between seamounts and tablemounts could have been produced. The theory explains why tablemounts are flat on top and also why the tops of some tablemounts have shallow-water deposits, despite being located in very deep water.

The origin of many seamounts and tablemounts is related to the volcanic activity occurring at hotspots; the origin of others is related to processes occurring at the mid-ocean ridge (**Figure 2.27**). Because of sea floor spreading, active volcanoes (seamounts) occur along the crest of the mid-ocean ridge. Some may be built up so high that they rise above sea level and become islands, at which point wave erosion becomes important. When sea floor spreading has moved the seamount off its source of magma (whether it is a mid-ocean ridge or a hotspot), the top of the seamount can be flattened by waves in just a few million years. This flattened seamount—now a tablemount—continues to be carried away from its source and, after millions of years, is submerged deeper into the ocean. Frequently, tops of tablemounts contain evidence of shallow-water conditions (such as ancient coral reef deposits) that were carried with them into deeper water.

Coral Reef Development

On his voyage aboard the HMS *Beagle*, the famous naturalist **Charles Darwin**[12] noticed a progression of stages in **coral reef** development. He hypothesized that the origin of coral reefs depended on the subsidence (sinking) of volcanic islands (**Figure 2.28**) and published the concept in *The Structure and Distribution of Coral Reefs* in 1842. What Darwin's hypothesis lacked was a mechanism for how volcanic islands subside. Much later, advances in plate tectonic theory and samples of the deep structure of coral reefs provided evidence to help support Darwin's hypothesis.

Reef-building corals are colonial animals that live in shallow, warm, tropical seawater and produce a hard skeleton of limestone. Once corals are established in an area that has the conditions necessary for their growth, they continue to grow upward layer by layer with each new generation attached to the skeletons of its predecessors. Over millions of years, a thick sequence of coral reef deposits may develop if conditions remain favorable.

The three stages of development in coral reefs are called fringing, barrier, and atoll. **Fringing reefs** (Figure 2.28a) initially develop along the margin of a landmass (an island or a continent), where the temperature, salinity, and turbidity (cloudiness) of the water are suitable for reef-building corals. Often, fringing reefs are associated with active volcanoes whose lava flows run down the flanks of the volcano and kill the coral. Thus, these fringing reefs are not very thick or well developed. Because of

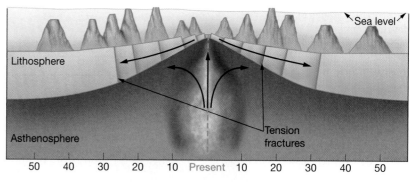

Age of ocean floor (millions of years)

Figure 2.27 Formation of seamounts and tablemounts at a mid-ocean ridge. Seamounts are tall volcanoes formed at volcanic centers such as the mid-ocean ridge. If they are tall enough to reach the surface, their tops are eroded flat by wave activity and become tablemounts. Through sea floor spreading, seamounts and tablemounts are transported into deeper water, sometimes carrying with them evidence of their tops once reaching shallow water.

ESSENTIAL CONCEPT 2.4A

Many independent lines of evidence, such as the detection of plate motion by satellites, provide strong support for the theory of plate tectonics.

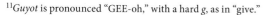

[11]*Guyot* is pronounced "GEE-oh," with a hard *g*, as in "give."

[12]For more information about Charles Darwin and the voyage of HMS *Beagle*, see Diving Deeper 1.2 in Chapter 1.

Figure 2.28 Stages of development in coral reefs. Cross-sectional view (*above*) and map view (*below*) of **(a)** a fringing reef, **(b)** a barrier reef, and **(c)** an atoll. With the right conditions for coral growth, a coral reef changes through time from fringing reef to barrier reef to atoll.

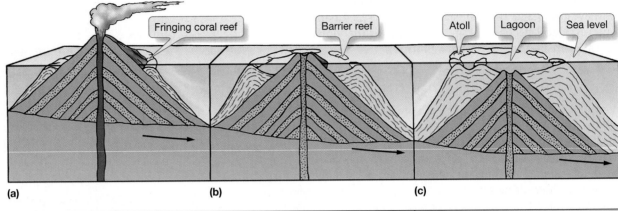

Web Animation
Seamounts/ Tablemounts and Coral Reef Stages

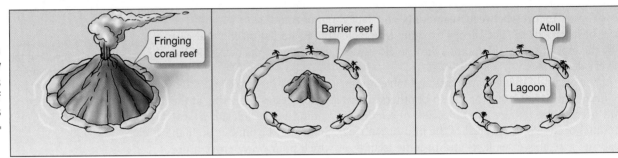

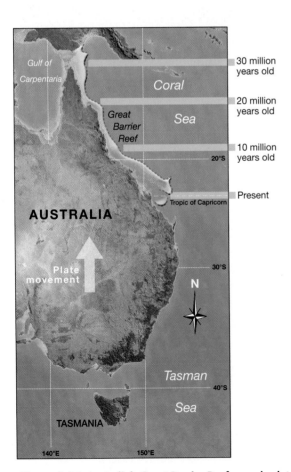

Figure 2.29 Australia's Great Barrier Reef records plate movement. About 30 million years ago, the Great Barrier Reef began to develop as northern Australia moved into tropical waters that allowed coral growth.

the close proximity of the landmass to the reef, runoff from the landmass can carry so much sediment that the reef is buried. The amount of living coral in a fringing reef at any given time is relatively small, with the greatest concentration in areas protected from sediment and salinity changes. If sea level does not rise or the land does not subside, the process stops at the fringing reef stage.

The **barrier reef** stage follows the fringing reef stage. Barrier reefs are linear or circular reefs separated from the landmass by a well-developed lagoon (Figure 2.28b). As the landmass subsides, the reef maintains its position close to sea level by growing upward. Studies of reef growth rates indicate that most have grown 3 to 5 meters (10 to 16 feet) per 1000 years during the recent geologic past. Evidence suggests that some fast-growing reefs in the Caribbean have grown more than 10 meters (33 feet) per 1000 years. Note that if the landmass subsides at a rate faster than coral can grow upward, the coral reef will be submerged in water too deep for it to live.

The largest reef system in the world is Australia's **Great Barrier Reef**, a series of more than 3000 individual reefs collectively in the barrier reef stage of development, home to hundreds of coral species and thousands of other reef-dwelling organisms. The Great Barrier Reef lies 40 kilometers (25 miles) or more offshore, averages 150 kilometers (90 miles) in width, and extends for more than 2000 kilometers (1200 miles) along Australia's shallow northeastern coast. The effects of the Indian–Australian Plate moving north toward the equator from colder Antarctic waters are clearly visible in the age and structure of the Great Barrier Reef (**Figure 2.29**). It is oldest (around 25 million years old) and thickest at its northern end because the northern part of Australia reached water warm enough to grow coral before the southern parts did. In other areas of the Pacific, Indian, and Atlantic Oceans, smaller barrier reefs are found around the tall volcanic peaks that form tropical islands.

The **atoll** (*atar* = crowded together) stage (Figure 2.28c) comes after the barrier reef stage. As a barrier reef around a volcano continues to subside, coral builds up toward the surface. After millions of years, the volcano becomes completely submerged, but the coral reef continues to grow. If the rate of subsidence is slow enough for the coral to keep up, a circular reef called an atoll is formed. The atoll encloses a lagoon usually not more than 30 to 50 meters (100 to 165 feet) deep. The reef generally has many channels that allow circulation between the lagoon and the open ocean. Buildups of crushed-coral debris often form narrow islands that encircle the central lagoon (**Figure 2.30**) and are large enough to allow human habitation.

Alternatively, a new theory has been put forward to explain the origin of coral atolls. The theory suggests that glacial cycles cause sea level to fluctuate, leading to episodes of reef exposure and dissolution when global sea level is lower during ice ages, alternating with coral reef submergence and deposition when sea level is higher during interglacial stages. Instead of the slow growth of ring-shaped coral above a sinking volcanic island, this alternating cycle may be responsible for the formation of coral atolls. Sea level change is discussed further in Chapter 10, "The Coast: Beaches and Shoreline Processes," and Chapter 16, "The Oceans and Climate Change."

Detecting Plate Motion with Satellites

Since the late 1970s, orbiting satellites have allowed the accurate positioning of locations on Earth. (This technique is also used for navigation by ships at sea; see Diving Deeper 1.1.) If the plates are moving, satellite positioning should show this movement over time. The map in **Figure 2.31** shows locations that have been measured in this manner and confirms that locations on Earth are indeed moving in good agreement with the direction and rate of motion predicted by plate tectonics.

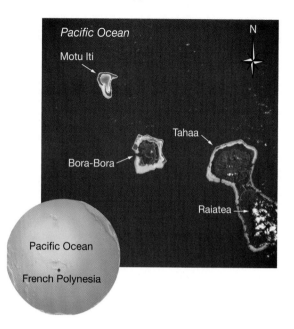

Figure 2.30 Barrier reefs and atoll. A portion of the Society Islands (French Polynesia) in the Pacific Ocean as photographed from the Space Shuttle. From lower right, the islands of Raiatea, Tahaa, and Bora-Bora are in the barrier reef stage of development, while Motu Iti is an atoll.

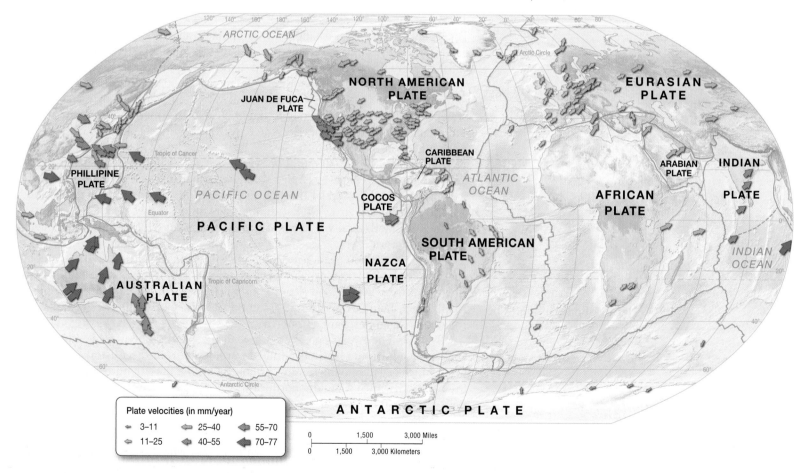

Figure 2.31 Satellite positioning of locations on Earth. Arrows show the direction of motion based on satellite measurement of positions on Earth. The rate of plate motion in millimeters per year is indicated with different-colored arrows (*see legend*). Plate boundaries are shown with blue lines and are dashed where uncertain or diffuse.

The successful prediction of how locations on Earth are moving with respect to one another very strongly supports plate tectonic theory.

CONCEPT CHECK 2.4

① How is the age distribution pattern of the Hawaiian Islands–Emperor Seamount chain explained by the position of the Hawaiian hotspot? What could have caused the curious bend in the chain?

② What are differences between a mid-ocean ridge and a hotspot?

③ How can plate tectonics be used to help explain the difference between a seamount and a tablemount?

④ Draw and describe each of the three stages of coral reef development. How does this sequence tie into the plate tectonic model?

2.5 How Has Earth Changed in the Past, and How Will it Look in the Future?

One of the most powerful features of any scientific theory is its ability to predict occurrences. Let's examine how plate tectonics can be used to determine the locations of the continents and oceans in the past, as well as what the continents and oceans will look like in the future.

The Past: Paleogeography

The study of historical changes of continental shapes and positions is called **paleogeography** (*paleo* = ancient, *geo* = earth, *graphy* = description of). As a result of paleogeographic changes, the size and shape of ocean basins have changed as well.

Figure 2.32 is a series of world maps showing the paleogeographic reconstructions of Earth at 60-million-year intervals. At 540 million years ago, many of the present-day continents are barely recognizable. North America was on the equator and rotated 90 degrees clockwise. Antarctica was on the equator and connected to many other continents.

Between 540 and 300 million years ago, the continents began to come together to form Pangaea. Notice that Alaska had not yet formed. Continents are thought to add material through the process of **continental accretion** (*ad* = toward, *crescere* = to grow). Like adding layers onto a snowball, bits and pieces of continents, islands, and volcanoes are added to the edges of continents and create larger landmasses.

From 180 million years ago to the present, Pangaea separated and the continents moved toward their present-day positions. North America and South America rifted away from Europe and Africa to produce the Atlantic Ocean. In the Southern Hemisphere, South America and a continent composed of India, Australia, and Antarctica began to separate from Africa.

By 120 million years ago, there was a clear separation between South America and Africa, and India had moved northward, away from the Australia–Antarctica mass, which began moving toward the South Pole. As the Atlantic Ocean continued to open, India moved rapidly northward and collided with Asia about 45 million years ago. Australia had also begun a rapid journey to the north since separating from Antarctica.

One major outcome of global plate tectonic events over the past 180 million years has been the creation of the Atlantic Ocean, which continues to grow as the sea floor spreads along the Mid-Atlantic Ridge. At the same time, the Pacific Ocean continues to shrink due to subduction along the many trenches that surround it and continental plates that bear in from both the east and west.

Web Animation
Plate Motions Through Time

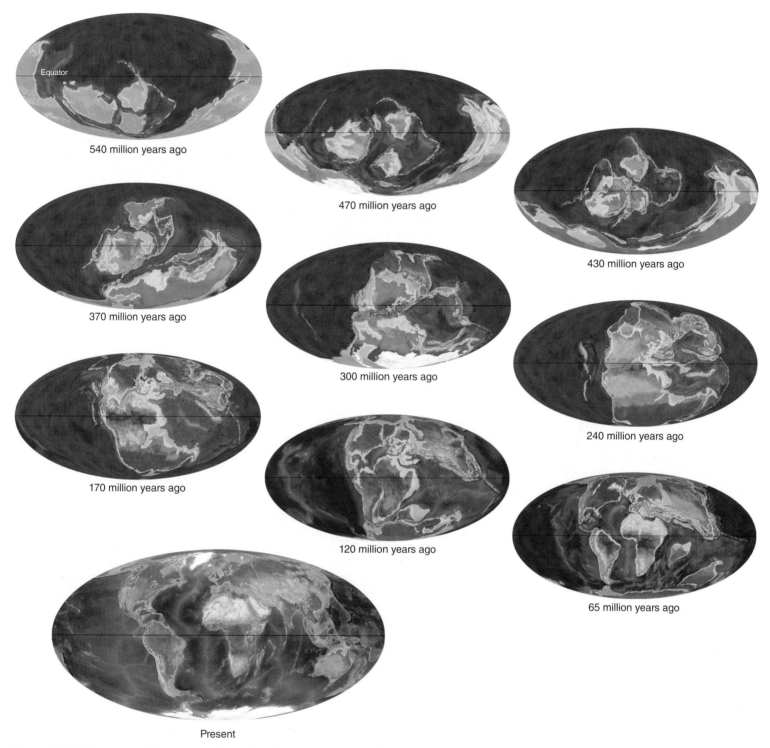

540 million years ago

470 million years ago

430 million years ago

370 million years ago

300 million years ago

240 million years ago

170 million years ago

120 million years ago

65 million years ago

Present

Figure 2.32 Paleogeographic reconstructions of Earth. The positions of the continents from 540 million years ago (*top*) to present (*large map at bottom*).

The Future: Some Bold Predictions

Using plate tectonics, a prediction of the future positions of features on Earth can be made based on the assumption that the rate and direction of plate motion will remain the same. Although these assumptions may not be entirely valid, they do provide a framework for the prediction of the positions of continents and other Earth features in the future.

STUDENTS SOMETIMES ASK . . .

> *Will the continents come back together and form a single landmass anytime soon?*

Yes, it is very likely that the continents will come back together, but not anytime soon. Because all the continents are on the same planetary body, a continent can travel only so far before it collides with other continents. Recent research suggests that the continents may form a supercontinent once every 500 million years or so. It has been 200 million years since Pangaea split up, so we only have about 300 million years to establish world peace! Even though that's a long time from now, researchers have already dubbed the new supercontinent "Amasia."

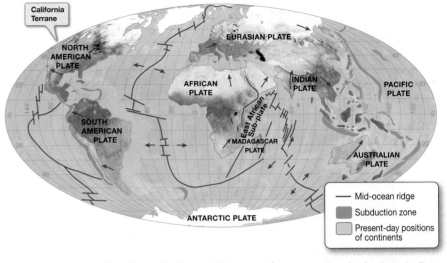

Figure 2.33 **The world as it may look 50 million years from now.** Purple shadows indicate the present-day positions of continents, and colored landmasses indicate positions of the continents in 50 million years. Red arrows indicate the direction of plate motion.

Figure 2.33 is a map of what the world may look like 50 million years from now, showing many notable differences from today. For instance, the East Africa Rift Valleys may enlarge to form a new linear sea, and the Red Sea may be greatly enlarged because of rifting there. India may continue to plow into Asia, further uplifting the Himalaya Mountains. As Australia moves north toward Asia, it may use New Guinea like a snowplow to accrete various islands. North America and South America may continue to move west, enlarging the Atlantic Ocean and decreasing the size of the Pacific Ocean. The land bridge of Central America may no longer connect North and South America; this would dramatically alter ocean circulation. Finally, the thin sliver of land that lies west of the San Andreas Fault may become an island in the North Pacific, soon to be accreted onto southern Alaska.

ESSENTIAL CONCEPT 2.5

The geographic positions of the continents and ocean basins are not fixed in time or place. Rather, they have changed in the past and will continue to change in the future.

CONCEPT CHECK 2.5

1 Using the paleogeographic reconstructions shown in Figure 2.32, determine when the following events first appear in the geologic record:

a. North America lies on the equator.

b. The continents come together as Pangaea.

c. The North Atlantic Ocean opens.

d. India separates from Antarctica.

2.6 How Can Plate Tectonics Be Used As A Working Model?

Since its inception by Alfred Wegener nearly 100 years ago, plate tectonics has been supported by a wealth of scientific evidence—some of which is presented in this chapter. Although there are still details to be worked out (such as the exact driving mechanism), the theory of plate tectonics has been universally accepted by Earth scientists today because it helps explain so many features and processes that are observed on Earth (**Diving Deeper 2.2**). Further, it has led to predictive models that have been used to successfully understand Earth behavior. One such example is the **Wilson cycle** (**Figure 2.34**), named in honor of geophysicist John Tuzo Wilson for his contribution to the early ideas of plate tectonics. The Wilson cycle uses plate tectonic processes to show the distinctive life cycle of ocean basins during their formation, growth, and destruction over many millions of years.

Stage, showing cross-sectional view	Motion	Physiography	Example
EMBRYONIC 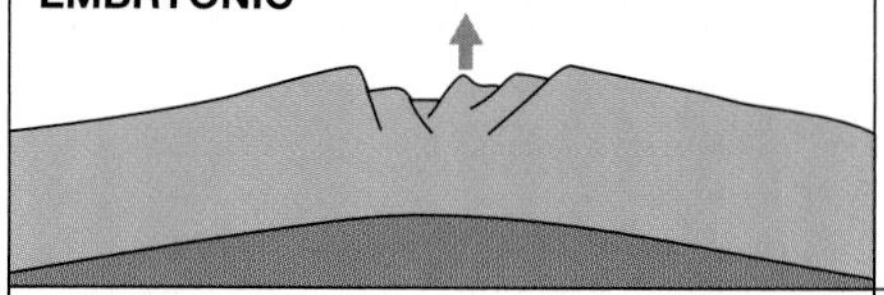	Uplift	Complex system of linear rift valleys on continent	East Africa rift valleys
JUVENILE 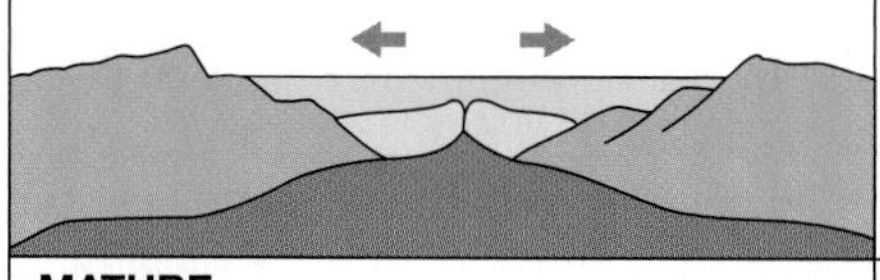	Divergence (spreading)	Narrow seas with matching coasts	Red Sea
MATURE 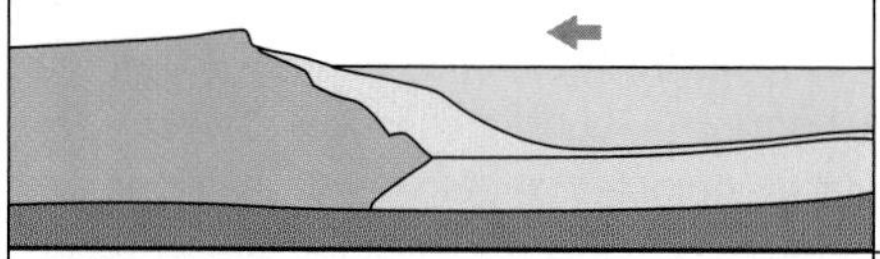	Divergence (spreading)	Ocean basin with continental margins	Atlantic and Arctic Oceans
DECLINING 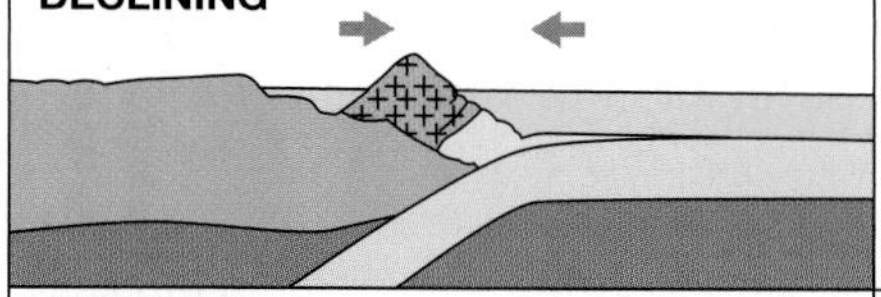	Convergence (subduction)	Island arcs and trenches around basin edge	Pacific Ocean
TERMINAL 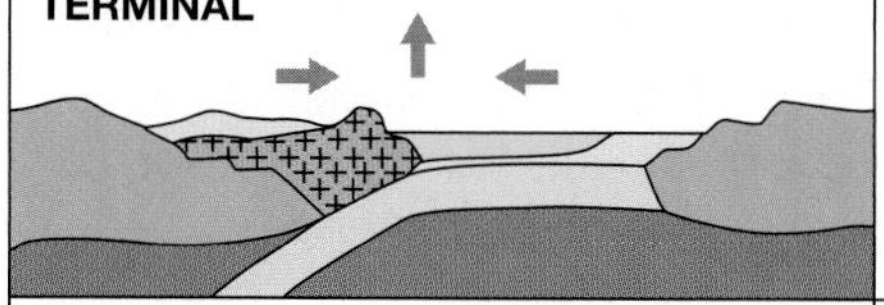	Convergence (collision) and uplift	Narrow, irregular seas with young mountains	Mediterranean Sea
SUTURING 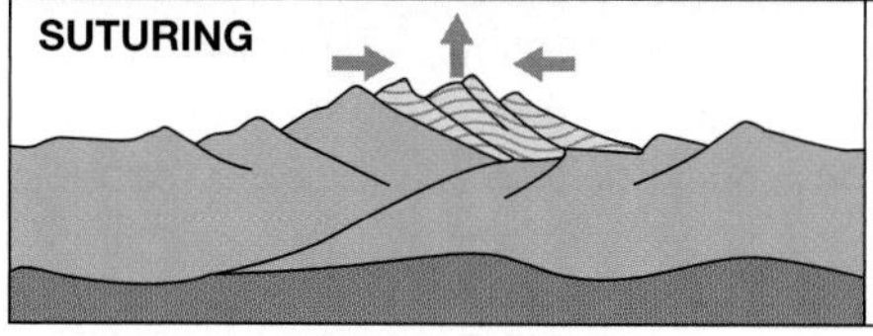	Convergence and uplift	Young to mature mountain belts	Himalaya Mountains

Figure 2.34 The Wilson cycle of ocean basin evolution. The Wilson cycle depicts the stages of ocean basin development, from the initial embryonic stage of formation to the destruction of the basin as continental masses collide and undergo suturing.

In the embryonic stage of the Wilson cycle, a heat source beneath the lithosphere creates uplift and begins to split a continent apart. The juvenile stage is characterized by further spreading, downdropping, and the formation of a narrow, linear sea. In the mature stage, an ocean basin is fully developed and a mid-ocean ridge runs down the middle of it. Eventually, a subduction zone occurs along the continental margin, and the plates come back together, producing the declining stage where the ocean basin shrinks. The terminal stage is marked by the plates coming back together, creating a progressively narrower ocean. Finally, in the suturing stage, the ocean disappears, the continents collide, and tall uplifted mountains are created. Over time, the uplifted mountains erode and the stage is set for the cycle to repeat.

Not only is plate tectonic activity primarily responsible for the creation of landforms, it also plays a prominent role in the development of ocean floor features—which is the topic of the next chapter. Armed with the knowledge of plate tectonic processes you've gained from this chapter, understanding the history and development of ocean floor features in various marine provinces will be a much simpler task.

ESSENTIAL CONCEPT 2.6

Working models of plate tectonics include the Wilson cycle, which describes the life cycle of ocean basins during their formation, growth, and destruction over millions of years.

CONCEPT CHECK 2.6

1 Determine the Wilson cycle stage for the following present-day locations, noting the features and processes that support your answers:

a. The Atlantic Ocean
b. The Pacific Ocean
c. The Red Sea
d. The Alps

Diving Deeper 2.2 Historical Feature

OPHIOLITES: A GIFT FROM THE SEA FLOOR TO THE BRONZE AGE

As a result of plate tectonic processes, most oceanic lithosphere that is created at spreading centers is ultimately subducted and returned to the underlying mantle. However, a very small amount—only about 0.001%—of oceanic lithosphere does not get subducted; instead, it is uplifted onto the edge of a continent in a process called *obduction* (*ob* = against, *duct* = lead). Fragments of the sea floor that are hoisted onto land are composed of a unique sequence of rocks called **ophiolites** (*ophis* = snake, *lite* = stone), which are so named because they are typically composed of mottled green-colored rocks that resemble snake skins. Ophiolite sequences on land have long been recognized (**Figure 2B**) but have only recently been connected to their environment of formation on the sea floor.

Figure 2C (*part a, left*) shows an idealized ophiolite sequence, which contains the following rock types vertically from top to bottom:

1. Layered *marine sediments*—such as chert, limestone, or shale—which get progressively older with depth below the sea floor.
2. *Pillow basalts*, which indicate the lava flows that produced them flowed out on the ocean floor and cooled quickly during contact with water.

3. *Vertical basalt dikes*, which form by being injected into cracks in oceanic crust as lithospheric plates pull apart and move away from the spreading center. Because they are essentially thin planar features, they are often called *sheet dikes*.
4. *Massive gabbro*, a rock similar to basalt in composition but of a coarser texture due to slower cooling. This gabbro may represent the cooling of magma near the roof of the magma chamber that was the source of the overlying basalts.
5. *Layered gabbro* that may have crystallized within the upper region of the magma chamber.
6. *Peridotite/serpentinite*: *Peridotite* is a mantle rock composed mostly of iron- and magnesium-rich minerals called pyroxene and olivine; *serpentinite* (*serpens* = snake) is hydrothermally altered peridotite. The *Moho* is the boundary that separates the layered peridotite/serpentinite zone with layered gabbro above; it represents the contact between oceanic crust and the underlying mantle.

The main features of an idealized ophiolite sequence can also be observed in cores drilled into the sea floor near mid-ocean ridges. Such is the case for cores recovered from the Deep Sea Drilling Project (DSDP) Drill Hole No. 504B,

which is located on the southern flank of the Costa Rica Rift (Figure 2C, *part c*). The hole was drilled to a depth of 1350 meters (4429 feet) beneath the ocean floor, at that time the deepest hole ever drilled into ocean crust. Although it did not penetrate through the complete ophiolite sequence,[1] the rocks it did penetrate show how well this sequence resembles those observed in ophiolites studied on land. (Figure 2C, *part a, right*).

Many ophiolites contain rich metallic ores with patterns of enrichment similar to those occurring in the sediments and oceanic crust near deep-sea hydrothermal vents on the sea floor (Figure 2C, *part b*). However, there is increasing evidence that many ophiolites may have formed at back-arc spreading centers associated with subduction zones rather than at oceanic ridges and rises.

One of the best-studied ophiolite sequences is the Troodos Massif on the island of Cyprus in the eastern Mediterranean Sea (Figure 2B, *part b*). Troodos Massif is also one of the oldest-known ore deposits, having been mined for copper[2] since the time of the Phoenicians. In fact, this copper deposit was largely responsible for helping civilization advance from the Stone

Figure 2B Distribution and mining of ophiolites. (a) Map showing the locations of major ophiolites. Ophiolites found near the margins of continents where subduction occurs are generally younger (less than 200 million years old) than those embedded deep within continents, some of which are as much as 1.2 billion years old. The location of DSDP Drill Hole No. 504B—the deepest hole yet drilled through ocean crust—is also shown. **(b)** Location of the Troodos Massif ophiolite on the island of Cyprus in the eastern Mediterranean Sea. **(c)** Mining of the Troodos Massif sulfide ore, which is the dark green material in the bottom of the pit.

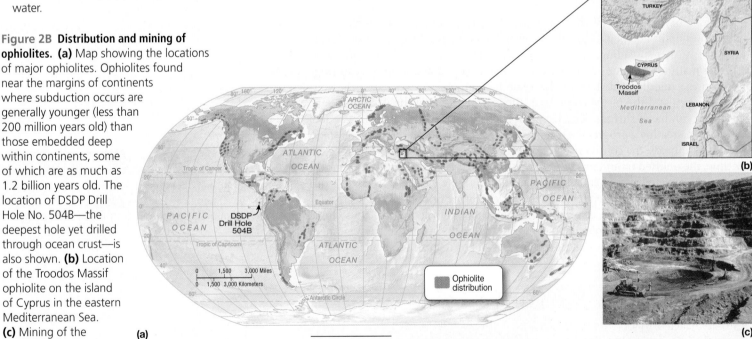

---

[1]A new series of deep-drilling expeditions—which began in 2002 with IODP Expedition 206—aim to drill completely though a sea floor ophiolite sequence.

[2]*Cyprus* means "copper"; even today, it is unclear if the island is named for the metal, or vice versa.

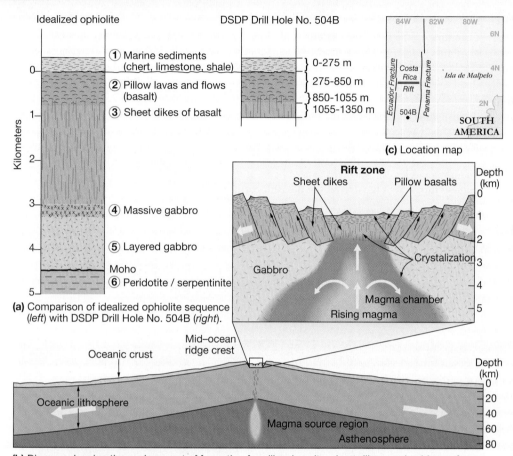

Idealized ophiolite

① Marine sediments (chert, limestone, shale)
② Pillow lavas and flows (basalt)
③ Sheet dikes of basalt
④ Massive gabbro
⑤ Layered gabbro
Moho
⑥ Peridotite / serpentinite

DSDP Drill Hole No. 504B

0–275 m
275–850 m
850–1055 m
1055–1350 m

(c) Location map

(a) Comparison of idealized ophiolite sequence (*left*) with DSDP Drill Hole No. 504B (*right*).

Rift zone

Sheet dikes   Pillow basalts
Gabbro
Crystalization
Magma chamber
Rising magma

Mid–ocean ridge crest

Oceanic crust
Oceanic lithosphere
Magma source region
Asthenosphere

Depth (km)

(b) Diagram showing the environment of formation for pillow basalts, sheet dikes, and gabbros of an ophiolite complex.

Figure 2C   Ophiolite rock types and environment of formation for an ophiolite complex.

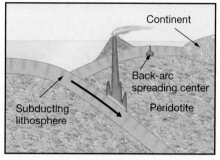

(a) Subducting lithosphere contains water-bearing minerals.

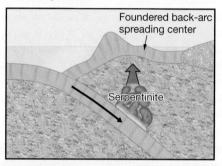

(b) Water-bearing minerals are heated, and hot water is released into the peridotite, which is altered to lower-density serpentinite. The expansion of the serpentinite lifts the overlying mantle and oceanic crust.

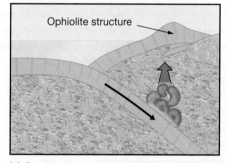

(c) Continued compressional forces break the uplifted segments of oceanic crust and wedge them onto the margin of a continent as ophiolites.

Figure 2D   Diagrammatic sequence showing the obduction of an ophiolite onto a continent.

Age to the Bronze Age. Archeological evidence indicates in about 2760 B.C., people living on Cyprus discovered that copper could be hardened with the addition of tin during smelting. The smelted bronze could then be fashioned into various weapons and tools that, at the time, had no equal. If the copper deposits of Troodos Massif were not so readily available, humankind might have remained in the Stone Age much longer!

How do ophiolites—which have high density—become obducted onto continents? **Figure 2D** shows the probable sequence of events (see also the Web Animation "Terrane Formation"). Although some serpentinite in ophiolites forms by hydrothermal alteration of peridotite at the mid-ocean ridge, serpentinite also forms when a plate is subducted.

Moreover, the subducting oceanic plate contains water-bearing minerals that release their water when heated. This heated water converts peridotite into serpentinite, which is much less dense than the surrounding peridotite. Because of compressive forces, ophiolite sequences including serpentinite are forced up above sea level along fractures and other cracks during subduction. Note that this sequence is responsible for the emplacement of ophiolites in locations such as the Franciscan Formation of coastal California.

Further exploration of the deep sea floor is needed to answer some of the details that remain about the origin and emplacement of ophiolites, but ophiolites do indeed fit well into the plate tectonic model.

Web Animation
Terrane Formation

Essential Concepts Review

2.1 What evidence supports continental drift?

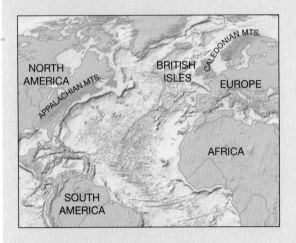

▶ According to the theory of plate tectonics, *the outermost portion of Earth is composed of a patchwork of thin, rigid lithospheric plates that move horizontally* with respect to one another. The idea began as a hypothesis called continental drift proposed by *Alfred Wegener* at the start of the 20th century. He suggested that *about 200 million years ago, all the continents were combined* into one large continent (*Pangaea*) surrounded by a single large ocean (*Panthalassa*).

▶ *Many lines of evidence were used to support the idea of continental drift*, including the similar shape of nearby continents, matching sequences of rocks and mountain chains, glacial ages and other climate evidence, and the distribution of fossil and present-day organisms. Although this evidence suggested that continents have drifted, other incorrect assumptions about the mechanism involved caused *many geologists and geophysicists to discount this hypothesis* throughout the first half of the 20th century.

Study Resources
Online Study Guide Quizzes, Web Animations

Critical Thinking Question

If you could travel back in time with three illustrations from this chapter to help Alfred Wegener convince the scientists of his day that continental drift does indeed exist, what would they be, and why?

2.2 What evidence supports plate tectonics?

▶ *More convincing evidence for drifting continents was introduced in the 1960s, when paleomagnetism*—the study of Earth's ancient magnetic field—*was developed* and the significance of features of the ocean floor became better known. The paleomagnetism of the ocean floor is permanently recorded in oceanic crust and reveals stripes of normal and reverse magnetic polarity in a symmetric pattern relative to the mid-ocean ridge.

▶ *Harry Hess advanced the idea of sea floor spreading. New sea floor is created at the crest of the mid-ocean ridge* and moves apart in opposite directions and is *eventually destroyed by subduction into an ocean trench*. This helps explain the pattern of magnetic stripes on the sea floor and why sea floor rocks increase linearly in age in either direction from the axis of the mid-ocean ridge. Other supporting evidence for plate tectonics includes oceanic heat flow measurements and the pattern of worldwide earthquakes. The combination of evidence *convinced geologists of Earth's dynamic nature* and helped advance the idea of continental drift into the more encompassing plate tectonic theory.

Study Resources
Online Study Guide Quizzes, Web Table 2.1, Web Animations

Critical Thinking Question

Why was the pattern of alternating reversals of Earth's magnetic field, as recorded in sea floor rocks, such an important piece of evidence for advancing the theory of plate tectonics?

2.3 What features occur at plate boundaries?

▶ *As new crust is added to the lithosphere at the mid-ocean ridge (divergent boundaries where plates move apart)*, the opposite ends of the *plates are subducted into the mantle at ocean trenches* or beneath continental mountain ranges such as the Himalayas (*convergent boundaries where plates come together*). In addition, oceanic ridges and rises are offset, and *plates slide past one another along transform faults (transform boundaries where plates slowly grind past one another).*

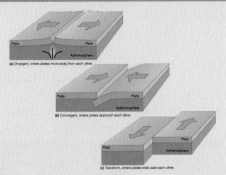

Study Resources
Online Study Guide Quizzes, Web Diving Deeper 2.1, Web Animations

Critical Thinking Question

List and describe the three types of plate boundaries. Include in your discussion any sea floor features that are related to these plate boundaries and include a real-world example of each. Construct a map view and cross section showing each of the three types of plate boundaries, including the direction of plate movement and associated features.

2.4 Testing the model: What are some applications of plate tectonics?

▶ *Tests of the plate tectonic model indicate that many features and phenomena provide support for shift-ing plates.* These include mantle plumes and their associated hotspots that record the motion of plates past them, the origin of flat-topped tablemounts, stages of coral reef development, and detection of plate motion by accurate positioning of locations on Earth using satellites.

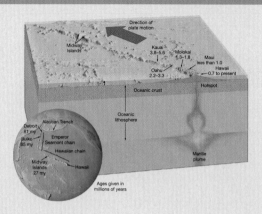

Study Resources
Online Study Guide Quizzes, Web Animations

Critical Thinking Question

Describe the differences in origin between the Aleutian Islands and the Hawaiian Islands. Provide evidence to support your explanation.

2.5 How has Earth changed in the past, and how will it look in the future?

▶ *The positions of various sea floor and continental features have changed in the past, continue to change today, and will look very different in the future.*

Study Resources
Online Study Guide Quizzes, Web Animation

Critical Thinking Question

Assume that you travel at the same rate as a fast-moving continent—at a rate of 10 centimeters (2.5 inches) per year. Calculate how long it would take you to travel from your present location to a nearby large city. Also, calculate how long it would take you to travel across the United States from the East Coast to the West Coast.

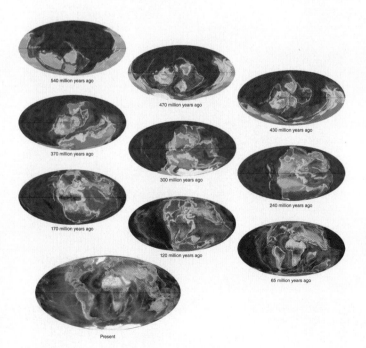

2.6 How can plate tectonics be used as a working model?

▶ Working models of plate tectonics include the *Wilson cycle*, which describes the *life cycle of ocean basins during their formation, growth, and destruction* over millions of years.

Study Resources
Online Study Guide Quizzes, Web Animation

Critical Thinking Question

Using the Wilson cycle as a predictive model, describe the sequence of events that will happen in the future to the East Africa Rift Valleys. Also, do the same for the Atlantic Ocean.

Stage, showing cross-sectional view	Motion	Physiography	Example
EMBRYONIC	Uplift	Complex system of linear rift valleys on continent	East Africa rift valleys
JUVENILE	Divergence (spreading)	Narrow seas with matching coasts	Red Sea
MATURE	Divergence (spreading)	Ocean basin with continental margins	Atlantic and Arctic Oceans
DECLINING	Convergence (subduction)	Island arcs and trenches around basin edge	Pacific Ocean
TERMINAL	Convergence (collision) and uplift	Narrow, irregular seas with young mountains	Mediterranean Sea
SUTURING	Convergence and uplift	Young to mature mountain belts	Himalaya Mountains

MasteringOceanography™

www.masteringoceanography.com

Looking for additional review and test prep materials? Visit the Study Area in MasteringOceanography to enhance your understanding of this chapter's content by accessing a variety of resources, including Self-Study Quizzes, Geoscience Animations, RSS feeds, flashcards, Web links, and an optional Pearson eText.

The North Atlantic sea floor. The sea floor has many interesting features, some of which are completely different from those on land. Recent improvements in technology have aided exploration of the sea floor and given scientists the ability to create stunning high-resolution images like this one that shows the Mid-Atlantic Ridge.

Before you begin reading this chapter, use the glossary at the end of this book to discover the meanings of any of these words you don't already know:

Volcanic arc · Seamount · Active margin · Continental shelf · Submarine fan · Sonar · fracture zone · Black smoker · Ocean trench · Mid-ocean ridge · Tablemount · Passive margin · Rift valley · Submarine canyon · Graded bedding · Pacific Ring of Fire · Island arc · Continental arc · Turbidity current · Continental rise · Continental slope · Pillow lava · Abyssal plain · Hydrothermal vent · Transform fault

3 MARINE PROVINCES

ESSENTIAL CONCEPTS

At the end of this chapter, you should be able to:

3.1 Discuss the techniques that are used to determine ocean bathymetry

3.2 Describe the sea floor features that exist on continental margins

3.3 Describe the sea floor features that exist in the deep-ocean basins

3.4 Describe the sea floor features that exist along the mid-ocean ridge

"Could the waters of the Atlantic be drawn off so as to expose to view this great sea-gash which separates the continents, and extends from the Arctic to the Antarctic, it would present a scene most rugged, grand, and imposing."

—Matthew Fontaine Maury (1854), the "father of oceanography," commenting about the Mid-Atlantic Ridge

What does the shape of the ocean floor look like? Over a century and a half ago, most scientists believed that the ocean floor was completely flat and carpeted with a thick layer of muddy sediment containing little of scientific interest. Further, it was believed that the deepest parts were somewhere in the middle of the ocean basins. However, as more and more vessels crisscrossed the seas to map the ocean floor and to lay transoceanic cables, scientists found that the terrain of the sea floor is highly varied and includes deep troughs, ancient volcanoes, submarine canyons, and great mountain chains. It is unlike anything on land and, as it turns out, some of the deepest parts of the oceans are actually very close to land!

As marine geologists and oceanographers began to analyze the features of the ocean floor, they realized that certain features had profound implications not only for the history of the ocean floor but also for the history of Earth. How could all these remarkable features have formed, and how can their origin be explained? Over long periods of time, the shape of the ocean basins has changed as continents have ponderously migrated across Earth's surface in response to forces within Earth's interior. The ocean basins as they presently exist reflect the processes of plate tectonics (the topic of the previous chapter), which help explain the origin of sea floor features.

3.1 What Techniques Are Used to Determine Ocean Bathymetry?

Bathymetry (*bathos* = depth, *metry* = measurement) is the measurement of ocean depths and the charting of the shape, or *topography* (*topos* = place, *graphy* = description of) of the ocean floor. Determining bathymetry involves measuring the vertical distance from the ocean surface down to the mountains, valleys, and plains of the sea floor.

Soundings

The first recorded attempt to measure the ocean's depth was conducted in the Mediterranean Sea in about 85 B.C. by a Greek named Posidonius. His mission was to answer an age-old question: How deep is the ocean? Posidonius's crew made a **sounding**[1] by letting out nearly 2 kilometers (1.2 miles) of line before the heavy weight on the end of the line touched bottom. For the next 2000 years, voyagers used sounding lines to probe the ocean's depths. The standard unit of ocean depth is the **fathom** (*fathme* = outstretched arms[2]), which is equal to 1.8 meters (6 feet).

The first systematic bathymetric measurements of the oceans were made in 1872 aboard the HMS *Challenger*, during its historic three-and-a-half-year voyage.[3] Every so often, *Challenger*'s crew stopped and measured the depth, along with many other ocean properties. These measurements indicated that the deep-ocean floor was not flat but had significant *relief* (variations in elevation), just as dry land does. However, determining bathymetry by making occasional soundings rarely gives a complete picture of the ocean floor. For instance, imagine trying to determine what the surface features on land look like while flying in a blimp at an altitude of several kilometers on a foggy night, using only a long weighted rope to determine your height above the surface. This is similar to how bathymetric measurements were collected from ships using sounding lines.

Echo Soundings

The presence of mid-ocean undersea mountains had long been known, but recognition of their full extent into a connected worldwide system had to await the invention and use of the **echo sounder**, or *fathometer*, in the early 1900s. An echo sounder sends a sound signal (called a **ping**) from the ship downward into the ocean, where it produces echoes when it bounces off any density difference, such as marine organisms or the ocean floor (**Figure 3.1**). Water is a good transmitter of sound, so the time it takes for the echoes to return[4] is used to determine the depth and corresponding shape of the ocean floor. In 1925, for example, the German vessel *Meteor* used echo sounding to identify the underwater mountain range running through the center of the South Atlantic Ocean.

Echo sounding, however, lacks detail and often gives an inaccurate view of the relief of the sea floor. For instance, the sound beam emitted from a ship 4000 meters (13,100 feet) above the ocean floor widens to a diameter of about 4600 meters (15,000 feet) at the bottom. Consequently, the first echoes to return from the bottom are usually from the closest (highest) peak within this broad area. Nonetheless, most of our knowledge of ocean bathymetry has been provided by the echo sounder.

Because sound from echo sounders bounces off any density difference, it was discovered that echo sounders could detect and track submarines. During World War II, antisubmarine warfare inspired many improvements in the technology of "seeing" into the ocean using sound.

During and after World War II, there was great improvement in sonar technology. For example, the **precision depth recorder (PDR)**, which was developed in the 1950s, uses a focused high-frequency sound beam to measure depths to a resolution

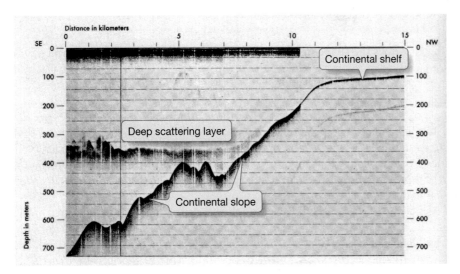

Figure 3.1 An echo sounder record. An echo sounder record of the East Coast U.S. offshore region shows the provinces of the sea floor. Vertical exaggeration (the amount of expansion of the vertical scale) is 12 times. The scattering layer probably represents a concentration of marine organisms.

---

[1]A *sounding* refers to a probe of the environment for scientific observation and was borrowed from atmospheric scientists, who released probes called soundings into the atmosphere. Ironically, the term does not actually refer to sound; the use of sound to measure ocean depths came later.

[2]This term is derived from the method used to bring depth sounding lines back on board a vessel by hand. While hauling in the line, workers counted the number of arm-lengths collected. By measuring the length of the person's outstretched arms, the amount of line taken in could be calculated. Much later, the distance of 1 fathom was standardized to equal exactly 6 feet.

[3]For more information about the accomplishments of the *Challenger* expedition, see Web Diving Deeper 5.2.

[4]This technique uses the speed of sound in seawater, which varies with salinity, pressure, and temperature but averages about 1507 meters (4945 feet) per second.

of about 1 meter (3.3 feet). Throughout the 1960s, PDRs were used extensively and provided a reasonably good representation of the ocean floor. From thousands of research vessel tracks, the first reliable global maps of sea floor bathymetry were produced. These maps helped confirm the ideas of sea floor spreading and plate tectonics.

Modern *acoustic* (*akouein* = to hear) instruments that use sound to map the sea floor include *multibeam echo sounders* (which use multiple frequencies of sound simultaneously) and side-scan **sonar** (an acronym for *so*und *na*vigation *a*nd *r*anging). **Seabeam**—the first multibeam echo sounder—made it possible for a survey ship to map the features of the ocean floor along a strip up to 60 kilometers (37 miles) wide. Multibeam systems use sound emitters directed away from both sides of a survey ship, with receivers permanently mounted on the ship's hull. Multibeam instruments emit multiple beams of sound waves, which are reflected off the ocean floor. As the sound waves bounce back with different strengths and timing, computers analyze these differences to determine the depth and shape of the sea floor and whether the bottom is rock, sand, or mud (**Figure 3.2**). In this way, multibeam surveying provides incredibly detailed imagery of the seabed. Because its beams of sound spread out with depth, multibeam systems have resolution limitations in deep water.

In deep water or where a detailed survey is required, side-scan sonar can provide enhanced views of the sea floor. Side-scan sonar systems such as **Sea MARC** (*Sea M*apping *a*nd *R*emote *C*haracterization) and **GLORIA** (*G*eological *L*ong-*R*ange *I*nclined *A*coustical instrument) can be towed behind a survey ship to produce a detailed strip map of ocean floor bathymetry (**Figure 3.3**). To maximize its resolution, a side-scan instrument can be towed behind a ship on a cable so that it "flies" just above the ocean floor.

Figure 3.2 Multibeam sonar. An artist's depiction of how a survey vessel uses multibeam sonar to map the ocean floor. Hull-mounted multibeam instruments emit multiple beams of sound waves, which are reflected off the ocean floor. Receivers collect data that allow oceanographers to determine the depth, shape, and even composition of the sea floor. As a ship travels back and forth throughout an area, it can produce a detailed image of sea floor bathymetry.

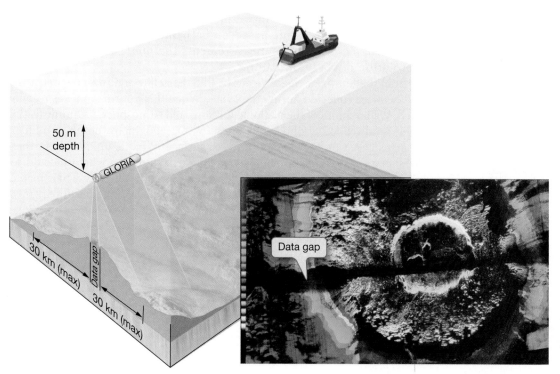

Figure 3.3 Side-scanning sonar. The side-scan sonar system GLORIA (*left*) is towed behind a survey ship and can map a strip of ocean floor (a swath) with a gap in data directly below the instrument. Side-scan sonar image of a volcano (*right*) with a summit crater about 2 kilometers (1.2 miles) in diameter in the Pacific Ocean. The black stripe through the middle of the image is the data gap.

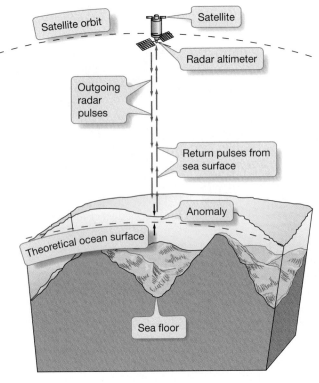

Figure 3.4 Satellite measurements of the ocean surface. A satellite measures the variation of ocean surface elevation, which is caused by gravitational attraction and mimics the shape of the sea floor. The sea surface *anomaly* is the difference between the measured and theoretical ocean surface.

Using Satellites to Map Ocean Properties from Space

Although multibeam and side-scan sonar produce very detailed bathymetric maps, mapping the sea floor by ship is an expensive and time-consuming process. A research vessel must tediously travel back and forth throughout an area (a process called "mowing the lawn") to produce an accurate map of bathymetric features (see Figure 3.2). Unfortunately, only a small percentage of the ocean floor has been mapped in this way.

An Earth-orbiting satellite, on the other hand, can observe large areas of the ocean at one time. Consequently, satellites are increasingly used to determine ocean properties. A list of recent U.S. oceanographic satellite missions and their objectives are shown in **Web Table 3.1**. Remarkably, recent satellite measurements of the ocean surface have been used to make maps of the sea floor. How does a satellite—which orbits at a great distance above the planet and can view only the ocean's *surface*—obtain a picture of the sea *floor*?

The answer lies in the fact that sea floor features directly influence Earth's gravitational field. Deep areas such as trenches correspond to a lower gravitational attraction, and large undersea objects such as seamounts exert an extra gravitational pull. These differences affect sea level, causing the ocean surface to bulge outward and sink inward, mimicking the relief of the ocean floor. A 2000-meter (6500-foot)-high seamount, for example, exerts a small but measurable gravitational pull on the water around it, creating a bulge 2 meters (7 feet) high on the ocean surface. These irregularities are easily detectable by satellites, which use microwave beams to measure sea level to an accuracy of 4 centimeters (1.5 inches). After corrections are made for waves, tides, currents, and atmospheric effects, the resulting pattern of dips and bulges at the ocean surface can be used to indirectly reveal ocean floor bathymetry (**Figure 3.4**). For example, **Figure 3.5** compares two different maps of the same area: one based on bathymetric data from ships (*top*) and the other based on satellite measurements (*bottom*), which shows much higher resolution of sea floor features.

Data from the European Space Agency's ERS-1 satellite and from Geosat, a U.S. Navy satellite, were collected during the 1980s. When this information was recently declassified, Walter Smith of the National Oceanic and Atmospheric Administration and David Sandwell of Scripps Institution of Oceanography began producing sea floor maps based on the shape of the sea surface. What is unique about these researchers' maps is that they provide a view of Earth similar to what we could see if we were able to drain the oceans and view the ocean floor directly. Their map of ocean surface gravity (**Figure 3.6**) uses depth soundings to calibrate the gravity measurements. Although gravity is not exactly bathymetry, this new map of the ocean

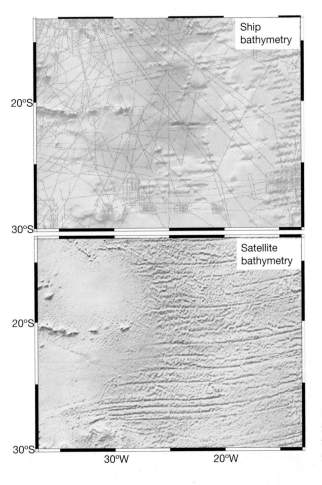

Figure 3.5 Comparing bathymetric maps of the sea floor. Both bathymetric maps show the same portion of the Brazil Basin in the South Atlantic Ocean. *Top:* A map made using conventional echo sounder records from ships (ship tracks shown by thin lines). *Bottom:* A map from satellite data made using measurements of the ocean surface.

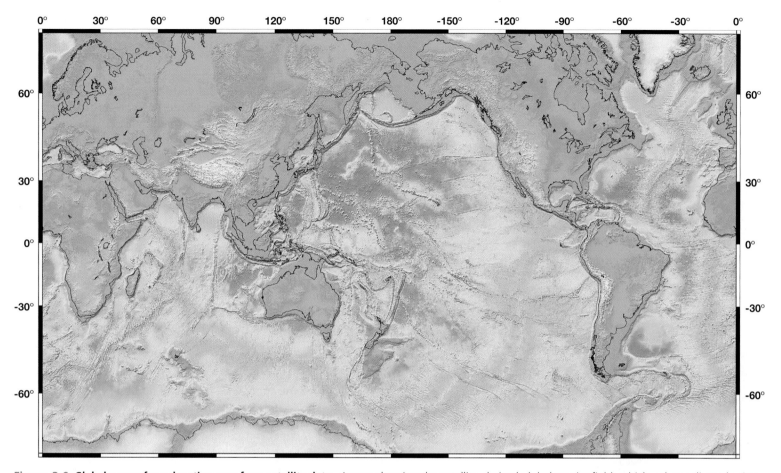

Figure 3.6 Global sea surface elevation map from satellite data. A map showing the satellite-derived global gravity field, which, when adjusted using measured depths, closely corresponds to ocean depth. Purple indicates deep water; the mid-ocean ridge (intermediate water depths) is mostly light green and yellow; pink indicates the shallowest water. The map also shows land surface elevations, with dark green indicating low elevations and white indicating high elevations.

floor clearly delineates many ocean floor features, such as the mid-ocean ridge, trenches, seamounts, and nemataths (island chains). In addition, this new mapping technique has revealed sea floor bathymetry in areas where research vessels have not conducted surveys.

Seismic Reflection Profiles

Oceanographers who want to know about ocean structure beneath the sea floor use strong low-frequency sounds produced by explosions or air guns, as shown in **Figure 3.7**. These sounds penetrate beneath the sea floor and reflect off the boundaries between different rock or sediment layers, producing **seismic reflection profiles**, which have applications in mineral and petroleum exploration.

ESSENTIAL CONCEPT 3.1

Sending pings of sound into the ocean (echo sounding) is a commonly used technique for determining ocean bathymetry. More recently, satellites are being used to map sea floor features.

CONCEPT CHECK 3.1

1 What is bathymetry?

2 Describe how an echo sounder works.

3 Discuss the development of bathymetric techniques, indicating significant advancements in technology.

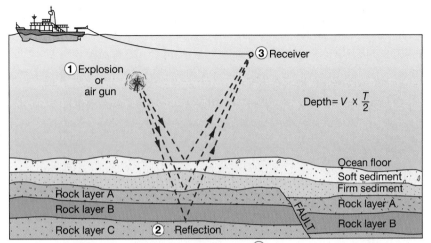

(a) An air gun explosion emits low frequency sounds ① that can penetrate bottom sediments and rock layers. The sound reflects off the boundaries between these layers ② and returns to the receiver ③.

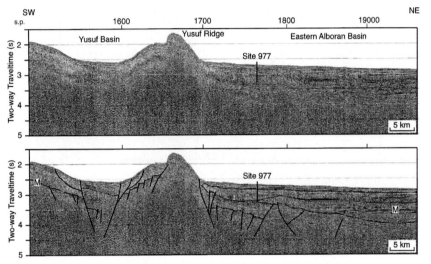

(b) *Top*: Seismic reflection profile of the western Mediterranean, showing the location of *JOIDES Resolution* Drill Site 977. *Bottom*: An interpretation of the same seismic profile showing faults (*black lines*). M = M-reflector, which was created during the drying up of the Mediterranean Sea approximately 5.5 million years ago.

Figure 3.7 Seismic profiling.

Figure 3.8 Major regions of the North Atlantic Ocean floor. Map view (*bottom*) and profile view (*top*), showing that the ocean floor can be divided into three major provinces: continental margins, deep-ocean basins, and the mid-ocean ridge.

3.2 What Features Exist on Continental Margins?

The ocean floor can be divided into three major provinces (**Figure 3.8**): (1) **continental margins**, which are shallow-water areas close to continents, (2) **deep-ocean basins**, which are deep-water areas farther from land, and (3) the **mid-ocean ridge**, which is composed of shallower areas near the middle of an ocean. Plate tectonic processes (discussed in the previous chapters) are integral to the formation of these provinces. Through the process of sea floor spreading, the mid-ocean ridge and deep-ocean basins are created. Elsewhere, as a continent is split apart, new continental margins are formed.

Passive versus Active Continental Margins

Continental margins can be classified as either passive or active, depending on their proximity to plate boundaries. **Passive margins** (**Figure 3.9**, *left*) are embedded within the interior of lithospheric plates and are therefore not in close proximity to any plate boundary. Thus, passive margins usually lack major tectonic activity (such as large earthquakes, eruptive volcanoes, and mountain building).

The East Coast of the United States, where there is no plate boundary, is an example of a passive continental margin. Passive margins are usually produced by rifting of continental landmasses and continued sea floor spreading over geologic time. Features of passive continental margins include the continental shelf, the continental slope, and the continental rise that extends toward the deep-ocean basins (Figures 3.9 and 3.10).

Active margins (Figure 3.9, *right*) are associated with lithospheric plate boundaries and are marked by a high degree of tectonic activity. Two types of active margins exist. **Convergent active margins** are associated with oceanic–continental

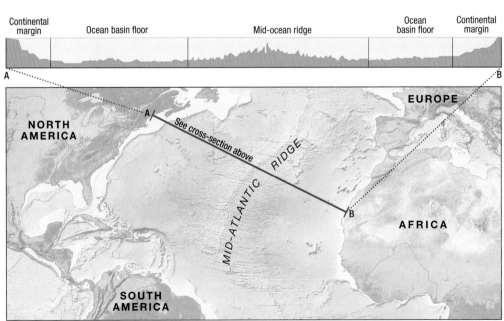

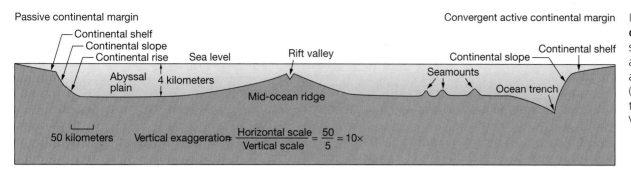

Passive continental margin

Convergent active continental margin

Figure 3.9 Passive and active continental margins. Cross-sectional view of typical features across an ocean basin, including a passive continental margin (*left*) and a convergent active continental margin (*right*). Vertical exaggeration is 10 times.

convergent plate boundaries. From the land to the ocean, features include an onshore arc-shaped row of active volcanoes, then a narrow shelf, a steep slope, and an offshore trench that delineates the plate boundary. Western South America, where the Nazca Plate is being subducted beneath the South American Plate, is an example of a convergent active margin. **Transform active margins** are less common and are associated with transform plate boundaries. At these locations, offshore faults usually parallel the main transform plate boundary fault and create linear islands, banks (shallowly submerged areas), and deep basins close to shore. Coastal California along the San Andreas Fault is an example of a transform active margin.

> **ESSENTIAL CONCEPT 3.2A**
>
> Passive continental margins lack a plate boundary and have different features than active continental margins, which include a plate boundary (either convergent or transform).

Continental Shelf

The **continental shelf** is defined as a generally flat zone extending from the shore beneath the ocean surface to a point at which a marked increase in slope angle occurs, called the **shelf break** (Figure 3.10). It is usually flat and relatively featureless because of marine sediment deposits but can contain coastal islands, reefs, and raised banks. The underlying rock is granitic continental crust, so the continental shelf is geologically part of the continent. Accurate sea floor mapping is essential for determining the extent of the continental shelf, which has come into question recently in the Arctic Ocean. The general bathymetry of the continental shelf can usually be predicted by examining the topography of the adjacent land. With few exceptions, the coastal topography extends beyond the shore and onto the continental shelf.

Continental shelves vary depending on the local geology and topography. For example, the average width of the continental shelf is about 70 kilometers (43 miles), but it varies from a few tens of meters to 1500 kilometers (930 miles). The broadest shelves occur off the northern coasts of Siberia and North America in the Arctic Ocean. Worldwide, the average depth at which the shelf break occurs is about 135 meters (443 feet). Around the continent of Antarctica, however, the shelf break occurs

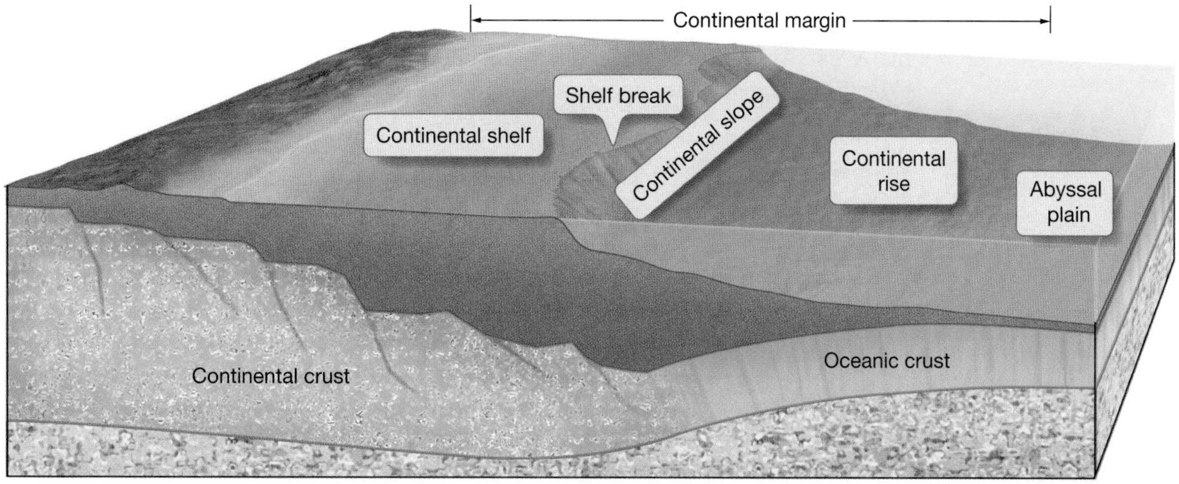

Figure 3.10 Features of a passive continental margin. Schematic view showing the main features of a passive continental margin.

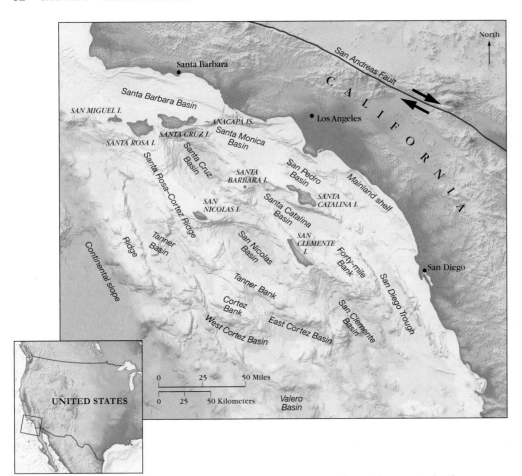

Figure 3.11 Continental borderland. A continental borderland, like the one offshore Southern California, consists of a
series of islands, shallow banks, and deep basins. Southern
California's continental borderland is a result of its proximity
to the San Andreas Fault, a major transform plate boundary.

Web Video
Virtual Field Trip: Underwater
Flyby of Southern California's
Offshore Sea Floor

at 350 meters (2200 feet). The average slope of the
continental shelf is only about a tenth of a degree,
which is similar to the slope given to a large parking lot for drainage purposes.

Sea level has fluctuated over the history of
Earth, causing the shoreline to migrate back
and forth across the continental shelf. When
colder climates prevailed during the most recent
ice age, for example, more of Earth's water was
frozen as glaciers on land, so sea level was lower
than it is today. During that time, more of the
continental shelf was exposed.

The type of continental margin determines
the shape and features associated with the continental shelf. For example, the east coast of South
America has a broader continental shelf than its
west coast. The east coast is a passive margin,
which typically has a wider shelf. In contrast,
the convergent active margin present along the
west coast of South America is characterized
by a narrow continental shelf and a shelf break
close to shore. For transform active margins
such as along California, the presence of offshore faults produces a continental shelf that is
not flat. Rather, it is marked by a high degree of
relief (islands, shallow banks, and deep basins)
called a **continental borderland** (Figure 3.11).

Continental Slope

The **continental slope**, which lies beyond the shelf break, is where the deep-ocean
basins begin. Total relief in this region is similar to that found in mountain ranges
on the continents. The break at the top of the slope may be from 1 to 5 kilometers (0.6 to 3 miles) above the deep-ocean basin at its base. Along convergent active
margins where the slope descends into submarine trenches, even greater vertical
relief is measured. Off the west coast of South America, for instance, the total relief
from the top of the Andes Mountains to the bottom of the Peru–Chile Trench is
about 15 kilometers (9.3 miles).

Worldwide, the slope of the continental slopes averages about 4 degrees but
varies from 1 to 25 degrees.[5] A study that compared different continental slopes in
the United States revealed that the average slope is just over 2 degrees. Around the
margin of the Pacific Ocean, the continental slopes average more than 5 degrees
because of the presence of convergent active margins that drop directly into deep
offshore trenches. The Atlantic and Indian Oceans, on the other hand, contain
many passive margins, which lack plate boundaries. Thus, the amount of relief is
lower and slopes in these oceans average about 3 degrees.

Submarine Canyons and Turbidity Currents

The continental slope and, to a lesser extent, the continental shelf exhibit **submarine
canyons**, which are narrow but deep submarine valleys that are V-shaped in profile
view and have branches or tributaries with steep to overhanging walls (Figure 3.12).
They resemble canyons formed on land that are carved by rivers and can be quite

---

[5]For comparison, the windshield of an aerodynamically designed car has a slope of about 25 degrees.

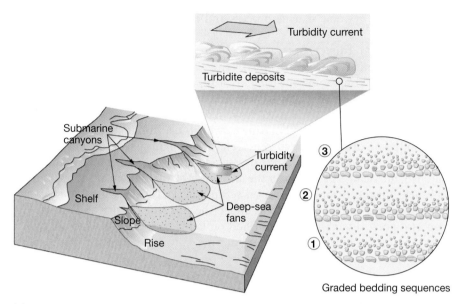

Graded bedding sequences

(a) Turbidity currents move downslope, eroding the continental margin to enlarge submarine canyons. Deep-sea fans are composed of turbidite deposits, which consist of sequences of graded bedding (*inset*).

(b) A diver descends into La Jolia Submarine Canyon, offshore California.

(c) Outcrop of layered turbidite deposits that have been tilted and uplifted onto land in California. Each light-colored layer is sandstone that marks the coarser bottom of a graded bedding sequence.

Figure 3.12 **Submarine canyons and turbidity currents.**

large. In fact, the Monterey Canyon off California is comparable in size to Arizona's Grand Canyon (**Figure 3.13**).

How are submarine canyons formed? Initially it was thought that submarine canyons were ancient river valleys created by the erosive power of rivers when sea level was lower and the continental shelf was exposed. Although some canyons are directly offshore from where rivers enter the sea, the majority of them are not. Many, in fact, are confined exclusively to the continental slope. In addition, submarine canyons continue to the base of the continental slope, which averages some 3500 meters (11,500 feet) below sea level. There is no evidence, however, that sea level has ever been lowered by that much.

Side-scan sonar surveys along the Atlantic coast indicate that the continental slope is dominated by submarine canyons from Hudson Canyon near New York City to Baltimore Canyon in Maryland. Canyons confined to the continental slope are straighter and have steeper canyon floor gradients than those that extend into the continental shelf. These characteristics suggest the canyons are created on the

STUDENTS SOMETIMES ASK . . .

Has anyone ever been caught and killed by a turbidity current?

Actually, no. This is mostly because turbidity currents occur in submarine canyons on the continental slope, which are normally so deep that even deep divers don't venture there. However, oceanographic equipment left on the floor of a submarine canyon has frequently been mangled or destroyed by the highly erosive power of turbidity currents. Other times, the equipment is simply swept away by a turbidity current and is never seen again.

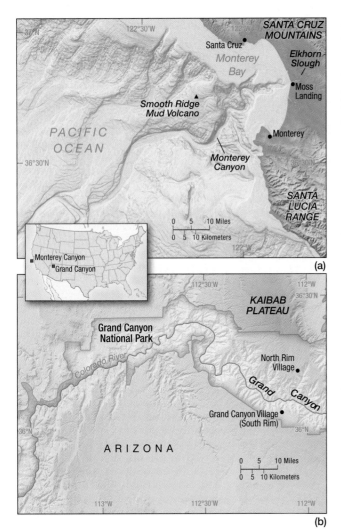

Figure 3.13 Comparison of the Monterey Submarine Canyon and Arizona's Grand Canyon. In these same-scale maps, it can be seen that the Monterey Submarine Canyon **(a)** is comparable to Arizona's Grand Canyon **(b)** in terms of length, depth, width, and steepness.

> ESSENTIAL CONCEPT 3.2B
>
> Turbidity currents are underwater avalanches of muddy water mixed with sediment that move down the continental slope and are responsible for carving submarine canyons.

continental slope by some marine process and enlarge into the continental shelf through time.

Both indirect and direct observation of the erosive power of **turbidity** (*turbidus* = disordered) **currents** (see Web Diving Deeper 3.3) have suggested that they are responsible for carving submarine canyons. Turbidity currents are underwater avalanches of muddy water mixed with rocks and other debris. The sediment portion of turbidity currents comes from sea floor materials that move across the continental shelf into the head of a submarine canyon and accumulate there, setting the stage for initiation of a turbidity current. Trigger mechanisms for turbidity currents include shaking by an earthquake, the oversteepening of sediment that accumulates on the shelf, hurricanes passing over the area, and the rapid input of sediment from flood waters. Once a turbidity current is set in motion, the dense mixture of water and debris moves rapidly downslope under the force of gravity and carves the canyon as it goes, resembling a flash flood on land. Turbidity currents are strong enough to transport huge rocks down submarine canyons and cause a considerable amount of erosion over time.

Continental Rise

The **continental rise** is a transition zone between the continental margin and the deep-ocean floor comprised of a huge submerged pile of debris. Where did all this debris come from, and how did it get there?

The existence of turbidity currents suggests that the material transported by these currents is responsible for the creation of continental rises. When a turbidity current moves through and erodes a submarine canyon, it exits through the mouth of the canyon. The slope angle decreases and the turbidity current slows, causing suspended material to settle out in a distinctive type of layering called **graded bedding** that *grades in size upward* (Figure 3.12a, *inset*). As the energy of the turbidity current dissipates, larger pieces settle first, then progressively smaller pieces settle, and eventually even very fine pieces settle out, which may occur weeks or months later.

An individual turbidity current deposits one graded bedding sequence. The next turbidity current may partially erode the previous deposit and then deposit another graded bedding sequence on top of the previous one. After some time, a thick sequence of graded bedding deposits can develop one on top of another. These stacks of graded bedding, which make up the continental rise, are called **turbidite deposits** (Figure 3.12c).

As viewed from above, the deposits at the mouths of submarine canyons are fan, lobate, or apron shaped (Figures 3.12a and 3.14). Consequently, these deposits are called **deep-sea fans**, or **submarine fans**. Deep-sea fans create the continental rise when they merge together along the base of the continental slope. Along convergent active margins, however, the steep continental slope leads directly into a deep-ocean trench. Sediment from turbidity currents accumulates in the trench and there is no continental rise.

One of the largest deep-sea fans in the world is the Indus Fan, a passive margin fan that extends 1800 kilometers (1100 miles) south of Pakistan (**Figure 3.14a**). The Indus River carries extensive amounts of sediment from the Himalaya Mountains to the coast. This sediment eventually makes its way down the submarine canyon and builds the fan, which, in some areas, has sediment that is more than 10 kilometers (6.2 miles) thick. The Indus Fan has a main submarine canyon channel that extends seaward onto the fan but soon divides into several branching distributary channels. These distributary channels are similar to those found on deltas, which form at the mouths of streams. On the lower fan, the surface has a very low slope, and the flow is no longer confined to channels, so it spreads out and forms layers of fine sediment across the fan surface. The Indus Fan has so much sediment, in fact, that it partially buries an active mid-ocean ridge, the Carlsberg Ridge!

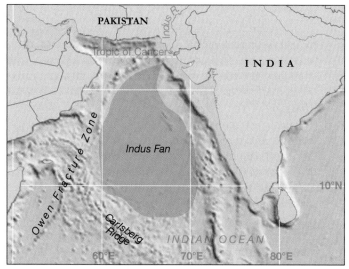

(a) Map of the Indus Fan, a large but otherwise typical example of a passive margin fan.

(b) Sonar perspective view of southeast Alaska's Chatham Fan, which rises 450 meters (1500 feet) above the surrounding sea floor. Vertical exaggeration is 20 times; view looking northeast.

Figure 3.14 **Examples of deep-sea (submarine) fans.**

CONCEPT CHECK 3.2

1. Describe the major features of a passive continental margin: continental shelf, continental slope, continental rise, submarine canyon, and deep-sea fans.

2. Explain how submarine canyons are created.

3. Explain what graded bedding is and how it forms.

3.3 What Features Exist in the Deep-Ocean Basins?

The deep-ocean floor lies beyond the continental margin province (the shelf, slope, and the rise) and contains a variety of features.

Abyssal Plains

Extending from the base of the continental rise into the deep-ocean basins are flat depositional surfaces with slopes of less than a fraction of a degree that cover extensive portions of the deep-ocean basins. These **abyssal** (*a* = without, *byssus* = bottom) **plains** average between 4500 meters (15,000 feet) and 6000 meters (20,000 feet) deep. They are not literally bottomless, but they are some of the deepest (and flattest) regions on Earth.

Abyssal plains are formed by fine particles of sediment slowly drifting onto the deep-ocean floor. Over millions of years, a thick blanket of sediment is produced by **suspension settling** as fine particles (analogous to "marine dust") accumulate on the ocean floor. With enough time, these deposits cover most irregularities of the deep ocean, as shown in **Figure 3.15**. In addition, sediment traveling in turbidity currents from land adds to the sediment load.

The type of continental margin determines the distribution of abyssal plains. For instance, few abyssal plains are located in the Pacific Ocean; instead,

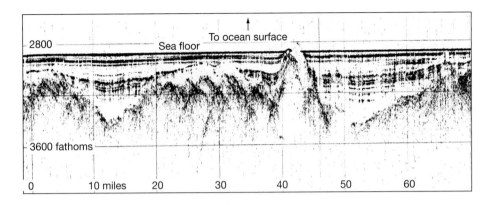

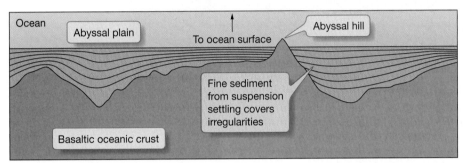

Figure 3.15 Abyssal plain formed by suspension settling. Seismic cross section (*above*) and matching drawing (*below*) for a portion of the deep Madeira Abyssal Plain in the eastern Atlantic Ocean, showing the irregular volcanic terrain buried by sediments.

most occur in the Atlantic and Indian Oceans. The deep-ocean trenches found on the convergent active margins of the Pacific Ocean prevent sediment from moving past the continental slope. In essence, the trenches act like a gutter that traps sediment transported off the land by turbidity currents. On the passive margins of the Atlantic and Indian Oceans, however, turbidity currents travel directly down the continental margin and deposit sediment on the abyssal plains. In addition, the distance from the continental margin to the floor of the deep-ocean basins in the Pacific Ocean is so great that most of the suspended sediment settles out before it reaches these distant regions. Conversely, the smaller size of the Atlantic and Indian Oceans does not prevent suspended sediment from reaching their deep-ocean basins.

Volcanic Peaks of the Abyssal Plains

Poking through the sediment cover of the abyssal plains are a variety of volcanic peaks, which extend to various elevations above the ocean floor. Some extend above sea level to form islands, while others are just below the surface (see **Web Diving Deeper 3.2**). Those that are below sea level but rise more than 1 kilometer (0.6 mile) above the deep-ocean floor and have a pointy top like an upside-down ice cream cone are called **seamounts**. Worldwide, scientists estimate that there are about 125,000 known seamounts, many of which originated at volcanic centers such as hotspots or the mid-ocean ridge. On the other hand, if a volcano has a flattened top, it is called a **tablemount**, or *guyot*. The origin of seamounts and tablemounts was discussed in Chapter 2 (refer to Figure 2.28).

Volcanic features on the ocean floor that are less than 1000 meters (0.6 mile) tall—the minimum height of a seamount—are called **abyssal hills, or seaknolls**. Abyssal hills are one of the most abundant features on the planet (several hundred thousand have been identified) and cover a large percentage of the entire ocean basin floor. Many are gently rounded in shape, and they have an average height of about 200 meters (650 feet). Most abyssal hills are created by stretching of crust during the creation of new sea floor at the mid-ocean ridge.

In the Atlantic and Pacific Oceans, many abyssal hills are found buried beneath abyssal plain sediment. In the Pacific Ocean, the abundance of active margins traps land-derived sediment, and so the rate of sediment deposition is lower. Consequently, extensive regions dominated by abyssal hills have resulted; these are called **abyssal hill provinces**. The evidence of volcanic activity on the bottom of the Pacific Ocean is particularly widespread. In fact, more than 20,000 volcanic peaks are known to exist on the Pacific sea floor.

Ocean Trenches and Volcanic Arcs

Along passive margins, the continental rise commonly occurs at the base of the continental slope and merges smoothly into the abyssal plain. In convergent active margins, however, the slope descends into a long, narrow, steep-sided **ocean trench**. Ocean trenches are deep linear scars in the ocean floor, caused by the collision of two plates along convergent plate margins (as discussed in Chapter 2). The landward side of the trench rises as a **volcanic arc** that may produce islands (such as the islands of Japan, an **island arc**) or a volcanic mountain range along the margin of a continent (such as the Andes Mountains, a **continental arc**).

Selected Pacific Ocean Trenches

Name	Depth (km)	Width (km)	Length (km)
Middle America	6.7	40	2800
Aleutian	7.7	50	3700
Peru-Chile	8.0	100	5900
Kermadec-Tonga	10.0	50	2900
Kuril	10.5	120	2200
Mariana	11.0	70	2550

Atlantic Ocean Trenches

Name	Depth (km)	Width (km)	Length (km)
South Sandwich	8.4	90	1450
Puerto Rico	8.4	120	1550

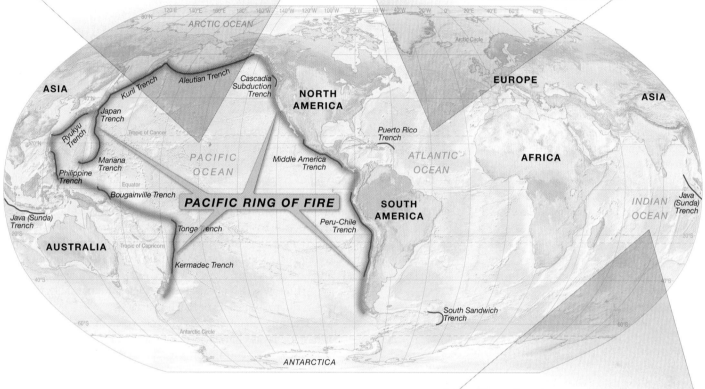

Indian Ocean Trenches

Name	Depth (km)	Width (km)	Length (km)
Java (Sunda)	7.5	80	4500

Figure 3.16 Location and dimensions of ocean trenches. The majority of ocean trenches *(red lines)* are along the margins of the Pacific Ocean, where plates are being subducted. Most of the world's large earthquakes (due to subduction) and active volcanoes (as volcanic arcs) occur around the Pacific Rim, which is why the area is also called the Pacific Ring of Fire *(red shading)*.

Figure 3.17 Profile across the Peru–Chile Trench and the Andes Mountains. Over a distance of 200 kilometers (125 miles), there is a change in elevation of more than 14,900 meters (49,000 feet) from the Peru–Chile Trench to the Andes Mountains. This dramatic relief is a result of plate interactions at a convergent active margin, producing a deep-ocean trench and associated continental volcanic arc. Vertical scale is exaggerated 10 times.

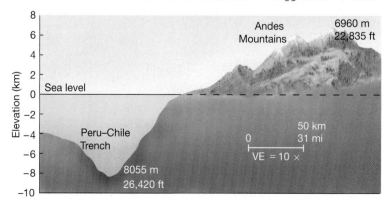

The deepest portions of the world's oceans are found in these trenches. In fact, the deepest point on Earth's surface—11,022 meters (36,161 feet)—is found in the Challenger Deep area of the Mariana Trench. The majority of ocean trenches are found along the margins of the Pacific Ocean (**Figure 3.16**), and only a few exist in the Atlantic and Indian Oceans.

THE PACIFIC RING OF FIRE The **Pacific Ring of Fire** occurs along the margins of the Pacific Ocean. It is home to the majority of Earth's active volcanoes and large earthquakes because of the prevalence of convergent plate boundaries along the Pacific Rim. A part of the Pacific Ring of Fire is South America's western coast, including the Andes Mountains and the associated Peru–Chile Trench. **Figure 3.17** shows a

Deep-ocean trenches and volcanic arcs result from the collision of two plates at convergent plate boundaries and mostly occur along the margins of the Pacific Ocean (the Pacific Ring of Fire).

STUDENTS SOMETIMES ASK . . .

What effect does all this volcanic activity along the mid-ocean ridge have at the ocean's surface?

Sometimes an underwater volcanic eruption is large enough to create what is called a *megaplume* of warm, mineral-rich water that is lower in density than the surrounding seawater and thus rises to the surface. Remarkably, a few research vessels have reported experiencing the effects of a megaplume at the surface while directly above an erupting sea floor volcano! Researchers on board describe bubbles of gas and steam at the surface, a marked increase in water temperature, and the presence of enough volcanic material to turn the water cloudy. In terms of warming the ocean, the heat released into the ocean at mid-ocean ridges is probably not very significant, mostly because the ocean is so good at absorbing and redistributing heat.

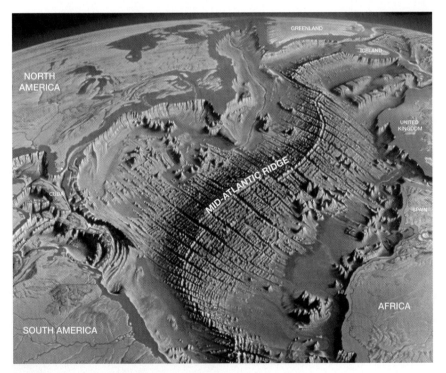

Figure 3.18 Floor of the North Atlantic Ocean. The global mid-ocean ridge cuts through the center of the Atlantic Ocean, where it is called the Mid-Atlantic Ridge.

cross-sectional view of this area and illustrates the tremendous amount of relief at convergent plate boundaries where deep-ocean trenches are associated with tall volcanic arcs.

CONCEPT CHECK 3.3

1 Describe the process by which abyssal plains are created.

2 Discuss the origin of the various volcanic peaks of the abyssal plains: seamounts, tablemounts, and abyssal hills.

3 What are some differences between a submarine canyon and an ocean trench?

3.4 What Features Exist Along the Mid-Ocean Ridge?

The global mid-ocean ridge is a continuous, fractured-looking mountain ridge that extends through all the ocean basins. The portion of the mid-ocean ridge found in the North Atlantic Ocean is called the Mid-Atlantic Ridge (**Figure 3.18**), and it dwarfs all mountain ranges on land. As discussed in Chapter 2, the mid-ocean ridge results from sea floor spreading along divergent plate boundaries. The enormous mid-ocean ridge forms Earth's longest mountain chain, extending across some 75,000 kilometers (46,600 miles) of the deep-ocean basin. The width of the mid-ocean ridge averages about 1000 kilometers (620 miles). The mid-ocean ridge is a topographically high feature, extending an average of 2.5 kilometers (1.5 miles) above the surrounding sea floor. The mid-ocean ridge contains only a few scattered islands, such as Iceland and the Azores, where it peeks above sea level. Remarkably, the mid-ocean ridge covers 23% of Earth's surface.

The mid-ocean ridge is entirely volcanic and is composed of basaltic lavas characteristic of the oceanic crust. Along most of its crest is a central downdropped **rift valley** created by sea floor spreading (rifting) where two plates diverge (see, for example, Figures 2.15 and 2.16). Along the Mid-Atlantic Ridge, for example, is a central rift valley that is as much as 30 kilometers (20 miles) wide and 3 kilometers (2 miles) deep. Here, molten rock presses upward toward the sea floor, setting off earthquakes, creating jets of superheated seawater, and eventually solidifying to form new oceanic crust. Cracks called *fissures* (*fissus* = split) and faults are commonly observed in the central rift valley. Swarms of small earthquakes occur along the central rift valley caused by the injection of magma into the sea floor or rifting along faults.

Segments of the mid-ocean ridge called **oceanic ridges** have a prominent rift valley and steep, rugged slopes, and **oceanic rises** have slopes that are gentler and less rugged. As explained in Chapter 2, the differences in overall shape are caused by the fact that oceanic ridges (such as the Mid-Atlantic Ridge) spread more slowly than oceanic rises (such as the East Pacific Rise).

Volcanic Features

Volcanic features associated with the mid-ocean ridge include tall volcanoes called *seamounts*[6] (**Figure 3.19a**) and

---

[6]In a number of cases, researchers have discovered seamounts that initially formed along the crest of the mid-ocean ridge and have split in two as the plates have spread apart.

EARTH'S HYPSOGRAPHIC CURVE: NEARLY EVERYTHING YOU NEED TO KNOW ABOUT EARTH'S OCEANS AND LANDMASSES IN ONE GRAPH

Earth's **hypsographic** (*hypos* = height, *graphic* = drawn) **curve** (**Figure 3A**), which shows the relationship between the height of the land and the depth of the oceans, is useful for illustrating many things. For example, the bar graph (Figure 3A, *left*) gives the percentage of Earth's surface area at various ranges of elevation and depth. The cumulative curve (Figure 3A, *right*) gives the percentage of surface area from the highest peaks to the deepest depths of the oceans. Together, they show that 70.8% of Earth's surface is covered by oceans and that the average depth of the ocean is 3729 meters (12,234 feet) while the average height of the land is only 840 meters (2756 feet). The difference, recalling our discussion of isostasy in Chapter 1, results from the greater density and lesser thickness of oceanic crust as compared to continental crust.

Earth's cumulative hypsographic curve (Figure 3A, *right*) shows five differently sloped segments. On land, the first steep segment of the curve represents tall mountains, while the gentle slope represents low coastal plains (and continues just offshore, representing the shallow parts of the continental margin). The first slope below sea level represents steep areas of the continental margins and also includes the mountainous mid-ocean ridge. Further offshore, the longest, flattest part of the whole curve represents the deep-ocean basins, followed by the last steep part, which represents ocean trenches.

Interestingly, the shape of Earth's hypsographic curve can be used to support the existence of plate tectonics on Earth. Specifically, the two flat areas and three sloped areas of the curve show that there is a very uneven

distribution of area at different depths and elevations. If there were no active mechanism involved in creating such features on Earth, the bar graph portions would all be about the same length, and the cumulative curve would be a straight line. Instead, the variations in the curve suggest that plate tectonics is actively working to modify Earth's surface. The flat portions of the curve represent various intraplate elevations both on land and underwater, while the slopes of the curve represent mountains, continental slopes, the mid-ocean ridge, and deep-ocean trenches, all of which are created by plate tectonic processes. Interestingly, analysis of hypsographic curves constructed for other planets and moons using satellite data have been used to show if plate tectonic processes are actively modifying the surfaces of those worlds.

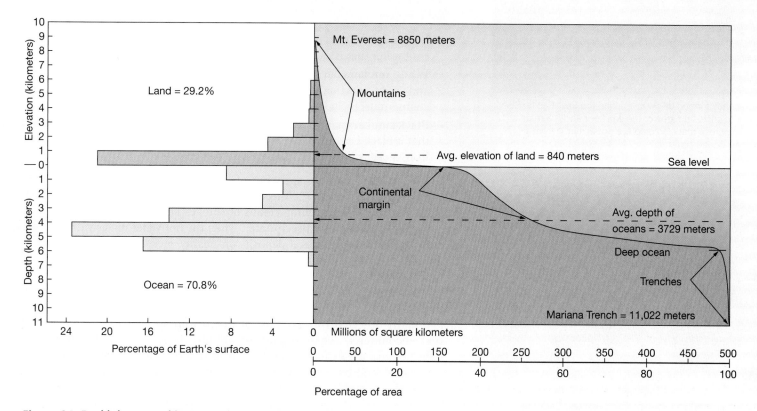

Figure 3A   Earth's hypsographic curve. The bar graph (*left*) gives the percentage of Earth's surface area at various ranges of elevation and depth. The cumulative hypsographic curve (*right*) gives the percentage of surface area from the highest peaks to the deepest depths of the oceans. Also shown are the average ocean depth and land elevation.

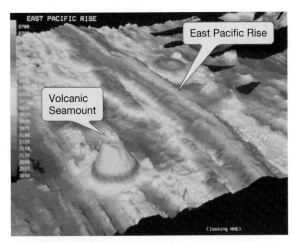

EAST PACIFIC RISE

East Pacific Rise

Volcanic
Seamount

(looking NNE)

(a) False-color perspective view based on sonar mapping of a portion of the East Pacific Rise (*center*) showing volcanic seamount (*left*). The depth, in meters, is indicated by the color scale along the left margin; vertical exaggeration is six times.

(b) Recently formed pillow lava along the East Pacific Rise. Photo shows an area of the sea floor about 3 meters (10 feet) across that also displays ripple marks from deep-ocean currents.

(c) Pillow lava that was once on the sea floor but has since been uplifted onto land at Port San Luis, California. Maximum width of an individual pillow is 1 meter (3.3 feet).

Figure 3.19 **Mid-ocean ridge volcanoes and pillow lava.**

recent underwater lava flows. When hot basaltic lava spills onto the sea floor, it is exposed to cold seawater that chills the margins of the lava. This creates **pillow lavas** or **pillow basalts**, which are smooth, rounded lobes of rock that resemble a stack of bed pillows (**Figures 3.19b** and **3.19c**).

Although most people are not aware of it, frequent volcanic activity is common along the mid-ocean ridge. In fact, 80% of Earth's volcanic activity takes place on the sea floor, and every year about 12 cubic kilometers (3 cubic miles) of molten rock erupts underwater. The amount of erupted lava along the mid-ocean ridge is large enough to fill an Olympic-sized swimming pool every three seconds! Bathymetric studies along the Juan de Fuca Ridge off Washington and Oregon, for example, revealed that 50 million cubic meters (1800 million cubic feet) of new lava had erupted between 1981 and 1987. Subsequent surveys of the area indicated many changes along the mid-ocean ridge, including new volcanic features, recent lava flows, and depth changes of up to 37 meters (121 feet). Interest in the continuing volcanic activity along the Juan de Fuca Ridge has led to the development of a permanent sea floor observation system there (see **Web Diving Deeper 2.1**). Other parts of the mid-ocean ridge, such as East Pacific Rise, also experience frequent volcanic activity (**Diving Deeper 3.2**).

Hydrothermal Vents

Other features in the central rift valley include **hydrothermal** (*hydro* = water, *thermo* = heat) **vents**. Hydrothermal vents are sea floor hot springs created when cold seawater seeps down along cracks and fractures in the ocean crust and approaches an underground magma chamber (**Figure 3.20**). The water picks up heat and dissolved substances and then works its way back toward the surface through a complex plumbing system, exiting through the sea floor. The temperature of the water that rushes out of a particular hydrothermal vent determines its appearance:

- **Warm-water vents** have water temperatures below 30°C (86°F) and generally emit water that is clear in color.
- **White smokers** have water temperatures from 30° to 350°C (86° to 662°F) and emit water that is white because of the presence of various light-colored compounds, including barium sulfide.
- **Black smokers** have water temperatures above 350°C (662°F) and emit water that is black because of the presence of dark-colored **metal sulfides**, including iron, nickel, copper, and zinc.

Many black smokers spew out of chimney-like structures (Figure 3.20b) that can be up to 60 meters (200 feet) high and were named for their resemblance to factory smokestacks belching clouds of smoke. The dissolved metal particles often come out of solution, or **precipitate**,[7] when the hot water mixes with cold seawater, creating coatings of mineral deposits on nearby rocks. Chemical analyses of these deposits reveal that they are composed of various metal sulfides and sometimes even silver and gold.

In addition, most hydrothermal vents foster unusual deep-ocean ecosystems that include organisms such as giant tubeworms, large clams, beds of mussels, and many other creatures—most of which were new to science when they were first encountered. These organisms are able to survive in the absence of sunlight because the vents discharge hydrogen sulfide gas, which is metabolized by archaeons[8] and bacteria and provides a food source for other organisms in the

---

[7]A *chemical precipitate* is formed whenever dissolved materials change from existing in the dissolved state to existing in the solid state.

[8]Archaeons are microscopic bacteria-like organisms—a newly discovered domain of life.

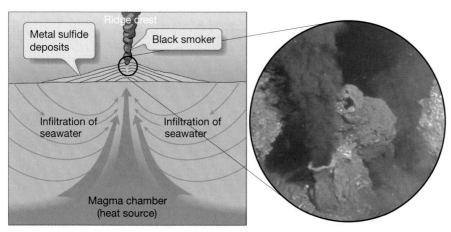

(a) Diagram showing hydrothermal circulation along the mid-ocean ridge and the creation of black smokers; photo (*inset*) showing a close-up view of a black smoker along the East Pacific Rise.

(b) Black smoker chimney and fissure at Susu north active site, Manus Basin, western Pacific Ocean. Chimney is about 3 meters (10 feet) tall.

Figure 3.20 Hydrothermal vents.

STUDENTS SOMETIMES ASK . . .

Has anyone seen pillow lava forming?

Amazingly, yes! In the 1960s, an underwater film crew ventured to Hawaii during an eruption of the volcano Kilauea, where lava spilled into the sea. They braved high water temperatures and risked being burned on the red-hot lava but filmed some incredible footage. Underwater, the formation of pillow lava occurs where a tube emits molten lava directly into the ocean. When hot lava comes into contact with cold seawater, it forms the characteristic smooth and rounded margins of pillow basalt. The divers also experimented with a hammer on newly formed pillows and were able to initiate new lava outpourings.

Web Video
Formation of Pillow Lava

STUDENTS SOMETIMES ASK . . .

If black smokers are so hot, why isn't there steam coming out of them instead of hot water?

Indeed, black smokers emit water that can be up to four times the boiling point of water at the ocean's surface and hot enough to melt lead. However, the depth where black smokers are found results in much higher pressure than at the surface. At these higher pressures, water has a much higher boiling point. Thus, water from hydrothermal vents remains in the liquid state instead of turning into water vapor (steam).

Web Video
Black Smoker Venting Fluid

ESSENTIAL CONCEPT 3.4A

The mid-ocean ridge is created by plate divergence and typically includes a central rift valley, faults and fissures, seamounts, pillow basalts, hydrothermal vents, and metal sulfide deposits.

community. Recent studies of active hydrothermal vent fields indicate that vents have short life spans of only a few years to several decades, which has important implications for the organisms that depend on hydrothermal vents. These interesting biocommunities are discussed in Chapter 15, "Animals of the Benthic Environment."

Fracture Zones and Transform Faults

The mid-ocean ridge is cut by a number of **transform faults**, which offset spreading zones. Oriented perpendicular to the spreading zones, transform faults give the mid-ocean ridge the zigzag appearance shown in Figure 3.18. As described in Chapter 2, transform faults occur to accommodate spreading of a linear ridge system on a spherical Earth and because different segments of the mid-ocean ridge spread apart at different rates.

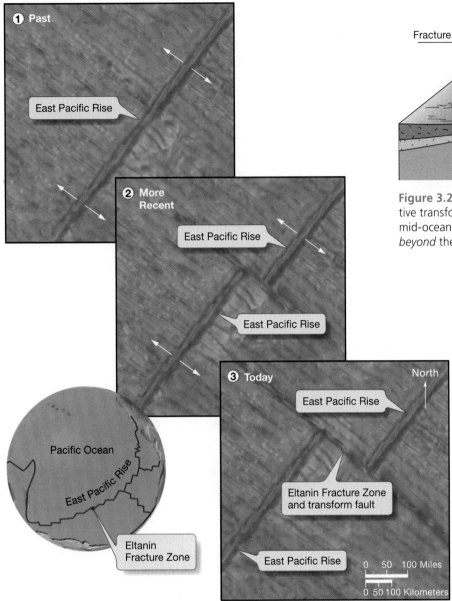

Figure 3.21 The Eltanin Fracture Zone. Enlargement of the Eltanin Fracture Zone in the South Pacific Ocean, showing its relationship to the East Pacific Rise. The Eltanin Fracture Zone is actually both a fracture zone and a transform fault; the name was given to it before the modern understanding of plate tectonic processes.

Web Animation
Transform Faults

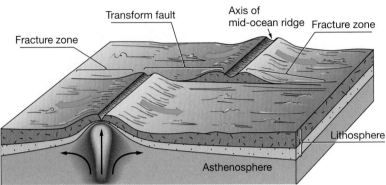

Figure 3.22 Transform faults and fracture zones. Transform faults are active transform plate boundaries that occur *between* the segments of the mid-ocean ridge. Fracture zones are inactive intraplate features that occur *beyond* the segments of the mid-ocean ridge.

On the Pacific Ocean sea floor, where scars are less rapidly covered by sediment than in other ocean basins, transform faults are prominently displayed (Figure 3.21). Here, they extend for thousands of kilometers away from the mid-ocean ridge and have widths of up to 200 kilometers (120 miles). These extensions, however, are not transform faults. Instead, they are **fracture zones**.

What is the difference between a transform fault and a fracture zone? Figure 3.22 shows that both run along the same long linear zone of weakness in Earth's crust. In fact, by following the same zone of weakness from one end to the other, it changes from a fracture zone to a transform fault and back again to a fracture zone. A transform fault is a seismically active area that offsets the axis of a mid-ocean ridge. A fracture zone, on the other hand, is a seismically inactive area that shows evidence of past transform fault activity. A helpful way to visualize the difference is that transform faults occur *between* offset segments of the mid-ocean ridge, while fracture zones occur *beyond* the offset segments of the mid-ocean ridge.

The relative direction of plate motion across transform faults and fracture zones further differentiates these two features. Across a transform fault, two lithospheric plates are moving in opposite directions. Across a fracture zone (which occurs entirely within a plate), there is no relative motion because the parts of the lithospheric plate cut by a fracture zone are moving in the same direction (Figure 3.22). Transform faults are actual plate boundaries, whereas fracture zones are not. Rather, fracture zones are ancient, inactive fault scars embedded within a plate.

Earthquake activity also differentiates transform faults from fracture zones. Earthquakes shallower than 10 kilometers (6 miles) are common when plates move in opposite directions along transform faults. Along fracture zones, where plate motion is in the same direction, seismic activity is almost completely absent. Table 3.1 summarizes the differences between transform faults and fracture zones.

NOW YOU SEE IT, NOW YOU DON'T: RECOVERING OCEANOGRAPHIC EQUIPMENT STUCK IN LAVA

Although the mid-ocean ridge is one of the most active features on the planet and experiences an abundance of volcanic activity, nobody has ever directly observed an undersea volcanic eruption there. However, a team of oceanographers on a research cruise to the East Pacific Rise in 2006 came close to this remarkable feat.

The story starts a year earlier, when scientists deployed 12 ocean-bottom seismometers (OBSs) over a few square kilometers of sea floor along an unusually active portion of the East Pacific Rise that is about 725 kilometers (450 miles) south of Acapulco, Mexico, and 2.5 kilometers (1.6 miles) deep. The OBSs—each about the size and weight of a small refrigerator—are designed to stay on the sea floor for up to a year and collect seismic data. Researchers returned in 2006, thinking that they would simply recover the instruments and send down others. When the research vessel sent a sonar signal to the OBSs to release their weights and use

their floats to return to the surface, only four came bobbing up. That's when the scientists suspected that a volcanic eruption had occurred. Three other OBSs responded to the signal but did not come to the surface, and five other instruments were not heard from, presumably because they were buried in lava.

Two months later, scientists returned with a camera-equipped sled that is towed behind a ship and were able to locate the three OBSs embedded in recent lava. Although they tried to nudge and pry them loose with the sled, the OBSs were thoroughly stuck. Wanting to retrieve the stuck OBSs with the hope that they had recorded data while riding out an active sea floor lava flow, the scientists had to wait until a year later, when the tethered robotic vehicle *Jason* was sent down to try to free the instruments. Using *Jason*'s video camera and its mechanical arms controlled remotely from a command center on the ship, the crew was able to pry away large chunks

of lava that locked the instruments in place. After much yanking, two of the OBSs finally broke free and rose to the surface, with help from attached floats. Although the researchers attempted to free the third OBS, it was never recovered because it was stuck too tightly in lava.

The recovered OBS instruments—although badly scorched from the hot lava (**Figure 3B**)—provided usable data that have given researchers new information about the volcanic processes that occur at the mid-ocean ridge. This and other evidence suggests that the fresh lava had erupted for six hours straight, heating and darkening the water above it and spreading along the ridge for more than 16 kilometers (10 miles). The researchers consider themselves lucky to have fortuitously caught Earth's crust in the very act of ripping itself apart, documenting swarms of undersea earthquakes and culminating in a volcanic eruption that buried their instruments in lava.

Figure 3B An ocean-bottom seismometer (OBS) stuck in lava. A 2006 sea floor eruption along the East Pacific Rise trapped this and several other OBS instruments in lava. The yellow plastic covering protects glass ball floats that are normally used to raise the instrument to the surface; additional attached floats are shown above the OBS. Scientists freed the device by using a robotic vehicle to remove chunks of lava that were embedded into the instrument and singed its outer casing. Inset (*right*) shows marine geologist Dan Fornari prying off chunks of recently erupted sea floor lava from the recovered instrument.

Web Video
Recovering
Oceanographic
Equipment Stuck in Lava

TABLE 3.1 COMPARISON BETWEEN TRANSFORM FAULTS AND FRACTURE ZONES

	Transform faults	Fracture zones
Plate boundary?	Yes—a transform plate boundary	No—an intraplate feature
Relative movement across feature	Movement in opposite directions	Movement in the same direction
	← ——— →	← ——— ←
Earthquakes?	Many	Few
Relationship to mid-ocean ridge	Occur *between* offset mid-ocean ridge segments	Occur *beyond* offset mid-ocean ridge segments
Geographic examples	San Andreas Fault, Alpine Fault, Dead Sea Fault	Mendocino Fracture Zone, Molokai Fracture Zone

Web Video
Transform Faults Versus
Fracture Zones

ESSENTIAL CONCEPT 3.4C

Plate tectonic processes create most ocean floor features.

Oceanic Islands

Some of the most interesting features of ocean basins are islands, which are unusually tall features that reach from the sea floor all the way above sea level. There are three basic types of oceanic islands: (1) islands associated with volcanic activity along the mid-ocean ridge (such as Ascension Island along the Mid-Atlantic Ridge); (2) islands associated with hotspots (such as the Hawaiian Islands in the Pacific Ocean); and (3) islands that are island arcs and associated with convergent plate boundaries (such as the Aleutian Islands in the Pacific Ocean). Note that all three types are volcanic in origin. In addition, there is a fourth type of island: islands that are parts of continents (such as the British Isles off Europe), but these occur close to shore and thus do not occur in the deep ocean.

CONCEPT CHECK 3.4

❶ Describe characteristics and features of the mid-ocean ridge, including the difference between oceanic ridges and oceanic rises.

❷ List and describe the different types of hydrothermal vents.

❸ What kinds of unusual life can be found associated with hydrothermal vents? How do these organisms survive?

❹ Describe the origin of the three basic types of oceanic islands.

❺ Describe differences between transform faults and fracture zones.

Essential Concepts Review

3.1 What techniques are used to determine ocean bathymetry?

▶ *Bathymetry is the measurement of ocean depths and the charting of ocean floor topography.* The varied bathymetry of the ocean floor was first determined using *soundings* to measure water depth. Later, the development of the *echo sounder* gave ocean scientists a more detailed representation of the sea floor.

▶ Today, much of our knowledge of the ocean floor has been obtained using various *multibeam echo sounders* or *side-scan sonar instruments* (to make detailed bathymetric maps of a small area of the ocean floor), *satellite measurement* of the ocean surface (to produce maps of the world ocean floor), and *seismic reflection profiles* (to examine Earth structure beneath the sea floor).

Study Resources
Online Study Guide Quizzes, Web Table 3.1, Web Animation, Web Video

Critical Thinking Question

Describe how satellite measurements of the ocean *surface* allow oceanographers to create a map of the sea *floor*.

3.2 What features exist on continental margins?

▸ *Passive continental margins are not associated with any plate boundaries.* Features of passive margins include, from the shore outward, *the continental shelf, the continental slope, and the continental rise.* The continental shelf is generally shallow, low relief, and gently sloping; it can also contain various features such as coastal islands, reefs, and banks. The boundary between the continental slope and the continental shelf is marked by an increase in slope that occurs at the *shelf break.*

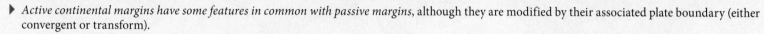

▸ Cutting deep into the continental slopes are *submarine canyons,* which resemble canyons on land but are created by erosive turbidity currents. *Turbidity currents* deposit their sediment load at the base of the continental slope, creating deep-sea fans that merge to produce a gently sloping continental rise. The deposits from turbidity currents (called *turbidite deposits*) have characteristic sequences of graded bedding. Active margins have similar features although they are modified by their associated plate boundary.

▸ *Active continental margins have some features in common with passive margins,* although they are modified by their associated plate boundary (either convergent or transform).

Study Resources
Online Study Guide Quizzes, Web Diving Deeper 3.3, Web Animation, Web Video

Critical Thinking Question

To help reinforce your knowledge of continental margins, draw and describe differences between passive and active continental margins from memory. Be sure to include a real-world example of each type, associated features, and how these features relate to plate tectonics.

3.3 What features exist in the deep-ocean basins?

▸ The *continental rises* gradually become flat, extensive, deep-ocean *abyssal plains,* which form by *suspension settling* of fine sediment. Poking through the sediment cover of the abyssal plains are numerous *volcanic peaks,* including volcanic islands, seamounts, tablemounts, and abyssal hills. In the Pacific Ocean, where sedimentation rates are low, abyssal plains are not extensively developed, and abyssal hill provinces cover broad expanses of ocean floor.

▸ *Along the margins of many continents*—especially those around the *Pacific Ring of Fire*—are *deep linear scars called ocean trenches* that are associated with convergent plate boundaries and volcanic arcs.

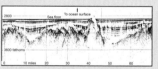

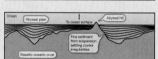

Study Resources
Online Study Guide Quizzes, Web Diving Deeper 3.2

Critical Thinking Question

In which ocean basin are most ocean trenches found? Use plate tectonic processes to help explain why.

3.4 What features exist along the mid-ocean ridge?

▸ The *mid-ocean ridge is a continuous mountain range* that winds through all ocean basins and is entirely *volcanic in origin.* Common features associated with the mid-ocean ridge include a *central rift valley, faults and fissures, seamounts, pillow basalts, hydrothermal vents, deposits of metal sulfides,* and *unusual life forms.* Segments of the mid-ocean ridge are either *oceanic ridges* if steep with rugged slopes (indicative of slow sea floor spreading) or *oceanic rises* if sloped gently and less rugged (indicative of fast spreading).

▸ *Long linear zones of weakness—fracture zones and transform faults—* cut across vast distances of ocean floor and *offset the axes of the mid-ocean ridge.* Fracture zones and transform faults are differentiated from one another based on the direction of movement across the feature. *Fracture zones (an intraplate feature) have movement in the same direction,* while *transform faults (a transform plate boundary) have movement in opposite directions.*

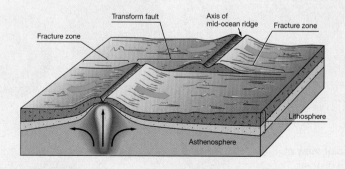

Study Resources
Online Study Guide Quizzes, Web Animation, Web Videos

Critical Thinking Question

Use pictures and words to describe the differences between fracture zones and transform faults.

MasteringOceanography™

www.masteringoceanography.com

Looking for additional review and test prep materials? Visit the Study Area in MasteringOceanography to enhance your understanding of this chapter's content by accessing a variety of resources, including Self-Study Quizzes, Geoscience Animations, RSS feeds, flashcards, Web links, and an optional Pearson eText.

Microscopic view of arranged diatoms. The objects in this photomicrograph are diatoms, a type of microscopic marine algae that exists in incredible abundance in the ocean. This image shows various species of diatoms magnified several hundred times and was made by carefully arranging them under a microscope.

4

MARINE SEDIMENTS

Before you begin reading this chapter, use the glossary at the end of this book to discover the meanings of any of these words you don't already know:

Coccolithophore **Calcareous ooze**
Radiolarian · Foraminifer
Precipitate **Biogenous sediment**
Test Texture
Calcite compensation depth (CCD) Metal sulfide
Neritic deposit **Ooze**
Lithogenous sediment Siliceous ooze Diatomaceous earth
Rotary drilling
Pelagic deposit Methane hydrate **Diatom Cosmogenous sediment**
Evaporite
Abyssal clay
Upwelling **Hydrogenous sediment** Manganese nodule

"From the sediments the history of the ocean emerged with all its wonders...."

— *Wolf H. Berger*, Oceans: Reflections on a Century of Exploration. University of California Press. 2009. Used by permission.

Why are **sediments** (*sedimentum* = settling) interesting to oceanographers? Although ocean sediments appear to be little more than eroded particles and fragments of dirt, dust, and other debris that have settled out of the water and accumulated on the ocean floor (**Figure 4.1**), they reveal much about Earth's history. As we'll see, sediments are quite useful for providing clues to past geographic distributions of marine organisms, movements of the ocean floor, ocean circulation patterns, climate changes on Earth, and even global extinction events. Remarkably, sediments that accumulate over time on the sea floor comprise a nearly continuous, undisturbed record of Earth history unlike anything on land. In essence, marine sediments represent Earth's largest museum with displays of Earth history dating back millions of years.

Over time, sediments can become *lithified* (*lithos* = stone, *fic* = making)—turned to rock—and form *sedimentary rock*. More than half of the rocks exposed on the continents are sedimentary rocks deposited in ancient ocean environments and uplifted onto land by plate tectonic processes. Perhaps surprisingly, even the tallest mountains on the continents—far from any ocean—contain telltale marine fossils, which indicate that these rocks originated on the ocean floor in the geologic past. For example, the summit of the world's tallest mountain (Mount Everest in the Himalaya Mountains) consists of limestone, which is a type of rock that originated as sea floor deposits.

Particles of sediment come from worn pieces of rocks, as well as living organisms, minerals dissolved in water, and outer space. Clues to sediment origin are found in its mineral composition and its **texture** (the size and shape of its particles).

This chapter is organized around **Table 4.1**, which shows the classification of marine sediments according to type, composition, sources, and main locations found. The chapter begins with a brief discussion about how sediments are collected and what sediments reveal about Earth history. As you work through the remainder of this chapter, note that each of the four main types of sediment are discussed (Table 4.1, *first column*). Then, within each type of sediment shown in Table 4.1, the composition (*second column*), origin (*third column*), and distribution (*fourth column*) are examined. Mixtures of marine sediment and sediment distribution are also considered. Finally, the chapter concludes with a discussion of the resources that marine sediments provide.

4.1 How Are Marine Sediments Collected, and What Historical Events Do they Reveal?

One of the difficulties of studying marine sediments is collecting adequate samples from the deep-ocean floor. Until relatively recently, the inaccessibility of the deep ocean has hindered the collection of marine sediments, especially those beneath the surface of the sea floor.

TABLE 4.1   CLASSIFICATION OF MARINE SEDIMENTS

Type	Composition		Sources		Main locations found
Lithogenous	Continental margin	Rock fragments, Quartz sand, Quartz silt, Clay		Rivers; coastal erosion; landslides	Continental shelf
				Glaciers	Continental shelf in high latitudes
				Turbidity currents	Continental slope and rise; ocean basin margins
	Oceanic	Quartz silt, Clay		Wind-blown dust; rivers	Abyssal plains and other regions of the deep-ocean basins
		Volcanic ash		Volcanic eruptions	
Biogenous	Calcium carbonate ($CaCO_3$)	Calcareous ooze (microscopic)	Warm surface waters	Coccolithophores (algae), Foraminifers (protozoans)	Low-latitude regions; sea floor above CCD; along mid-ocean ridges and the tops of volcanic peaks
		Shells and coral fragments (macroscopic)		Macroscopic shell-producing organisms	Continental shelf; beaches
				Coral reefs	Shallow low-latitude regions
	Silica ($SiO_2 \cdot nH_2O$)	Siliceous ooze	Cold surface waters	Diatoms (algae), Radiolarians (protozoans)	High-latitude regions; sea floor below CCD; upwelling areas where cold, deep water rises to the surface, especially that caused by surface current divergence near the equator
Hydrogenous	Manganese nodules (manganese, iron, copper, nickel, cobalt)			Precipitation of dissolved materials directly from seawater due to chemical reactions	Abyssal plain
	Phosphorite (phosphorous)				Continental shelf
	Oolites ($CaCO_3$)				Shallow shelf in low-latitude regions
	Metal sulfides (iron, nickel, copper, zinc, silver)				Hydrothermal vents at mid-ocean ridges
	Evaporites (gypsum, halite, other salts)				Shallow restricted basins where evaporation is high in low-latitude regions
Cosmogenous	Iron–nickel spherules, Tektites (silica glass)			Space dust	In very small proportions mixed with all types of sediment and in all marine environments
	Iron–nickel meteorites			Meteors	Localized near meteor impact structures

Figure 4.1 **Oceanic sediment.** View of the deep-ocean floor from a submersible. Most of the deep-ocean floor is covered with particles of material that have settled out through the water.

Collecting Marine Sediments

Collecting sediments suitable for analysis from the deep ocean is an arduous process. During early exploration of the oceans, a bucket-like device called a *dredge* was used to scoop up sediment from the deep-ocean floor for analysis. This technique, however, was tedious and had many limitations. For example, it often didn't work right and the dredge came up empty. It also disturbed the sediment and could only gather samples from the surface of the ocean floor. Later, the *gravity corer*—a hollow steel tube with a heavy weight on top—was thrust into the sea floor to collect the first **cores** (cylinders of sediment and rock). Although the gravity corer could sample below the surface, its depth of penetration was limited. Today, specially designed ships perform **rotary drilling** to collect cores from the deep ocean.

In 1963, the U.S. National Science Foundation funded a program that borrowed drilling technology from the offshore oil industry to obtain long sections of core from deep below the surface of the ocean floor. The program united four leading oceanographic institutions (Scripps Institution of Oceanography in California; Rosenstiel School of Atmospheric and Oceanic Studies at the University of Miami, Florida; Lamont-Doherty Earth Observatory of Columbia University in New York; and the Woods Hole Oceanographic Institution in Massachusetts) to form the *Joint Oceanographic Institutions for Deep Earth Sampling (JOIDES)*. The oceanography departments of several other leading universities later joined JOIDES.

The first phase of the **Deep Sea Drilling Project (DSDP)** was initiated in 1968, when the specially designed drill ship *Glomar Challenger* was launched. It had a tall drilling rig resembling a steel tower. Cores could be collected by drilling into the ocean floor in water up to 6000 meters (3.7 miles) deep. From the initial cores collected, scientists confirmed the existence of sea floor spreading by documenting that (1) the age of the ocean floor increased progressively with distance from the mid-ocean ridge; (2) sediment thickness increased progressively with distance from the mid-ocean ridge; and (3) Earth's magnetic field polarity reversals were recorded in ocean floor rocks.

Although the oceanographic research program was initially financed by the U.S. government, it became international in 1975, when West Germany, France, Japan, the United Kingdom, and the Soviet Union also provided financial and scientific support. In 1983, the Deep Sea Drilling Project became the **Ocean Drilling Program (ODP)**, with 20 participating countries under the supervision of Texas A&M University and a broader objective of drilling the thick sediment layers near the continental margins.

In 1985, the *Glomar Challenger* was decommissioned and replaced by the drill ship **JOIDES Resolution** (**Figure 4.2**). The new ship also has a tall metal drilling rig to conduct *rotary drilling*. The drill pipe is in individual sections of 9.5 meters (31 feet), and sections can be screwed together to make a single string of pipe up to 8200 meters (27,000 feet) long (**Figure 4.3**). The drill bit, located at the end of

Figure 4.2 **The drill ship JOIDES Resolution.** Aerial view of the *JOIDES Resolution*, which has a tall metal derrick in the middle of the ship for rotary drilling. The helipad is also visible at the stern of the vessel.

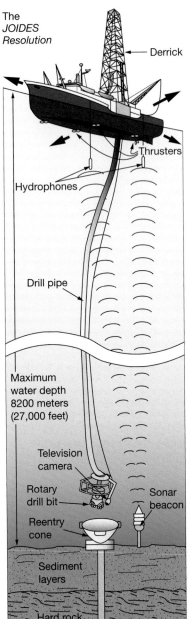

The *JOIDES Resolution*

Derrick

Thrusters

Hydrophones

Drill pipe

Maximum water depth 8200 meters (27,000 feet)

Television camera

Rotary drill bit

Reentry cone

Sonar beacon

Sediment layers

Hard rock

Figure 4.3 Rotary drilling from the *JOIDES Resolution*. Using its array of thrusters, the *JOIDES Resolution* can remain in one place at the surface while performing rotary drilling, which is shown diagrammatically.

the pipe string, rotates as it is pressed against the ocean bottom and can drill up to 2100 meters (6900 feet) into the sea floor. Like twirling a soda straw into a layer cake, the drilling operation crushes the rock around the outside and retains a cylinder of rock (a *core sample*) on the inside of the hollow pipe. A core can then be raised to the surface from inside the pipe and analyzed using state-of-the-art laboratory facilities on board the *Resolution*. Worldwide, more than 2000 holes have been drilled into the sea floor using this method, allowing the collection of cores that provide scientists with valuable information about Earth history as recorded in sea floor sediments.

In 2003, the ODP was replaced by the **Integrated Ocean Drilling Program (IODP)**, whose main participants are Japan, the United States, and the European Union. This new international effort does not rely on just one drill ship but uses multiple vessels for exploration. One of the new vessels that began operations in 2007 is a state-of-the-art drill ship named *Chikyu* (which means "Planet Earth" in Japanese) that can drill up to 7000 meters (23,000 feet) into the sea floor, with plans to upgrade the vessel with new drilling technology to allow it to drill even deeper, perhaps as deep as through Earth's crust into the mantle. The program's primary objective is to collect cores that will allow scientists to better understand Earth history and Earth system processes, including the properties of the deep crust, climate change patterns, the microbiology of the deep-ocean floor, and Earthquake mechanisms. For example, shortly after the devastating 2011 Tohoku-Oki Earthquake and resulting tsunami, the *Chikyu* drillship began an expedition to the Japan Trench to drill into the fault zone near the site of the earthquake and study friction by taking detailed temperature measurements beneath the sea floor.

Environmental Conditions Revealed by Marine Sediments

Marine sediments provide a wealth of information about past conditions on Earth. As sediment accumulates on the ocean floor, it preserves the materials—and the

conditions of the environment—that existed in the overlying water column. By carefully analyzing cylindrical cores of sediment collected from the sea floor and interpreting them (**Figure 4.4**), Earth scientists can infer past environmental conditions such as sea surface temperature, nutrient supply, abundance of marine life, atmospheric winds, ocean current patterns, volcanic eruptions, major extinction events, changes in Earth's climate, and the movement of tectonic plates. In fact, most of what is known of Earth's past geology, climate, and biology has been learned through studying ancient marine sediments.

Paleoceanography

The study of how the ocean, atmosphere, and land have interacted to produce changes in ocean chemistry, circulation, biology, and climate is called **paleoceanography** (*paleo* = ancient, *ocean* = the marine environment, *graphy* = description of), which relies on sea floor sediments to gain insight into these past changes. Recent paleoceanographic studies, for example, have linked changes in

Web Video
Rotary Drilling

Figure 4.4 Examining deep-ocean sediment cores. Long cylinders of sediment and rock called *cores* are cut in half and examined, revealing interesting aspects of Earth history.

deep-ocean circulation with rapid climate change. In the North Atlantic Ocean, cold, relatively salty water sinks and forms a body of water called *North Atlantic Deep Water*. Water in this deep current circulates through the global ocean, driving deep-ocean circulation and global heat transport and, thus, impacting global climate. This is widely viewed as one of the most climatically sensitive regions on Earth, and North Atlantic sea floor sediments from the past several million years have revealed that the region has experienced abrupt changes to its ocean–atmosphere system, triggered by fluctuations of freshwater from melting glaciers. Understanding the timing, mechanisms, and causes of this abrupt climate change is one of the major challenges facing paleoceanography today.

ESSENTIAL CONCEPT 4.1

Marine sediments accumulate on the ocean floor and contain a record of Earth history, including past environmental conditions.

CONCEPT CHECK 4.1

1. Using Table 4.1, list and describe the characteristics of the four basic types of marine sediment.
2. Describe the process of how a drill ship like the *JOIDES Resolution* obtains core samples from the deep-ocean floor.
3. What types of past environmental conditions can be inferred by studying cores of sediment?

4.2 What Are the Characteristics Of Lithogenous Sediment?

Lithogenous (*lithos* = stone, *generare* = to produce) **sediment** is derived from pre-existing rock material that originates on the continents or islands from erosion, volcanic eruptions, or blown dust. Note that lithogenous sediment is sometimes referred to as **terrigenous** (*terra* = land, *generare* = to produce) **sediment**.

Origin Of Lithogenous Sediment

Lithogenous sediment begins as rocks on continents or islands. Over time, **weathering** agents such as water, temperature extremes, and chemical effects break rocks into smaller pieces, as shown in **Figure 4.5**. When rocks are in smaller pieces, they can be more easily **eroded** (picked up) and transported. This eroded material is the basic component of which all lithogenous sediment is composed.

Eroded material from the continents is carried to the oceans by streams, wind, glaciers, and gravity (**Figure 4.6**). Each year, stream flow alone carries about 20 billion metric tons (22 billion short tons) of sediment to Earth's continental margins; almost 40% is provided by runoff from Asia.

Transported sediment can be deposited in many environments, including bays or lagoons near the ocean, as deltas at the mouths of rivers, along beaches at the shoreline,

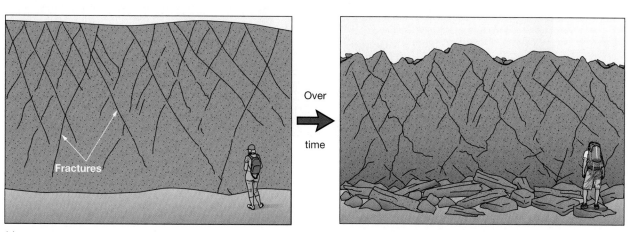

Over time

(a)   (b)

Figure 4.5 Weathering. Over time, weathering occurs along fractures and breaks rock into smaller fragments, which are much easier to transport.

Figure 4.6 Sediment transporting media. Photos showing examples of various sediment transporting media, including **(a)** stream, **(b)** wind, **(c)** glacier, and **(d)** gravity.

(a) Stream: Po River, Italy, which displays a prominent delta and a visible sediment plume in the water.

(b) Wind: Dust storm approaching a military base, Australia.

(c) Glacier: Riggs Glacier, Glacier Bay National Park, Alaska, which displays a dark stripe of sediment along its length called a *medial moraine*.

(d) Gravity, which creates landslides: Del Mar, California.

or further offshore across the continental margin. It can also be carried beyond the continental margin to the deep-ocean basin by turbidity currents, as discussed in Chapter 3.

The greatest quantity of lithogenous material is found around the margins of the continents, where it is constantly moved by high-energy currents along the shoreline and in deeper turbidity currents. Lower-energy currents distribute finer components that settle out onto the deep-ocean basins. Microscopic particles from wind-blown dust or volcanic eruptions can even be carried far out over the open ocean by prevailing winds. These particles either settle into fine layers as the velocity of the wind decreases or disperse into the ocean when they serve as nuclei around which raindrops and snowflakes form.

Composition Of Lithogenous Sediment

The composition of lithogenous sediment reflects the material from which it was derived. All rocks are composed of discrete crystals of naturally occurring compounds called *minerals*. One of the most abundant, chemically stable, and durable minerals in Earth's crust is **quartz**, composed of silicon and oxygen in the form of SiO_2—the same composition as ordinary glass. Quartz is a major component of most rocks. Because quartz is resistant to abrasion, it can be transported long distances and deposited far from its source area. The majority of lithogenous deposits—such as beach sands—are composed primarily of quartz (**Figure 4.7**).

A large percentage of lithogenous particles that find their way into deep-ocean sediments far from continents are transported by prevailing winds that remove small particles from the continents' subtropical desert regions. The map in **Figure 4.8** shows a close relationship between the location of microscopic fragments of lithogenous quartz in the surface sediments of the ocean floor and the strong prevailing winds in the desert regions of Africa, Asia, and Australia. Satellite observations of dust storms (Figure 4.8, *inset*) confirm this relationship. Sediment is not the only item transported by wind. In fact, there has been recent documentation of the transportation of a variety of airborne substances—including viruses, pollutants, and even living insects—from Africa all the way across the Atlantic Ocean to North America.

Sediment Texture

One of the most important properties of lithogenous sediment is its texture, including its **grain size**.[1] The **Wentworth scale of grain size** (**Table 4.2**) indicates that particles can be classified as boulders (largest), cobbles, pebbles, granules, sand, silt, or clay (smallest). Sediment size is proportional to the energy needed to lay down a deposit. Deposits laid down where wave action is strong (areas of high energy) may be composed primarily of larger particles—cobbles and boulders. Fine-grained particles, on the other hand, are deposited where the energy level is low and the current speed is minimal. When clay-sized particles—many of which are flat—are deposited, they tend to stick together by cohesive forces. Consequently, higher-energy conditions than what would be expected based on grain size alone are required to erode and transport clays. In general, however, lithogenous sediment tends to become finer with increasing distance from shore. This relationship is mostly because high-energy transporting media predominate close to shore and lower-energy conditions exist in the deep-ocean basins.

The texture of lithogenous sediment also depends on its **sorting**. Sorting is a measure of the uniformity of grain sizes and indicates the selectivity of the transportation process. For example, sediments composed of particles that are primarily the same size are well sorted—such as in coastal sand dunes, where winds can only pick up a certain size particle. Poorly sorted deposits, on the other hand, contain a variety of different sized particles and indicate a transportation process capable of picking up clay- to boulder-sized particles. An example of poorly sorted sediment is that which is carried by a glacier and left behind when the glacier melts.

Figure 4.7 Lithogenous beach sand. Lithogenous beach sand is composed mostly of particles of white quartz, plus small amounts of other minerals. This sand, from North Beach, Hampton, New Hampshire, is magnified approximately 23 times.

STUDENTS SOMETIMES ASK . . .

How effective is wind as a transporting agent?

Any material that gets into the atmosphere—including dust from dust storms, soot from forest fires, specks of pollution, and ash from volcanic eruptions—is transported by wind and can be found as deposits on the ocean floor. Every year, wind storms lift an estimated 3 billion metric tons (3.3 billion short tons) of this material into the atmosphere, where it gets transported around the globe. As much as three-quarters of these particles—mostly dust—come from Africa's Sahara Desert; once airborne, they are carried out across the Atlantic Ocean (see Figure 4.8). Much of this dust falls in the Atlantic, and that's why ships traveling downwind from the Sahara Desert often arrive at their destinations quite dusty. Some of it falls in the Caribbean (where the pathogens it contains have been linked to stress and disease among coral reefs), the Amazon (where its iron and phosphorus fertilize nutrient-poor soil), and across the southern United States as far west as New Mexico. The dust also contains bacteria and pesticides—even African desert locusts have been transported alive across the Atlantic Ocean during strong wind storms!

[1]Sediment grains are also known as particles, fragments, or clasts.

Figure 4.8 Lithogenous quartz in surface sediments of the world's oceans and transport by wind. High concentrations of microscopic lithogenous quartz in deep-sea sediment (*map*) match prevailing winds from land (*green arrows*). SeaStar SeaWiFS satellite image (*photo*) on February 26, 2000, shows a wind storm that has blown dust from the Sahara Desert off the northwest coast of Africa. Some of this dust is transported across the Atlantic Ocean to South America, the Caribbean, and North America.

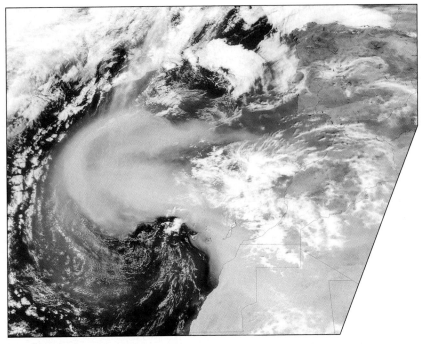

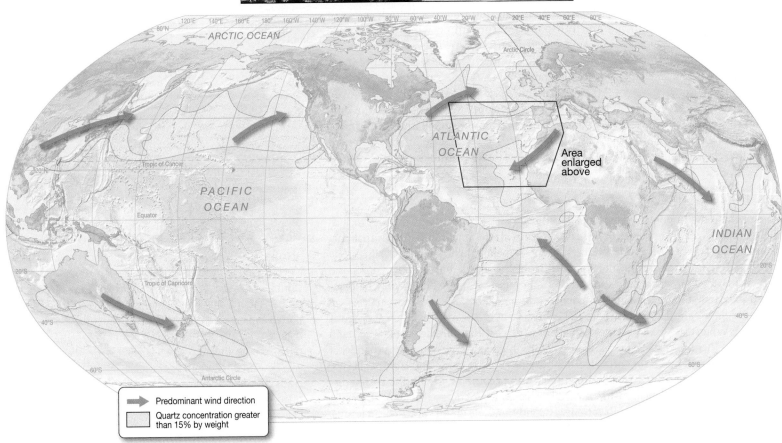

Distribution Of Lithogenous Sediment

Marine sedimentary deposits can be categorized as either neritic or pelagic. **Neritic** (*neritos* = of the coast) **deposits** are found on continental shelves and in shallow water near islands; these deposits are generally coarse grained. Alternatively, **pelagic** (*pelagios* = of the sea) **deposits** are found in the deep-ocean basins and are typically fine grained. Moreover, lithogenous sediment in the ocean is ubiquitous: At least a small percentage of lithogenous sediment is found nearly everywhere on the ocean floor.

TABLE 4.2   WENTWORTH SCALE OF GRAIN SIZE FOR SEDIMENTS

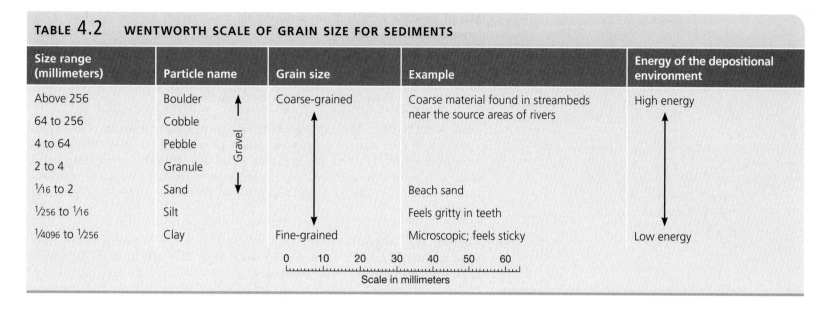

Size range (millimeters)	Particle name	Grain size	Example	Energy of the depositional environment
Above 256	Boulder	Coarse-grained	Coarse material found in streambeds near the source areas of rivers	High energy
64 to 256	Cobble			
4 to 64	Pebble			
2 to 4	Granule			
1/16 to 2	Sand		Beach sand	
1/256 to 1/16	Silt		Feels gritty in teeth	
1/4096 to 1/256	Clay	Fine-grained	Microscopic; feels sticky	Low energy

Scale in millimeters: 0 10 20 30 40 50 60

NERITIC DEPOSITS   Lithogenous sediment dominates most neritic deposits. Lithogenous sediment is derived from rocks on nearby landmasses, consists of coarse-grained deposits, and accumulates rapidly on the continental shelf, slope, and rise. Examples of lithogenous neritic deposits include beach deposits, continental shelf deposits, turbidite deposits, and glacial deposits.

Beach Deposits   Beaches are made of whatever materials are locally available. Beach materials are composed mostly of quartz-rich sand that is washed down to the coast by rivers but can also be composed of a wide variety of sizes and compositions. This material is transported by waves that crash against the shoreline, especially during storms.

Continental Shelf Deposits   At the end of the last ice age (about 18,000 years ago), glaciers melted and sea level rose. As a result, many rivers of the world today deposit their sediment in drowned river mouths rather than carry it onto the continental shelf as they did during the geologic past. In many areas, the sediments that cover the continental shelf—called *relict* (relict = left behind) *sediments*—were deposited from 3000 to 7000 years ago and have not yet been covered by more recent deposits. These sediments presently cover about 70% of the world's continental shelves. In other areas, deposits of sand ridges on the continental shelves appear to have been formed more recently than the most recent ice age and at present water depths.

Turbidite Deposits   As discussed in Chapter 3, **turbidity currents** are underwater avalanches that periodically move down the continental slopes and carve submarine canyons. Turbidity currents also carry vast amounts of neritic material. This material spreads out as deep-sea fans, comprises the continental rise, and gradually thins toward the abyssal plains. These deposits are called **turbidite deposits** and are composed of characteristic layering called *graded bedding* (see Figure 3.12).

Glacial Deposits   Poorly sorted deposits containing particles ranging from boulders to clays may be found in the high-latitude[2] portions of the continental shelf. These

[2]High-latitude regions are those far from the equator (either north or south); low latitudes are areas close to the equator.

glacial deposits were laid down during the most recent ice age by glaciers that covered the continental shelf and eventually melted. Glacial deposits are currently forming around the continent of Antarctica and around Greenland by **ice rafting**. In this process, rock particles trapped in glacial ice are carried out to sea by icebergs that break away from coastal glaciers. As the icebergs melt, lithogenous particles of many sizes are released and settle onto the ocean floor.

PELAGIC DEPOSITS Turbidite deposits of neritic sediment on the continental rise can spill over into the deep-ocean basin. However, most pelagic deposits are composed of fine-grained material that accumulates slowly on the deep-ocean floor. Pelagic lithogenous sediment includes particles that have come from volcanic eruptions, windblown dust, and fine material that is carried by deep-ocean currents.

Abyssal Clay **Abyssal clay** is composed of at least 70% (by weight) fine, clay-sized particles from the continents. Even though they are far from land, deep abyssal plains contain thick sequences of abyssal clay deposits composed of particles transported great distances by winds or ocean currents and deposited on the deep-ocean floor. Because abyssal clays contain oxidized iron, they are commonly red-brown or buff in color and are sometimes referred to as **red clays**. The predominance of abyssal clay on abyssal plains is caused not by an abundance of clay settling on the ocean floor but by the absence of other material that would otherwise dilute it.

CONCEPT CHECK 4.2

❶ Describe the origin, composition, texture, and distribution of lithogenous sediment.

❷ Why is most lithogenous sediment composed of quartz grains? What is the chemical composition of quartz?

❸ What is the difference between neritic and pelagic deposits? Give examples of lithogenous sediment found in each.

4.3 What Are the Characteristics Of Biogenous Sediment?

Biogenous (*bio* = life, *generare* = to produce) **sediment** (also called *biogenic sediment*) is derived from the remains of hard parts of once-living organisms.

Origin Of Biogenous Sediment

Biogenous sediment begins as the hard parts (shells, bones, and teeth) of living organisms ranging from minute algae and protozoans to fish and whales. When organisms that produce hard parts die, their remains settle onto the ocean floor and can accumulate as biogenous sediment.

Biogenous sediment can be classified as either macroscopic or microscopic. **Macroscopic biogenous sediment** is large enough to be seen without the aid of a microscope and includes shells, bones, and teeth of large organisms. Except in certain tropical beach localities where shells and coral fragments are numerous, this type of sediment is relatively rare in the marine environment, especially in deep water where fewer organisms live. Much more abundant is **microscopic biogenous sediment**, which contains particles so small they can be seen well only through a microscope. Microscopic organisms produce tiny shells called **tests** (*testa* = shell) that begin to sink after the organisms die and continually rain down in great numbers onto the ocean floor. These microscopic tests can accumulate on the deep-ocean floor and form deposits called **ooze** (*wose* = juice). As its name implies, ooze resembles very fine-grained, mushy material.[3] Technically, biogenous ooze must contain at least 30% biogenous test material by weight. What comprises the other part—up to 70%—of an ooze? Commonly, it is fine-grained lithogenous clay that is deposited along with biogenous tests in the deep ocean. By volume, much more microscopic ooze than macroscopic biogenous sediment exists on the ocean floor.

The organisms that contribute to biogenous sediment are chiefly **algae** (*alga* = seaweed) and **protozoans** (*proto* = first, *zoa* = animal). Algae are primarily aquatic, eukaryotic,[4] photosynthetic organisms, ranging in size from microscopic single cells to large organisms like giant kelp. Protozoans are any of a large group of single-celled, eukaryotic, usually microscopic organisms that are generally not photosynthetic.

Composition Of Biogenous Sediment

The two most common chemical compounds in biogenous sediment are **calcium carbonate** ($CaCO_3$, which forms the mineral **calcite**) and **silica** (SiO_2). Often, the silica is chemically combined with water to produce $SiO_2 \cdot nH_2O$, the hydrated form of silica, which is called *opal*.

SILICA Most of the silica in biogenous ooze comes from microscopic algae called **diatoms** (*diatoma* = cut in half) and protozoans called **radiolarians** (*radio* = a spoke or ray).

Because diatoms photosynthesize, they need strong sunlight and are found only within the upper, sunlit surface waters of the ocean. Most diatoms are free-floating, or **planktonic** (*planktos* = wandering). The living organism builds a glass greenhouse out of silica as a protective covering and lives inside. Most species have two parts to their test that fit together like a petri dish or pillbox (**Figure 4.9a**). The tiny tests are perforated with small holes in intricate patterns to allow nutrients to pass in and waste products to pass out. Where diatoms are abundant at the ocean surface, thick deposits of diatom-rich ooze can accumulate below on the ocean floor. When this ooze lithifies, it becomes **diatomaceous earth**,[5] a lightweight white rock composed of diatom tests and clay (**Diving Deeper 4.1**).

Radiolarians are microscopic single-celled protozoans, most of which are also planktonic. As their name implies, they often have long spikes or rays of silica protruding from their siliceous shell (**Figure 4.9b**). They do not photosynthesize but rely on external food sources such as bacteria and other plankton. Radiolarians typically display well-developed symmetry, which is why they have been described as the "living snowflakes of the sea."

The accumulation of siliceous tests of diatoms, radiolarians, and other silica-secreting organisms produces **siliceous ooze** (**Figure 4.9c**).

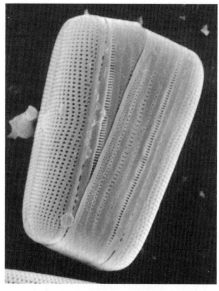

(a) Diatom (length = 30 micrometers, equal to 30 millionths of a meter), showing how the two parts of the diatom's test fit together.

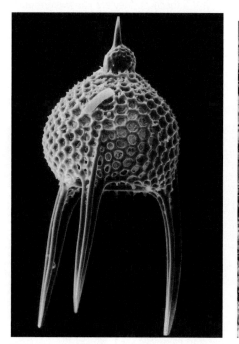

(b) Radiolarian (length = 100 micrometers).

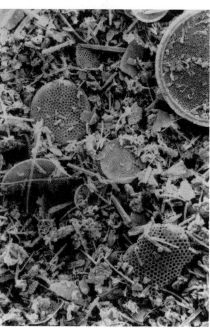

(c) Siliceous ooze, showing mostly fragments of diatom tests (magnified 250 times).

Figure 4.9 Microscopic siliceous tests. Scanning electron micrographs of various siliceous tests.

---

[3]Ooze has the consistency of toothpaste mixed about half and half with water. As a way to remember this term, imagine walking barefoot across the deep-ocean floor and having the fine sediment there *ooze* between your toes.

[4]Eukaryotic (eu = good, *karyo* = the nucleus) cells contain a distinct membrane-bound nucleus.

[5]Diatomaceous earth is also called diatomite, tripolite, or kieselguhr.

DIATOMS: THE MOST IMPORTANT THINGS YOU HAVE (PROBABLY) NEVER HEARD OF

"Few objects are more beautiful than the minute siliceous cases of the diatomaceae: were these created that they might be examined and admired under the higher powers of the microscope?"

—*Charles Darwin (1872)*

Diatoms are microscopic single-celled photosynthetic organisms. Each one lives inside a protective silica test, most of which contain two halves that fit together like a shoebox and its lid. First described with the aid of a microscope in 1702, their tests are exquisitely ornamented with holes, ribs, and radiating spines unique to individual species. The fossil record indicates that diatoms have been on Earth since the Jurassic Period (180 million years ago), and more than 70,000 species of diatoms have been identified.

Diatoms live for a few days to as much as a week, can reproduce sexually or asexually, and occur individually or linked together into long communities. They are found in great abundance floating in the ocean and in certain freshwater lakes but can also be found in many diverse environments, such as on the undersides of polar ice, on the skins of whales, in soil, in thermal springs, and even on brick walls.

When marine diatoms die, their tests rain down and accumulate on the sea floor as siliceous ooze. Hardened deposits of siliceous ooze, called *diatomaceous earth*, can be as much as 900 meters (3000 feet) thick. Diatomaceous earth consists of billions of minute silica tests and has many unusual properties: It is lightweight, has an inert chemical composition, is resistant to high temperatures, and has excellent filtering properties. Diatomaceous earth is used to produce a variety of common products (**Figure 4A**). The main uses of diatomaceous earth include:

- Filters (for refining sugar, separating impurities from wine, straining yeast from beer, and filtering swimming pool water)
- Mild abrasives (in toothpaste, facial scrubs, matches, and household cleaning and polishing compounds)
- Absorbents (for chemical spills, in cat litter, and as a soil conditioner)
- Chemical carriers (in pharmaceuticals, paint, and even dynamite)

Other products from diatomaceous earth include optical-quality glass (because of the pure silica content of diatoms) and space shuttle tiles (because they are lightweight and provide good insulation). Diatomaceous earth is also used as an additive in concrete, a filler in tires, an anticaking agent, a natural pesticide, and even a building stone in the construction of houses.

Further, the vast majority of oxygen that all animals breathe is a by-product of photosynthesis by diatoms. In addition, each living diatom contains a tiny droplet of oil. When diatoms die, their tests containing droplets of oil accumulate on the sea floor and are the beginnings of petroleum deposits, such as those found offshore of California.

Given their many practical applications, it is difficult to imagine how different our lives would be without diatoms!

Figure 4A **Products containing or produced using diatomaceous earth (diatom *Thalassiosira eccentrica*, *inset*).**

CALCIUM CARBONATE Two significant sources of calcium carbonate biogenous ooze are the **foraminifers** (*foramen* = an opening)—close relatives of radiolarians—and microscopic algae called **coccolithophores** (*coccus* = berry, *lithos* = stone, *phorid* = carrying).

Coccolithophores are single-celled algae, most of which are planktonic. Coccolithophores produce thin plates or shields made of calcium carbonate, 20 or 30 of which overlap to produce a spherical test (Figure 4.10a). Like diatoms, coccolithophores photosynthesize, so they need sunlight to live. Coccolithophores are about 10 to 100 times smaller than most diatoms (Figure 4.10b), which is why coccolithophores are often called **nannoplankton** (*nanno* = dwarf, *planktos* = wandering).

When the organism dies, the individual plates (called **coccoliths**) disaggregate and can accumulate on the ocean floor as coccolith-rich ooze. When this ooze lithifies over time, it forms a white deposit called **chalk**, which is used for a variety of purposes (including writing on chalkboards). The White Cliffs of southern England

(a) Coccolithophores (diameter of individual coccolithophores = 20 micrometers, equal to 20 millionths of a meter).

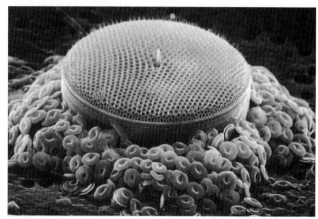

(b) Diatom (siliceous) and coccoliths (diameter of diatom = 70 micrometers).

(c) Foraminifers (most species 400 micrometers in diameter).

(d) Calcareous ooze, which also includes some siliceous radiolarian tests (magnified 160 times).

Figure 4.10 **Microscopic calcareous tests.** Scanning electron micrographs (*above*) and photomicrographs (*below*) of various calcareous tests.

Figure 4.11 The White Cliffs of southern England. The White Cliffs near Dover in southern England are composed of chalk, which is hardened coccolith-rich calcareous ooze. Inset shows a colored image of the coccolithophore *Emiliana huxleyi* (diameter = 20 micrometers, equal to 20 millionths of a meter).

are composed of hardened, coccolith-rich calcium carbonate ooze, which was deposited on the ocean floor and has been uplifted onto land (**Figure 4.11**). Deposits of chalk the same age as the White Cliffs are so common throughout Europe, North America, Australia, and the Middle East that the geologic period in which these deposits formed is named the Cretaceous (*creta* = chalk) Period.

Foraminifers are single-celled protozoans, many of which are planktonic, ranging in size from microscopic to macroscopic. They do not photosynthesize, so they must ingest other organisms for food. Foraminifers produce a hard calcium carbonate test in which the organism lives (**Figure 4.10c**). Most foraminifers produce a segmented or chambered test, and all tests have a prominent opening in one end. Although very small in size, the tests of foraminifers resemble the large shells that one might find at a beach.

Deposits comprised primarily of tests of foraminifers, coccoliths, and other calcareous-secreting organisms are called **calcareous ooze** (**Figure 4.10d**).

Distribution Of Biogenous Sediment

Biogenous sediment is one of the most common types of pelagic deposits. The distribution of biogenous sediment on the ocean floor depends on three fundamental processes: (1) productivity, (2) destruction, and (3) dilution.

Productivity is the number of organisms present in the surface water above the ocean floor. Surface waters with high biologic productivity contain many living and reproducing organisms—conditions that are likely to produce biogenous sediments. Conversely, surface waters with low biologic productivity contain too few organisms to produce biogenous oozes on the ocean floor.

Destruction occurs when skeletal remains (tests) dissolve in seawater at depth. In some cases, biogenous sediment dissolves before ever reaching the sea floor; in other cases, it is dissolved before it has a chance to accumulate into deposits on the sea floor.

Dilution occurs when the deposition of other sediments decreases the percentage of the biogenous sediment found in marine deposits. For example, other types of sediments can dilute biogenous test material below the 30% necessary to classify it as ooze. Dilution occurs most often because of the abundance of coarse-grained lithogenous material in neritic environments, so biogenous oozes are uncommon along continental margins.

NERITIC DEPOSITS Although neritic deposits are dominated by lithogenous sediment, both microscopic and macroscopic biogenous material may be incorporated into lithogenous sediment in neritic deposits. In addition, biogenous carbonate deposits are common in some areas.

Carbonate Deposits **Carbonate** minerals are those that contain CO_3 in their chemical formula—such as calcium carbonate, $CaCO_3$. Rocks from the marine environment composed primarily of calcium carbonate are called **limestones**. Most limestones contain fossil marine shells, suggesting a biogenous origin, while others appear to have formed directly from seawater without the help of any marine organism. Modern environments where calcium carbonate is currently forming (such as in the Bahama Banks, Australia's Great Barrier Reef, and the Persian Gulf) suggest that these carbonate deposits occurred in shallow, warm-water shelves and around tropical islands as coral reefs and beaches.

Ancient marine carbonate deposits constitute 2% of Earth's crust and 25% of all sedimentary rocks on Earth. In fact, these marine limestone deposits form the underlying bedrock of Florida and many midwestern states, from Kentucky to Michigan and from Pennsylvania to Colorado. Percolation of groundwater through

Figure 4.12 Stromatolites. Stromatolites are bulbous algal mats that grow in warm, shallow, high-salinity water such as in Shark Bay, Australia.

(a) Location map of Shark Bay, Australia.

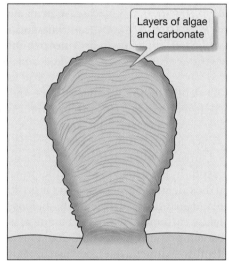

(b) Shark Bay stromatolites, which form in high-salinity tidal pools and reach a maximum height of about 1 meter (3.3 feet).

Layers of algae and carbonate

(c) Diagrammatic cross section through a stromatolite, showing internal fine layering.

these deposits has dissolved the limestone to produce sinkholes and, in some cases, spectacular caverns.

Stromatolites Stromatolites are lobate structures consisting of fine layers of carbonate that form in specific warm, shallow-water environments such as the high salinity tidal pools in Shark Bay, Western Australia (**Figure 4.12**). Cyanobacteria[6] produce these deposits by trapping fine sediment in mucous mats. Other types of algae produce long filaments that bind carbonate particles together. As layer upon layer of these algae colonize the surface, a bulbous structure is formed. In the geologic past—particularly from about 1 to 3 billion years ago—conditions were ideal for the development of stromatolites, so stromatolite structures hundreds of meters high can be found in rocks from these ages.

PELAGIC DEPOSITS Microscopic biogenous sediment (ooze) is common on the deep-ocean floor because there is so little lithogenous sediment deposited at great distances from the continents that could dilute the biogenous material.

Siliceous Ooze Siliceous ooze contains at least 30% of the hard remains of silica-secreting organisms. When the siliceous ooze consists mostly of diatoms, it is called *diatomaceous ooze*. When it consists mostly of radiolarians, it is called *radiolarian ooze*. When it consists mostly of single-celled silicoflagellates—another type of protozoan—it is called *silicoflagellate ooze*.

The ocean is undersaturated with silica at all depths, so the destruction of siliceous biogenous particles by dissolving in seawater occurs continuously and slowly at all depths. How can siliceous ooze accumulate on the ocean floor if it is being dissolved? One way is to accumulate the siliceous tests faster than seawater can dissolve them. For instance, many tests sinking at the same time will create a

---

[6]Cyanobacteria (*kuanos* = dark blue) are simple, ancient creatures whose ancestry can be traced back to some of the first photosynthetic organisms.

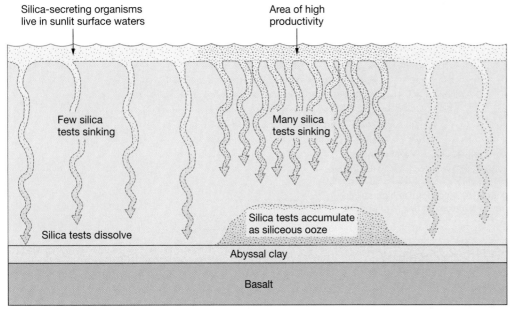

Silica-secreting organisms live in sunlit surface waters

Area of high productivity

Few silica tests sinking

Many silica tests sinking

Silica tests accumulate as siliceous ooze

Silica tests dissolve

Abyssal clay

Basalt

Figure 4.13 Accumulation of siliceous ooze. Siliceous ooze accumulates on the ocean floor beneath areas of high productivity, where the rate of accumulation of siliceous tests is greater than the rate at which silica is being dissolved.

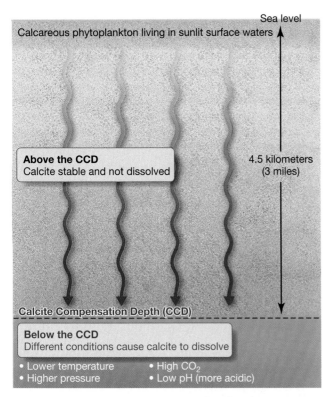

Sea level

Calcareous phytoplankton living in sunlit surface waters

Above the CCD
Calcite stable and not dissolved

4.5 kilometers (3 miles)

Calcite Compensation Depth (CCD)

Below the CCD
Different conditions cause calcite to dissolve

• Lower temperature   • High CO_2
• Higher pressure   • Low pH (more acidic)

Figure 4.14 Characteristics of water above and below the calcite compensation depth (CCD). Schematic diagram showing the calcite compensation depth (CCD). Above the CCD, calcite is stable and does not dissolve; below the CCD, ocean conditions cause calcite to dissolve rapidly.

deposit of siliceous ooze on the sea floor below (**Figure 4.13**).[7] Once buried beneath other siliceous tests, they are no longer exposed to the dissolving effects of seawater. Thus, siliceous ooze is commonly found in areas below surface waters with high biologic productivity of silica-secreting organisms.

Calcareous Ooze and the CCD Calcareous ooze contains at least 30% of the hard remains of calcareous-secreting organisms. When it consists mostly of coccolithophores, it is called *coccolith ooze*. When it consists mostly of foraminifers, it is called *foraminifer ooze*. One of the most common types of foraminifer ooze is *Globigerina ooze*, named for a foraminifer that is especially widespread in the Atlantic and South Pacific oceans. Other calcareous oozes include *pteropod oozes* and *ostracod oozes*.

The destruction of calcium carbonate varies with depth. At the warmer surface and in the shallow parts of the ocean, seawater is generally saturated with calcium carbonate, so calcite does not dissolve. In the deep ocean, however, the colder water contains greater amounts of carbon dioxide, which forms carbonic acid and causes calcareous material to dissolve. The higher pressure at depth also helps speed the dissolution of calcium carbonate.

The depth in the ocean at which the pressure is high enough and the amount of carbon dioxide in deep-ocean waters is great enough to begin dissolving calcium carbonate is called the **lysocline** (*lusis* = a loosening, *cline* = slope). Below the lysocline, calcium carbonate dissolves at an increasing rate with increasing depth until the **calcite compensation depth (CCD)**[8] is reached (**Figure 4.14**). At the CCD and greater depths, sediment does not usually contain much calcite because it readily dissolves; even the thick tests of foraminifers dissolve within a day or two. The CCD, on average, is 4500 meters (15,000 feet) below sea level, but, depending on the chemistry of the deep ocean, it may be as deep as 6000 meters (20,000 feet) in portions of the Atlantic Ocean or as shallow as 3500 meters (11,500 feet) in the Pacific Ocean. The depth of the lysocline also varies from ocean to ocean but averages about 4000 meters (13,100 feet).

In the geologic past, higher concentrations of carbon dioxide in the atmosphere have led to increased amounts of dissolved carbon dioxide in the ocean, thereby making the ocean more acidic and causing the CCD to rise. Currently, scientists have documented an increase in ocean acidity due to higher levels of atmospheric carbon dioxide caused by human-caused emissions. Increased ocean acidity and its effect on marine life are discussed in Chapter 16, "The Oceans and Climate Change."

Because of the CCD, modern carbonate oozes are generally rare below 5000 meters (16,400 feet). Still, buried deposits of ancient calcareous ooze are found beneath the CCD. How can calcareous ooze exist below the CCD? The necessary conditions are shown in **Figure 4.15**. The mid-ocean ridge is a topographically high feature that rises above the sea floor. It often pokes up above the CCD, even though the surrounding deep-ocean floor is below the CCD. Thus, calcareous ooze deposited on top of the mid-ocean ridge will not be dissolved. Sea floor spreading causes

---

[7]An analogy to this is trying to get a layer of sugar to form on the bottom of a cup of hot coffee. If a few grains of sugar are slowly dropped into the cup, a layer of sugar won't accumulate. If a whole bowl full of sugar is dumped into the coffee, however, a thick layer of sugar will form on the bottom of the cup.

[8]Because the mineral calcite is composed of calcium carbonate, the *calcite compensation depth* is also known as the *calcium carbonate compensation depth or the carbonate compensation depth*. All go by the handy abbreviation CCD.

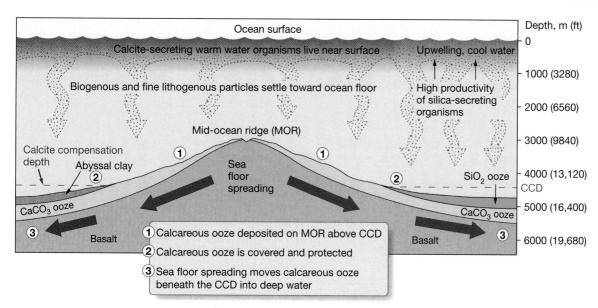

Figure 4.15 Sea floor spreading and sediment accumulation. Relationships among carbonate compensation depth, the mid-ocean ridge, sea floor spreading, productivity, and destruction that allow calcareous ooze to be preserved below the CCD.

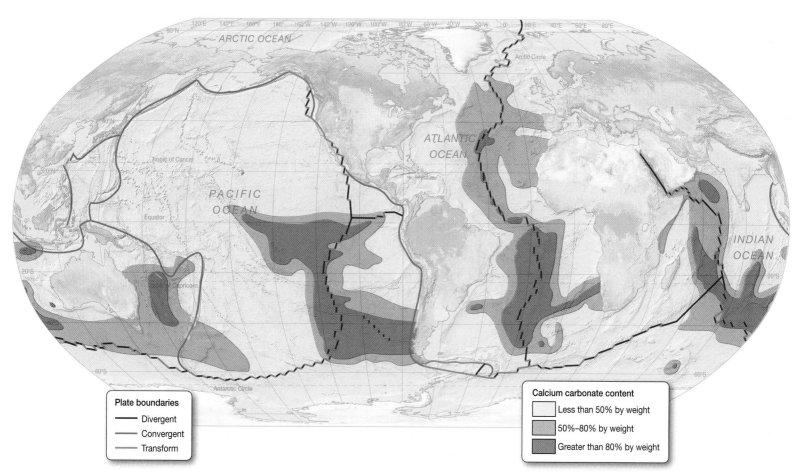

Figure 4.16 Distribution of calcium carbonate in modern surface sediments. High percentages of calcareous ooze closely follow the mid-ocean ridge, most of which is above the CCD.

the newly created sea floor and the calcareous sediment on top of it to move into deeper water away from the ridge, eventually being transported below the CCD. This calcareous sediment will dissolve below the CCD unless it is covered by a deposit that is unaffected by the CCD (such as siliceous ooze or abyssal clay).

The map in **Figure 4.16** shows the percentage (by weight) of calcium carbonate in the modern surface sediments of the ocean basins. High concentrations of

Web Animation
How Calcareous Ooze Can Be
Found Beneath the CCD

TABLE 4.3 COMPARISON OF ENVIRONMENTS INTERPRETED FROM DEPOSITS OF SILICEOUS AND CALCAREOUS OOZE IN SURFACE SEDIMENTS

	Siliceous ooze	Calcareous ooze
Surface water temperature above sea floor deposits	Cool	Warm
Main location found	Sea floor beneath cool surface water in high latitudes	Sea floor beneath warm surface water in low latitudes
Other factors	Upwelling brings deep, cold, nutrient-rich water to the surface	Calcareous ooze dissolves below the CCD
Other locations found	Sea floor beneath areas of upwelling, including along the equator	Sea floor beneath warm surface water in low latitudes along the mid-ocean ridge

calcareous ooze (sometimes exceeding 80%) are found along segments of the mid-ocean ridge, but little is found in deep-ocean basins below the CCD. For example, in the northern Pacific Ocean—one of the deepest parts of the world ocean—there is very little calcium carbonate in the sediment. Calcium carbonate is also rare in sediments accumulating beneath cold, high-latitude waters where calcareous-secreting organisms are relatively uncommon.

Table 4.3 compares the environmental conditions that can be inferred from siliceous and calcareous oozes. It shows that siliceous ooze typically forms below cool surface water regions, including areas of **upwelling** where deep-ocean water comes to the surface and supplies nutrients that stimulate high rates of biological productivity. Calcareous ooze, on the other hand, is found on the shallower areas of the ocean floor beneath warmer surface water.

ESSENTIAL CONCEPT 4.3

Biogenous sediment is produced from the hard remains of once-living organisms. Microscopic biogenous sediment is especially widespread and forms deposits of ooze on the ocean floor.

CONCEPT CHECK 4.3

1 Describe the origin, composition, and distribution of biogenous sediment.

2 List the two major chemical compounds of which most biogenous sediment is composed and examples of the organisms that produce them. Sketch these organisms.

3 What are several reasons diatoms are so remarkable? List products

that contain or are produced using diatomaceous earth.

4 If siliceous ooze is slowly but constantly dissolving in seawater, how can deposits of siliceous ooze accumulate on the ocean floor?

5 Explain the stages of progression that result in calcareous ooze existing below the CCD.

4.4 What Are the Characteristics Of Hydrogenous Sediment?

Hydrogenous (*hydro* = water, *generare* = to produce) **sediment** is derived from the dissolved material in water.

Origin Of Hydrogenous Sediment

Seawater contains many dissolved materials. Chemical reactions within seawater cause certain minerals to come out of solution, or **precipitate** (change from the dissolved to the solid state). Precipitation usually occurs when there is a *change in conditions*, such as a change in temperature or pressure or the addition of chemically active fluids. To make rock candy, for instance, a pan of water is heated and sugar is added. When the water is hot and the sugar dissolved, the pan is removed from the heat, and the sugar water is allowed to cool. The *change in temperature* causes the sugar to become oversaturated, which causes it to precipitate. As the water cools, the sugar precipitates on anything that is put in the pan, such as pieces of string or kitchen utensils.

Composition and Distribution Of Hydrogenous Sediment

Although hydrogenous sediments represent a relatively small portion of the overall sediment in the ocean, they have many different compositions and are distributed in diverse environments of deposition.

MANGANESE NODULES   Manganese nodules are rounded, hard lumps of manganese, iron, and other metals typically 5 centimeters (2 inches) in diameter up to a maximum of about 20 centimeters (8 inches). When cut in half, they often reveal a layered structure formed by precipitation around a central nucleation object (**Figure 4.17a**). The nucleation object may be a piece of lithogenous sediment, coral, volcanic rock, a fish bone, or a shark's tooth. Manganese nodules are found on the deep-ocean floor at concentrations of about 100 nodules per square meter (square yard). In some areas, they occur in even greater abundance (**Figure 4.17b**), resembling a scattered field of baseball-sized nodules. The formation of manganese nodules requires extremely low rates of lithogenous or biogenous input so that the nodules are not buried.

The major components of these nodules are manganese dioxide (around 30% by weight) and iron oxide (around 20%). The element manganese is important for making high-strength steel alloys. Other accessory metals present in manganese nodules include copper (used in electrical wiring, in pipes, and to make brass and bronze), nickel (used to make stainless steel), and cobalt (used as an alloy with iron to make strong magnets and steel tools). Although the concentration of these accessory metals is usually less than 1%, they can exceed 2% by weight, which may make them attractive exploration targets in the future.

The origin of manganese nodules has puzzled oceanographers since manganese nodules were first discovered in 1872 during the voyage of HMS *Challenger*.[9] If manganese nodules are truly hydrogenous and precipitate from seawater, then how can they have such high concentrations of manganese (which occurs in seawater at concentrations often too small to measure accurately)? Furthermore, why are the nodules on *top* of ocean floor sediment and not buried by the constant rain of sedimentary particles?

Unfortunately, nobody has definitive answers to these questions. Perhaps manganese nodules are created by one of the slowest chemical reactions known—on average, they grow at a rate of about 5 millimeters (0.2 inch) per *million years*. Recent research suggests that the formation of manganese nodules

(a) Manganese nodules, including some that are cut in half.

(b) Close-up of a sliced manganese nodule, revealing its central nucleation object and layered internal structure.

(c) An abundance of manganese nodules on a portion of the deep South Pacific Ocean floor about 4 meters (13 feet) across.

Figure 4.17   Manganese nodules.

[9]For more information about the accomplishments of the *Challenger* expedition, see Web Diving Deeper 5.2.

may be aided by bacteria and an as-yet-unidentified marine organism that intermittently lifts and rotates them. Other studies reveal that the nodules don't form continuously over time but in spurts that are related to specific conditions such as a low sedimentation rate of lithogenous clay and strong deep-water currents. Remarkably, the larger the nodules are, the faster they grow. The origin of manganese nodules is widely considered the most interesting unresolved problem in marine chemistry.

PHOSPHATES Phosphorus-bearing compounds (**phosphates**) occur abundantly as coatings on rocks and as nodules on the continental shelf and on banks at depths shallower than 1000 meters (3300 feet). Concentrations of phosphates in such deposits commonly reach 30% by weight and indicate abundant biological activity in surface water above where they accumulate. Because phosphates are valuable as fertilizers, ancient marine phosphate deposits that have been uplifted onto land are extensively mined to supply agricultural needs.

CARBONATES The two most important carbonate minerals in marine sediment are **aragonite** and calcite. Both are composed of calcium carbonate ($CaCO_3$), but aragonite has a different crystalline structure that is less stable and changes into calcite over time. Carbonates are widely used in the construction industry, in the production of cement, and they are commonly used medicinally as calcium supplements or antacids.

As previously discussed, most carbonate deposits are biogenous in origin. However, hydrogenous carbonate deposits can precipitate directly from seawater in tropical climates to form aragonite crystals less than 2 millimeters (0.08 inch) long. In addition, **oolites** (*oo* = egg, *lithos* = rock) are small calcite spheres 2 millimeters (0.08 inch) or less in diameter that have layers like an onion and form in some shallow tropical waters where concentrations of $CaCO_3$ are high. Oolites are thought to precipitate around a nucleus and grow larger as they roll back and forth on beaches by wave action, but some evidence suggests that a type of algae may aid their formation.

METAL SULFIDES Deposits of **metal sulfides** are associated with hydrothermal vents and black smokers along the mid-ocean ridge. These deposits contain iron, nickel, copper, zinc, silver, and other metals in varying proportions. Transported away from the mid-ocean ridge by sea floor spreading, these deposits can be found throughout the ocean floor and can even be uplifted onto continents (see Diving Deeper 2.2).

EVAPORITES Generally, **evaporite minerals** form wherever there are high evaporation rates (dry climates) accompanied by restricted open ocean circulation. One such example is the Mediterranean Sea, which contains thick deposits of evaporites on its floor that suggest the sea completely dried up in the geologic past (see Web Diving Deeper 4.1). As water evaporates in these dry areas, the remaining seawater becomes saturated with dissolved minerals, which then begin to precipitate (form a solid). Because they are heavier than seawater, the minerals sink to the bottom or form a white crust of evaporite minerals around the edges of these areas (**Figure 4.18**). Collectively termed "salts," some

Figure 4.18 Evaporative salts. Due to a high evaporation rate, salts (white material) precipitate onto the floor of Death Valley, California.

evaporite minerals, such as *halite* (common table salt, NaCl), taste salty, and some, such as the calcium sulfate minerals *anhydrite* ($CaSO_4$) and *gypsum* ($CaSO_4 \cdot H_2O$) do not.

4.5 What Are the Characteristics Of Cosmogenous Sediment?

Cosmogenous (*cosmos* = universe, *generare* = to produce) **sediment** is derived from extraterrestrial sources.

Origin, Composition, and Distribution Of Cosmogenous Sediment

Forming an insignificant portion of the overall sediment on the ocean floor, cosmogenous sediment consists of two main types: microscopic **spherules** and macroscopic **meteor** debris.

Microscopic spherules are small globular masses. Some spherules are composed of silicate rock material and show evidence of being formed by extraterrestrial impact events on Earth or other planets that eject small molten pieces of crust into space. These **tektites** (*tektos* = molten) then rain down on Earth and can form *tektite fields*. Other spherules are composed mostly of iron and nickel (**Figure 4.19**) that form in the asteroid belt between the orbits of Mars and Jupiter and are produced when asteroids collide. This material constantly rains down on Earth as a general component of *space dust* or *micrometeorites* that float harmlessly through the atmosphere. Although about 90% of micrometeorites are destroyed by frictional heating as they enter the atmosphere, it has been estimated that as much as 300,000 metric tons (330,000 short tons) of space dust reach Earth's surface each year, which equates to about 10 kilograms (22 pounds) every second of every day! The iron-rich space dust that lands in the oceans often dissolves in seawater. Glassy tektites, however, do not dissolve as easily and sometimes comprise minute proportions of various marine sediments.

Macroscopic meteor debris is rare on Earth but can be found associated with meteor impact sites. Evidence suggests that throughout time meteors have collided with Earth at great speeds and that some larger ones have released energy equivalent to the explosion of multiple large nuclear bombs. To date, nearly 200 meteorite impact structures have been identified on Earth, most of them on land but new ones are being discovered on the ocean floor (see Web Diving Deeper 4.2). The debris from meteors—called **meteorite** material—settles out around the impact site and is either composed of silicate rock material (called *chondrites*) or iron and nickel (called *irons*).

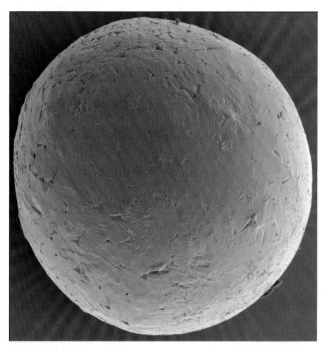

Figure 4.19 Microscopic cosmogenous spherule. Scanning electron micrograph of an iron-rich spherule of cosmic dust. Bar scale of 75 micrometers is equal to 75 millionths of a meter.

STUDENTS SOMETIMES ASK ...

How are scientists able to identify cosmogenous sediment? I mean, how can they tell that it's extraterrestrial?

Cosmogenous sediment can be differentiated from other sediment types primarily by its structure but also by its composition. Cosmogenous sediment can be either silicate rock or rich in iron—both of which are common compositions of lithogenous sediment. However, glassy fragments indicative of melting (called tektites) are uniquely cosmogenous, as are iron-rich spherules (see Figure 4.19). Compositionally, cosmogenous particles from outer space typically contain more nickel than those that originate in other ways; most of the nickel in Earth's crust sank below the surface during density stratification early in Earth's history.

4.6 How Are Pelagic and Neritic Deposits Distributed?

The ocean is a messy place. Lithogenous and biogenous sediment rarely occur as absolutely pure deposits that do not contain other types of sediment. As a result, most marine sediments occur as mixtures.

Mixtures Of Marine Sediment

There are many types of mixtures of marine sediment. Consider the following examples:

- Most calcareous oozes contain some siliceous material and vice versa (see, for example, Figure 4.10d).
- The abundance of clay-sized lithogenous particles throughout the world and the ease with which they are transported by winds and currents means that these particles are incorporated into every sediment type.
- The composition of biogenous ooze includes up to 70% fine-grained lithogenous clays.
- Most lithogenous sediment contains small percentages of biogenous particles.
- There are many types of hydrogenous sediment.
- Tiny amounts of cosmogenous sediment are mixed in with all other sediment types.

Deposits of sediment on the ocean floor are usually a mixture of different sediment types. **Figure 4.20** shows the distribution of sediment across a passive continental margin and illustrates how mixtures can occur. Typically, however, one type of sediment dominates, which allows the deposit to be classified as primarily lithogenous, biogenous, hydrogenous, or cosmogenous.

ESSENTIAL CONCEPT 4.6A

Although most ocean sediment is a mixture of various sediment types, it is usually dominated by lithogenous, biogenous, hydrogenous, or cosmogenous material.

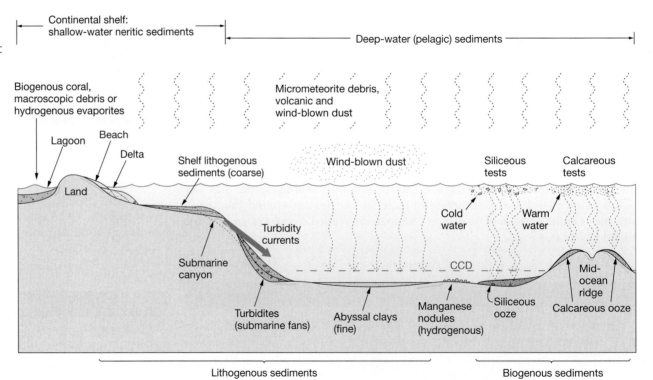

Figure 4.20 Distribution of sediment across a passive continental margin. Schematic cross-sectional view of various sediment types and their distribution across an idealized passive margin.

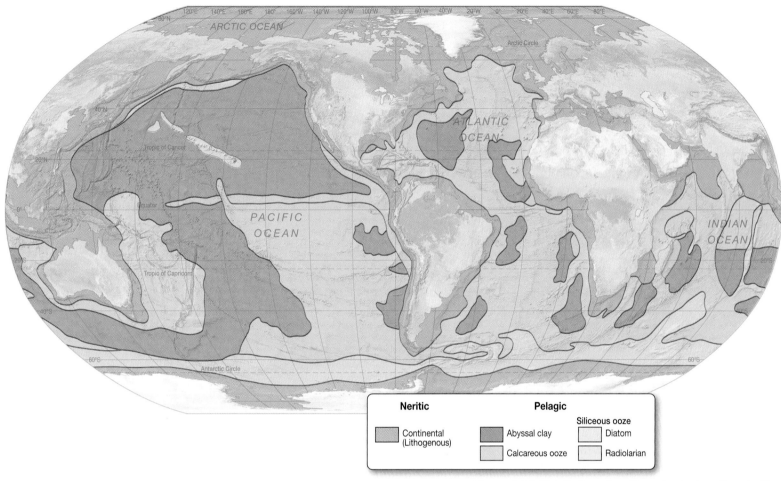

Figure 4.21   **Distribution of neritic and pelagic sediments.** The worldwide distribution of neritic (nearshore) and pelagic (open ocean) sediments shows that neritic deposits are dominated by lithogenous materials while pelagic deposits are dominated by various types of biogenous oozes and lithogenous abyssal clay.

Neritic Deposits

Neritic (nearshore) deposits cover about one-quarter of the ocean floor, and pelagic (deep-ocean basin) deposits cover the other three-quarters. The map in **Figure 4.21** shows the distribution of neritic and pelagic deposits in the world's oceans. Coarse-grained lithogenous neritic deposits dominate continental margin areas (*brown shading*), which is not surprising because lithogenous sediment is derived from nearby continents. Although neritic deposits usually contain biogenous, hydrogenous, and cosmogenous particles, these constitute only a minor percentage of the total sediment mass.

Pelagic Deposits

Figure 4.21 shows that pelagic deposits are dominated by biogenous calcareous oozes (*light purple shading*), which are found on the relatively shallow deep-ocean areas along the mid-ocean ridge. Biogenous siliceous oozes are found beneath areas of unusually high biological productivity such as the North Pacific, the Antarctic (*light yellow shading*, where diatomaceous ooze occurs), and the equatorial Pacific (*dark yellow shading*, where radiolarian ooze occurs). Fine lithogenous pelagic deposits of abyssal clays (*dark purple shading*) are common in deeper areas of the ocean basins. Hydrogenous and cosmogenous sediment comprise only a small proportion of pelagic deposits in the ocean.

 Figure 4.22 shows the proportion of each ocean floor that is covered by pelagic calcareous ooze, siliceous ooze, or abyssal clay. Calcareous oozes predominate, covering almost 48% of the world's deep-ocean floor. Abyssal clay covers 38% and siliceous oozes 14% of the world ocean floor area. The pie charts also

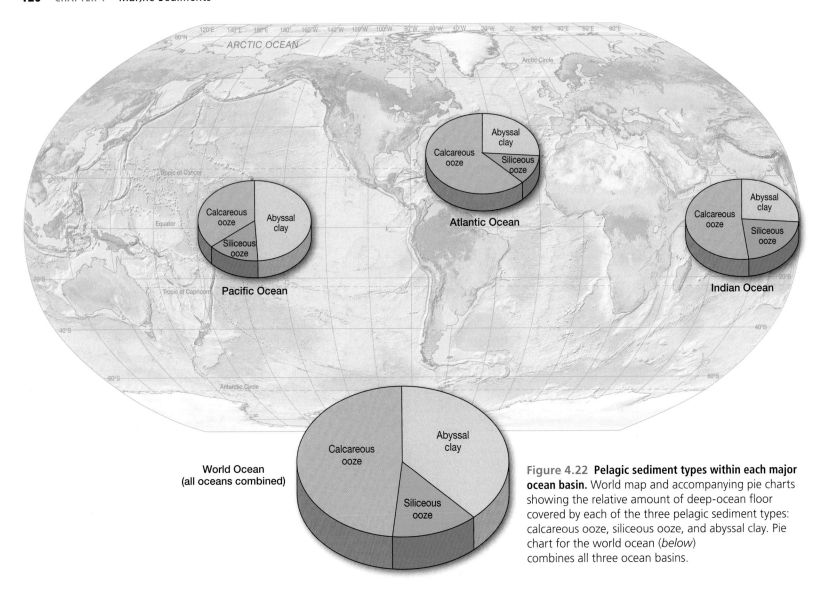

Figure 4.22 Pelagic sediment types within each major ocean basin. World map and accompanying pie charts showing the relative amount of deep-ocean floor covered by each of the three pelagic sediment types: calcareous ooze, siliceous ooze, and abyssal clay. Pie chart for the world ocean (*below*) combines all three ocean basins.

show that the amount of ocean basin floor covered by calcareous ooze decreases in deeper basins because they generally lie beneath the CCD. The dominant oceanic sediment in the deepest basin—the Pacific—is abyssal clay (see also Figure 4.21). Calcareous ooze is the most widely deposited sediment in the shallower Atlantic and Indian Oceans. Siliceous oozes cover a smaller percentage of the ocean bottom in all the oceans because regions of high productivity of siliceous-secreting organisms are generally restricted to the equatorial region (for radiolarians) and high latitudes (for diatoms). Table 4.4 shows the average rates of deposition of selected marine sediments in neritic and pelagic deposits.

ESSENTIAL CONCEPT 4.6B

Neritic deposits occur close to shore and are dominated by coarse lithogenous material. Pelagic deposits occur in the deep ocean and are dominated by biogenous oozes and fine lithogenous clay.

How Sea Floor Sediments Represent Surface Conditions

Microscopic biogenous tests should take from 10 to 50 years to sink from the ocean surface where the organisms lived to the abyssal depths where biogenous ooze accumulates. During this time, even a sluggish horizontal ocean current

TABLE 4.4   AVERAGE RATES OF DEPOSITION OF SELECTED MARINE SEDIMENTS

Type of sediment/deposit	Average rate of deposition (per 1000 years)	Thickness of deposit after 1000 years equivalent to . . .
Coarse lithogenous sediment, neritic deposit	1 meter (3.3 feet)	A meter stick
Biogenous ooze, pelagic deposit	1 centimeter (0.4 inch)	The diameter of a dime
Abyssal clay, pelagic deposit	1 millimeter (0.04 inch)	The thickness of a dime
Manganese nodule, pelagic deposit	0.001 millimeter (0.00004 inch)	A microscopic dust particle

of only 0.05 kilometer (0.03 mile) per hour could carry tests as much as 22,000 kilometers (13,700 miles) before they settled onto the deep-ocean floor. Why, then, do biogenous tests on the deep-ocean floor closely reflect the population of organisms living in the surface water directly above? Remarkably, about 99% of the particles that fall to the ocean floor do so as part of *fecal pellets*, which are produced by tiny animals that eat algae and protozoans living in the water column, digest their tissues, and excrete their hard parts. These pellets are full of the remains of algae and protozoans from the surface waters (**Figure 4.23**) and, though still small, are large enough to sink to the deep-ocean floor in only 10 to 15 days.

Worldwide Thickness Of Marine Sediments

Figure 4.24 is a map of marine sediment thickness. The map shows that areas of thick sediment accumulation occur on the continental shelves and rises, especially near the mouths of major rivers. The reason sediments in these locations are so thick is because they are close to major sources of lithogenous sediment. Conversely,

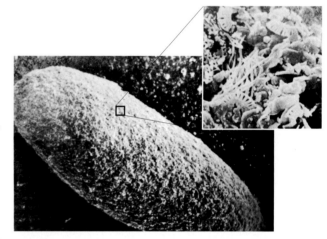

Figure 4.23 Fecal pellet. A 200-micrometer (0.008-inch)-long fecal pellet, which is large enough to sink rapidly from the surface to the ocean floor. Close-up of the surface of a fecal pellet (*inset*) shows the remains of coccoliths and other debris.

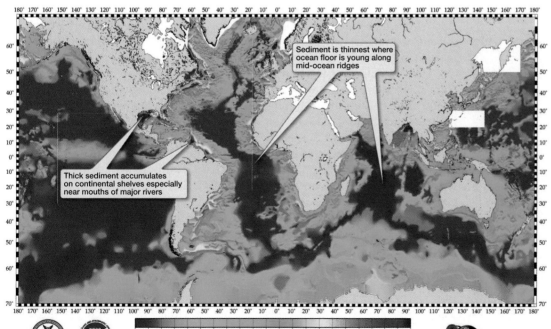

Figure 4.24 Marine sediment thickness. Map showing the thickness of sediments in the oceans and marginal seas. Thickness shown in meters; dark blue color represents thinnest sediments, and red represents thickest sediment accumulations. White color indicates no available data.

Sediment is thinnest where ocean floor is young along mid-ocean ridges

Thick sediment accumulates on continental shelves especially near mouths of major rivers

Thickness in Meters
0   500   1000   5000   10000   20000

areas where marine sediments are thinnest are where the ocean floor is young, such as along the crest of the mid-ocean ridge. Since sediments accumulate slowly in the deep ocean and the sea floor is continually being created here, there hasn't been enough time for much sediment to accumulate. However, as the sea floor moves away from the mid-ocean ridge, it gets progressively older and carries a thicker pile of sediments.

STUDENTS SOMETIMES ASK ...

Are there any areas of the ocean floor where no sediment is being deposited?

Various types of sediment accumulate on nearly all areas of the ocean floor in the same way dust accumulates in all parts of your home (which is why marine sediment is often referred to as "marine dust"). Even the deep-ocean floor far from land receives small amounts of windblown material, microscopic biogenous particles, and space dust. There are some places in the ocean, however, where very little sediment accumulates. A few such places include the South Pacific Bare Zone east of New Zealand, where a combination of factors limit sediment accumulation; along the continental slope, where there is active erosion by turbidity and other deep-ocean currents; and along the mid-ocean ridge, where the age of the sea floor is so young (because of sea floor spreading) and the rates of sediment accumulation far from land are so slow that there hasn't been enough time for sediments to accumulate.

CONCEPT CHECK 4.6

① Why is it so rare to find a pure marine sediment type? Give some examples of mixtures of sediment.

② Why is lithogenous sediment the most common neritic deposit? Why are biogenous oozes the most common pelagic deposits?

③ How do fecal pellets help explain why the particles found in the ocean surface waters are closely reflected in the particle composition of the sediment directly beneath? Why is this unexpected?

4.7 What Resources Do Marine Sediments Provide?

The sea floor is rich in potential mineral and organic resources. Much of these resources, however, are not easily accessible, so their recovery involves technological challenges and high cost. Nevertheless, let's examine some of the most appealing exploration targets.

Energy Resources

The main energy resources associated with marine sediments are *petroleum* and *gas hydrates*.

PETROLEUM The ancient remains of microscopic organisms, buried within marine sediments before they could decompose, are the source of today's **petroleum** (oil and natural gas) deposits. Of the nonliving resources extracted from the oceans, more than 95% of the economic value is in petroleum products.

The percentage of world oil produced from offshore regions has increased from small amounts in the 1930s to more than 30% today. Most of this increase results from continuing technological advancements employed by offshore drilling platforms (**Figure 4.25**). Major offshore reserves exist in the Persian Gulf, in the Gulf of Mexico, off Southern California, in the North Sea, and in the East Indies. Additional reserves are probably located off the north coast of Alaska and in the Canadian Arctic, Asian seas, Africa, and Brazil. With almost no likelihood of finding major new reserves on land, future offshore petroleum exploration will continue to be intense, especially in deeper waters of the continental margins. However, a major drawback to offshore petroleum exploration is the inevitable oil spills caused by inadvertent leaks or blowouts during the drilling process.

GAS HYDRATES **Gas hydrates**, which are also known as *clathrates* (*clathri* = a lattice) are unusually compact chemical structures made of water and natural gas. They form only when high pressures squeeze chilled water and gas molecules into an icelike solid. Although hydrates can contain a variety of gases—including carbon dioxide, hydrogen sulfide, and larger hydrocarbons such as ethane and propane—**methane hydrates** are by far the most common hydrates in nature. Gas

Figure 4.25 Offshore oil-drilling platform. Constructed on tall stilts, drilling platforms are important for extracting petroleum reserves from beneath the continental shelves.

hydrates occur beneath Arctic permafrost areas on land and under the ocean floor, where they were discovered in 1976.

In deep-ocean sediments, where pressures are high and temperatures are low, water and natural gas combine in such a way that the gas is trapped inside a lattice-like cage of water molecules. Vessels that have drilled into gas hydrates have retrieved cores of mud mixed with chunks or layers of gas hydrates that fizzle and decompose quickly when they are exposed to the relatively warm, low-pressure conditions at the ocean surface. Gas hydrates resemble chunks of ice but ignite when lit by a flame because methane and other flammable gases are released as gas hydrates vaporize (**Figure 4.26**).

Most oceanic gas hydrates are created when bacteria break down organic matter trapped in sea floor sediments, producing methane gas with minor amounts of ethane and propane. These gases can be incorporated into gas hydrates under high-pressure and low-temperature conditions. Most ocean floor areas below 525 meters (1720 feet) provide these conditions, but gas hydrates seem to be confined to continental margin areas, where high productivity surface waters enrich ocean floor sediments below with organic matter.

Studies of the deep-ocean floor reveal that at least 50 sites worldwide may contain extensive gas hydrate deposits. Interestingly, sea floor methane seeps support a rich community of organisms, many of which are species new to science.

The release of methane from the sea floor to the atmosphere can have dramatic effects on global climate. Research suggests that at various times in the geologic past, changes in sea level or sea floor instability have released large quantities of methane, which is the third-most-important greenhouse gas after water vapor and carbon dioxide. In fact, recent analysis of sea floor sediments off Norway suggests that an abrupt increase in global temperature about 55 million years ago was driven by an explosive release of gas hydrates from the sea floor. Today, a major concern is that recent climate changes could warm ocean waters enough to release additional methane that is trapped beneath the seabed, causing even more warming. Sudden releases of methane hydrates have also been linked to underwater slope failure, which can cause *seismic sea waves*, or *tsunami* (see Chapter 8, "Waves and Water Dynamics").

Some estimates indicate that as much as 20 quadrillion cubic meters (700 quadrillion cubic feet) of methane are locked up in sediments containing gas hydrates. This is equivalent to about *twice* as much carbon as Earth's coal, oil, and conventional gas reserves combined (**Figure 4.27**), so gas hydrates may potentially be the world's largest source of usable energy. A major drawback in exploiting reserves of gas hydrate is that they rapidly decompose at surface temperatures and pressures.

Another problem is that they are typically too dispersed within the sea floor to make collecting them economically feasible. An additional concern is that commercial extraction of methane hydrates might exacerbate fossil fuel–driven climate changes. Nonetheless, a Japanese research team is currently evaluating the economic potential of collecting methane hydrates in the Nankai Trough off Japan and could begin producing methane as early as 2016.

Other Resources

Other resources associated with marine sediments include *sand and gravel, evaporative salts, phosphorite, manganese nodules and crusts,* and *rare-earth elements.*

SAND AND GRAVEL Sand and gravel, which includes rock fragments that are washed out to sea and shells of marine organisms, is mined by offshore barges using suction dredges. This material is primarily used as aggregate in concrete, as fill material in grading projects, and on recreational beaches. In terms of economic value, offshore sand and gravel is the second largest sea floor deposit behind petroleum.

(a) A sample retrieved from the ocean floor shows layers of white icelike gas hydrate mixed with mud.

(b) Gas hydrates decompose when exposed to surface conditions and release natural gas, which can be ignited.

Figure 4.26 Gas hydrates. Gas hydrates are icelike substances that form in deep-ocean sediments and are composed of natural gas combined with frozen water.

STUDENTS SOMETIMES ASK …

When will we run out of oil?

Not anytime soon. However, from an economic perspective, when the world runs completely out of oil—a finite resource—is not as relevant as when production begins to taper off. When this happens, we will run out of the *abundant* and *cheap* oil on which all industrialized nations depend. Several oil-producing countries are already past the peak of their production—including the United States and Canada, both of which peaked in 1972. Current estimates indicate that sometime within the next few decades (and according to some projections, as early as 2014), more than half of all known and likely-to-be-discovered oil will be gone. Other experts have suggested that petroleum production has already reached its plateau. Once the decline begins, it will be increasingly costly to produce oil, and prices will rise dramatically—unless demand declines proportionally or other sources such as coal, extra-heavy oil, tar sands, or gas hydrates become readily available.

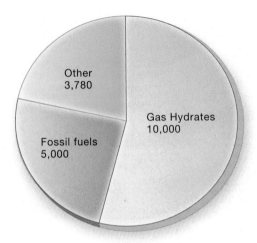

Figure 4.27 Organic carbon in Earth reservoirs. Pie diagram of organic carbon in Earth reservoirs, showing that gas hydrates contain twice as much organic carbon as all fossil fuels combined. Values in billions of tons of carbon; "other" includes sources such as soil, peat, and living organisms.

Figure 4.28 Mining sea salt. A salt mining operation at Scammon's Lagoon, Baja California, Mexico. Low-lying areas near the lagoon are allowed to flood with seawater, which evaporates in the arid climate and leaves deposits of salt that are then collected.

Offshore deposits are a major source of sand and gravel in New England, New York, and throughout the Gulf Coast. Many European countries, Iceland, Israel, and Lebanon also depend heavily on such deposits.

Some offshore sand and gravel deposits are rich in valuable minerals. Gem-quality diamonds, for example, are recovered from gravel deposits on the continental shelf offshore of South Africa and Australia, where waves rework them during low stands of the sea. Sediments rich in tin have been mined offshore of southeast Asia from Thailand to Indonesia. Platinum and gold have been found in deposits offshore of gold mining areas throughout the world, and some Florida beach sands are rich in titanium. The largest unexplored potential for metallic minerals in offshore sand deposits may exist along the west coast of South America, where rivers have transported Andean metallic minerals.

EVAPORATIVE SALTS When seawater evaporates, the salts increase in concentration until they can no longer remain dissolved, so they precipitate out of solution and form **salt deposits** (Figure 4.28). Extensive sea floor salt deposits indicate that entire seas such as the Mediterranean Sea completely dried up in the geologic past (see Web Diving Deeper 4.1).

The most economically useful salts are *gypsum* and *halite*. Gypsum is used in plaster of Paris to make casts and molds and is the main component in gypsum board (wallboard or sheet rock). Halite—common table salt—is widely used for seasoning, curing, and preserving foods. It is also used to de-ice roads, in water conditioners, in agriculture, and in the clothing industry for dying fabric.

In addition, halite is used in the production of chemicals such as sodium hydroxide (to make soap products), sodium hypochlorite (for disinfectants, bleaching agents, and PVC piping), sodium chlorate (for herbicides, matches, and fireworks), and hydrochloric acid (for use in chemical applications and for cleaning scaled pipes). The manufacture and use of salt is one of the oldest chemical industries.[10]

PHOSPHORITE (PHOSPHATE MINERALS) **Phosphorite** is a sedimentary rock consisting of various phosphate minerals containing the element phosphorus, an important plant nutrient. Consequently, phosphate deposits can be used to produce phosphate fertilizer. Although there is currently no commercial phosphorite mining occurring in the oceans, the marine reserve is estimated to exceed 45 billion metric tons (50 billion short tons). Phosphorite occurs in the ocean at depths of less than 300 meters (1000 feet) on the continental shelf and slope in regions of upwelling and high productivity.

Some shallow sand and mud deposits contain up to 18% phosphate. Many phosphorite deposits occur as nodules, with a hard crust formed around a nucleus. The nodules may be as small as a sand grain or as large as 1 meter (3.3 feet) in diameter and may contain more than 25% phosphate. For comparison, most land sources of phosphate have been enriched to more than 31% by groundwater leaching. Florida, for example, has large phosphorite deposits and supplies about one-quarter of the world's phosphates.

MANGANESE NODULES AND CRUSTS *Manganese nodules* are rounded, hard, golf- to tennis-ball-sized lumps of metals that contain significant concentrations of manganese, iron, and smaller concentrations of copper, nickel, and cobalt, all of which have a variety of economic uses. In the 1960s, mining companies began to assess the feasibility of mining manganese nodules from the deep-ocean floor

---

[10]An interesting historical note about salt is that part of a Roman soldier's pay was in salt. That portion was called the *salarium*, from which the word *salary* is derived. If a soldier did not earn it, he was not worth his salt.

(**Figure 4.29**). The map in **Figure 4.30** shows that vast areas of the sea floor contain manganese nodules, particularly in the Pacific Ocean.

Technologically, mining the deep-ocean floor for manganese nodules is possible. However, the political issue of determining international mining rights at great distances from land has hindered exploitation of this resource. In addition, environmental concerns about mining the deep-ocean floor have not been fully addressed. Evidence suggests that it takes at least several million years for manganese nodules to form and that their formation depends on a particular set of physical and chemical conditions that probably do not last long at any location. In essence, they are a nonrenewable resource that will not be replaced for a very long time once they are mined.

Of the five metals commonly found in manganese nodules, cobalt is the only metal deemed "strategic" (essential to national security) for the United States. It is required to produce dense, strong alloys with other metals for use in high-speed cutting tools, powerful permanent magnets, and jet engine parts. Currently, the United States must import all of its cobalt from large deposits in southern Africa. However, the United States has considered deep-ocean nodules and **crusts** (hard coatings on other rocks) as a more reliable source of cobalt.

In the 1980s cobalt-rich manganese crusts were discovered on the upper slopes of islands and seamounts that lie relatively close to shore and within the jurisdiction of the United States and its territories. The cobalt concentrations in these crusts are about one-and-a-half times as rich as the best African ores and at least twice as rich as deep-sea manganese nodules. However, interest in mining these deposits has faded because of lower metal prices from land-based sources.

RARE-EARTH ELEMENTS *Rare-earth elements*—an assortment of 17 chemically similar metallic elements such as lanthanum and neodymium—are used in a variety of electronic, optical, magnetic, and catalytic applications. For example, rare-earth elements are used in a host of technological gadgets from cell phones and television screens to fluorescent lightbulbs and batteries in electric cars. Demand for rare-earth elements has skyrocketed in recent years, with China supplying about 90% in 2012.

Over millions of years, deep-sea hot springs associated with the mid-ocean ridge pulled rare-earth elements out of seawater and enriched them in sea floor muds. A recent study of rare-earth elements on the floor of the Pacific Ocean indicated that some locations are particularly enriched. For example, an area of the sea floor near Hawaii measuring 1 square kilometer (0.4 square mile) holds as much as 25,000 metric tons (27,500 short tons) of rare-earth elements. Overall, estimates suggest that the ocean floor might hold more rare-earth elements than all the known deposits on land.

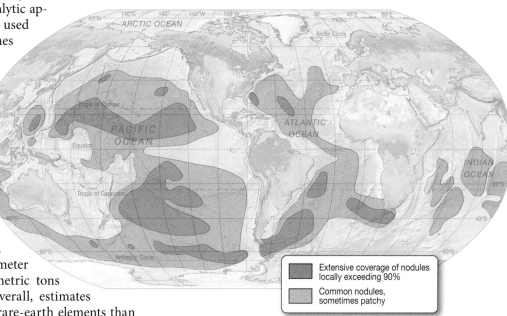

Figure 4.29 Mining manganese nodules. Manganese nodules can be collected by dredging the ocean floor. This metal dredge is shown unloading nodules onto the deck of a ship.

Figure 4.30 Distribution of manganese nodules on the sea floor. High concentrations of manganese nodules are located on some regions of the deep-sea floor, particularly in the Pacific and Atlantic Oceans.

Legend:
- Extensive coverage of nodules locally exceeding 90%
- Common nodules, sometimes patchy

CONCEPT CHECK 4.7

1. Discuss the present importance and the future prospects for the production of petroleum, sand and gravel, phosphorite, manganese nodules and crusts, and rare-earth elements.

2. What are gas hydrates, where are they found, and why are they important?

ESSENTIAL CONCEPT 4.7

Ocean sediments contain many important resources, including petroleum, gas hydrates, sand and gravel, evaporative salts, phosphorite, manganese nodules and crusts, and rare-earth elements.

Essential Concepts Review

4.1 How are marine sediments collected, and what historical events do they reveal?

The JOIDES Resolution

Derrick

Hydrophones

Drill pipe

Maximum water depth 8200 meters (27,000 feet)

Television camera

Rotary drill bit

Sonar beacon

Reentry cone

Sediment layers

Hard rock

Thrusters

▶ Sediments that accumulate on the ocean floor are *classified by origin* as *lithogenous* (derived from rock), *biogenous* (derived from organisms), *hydrogenous* (derived from water), or *cosmogenous* (derived from outer space).

▶ The existence of *sea floor spreading was confirmed when the* Glomar Challenger *began the Deep Sea Drilling Project* to sample ocean sediments and the underlying crust, which was continued by the *Ocean Drilling Program's* JOIDES Resolution. Today, the *Integrated Ocean Drilling Program* continues the important work of retrieving sediments from the deep-ocean floor.

▶ Analysis and interpretation of marine sediments reveal that *Earth has had an interesting and complex history* including *mass extinctions*, the *drying of entire seas*, *global climate change*, and the *movement of tectonic plates*.

Study Resources
Online Study Guide Quizzes, Web Video

Critical Thinking Question

It has been said that in the early days of oceanography, collecting marine sediments using a dredge was akin to collecting land samples from a hot air balloon using a bucket—at an a height above the ground of several kilometers (a few miles) and at night. Evaluate the effectiveness of this type of sample collecting (for example, is it representative of the environment being sampled?).

4.2 What are the characteristics of lithogenous sediment?

▶ *Lithogenous sediments reflect the composition of the rock from which they were derived.* Sediment *texture*—determined in part by the size, sorting, and rounding of particles—is affected greatly by how the particles were transported (by water, wind, ice, or gravity) and the energy conditions under which they were deposited. *Coarse lithogenous material dominates neritic deposits* that accumulate rapidly along the margins of continents, while *fine abyssal clays are found in pelagic deposits*.

Study Resources
Online Study Guide Quizzes

Critical Thinking Question

If a deposit has a coarse grain size, what does this indicate about the energy of the transporting medium? Give several examples of various transporting media that would produce such a deposit.

4.3 What are the characteristics of biogenous sediment?

▶ *Biogenous sediment consists of the hard remains* (shells, bones, and teeth) *of organisms.* These are composed of either *silica* (SiO_2) from diatoms and radiolarians or *calcium carbonate* ($CaCO_3$) from foraminifers and coccolithophores. *Accumulations of microscopic shells* (tests) of organisms must comprise at least 30% of the deposit for it to be classified as *biogenic ooze*.

▶ *Biogenous oozes are the most common type of pelagic deposits.* The rate of biological productivity relative to the rates of destruction and dilution of biogenous sediment determines whether abyssal clay or oozes will form on the ocean floor. *Siliceous ooze will form only below areas of high biologic productivity of silica-secreting organisms at the surface. Calcareous ooze will form only above the calcite compensation depth (CCD)*—the depth where seawater dissolves calcium carbonate—although it can be covered and transported into deeper water through sea floor spreading.

Study Resources
Online Study Guide Quizzes, Web Animation, Web Video

Critical Thinking Question

How do oozes differ from abyssal clay? Discuss how productivity, destruction, and dilution combine to determine whether an ooze or abyssal clay will form on the deep-ocean floor.

4.4 What are the characteristics of hydrogenous sediment?

▶ *Hydrogenous sediment* includes manganese nodules, phosphates, carbonates, metal sulfides, and evaporites that *precipitate directly from water* or are formed by the interaction of substances dissolved in water with materials on the ocean floor. Hydrogenous sediments represent a relatively small proportion of marine sediment and are distributed in many diverse environments.

Study Resources
Online Study Guide Quizzes, Web Diving Deeper 4.1

Critical Thinking Question

Construct a table that shows the various types of hydrogenous sediment and list both their origin and how they are used by humans.

4.5 What are the characteristics of cosmogenous sediment?

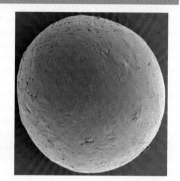

▶ *Cosmogenous sediment is composed of either macroscopic meteor debris* or *microscopic iron–nickel and silicate spherules* that result from asteroid collisions or extraterrestrial impacts. Minute amounts of cosmogenous sediment are mixed into most other types of ocean sediment.

Study Resources
Online Study Guide Quizzes, Web Diving Deeper 4.2

Critical Thinking Question

Using Web Diving Deeper 4.2, describe what happened on Earth at the Cretaceous–Tertiary (K–T) boundary. What evidence was used to confirm the event, and what environmental effects did it cause?

4.6 How are pelagic and neritic deposits distributed?

▶ Although *most ocean sediment is a mixture of various sediment types*, it is usually dominated by lithogenous, biogenous, hydrogenous, or cosmogenous material.

▶ *The distribution of neritic and pelagic sediment is influenced by many factors*, including proximity to sources of lithogenous sediment, productivity of microscopic marine organisms, the depth of the ocean floor, and the distribution of various sea floor features. *Fecal pellets* rapidly transport biogenous particles to the deep-ocean floor and cause the composition of sea floor deposits to match the organisms living in surface waters immediately above them.

Study Resources
Online Study Guide Quizzes

Critical Thinking Question

Using the depositional rates shown in Table 4.4, how long would it take to make a deposit 1 meter (3.3 feet) thick of biogenous ooze? A deposit 1 meter (3.3 feet) thick of abyssal clay?

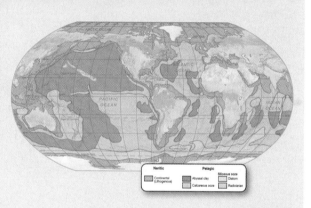

4.7 What resources do marine sediments provide?

▶ *The most valuable nonliving resource from the ocean today is petroleum*, which is recovered from below the continental shelves and used as a source of energy. *Gas hydrates* include vast deposits of ice-like material that may someday be used as a source of energy. Other important resources include *sand and gravel* (including deposits of valuable minerals), *evaporative salts, phosphorite, manganese nodules and crusts*, and *rare-earth elements*.

Study Resources
Online Study Guide Quizzes, Web Video

Critical Thinking Question

A company wants to mine sea floor minerals. What technological issues would there be for developing a mining operation on the sea floor? Also, evaluate the environmental factors that should be considered before mining materials on the sea floor.

MasteringOceanography™
www.masteringoceanography.com

Looking for additional review and test prep materials? Visit the Study Area in MasteringOceanography to enhance your understanding of this chapter's content by accessing a variety of resources, including Self-Study Quizzes, Geoscience Animations, RSS feeds, flashcards, Web links, and an optional Pearson eText.

Water molecules and the ocean.
The objects shown on this image are water molecules, magnified by many orders of magnitude. Most surface water on Earth is in the ocean; a single droplet of water contains more water molecules than there are grains of sand on a large beach.

5

Before you begin reading this chapter, use the glossary at the end of this book to discover the meanings of any of these words you don't already know:

Polarity **Halocline Heat capacity**
Parts per thousand (‰) Thermocline
Evaporation**Molecule**Buffering **Latent heat** Isopycnal
Brackish pH scale
Covalent bond Principle of constant proportions**Pycnocline**Hydrologic cycle
Salinometer **Desalination** Isothermal
Hypersaline Ionic bond **Salinity**Specific heat
Precipitation
Hydrogen bond

WATER AND SEAWATER

"Chemistry . . . is one of the broadest branches of science, if for no other reason that, when we think about it, everything is chemistry."

—Luciano Caglioti,
The Two Faces of Chemistry (1985)

Why are temperature extremes found at places far from the ocean, while areas close to the ocean rarely experience severe temperature variations? The mild climates found in coastal regions are made possible by the unique thermal properties of water. These and other properties of water, which stem from the arrangement of its atoms and how its molecules stick together, give water the ability to store vast quantities of heat and to dissolve almost everything.

Water is so common that we often take it for granted, yet it is one of the most peculiar substances on Earth. For example, almost every other liquid contracts as it approaches its freezing point, but water actually *expands* as it freezes. Thus water stays at the surface as it starts to freeze, and ice floats—a rare property shared by very few other substances. If its nature were otherwise, all temperate-zone lakes, ponds, rivers, and even oceans would eventually freeze solid from the bottom up, and life as we know it could not exist. Instead, a floating skin of ice forms at the surface and acts as an insulative cover to protect the marine organisms that live in the liquid water below.

The chemical properties of water are also essential for sustaining all forms of life. In fact, the primary component of all living organisms is water. The water content of organisms, for instance, ranges from about 65% (humans) to 95% (most plants) to 99% in some jellyfish. Water is the ideal medium to have within our bodies because it facilitates chemical reactions. Our blood, which serves to transport nutrients and remove wastes within our bodies, is 83% water. The very presence of water on our planet makes life possible, and its remarkable properties make our planet livable.

5.1 Why Does Water Have Such Unusual Chemical Properties?

To understand why water has such unusual properties, let's examine its chemical structure.

Atomic Structure

Atoms (*a* = not, *tomos* = cut) are the basic building blocks of all matter. Every physical substance in our world—chairs, tables, books, people, the air we breathe—is composed of atoms. An atom resembles a microscopic sphere (**Figure 5.1**) and was originally thought to be the smallest form of matter. Additional study has revealed that atoms are composed of even smaller particles, called subatomic

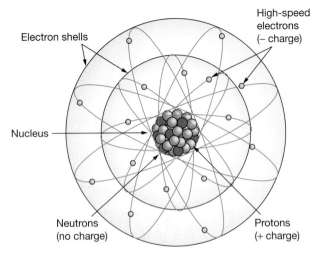

Electron shells

High-speed electrons (– charge)

Nucleus

Neutrons (no charge)

Protons (+ charge)

Figure 5.1 Simplified model of an atom. An atom consists of a central nucleus composed of protons and neutrons that is encircled by electrons.

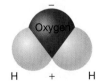

Web Video

Just How Small Is an Atom?

Shared electrons

Oxygen

Shared electrons

8+

Hydrogen

Hydrogen

+   +

105°

(a) Geometry of a water molecule. The oxygen end of the molecule is negatively charged, and the hydrogen regions exhibit a positive charge. Covalent bonds occur between the oxygen and the two hydrogen atoms as electrons are shared.

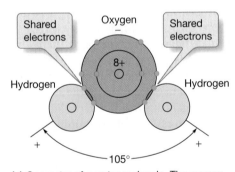

Oxygen

H   +   H

(b) A three-dimensional representation of the water molecule.

$$\overset{-}{O}$$
$$\underset{H \;\; + \;\; H}{}$$

(c) The water molecule represented by letters (*H* = hydrogen, *O* = oxygen).

Figure 5.2 Representations of the water molecule.

particles.[1] As shown in Figure 5.1, the **nucleus** (*nucleos* = a little nut) of an atom is composed of **protons** (*protos* = first) and **neutrons** (*neutr* = neutral) that are bound together by strong forces. Protons have a positive electrical charge, whereas neutrons have no electrical charge. Both protons and neutrons have about the same mass, which is extremely small. Surrounding the nucleus are particles called **electrons** (*electro* = electricity), which have about $\frac{1}{2000}$ the mass of either protons or neutrons. Electrical attraction between positively charged protons and negatively charged electrons holds electrons in layers, or shells, around the nucleus.

The overall electrical charge of most atoms is balanced because each atom contains an equal number of protons and electrons. An oxygen atom, for example, has eight protons and eight electrons. Most oxygen atoms also have eight neutrons, which do not affect the overall electrical charge because neutrons are electrically neutral. The number of protons is what distinguishes atoms of the 118 known chemical elements from one another. For example, an oxygen atom (and only an oxygen atom) has eight protons. Similarly, a hydrogen atom (and only a hydrogen atom) has one proton, a helium atom has two protons, and so on (for more details, see Appendix IV, "A Chemical Background: Why Water Has 2 H's and 1 O"). In some cases, an atom will lose or gain one or more electrons and thus have an overall electrical charge. These atoms are called **ions** (*ienai* = to go).

The Water Molecule

A **molecule** (*molecula* = a mass) is a group of two or more atoms held together by mutually shared electrons. It is the smallest form of a substance that can exist yet still retain the original properties of that substance. When atoms combine with other atoms to form molecules, they share or trade electrons and establish chemical bonds. For instance, the chemical formula for water—H_2O—indicates that a water molecule is composed of two hydrogen atoms chemically bonded to one oxygen atom.

GEOMETRY   Atoms can be represented as spheres of various sizes, and the more electrons an atom contains, the larger its sphere. It turns out that an oxygen atom (with eight electrons) is about twice the size of a hydrogen atom (with one electron). A water molecule consists of a central oxygen atom covalently bonded to the two hydrogen atoms, which are separated by an angle of about 105 degrees (**Figure 5.2a**). The **covalent** (*co* = with, *valere* = to be strong) **bonds** in a water molecule are due to the sharing of electrons between oxygen and each hydrogen atom. They are relatively strong chemical bonds, so a lot of energy is needed to break them.

Figure 5.2b shows a water molecule in a more compact representation, and in **Figure 5.2c** letter symbols are used to represent the atoms in water (*O* for oxygen, *H* for hydrogen). Instead of water's atoms being in a straight line, *both hydrogen atoms are on the same side of the oxygen atom*. This curious bend in the geometry of the water molecule is the underlying cause of most of the unique properties of water.

POLARITY   The bent geometry of the water molecule gives a slight overall negative charge to the side of the oxygen atom and a slight overall positive charge to the side of the hydrogen atoms (Figure 5.2a). This slight separation of charges gives the entire molecule an electrical **polarity** (*polus* = pole, *ity* = having the quality of), so water molecules are **dipolar** (*di* = two, *polus* = pole). Other common dipolar objects are flashlight batteries, car batteries, and bar magnets. In fact, a good way to visualize the polarity of water molecules is to view them as if they contain a tiny, weak bar magnet.

INTERCONNECTIONS OF MOLECULES   If you've ever experimented with bar magnets, you know they have polarity and orient themselves relative to one another such

---

[1]It has been discovered that subatomic particles themselves are composed of a variety of even smaller particles, such as *quarks*, *leptons*, and *bosons*.

Why does a water molecule have the unusual shape that it does?

Based on simple symmetry considerations and charge separations, a water molecule should have its two hydrogen atoms on opposite sides of the oxygen atom, thus producing a linear shape, like the shape of many other molecules. But water's odd shape where both hydrogen atoms are on the same side of the oxygen atom stems from the fact that oxygen has four bonding sites, which are evenly spaced around the oxygen atom. No matter which two bonding sites are occupied by hydrogen atoms, there is a curious bend in each water molecule.

that the positive end of one bar magnet is attracted to the negative end of another. Water molecules have polarity, too, and as a result they orient themselves relative to one another. In water, the positively charged hydrogen area of one water molecule interacts with the negatively charged oxygen end of an adjacent water molecule, forming a **hydrogen bond** (**Figure 5.3**). The hydrogen bonds between water molecules are much weaker than the covalent bonds that hold individual water molecules together. In essence, weaker hydrogen bonds form *between* adjacent water molecules, and stronger covalent bonds occur *within* water molecules.

Even though hydrogen bonds are weaker than covalent bonds, they are strong enough to cause water molecules to stick to one another and exhibit **cohesion** (*cohaesus* = a clinging together). The cohesive properties of water cause it to "bead up" on a waxed surface, such as a freshly waxed car. They also give water its **surface tension**. Water's surface has a thin "skin" that allows a glass to be filled just above the brim without spilling any of the water. Surface tension results from the formation of hydrogen bonds between the outermost layer of water molecules and the underlying molecules. Water's ability to form hydrogen bonds causes it to have the highest surface tension of any liquid except the element mercury.[2]

WATER: THE UNIVERSAL SOLVENT   Water molecules stick not only to other water molecules, but also to other polar chemical compounds. In doing so, water molecules can reduce the attraction between ions of opposite charges by as much as 80 times. For instance, ordinary table salt—sodium[3] chloride, NaCl—consists of an alternating array of positively charged sodium ions and negatively charged chloride ions (**Figure 5.4a**). The **electrostatic** (*electro* = electricity, *stasis* = standing) **attraction** between oppositely charged ions produces an **ionic** (*ienai* = to go) **bond**. When solid NaCl is placed in water, the electrostatic attraction (ionic bonding) between the sodium and chloride ions is reduced by 80 times. This, in turn, makes it much easier for the sodium ions and chloride ions to separate. When the ions separate, the positively charged sodium ions become attracted to the negative ends of the water molecules, the negatively charged chloride ions become attracted to the positive ends of the water molecules (**Figure 5.4b**), and the salt is dissolved in water. The process by which water molecules completely surround ions is called *hydration* (*hydra* = water, *ation* = action or process).

Because water molecules interact with other water molecules and other polar molecules, water is able to dissolve nearly everything.[4] Given enough time, water can dissolve more substances and in greater quantity than any other known substance. This is why water is called "the universal solvent." It is also why the ocean

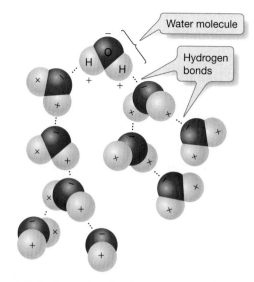

Figure 5.3 Hydrogen bonding in water. Dashed lines indicate locations of hydrogen bonds, which occur between water molecules.

A water molecule has a bend in its geometry, with the two hydrogen atoms on the same side of the oxygen atom. This property gives water its polarity and ability to form hydrogen bonds.

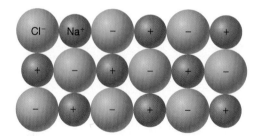

(a) Atomic structure of table salt, which is composed of sodium chloride (Na^+ = sodium ion, Cl^- = chlorine ion).

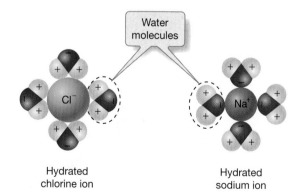

(b) As sodium chloride is dissolved, the positively charged ends of water molecules are attracted to the negatively charged Cl^- ion, while the negatively charged ends are attracted to the positively charged Na^+ ion.

Figure 5.4 Water as a solvent.

---

[2]Mercury is the only metal that is a liquid at normal surface temperatures, which is why it was commonly used in older thermometers. Today, mercury thermometers have been replaced by digital thermometers, which do not contain toxic mercury.

[3]Sodium is represented by the letters Na because the Latin term for sodium is *natrium*.

[4]If water is such a good solvent, why doesn't oil dissolve in water? As you might have guessed, the chemical structure of oil is remarkably nonpolar. With no positive or negative ends to attract the polar water molecule, oil will not dissolve in water.

contains so much dissolved material—an estimated 50 quadrillion tons (50 million billion tons) of salt—which makes seawater taste "salty."

CONCEPT CHECK 5.1

1. Sketch a model of an atom, showing the positions of the subatomic particles protons, neutrons, and electrons.
2. Describe what condition exists in water molecules to make them dipolar.
3. Sketch several water molecules, showing all covalent and hydrogen bonds. Be sure to indicate the polarity of each water molecule.
4. How does hydrogen bonding produce the surface tension phenomenon of water?
5. Discuss how the dipolar nature of a water molecule makes it such an effective solvent of ionic compounds.

5.2 What Other Important Properties Does Water Possess?

Water's other important properties include its thermal properties (such as water's freezing and boiling points, heat capacity, and latent heats) and its density.

Water's Thermal Properties

Water exists on Earth as a solid, a liquid, and a gas and has the ability to store and release great amounts of heat. Water's thermal properties influence the world's heat budget and are in part responsible for the development of tropical cyclones, worldwide wind belts, and ocean surface currents.

HEAT, TEMPERATURE, AND CHANGES OF STATE Matter around us is usually in one of the three common states: solid, liquid, or gas.[5] What must happen to change the state of a compound? The attractive forces between molecules or ions in a substance must be overcome if the state of the substance is to be changed from solid to liquid or from liquid to gas. These attractive forces include hydrogen bonds and van der Waals forces. The **van der Waals forces**—named for Dutch physicist Johannes Diderik van der Waals (1837–1923)—are relatively weak interactions that become significant only when molecules are very close together, as in the solid and liquid states (but not the gaseous state). Energy must be added to the molecules or ions so they can move fast enough to overcome these attractions.

What form of energy changes the state of matter? Very simply, adding or removing heat causes a substance to change its state of matter. For instance, adding heat to ice cubes causes them to melt, and removing heat from water causes ice to form. Before proceeding, let's clarify the difference between heat and temperature:

- **Heat** is the *energy transfer from one body to another due to a difference in temperature*. Heat is proportional to the energy level of moving molecules and thus is the total internal energy—both **kinetic** (*kinetos* = moving) **energy** and **potential** (*potentia* = power) **energy**—transferred from one body to another. For example, water can exist as a solid, liquid, or gas, depending on the amount of heat added. Heat may be generated by combustion (a chemical reaction commonly called "burning"), through other chemical reactions, by friction, or from radioactivity; it can be transferred by conduction, by convection, or by radiation. A **calorie** (*calor* = heat) is the amount of heat required to raise the temperature of 1 gram of water[6] by 1 degree centigrade. The familiar "calories" used to measure

---

[5]*Plasma* is widely recognized as a fourth state of matter distinct from solids, liquids, and normal gases. Plasma is a gaseous substance in which atoms have been ionized—that is to say, stripped of electrons. Plasma television screens take advantage of the fact that plasmas are strongly influenced by electric currents.

[6]1 gram (0.035 ounce) of water is equal to about 10 drops.

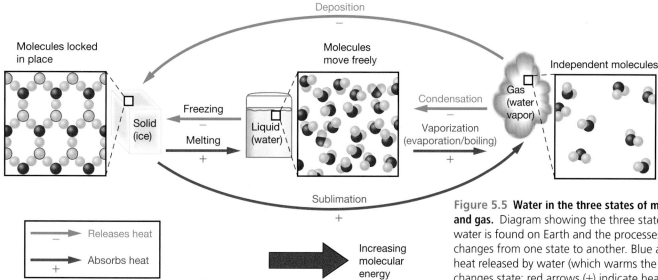

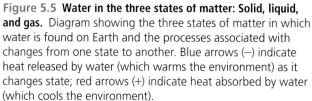

Figure 5.5 **Water in the three states of matter: Solid, liquid, and gas.** Diagram showing the three states of matter in which water is found on Earth and the processes associated with changes from one state to another. Blue arrows (−) indicate heat released by water (which warms the environment) as it changes state; red arrows (+) indicate heat absorbed by water (which cools the environment).

Web Animation
Phase Changes of Water

the energy content of foods is actually a *kilocalorie*, or 1000 calories. Although the metric unit for thermal energy is the *joule*, calories are directly tied to some of water's thermal properties, as will be discussed in the next section.

- **Temperature** is the *direct measure of the average kinetic energy of the molecules that make up a substance.* The greater the temperature, the greater the kinetic energy of the substance. Temperature changes when heat energy is added to or removed from a substance. Temperature is usually measured in degrees centigrade (°C) or degrees Fahrenheit (°F).

Figure 5.5 shows water molecules in the solid, liquid, and gaseous states. In the *solid state* (ice), water has a rigid structure and does not normally flow over short time scales. Intermolecular bonds are constantly being broken and reformed, but the molecules remain firmly attached. That is, the molecules vibrate with energy but remain in relatively fixed positions. As a result, solids do not conform to the shape of their container.

In the *liquid state* (water), water molecules still interact with each other, but they have enough kinetic energy to flow past each other and take the shape of their container. Intermolecular bonds are being formed and broken at a much greater rate than in the solid state.

In the *gaseous state* (water **vapor**), water molecules no longer interact with one another except during random collisions. Water vapor molecules flow very freely, filling the volume of whatever container they are placed in.

WATER'S FREEZING AND BOILING POINTS If enough heat energy is added to a solid, it melts to a liquid. The temperature at which melting occurs is the substance's **melting point**. If enough heat energy is removed from a liquid, it freezes to a solid. The temperature at which freezing occurs is the substance's **freezing point**, which is the same temperature as the melting point (Figure 5.5). For pure water, melting and freezing occur at 0°C (32°F).[7]

If enough heat energy is added to a liquid, it converts to a gas. The temperature at which boiling occurs is the substance's **boiling point**. If enough heat energy is removed from a gas, it **condenses** to a liquid. The highest temperature at which condensation occurs is the substance's **condensation point**, which is the same temperature as the boiling point (Figure 5.5). For pure water, boiling and condensation occur at 100°C (212°F).

Both the freezing and boiling points of water are unusually high compared to those of similar chemical substances. As shown in **Figure 5.6**, if water followed the pattern of other chemical compounds with molecules of similar mass, it should melt

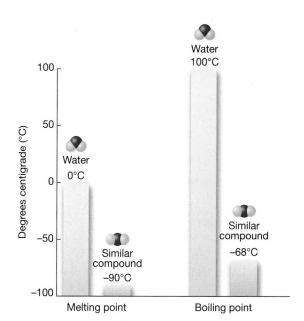

Figure 5.6 **Comparison of melting and boiling points of water with similar chemical compounds.** Bar graph showing the melting and boiling points of water compared to the melting and boiling points of similar chemical compounds. Water would have properties like those of similar chemical compounds if it did not have its unique geometry and resulting polarity.

---

[7]All melting/freezing/boiling points discussed in this chapter assume a standard sea level pressure of 1 atmosphere (14.7 pounds per square inch).

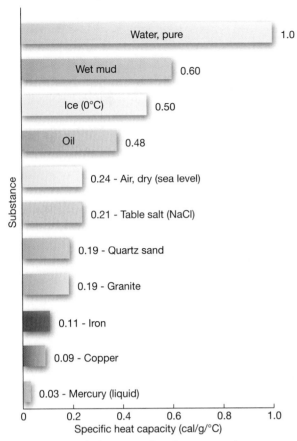

Figure 5.7 Specific heat capacity of common substances. Graph showing the specific heat capacity of common substances at 20°C (68°F). Note that water has a very high specific heat capacity, which means it takes a lot of energy to increase water temperature.

at −90°C (−130°F) and boil at −68°C (−90°F). If that were the case, all water on Earth would be in the gaseous state. Instead, water melts and boils at the relatively high temperatures of 0°C (32°F) and 100°C (212°F),[8] respectively, because additional heat energy is required to overcome its hydrogen bonds and van der Waals forces. Thus, if not for the unusual geometry and resulting polarity of the water molecule, all water on Earth would be boiled away, and life as we know it would not exist.

WATER'S HEAT CAPACITY AND SPECIFIC HEAT Heat capacity is *the amount of heat energy required to raise the temperature of a substance by 1 degree centigrade.* Substances that have high heat capacity can absorb (or lose) large quantities of heat with only a small change in temperature. Conversely, substances that change temperature rapidly when heat is applied—such as oil or metals—have lower heat capacity.

The heat capacity per unit mass of a body, called *specific heat capacity* or, more simply, **specific heat**, is used to more directly compare the heat capacity of substances. For example, as shown in **Figure 5.7**, pure water has a high specific heat capacity that is exactly 1 calorie per gram,[9] whereas other common substances have much lower specific heats. Note that metals such as iron and copper—which heat up rapidly when heat is applied—have capacity values that are about 10 times lower than that of water.

Why does water have such high heat capacity? The reason is because it takes more energy to increase the kinetic energy of hydrogen-bonded water molecules than it does for substances in which the dominant intermolecular interaction is the much weaker van der Waals force. As a result, water gains or loses much more heat than other common substances while undergoing an equal temperature change. In addition, water resists any change in temperature, as you may have observed when heating a large pot of water. When heat is applied to the pot, which is made of metal that has a low heat capacity, the pot heats up quickly. The water *inside* the pot, however, takes a long time to heat up (hence, the saying that a watched pot never boils but an unwatched pot boils over!). Making the water boil takes even more heat because all the hydrogen bonds must be broken. The exceptional capacity of water to absorb large quantities of heat helps explain why water is used in home heating, industrial and automobile cooling systems, and home cooking applications.

WATER'S LATENT HEATS When water undergoes a change of state—that is, when ice melts or water freezes, or when water boils or water vapor condenses—a large amount of heat is absorbed or released. The amount of heat absorbed or released is due to water's high latent (*latent* = hidden) heats and is closely related to water's unusually high heat capacity. As water evaporates from your skin, it cools your body by absorbing heat (this is why sweating cools your body). Conversely, if you have ever been scalded by water vapor—steam—you know that steam releases an enormous amount of latent heat when it condenses to a liquid.

Latent Heat of Melting The graph in **Figure 5.8** shows how latent heat affects the amount of energy needed to increase water temperature and change the state of water. Beginning with 1 gram of ice (*lower left*), the addition of 20 calories of heat

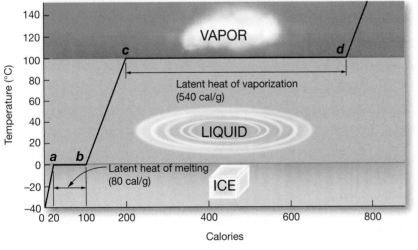

Figure 5.8 Latent heats and changes of state of water. The latent heat of melting (80 calories per gram) is much less than the latent heat of vaporization (540 calories per gram). See text for description of points *a, b, c,* and *d.*

---

[8]Note that the temperature scale *centigrade* (*centi* = a hundred, *grad* = step) is based on 100 even divisions between the melting and boiling points of pure water. It is also called the Celsius scale, after its founder (see Appendix I, "Metric and English Units Compared").

[9]Note that the specific heat capacity of water is used as the unit of heat quantity, the calorie. Thus, water is the standard against which the specific heats of other substances are compared.

raises the temperature of the ice by 40 degrees, from −40°C to 0°C (point *a* on the graph). The temperature remains at 0°C (32°F) even though more heat is being added, as shown by the plateau on the graph between points *a* and *b*. The temperature of the water does not change until 80 more calories of heat energy have been added. The **latent heat of melting** is the energy needed to break the intermolecular bonds that hold water molecules rigidly in place in ice crystals. The temperature remains unchanged until most of the bonds are broken and the mixture of ice and water has changed completely to 1 gram of water.

After the change from ice to liquid water has occurred at 0°C (32°F), additional heat raises the water temperature between points *b* and *c* in Figure 5.8. As it does, it takes 1 calorie of heat to raise the temperature of the gram of water 1°C (or 1.8°F). Therefore, another 100 calories must be added before the gram of water reaches the boiling point of 100°C (212°F). So far, a total of 200 calories has been added to reach point *c*.

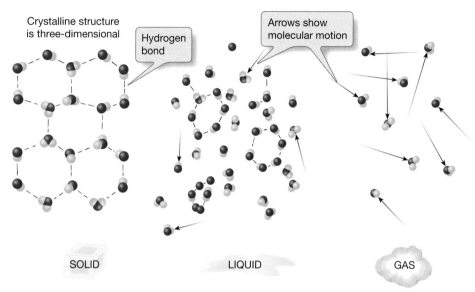

(a) In the solid state, water exists as ice, and there are hydrogen bonds between all water molecules.

(b) In the liquid state, there are some hydrogen bonds.

(c) In the gaseous state, there are no hydrogen bonds, and the water molecules move rapidly and independently.

Figure 5.9 Hydrogen bonds in H_2O and the three states of matter.

Latent Heat of Vaporization   The graph in Figure 5.8 flattens out again at 100°C (212°F), between points *c* and *d*. This plateau represents the **latent heat of vaporization**, which is 540 calories per gram for water. This is the amount of heat that must be added to 1 gram of a substance at its boiling point to break the intermolecular bonds and complete the change of state from liquid to vapor (gas).

The drawings in **Figure 5.9**, which show the structure of water molecules in the solid, liquid, and gaseous states, help explain why the latent heat of vaporization is so much greater than the latent heat of melting. To go from a solid to a liquid, just enough hydrogen bonds must be broken to allow water molecules to slide past one another. To go from a liquid to a gas, however, all of the hydrogen bonds must be completely broken so that individual water molecules can move about freely.

Latent Heat of Evaporation   Sea surface temperatures average 20°C (68°F) or less. How, then, does liquid water convert to vapor at the surface of the ocean? The conversion of a liquid to a gas below the boiling point is called **evaporation**. At ocean surface temperatures, individual molecules converted from the liquid to the gaseous state have less energy than do water molecules at 100°C (212°F). To gain the additional energy necessary to break free of the surrounding ocean water molecules, an individual molecule must capture heat energy from its neighbors. In other words, the molecules left behind have lost heat energy to those that evaporate, which explains the cooling effect of evaporation.

It takes more than 540 calories of heat to produce 1 gram of water vapor from the ocean surface at temperatures less than 100°C (212°F). At 20°C (68°F), for instance, the **latent heat of evaporation** is 585 calories per gram. More heat is required because more hydrogen bonds must be broken. At higher temperatures, liquid water has fewer hydrogen bonds because the molecules are vibrating and jostling about more.

Latent Heat of Condensation   When water vapor is cooled sufficiently, it condenses to a liquid and releases its **latent heat of condensation** into the surrounding air. On a small scale, the heat released is enough to cook food; this is how a steamer works. On a large scale, the heat released is sufficient to power large thunderstorms and even hurricanes (see Chapter 6, "Air–Sea Interaction").

Latent Heat of Freezing Heat is also released when water freezes. The amount of heat released when water freezes is the same amount that was absorbed when the water was melted in the first place. Thus, the **latent heat of freezing** is identical to the latent heat of melting. Similarly, the latent heats of vaporization and condensation are identical.

GLOBAL THERMOSTATIC EFFECTS The **thermostatic** (*thermos* = heat, *stasis* = standing) **effects** of water are those properties that act to moderate changes in temperature, which in turn affect Earth's climate. For example, the huge amount of heat energy exchanged in the evaporation–condensation cycle helps make life possible on Earth. The Sun radiates energy to Earth, where some is stored in the oceans. Evaporation removes this heat energy from the oceans and carries it high into the atmosphere. In the cooler upper atmosphere, water vapor condenses into clouds, which are the basis of **precipitation** (mostly rain and snow) that releases latent heat of condensation. The map in **Figure 5.10** shows how this cycle of evaporation and condensation removes huge amounts of heat energy from the low-latitude oceans and adds huge amounts of heat energy to the heat-deficient higher latitudes. In addition, the heat released when sea ice forms further moderates Earth's high-latitude regions near the poles.

The exchange of latent heat between ocean and atmosphere is very efficient. For every gram of water that condenses in cooler latitudes, the amount of heat released to warm these regions equals the amount of heat removed from the tropical ocean when that gram of water was evaporated initially. The end result is that the thermal properties of water have prevented wide variations in Earth's temperature, thus moderating Earth's climate. Because rapid environmental changes can often result in the death of many life-forms, our planet's moderated climate is one of the main reasons life exists on Earth.

Figure 5.10 Atmospheric transport of surplus heat from low latitudes into heat-deficient high latitudes. The heat removed from the tropical ocean (*evaporation latitudes*) is carried toward the poles (*red arrows*) and is released at higher latitudes through precipitation (*precipitation latitudes*), thus moderating Earth's climate.

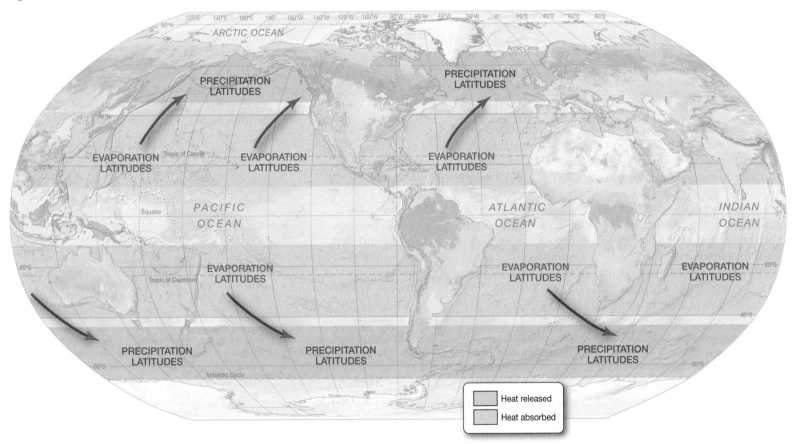

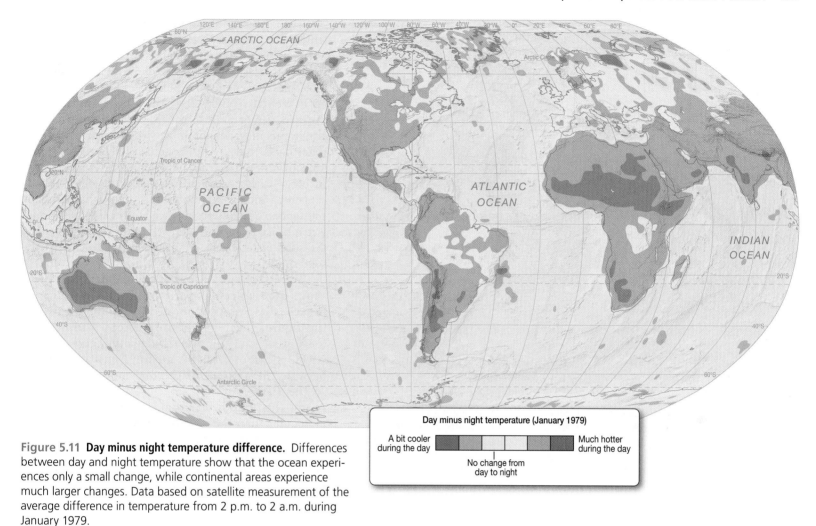

Figure 5.11 Day minus night temperature difference. Differences between day and night temperature show that the ocean experiences only a small change, while continental areas experience much larger changes. Data based on satellite measurement of the average difference in temperature from 2 p.m. to 2 a.m. during January 1979.

Another thermostatic effect of the ocean can be seen in **Figure 5.11**, which shows the temperature difference between day and night. The map shows that in the ocean, there is only a small difference in temperature between day and night, while the land experiences a much greater variation. This difference between ocean and land is due to the higher heat capacity of water, which gives it the ability to absorb the daily gains and minimize the daily losses of heat energy much more easily than land materials. The term **marine effect** describes locations that experience the moderating influences of the ocean, usually along coastlines or islands. **Continental effect**—a condition of *continentality*—refers to areas less affected by the sea and therefore having a greater range of temperature differences (both daily and yearly).

Water Density

Recall from Chapter 1 that density is mass per unit volume and can be thought of as *how heavy something is for its size*. Ultimately, density is related to how tightly the molecules or ions of a substance are packed together. Typical units of density are grams per cubic centimeter (g/cm^3). Pure water, for example, has a density of 1.0 g/cm^3. Note that temperature, salinity, and pressure all affect water density.

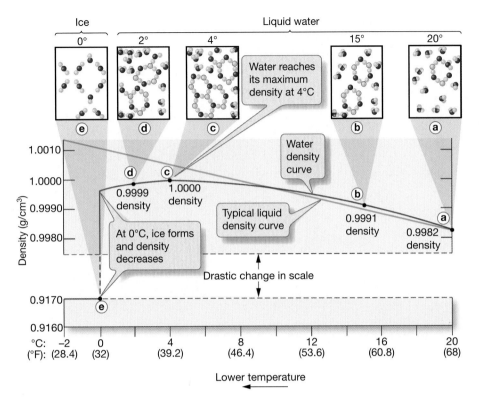

Figure 5.12 **Water density as a function of temperature and the formation of ice.** Curves showing the density of freshwater (*red curve*) as it freezes (*right to left*) and the density of a typical liquid (*green curve*). Water reaches its maximum density at 4°C (39°F), but below that water becomes less dense as ice begins to form. At 0°C (32°F), ice forms, its crystal structure expands dramatically, and density decreases. As a result, ice floats.

Web Animation
The Formation of Ice

Figure 5.13 **Snowflakes.** Scanning electron photomicrograph of actual snowflakes magnified about 500 times. Hexagonal snowflakes indicate the internal structure of water molecules held together by hydrogen bonds.

The density of most substances increases as temperature decreases. For example, cold air sinks and warm air rises because cold air is denser than warm air. Density increases as temperature decreases because the molecules lose energy and slow down, so the same number of molecules occupy less space. This shrinkage caused by cold temperatures, called **thermal contraction**, also occurs in water, but only to a certain point. As pure water cools to 4°C (39°F), its density increases. From 4°C down to 0°C (32°F), however, its density *decreases*. In other words, water stops contracting and actually expands, which is highly unusual among Earth's many substances. The result is that ice is less dense than liquid water, so ice floats on water. For most other substances, the solid state is denser than the liquid state, so the solid sinks.

Why is ice less dense than liquid water? **Figure 5.12** shows how molecular packing changes as water approaches its freezing point. From points *a* to *c* in the figure, the temperature decreases from 20°C (68°F) to 4°C (39°F) and the density increases from 0.9982 g/cm³ to 1.0000 g/cm³. Density increases because the amount of thermal motion decreases, so the water molecules occupy less volume. As a result, the window at point *c* contains more water molecules than the windows at points *a* or *b*. When the temperature is lowered below 4°C (39°F), the overall volume increases again because water molecules begin to line up to form ice crystals. Ice crystals are bulky, open, six-sided structures in which water molecules are widely spaced. Their characteristic hexagonal shape (**Figure 5.13**) mimics the hexagonal molecular structure resulting from hydrogen bonding between water molecules (see Figure 5.9a). By the time water fully freezes (point *e*), the density of the ice is much less than that of water at 4°C (39°F), the temperature at which water achieves its maximum density.

When water freezes, its volume increases by about 9%. Anyone who has put a beverage in a freezer for "just a few minutes" to cool it down and inadvertently forgotten about it has experienced the volume increase associated with water's expansion as it freezes—usually resulting in a burst beverage container (**Figure 5.14**). The force exerted when ice expands is powerful enough to break apart rocks, split pavement on roads and sidewalks, and crack water pipes.

Increasing the pressure or adding dissolved substances decreases the temperature of maximum density for freshwater because the formation of bulky ice crystals is inhibited. Increasing pressure increases the number of water molecules in a given volume and inhibits the number of ice crystals that can be created. Increasing amounts of dissolved substances inhibits the formation of hydrogen bonds, which further restricts the number of ice crystals that can form. To produce ice crystals equal in volume to those that could be produced at 4°C (39°F) in freshwater, more energy must be removed, causing a reduction in the temperature of maximum density.

Dissolved solids reduce the freezing point of water, too. This is one of the reasons most seawater never freezes, except near Earth's frigid poles (and even then, only at the surface). It's also why salt is spread on roads and sidewalks during the winter in cold climates. The salt lowers the freezing point of water, allowing ice-free roads and sidewalks at temperatures that are several degrees below freezing.

For a summary of the physical and biological significance of the unusual properties of seawater, see **Web Table 5.1**.

Figure 5.14 Glass bottle shattered by frozen water. This glass bottle was filled with water, sealed, and put into a freezer. As water freezes, it expands by 9% as it forms hydrogen bonds and forms an open lattice structure, which increased the pressure and caused the bottle to fracture.

5.3 How Salty Is Seawater?

What is the difference between pure water and seawater? One of the most obvious differences is that seawater contains dissolved substances that give it a distinctly salty taste. These dissolved substances are not simply sodium chloride (table salt); they include various other salts, metals, and dissolved gases. The oceans, in fact, contain enough salt to cover the entire planet with a layer more than 150 meters (500 feet) thick (about the height of a 50-story skyscraper). Unfortunately, the salt content of seawater makes it unsuitable for drinking or irrigating most crops and causes it to be highly corrosive to many materials.

Salinity

Salinity (*salinus* = salt) is the total amount of solid material dissolved in water, including dissolved gases (because even gases become solids at low enough temperatures) but excluding dissolved organic substances. Salinity does *not* include fine particles being held in suspension (turbidity) or solid material in contact with water because these materials are not dissolved. Salinity is the ratio of the mass of dissolved substances to the mass of the water sample.

The salinity of seawater is typically about 3.5%, about 220 times saltier than freshwater. Seawater with a salinity of 3.5% indicates that it also contains 96.5% pure water, as shown in **Figure 5.15**. Because seawater is mostly pure water, its physical properties are very similar to those of pure water, with only slight variations.

Figure 5.15 and **Table 5.1** show that the elements chlorine, sodium, sulfur (as the sulfate ion), magnesium, calcium, and potassium account for over 99% of the dissolved solids in seawater. More than 80 other chemical elements have been identified in seawater—most in extremely small amounts—and probably all of Earth's naturally occurring elements exist in the sea. Remarkably, trace amounts of dissolved components in seawater are vital for human survival (Diving Deeper 5.1).

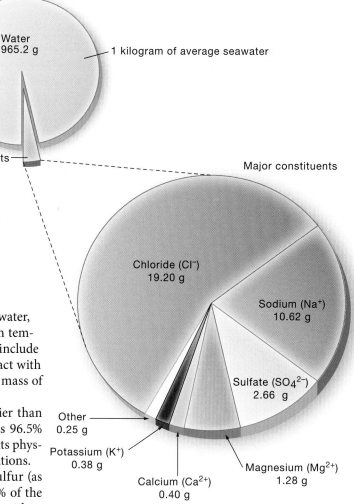

Figure 5.15 Major dissolved components in seawater. Diagrammatic representation of the most abundant components in a kilogram of 35‰ salinity seawater. Constituents are listed in grams per kilogram, which is equivalent to parts per thousand (‰).

TABLE 5.1 SELECTED DISSOLVED MATERIALS IN 35‰ SEAWATER

1. Major constituents (in parts per thousand by weight, ‰)

Constituent	Concentration (‰)	Ratio of constituent/total salts (%)
Chloride (Cl^-)	19.2	55.04
Sodium (Na^+)	10.6	30.61
Sulfate (SO_4^{2-})	2.7	7.68
Magnesium (Mg^{2+})	1.3	3.69
Calcium (Ca^{2+})	0.40	1.16
Potassium (K^+)	0.38	1.10
Total	**34.58‰**	**99.28%**

2. Minor constituents (in parts per million by weight, ppm[a])

Gases		Nutrients		Others	
Constituent	Concentration (ppm)	Constituent	Concentration (ppm)	Constituent	Concentration (ppm)
Carbon dioxide (CO_2)	90	Silicon (Si)	3.0	Bromide (Br^-)	65.0
Nitrogen (N_2)	14	Nitrogen (N)	0.5	Carbon (C)	28.0
Oxygen (O_2)	6	Phosphorus (P)	0.07	Strontium (Sr)	8.0
		Iron (Fe)	0.002	Boron (B)	4.6

3. Trace constituents (in parts per billion by weight, ppb[b])

Constituent	Concentration (ppb)	Constituent	Concentration (ppb)	Constituent	Concentration (ppb)
Lithium (Li)	185	Zinc (Zn)	10	Lead (Pb)	0.03
Rubidium (Rb)	120	Aluminum (Al)	2	Mercury (Hg)	0.03
Iodine (I)	60	Manganese (Mn)	2	Gold (Au)	0.005

[a]Note that 1000 ppm = 1‰.
[b]Note that 1000 ppb = 1 ppm.

Salinity is often expressed in **parts per thousand (‰).** For example, as 1% is 1 part in 100, 1‰ is 1 part in 1000. When converting from percent to parts per thousand, the decimal is simply moved one place to the right. For instance, typical seawater salinity of 3.5% is the same as 35‰. Advantages of expressing salinity in parts per thousand are that decimals are often avoided and values convert directly to grams of salt per kilogram of seawater. For example, 35‰ seawater has 35 grams of salt in every 1000 grams of seawater.[10]

Determining Salinity

Early methods of determining seawater salinity involved evaporating a carefully weighed amount of seawater and weighing the salts that precipitated from it. However, the accuracy of this time-consuming method is limited because some water can remain bonded to salts that precipitate and some substances can evaporate along with the water.

[10]Note that the units "parts per thousand" are effectively parts per thousand by weight. Salinity values, however, lack units because the salinity of a water sample is determined as the *ratio* of the electrical conductivity of the sample to the electrical conductivity of a standard. Thus, salinity values are sometimes reported in *p.s.u.*, or *practical salinity units*, which are equivalent to parts per thousand.

Diving Deeper 5.1 Oceans and People

HOW TO AVOID GOITERS

The nutritional label on containers of salt usually proclaims "this product contains iodine, a necessary nutrient." Why is iodine necessary in our diets? It turns out that if a person's diet contains an insufficient amount of iodine, a potentially life-threatening affliction called **goiters** (*guttur* = throat) may result (Figure 5A).

Iodine is used by the thyroid gland, which is a butterfly-shaped organ located in the neck in front of and on either side of the trachea (windpipe). The thyroid gland manufactures hormones that regulate cellular metabolism essential for mental development and physical growth. If people lack iodine in their diet, their thyroid glands cannot function properly. Often, this results in the enlargement or swelling of the thyroid gland. Severe symptoms include dry skin, loss of hair, puffy face, weakness of muscles, weight increase, diminished vigor, mental sluggishness, and a large nodular growth on the neck called a goiter. If proper steps are not taken to correct this disease, it can lead to cancer. Iodine ingested regularly often begins to reverse

Figure 5A A woman with goiters.

the effects. In advanced stages, surgery to remove the goiter or exposure to radioactivity is the only course of action.

How can you avoid goiters? Fortunately, goiters can be prevented with a diet that contains just *trace amounts* of iodine. Where can you get iodine in your diet? All products from the sea contain trace amounts of iodine because iodine is one of the many elements dissolved in seawater. Sea salt, seafood, seaweed, and other sea products contain plenty of iodine to help prevent goiters. Although goiters are rarely a problem in developed nations like the United States, goiters pose a serious health hazard in many underdeveloped nations, especially those far from the sea. In the United States, however, many people get too much iodine in their diet, leading to the overproduction of hormones by the thyroid gland. That's why most stores that sell iodized salt also carry noniodized salt for those people who have a *hyperthyroid* (*hyper* = excessive, *thyroid* = the thyroid gland) condition and must restrict their intake of iodine.

Another way to measure salinity is to use the *principle of constant proportions*, which was firmly established by chemist William Dittmar (1859–1951) when he analyzed the water samples collected during the *Challenger* Expedition (see Web Diving Deeper 5.2). The **principle of constant proportions** states that the major dissolved constituents responsible for the salinity of seawater occur nearly everywhere in the ocean in exactly the same proportions, independent of salinity. The ocean, therefore, is well mixed. When salinity changes, moreover, the salts don't leave (or enter) the ocean, but water molecules do. Seawater has *constancy of composition*, so the concentration of a single major constituent can be measured to determine the total salinity of a given water sample. The constituent that occurs in the greatest abundance and is the easiest to measure accurately is the chloride ion, Cl^-. The weight of this ion in a water sample is its **chlorinity**.

In any sample of ocean water worldwide, the chloride ion accounts for 55.04% of the total proportion of dissolved solids (Figure 5.15 and Table 5.1). Therefore, by measuring only the chloride ion concentration, the total salinity of a seawater sample can be determined using the following relationship:

$$\text{Salinity (\textperthousand)} = 1.80655 \times \text{chlorinity (\textperthousand)}^* \tag{5.1}$$

---

*The number 1.80655 comes from dividing 1 by 0.5504 (the chloride ion's proportion in seawater of 55.04%). However, if you actually divide this, you will get 1.81686, which is different from the original value by 0.57%. Empirically, oceanographers have found that seawater's constancy of composition is an approximation and have agreed to use 1.80655 because it more accurately represents the total salinity of seawater.

Freshwater   Saltwater

Figure 5.16 Salinity affects water conductivity. Increasing the amount of dissolved substances increases the conductivity of the water. A light bulb with bare electrodes shows that the higher the salinity, the more electricity is transmitted and the brighter the bulb will be lit.

STUDENTS SOMETIMES ASK . . .

What is the strategy behind adding salt to a pot of water when making pasta? Does it make the water boil faster?

Adding salt to water will not make the water boil faster. It will, however, make the water boil at a slightly higher temperature because dissolved substances raise its boiling point (and *lower* its freezing point; see Table 5.2). Thus, the pasta will cook in slightly less time. In addition, the salt adds flavoring, so the pasta may taste better, too. Be sure to add the salt after the water has come to a boil, though, or it will take longer to reach a boil. This is a wonderful use of chemical principles—helping you to cook better!

For example, the average chlorinity of the ocean is 19.2‰, so the average salinity is 1.80655 × 19.2‰ which rounds to 34.7‰. In other words, on average there are 34.7 parts of dissolved material in every 1000 parts of seawater.

Standard seawater consists of ocean water analyzed for chloride ion content to the nearest ten-thousandth of a part per thousand by the Institute of Oceanographic Services in Wormly, England. It is then sealed in small glass vials called *ampules* and sent to laboratories throughout the world for use as a reference standard in calibrating analytical equipment.

Seawater salinity can be measured very accurately with modern oceanographic instruments such as a **salinometer** (*salinus* = salt, *meter* = measure). Most salinometers measure seawater's *electrical conductivity* (the ability of a substance to transmit electric current), which increases as more substances are dissolved in water (**Figure 5.16**). Salinometers can determine salinity to resolutions of better than 0.003%.

Comparing Pure Water and Seawater

Table 5.2 compares various properties of pure water and seawater. Because seawater is 96.5% water, most of its physical properties are very similar to those of pure water. For instance, the color of pure water and seawater is identical.

The dissolved substances in seawater, however, give it slightly different yet important physical properties as compared to pure water. For example, recall that dissolved substances interfere with pure water changing state. The freezing points and boiling points in Table 5.2 show that dissolved substances decrease the freezing point and increase the boiling point of water. Thus, seawater freezes at a temperature of −1.9°C (28.6°F), which is lower than the freezing point of pure water (0°C [32°F]). Similarly, seawater boils at a temperature of 100.6°C (213.1°F), which is higher than the freezing point of pure water (100°C [212°F]). In effect, the salts in seawater extend the range of temperatures in which water is a liquid. This same principle applies to antifreeze used in automobile radiators. Antifreeze lowers the freezing point of the water in a radiator and increases the boiling point, thus extending the range over which the water remains in the liquid state. Antifreeze, therefore, protects your radiator from freezing in the winter *and* from boiling over in the summer.

Seawater salinity can be measured using a salinometer and averages 35‰. The dissolved components in seawater give it different yet important physical properties as compared to pure water.

TABLE 5.2   COMPARISON OF SELECTED PROPERTIES OF PURE WATER AND SEAWATER

Property		Pure water	35‰ seawater
Color (light transmission)	Small quantities of water	Clear (high transparency)	Same as for pure water
	Large quantities of water	Blue-green because water molecules scatter blue and green wavelengths best	Same as for pure water
Odor		Odorless	Distinctly marine
Taste		Tasteless	Distinctly salty
pH		7.0 (neutral)	Surface waters, range = 8.0–8.3; average = 8.1 (slightly alkaline)
Density at 4°C (39°F)		1.000 g/cm^3	1.028 g/cm^3
Freezing point		0°C (32°F)	−1.9°C (28.6°F)
Boiling point		100°C (212°F)	100.6°C (213.1°F)

Other important properties of seawater (such as pH and density) are discussed later in this chapter.

CONCEPT CHECK 5.3

1 What is the average salinity of seawater? What units are normally used, and why are those units useful?

2 What condition of salinity makes it possible to determine the total salinity of ocean water by measuring the concentration of only one constituent, the chloride ion?

3 What are goiters? How can they be avoided?

4 How are seawater and pure water similar? How are the two different?

5.4 Why Does Seawater Salinity Vary?

Using salinometers and other techniques, oceanographers have determined that salinity varies from place to place in the oceans. What are the patterns of seawater salinity, and what causes them?

Salinity Variations

In the open ocean far from land, salinity varies between about 33 and 38‰. In coastal areas, salinity variations can be extreme. In the Baltic Sea, for example, salinity averages only 10‰ because physical conditions create **brackish** (*brak* = salt, *ish* = somewhat) water. Brackish water is produced in areas where freshwater (from rivers and high rainfall) and seawater mix. In the Red Sea, on the other hand, salinity averages 42‰ because physical conditions produce **hypersaline** (*hyper* = excessive, *salinus* = salt) water. Hypersaline water is typical of seas and inland bodies of water that experience high evaporation rates and limited open-ocean circulation.

Some of the most hypersaline water in the world is found in inland lakes, which are often called seas because they are so salty. The Great Salt Lake in Utah, for example, has a salinity of 280‰, and the Dead Sea on the border of Israel and Jordan has a salinity of 330‰. The water in the Dead Sea, therefore, contains 33% dissolved solids and is almost *10 times saltier than seawater*. As a result, hypersaline waters are so dense and buoyant that one can easily float (**Figure 5.17**), even with arms and legs sticking up above water level! Hypersaline waters also taste much saltier than seawater.

Figure 5.17 High-salinity water of the Dead Sea allows swimmers to easily float. The Dead Sea, which has 330‰ salinity (almost 10 times the salinity of seawater), has high density. As a result, it also has high buoyancy that allows swimmers to float easily.

Salinity of seawater in coastal areas also varies seasonally. For example, the salinity of seawater off Miami Beach, Florida, varies from about 34.8‰ in October to 36.4‰ in May and June, when evaporation is high. Offshore of Astoria, Oregon, seawater salinity is always relatively low because of the vast freshwater input from the Columbia River. Here, surface water salinity can be as low as 0.3‰ in April and May (when the Columbia River is at its maximum flow rate) and 2.6‰ in October (the dry season).

Other types of water have much lower salinity. Tap water, for instance, has salinity somewhere below 0.8‰, and good-tasting tap water is usually below 0.6‰. Salinity of premium bottled water is on the order of 0.3%, with the salinity often displayed prominently on its label, usually as total dissolved solids (TDS) in units of parts per million (ppm), where 1000 ppm equals 1‰.

Processes Affecting Seawater Salinity

Processes affecting seawater salinity change either the amount of water (H_2O molecules) or the amount of dissolved substances in the water. Adding more water, for instance, dilutes the dissolved component and lowers the salinity of the sample. Conversely, removing water increases salinity. Changing the salinity in these ways does not affect the *amount* or the *composition* of the dissolved components, which remain in constant proportions. Let's first examine processes that affect the amount of water in seawater before turning our attention to processes that influence dissolved components.

PROCESSES THAT DECREASE SEAWATER SALINITY Table 5.3 summarizes the processes that affect seawater salinity. Precipitation, **runoff** (stream discharge), melting icebergs, and melting sea ice *decrease* seawater salinity by adding more freshwater to the ocean. Precipitation is the way atmospheric water returns to Earth as rain, snow, sleet, and hail. Worldwide, about three-quarters of all precipitation falls directly back into the ocean and one-quarter falls onto land. Precipitation falling directly into the oceans adds freshwater, reducing seawater salinity.

TABLE 5.3   PROCESSES THAT AFFECT SEAWATER SALINITY

Process	How accomplished	Adds or removes	Effect on salt in seawater	Effect on H_2O in seawater	Salinity increase or decrease?	Source of freshwater from the sea?
Precipitation	Rain, sleet, hail, or snow falls directly on the ocean	Adds very fresh water	None	More H_2O	Decrease	N/A
Runoff	Streams carry water to the ocean	Adds mostly fresh water	Negligible addition of salt	More H_2O	Decrease	N/A
Icebergs melting	Glacial ice calves into the ocean and melts	Adds very fresh water	None	More H_2O	Decrease	Yes, icebergs from the Antarctic have been towed to South America
Sea ice melting	Sea ice melts in the ocean	Adds mostly fresh water and some salt	Adds a small amount of salt	More H_2O	Decrease	Yes, sea ice can be melted and is better than drinking seawater
Sea ice forming	Seawater freezes in cold ocean areas	Removes mostly freshwater	30% of salts in seawater are retained in ice	Less H_2O	Increase	Yes, through multiple freezings, called *freeze separation*
Evaporation	Seawater evaporates in hot climates	Removes very pure water	None (essentially all salts are left behind)	Less H_2O	Increase	Yes, through evaporation of seawater and condensation of water vapor, called *distillation*

Most of the precipitation that falls on land returns to the oceans indirectly as stream runoff. Even though this water dissolves minerals on land, the runoff is relatively pure water, as shown in Table 5.4. Runoff, therefore, adds mostly water to the ocean, causing seawater salinity to decrease.

Icebergs are chunks of ice that have broken free (*calved*) from a glacier when it flows into an ocean or marginal sea and begins to melt. Glacial ice originates as snowfall in high mountain areas, so icebergs are composed of freshwater. When icebergs melt in the ocean, they add freshwater, which is another way in which seawater salinity is reduced.

Sea ice forms when ocean water freezes in high-latitude regions and is composed primarily of freshwater. When warmer temperatures return to high-latitude regions in the summer, sea ice melts in the ocean, adding mostly freshwater with a small amount of salt to the ocean. Seawater salinity, therefore, is decreased.

PROCESSES THAT INCREASE SEAWATER SALINITY The formation of sea ice and evaporation *increase* seawater salinity by removing water from the ocean (Table 5.3). Sea ice forms when seawater freezes. Depending on the salinity of seawater and the rate of ice formation, about 30% of the dissolved components in seawater are retained in sea ice. This means that 35‰ seawater creates sea ice with about 10‰ salinity (30% of 35‰ is 10‰). Consequently, the formation of sea ice removes mostly freshwater from seawater, increasing the salinity of the remaining unfrozen water. High-salinity water also has a high density, so it sinks below the surface.

Recall that evaporation is the conversion of water molecules from the liquid state to the vapor state at temperatures below the boiling point. Evaporation removes water from the ocean, leaving its dissolved substances behind. Evaporation, therefore, increases seawater salinity. Worldwide, about 86% of all evaporation occurs in the oceans.

THE HYDROLOGIC CYCLE Figure 5.18 shows how the **hydrologic** (*hydro* = water, *logos* = study of) **cycle** relates the processes that affect seawater salinity. These processes recycle water among the ocean, the atmosphere, and the continents, so water is in continual motion between the different components (*reservoirs*) of the hydrologic cycle. Earth's water supply exists in the following proportions:

- 97.2% in the world ocean
- 2.15% frozen in glaciers and ice caps
- 0.62% in groundwater and soil moisture
- 0.02% in streams and lakes
- 0.001% as water vapor in the atmosphere

ESSENTIAL CONCEPT 5.4

Various surface processes either decrease seawater salinity (precipitation, runoff, icebergs melting, or sea ice melting) or increase seawater salinity (sea ice forming and evaporation).

TABLE 5.4 COMPARISON OF MAJOR DISSOLVED COMPONENTS IN STREAMS WITH THOSE IN SEAWATER

Constituent	Concentration in streams (parts per million by weight)	Concentration in seawater (parts per million by weight)
Bicarbonate ion (HCO_3^-)	58.4	trace
Calcium ion (Ca^{2+})	15.0	400
Silicate (SiO_2)	13.1	3
Sulfate ion (SO_4^{2-})	11.2	2700
Chloride ion (Cl^-)	7.8	19,200
Sodium ion (Na^+)	6.3	10,600
Magnesium ion (Mg^{2+})	4.1	1300
Potassium ion (K^+)	2.3	380
Total (parts per million)	119.2 ppm	34,793 ppm
Total (%)	0.1192%	34.8%

Figure 5.18 The hydrologic cycle. All water is in continual motion between the various components (reservoirs) of the hydrologic cycle. Volumes are Earth's average yearly amounts in cubic kilometers; table shows average yearly flux between reservoirs; ice not shown.

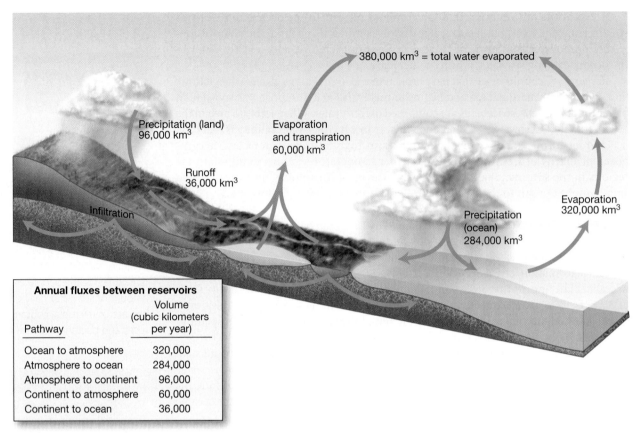

380,000 km³ = total water evaporated

Precipitation (land)
96,000 km³

Evaporation and transpiration
60,000 km³

Runoff
36,000 km³

Infiltration

Precipitation (ocean)
284,000 km³

Evaporation
320,000 km³

Annual fluxes between reservoirs	
Pathway	Volume (cubic kilometers per year)
Ocean to atmosphere	320,000
Atmosphere to ocean	284,000
Atmosphere to continent	96,000
Continent to atmosphere	60,000
Continent to ocean	36,000

Web Animation
Earth's Water and the Hydrologic Cycle

STUDENTS SOMETIMES ASK . . .

You mentioned that when seawater freezes, it produces ice with about 10% salinity. Once that ice melts, can a person drink it with no ill effects?

Early Arctic explorers found out the answer to your question by necessity. Some of these explorers who traveled by ship in high-latitude regions became inadvertently or purposely entrapped by sea ice (see, for example, **Web Diving Deeper 7.1**, which describes the remarkable voyage of the *Fram*). Lacking other water sources, they used melted sea ice. Although newly formed sea ice contains little salt, it does trap a significant amount of brine (drops of salty water). Depending on the rate of freezing, newly formed ice may have a total salinity from 4 to 15%. The more rapidly it forms, the more brine it captures and the higher the salinity. Melted sea ice with salinity this high doesn't taste very good, and it still causes dehydration, but not as quickly as drinking 35‰ seawater does. Over time, however, the brine will trickle down through the coarse structure of the sea ice, so its salinity decreases. By the time it is a year old, sea ice normally becomes relatively pure. Drinking melted sea ice enabled these early explorers to survive.

In addition, Figure 5.18 shows the average yearly amounts of transfer, or *flux*, of water between various reservoirs.

Dissolved Components Added to and Removed from Seawater

Seawater salinity is a function of the amount of dissolved components in seawater. Interestingly, dissolved substances do not remain in the ocean forever. Instead, they are cycled into and out of seawater by the processes shown in **Figure 5.19**. These processes include stream runoff, in which streams dissolve ions from continental rocks and carry them to the sea, and volcanic eruptions, both on the land and on the sea floor. Other sources include the atmosphere (which contributes gases) and biological interactions.

Stream runoff is the primary method by which dissolved substances are added to the oceans. Table 5.4 compares the major components dissolved in stream water with those in seawater. It shows that streams have far lower salinity and a vastly different composition of dissolved substances than seawater. For example, bicarbonate ion (HCO_3^-) is the most abundant dissolved constituent in stream water yet is found in only trace amounts in seawater. Conversely, the most abundant dissolved component in seawater is the chloride ion (Cl^-), which exists in very small concentrations in streams.

If stream water is the main source of dissolved substances in seawater, why do the components of the two not match each other more closely? One of the reasons is that some dissolved substances stay in the ocean and accumulate over time. **Residence time** is the average length of time that a substance resides in the ocean. Long residence times lead to higher concentrations of the dissolved substance. The sodium ion (Na^+) for instance, has a residence time of 260 million years, and, as a result, has a high concentration in the ocean. Other elements such as aluminum have a residence time of only 100 years and occur in seawater in much lower concentrations.

Are the oceans becoming saltier through time? This might seem logical since new dissolved components are constantly being added to the oceans and because most salts have long residence times. However, analysis of ancient marine organisms and sea floor sediments suggests that the oceans have not increased in salinity over time. This must be because the rate at which an element is added to the ocean equals the rate at which it is removed, so the average *amounts* of various elements remain constant (this is called a *steady-state* condition).

Materials added to the oceans are counteracted by several processes that cycle dissolved substances out of seawater. When waves break at sea, for example, sea spray releases tiny salt particles into the atmosphere, where they may be blown over land before being washed back to Earth. The amount of material leaving the ocean in this way is enor-

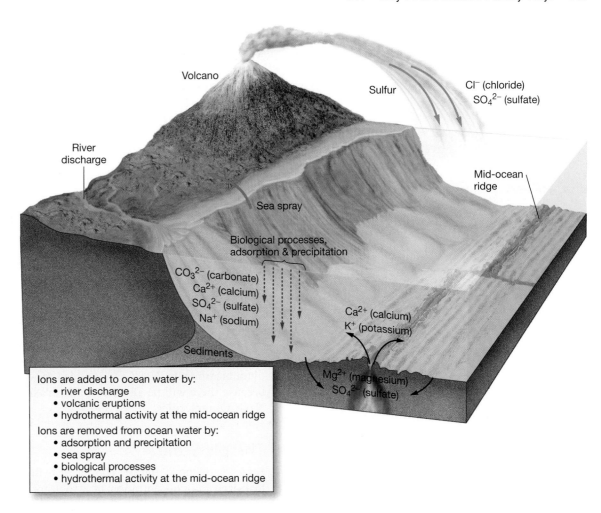

Ions are added to ocean water by:
• river discharge
• volcanic eruptions
• hydrothermal activity at the mid-ocean ridge

Ions are removed from ocean water by:
• adsorption and precipitation
• sea spray
• biological processes
• hydrothermal activity at the mid-ocean ridge

mous: According to a recent study, as much as 3.3 billion metric tons (3.6 billion short tons) of salt as sea spray enter the atmosphere each year. Another example is the infiltration of seawater along mid-ocean ridges near hydrothermal vents (see Figure 5.19), which incorporates magnesium and sulfate ions into sea floor mineral deposits. In fact, chemical studies of seawater indicate that the *entire volume of ocean water* is recycled through this hydrothermal circulation system at the mid-ocean ridge approximately every 3 million years. As a result, the chemical exchange between ocean water and the basaltic crust has a major influence on the composition of ocean water.

Dissolved substances are also removed from seawater in other ways. Calcium, carbonate, sulfate, sodium, and silicon are deposited in ocean sediments within the shells of dead microscopic organisms and animal feces. Vast amounts of dissolved substances can be removed when inland arms of seas dry up, leaving salt deposits called *evaporites* (such as those beneath the Mediterranean Sea; see Web Diving Deeper 4.1). In addition, ions dissolved in ocean water are removed by adsorption (physical attachment) to the surfaces of sinking clay and biological particles.

Figure 5.19 The cycling of dissolved components in seawater. Dissolved components are added to seawater primarily by river discharge and volcanic eruptions, while they are removed by adsorption, precipitation, ion entrapment in sea spray, and marine organisms that produce hard parts. Chemical reactions at the mid-ocean ridge add and remove various dissolved components.

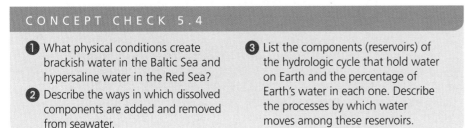

CONCEPT CHECK 5.4

❶ What physical conditions create brackish water in the Baltic Sea and hypersaline water in the Red Sea?

❷ Describe the ways in which dissolved components are added and removed from seawater.

❸ List the components (reservoirs) of the hydrologic cycle that hold water on Earth and the percentage of Earth's water in each one. Describe the processes by which water moves among these reservoirs.

5.5 Is Seawater Acidic or Basic?

An **acid** is a compound that releases hydrogen ions (H⁺) when dissolved in water. The resulting solution is said to be *acidic*. A strong acid readily and completely releases hydrogen ions when dissolved in water. An **alkaline**, or a **base**, is a compound that releases hydroxide ions (OH⁻) when dissolved in water. The resulting solution is said to be *alkaline*, or *basic*. A strong base readily and completely releases hydroxide ions when dissolved in water.

Both hydrogen ions and hydroxide ions are present in extremely small amounts at all times in water because water molecules dissociate and reform. Chemically, this is represented by the equation:

$$H_2O \overset{\text{dissociate}}{\underset{\text{reform}}{\rightleftharpoons}} H^+ + OH^- \tag{5.2}$$

Note that if the hydrogen ions and hydroxide ions in a solution are due only to the dissociation of water molecules, they are always found in equal concentrations, and the solution is consequently neutral.

When substances dissociate in water, they can make the solution acidic or basic. For example, if hydrochloric acid (HCl) is added to water, the resulting solution will be acidic because there will be a large excess of hydrogen ions from the dissociation of the HCl molecules. Conversely, if a base such as baking soda (sodium bicarbonate, $NaHCO_3$) is added to water, the resulting solution will be basic because there will be an excess of hydroxide ions (OH⁻) from the dissociation of $NaHCO_3$ molecules.

The pH Scale

Figure 5.20 shows the **pH** (power of hydrogen) **scale**, which is a measure of the hydrogen ion concentration of a solution. Values for pH range from 0 (strongly acidic) to 14 (strongly alkaline or basic), and the pH of a **neutral** solution such as pure water is 7.0. The pH scale is not linear: A decrease of 1.0 pH unit corresponds to a 10-fold increase in the concentration of hydrogen ions, making the water more acidic, whereas a change of 1.0 unit upward corresponds to a 10-fold decrease, making the water more alkaline.

Figure 5.20 The pH scale. The pH scale ranges from 0 (highly acidic) to 14 (highly alkaline). A pH of 7 is neutral; the pH values of common substances are also shown. Because pH is not a linear scale, a 1-unit change in pH represents a tenfold change in hydrogen ion concentration.

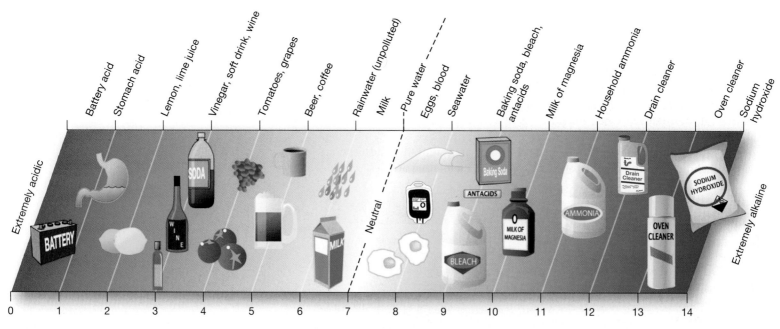

pH Values of Common Substances

Ocean surface waters have a pH that averages about 8.1 and ranges from about 8.0 to 8.3, so seawater is slightly alkaline. At depth, seawater pH is generally lower than surface waters (**Figure 5.21**). Water in the ocean combines with carbon dioxide to form a weak acid, called carbonic acid (H_2CO_3), which dissociates and releases hydrogen ions (H^+):

$$H_2O + CO_2 \longrightarrow H_2CO_3 \longrightarrow H^+ + HCO_3^- \tag{5.3}$$

This reaction would seem to make the ocean slightly acidic. Carbonic acid, however, keeps the ocean slightly alkaline through the process of *buffering*.

The Carbonate Buffering System

The chemical reactions in **Figure 5.22** show that carbon dioxide (CO_2) combines with water (H_2O) to form carbonic acid (H_2CO_3). Carbonic acid can then lose a hydrogen ion (H^+) to form the negatively charged bicarbonate ion (HCO_3^-). The bicarbonate ion can lose its hydrogen ion, too, though it does so less readily than carbonic acid. When the bicarbonate ion loses its hydrogen ion, it forms the double-charged negative carbonate ion (CO_3^{2-}), some of which combines with calcium ions to form calcium carbonate ($CaCO_3$). Some of the calcium carbonate is precipitated by various inorganic and organic means, and then it sinks and cycles back into the ocean by dissolving at depth.

The equations below Figure 5.22 show how these chemical reactions involving carbonate minimize changes in the pH of the ocean in a process called **buffering**. Buffering protects the ocean from getting too acidic or too basic, similarly to how buffered aspirin protects sensitive stomachs. For example, if the pH

STUDENTS SOMETIMES ASK . . .

If water molecules are so good at dissolving almost everything, then why does pure water have a neutral pH of 7.0?

Indeed, pure water's neutral pH might seem surprising in light of its tremendous ability to dissolve substances. Intuitively, it seems like water should be acidic and thus have a low pH. However, pH measures the amount of hydrogen ions (H^+) in solution, not the ability of a substance to dissolve by forming hydrogen bonds (as water molecules do).

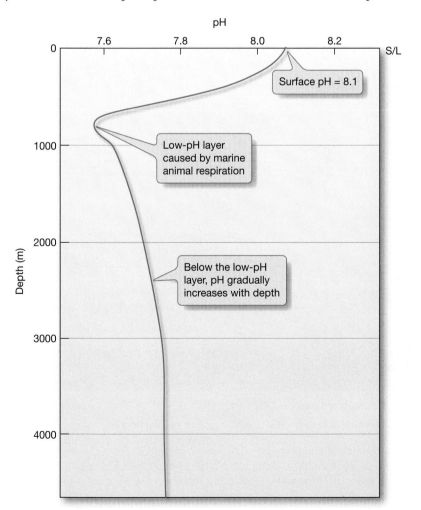

Figure 5.21 **Seawater pH varies with ocean depth.** The pH of ocean surface waters averages about 8.1, which is slightly alkaline. The low-pH layer is caused by marine animal respiration, which releases carbon dioxide into the water, making the water more acidic. Below the low-pH layer, pH gradually increases with depth.

Web Animation
The Carbonate Buffering System

of the ocean increases (becomes too basic), it causes H_2CO_3 to release H^+, and pH drops. Conversely, if the pH of the ocean decreases (becomes too acidic), HCO_3^- combines with H^+ to remove it, causing pH to rise. In this way, buffering prevents large swings of ocean water pH and allows the ocean to stay within a limited range of pH values. Recently, however, increasing amounts of carbon dioxide from human emissions are beginning to enter the ocean and change the ocean's pH, making it more acidic. For more details on this process, see Chapter 16, "The Oceans and Climate Change."

Deep-ocean water contains more carbon dioxide than surface water because deep water is cold and has the ability to dissolve more gases. Also, the higher pressures of the deep ocean further aid the dissolution of gases in seawater. If carbon dioxide combines with water to form carbonic acid, why isn't the cold water of the deep ocean highly acidic? When microscopic marine organisms that make their shells out of calcium carbonate (calcite) die and sink into the deep ocean, they neutralize the acid through buffering. In essence, these organisms act as an "antacid" for the deep ocean, analogous to the way commercial antacids use calcium carbonate to neutralize excess stomach acid. As explained in Chapter 4, these shells are readily dissolved below the calcite (calcium carbonate) compensation depth (CCD).

ESSENTIAL CONCEPT 5.5

Reactions involving carbonate chemicals serve to buffer the ocean and help maintain its average pH at 8.1 (slightly alkaline, or basic).

CONCEPT CHECK 5.5

1. Explain the difference between an acidic substance and an alkaline (basic) substance.

2. How does the ocean's buffering system work?

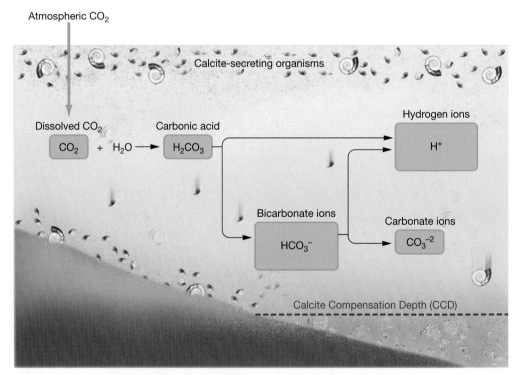

If seawater too basic: $H_2CO_3 \longrightarrow HCO_3^- + H^+$; pH drops
If seawater too acidic: $HCO_3^- + H^+ \longrightarrow H_2CO_3$; pH rises

Figure 5.22 The carbonate buffering system. Atmospheric carbon dioxide (CO_2) enters the ocean and undergoes chemical reactions. If seawater is too basic, chemical reactions occur that release H^+ into seawater and lower pH. If seawater is too acidic, chemical reactions occur that remove H^+ from seawater and cause pH to rise. Thus, buffering keeps the pH of seawater constant.

5.6　How Does Seawater Salinity Vary at the Surface and With Depth?

Average seawater salinity is 35%, but it varies significantly from place to place at the surface and also with depth.

Surface Salinity Variation

Figure 5.23 shows how salinity varies at the surface with latitude. The red curve shows temperature, which decreases at high latitudes and increases near the equator. The green curve shows salinity, which is lowest at high latitudes, peaks near the Tropics of Cancer and Capricorn, and dips near the equator.

Why does surface salinity vary in the pattern shown in Figure 5.23? At high latitudes, abundant precipitation and runoff and the melting of freshwater icebergs all decrease salinity. In addition, cool temperatures limit the amount of evaporation that takes place (which would increase salinity). The formation and melting of sea ice balance each other out in the course of a year and are not a factor in changes in salinity.

The pattern of Earth's atmospheric circulation (see Chapter 6, "Air–Sea Interaction") causes warm, dry air to descend at lower latitudes near the Tropics of Cancer and Capricorn, so evaporation rates are high and salinity increases. In addition, little precipitation and runoff occur to decrease salinity. As a result, the regions near the Tropics of Cancer and Capricorn are the continental *and* maritime deserts of the world.

Temperatures are warm near the equator, so evaporation rates are high enough to increase salinity. Increased precipitation and runoff partially offsets the high salinity, though. For example, daily rain showers are common along the equator, adding water to the ocean and lowering its salinity.

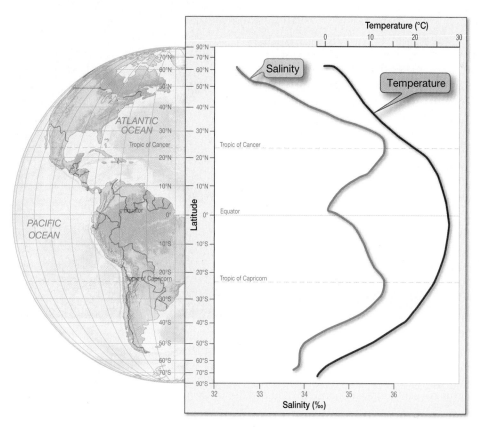

Figure 5.23　Surface salinity variation. Sea surface temperature (*red curve*) is lowest at the poles and highest at the equator. Surface seawater salinity (*green curve*) is lowest at the poles, peaks at the Tropics of Cancer and Capricorn, and dips near the equator. The presence of large amounts of runoff from land in far northern latitudes causes salinity to be lower there as compared to equivalent latitudes in the Southern Hemisphere.

STUDENTS SOMETIMES ASK . . .

Why do carbonated beverages burn my throat when I drink them?

When carbon dioxide gas (CO_2) dissolves in water (H_2O), its molecules often cling to water molecules and form carbonic acid (H_2CO_3). Carbonic acid is a weak acid, in which most molecules are intact at any given moment. However, some of those molecules naturally break apart and exist as two fragments: a negatively charged $H_2CO_3^-$ ion and a positively charged H^+ ion. The H^+ ions are responsible for acidity—the higher their concentration in a solution, the more acidic that solution. The presence of carbonic acid in carbonated water makes that water acidic—the more carbonated, the more acidic. What you're feeling when you drink a carbonated beverage is the moderate acidity of that beverage irritating your throat.

Figure 5.24 is a map of satellite-collected data that shows how ocean surface salinity varies worldwide. Notice how the overall pattern of the satellite image matches the graph in Figure 5.23. For example, both the graph and the satellite image show high salinity in the subtropics (Figure 5.24, *orange*) and lower salinity in rainy polar regions and equatorial belts (Figure 5.24, *blue*). In addition, notice that the Atlantic Ocean has higher salinity values than the Pacific. The Atlantic Ocean's higher overall salinity is caused by its proximity to land, which experiences continental effect. This causes high rates of evaporation in the narrower Atlantic Ocean, particularly in the tropics. The satellite image also reveals an expanse of low-salinity water from the Amazon River's outflow (Figure 5.24, *purple*).

Salinity Variation with Depth

Figure 5.25 shows how seawater salinity varies with depth. The graph displays data for the open ocean far from land and shows one curve for high-latitude regions and one for low-latitude regions.

For low-latitude regions (such as in the tropics), the curve begins at the surface, with relatively high salinity (as discussed in the preceding section). Note that even at the equator where salinity dips, surface salinity is still relatively high. Then with increasing depth, the curve swings toward an intermediate salinity value.

For high-latitude regions (such as near Antarctica or in the Gulf of Alaska), the curve begins at the surface, with relatively low salinity (again, see the discussion in the preceding section). With increasing depth, the curve also swings toward an intermediate salinity value that approaches the value of the low-latitude salinity curve at the same depth.

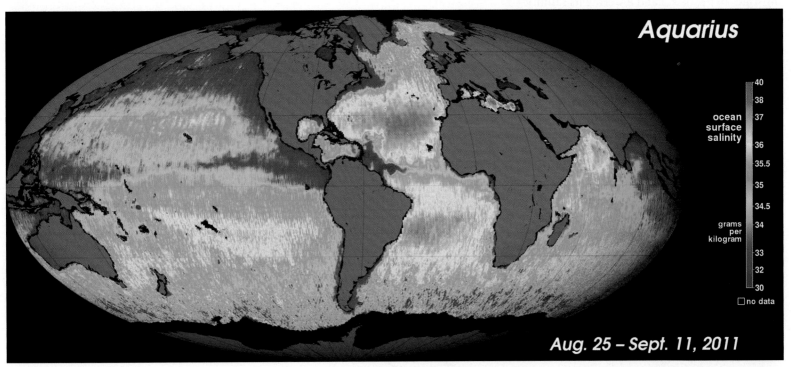

Figure 5.24 Surface salinity of the oceans. Satellite-derived ocean surface salinity map from data collected by the Aquarius satellite during August 25 to September 11, 2011. The map shows that the lowest salinities (*purple*, *blue*) occur near land, in high latitudes, and along the equatorial belt, while the highest salinities (*red*, *orange*) occur in the subtropics. Values in grams per kilogram or parts per thousand (‰) by weight; black regions indicate no data.

These two curves, which together resemble the outline of a wide Champagne glass, show that salinity varies widely at the surface but hardly varies at all in the deep ocean. Why is this so? It occurs because all the processes that affect seawater salinity (precipitation, runoff, melting icebergs, melting sea ice, sea ice forming, and evaporation) occur at the *surface* and thus have no effect on deep water below.

Halocline

Both curves in Figure 5.25 show a rapid change in salinity between the depths of about 300 meters (980 feet) and 1000 meters (3300 feet). For the low-latitude curve, the change is a *decrease* in salinity. For the high-latitude curve, the change is an *increase* in salinity. In both cases, this layer of rapidly changing salinity with depth is called a **halocline** (*halo* = salt, *cline* = slope). Haloclines separate layers of different salinity in the ocean.

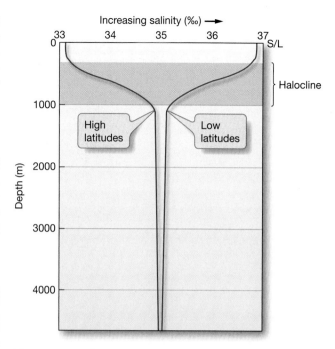

Figure 5.25 Salinity variation with depth. Vertical open-ocean profile showing high- and low-latitude salinity variation (horizontal scale in ‰) with depth (vertical scale in meters, with sea level at the top). The layer of rapidly changing salinity with depth is the halocline.

CONCEPT CHECK 5.6

1 Why is there low surface salinity in the high latitudes, and why is there higher surface salinity in the low latitudes?

2 Explain the dip in surface salinity values near the equator that is shown in Figure 5.23.

3 Describe the halocline, including where it occurs in the ocean.

ESSENTIAL CONCEPT 5.6

A halocline is a layer of rapidly changing salinity that occurs in both high- and low-latitude regions.

5.7 How Does Seawater Density Vary With Depth?

The density of pure water is 1.000 gram per cubic centimeter (g/cm³) at 4°C (39°F). This value serves as a standard against which the density of all other substances can be measured. Seawater contains various dissolved substances that increase its density. In the open ocean, seawater density averages between 1.022 and 1.030 g/cm³ (depending on its salinity). Thus, the density of seawater is 2 to 3% greater than that of pure water. Unlike freshwater, seawater continues to increase in density until it freezes at a temperature of −1.9°C (28.6°F). (Recall that below 4°C [39°F], the density of freshwater actually *decreases*; see Figure 5.12.) At its freezing point, however, seawater behaves in a similar fashion to freshwater: Its density decreases dramatically, which is why sea ice floats, too.

Density is an important property of ocean water because density differences determine the vertical position of ocean water and cause water masses to float or sink, thereby creating deep-ocean currents. For example, if seawater with a density of 1.030 g/cm³ were added to freshwater with a density of 1.000 g/cm³, the denser seawater would sink below the freshwater, initiating a deep current.

Factors Affecting Seawater Density

The ocean, like Earth's interior, is layered according to density. Low-density water exists near the surface, and higher-density water occurs below. Except for some shallow inland seas with high rates of evaporation that create high-salinity water, the highest-density water is found at the deepest ocean depths. Let's examine how temperature, salinity, and pressure influence seawater density by expressing the relationships using arrows (up arrow = increase; down arrow = decrease):

- As temperature increases (↑), seawater density decreases (↓) [11] (due to thermal expansion).
- As salinity increases (↑), seawater density increases (↑) (due to the addition of more dissolved material).
- As pressure increases (↑), seawater density increases (↑) (due to the compressive effects of pressure).

Of these three factors, only temperature and salinity influence the density of surface water. Pressure influences seawater density only when very high pressures are encountered, such as in deep-ocean trenches. Still, the density of seawater in the deep ocean is only about 5% greater than at the ocean surface, showing that despite tons of pressure per square centimeter, water is nearly incompressible. Unlike air, which can be compressed and put in a tank for use in scuba diving, the molecules in liquid water are already close together and cannot be compressed much more. Therefore, pressure has the least effect on influencing the density of surface water and can largely be ignored.

Temperature, on the other hand, has the greatest influence on surface seawater density because the range of surface seawater temperature is greater than that of salinity. In fact, only in the extreme polar areas of the ocean, where temperatures are low and remain relatively constant, does salinity significantly affect density. Cold water that also has high salinity is some of the highest-density water in the world. The density of seawater—the result of its salinity and temperature—influences currents in the deep ocean because high-density water sinks below less-dense water.

Temperature and Density Variation with Depth

The four graphs in **Figure 5.26** show how seawater temperature and density vary with depth in both low-latitude and high-latitude regions. Let's examine each graph individually.

Figure 5.26a shows how temperature varies with depth in low-latitude regions, where surface waters are warmed by high Sun angles and constant length of days. However, the Sun's energy does not penetrate very far into the ocean. Surface water temperatures remain relatively constant until a depth of about 300 meters (980 feet) because of good surface mixing mechanisms such as surface currents, waves, and tides. Below 300 meters (980 feet), the temperature decreases rapidly until a depth of about 1000 meters (3300 feet). Below 1000 meters, the water's low temperature again remains constant down to the ocean floor.

Figure 5.26b shows how temperature varies with depth in high-latitude regions, where surface waters remain cool year-round and deep-water temperatures are about the same as the surface. The temperature curve for high-latitude regions, therefore, is a straight vertical line, which indicates uniform conditions at the surface and at depth.

The density curve for low-latitude regions in Figure 5.26c shows that density is relatively low at the surface. Density is low because surface water temperatures are high. (Remember that temperature has the greatest influence on density and temperature is inversely proportional to density.) Below the surface, density remains constant also until a depth of about 300 meters (980 feet) because of good surface mixing. Below 300 meters (980 feet), the density increases rapidly until a depth of about 1000 meters (3300 feet). Below 1000 meters, the water's low density again remains constant down to the ocean floor.

---

[11] A relationship where one variable *decreases* as a result of another variable's *increase* is known as an inverse relationship, in which the two variables are *inversely proportional*.

The density curve for high-latitude regions (Figure 5.26d) shows very little variation with depth. Density is relatively high at the surface because surface water temperatures are low. Density is high below the surface, too, because water temperature is also low. The density curve for high-latitude regions, therefore, is a straight vertical line, which indicates uniform conditions at the surface and at depth. These conditions allow cold high-density water to form at the surface, sink, and initiate deep-ocean currents.

Temperature is the most important factor influencing seawater density, so the temperature graphs (Figure 5.26a and Figure 5.26b) strongly resemble the corresponding density graphs (Figure 5.26c and Figure 5.26d, respectively). The only difference is that they are *a mirror image of each other*, illustrating that temperature and density are inversely proportional to one another.

Thermocline and Pycnocline

Analogous to the halocline (the layer of rapidly changing salinity shown in Figure 5.25), the low-latitude temperature graph in Figure 5.26a displays a curving line that indicates a layer of rapidly changing temperature called a **thermocline** (*thermo* = heat, *cline* = slope). Similarly, the low-latitude density graph in Figure 5.26c displays a curving line that indicates a **pycnocline** (*pycno* = density, *cline* = slope), which is a layer of rapidly changing density. Note that the high-latitude graphs of temperature (Figure 5.26b) and density (Figure 5.26d) lack both a thermocline and a pycnocline, respectively, because these lines show a constant value with depth (they are straight lines and don't curve). Like a halocline, a thermocline and a pycnocline typically occur between about 300 meters (980 feet) and 1000 meters (3300 feet) below the surface. The temperature difference between water above and below the thermocline can be used to generate electricity (see **Web Diving Deeper 5.1**).

When a pycnocline is established in an area, it presents an incredible barrier to mixing between low-density water above and high-density water below. A pycnocline has a high gravitational stability and thus physically isolates adjacent layers of water.[12] The pycnocline results from the combined effect of the thermocline and the halocline because temperature and salinity influence density. The interrelationship

Figure 5.26 Temperature and density variations with depth. Graphs showing temperature and density profiles in low and high latitudes. Note the inverse relationship between temperature and density by comparing graph *a* with graph *c* (low latitudes) and graph *b* with graph *d* (high latitudes).

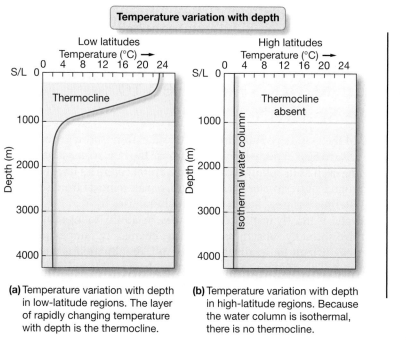

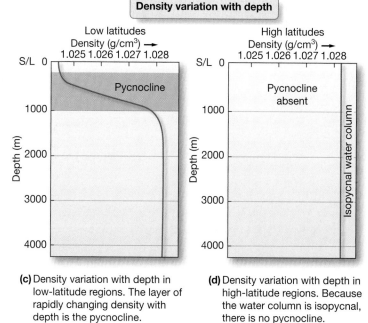

(a) Temperature variation with depth in low-latitude regions. The layer of rapidly changing temperature with depth is the thermocline.

(b) Temperature variation with depth in high-latitude regions. Because the water column is isothermal, there is no thermocline.

(c) Density variation with depth in low-latitude regions. The layer of rapidly changing density with depth is the pycnocline.

(d) Density variation with depth in high-latitude regions. Because the water column is isopycnal, there is no pycnocline.

[12]This is similar to a temperature inversion in the atmosphere, which traps cold (high-density) air underneath warm (low-density) air.

of these three layers determines the degree of separation between the upper-water and deep-water masses.

There are essentially three distinct water masses based on density. The **mixed surface layer** occurs above a strong permanent thermocline (and corresponding pycnocline). The water is uniform because it is well mixed by surface currents, waves, and tides. The thermocline and pycnocline occur in a relatively low-density layer called the **upper water**, which is well developed throughout the low and middle latitudes. Denser and colder **deep water** extends from below the thermocline/pycnocline to the deep-ocean floor.

At depths above the main thermocline, divers often experience lesser thermoclines (and corresponding pycnoclines) when descending into the ocean. Thermoclines can develop in swimming pools, ponds, and lakes, too. During the spring and fall, when nights are cool but days can be quite warm, the Sun heats the surface water of the pool, yet the water below the surface can be quite cold. If the pool has not been mixed, a thermocline isolates the warm surface layer from the deeper cold water. The cold water below the thermocline can be quite a surprise for anyone who dives into the pool!

In high-latitude regions, the temperature of the surface water remains cold year round, so there is very little difference between the temperature at the surface and in deep water below. Thus, a thermocline and corresponding pycnocline rarely develop in high-latitude regions. Only during the short summer when the days are long does the Sun begin to heat surface waters. Even then, the water does not heat up very much. Nearly all year, then, the water column in high latitudes is **isothermal** (*iso* = same, *thermo* = heat) and **isopycnal** (*iso* = same, *pycno* = density), allowing good vertical mixing between surface and deeper waters.

CONCEPT CHECK 5.7

1. What are the three factors that affect seawater density? Describe how each factor influences seawater density, including which one is the most important.

2. Describe the thermocline, including where it occurs in the ocean.

3. Describe the pycnocline, including where it occurs in the ocean.

4. Why is there such a close association between (a) the curve showing seawater density variation with ocean depth and (b) the curve showing seawater temperature variation with ocean depth?

5.8 What Methods Are Used to Desalinate Seawater?

More than a third of the world's population already suffers from shortages of drinkable water—with a rise to 50% expected by 2025. Because the human consumption of freshwater is growing even as its supply is dwindling, several countries have begun to use the ocean as a source of freshwater. **Desalination**, or salt removal from seawater, can provide freshwater for business, home, and agricultural use.

Although seawater is mostly just water molecules, its ability to form hydrogen bonds, easily dissolve so many substances, and resist changes in temperature and state makes seawater difficult to desalinate. As a result, desalination is energy intensive and expensive. The high cost of desalination, however, is only one issue. Recent studies, for example, indicate that desalination can negatively affect marine life by entrapping them in intake pipes and by releasing highly salty leftover brine back to the ocean. Still, using the sea as a source of freshwater is attractive to many coastal communities that have few other sources.

Currently, there are more than 13,000 desalination plants worldwide, the majority of them very small and located in arid regions of the Middle East, Caribbean, and Mediterranean. These plants produce more than 45 billion liters (12 billion gallons) of freshwater daily. The United States produces only about 10% of the world's desalted water, primarily in Florida. To date, only a limited number of desalination plants have been built along the California coast, primarily because the cost of desalination is generally higher than the costs of other water supply alternatives available in California (such as water transfers and groundwater pumping) but also because the extensive permitting process is an impediment to building desalination facilities. However, as drought conditions occur and concern over water availability increases, desalination projects are being proposed at numerous locations in the state.

Because desalinated water requires a lot of energy and thus is expensive to produce, most desalination plants are small-scale operations. In fact, desalination plants provide less than 0.5% of human water needs. More than half of the world's desalination plants use *distillation* to purify water, while most of the remaining plants use *membrane processes*.

Distillation

The process of **distillation** (*distillare* = to trickle) is shown schematically in **Figure 5.27**. In distillation, saltwater is boiled, and the resulting water vapor is passed through a cooling condenser, where it condenses and is collected as freshwater. This simple procedure is very efficient at purifying seawater. For instance, distillation of 35% seawater produces freshwater with a salinity of only 0.03%, which is about 10 times fresher than bottled water, so it needs to be mixed with less pure water to make it taste better. Distillation is expensive, however, because it requires large amounts of heat energy to boil the saltwater. Because of water's high latent heat of vaporization, it takes 540 calories to convert only 1 gram (0.035 ounce) of water at the boiling point to the vapor state.[13] Increased efficiency, such as using the waste heat from a power plant, is required to make distillation practical on a large scale.

Solar distillation, which is also known as **solar humidification**, does not require supplemental heating and has been used successfully in small-scale agricultural experiments in arid regions such as Israel, West Africa, and Peru. Solar humidification is similar to distillation in that saltwater is evaporated in a covered container, but the water is heated by direct sunlight instead (Figure 5.27). Saltwater in the container evaporates, and the water vapor that condenses on the cover runs into collection trays. The major difficulty lies in effectively concentrating the energy of sunlight into a small area to speed evaporation.

Membrane Processes

Electrolysis can be used to desalinate seawater, too. In this method, two volumes of freshwater—one containing a positive electrode and the other a negative electrode—are placed on either side of a volume of seawater. The seawater is separated from each of the freshwater reservoirs by semipermeable membranes. These membranes are permeable to salt ions but not to water molecules. When an electrical current is applied, positive ions such as sodium ions are attracted to the negative electrode, and negative ions such as chloride ions are attracted to the positive electrode. In time, enough ions are removed through the membranes to convert the seawater to freshwater. The major drawback to electrolysis is that it requires large amounts of energy.

Reverse osmosis (*osmos* = to push) may have potential for large-scale desalination. In osmosis, water molecules naturally pass through a thin, semipermeable membrane from a freshwater solution to a saltwater solution. In reverse osmosis, water on the salty side is highly pressurized to drive water molecules—but not salt and other impurities—through the membrane to the freshwater side (**Figure 5.28**). A significant problem with reverse osmosis is that the membranes are flimsy, become clogged, and must be replaced frequently. Advanced composite materials may help eliminate these problems because they are sturdier, provide better filtration, and last up to 10 years.

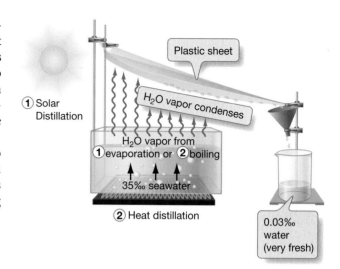

Figure 5.27 Distillation. The process of distillation requires boiling saltwater (*heat distillation*) or using the Sun's energy to evaporate seawater (*solar distillation*). In either case, the water vapor is captured and condensed in a process that produces very pure water.

---

[13]Even at 100% efficiency, it still requires a whopping *540,000 calories* of heat energy to produce 1 liter (about 1 quart) of distilled water.

Worldwide, at least 30 countries are operating reverse osmosis units. Saudi Arabia—where energy from oil is cheap but water is scarce—has the world's largest reverse osmosis plant, which produces 485 million liters (128 million gallons) of desalinated water daily. The largest plant in the United States opened in 2008 in Tampa Bay, Florida, and produces up to 95 million liters (25 million gallons) of freshwater per day, which provides about 10% of the drinking water supply of the Tampa Bay region. Once permits are obtained, a new facility in Carlsbad, California, is designed to produce twice as much freshwater as the Tampa Bay plant. Reverse osmosis is also used in many household water purification units and aquariums.

Other Methods Of Desalination

Seawater selectively excludes dissolved substances as it freezes—a process called **freeze separation**. As a result, the salinity of sea ice (once it is melted) is typically 70% lower than seawater. To make this an effective desalination technique, though, the water must be frozen and thawed multiple times, with the salts washed from the ice between each thawing. Like electrolysis, freeze separation requires large amounts of energy, so it may be impractical except on a small scale.

Yet another way to obtain freshwater is to melt naturally formed ice. Imaginative thinkers have proposed towing large icebergs to coastal waters off countries that need freshwater. Once there, the freshwater produced as the icebergs melt could be captured and pumped ashore. Studies have shown that towing large Antarctic icebergs to arid regions would be technologically feasible and, for certain Southern Hemisphere locations, economically feasible, too.

Other novel approaches to desalination include crystallization of dissolved components directly from seawater, solvent demineralization using chemical catalysts, and even making use of salt-eating bacteria!

ESSENTIAL CONCEPT 5.8

Although desalination of seawater is costly, desalination plants use distillation, solar humidification, electrolysis, freeze separation, and reverse osmosis to purify seawater for domestic use.

CONCEPT CHECK 5.8

❶ Why is seawater desalination so expensive?

❷ List and describe the two main methods of seawater desalination.

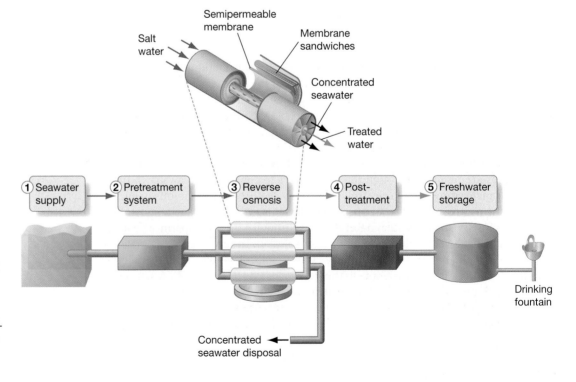

Figure 5.28 Reverse osmosis. Diagrammatic view of the steps involved in using reverse osmosis to produce freshwater. The enlargement shows that seawater is forced through a semipermeable membrane with microscopically fine holes that allow water molecules to pass but trap the salts.

Essential Concepts Review

5.1 Why does water have such unusual chemical properties?

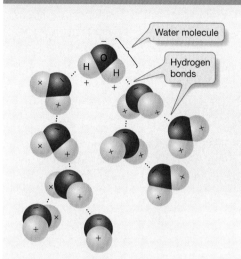

Water molecule

Hydrogen bonds

▶ *Water's remarkable properties help make life as we know it possible on Earth.* These properties include the arrangement of its atoms, how its molecules stick together, its ability to dissolve almost everything, and its heat storage capacity.

▶ *The water molecule is composed of one atom of oxygen and two atoms of hydrogen* (H_2O). The two hydrogen atoms, which are covalently bonded to the oxygen atom, are *attached to the same side of the oxygen atom* and produce a bend in the geometry of a water molecule. This geometry makes water molecules *polar*, which allows them to form hydrogen bonds with other water molecules or other substances and gives water its remarkable properties. Water, for example, is *"the universal solvent"* because it can hydrate charged particles (ions), thereby dissolving them.

Study Resources
Online Study Guide Quizzes, Web Video

Critical Thinking Question

Using chemical principles, explain why water is considered "the universal solvent."

5.2 What other important properties does water possess?

▶ *Water is one of the few substances that exists naturally on Earth in all three states of matter (solid, liquid, and gas).* Hydrogen bonding gives water *unusual thermal properties*, such as a high freezing point (0°C [32°F]) and boiling point (100°C [212°F]), a high heat capacity and high specific heat (1 calorie per gram), a high latent heat of melting (80 calories per gram), and a high latent heat of vaporization (540 calories per gram). Water's high heat capacity and latent heats have important implications in regulating global thermostatic effects.

▶ Like most other chemical substances, *the density of water increases as temperature decreases* and reaches a maximum density at 4°C (39°F). *Below 4°C, however, water density decreases with temperature*, due to the formation of bulky ice crystals. *As water freezes, it expands by about 9% in volume, so ice floats* on water.

Study Resources
Online Study Guide Quizzes, Web Table 5.1, Web Animations

Critical Thinking Question

Describe the differences between the three states of matter, using the arrangement of molecules in your explanation.

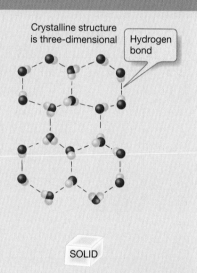

Crystalline structure is three-dimensional

Hydrogen bond

SOLID

5.3 How salty is seawater?

BATTERY

Freshwater

Saltwater

▶ *Salinity is the amount of dissolved solids in ocean water.* It averages about 35 grams of dissolved solids per kilogram of ocean water (*35 parts per thousand* [‰]) but ranges from brackish to hypersaline. Six ions—chloride, sodium, sulfate, magnesium, calcium, and potassium—account for over 99% of the dissolved solids in ocean water. These ions always occur in a *constant proportion* in any seawater sample, so salinity can be determined by measuring the concentration of only one—typically, the chloride ion.

▶ *The physical properties of pure water and seawater are remarkably similar,* with a few notable exceptions. Compared to pure water, seawater has a *higher pH, density, and boiling point* (but a *lower freezing point*).

Study Resources
Online Study Guide Quizzes, Web Diving Deeper 5.2, Web Video

Critical Thinking Question

What is your state sales tax, in parts per thousand?

Essential Concepts Review (cont.)

5.4 Why does seawater salinity vary?

▶ *Dissolved components in seawater are added and removed by a variety of processes.* Precipitation, runoff, and the melting of icebergs and sea ice add freshwater to seawater and decrease its salinity. The formation of sea ice and evaporation remove freshwater from seawater and increase its salinity. The hydrologic cycle includes all the reservoirs of water on Earth, including the *oceans*, which contain *97% of Earth's water.* The residence time of various elements indicates how long they stay in the ocean and implies that *ocean salinity has remained constant through time.*

Study Resources
Online Study Guide Quizzes, Web Animation

Critical Thinking Question

What evidence is used to support the hypothesis that ocean salinity has remained constant through time?

5.5 Is seawater acidic or basic?

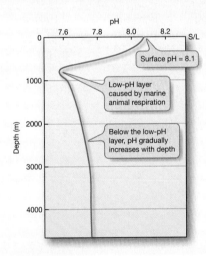

▶ Ocean surface water has an *average pH of 8.1, making it slightly alkaline,* but the pH of seawater varies both at the surface and at depth. *A natural buffering system* based on the chemical reaction of carbon dioxide in water *exists in the ocean.* This buffering system *regulates any changes in pH,* creating a stable ocean environment.

Study Resources
Online Study Guide Quizzes, Web Animation

Critical Thinking Question

What is the pH scale? Use examples of common household items and include their pH values.

5.6 How does seawater salinity vary at the surface and with depth?

▶ *The salinity of surface water varies considerably due to surface processes,* with the maximum salinity found near the Tropics of Cancer and Capricorn and the minimum salinity found in high-latitude regions. Salinity also varies with depth down to about 1000 meters (3300 feet), but below that the salinity of deep water is very consistent. *A halocline is a layer of rapidly changing salinity.*

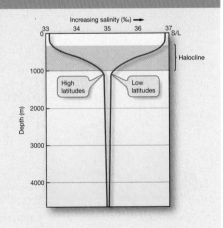

Study Resources
Online Study Guide Quizzes

Critical Thinking Question

Using the processes that affect seawater salinity, explain why there is such a large range of salinity variation at the surface but such a narrow range of salinity at depth.

5.7 How does seawater density vary with depth?

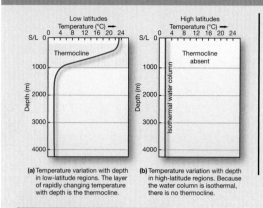

(a) Temperature variation with depth in low-latitude regions. The layer of rapidly changing temperature with depth is the thermocline.

(b) Temperature variation with depth in high-latitude regions. Because the water column is isothermal, there is no thermocline.

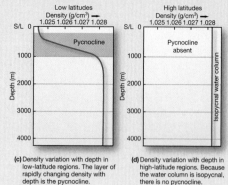

(c) Density variation with depth in low-latitude regions. The layer of rapidly changing density with depth is the pycnocline.

(d) Density variation with depth in high-latitude regions. Because the water column is isopycnal, there is no pycnocline.

▶ *Seawater density increases as temperature decreases and salinity increases*, though temperature influences surface seawater density more strongly than salinity (the influence of pressure is negligible). Temperature and density vary considerably with depth in low-latitude regions, creating a *thermocline* (*layer of rapidly changing temperature*) and corresponding *pycnocline* (*layer of rapidly changing density*), both of which are generally absent in high latitudes.

Study Resources
Online Study Guide Quizzes

Critical Thinking Question

Seawater density is inversely proportional to and largely controlled by its temperature. Using plain English, explain how an inversely proportional relationship works.

5.8 What methods are used to desalinate seawater?

▶ Although *desalination of seawater* is costly, it *provides freshwater* for business, home, and agricultural use. *Distillation, solar humidification, electrolysis, reverse osmosis,* and *freeze separation* are methods currently used to desalinate seawater.

Study Resources
Online Study Guide Quizzes,
Web Diving Deeper 5.1

Critical Thinking Question

Compare and contrast the following seawater desalination methods: distillation, solar humidification, electrolysis, and reverse osmosis.

MasteringOceanography™
www.masteringoceanography.com

Looking for additional review and test prep materials? Visit the Study Area in MasteringOceanography to enhance your understanding of this chapter's content by accessing a variety of resources, including Self-Study Quizzes, Geoscience Animations, RSS feeds, flashcards, Web links, and an optional Pearson eText.

Iceberg above and below water. Composite image of an iceberg, showing that 90% of an iceberg's mass is below water. Interactions between sea ice, the ocean, and the atmosphere help regulate Earth's climate.

6

Before you begin reading this chapter, use the glossary at the end of this book to discover the meanings of any of these words you don't already know:

Trade winds **Weather** Solstice Equinox
Horse latitudes Polar front Eye of the hurricane Hadley cell
Prevailing westerly wind belt **Air mass** Seasons
Tropical cyclone **Doldrums** **Polar easterly wind belt** Sea ice
Climate Hurricane Storm surge **Convection cell Coriolis effect**
Troposphere Intertropical Convergence Zone (ITCZ) ferrel cell Iceberg
Polar high pressure

AIR–SEA INTERACTION

ESSENTIAL CONCEPTS

At the end of this chapter, you should be able to:

6.1 Explain variations in solar radiation on Earth, including the cause of Earth's seasons

6.2 Describe the physical properties of the atmosphere

6.3 Demonstrate an understanding of the Coriolis effect

6.4 Explain global atmospheric circulation patterns

6.5 Describe oceanic weather and climate patterns, including hurricanes

6.6 Describe how sea ice and icebergs are formed

6.7 Discuss advantages and disadvantages of harnessing winds as a source of energy

Down dropt the breeze, the sails dropt down,
'Twas sad as sad could be;
And we did speak only to break
The silence of the sea!

Day after day, day after day,
We stuck, nor breath nor motion;
As idle as a painted ship
Upon a painted ocean.

> *—Samuel Taylor Coleridge, about ships*
> *getting stuck in the horse latitudes,*
> Rime of the Ancient Mariner *(1798)*

One of the most remarkable things about our planet is that the atmosphere and the ocean act as one interdependent system. Observations of the atmosphere–ocean system show that what happens in one causes changes in the other. Further, the two parts of this system are linked by complex feedback loops, some of which reinforce a change and others that nullify any changes. Surface currents in the oceans, for instance, are a direct result of Earth's atmospheric wind belts. Conversely, certain atmospheric weather phenomena are manifested in the oceans. In order to understand the behavior of the atmosphere and the oceans, their mutual interactions and relationships must be examined.

Solar energy heats the surface of Earth and creates atmospheric winds, which, in turn, drive most of the surface currents and waves in the ocean. Radiant energy from the Sun, therefore, is responsible for motion in the atmosphere and the ocean. In fact, variations in solar radiation drive the global ocean–atmosphere engine, creating pressure and density differences that stir currents and waves in both the atmosphere and the ocean. Recall from Chapter 5 that the atmosphere and ocean use the high heat capacity of water to constantly exchange this energy, shaping Earth's global weather patterns in the process.

Periodic extremes of atmospheric weather, such as droughts and profuse precipitation, are related to periodic changes in oceanic conditions. For instance, it was recognized as far back as the 1920s that El Niño—an ocean event—is tied to catastrophic weather events worldwide. What is as yet unclear, however, is if changes in the ocean produce changes in the atmosphere that lead to El Niño conditions—or vice versa. El Niño–Southern Oscillation events are discussed in Chapter 7, "Ocean Circulation."

Air–sea interactions have important implications in global warming, too. A multitude of recent studies have confirmed that the atmosphere is experiencing unprecedented warming as a result of human-caused emissions of carbon dioxide and other gases that absorb and trap heat in the atmosphere. This atmospheric heat is being transferred to the oceans and has the potential to cause widespread marine ecosystem changes. This issue is discussed in Chapter 16, "The Oceans and Climate Change."

In this chapter, we'll examine the redistribution of solar heat by the atmosphere and its influence on oceanic conditions. First, large-scale phenomena that influence air–sea interactions are studied, and then smaller-scale phenomena are examined.

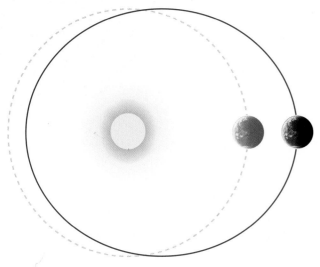

Figure 6.1 Earth's elliptical orbit. As Earth orbits the Sun, it follows a slightly elliptical path (*red line*) that is not perfectly circular (*yellow dashed line*). Earth's elliptical orbit is not the cause of Earth's seasons (see Figure 6.2). View is directly above Earth's plane of the ecliptic; Earth's elliptical orbit is exaggerated for clarity (not to scale).

Web Animation
Earth-Sun Relations

6.1 What Causes Variations in Solar Radiation on Earth?

A variety of factors cause changes in the amount of solar radiation (solar energy) that Earth receives. One of the most striking examples is the daytime–nighttime cycle: The side of Earth facing the Sun (the daytime side) receives a tremendous dose of intense solar radiation, while the nighttime side receives none. An example of a longer-term cycle is the change in seasons.

What Causes Earth's Seasons?

This seemingly simple question is the source of a common misconception: Even though Earth does revolve around the Sun in an elliptical orbit that varies only slightly from a perfect circle (**Figure 6.1**), Earth's seasons are not caused by Earth's changing distance from the Sun. The surface connecting all points in Earth's orbit is called the **plane of the ecliptic** (**Figure 6.2**). More importantly, Figure 6.2 shows that Earth's axis of rotation is not perpendicular ("upright") relative to the plane of the ecliptic; rather, it tilts at an angle of 23.5 degrees. As a result, *different hemispheres on Earth are tilted more directly toward or away from the Sun* during Earth's yearly orbit (Figure 6.2, *insets*), which is the cause of Earth's seasons (not Earth's elliptical orbit). An interesting consequence of Earth's tilt is that throughout its yearly cycle, Earth's axis *always points in the same direction*, which is toward Polaris, the North Star.

So, the tilt of Earth's rotational axis—not its elliptical orbit—is what causes Earth to have seasons. Let's examine a yearly progression of the seasons from spring, summer, fall, to winter:

- At the **vernal equinox** (*vernus* = spring; *equi* = equal, *noct* = night), which occurs on or about March 21, the Sun is directly overhead along the equator.

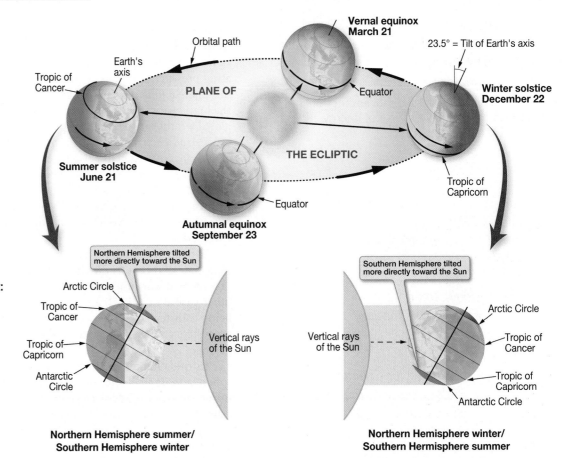

Figure 6.2 Perspective view of Earth's orbit: Why Earth has seasons. As Earth orbits the Sun during a year, its axis of rotation constantly tilts 23.5 degrees from perpendicular (relative to the plane of the ecliptic) and always points in the same direction, causing different areas to experience vertical rays of the Sun. This tilt causes Earth to have seasons, such as when the Northern Hemisphere is tilted toward the Sun during the summer (*left inset*) and away from the Sun during the winter (*right inset*).

During this time, all places in the world experience equal lengths of night and day (hence the name *equinox*). In the Northern Hemisphere, the vernal equinox is also known as the *spring equinox*.

- At the **summer solstice** (*sol* = the Sun, *stitium* = a stoppage), which occurs on or about June 21, the Sun reaches its most northerly point in the sky, directly overhead along the **Tropic of Cancer**, at 23.5 degrees north latitude (Figure 6.2, *left inset*). To an observer on Earth, the noonday Sun reaches its northernmost or southernmost position in the sky at this time and appears to pause—hence the term *solstice*—before beginning its next six-month cycle.

- At the **autumnal** (*autumnus* = fall) **equinox**, which occurs on or about September 23, the Sun is directly overhead along the equator again. In the Northern Hemisphere, the autumnal equinox is also known as the *fall equinox*.

- At the **winter solstice**, which occurs on or about December 22, the Sun is directly overhead along the **Tropic of Capricorn**, at 23.5 degrees south latitude (Figure 6.2, *right inset*). In the Southern Hemisphere, the seasons are reversed. Thus, the winter solstice is the time when the Southern Hemisphere is most directly facing the Sun, which is the beginning of the Southern Hemisphere summer.

Because Earth's rotational axis is tilted 23.5 degrees, the Sun's **declination** (angular distance from the equatorial plane) varies between 23.5 degrees north and 23.5 degrees south of the equator on a yearly cycle. As a result, the region between these two latitudes (called the **tropics**) receives much greater annual radiation than polar areas.

Seasonal changes in the angle of the Sun and the length of day profoundly influence Earth's climate. In the Northern Hemisphere, for example, the longest day occurs on the summer solstice and the shortest day on the winter solstice.

Daily heating of Earth also influences climate in most locations. Exceptions to this pattern occur north of the **Arctic Circle** (66.5 degrees north latitude) and south of the **Antarctic Circle** (66.5 degrees south latitude), which at certain times of the year do not experience daily cycles of daylight and darkness. For instance, during the Northern Hemisphere winter, the area north of the Arctic Circle receives no direct solar radiation at all and experiences up to six months of darkness. At the same time, the area south of the Antarctic Circle receives continuous radiation ("midnight Sun"), so it experiences up to six months of light. Half a year later, during the Northern Hemisphere summer (the Southern Hemisphere winter), the situation is reversed.

ESSENTIAL CONCEPT 6.1A

Earth's axis is tilted at an angle of 23.5 degrees, which causes the Northern and Southern Hemispheres to take turns "leaning toward" the Sun every six months and results in the change of seasons.

How Latitude Affects the Distribution Of Solar Radiation

If Earth were a flat plate in space, with its flat side directly facing the Sun, sunlight would fall equally on all parts of Earth. Earth is spherical, however, so the amount and intensity of solar radiation received at higher latitudes are much less than at lower latitudes. The following factors influence the amount of radiation received at low and high latitudes:

- Sunlight strikes low latitudes at a high angle, so the radiation is concentrated in a relatively small area (area *A* in **Figure 6.3**). Sunlight strikes high latitudes at a low angle, so the same amount of radiation is spread over a larger area (area *B* in Figure 6.3).

- Earth's atmosphere absorbs some radiation, so less radiation strikes Earth at high latitudes than at low latitudes because sunlight passes through more of the atmosphere at high latitudes.

- The **albedo** (*albus* = white) of various Earth materials is the percentage of incident radiation that is reflected back to space. The average albedo of Earth's surface is about 30%. More radiation is reflected back to space at high latitudes because ice has a much higher albedo than soil, vegetation, or water.

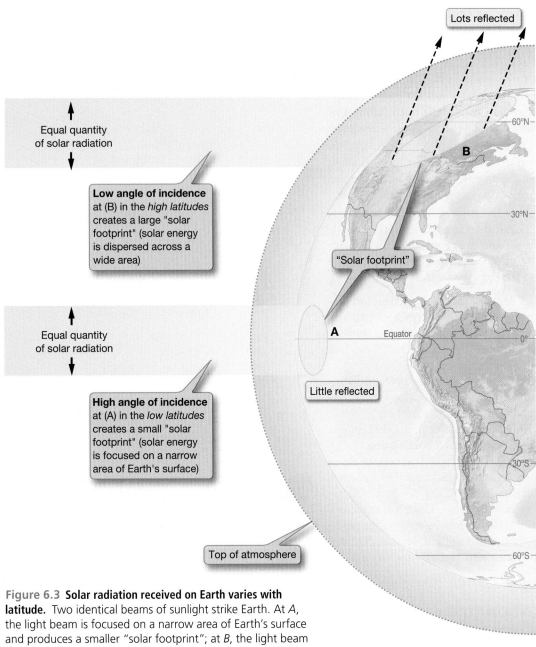

- The angle at which sunlight strikes the ocean surface determines how much is absorbed and how much is reflected. If the Sun shines down on a smooth sea from directly overhead, only 2% of the radiation is reflected, but if the Sun is only 5 degrees above the horizon, 40% is reflected back into the atmosphere (Table 6.1). Thus, the ocean reflects more radiation at high latitudes than at low latitudes.

Because of all these reasons, the intensity of radiation at high latitudes is greatly decreased compared with the intensity of radiation received in equatorial regions.

Other factors influence the amount of solar energy that reaches Earth. For example, the amount of radiation received at Earth's surface varies *daily* because Earth rotates on its axis, so the surface experiences daylight and darkness each day. In addition, the amount of radiation varies *annually* due to Earth's seasons, as discussed in the previous section.

Oceanic Heat Flow

Close to the poles, much incoming solar radiation strikes Earth's surface at low angles. In addition, ice has a high albedo, so more energy is reflected back into space than is absorbed. In contrast, between about 35 degrees north latitude and 40 degrees south latitude,[1] sunlight strikes Earth at much higher angles, and more energy is absorbed than is reflected

Figure 6.3 Solar radiation received on Earth varies with latitude. Two identical beams of sunlight strike Earth. At *A*, the light beam is focused on a narrow area of Earth's surface and produces a smaller "solar footprint"; at *B*, the light beam is dispersed across a wide area and produces a larger "solar footprint." Additionally, more light is reflected at *B*. Thus, the amount of solar energy received at higher latitudes is much less than that at lower latitudes.

TABLE 6.1 REFLECTION AND ABSORPTION OF SOLAR ENERGY RELATIVE TO THE ANGLE OF INCIDENCE ON A FLAT SEA

Elevation of the Sun above the horizon	90°	60°	30°	15°	5°
Reflected radiation (%)	2	3	6	20	40
Absorbed radiation (%)	98	97	94	80	60

[1]Note that this latitudinal range extends farther into the Southern Hemisphere because the Southern Hemisphere has more ocean surface area in the middle latitudes than the Northern Hemisphere does.

back into space. The graph in **Figure 6.4** shows how incoming sunlight and outgoing heat combine on a daily basis for a net heat gain in low-latitude oceans and a net heat loss in high-latitude oceans.

Based on Figure 6.4, you might expect the equatorial zone to grow progressively warmer and the polar regions to grow progressively cooler. The polar regions are always considerably colder than the equatorial zone, but the temperature *difference* remains the same because excess heat is transferred from the equatorial zone to the poles. How is this accomplished? Circulation in both the oceans and the atmosphere transfers the heat.

ESSENTIAL CONCEPT 6.1B

Low-latitude regions receive more solar radiation than high-latitude regions, but oceanic and atmospheric circulation transfer heat around the globe.

CONCEPT CHECK 6.1

1. Sketch a labeled diagram to explain the cause of Earth's seasons.
2. Along the Arctic Circle, how would the Sun appear during the summer solstice? During the winter solstice?
3. If there is a net annual heat loss at high latitudes and a net annual heat gain at low latitudes, why does the temperature difference between these regions not increase?

6.2 What Physical Properties Does the Atmosphere Possess?

The atmosphere transfers heat and water vapor from place to place on Earth. Within the atmosphere, complex relationships exist among air composition, temperature, density, water vapor content, and pressure. Before we apply these relationships, let's examine the atmosphere's composition and some of its physical properties.

Composition Of the Atmosphere

Figure 6.5 lists the composition of dry air and shows that the atmosphere consists almost entirely of nitrogen and oxygen. Other gases include argon (an inert gas), carbon dioxide, and others in trace amounts. Although these gases are present in very small amounts, they can trap significant amounts of heat within the atmosphere. For more about how these gases trap heat in the atmosphere, see Chapter 16, "The Oceans and Climate Change."

Temperature Variation in the Atmosphere

Intuitively, it seems logical that the higher one goes in the atmosphere, the warmer it should be since it's closer to the Sun. However, as unusual as it seems, the atmosphere is actually heated from *below*. Moreover, the Sun's energy passes through the Earth's atmosphere and heats the Earth's surface (both land and water), which then reradiates this energy as heat into the atmosphere. This process is known as the *greenhouse effect* and will be discussed in more detail in Chapter 16, "The Oceans and Climate Change."

Figure 6.6 shows a temperature profile of the atmosphere. The lowermost portion of the atmosphere, which extends from the surface to about 12 kilometers (7 miles), is called the **troposphere** (*tropo* = turn, *sphere* = a ball) and is where all weather is produced. The troposphere gets its name because of the abundance of mixing that occurs within this layer of the atmosphere, mostly as a result of being heated from below. Within the troposphere, temperature

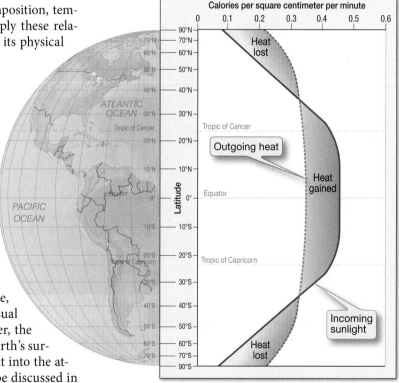

Figure 6.4 Heat gained and lost from the ocean varies with latitude. Heat gained by the oceans in equatorial latitudes (*orange shading*) equals heat lost in polar latitudes (*blue shading*). On average, the two balance each other, and the excess heat from equatorial latitudes is transferred to heat-deficient polar latitudes by both oceanic and atmospheric circulation.

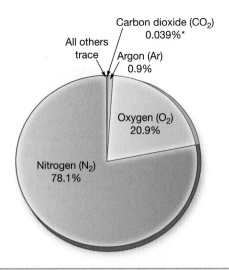

*Note that the concentration of carbon dioxide in the atmosphere is increasing by 0.5% per year due to human activities

Figure 6.5 Composition of dry air. Pie chart showing the composition of dry air (without any water vapor) by volume. Nitrogen and oxygen gas comprise 99% of the total, with several trace gases making up the rest; the most significant trace gas is carbon dioxide, an important greenhouse gas.

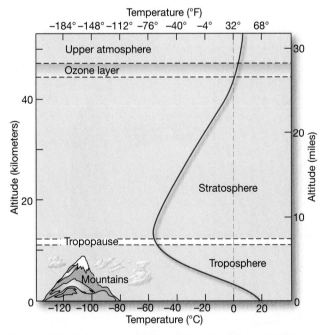

Figure 6.6 Temperature profile of the atmosphere. Within the troposphere, the atmosphere gets cooler with increasing altitude. Above the troposphere, the atmosphere generally warms.

gets cooler with altitude to the point that at high altitudes, the air temperature is well below freezing. If you have ever flown in a jet airplane, for instance, you may have noticed that any water on the wings or inside your window freezes during a high-altitude flight.

Density Variation in the Atmosphere

It may seem surprising that air has density, but since air is composed of molecules, it certainly does. Temperature has a dramatic effect on the density of air. At higher temperatures, for example, air molecules move more quickly, take up more space, and density is decreased. Thus, the general relationship between density and temperature is as follows:

- Warm air is less dense, so it rises; this is commonly expressed as "heat rises."
- Cool air is more dense, so it sinks.

Figure 6.7 shows how a radiator (heater) uses convection to heat a room. The heater warms the nearby air and causes it to expand. This expansion makes the air less dense, causing it to rise. Conversely, a cold window cools the nearby air and causes it to contract, thereby becoming more dense, which causes it to sink. A **convection** (*con* = with, *vect* = carried) **cell** forms, composed of the rising and sinking air moving in a circular fashion, similar to the convection in Earth's mantle discussed in Chapter 2.

Atmospheric Water Vapor Content

The amount of water vapor in air depends in part on the air's temperature. Warm air, for instance, can hold more water vapor than cold air because the air molecules are moving more quickly and come into contact with more water vapor. Thus, warm air is typically moist, and, conversely, cool air is typically dry. As a result, a warm, breezy day speeds evaporation when you hang your laundry outside to dry.

Water vapor influences the density of air. The addition of water vapor decreases the density of air because water vapor has a lower density than air. Thus, humid air is less dense than dry air.

Atmospheric Pressure

Atmospheric pressure is 1.0 atmosphere[2] (14.7 pounds per square inch) at sea level and decreases with increasing altitude. Atmospheric pressure depends on the weight of the column of air above. For instance, a thick column of air produces higher atmospheric pressure than a thin column of air. An analogy to this is water pressure in a swimming pool: The thicker the column of water above, the higher the water pressure. Thus, the highest pressure in a pool is at the bottom of the deep end.

Similarly, the thicker column of air at sea level means air pressure is high at sea level and decreases with increasing elevation. When sealed bags of potato chips or pretzels are taken to a high elevation, the pressure is much lower than where they were sealed, sometimes causing the bags to burst. You may also have experienced this change in pressure when your ears "popped" during the takeoff or landing of an airplane or while driving on steep mountain roads.

---

[2]The *atmosphere* is a unit of pressure; 1.0 atmosphere is the average pressure exerted by the overlying atmosphere at sea level and is equivalent to 760 millimeters of mercury, 101,300 Pascal, or 1013 millibars.

Changes in atmospheric pressure cause air movement as a result of changes in the molecular density of the air. The general relationship is shown in **Figure 6.8**, which indicates that:

- A column of cool, dense air causes high pressure at the surface, which will lead to sinking air (movement *toward* the surface and compression).
- A column of warm, less dense air causes low pressure at the surface, which will lead to rising air (movement *away from* the surface and expansion).

In addition, sinking air tends to warm because of its compression, while rising air tends to cool due to expansion. Note that there are complex relationships among air composition, temperature, density, water vapor content, and pressure.

Movement Of the Atmosphere

Air *always* moves from high-pressure regions toward low-pressure regions. This moving air is called **wind**. If a balloon is inflated and let go, what happens to the air inside the balloon? It rapidly escapes, moving from a high-pressure region inside the balloon (caused by the balloon pushing on the air inside) to the lower-pressure region outside the balloon.

An Example: A Nonspinning Earth

Imagine for a moment that Earth is not spinning on its axis but that the Sun rotates around Earth, with the Sun directly above Earth's equator at all times (**Figure 6.9**). Because more solar radiation is received along the equator than at the poles, the air at the equator in contact with Earth's surface is warmed. This warm, moist air rises, creating low pressure at the surface. This rising air cools (see Figure 6.6) and releases its moisture as rain. Thus, a zone of low pressure and much precipitation occurs along the equator.

As the air along the equator rises, it reaches the top of the troposphere and begins to move toward the poles. Because the temperature is much lower at high altitudes, the air cools, and its density increases. This cool, dense air sinks at the poles, creating high pressure at the surface. The sinking air is quite dry because cool air cannot hold much water vapor. Thus, the poles experience high pressure and clear, dry weather.

Which way will surface winds blow? Air always moves from high pressure to low pressure, so air travels from the high pressure at the poles toward the low

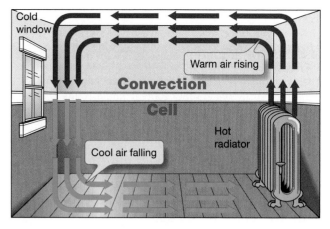

Figure 6.7 Convection in a room. A circular-moving loop of air (a convection cell) is caused by warm air rising and cool air sinking.

STUDENTS SOMETIMES ASK . . .

Why is there so much nitrogen in the atmosphere?

To understand the abundance of nitrogen in the atmosphere, it's useful to compare it to oxygen, the next most abundant element in the atmosphere. Figure 6.5, for example, shows that nitrogen is about four times as abundant in the atmosphere as oxygen. However, if we consider the relative abundances of oxygen and nitrogen over the entire Earth, oxygen is about 10,000 times more abundant. These earthly abundances overall reflect the composition of the material from which Earth originally formed and the process of Earth's accretion. Oxygen is a major component of the solid Earth, along with silicon and elements such as magnesium, calcium, and sodium. Nitrogen is not stable as part of a crystal lattice, so it is not incorporated into the solid Earth. This is one reason why nitrogen is so enriched in the atmosphere relative to oxygen. The other primary reason is that, unlike oxygen, nitrogen is very stable in the atmosphere and is not involved to a great extent in chemical reactions that occur there. Thus, over geologic time, it has built up in the atmosphere to a much greater extent than oxygen.

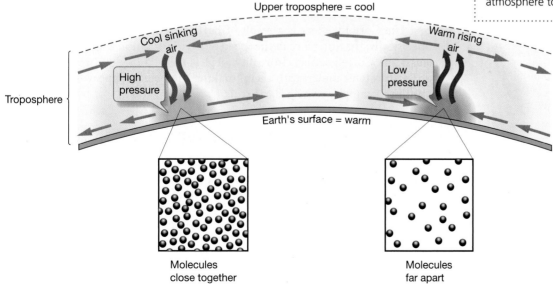

Figure 6.8 High and low atmospheric pressure zones. A column of cool, dense air causes high pressure at the surface (*left*), which will lead to sinking air. A column of warm, less dense air causes low pressure at the surface (*right*), which will lead to rising air.

pressure at the equator. Thus, there are strong northerly winds in the Northern Hemisphere and strong southerly winds in the Southern Hemisphere.[3] The air warms as it makes its way back to the equator, completing the loop (called a *convection cell* or *circulation cell*; see Figure 6.7).

Is this fictional case of a nonspinning Earth a good analogy for what is really happening on Earth? Actually, it is not, even though the *principles* that drive the physical movement of air remain the same whether Earth is spinning or not. Let's now examine how Earth's spin influences atmospheric circulation.

CONCEPT CHECK 6.2

❶ Describe the physical properties of the atmosphere, including its composition, temperature, density, water vapor content, pressure, and movement.

❷ Is Earth's atmosphere heated from above or below? Explain.

6.3 How Does the Coriolis Effect Influence Moving Objects?

Web Video
Coriolis Effect on a Merry-Go-Round

The **Coriolis effect** changes the intended path of a moving body. Named after Gaspard Gustave de Coriolis, the French engineer who first calculated its influence in 1835, it is often incorrectly called the Coriolis *force*. It does not accelerate the moving body, so it does not influence the body's speed. As a result, it is an effect and not a true force.

The Coriolis effect causes moving objects on Earth to follow curved paths. In the Northern Hemisphere, an object will follow a path to the *right* of its intended direction; in the Southern Hemisphere, an object will follow a path to the *left* of its intended direction. The directions right and left are the *viewer's perspective looking in the direction in which the object is traveling*. For example, the Coriolis effect very slightly influences the movement of a ball thrown between two people. In the Northern Hemisphere, the ball will veer slightly to its right *from the thrower's perspective*.

The Coriolis effect acts on all moving objects. However, it is much more pronounced on objects traveling long distances, especially north or south. This is why the Coriolis effect has a dramatic effect on atmospheric circulation and the movement of ocean currents.

The Coriolis effect is a result of Earth's rotation toward the east. More specifically, the *difference* in the speed of Earth's rotation at different latitudes causes the Coriolis effect. In reality, objects travel along straight-line paths,[4] but Earth rotates underneath them, making the objects appear to curve. Let's look at two examples to help clarify this.

Example 1: Perspectives and Frames Of Reference on a Merry-Go-Round

A merry-go-round is a useful experimental apparatus with which to test some of the concepts of the Coriolis effect. A merry-go-round is a large circular wheel that

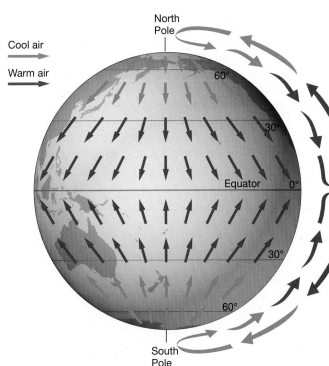

Figure 6.9 Atmospheric circulation on a nonspinning Earth. A fictional nonspinning Earth with the Sun rotating around Earth directly above Earth's equator at all times. Arrows show the pattern of winds that would develop due to uneven solar heating on Earth.

[3]Notice that winds are named based on the direction *from which they are moving.*
[4]Newton's first law of motion (the law of inertia) states that every body persists in its state of rest or of uniform motion in a straight line unless it is compelled to change that state by forces imposed upon it.

rotates around its center. It has bars that people hang onto while the merry-go-round spins, as shown in **Figure 6.10**.

Imagine that you are on a merry-go-round that is spinning counterclockwise, as viewed from above (Figure 6.10). As you are spinning, what will happen to you if you let go of the bar? If you guessed that you would fly off along a straight-line path perpendicular to the merry-go-round (Figure 6.10, *Path A*), that's not quite right. Your angular momentum would propel you in a straight line *tangent* to your circular path on the merry-go-round at the point where you let go (Figure 6.10, *Path B*). The law of inertia states that a moving object will follow a straight-line path until it is compelled to change that path by other forces. Thus, you would follow a straight-line path (*Path B*) until you collide with some object such as other playground equipment or the ground. From the perspective of another person on the merry-go-round, your departure along *Path B* would *appear* to curve to the right due to the merry-go-round's rotation.

Imagine that you are back on the merry-go-round, spinning counterclockwise, but you are now joined by another person who is facing you directly but on the opposite side of the merry-go-round. If you were to toss a ball to the other person, what path would it appear to follow? Even though you threw the ball straight at the other person, from *your perspective* the ball's path would appear to curve to the right (*dashed gray line*). That's because the frame of reference (in this example, the merry-go-round) has rotated during the time that it took the ball to reach where the other person had been (Figure 6.10). A person viewing the merry-go-round from directly overhead would observe that the ball did indeed travel along a straight-line path (Figure 6.10, *Path C*), just as your path was straight when you let go of the merry-go-round bar. Similarly, the perspective of being on the rotating Earth causes objects to appear to travel along curved paths. This is the Coriolis effect. The merry-go-round spinning in a counterclockwise direction is analogous to the Northern Hemisphere because, as viewed from above the North Pole, Earth is spinning counterclockwise. Thus, moving objects appear to follow curved paths to the *right* of their intended direction in the Northern Hemisphere.

If the other person on the merry-go-round had thrown a ball toward you, it would also appear to have curved. From the perspective of the other person, the ball would appear to curve to its right, just as the ball you threw curved. From your perspective, however, the ball thrown toward you would appear to curve to its *left*. The perspective to keep in mind when considering the Coriolis effect is the one *looking in the same direction that the object is moving*.

To simulate the Southern Hemisphere, the merry-go-round would need to rotate in a *clockwise* direction, which is analogous to Earth when viewed from above the South Pole. Thus, moving objects appear to follow curved paths to the *left* of their intended direction in the Southern Hemisphere.

Example 2: A Tale Of Two Missiles

The distance that a point on Earth has to travel in a day is shorter with increasing latitude. A location near the pole, for example, travels in a circle not nearly as far in a day as will an area near the equator. Because both areas travel their respective distances in one day, the velocity of the two areas must not be the same. **Figure 6.11a** shows that as Earth rotates on its axis, the velocity decreases with latitude, ranging from more than 1600 kilometers (1000 miles) per hour at the equator to 0 kilometers per hour at the poles. *This change in velocity with latitude is the true cause of the Coriolis effect.* The following example illustrates how velocity changes with latitude.

Imagine that we have two missiles that fly in straight lines toward their destinations. For simplicity, assume that the flight of each missile takes one hour regardless of the distance flown. The first missile is launched from the North Pole toward New Orleans, Louisiana, which is at 30 degrees north latitude (**Figure 6.11b**).

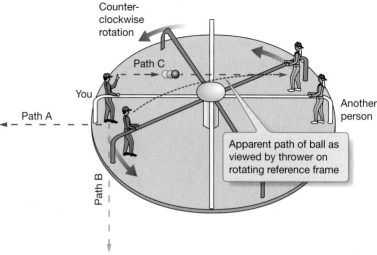

Figure 6.10 A merry-go-round spinning counterclockwise as viewed from above. See text for description of *Paths A*, *B*, and *C*. The curving dashed gray line shows the apparent path of the ball between the two people as viewed by the thrower on the moving merry-go-round.

STUDENTS SOMETIMES ASK . . .

Is it true that the Coriolis effect causes water to drain one way in the Northern Hemisphere and the other way in the Southern?

In most cases, no. Theoretically, the water moves too slowly and the distance across a basin in your home is too small to generate a Coriolis-induced whirlpool (vortex) in such a basin. If all other effects are nullified, however, the Coriolis effect comes into play and makes draining water spiral counterclockwise north of the equator and the other way in the Southern Hemisphere (the same direction that hurricanes spin). But the Coriolis effect is extremely weak on small systems like a basin of water. The shape and irregularities of the basin, local slopes, or any external movement can easily outweigh the Coriolis effect in determining the direction in which water drains.

STUDENTS SOMETIMES ASK . . .

If Earth is spinning so fast, why don't we feel it?

Despite Earth's constant rotation, we have the illusion that Earth is still. The reason that we don't feel the motion is because Earth rotates smoothly and quietly, and the atmosphere moves along with us. Thus, all sensations we receive tell us there is no motion and the ground is comfortably at rest—even though most of the United States is continually moving at speeds greater than 800 kilometers (500 miles) per hour!

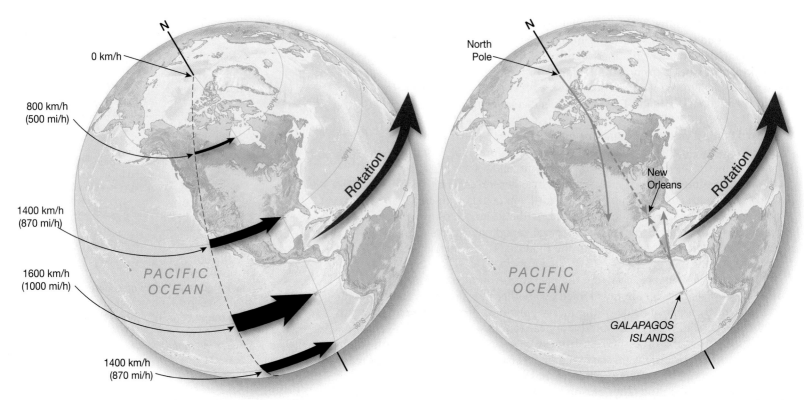

(a) The velocity of any point on Earth varies with latitude from about 1600 kilometers (1000 miles) per hour at the equator to 0 kilometers per hour at either pole.

Figure 6.11 The Coriolis effect and missile paths.

(b) The path of missiles shot towards New Orleans from the North Pole and from the Galápagos Islands on the equator. Dashed lines indicate intended paths; solid lines indicate paths that the missiles would travel as viewed from Earth's surface.

STUDENTS SOMETIMES ASK . . .

Why are space missions launched from low-latitude regions?

The reason that the United States launches its space missions from Florida is to take advantage of Earth's additional rotational speed at lower latitudes (note arrows in Figure 6.11a), thereby giving space vehicles more momentum once they get into space. And, the further south you go, the more momentum the rockets naturally obtain; that's why some countries (such as France) launch rockets from their territories in tropical islands. In fact, the multinational company Sea Launch currently operates a floating launching pad along the equator about 1600 kilometers (1000 miles) south of Hawaii.

Web Animation
Coriolis Effect

Does the missile land in New Orleans? Actually, no. Earth rotates eastward at 1400 kilometers (870 miles) per hour along the 30 degrees latitude line (Figure 6.11a), so the missile lands somewhere near El Paso, Texas, 1400 kilometers west of its target. From your perspective at the North Pole, the path of the missile appears to curve *to its right* in accordance with the Coriolis effect. In reality, New Orleans has moved out of the line of fire due to Earth's rotation.

The second missile is launched toward New Orleans from the Galápagos Islands, which are directly south of New Orleans along the equator (Figure 6.11b). From their position on the equator, the Galápagos Islands are moving east at 1600 kilometers (1000 miles) per hour, 200 kilometers (124 miles) per hour faster than New Orleans (Figure 6.11a). At takeoff, therefore, the missile is also moving toward the east 200 kilometers per hour faster than New Orleans. Thus, when the missile returns to Earth one hour later at the latitude of New Orleans, it will land offshore of Alabama, 200 kilometers east of New Orleans. Again, from your perspective on the Galápagos Islands, the missile appears to curve *to its right*. Keep in mind that both of these missile examples ignore friction, which would greatly reduce the amount the missiles deflect to the right of their intended courses.

Changes in the Coriolis Effect with Latitude

The first missile (shot from the North Pole) missed the target by 1600 kilometers (1000 miles), while the second missile (shot from the Galápagos Islands) missed its target by only 200 kilometers (124 miles). What was responsible for the difference? Not only does the rotational velocity of points on Earth range from 0 kilometers per hour at the poles to more than 1600 kilometers (1000 miles) per hour at the equator, but the *rate of change* of the rotational velocity (per degree of latitude) increases as the pole is approached from the equator.

For example, the rotational velocity differs by 200 kilometers (124 miles) per hour between the equator (0 degrees) and 30 degrees north latitude. From

30 degrees north latitude to 60 degrees north latitude, however, the rotational velocity differs by 600 kilometers (372 miles) per hour. Finally, from 60 degrees north latitude to the North Pole (where the rotational velocity is zero), the rotational velocity differs by more than 800 kilometers (500 miles) per hour.

Thus, the maximum Coriolis effect is at the poles, and there is no Coriolis effect at the equator. The magnitude of the Coriolis effect depends much more, however, on the length of time the object (such as an air mass or ocean current) is in motion. Even at low latitudes, where the Coriolis effect is small, a large Coriolis deflection is possible if an object is in motion for a long time. In addition, because the Coriolis effect is caused by the *difference* in velocity of different latitudes on Earth, there is no Coriolis effect for those objects moving due east or due west along the equator.

For a summary of the Coriolis effect, see Web Table 6.1.

ESSENTIAL CONCEPT 6.3

The Coriolis effect causes moving objects to curve to the right in the Northern Hemisphere and to the left in the Southern Hemisphere. It is at its maximum at the poles and is nonexistent at the equator

CONCEPT CHECK 6.3

1 Describe the Coriolis effect in both the Northern and Southern Hemisphere.

2 What is the underlying cause of the Coriolis effect?

3 Explain how the strength of the Coriolis effect changes with latitude.

6.4 What Global Atmospheric Circulation Patterns Exist?

Figure 6.12 shows atmospheric circulation and the corresponding wind belts on a spinning Earth, which presents a more complex pattern than that of the fictional nonspinning Earth (Figure 6.9).

Circulation Cells

The greater heating of the atmosphere over the equator causes the air to expand, to decrease in density, and to rise. As the air rises, it cools by expansion because the pressure is lower, and the water vapor it contains condenses and falls as rain in the equatorial zone. The resulting dry air mass travels north or south of the equator. Around 30 degrees north and south latitude, the air cools off enough to become denser than the surrounding air, so it begins to descend, completing the loop (Figure 6.12). These circulation cells are called **Hadley cells** after noted English meteorologist George Hadley (1685–1768).

In addition to Hadley cells, each hemisphere has a **Ferrel cell** between 30 and 60 degrees latitude and a **polar cell** between 60 and 90 degrees latitude. The Ferrel cell—named after American meteorologist William Ferrel (1817–1891), who invented the three-cells-per-hemisphere model for atmospheric circulation—is not driven solely by differences in solar heating; if it were, air within it would circulate in the opposite direction. Similarly to the movement of interlocking gears, the Ferrel cell moves in the direction that coincides with the movement of the two adjoining circulation cells.

Pressure

A column of cool, dense air moves *toward* the surface and creates high pressure. The descending air at about 30 degrees north and south latitude creates high-pressure zones called the **subtropical highs**. Similarly, descending air at the poles creates high-pressure regions called the **polar highs**.

What kind of weather is experienced in these high-pressure areas? Descending air is quite dry, and it tends to warm under its own weight, so these areas typically

STUDENTS SOMETIMES ASK . . .

I've heard that the Coriolis effect is really a force but it is often described as a fictitious force. What is a fictitious force?

The forces you feel in a moving car—those that push you back into your seat when the driver steps on the gas or throw you sideways when the car makes sharp turns—are everyday examples of fictitious forces. In general, these influences arise because the natural frame of reference for a given situation (such as the car) is itself accelerating.

A classic example of these types of apparent influences involves the Coriolis "force" and a pendulum. Consider a back-and-forth swinging pendulum that is suspended directly over the North Pole. To an earthly observer, it would appear to rotate 360 degrees every day and thus would seem to be acted upon by a sideways force (that is, perpendicular to the plane of swing). If you viewed this pendulum from a stationary point in outer space, however, it would appear to swing in a single, fixed plane while Earth turned underneath it. From this outer-space perspective, there is no sideways force deflecting the pendulum's sway. That is why the somewhat pejorative term *fictitious* is attached to this force and also why Coriolis is more properly termed an *effect* (not a true force). Similarly, in the car, no real force pushes you back into your seat, your senses notwithstanding; what you feel is the moving frame of reference caused by the car's acceleration.

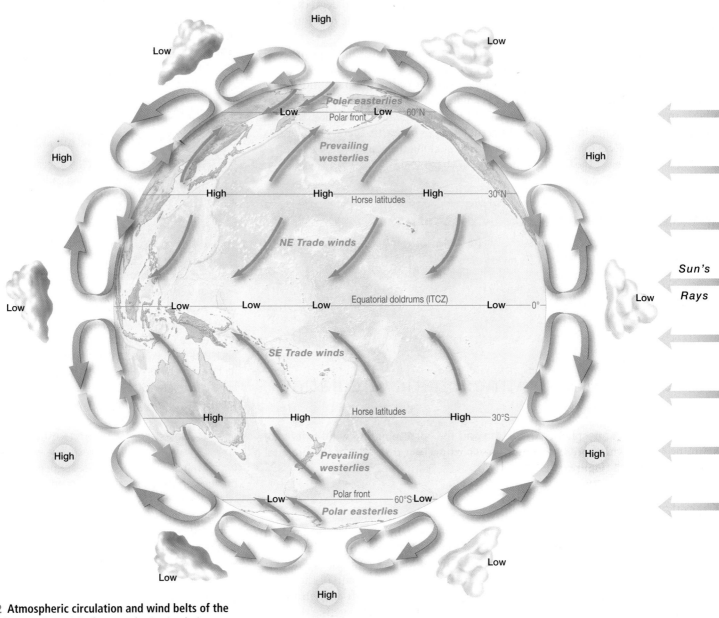

Figure 6.12 Atmospheric circulation and wind belts of the world. The three-cell model of atmospheric circulation creates the major wind belts of the world. Boundaries between wind belts and surface atmospheric pressures are also shown. The general pattern of wind belts is modified by seasonal changes and the distribution of continents.

Web Animation
Global Wind Patterns

experience dry, clear, fair conditions. The conditions are not necessarily warm (such as at the poles)—just dry and associated with clear skies.

A column of warm, low-density air rises *away* from the surface and creates low pressure. Thus, rising air creates a band of low pressure at the equator—the **equatorial low**—and at about 60 degrees north and south latitude—the **subpolar low**. The weather in areas of low pressure is dominated by cloudy conditions with lots of precipitation, because rising air cools and cannot hold its water vapor.

Wind Belts

The lowermost portion of the circulation cells—that is, the part that is closest to the surface—generates the major wind belts of the world. The masses of air that move across Earth's surface from the subtropical high-pressure belts toward the equatorial low-pressure belt constitute the **trade winds**. These steady winds are named from the term *to blow trade*, which means to blow in a regular course. If Earth did not rotate, these winds would blow in a north–south direction. In the Northern

Hemisphere, however, the **northeast trade winds** curve to the right due to the Coriolis effect and blow *from northeast to southwest*. In the Southern Hemisphere, on the other hand, the **southeast trade winds** curve to the left due to the Coriolis effect and blow *from southeast to northwest*.

Some of the air that descends in the subtropical regions moves along Earth's surface to higher latitudes as the **prevailing westerly wind belts**. Because of the Coriolis effect, the prevailing westerlies blow from southwest to northeast in the Northern Hemisphere and from northwest to southeast in the Southern Hemisphere.

Air moves away from the high pressure at the poles, too, producing the **polar easterly wind belts**. The Coriolis effect is maximized at high latitudes, so these winds are deflected strongly. The polar easterlies blow from the northeast in the Northern Hemisphere, and from the southeast in the Southern Hemisphere. When the polar easterlies come into contact with the prevailing westerlies near the subpolar low pressure belts (at 60 degrees north and south latitude), the warmer, less dense air of the prevailing westerlies rises above the colder, more dense air of the polar easterlies.

Boundaries

The boundary between the two trade wind belts along the equator is known as the **doldrums** (*doldrum* = dull) because, long ago, sailing ships were becalmed there by the lack of winds. Sometimes stranded for days or weeks, the situation was unfortunate but not life-threatening: Daily rain showers supplied sailors with plenty of freshwater. Today, meteorologists refer to this globe-circling region as the **Intertropical Convergence Zone (ITCZ)** because it is the region between the tropics where the Northern and Southern Hemisphere trade winds converge (Figure 6.12).

The boundary between the trade winds and the prevailing westerlies (centered at 30 degrees north or south latitude) is known as the **horse latitudes**. Sinking air in these regions causes high atmospheric pressure (associated with the *subtropical high pressure*) and results in clear, dry, and fair conditions. Because the air is sinking, the horse latitudes are known for surface winds that are light and variable.

The boundary between the prevailing westerlies and the polar easterlies at 60 degrees north or south latitude is known as the **polar front**. This is a battleground for different air masses, so cloudy conditions and lots of precipitation are common here.

Clear, dry, fair conditions are associated with the high pressure at the poles, so precipitation is minimal. The poles are often classified as cold deserts because the annual precipitation is so low.

Table 6.2 summarizes the characteristics of global wind belts and boundaries.

Web Video
Satellite Video of
Major Wind Belts

TABLE 6.2 CHARACTERISTICS OF WIND BELTS AND BOUNDARIES

Region (north or south latitude)	Name of wind belt or boundary	Atmospheric pressure	Characteristics
Equatorial (0–5 degrees)	Doldrums (boundary)	Low	Light, variable winds. Abundant cloudiness and much precipitation. Breeding ground for hurricanes.
5–30 degrees	Trade winds (wind belt)	—	Strong, steady winds, generally from the east.
30 degrees	Horse latitudes (boundary)	High	Light, variable winds. Dry, clear, fair weather with little precipitation. Major deserts of the world.
30–60 degrees	Prevailing westerlies (wind belt)	—	Winds generally from the west. Brings storms that influence weather across the United States.
60 degrees	Polar front (boundary)	Low	Variable winds. Stormy, cloudy weather year round.
60–90 degrees	Polar easterlies (wind belt)	—	Cold, dry winds generally from the east.
Poles (90 degrees)	Polar high pressure (boundary)	High	Variable winds. Clear, dry, fair conditions, cold temperatures, and minimal precipitation. Cold deserts.

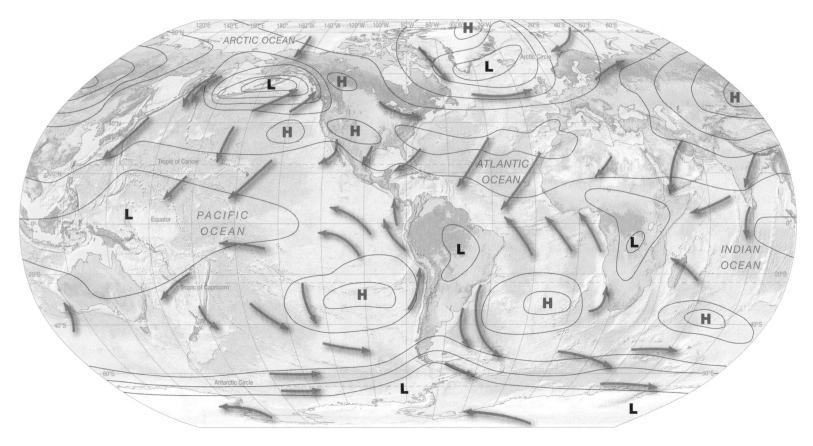

Figure 6.13 January sea-level atmospheric pressures and winds. Average atmospheric pressure pattern for January. High (*H*) and low (*L*) atmospheric pressure zones correspond closely to those shown in Figure 6.12 but are modified by the change of seasons and the distribution of continents. Arrows show direction of winds, which move from high- to low-pressure regions.

Web Animation
Seasonal Pressure and Precipitation Patterns

Circulation Cells: Idealized or Real?

The three-cell model of atmospheric circulation first proposed by Ferrel provides a simplified model of the general circulation pattern on Earth. This circulation model is idealized and does not always match the complexities observed in nature, particularly for the location and direction of motion of the Ferrel and polar cells. Nonetheless, it generally matches the pattern of major wind belts of the world and provides a general framework for understanding why they exist.

Further, the following factors significantly alter the idealized wind, pressure, and atmospheric circulation patterns illustrated in Figure 6.12:

1. The tilt of Earth's rotation axis, which produces seasons

2. The lower heat capacity of continental rock compared to seawater,[5] which makes the air over continents colder in winter and warmer in summer than the air over adjacent oceans

3. The uneven distribution of land and ocean over Earth's surface, which particularly affects patterns in the Northern Hemisphere

During winter, therefore, the continents usually develop atmospheric high-pressure cells from the weight of cold air centered over them and, during the summer, they usually develop low-pressure cells (**Figure 6.13**). In fact, such seasonal shifts in atmospheric pressure over Asia cause *monsoon winds*, which have a dramatic effect on Indian Ocean currents. and will be discussed in Chapter 7, "Ocean Circulation." In general, however, the patterns of atmospheric high- and low-pressure zones shown in Figure 6.13 corresponds closely to those shown in Figure 6.12.

---

[5]An object that has low heat capacity heats up quickly when heat energy is applied. Recall from Figure 5.7 that water has one of the highest specific heat capacities of common substances.

Global wind belts have had a profound effect on ocean explorations (**Diving Deeper 6.1**). The world's wind belts also closely match the pattern of ocean surface currents, which are discussed in Chapter 7, "Ocean Circulation."

CONCEPT CHECK 6.4

1 Sketch the pattern of surface wind belts on Earth, showing atmospheric circulation cells, zones of high and low pressure, the names of the wind belts, and the names of the boundaries between the wind belts.

2 Why are there high-pressure caps at each pole and a low-pressure belt in the equatorial region?

3 Discuss the patterns or trends that you notice about the global wind belts and boundaries as shown in Figure 6.12 and delineated in Table 6.2.

Web Animation
Cyclones and Anticyclones

6.5 What Weather and Climate Patterns Occur in the Oceans?

Because of the ocean's huge extent over Earth's surface and also because of water's unusual thermal properties, the ocean dramatically influences global weather and climate patterns.

Weather versus Climate

Weather describes the conditions of the atmosphere at a given time and place. **Climate** is the long-term average of weather. If we observe the weather conditions in an area over a long period, we can begin to draw some conclusions about its climate. For instance, if the weather in an area is dry over many years, we can say that the area has an arid climate.

Winds

Recall that air always moves from high pressure toward low pressure and that the movement of air is called *wind*. However, as air moves away from high-pressure regions and toward low-pressure regions, the Coriolis effect modifies its direction. In the Northern Hemisphere, for example, air moving from high to low pressure curves to the right and results in a counterclockwise[6] flow of air around low-pressure cells (called **cyclonic** [*kyklon* = moving in a circle] **flow**). Similarly, as the air leaves the high-pressure region and curves to the right, it establishes a clockwise flow of air around high-pressure cells (called **anticyclonic flow**). **Figure 6.14** shows how a screwdriver can help you remember how air moves around high- and low-pressure regions: High pressures are similar to a high screw that needs to be tightened, so a screwdriver would be turned clockwise; low pressures are similar to a tightened screw that needs to be loosened, so a screwdriver would be turned counterclockwise. Because winter high-pressure cells are replaced by summer low-pressure cells over the continents, wind patterns associated with continents often reverse themselves seasonally.

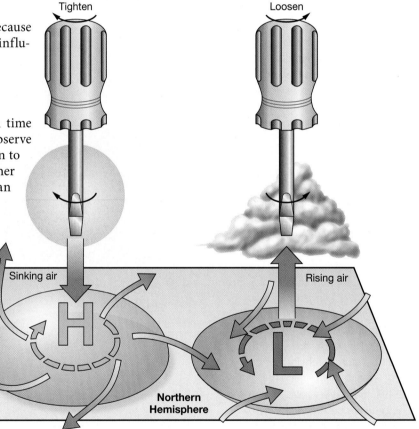

Figure 6.14 High- and low-pressure regions and air flow. As air moves away from a high-pressure region (*H*) toward a low-pressure region (*L*), the Coriolis effect causes the air to curve to the right in the Northern Hemisphere. This results in clockwise winds around high-pressure regions (anticyclonic flow) and counterclockwise winds around low-pressure regions (cyclonic flow). One way to remember this is to think of high pressures as being similar to a high screw that needs tightening (clockwise motion) and low pressures as being similar to a tightened screw that needs loosening (counterclockwise motion).

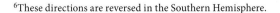

[6]These directions are reversed in the Southern Hemisphere.

WHY CHRISTOPHER COLUMBUS NEVER SET FOOT ON NORTH AMERICA

The Italian navigator and explorer **Christopher Columbus** is widely credited with discovering North America in the year 1492. However, America was already populated with many natives, and the Vikings made a voyage to North America that predated Columbus's by about 500 years. Moreover, the pattern of the major wind belts of the world prevented his sailing ships from reaching continental North America during his four voyages.

Rather than sailing east, Columbus was determined to reach the East Indies (today the country of Indonesia) by sailing west across the Atlantic Ocean. An astronomer in Florence, Italy, named Toscanelli was the first to suggest such a route in a letter to the king of Portugal. Columbus later contacted Toscanelli and was told how far he would have to sail west to reach India. Today, we know that this distance would have carried him just west of North America.

After years of difficulties in initiating the voyage, Columbus received the financial backing of the Spanish monarchs Ferdinand V and Isabella I. He set sail with 88 men and three ships (the *Niña*, the *Pinta*, and the *Santa María*) on August 3, 1492, from the Canary Islands off Africa (**Figure 6A**). The Canary Islands are located at 28 degrees north latitude and are within the northeast trade winds, which blow steadily from the northeast to the southwest. Instead of sailing directly west, which would have allowed Columbus to reach central Florida, the map in Figure 6A shows that Columbus sailed a more southerly route.

During the morning of October 12, 1492, the first land was sighted; this is generally believed to have been Watling Island in The Bahamas, southeast of Florida. Based on the inaccurate information he had been given, Columbus was convinced that he had arrived in the East Indies and was somewhere near India. Consequently,

he called the inhabitants "Indians," and the area is known today as the *West Indies*. Later during this voyage, he explored the coasts of Cuba and Hispaniola (the island comprising modern-day Haiti and the Dominican Republic).

On his return journey, Columbus sailed to the northeast and picked up the prevailing westerlies, which transported him away from North America and toward Spain. Upon his return to Spain and the announcement of his discovery, additional voyages were planned. Columbus made three more trips across the Atlantic Ocean, following similar paths through the Atlantic. Thus, his ships were controlled by the trade winds on the outbound voyage and the prevailing westerlies on

the return trip. During his next voyage, in 1493, Columbus explored Puerto Rico and the Leeward Islands and established a colony on Hispaniola. In 1498, he explored Venezuela and landed on South America, unaware that it was a new continent to Europeans. On his last voyage in 1502, he reached Central America.

Although he is today considered a master mariner, Columbus died in neglect in 1506, still convinced that he had explored islands near India. Even though he never set foot on the North American mainland, his journeys inspired other Spanish and Portuguese navigators to explore the "New World," including the coasts of North and South America.

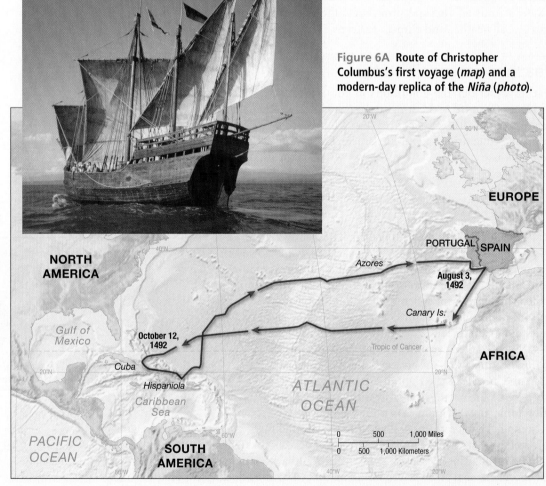

Figure 6A Route of Christopher Columbus's first voyage (*map*) and a modern-day replica of the *Niña* (*photo*).

SEA AND LAND BREEZES   Other factors that influence regional winds, especially in coastal areas, are **sea breezes** and **land breezes** (**Figure 6.15**). When an equal amount of solar energy is applied to both land and ocean, the land heats up about five times more due to its lower heat capacity. The land heats the air around it and, during the afternoon, the warm, low-density air over the land rises. Rising air creates a low-pressure region over the land, pulling the cooler air over the ocean toward land, creating what is known as a *sea breeze*. At night, the land surface cools about five times more rapidly than the ocean and cools the air around it. This cool, high-density air sinks, creating a high-pressure region that causes the wind to blow from the land. This is known as a *land breeze*, and it is most prominent in the late evening and early morning hours.

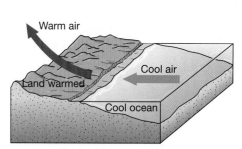

(a) Sea breezes occur when air warmed by the land rises and is replaced by cool air from the ocean.

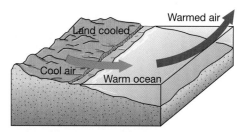

(b) Land breezes occur when the land has cooled, causing dense air to sink and flow toward the warmer ocean.

Figure 6.15 Sea and land breezes.   Diagrammatic view of the formation of a daytime sea breeze **(a)** and a nighttime land breeze **(b)**.

Storms and Fronts

At very high and very low latitudes, there is little daily and minor seasonal change in weather.[7] Equatorial regions are usually warm, damp, and typically calm because the dominant direction of air movement in the doldrums is upward. Midday rains are common, even during the supposedly "dry" season. It is within the *middle latitudes* between 30 and 60 degrees north or south latitude where storms are common.

Storms are atmospheric disturbances characterized by strong winds, precipitation, and often thunder and lightning. Due to the seasonal change of pressure systems over continents, air masses from the high and low latitudes may move into the middle latitudes, meet, and produce severe storms. **Air masses** are large volumes of air that have a definite area of origin and distinctive characteristics. Several air masses influence the United States, including polar air masses and tropical air masses (**Figure 6.16**). Some air masses originate over land (c = continental) and are therefore dryer, but most originate over the sea (m = maritime) and are moist. Some are colder (P = polar; A = Arctic) and some are warm (T = tropical). Typically, the United States is influenced more by polar air masses during the winter and more by tropical air masses during the summer.

As polar and tropical air masses move into the middle latitudes, they also move gradually in an easterly direction. A **warm front** is the contact between a warm air mass moving into an area occupied by cold air. A **cold front** is the contact between a cold air mass moving into an area occupied by warm air (**Figure 6.17**).

These confrontations are brought about by the movement of the **jet stream**, which is a narrow, fast-moving, easterly flowing air mass. It exists above the middle latitudes just below the top of the troposphere, centered at an altitude of about 10 kilometers (6 miles). It usually follows a wavy path and may cause unusual weather by steering a polar air mass far to the south or a tropical air mass far to the north.

Regardless of whether a warm front or cold front is produced, the warmer, less-dense air always rises above the denser cold air. The warm air cools as it rises, so its water vapor condenses as precipitation. A cold front is usually steeper, and the temperature difference across it is greater than a warm front. Therefore, rainfall along a cold front is usually heavier and briefer than rainfall along a warm front.

Web Animation
Cold Fronts and Warm Fronts

Tropical Cyclones (Hurricanes)

Tropical cyclones (*kyklon* = moving in a circle) are huge rotating masses of low pressure characterized by strong winds and torrential rain. They are the largest storm systems on Earth, though they are not associated with any fronts. In North and South America, tropical cyclones are called **hurricanes** (*Huracan* = Taino god

---

[7]In fact, in equatorial Indonesia, the vocabulary of Indonesians doesn't include the word *seasons*.

Figure 6.16 Air masses that affect U.S. weather. Polar air masses are shown in blue, and tropical air masses are shown in red. Air masses are classified based on their source region: The designation continental (c) or maritime (m) indicates moisture content, whereas polar (P), Arctic (A), and tropical (T) indicate temperature conditions.

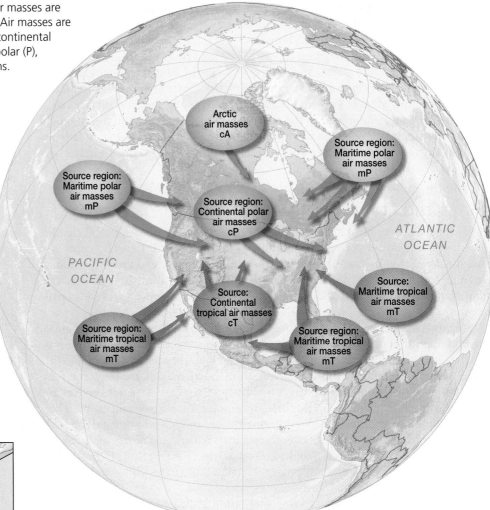

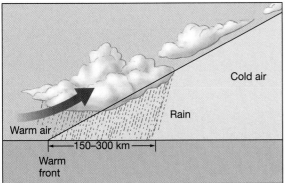

(a) Cross section through a gradually rising warm front

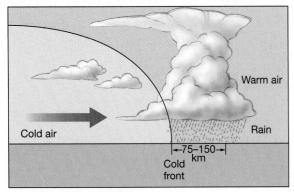

(b) Cross section through a steeper cold front

Figure 6.17 Warm and cold fronts. Cross sectional view through a gradually rising warm front **(a)** and a steeper cold front **(b)**. With both fronts, warm air rises and precipitation is produced.

of wind); in the western North Pacific Ocean, they are called **typhoons** (*tai-fung* = great wind); and in the Indian Ocean, they are called **cyclones**. No matter what they are called, tropical cyclones can be highly destructive. In fact, the energy contained in a *single* hurricane is greater than that generated by all energy sources in the United States over the past 20 years.

ORIGIN Remarkably, what powers tropical storms is the release of vast amounts of *latent heat of condensation*[8] that is carried within water vapor and is released as water condenses to form clouds in a hurricane. A tropical cyclone begins as a low-pressure cell that breaks away from the equatorial low-pressure belt and grows as it picks up heat energy in the following manner. Surface winds feed moisture (in the form of water vapor) into the storm. When water evaporates, it stores tremendous amounts of heat in the form of latent heat of evaporation. When water vapor condenses into a liquid (in this case, clouds and rain), it releases this stored heat—latent heat of condensation—into the surrounding atmosphere, which causes the atmosphere to warm and the air to rise. This rising air causes surface pressure to decrease, drawing additional warm moist surface air into the storm. This air, as it rises and cools, condenses into clouds and releases even more latent heat, further powering the storm and continuously repeating itself as a feedback loop, each time intensifying the storm.

---

[8]For a discussion of water's latent heats, see Chapter 5.

Tropical storms are classified according to their maximum sustained wind speed:

- If winds are less than 61 kilometers (38 miles) per hour, the storm is classified as a *tropical depression.*
- If winds are between 61 and 120 kilometers (38 and 74 miles) per hour, the storm is called a *tropical storm.*
- If winds exceed 120 kilometers (74 miles) per hour, the storm is a *tropical cyclone.*

The **Saffir-Simpson Scale** of hurricane intensity (Table 6.3) further divides tropical cyclones into categories based on wind speed and damage. In some cases, in fact, the wind in tropical cyclones attains speeds as high as 400 kilometers (250 miles) per hour!

Worldwide, about 100 storms grow to hurricane status each year. The conditions needed to create a hurricane are as follows:

Web Animation
Hurricanes

- Ocean water with a temperature greater than which provides an abundance of water vapor to the atmosphere through evaporation.
- Warm, moist air, which supplies vast amounts of latent heat as the water vapor in the air condenses and fuels the storm.
- The Coriolis effect, which causes the hurricane to spin counterclockwise in the Northern Hemisphere and clockwise in the Southern Hemisphere. No hurricanes occur directly on the equator because the Coriolis effect is nonexistent there.[9]

These conditions are found during the late summer and early fall, when the tropical and subtropical oceans are at their maximum temperature. Even though hurricanes sometimes form outside hurricane season, the official Atlantic basin hurricane season is from June 1 to November 30 each year. These dates conventionally delimit the period when most tropical cyclones form in the Atlantic basin.

MOVEMENT When hurricanes are initiated in the low latitudes, they are affected by the trade winds and generally move from east to west across ocean basins. Hurricanes typically last from 5 to 10 days and sometimes migrate into the middle latitudes

TABLE 6.3   THE SAFFIR-SIMPSON SCALE OF HURRICANE INTENSITY

| Category | Wind Speed | | Typical storm surge (sea level height above normal) | | Damage |
	km/hr	mi/hr	meters	feet	
1	120–153	74–95	1.2–1.5	4–5	Minimal: Minor damage to buildings
2	154–177	96–110	1.8–2.4	6–8	Moderate: Some roofing material, door, and window damage; some trees blown down
3	178–209	111–130	2.7–3.7	9–12	Extensive: Some structural damage and wall failures; foliage blown off trees and large trees blown down
4	210–249	131–155	4.0–5.5	13–18	Extreme: More extensive structural damage and wall failures; most shrubs, trees, and signs blown down
5	>250	>155	>5.8	>19	Catastrophic: Complete roof failures and entire building failures common; all shrubs, trees, and signs blown down; flooding of lower floors of coastal structures

[9]An unusual confluence of weather conditions in 2001 created the first-ever documented instance of a tropical cyclone almost directly over the equator. Statistical models indicate that such an event occurs only once every 300–400 years.

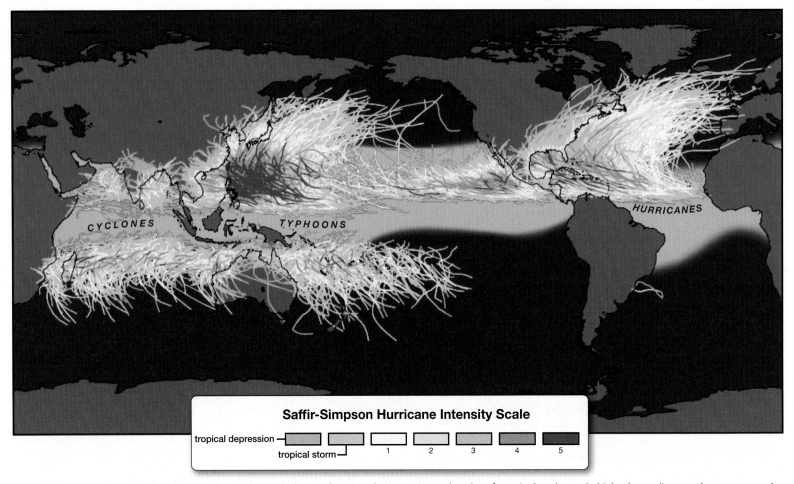

Figure 6.18 Historic tropical cyclone tracks. Color-coded map showing the intensity and paths of tropical cyclones (which, depending on the area, can also be called *hurricanes* or *typhoons*) over the past 150 years. Cyclones originate in low-latitude regions that have warm ocean surface temperatures (*red shading*). Once formed, cyclones are influenced by the trade winds, so generally travel from east to west. Note that cyclones curve to the right north of the equator and to the left south of the equator because of the Coriolis effect, which causes them to move into the middle latitudes where they are steered by the prevailing westerlies toward the east. Once cyclones track away from the tropics into cooler water or land, their energy source is cut off, which causes cyclones to dissipate.

(Figure 6.18). In rare cases, hurricanes have done considerable damage to the northeastern United States and have even affected Nova Scotia, Canada. Figure 6.18 also shows how hurricanes are affected by the Coriolis effect: In the Northern Hemisphere, they curve to the right and in the Southern Hemisphere, they curve to the left. Moreover, this serves to carry them out of the tropics and into the middle latitudes, where they are steered towards the east by the prevailing westerlies. Once hurricanes move over cooler water or land, their energy source is cut off, which causes hurricanes to dissipate.

The diameter of a typical hurricane is less than 200 kilometers (124 miles), although extremely large hurricanes can exceed diameters of 800 kilometers (500 miles). As air moves across the ocean surface toward the low-pressure center, it is drawn up around the **eye of the hurricane** (Figure 6.19). The air in the vicinity of the eye spirals upward, so horizontal wind speeds may be less than 15 kilometers (25 miles) per hour. The eye of the hurricane, therefore, is usually calm. Hurricanes are composed of spiral rain bands where intense rainfall caused by severe thunderstorms can produce tens of centimeters (several inches) of rainfall per hour.

IMPACT OF OTHER FACTORS A host of factors influences the development and strength of hurricanes. For example, warmer sea surface temperatures tend to favor the development of hurricanes, while strong wind shear in the upper atmosphere can ventilate heat away from a developing hurricane and thus interfere with hurricane formation. Other factors that can either enhance or disrupt the development and intensification of hurricanes include the amount of atmospheric convective instability,

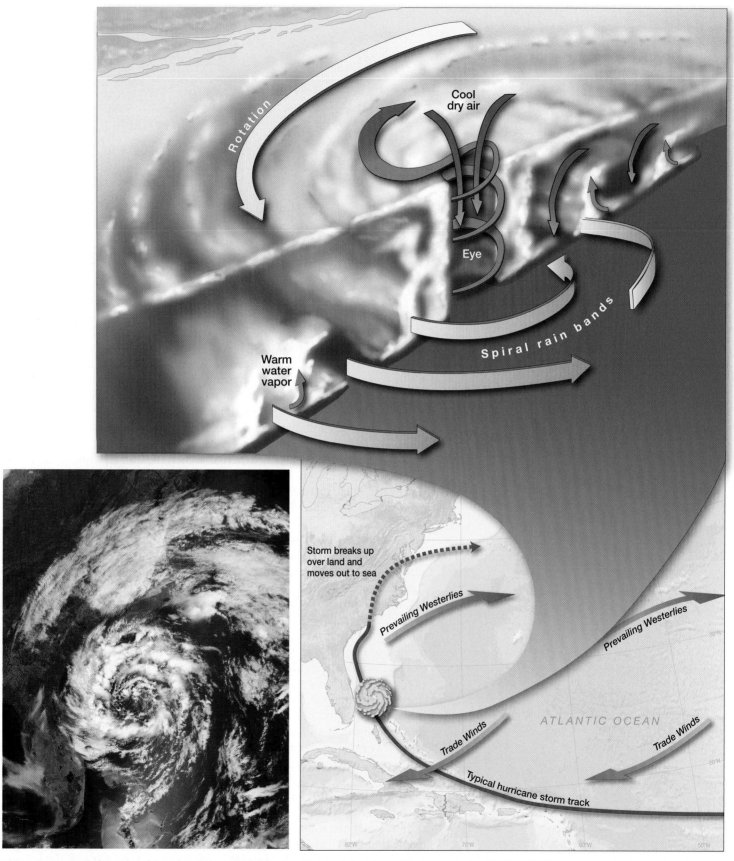

Figure 6.19 Typical North Atlantic hurricane storm track and internal structure. Hurricanes in the North Atlantic are blown by the trade winds toward North America. As they curve to the right because of the Coriolis effect, they move northward and often make landfall or pass close to shore. Eventually, they break up and are transported out to sea by the prevailing westerlies. Photo shows subtropical storm Andrea off the East Coast in 2007; the internal structure of a hurricane is also shown (*enlargement*).

Figure 6.20 Storm surge. As a tropical cyclone in the Northern Hemisphere moves ashore, the low-pressure center around which the storm winds blow combined with strong onshore winds produces a high-water storm surge that floods and batters the coast. The area of the coast that is hit with the right front quadrant of the hurricane—where onshore winds further pile up water—experiences the most severe storm surge. Photograph (*inset*) shows a storm surge in New Jersey caused by Hurricane Felix in August 1995.

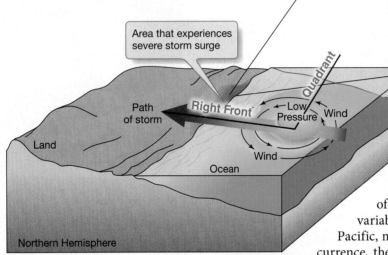

air humidity, vorticity of spinning winds, and even El Niño/La Niña events (which are discussed in Chapter 7, "Ocean Circulation").

New research combined with careful analysis of historical data shows an out-of-phase relationship between tropical cyclone variability in the North Atlantic and the eastern North Pacific, meaning that when one basin has high storm occurrence, the other has low storm occurrence. The recognition of this pattern has helped improve hurricane forecasting.

Human-caused climate change has been linked to the documented warming of ocean surface waters, which fuels hurricanes. As a result, several recent studies show that hurricane severity is expected to increase. In fact, climate models suggest that tropical storms will likely decrease in number, but the risk of Category 4 and 5 storms will likely increase. For more details about climate change and its impact on hurricanes, see Chapter 16, "The Oceans and Climate Change."

TYPES OF DESTRUCTION Destruction from hurricanes is caused by high winds and flooding from intense rainfall. **Storm surge**, however, causes the majority of a hurricane's coastal destruction. In fact, storm surge is responsible for 90% of the deaths associated with hurricanes.

When a hurricane develops over the ocean, its low-pressure center produces a low "hill" of water (**Figure 6.20**). As the hurricane migrates across the open ocean, the hill moves with it. As the hurricane approaches shallow water near shore, the portion of the hill over which the wind is blowing shoreward produces a mass of elevated, wind-driven water. This mass of water—the storm surge—can be as high as 12 meters (40 feet), resulting in a dramatic increase in sea level at the shore, large storm waves, and tremendous destruction to low-lying coastal areas (particularly if it occurs at high tide). In addition, the area of the coast that is hit with the right front quadrant of the hurricane—where onshore winds further pile up water—experiences the most severe storm surge (Figure 6.20). Table 6.3 shows typical storm surge heights associated with Saffir-Simpson hurricane intensities.

HISTORIC DESTRUCTION ON THE U.S. MAINLAND Periodic destruction from hurricanes occurs along the East Coast and the Gulf Coast regions of the United States. In fact, the most deadly natural disaster in U.S. history was caused by a hurricane that struck Galveston Island, Texas, in September 1900. Galveston Island is a thin strip of sand called a *barrier island* located in the Gulf of Mexico off Texas (**Figure 6.21**). In 1900, it was a popular beach resort that averaged only 1.5 meters (5 feet) above sea level. At least 6000 people in and around Galveston were killed when the hurricane's 6-meter (20-foot)-high storm surge completely submerged the island, accompanied by heavy rainfall and winds of 160 kilometers (100 miles) per hour.

A category 4 hurricane like the one in 1900 that devastated Galveston have been surpassed by Category 5 hurricanes that made landfall only three times in the United States: (1) in 1935, an unnamed hurricane[10] flattened the Florida Keys; (2) in 1969, Hurricane Camille struck Mississippi; and (3) in 1992, Hurricane Andrew came ashore in southern Florida, with winds as high as 258 kilometers (160 miles) per hour, ripping down every tree in its path as it crossed the Everglades. Hurricane Andrew did more than $26.5 billion of damage in Florida and along the Gulf Coast. In the aftermath of Hurricane Andrew, more than 250,000 people were left homeless and although most people heeded the warnings to evacuate, 54 were killed.

In October 1998, Hurricane Mitch proved to be one of the most devastating tropical cyclones to affect the Western Hemisphere. At its peak, it was estimated to have winds of 290 kilometers (180 miles) per hour—a strong Category 5 hurricane. It hit Central America with winds of 160 kilometers (100 miles) per hour and as much as 130 centimeters (51 inches) of total rainfall, causing widespread flooding and mudslides in Honduras and Nicaragua that destroyed entire towns. The hurricane resulted in more than 11,000 deaths, left more than 2 million homeless, and caused more than $10 billion in damage across the region.

In September 2008, Hurricane Ike reached Category 4 in the Gulf of Mexico and made landfall near Galveston in low-lying Gilchrist, Texas, as a Category 2 hurricane. Ike resulted in 146 deaths and $24 billion in damages, making it the third costliest U.S. hurricane of all time, behind only Hurricane Katrina (2005) and Hurricane Andrew (1992). In August 2011, Hurricane Irene achieved Category 3 status in the Caribbean and wreaked havoc as it moved along the eastern seaboard from Florida to New England. In all, Irene caused severe flooding that was responsible for 56 deaths and more than $10 billion in damages (**Figure 6.22**).

In October 2012, Hurricane Sandy, a large Category 1 storm, affected the eastern seaboard from Florida to Maine. Hurricane Sandy was the largest Atlantic hurricane on record, with winds spanning 1800 kilometers (1100 miles). When Hurricane Sandy came ashore in New Jersey, it caused flooding there and in many neighboring states, destroyed thousands of homes, left millions without electric service, and killed hundreds. It also caused at least $50 billion in damage, making it the second costliest hurricane in U.S. history, behind only Hurricane Katrina.

Figure 6.21   The Galveston hurricane of 1900. Destruction from the 1900 hurricane at Galveston and location map of Galveston, Texas. At least 6000 people died as a result of the Galveston hurricane, which completely submerged Galveston Island and still stands as the single deadliest U.S. natural disaster.

THE RECORD-BREAKING 2005 ATLANTIC HURRICANE SEASON: HURRICANES KATRINA, RITA, AND WILMA   Although the official Atlantic hurricane season extends each year from June 1 to November 30, the 2005 Atlantic hurricane season persisted into January 2006 and was the most active season on record, shattering numerous records. For example, a record 27 named tropical storms formed, of which a record 15 became hurricanes. Of these, seven strengthened into major hurricanes, a record-tying five became Category 4 hurricanes and a record four reached Category 5 strength, the highest categorization for hurricanes on the Saffir-Simpson Scale of hurricane intensity (see Table 6.3). For the first time ever, NOAA's National Hurricane Center, which oversees the naming of Atlantic Hurricanes, ran out of the usual names for storms and resorted to naming storms using the Greek alphabet.

---

[10]Prior to 1950, Atlantic hurricanes were not named, but this hurricane is often referred to as the "Labor Day Hurricane" because it came ashore then. Today, hurricanes are named by forecasters using an alphabetized list of female and male names.

Figure 6.22 Damage from Hurricane Irene in 2011. Hurricane Irene, which created a wide path of destruction along the East Coast from Florida to New England, caused intense flooding that resulted in 56 deaths and more than $10 billion in damage.

The most notable storms of the 2005 season were the five Category 4 and Category 5 hurricanes: Dennis, Emily, Katrina, Rita, and Wilma. These storms made a combined 12 landfalls as major hurricanes (Category 3 strength or higher) throughout Cuba, Mexico, and the Gulf Coast of the United States, causing over $100 billion in damage and more than 2000 deaths.

Hurricane Katrina, the sixth-strongest Atlantic hurricane ever recorded, was the costliest and one of the deadliest hurricanes in U.S. history. Katrina formed over The Bahamas on August 23 and crossed southern Florida as a moderate Category 1 hurricane before passing over the warm Loop Current and strengthening rapidly in the Gulf of Mexico, causing it to become one of the strongest hurricanes ever recorded in the Gulf. The storm weakened considerably before making its second landfall as a Category 3 storm on the morning of August 29 in southeast Louisiana (**Figure 6.23a**). Still, Katrina was the largest hurricane of its strength to make landfall in the United States in recorded history; its sheer size caused devastation over a radius of 370 kilometers (230 miles). Katrina's 9-meter (30-foot) storm surge—the highest ever recorded in the United States—caused severe damage along the coasts of Mississippi, Louisiana, and Alabama.

Worse yet, Katrina was on a collision course with New Orleans. This scenario was considered a potential catastrophe because nearly all of the New Orleans metropolitan area is below sea level along Lake Pontchartrain. Even without a direct hit, the storm surge from Katrina was forecast to be greater than the height of the levees protecting New Orleans. This risk of devastation was well known; several previous studies warned that a direct hurricane strike on New Orleans could lead to massive flooding, which would lead to thousands of drowning deaths, as well as many more suffering from disease and dehydration after the hurricane passed. Although Katrina passed to the east of New Orleans, levees separating Lake Pontchartrain from New Orleans were breached by Katrina's high winds, storm surge, and heavy rains, ultimately flooding roughly 80% of the city and many neighboring areas (**Figure 6.23b**). Damages from Katrina are estimated to have been $75 billion, easily making it the costliest hurricane in U.S. history. The storm also left hundreds of thousands homeless and killed at least 1600 people, making it the deadliest U.S. hurricane since the 1928 Okeechobee Hurricane. The lack of adequate disaster response by the Federal Emergency Management Agency (FEMA) led to a U.S. Senate investigation in 2006 that recommended disbanding the agency and creating a new National Preparedness and Response Agency.

Hurricane Rita set records as the fourth most intense Atlantic hurricane ever recorded and the most intense tropical cyclone observed in the Gulf of Mexico, breaking the record set by Katrina just three weeks earlier. Rita reached its maximum intensity on September 21, with sustained winds of 290 kilometers (180 miles) per hour and an estimated minimum pressure of 89,500 Pascal (895 millibars, or 0.884 atmosphere). Hurricane Rita's unusually rapid intensification in the Gulf can likely be attributed to its passage over the warm Loop Current as well as higher-than-normal sea surface temperatures in the Gulf. Rita made landfall on September 24 near the Texas–Louisiana border as a Category 3 hurricane. Rita's 6-meter (20-foot) storm surge caused extensive damage along the coasts of Louisiana and extreme southeastern Texas, completely destroying some coastal communities and causing $10 billion in damage.

Later during the same season, Hurricane Wilma set numerous records for both strength and seasonal activity. Wilma was only the third Category 5 ever to develop during the month of October, and its pressure of 88,200 Pascal (882 millibars, or 0.871 atmosphere) ranked it as the most intense hurricane ever recorded in the Atlantic basin. Its maximum sustained near-surface wind speed reached 282 kilometers (175 miles) per hour, with gusts up to 320 kilometers (200 miles) per hour. Wilma made several landfalls, with the most destructive effects felt in the Yucatán Peninsula of Mexico, Cuba, and southern Florida. At least 62 deaths were reported, and damage was estimated at $16 billion to $20 billion ($12.2 billion in the U.S.), ranking Wilma among the top 10 costliest hurricanes ever recorded in

the Atlantic and the sixth costliest storm in U.S. history. Wilma also affected 11 countries with winds or rainfall, more than any other hurricane
in recent history.

HISTORIC DESTRUCTION IN OTHER REGIONS   The majority of the world's
tropical cyclones are formed in the waters north of the equator in the western Pacific Ocean. These storms, called *typhoons*, do enormous damage to
coastal areas and islands in Southeast Asia (see Figure 6.18).

Other areas of the world such as Bangladesh experience tropical cyclones on a regular basis. Bangladesh borders the Indian Ocean and is particularly vulnerable because it is a highly populated and low-lying country,
much of it only 3 meters (10 feet) above sea level. In 1970, a 12-meter
(40-foot)-high storm surge from a tropical cyclone killed an estimated
1 million people. Another tropical cyclone hit the area in 1972 and caused
up to 500,000 deaths. In 1991, Hurricane Gorky's winds of 233 kilometers
(145 miles) per hour and large storm surge caused extensive damage and
killed 200,000 people.

Even islands near the centers of ocean basins can be struck by hurricanes. The Hawaiian Islands, for example, were hit hard by Hurricane Dot
in August 1959 and by Hurricane Iwa in November 1982. Hurricane Iwa hit
very late in the hurricane season and produced winds up to 130 kilometers
(81 miles) per hour. Damage of more than $100 million occurred on the
islands of Kauai and Oahu. Niihau, a small island that is inhabited by only
a few hundred native Hawaiians, was directly in the path of the storm
and suffered severe property damage but no serious injuries. Hurricane
Iniki roared across the islands of Kauai and Niihau in September 1992,
with 210-kilometer (130-mile)-per-hour winds. It was the most powerful
hurricane to hit the Hawaiian Islands in the past 100 years, with property
damage that approached $1 billion.

FUTURE THREAT TO LIFE AND PROPERTY   Each year, tropical cyclones and
hurricanes leave millions homeless worldwide and account for, on average,
over $100 billion of damage in the United States alone. Hurricanes will
continue to be a threat to life and property around the globe. Because of
increasingly accurate forecasts and prompt evacuation, however, the loss of
life has been decreasing. Property damage, on the other hand, has been increasing because increasing coastal populations have resulted in more and
more construction along the coast. Inhabitants of areas subject to a hurricane's destructive force must be made aware of the danger so that they can
be prepared for its eventuality.

The Ocean's Climate Patterns

Just as land areas have climate patterns, so do regions of the oceans. The open
ocean is divided into climatic regions that run generally east–west (parallel
to lines of latitude) and have relatively stable boundaries that are somewhat
modified by ocean surface currents (**Figure 6.24**).

The **equatorial** region spans the equator, which gets an abundance of solar radiation. As a result, the major air movement is upward because heated air rises.
Surface winds, therefore, are weak and variable, which is why this region is called
the *doldrums*. Surface waters are warm and the air is saturated with water vapor.
Daily rain showers are common, which keeps surface salinity relatively low. The
equatorial regions just north or south of the equator are also the breeding grounds
for tropical cyclones.

Tropical regions extend north or south of the equatorial region up to the
Tropic of Cancer and the Tropic of Capricorn, respectively. They are characterized
by strong trade winds, which blow from the northeast in the Northern Hemisphere

(a) Satellite view of Hurricane Katrina coming ashore along the Gulf Coast
on August 29, 2005, showing the hurricane's counterclockwise direction
of spin and prominent central eye. Hurricane Katrina, which had a
diameter of about 670 kilometers (415 miles), was the largest hurricane
of its strength to make landfall in the United States in recorded history.

(b) Hurricane Katrina breached levees and flooded New Orleans,
Louisiana, causing damages of more than $75 billion and claiming
at least 1600 lives.

**Figure 6.23 Hurricane Katrina, the most destructive hurricane
in U.S. history.**

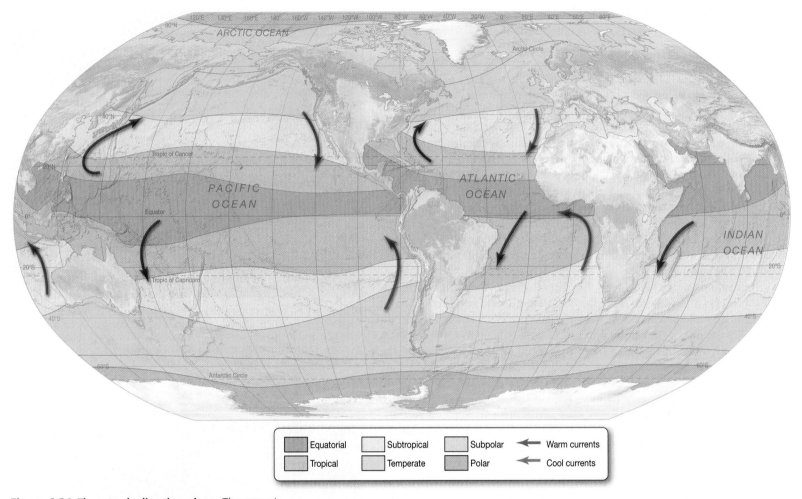

Figure 6.24 The ocean's climatic regions. The ocean's climatic regions are defined primarily by latitude but are modified by ocean currents and wind belts. Red arrows indicate warm surface currents; blue arrows indicate cool surface currents.

and from the southeast in the Southern Hemisphere. These winds push the equatorial currents and create moderately rough seas. Relatively little precipitation falls at higher latitudes within tropical regions, but precipitation increases toward the equator. Once tropical cyclones form, they gain energy here as large quantities of heat are transferred from the ocean to the atmosphere.

Beyond the tropics are the **subtropical** regions. Belts of high pressure are centered there, so the dry, descending air produces little precipitation and a high rate of evaporation, resulting in the highest surface salinities in the open ocean (see the surface salinity map in Chapter 5, Figure 5.24). Winds are weak and currents are sluggish, typical of the horse latitudes. However, strong boundary currents (along the boundaries of continents) flow north and south, particularly along the western margins of the subtropical oceans.

The **temperate** regions (also called the *middle latitudes* or *midlatitudes*) are characterized by strong westerly winds (the prevailing westerlies) that blow from the southwest in the Northern Hemisphere and from the northwest in the Southern Hemisphere (see Figure 6.12). Severe storms are common, especially during winter, and precipitation is heavy. In fact, the North Atlantic is noted for fierce storms, which have claimed many ships and numerous lives over the centuries.

The **subpolar** region experiences extensive precipitation due to the subpolar low. Sea ice covers the subpolar ocean in winter, but it melts away, for the most part, in summer. Icebergs are common, and the surface temperature seldom exceeds 5°C (41°F) in the summer months.

Surface temperatures remain at or near freezing in the **polar** regions, which are covered with ice throughout most of the year. The polar high pressure dominates the area, which includes the Arctic Ocean and the ocean adjacent to Antarctica. There is no sunlight during the winter and constant daylight during the summer.

6.6 How Do Sea Ice and Icebergs Form?

Low temperatures in high-latitude regions cause a permanent or nearly permanent ice cover on the sea surface. The term **sea ice** is used to distinguish such masses of frozen seawater from **icebergs**, which are also found at sea but originate by breaking off (*calving*) from glaciers that originate on land. Sea ice is found throughout the year around the margin of Antarctica, within the Arctic Ocean, and in the extreme high-latitude region of the North Atlantic Ocean.

Formation Of Sea Ice

Sea ice is ice that forms directly from seawater (**Figure 6.25**). It begins as small, needle-like, hexagonal (six-sided) crystals, which eventually become so numerous that a *slush* develops. As the slush begins to form into a thin sheet, it is broken by wind stress and wave action into disk-shaped pieces called **pancake ice** (Figure 6.25a). As further freezing occurs, the pancakes coalesce to form **ice floes** (*flo* = layer).

The rate at which sea ice forms is closely tied to temperature conditions. Large quantities of ice form in relatively short periods when the temperature falls to extremely low levels (such as temperatures below –30°C [–22°F]). Even at these low temperatures, the rate of ice formation slows as sea ice thickens because the ice (which has poor heat conduction) effectively insulates the underlying water from freezing. In addition, calm water enables pancake ice to join together more easily, which aids the formation of sea ice.

The process of sea ice formation tends to be a self-perpetuating process. As sea ice forms at the surface, only a small percentage of the dissolved components can be accommodated into the crystalline structure of ice. As a result, most of the dissolved substances remain in the surrounding seawater, which causes its salinity to increase. Recall from Chapter 5 that increasing the amount of dissolved materials decreases the freezing point of water, which doesn't appear to enhance ice formation. However, also recall that increasing the salinity of water increases its density and its tendency to sink. As it sinks below the surface, it is replaced by lower-salinity (and lower-density) water from below, which will freeze more readily than the high-salinity water it replaced, thereby establishing a circulation pattern that enhances the formation of sea ice.

Recent satellite analyses of the extent of Arctic Ocean sea ice shows that it has decreased dramatically in the past few decades. This accelerated melting appears to

Figure 6.25 Sea ice.
Sea ice forms directly from freezing seawater. Various features of sea ice include pancake ice **(a)**, sea ice ribbons **(b)**, and structures associated with rafted ice **(c)**.

(a) Pancake ice, which is frozen slush that is broken by wind stress and wave action into disk-shaped pieces.

(b) High-altitude view of the Larsen Ice Shelf on the Antarctic Peninsula, where ribbons of sea ice (*top*) remain seaward of the shelf ice during September (the beginning of the spring season in the Southern Hemisphere). Width of view is approximately 100 kilometers (62 miles).

(c) Ice stuctures associated with rafted ice, which is created as ice floes expand and raft onto one another.

be linked to shifts in Northern Hemisphere atmospheric circulation patterns that have caused the region to experience anomalous warming. For more on this topic, see Chapter 16, "The Oceans and Climate Change."

Formation Of Icebergs

An *iceberg* is a body of floating ice that has broken away from a glacier (**Figure 6.26a**) and so is quite distinct from sea ice. Icebergs are formed by vast ice sheets on land, which grow from the accumulation of snow and slowly flow outward to the sea. Once at sea, the ice either breaks up and produces icebergs there or, because it is less dense than water, floats on top of the water, often extending a great distance away from shore before breaking up under the stress of current, wind, and wave action. Most calving occurs during the summer months, when temperatures are highest.

In the Arctic, icebergs originate primarily by calving from glaciers that extend to the ocean along the western coast of Greenland (**Figure 6.26b**). Icebergs are also produced by glaciers along the eastern coasts of Greenland, Ellesmere Island, and other Arctic islands. In all, about 10,000 or so icebergs are calved off these glaciers each year. Many of these icebergs are carried by the East Greenland Current and the West Greenland Current (Figure 6.26b, *arrows*) into North Atlantic shipping lanes, where they become navigational hazards. In recognition of this fact, the area is called *Iceberg*

Alley; it is here that the luxury liner RMS *Titanic* hit an iceberg and sank (see **Web Diving Deeper 6.1**). Because of their large size, some of these icebergs take several years to melt, and, in that time, they may be carried as far south as 40 degrees north latitude, which is the same latitude as Philadelphia, Pennsylvania.

SHELF ICE In Antarctica, where glaciers cover nearly the entire continent, the edges of glaciers form thick floating sheets of ice called **shelf ice** that break off and produce vast plate-like icebergs (**Figures 6.26c** and **6.26d**). In March 2000, for example, a Connecticut-sized iceberg (11,000 square kilometers [4250 square miles]) known as *B-15* and nicknamed "Godzilla" broke loose from the Ross Ice Shelf into the Ross Sea. Since its origin more than a decade ago, B-15 has broken up into smaller icebergs, but many of its parts still exist. Even larger icebergs have also been observed in Antarctic waters. For instance, the largest iceberg ever recorded measured an incredible 32,500 square kilometers (12,500 square miles)—nearly three times the size of B-15 or about the same size as Connecticut and Massachusetts combined.

Icebergs from shelf ice have flat tops that may stand as much as 200 meters (650 feet) above the ocean surface, although most rise less than 100 meters (330 feet) above sea level, and as much as 90% of their mass is below the waterline. Once icebergs are created, ocean currents driven by strong winds carry the icebergs north, where they eventually melt. Because this region is not a major shipping route, the icebergs pose little serious navigation hazard except to supply ships traveling to Antarctica. Officers aboard ships sighting these gigantic bergs have, in some cases, mistaken them for land!

The rate at which Antarctica is producing icebergs—especially large icebergs—has recently increased, most likely as a result of Antarctic warming. For more information about Antarctic warming and its relationship to climate change, see Chapter 16, "The Oceans and Climate Change."

(a) Icebergs, such as this small North Atlantic berg, are formed when pieces of ice calve from glaciers that extend to the sea.

(b) Map showing North Atlantic currents (*blue arrows*), typical iceberg distribution (*white triangles*), and the site of the 1912 Titanic sinking (*black X*).

(c) Aerial view of part of a large tabular Antarctic iceberg.

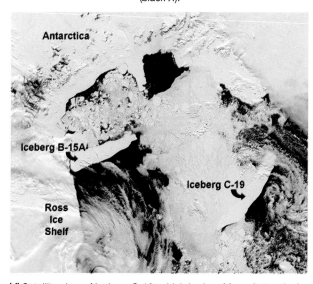
(d) Satellite view of iceberg C-19, which broke of from Antarctica's Ross Ice Shelf in May 2002. Also shown is iceberg B-15A, which is part of the larger B-15 iceberg that was the size of Connecticut when it calved in March 2000.

Figure 6.26 Icebergs. Icebergs form when land-based freshwater glacial ice breaks off into the sea. The world's two main iceberg-producing regions are Greenland (*a* and *b*) and Antarctica (*c* and *d*).

CONCEPT CHECK 6.6

1 Why does the formation of sea ice tend to be a self-perpetuating process?

2 Describe differences between sea ice, icebergs, and shelf ice, including how each is formed.

ESSENTIAL CONCEPT 6.6

Sea ice is created when seawater freezes; icebergs form when chunks of ice break off from coastal glaciers that reach the sea.

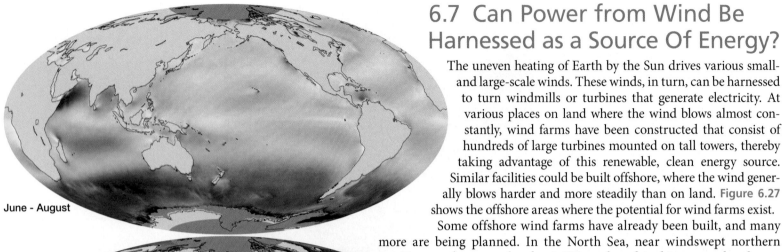

June - August

December - February

Wind Power Density (W/m²)

| 0 | 250 | 500 | 1000 | 2000 |

Figure 6.27 Global ocean wind energy potential. Average ocean wind intensity maps during 2000–2007 for June–August (*top*) and December–February (*bottom*). Areas of high wind power density, where winds are strongest and the potential for wind farms is greatest, are shown in purple, while low power density regions where winds are light are shown in light blue and white.

6.7 Can Power from Wind Be Harnessed as a Source Of Energy?

The uneven heating of Earth by the Sun drives various small- and large-scale winds. These winds, in turn, can be harnessed to turn windmills or turbines that generate electricity. At various places on land where the wind blows almost constantly, wind farms have been constructed that consist of hundreds of large turbines mounted on tall towers, thereby taking advantage of this renewable, clean energy source. Similar facilities could be built offshore, where the wind generally blows harder and more steadily than on land. **Figure 6.27** shows the offshore areas where the potential for wind farms exist.

Some offshore wind farms have already been built, and many more are being planned. In the North Sea, near windswept northern Europe, for example, about 100 sea-based turbines are already operating (**Figure 6.28**), with hundreds more planned. In fact, Denmark generates 18% of its power by wind—more than any other country—and hopes to increase its proportion of wind power to 50% by 2030. In the United States, America's first offshore wind farm, Cape Wind, is scheduled to be built on Horseshoe Shoal in Nantucket Sound, Massachusetts, 8 kilometers (5 miles) south of Cape Cod. Construction of the wind farm is expected to begin in 2013, with a goal of completing 130 wind turbines capable of producing 420 megawatts of power, enough to supply the energy needs of nearly 350,000 average U.S. homes.

One major disadvantage of wind power is that wind strength varies and sometimes it's not windy at all, which is particularly problematic when a high demand for electricity exists. There is also the problem of getting the energy to viable markets such as population centers. On paper, wind and solar power could supply the U.S. and some other countries with all the electricity they require. In practice, however, both sources are too erratic to supply more than about 20% of a region's total energy capacity, according to the U.S. Department of Energy. What are needed are cheap and efficient ways of storing the generated power, then tapping those

Figure 6.28 Offshore wind farm. Offshore wind turbines form a part of a wind farm in the North Sea off the coast of Blyth in the United Kingdom.

supplies when needed. Some of the best solutions to the storage problem include pumping water to high areas and using the water to turn turbines later, using pumps to store underground air at high pressures, and storing energy in advanced batteries.

ESSENTIAL CONCEPT 6.7

There is vast potential for developing wind power as a renewable source of energy, although the erratic nature of wind energy is problematic. Several offshore wind farm systems currently exist.

CONCEPT CHECK 6.7

1 Discuss the advantages and disadvantages of building an offshore wind farm.

2 Describe the location, timeframe, and power generating capability of America's first offshore wind farm.

Essential Concepts Review

6.1 What causes variations in solar radiation on Earth?

▶ *The atmosphere and the ocean act as one interdependent system,* linked by complex feedback loops. There is a close association between most atmospheric and oceanic phenomena.

▶ *The Sun heats Earth's surface unevenly due to the change of seasons (caused by the tilt of Earth's rotational axis,* which is 23.5 degrees from vertical) and the daily cycle of sunlight and darkness (Earth's rotation on its axis). *Differences in latitude also cause changes in the amount of solar radiation* received on Earth.

Study Resources
Online Study Guide Quizzes, Web Animation

Critical Thinking Question

Describe the effect on Earth as a result of Earth's axis of rotation being angled 23.5 degrees from perpendicular relative to the plane of the ecliptic. What would happen if Earth were not tilted on its axis?

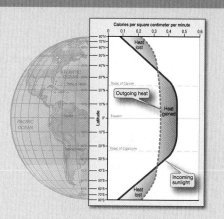

6.2 What physical properties does the atmosphere possess?

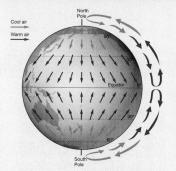

▶ *The uneven distribution of solar energy on Earth* influences most of the physical properties of the atmosphere (such as *temperature, density, water vapor content,* and *pressure differences*) that produce *atmospheric movement.*

Study Resources
Online Study Guide Quizzes

Critical Thinking Question

In a nonspinning Earth, describe the basic atmospheric circulation pattern that would exist.

6.3 How does the Coriolis effect influence moving objects?

▶ *The Coriolis effect influences the paths of moving objects on Earth and is caused by Earth's rotation.* Because Earth's surface rotates at different velocities at different latitudes, *objects in motion tend to veer to the right in the Northern Hemisphere and to the left in the Southern Hemisphere.* The Coriolis effect is *nonexistent at the equator but increases with latitude,* reaching a maximum at the poles.

Study Resources
Online Study Guide Quizzes, Web Animation, Web Table 6.1, Web Video

Critical Thinking Question

How does the Coriolis effect influence the direction of moving objects? How does it affect the speed of moving objects? Explain.

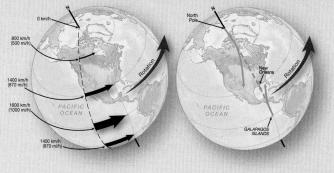

(a)   (b)

Essential Concepts Review (cont.)

6.4 What global atmospheric circulation patterns exist?

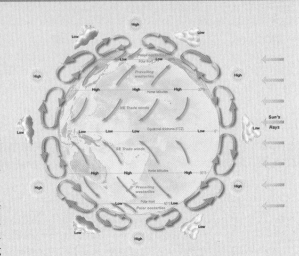

▶ *More solar energy is received than is radiated back into space at low latitudes than at high latitudes.* On the spinning Earth, this creates *three circulation cells in each hemisphere*: a *Hadley cell* between 0 and 30 degrees latitude, a *Ferrel cell* between 30 and 60 degrees latitude, and a *polar cell* between 60 and 90 degrees latitude. High-pressure regions, where dense air descends, are located at about 30 degrees north or south latitude and at the poles. Belts of low pressure, where air rises, are generally found at the equator and at about 60 degrees latitude.

▶ *The movement of air within the circulation cells produces the major wind belts of the world.* The air at Earth's surface that is moving away from the subtropical highs produces *trade winds* moving toward the equator and *prevailing westerlies* moving toward higher latitudes. The air moving along Earth's surface from the polar high to the subpolar low creates the *polar easterlies*.

▶ *Calm winds characterize the boundaries between the major wind belts of the world.* The boundary between the two trade wind belts is called the *doldrums*, which coincides with the Intertropical Convergence Zone (ITCZ). The boundary between the trade winds and the prevailing westerlies is called the *horse latitudes*. The boundary between the prevailing westerlies and the polar easterlies is called the *polar front*.

▶ The *tilt of Earth's axis of rotation*, the *lower heat capacity of rock material* compared to seawater, and the *distribution of continents modify the wind and pressure belts of the idealized three-cell model.* However, the three-cell model closely matches the pattern of the major wind belts of the world.

Study Resources
Online Study Guide Quizzes, Web Animations, Web Video

Critical Thinking Question

To help reinforce your knowledge of atmospheric circulation patterns, draw the pattern of surface wind belts on Earth, showing atmospheric circulation cells, zones of high and low pressure, the names of the wind belts, and the names of the boundaries between the wind belts. Discuss why the idealized belts of high and low atmospheric pressure shown in Figure 6.12 are modified (see Figure 6.13).

6.5 What weather and climate patterns occur in the oceans?

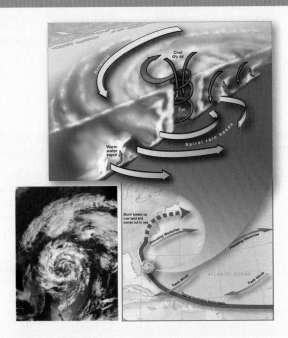

▶ *Weather describes the conditions of the atmosphere at a given place and time, while climate is the long-term average of weather.* Atmospheric motion (*wind*) *is always from high-pressure regions toward low-pressure regions.* In the Northern Hemisphere, therefore, there is a *counterclockwise cyclonic movement* of air around low-pressure cells and a *clockwise anticyclonic movement* around high-pressure cells. Coastal regions commonly experience *sea and land breezes*, due to the daily cycle of heating and cooling.

▶ *Many storms are due to the movement of air masses.* In the middle latitudes, cold air masses from higher latitudes meet warm air masses from lower latitudes and create *cold and warm fronts* that move from west to east across Earth's surface. *Tropical cyclones (hurricanes) are large, powerful storms that mostly affect tropical regions of the world.* Destruction caused by hurricanes is caused by storm surge, high winds, and intense rainfall.

▶ *The ocean's climate patterns are closely related to the distribution of solar energy and the wind belts of the world.* Ocean surface currents somewhat modify oceanic climate patterns.

Study Resources
Online Study Guide Quizzes, Web Animations, Web Video

Critical Thinking Question

What is the difference between weather and climate? If it rains in a particular area during a day, does that mean that the area has a wet climate? Explain.

6.6 How do sea ice and icebergs form?

▶ *In high latitudes, low temperatures freeze seawater and produce sea ice*, which forms as a slush and breaks into pancakes that ultimately grow into ice floes. *Icebergs form when chunks of ice break off glaciers* that form on Antarctica, Greenland, and some Arctic islands. Floating sheets of ice called *shelf ice* near Antarctica produce the largest icebergs.

(a)

(b)

(c)

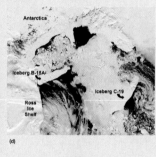
(d)

Study Resources
Online Study Guide Quizzes, Web Diving Deeper 6.1

Critical Thinking Question

What places in the world do most icebergs form? What hazards do they present? Explain.

6.7 Can power from wind be harnessed as a source of energy?

▶ *Winds can be harnessed as a source of power.* There is vast potential for developing this clean, renewable resource but *problems exist* related to generating power when needed, supplying it to consumers, and storing it. Several offshore wind farm systems currently exist.

Study Resources
Online Study Guide Quizzes

Critical Thinking Question

What are some negative environmental factors that could inhibit the development of offshore wind farms?

MasteringOceanography™
www.masteringoceanography.com

Looking for additional review and test prep materials? Visit the Study Area in MasteringOceanography to enhance your understanding of this chapter's content by accessing a variety of resources, including Self-Study Quizzes, Geoscience Animations, RSS feeds, flashcards, Web links, and an optional Pearson eText.

Detection of ocean circulation processes from space.
This composite SeaWiFS/SeaStar satellite view during
the austral summer highlights ocean circulation patterns. The deep blue color represents low chlorophyll
(phytoplankton) concentrations, and the orange and
red colors represent high chlorophyll (phytoplankton)
concentrations. Note the wavy pattern of eddies
between Africa and Antarctica, where the Agulhas
Current meets the Antarctic Circumpolar Current
and is turned to the east, creating the Agulhas
Retroflection. Off the west coast of Africa, coastal
upwelling is shown in bright red colors.

Before you begin reading this chapter, use the glossary at the end of this book to discover the meanings of any of these words you don't already know:

Indian Ocean Subtropical Gyre
Monsoon
West Wind Drift North Atlantic Subtropical Gyre Gyre
South Atlantic Subtropical Gyre Thermohaline circulation Sverdrup (Sv)
Conveyer-belt circulation Antarctic Bottom Water
Western intensification Upwelling El Niño—Southern Oscillation (ENSO)
Equatorial countercurrent Ekman transport Ocean current
Pacific warm pool
Geostrophic current South Pacific Subtropical Gyre
La Niña Productivity
Downwelling North Pacific Subtropical Gyre
Subpolar gyre Ekman spiral

OCEAN CIRCULATION

ESSENTIAL CONCEPTS

At the end of this chapter, you should be able to:

7.1 Demonstrate an understanding of how ocean currents are measured

7.2 Explain how ocean surface currents are organized

7.3 Describe upwelling and the conditions that produce it

7.4 Describe the main surface circulation patterns in each ocean basin

7.5 Discuss the origin and characteristics of deep-ocean currents

7.6 Discuss advantages and disadvantages of harnessing currents as a source of energy

"The coldest winter I ever spent was a summer in San Francisco."

—Anonymous, but often attributed to Mark Twain; said in reference to San Francisco's cool summer weather caused by coastal upwelling

Ocean currents are masses of ocean water that flow from one place to another. The amount of water can be large or small, currents can be at the surface or deep below, and the phenomena that create them can be simple or quite complex. Simply put, currents are *water masses in motion*.

Huge current systems dominate the surfaces of the major oceans. These currents transfer heat from warmer to cooler areas on Earth, just as the major wind belts of the world do. Wind belts transfer about two-thirds of the total amount of heat from the tropics to the poles; ocean surface currents transfer the other third. Ultimately, energy from the Sun drives surface currents, and they closely follow the pattern of the world's major wind belts. As a result, the movement of currents has aided the travel of prehistoric people across ocean basins. Ocean currents also influence the abundance of life in surface waters by affecting the growth of microscopic algae, which are the basis of most oceanic food webs.

More locally, surface currents affect the climates of coastal continental regions. Cold currents flowing toward the equator on the western sides of continents produce arid conditions. Conversely, warm currents flowing poleward on the eastern sides of continents produce warm, humid conditions. Ocean currents, for example, contribute to the mild climate of northern Europe and Iceland, whereas conditions at similar latitudes along the Atlantic coast of North America (such as Labrador, Canada) are much colder. In addition, water sinks in high-latitude regions, initiating deep currents that help regulate the planet's climate.

7.1 How Are Ocean Currents Measured?

Ocean currents are either *wind driven* or *density driven*. Moving air masses—particularly the major wind belts of the world—set wind-driven currents in motion. Wind-driven currents move water horizontally and occur primarily in the ocean's surface waters, so these currents are called **surface currents**. Density-driven circulation, on the other hand, moves water vertically and accounts for the thorough mixing of the deep masses of ocean water. Some surface waters become high in density—through low temperature and/or high salinity—and so sink beneath the surface. This dense water sinks and spreads slowly beneath the surface, so these currents are called **deep currents**.

197

(a) Drift current meter. Depth of metal vanes is 1 meter (3.3 feet).

Figure 7.1 Examples of current-measuring devices. Pictures showing a drift current meter **(a)** and a propellor-type flow meter **(b)**.

(b) A propellor-type flow meter being brought back aboard a research vessel.

Surface Current Measurement

Surface currents rarely flow in the same direction and at the same rate for very long, so measuring average flow rates can be difficult. Some consistency, however, exists in the *overall* surface current pattern worldwide. Surface currents can be measured directly or indirectly.

DIRECT METHODS Two main methods are used to measure currents *directly*. In one, a floating device is released into the current and tracked through time. Typically, radio-transmitting float bottles or other devices are used (**Figure 7.1a**), but other accidentally released items also make good drift meters (**Diving Deeper 7.1**). The other method uses a current-measuring device such as the propeller flow meter shown in **Figure 7.1b** that is deployed from a fixed position such as a pier or a stationary ship. A propeller device can also be towed behind a ship, with the ship's speed subtracted to determine the current's true flow rate.

INDIRECT METHODS Three different methods can be used to measure surface currents *indirectly*. Water flows parallel to a pressure gradient, so one method is to determine the internal distribution of density and the corresponding pressure gradient across an area of the ocean. A second method uses radar altimeters—such as those launched aboard Earth-observing satellites today—to determine the lumps and bulges at the ocean surface, which are a result of the shape of the underlying sea floor[1] as well as current flow. From these data, *dynamic topography* maps can be produced that show the speed and direction of surface currents (**Figure 7.2**). A third method uses a *Doppler flow*

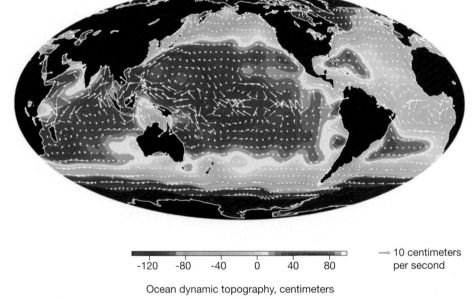

Figure 7.2 Satellite view of ocean dynamic topography. Map showing TOPEX/Poseidon radar altimeter data, in centimeters, from September 1992 to September 1993. Red colors are areas that have higher–than-normal sea level; blue colors are areas that are lower than normal. White arrows indicate the flow direction of currents, with longer arrows indicating faster flow rates.

→ 10 centimeters per second

-120   -80   -40   0   40   80

Ocean dynamic topography, centimeters

[1]Note that this technique is described in Chapter 3, Section 3.1: "Using Satellites to Map Ocean Properties from Space."

Diving Deeper 7.1 Oceans and People

RUNNING SHOES AS DRIFT METERS: JUST DO IT

Any floating object can serve as a makeshift drift meter, as long as it is known where the object entered the ocean and where it was retrieved. The path of the object can then be inferred, providing information about the movement of surface currents. If the time of release and retrieval are known, the speed of currents can also be determined. Oceanographers have long used *drift bottles* (floating "messages in a bottle" or radio-transmitting devices set adrift in the ocean) to track the movement of currents.

Many objects have inadvertently become drift meters when ships lose cargo at sea. Worldwide, in fact, as many as 10,000 shipping containers are lost overboard each year (**Figure 7A**, *top photo*). In this way, Nike athletic shoes and colorful floating bathtub toys (Figure 7A, *middle photo*) have advanced the understanding of current movement in the North Pacific Ocean.

In May 1990, the container vessel *Hansa Carrier* was en route from Korea to Seattle, Washington, when it encountered a severe North Pacific storm. The ship was transporting 12.2-meter (40-foot)-long rectangular metal shipping containers, many of which were lashed to the ship's deck for the voyage. During the storm, the ship lost 21 deck containers overboard, including 5 that held Nike athletic shoes. The shoes floated, and those that were released from their containers were carried east by the North Pacific Current. Within six months, thousands of the shoes began to wash up along the beaches of Alaska, Canada, Washington, and Oregon (Figure 7A, *map*), more than 2400 kilometers (1500 miles) from the site of the spill.

Continued...

Figure 7A Map showing path of drifting shoes and recovery locations from the 1990 spill (*map*), spilled shipping containers from a cargo ship (*top photo*), recovered shoes and plastic bathtub toys (*middle photo*), and oceanographer Curtis Ebbesmeyer displaying floating bathtub toys (*bottom photo*).

Diving Deeper 7.1 (*Continued*)

A few shoes were found on beaches in northern California, and over two years later, shoes from the spill were even recovered from the north end of the Big Island of Hawaii!

Even though the shoes had spent considerable time drifting in the ocean, they were in good shape and wearable (after barnacles and oil were removed). Because the shoes were not tied together, many beachcombers found individual shoes or pairs that did not match. Many of the shoes retailed for around $100, so people interested in finding matching pairs placed ads in newspapers or attended local swap meets.

With help from the beachcombing public (as well as lighthouse operators), information on the location and number of shoes collected was compiled during the months following the spill. Serial numbers inside the shoes were traced to individual containers, and they indicated that only four of the five containers had released their shoes; evidently, one entire container sank without opening. Thus, a maximum of 30,910 pairs of shoes (61,820

individual shoes) were released. The almost instantaneous release of such a large number of drift items helped oceanographers refine computer models of North Pacific circulation. Before the shoe spill, the largest number of drift bottles purposefully released at one time by oceanographers had been about 30,000. Although only 2.6% of the shoes were recovered, this compares favorably with the 2.4% recovery rate of drift bottles released by oceanographers conducting research.

In January 1992, another cargo ship lost 12 containers during a storm to the north of where the shoes had previously spilled. One of these containers held 29,000 packages of small, floatable, colorful plastic bathtub toys in the shapes of blue turtles, yellow ducks, red beavers, and green frogs (Figure 7A, *middle*). Even though the toys were housed in plastic packaging glued to a cardboard backing, studies showed that after 24 hours in seawater, the glue deteriorated, and more than 100,000 of the toys were released.

The floating bathtub toys began to come ashore in southeastern Alaska 10 months later, verifying the computer models. Models suggest that as long as the bathtub toys are able to float, they will continue to be carried by the Alaska Current, eventually dispersing throughout the North Pacific Ocean. For example, some of the toys have made it all the way to South America. Others have even found their way into the Arctic Ocean, where they became entrapped and transported within the Arctic Ocean's floating ice pack, later released into the northern Atlantic Ocean. Remarkably, the tub toys have been found along beaches in the United Kingdom and eastern North America entire ocean basins from where they were accidentally released into the ocean.

Oceanographers such as Curtis Ebbesmeyer (Figure 7A, *bottom photo*) continue to study ocean currents by tracking these and other floating items spilled by cargo ships (see **Web Table 7.1**).

---

meter to transmit low-frequency sound signals through the water. The flow meter measures the shift in frequency between the sound waves emitted and those backscattered by particles in the water to determine current movement.

Deep Current Measurement

The great depth at which deep currents exist makes them even more difficult to measure than surface currents. Most often, they are mapped using underwater floats that are carried within deep currents. One such unique oceanographic program that began in 2000, called **Argo**, is a global array of free-drifting profiling floats (**Figure 7.3b**) that move vertically and measure the temperature, salinity, and

Figure 7.3 **The Argo system of free-drifting submersible floats. (a)** Map showing the location of Argo floats and **(b)** a recently-deployed Argo float.

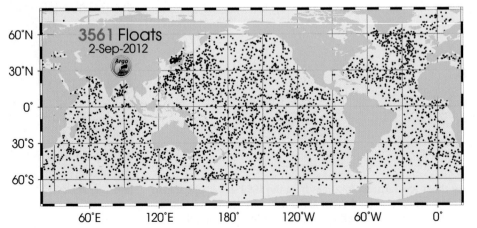

(a) Map showing the location of Argo floats, which can dive to 2000 meters (6600 feet) and collect data on the ocean properties before resurfacing and transmitting their data.

(b) Argo floats are deployed from research or cargo vessels.

other water characteristics of the upper 2000 meters (6600 feet) of the ocean. Once deployed, each float sinks to a particular depth, drifts for up to 10 days collecting data, then resurfaces and transmits data on its location and ocean variables, which are made publically available within hours. Each float then sinks back down to a programmed depth and drifts for up to another 10 days, collecting more data, before resurfacing and repeating the cycle. In 2007, the goal of the program was achieved with the launch of the 3000th Argo float; currently, more than 3500 floats are operating worldwide (**Figure 7.3a**). The program will allow oceanographers to develop a forecasting system for the oceans analogous to weather forecasting on land and also track changes in ocean properties as a result of human-caused climate change.

Other techniques used for measuring deep currents include identifying the distinctive temperature and salinity characteristics of a deep-water mass and tracking telltale chemical tracers. Some tracers are naturally absorbed into seawater, while others are intentionally added. Some useful tracers that have inadvertently been added to seawater include tritium (a radioactive isotope of hydrogen produced by nuclear bomb tests in the 1950s and early 1960s) and chlorofluorocarbons (Freon and other gases that deplete Earth's ozone layer).

ESSENTIAL CONCEPT 7.1

Wind-induced surface currents are measured with floating objects, by satellites, and by other techniques. Density-induced deep currents are measured using submerged floats, water properties, and chemical tracers.

CONCEPT CHECK 7.1

❶ Compare the forces that are directly responsible for creating horizontal and deep vertical circulation in the oceans. What is the ultimate source of energy that drives both circulation systems?

❷ Describe the different ways in which currents are measured.

7.2 How Are Ocean Surface Currents Organized?

Surface currents occur within and above the *pycnocline* (layer of rapidly changing density) to a depth of about 1 kilometer (0.6 mile) and affect only about 10% of the world's ocean water.

Origin Of Surface Currents

In a simplistic case, surface currents develop from friction between the ocean and the wind that blows across its surface. Only about 2% of the wind's energy is transferred to the ocean surface, so a 50-knot[2] wind will create a 1-knot current. You can simulate this on a tiny scale simply by blowing gently and steadily across a cup of coffee.

If there were no continents on Earth, the surface currents would generally follow the major wind belts of the world. In each hemisphere, therefore, a current would flow between 0 and 30 degrees latitude as a result of the trade winds, a second would flow between 30 and 60 degrees latitude as a result of the prevailing westerlies, and a third would flow between 60 and 90 degrees latitude as a result of the polar easterlies.

In reality, however, ocean surface currents are driven by more than just the wind belts of the world. The distribution of continents on Earth is one factor that influences the nature and direction of flow of surface currents in each ocean basin.

[2]A *knot* is a speed of 1 nautical mile per hour. A nautical mile is defined as the distance of 1 minute of latitude and is equivalent to 1.15 statute (land) miles, or 1.85 kilometers.

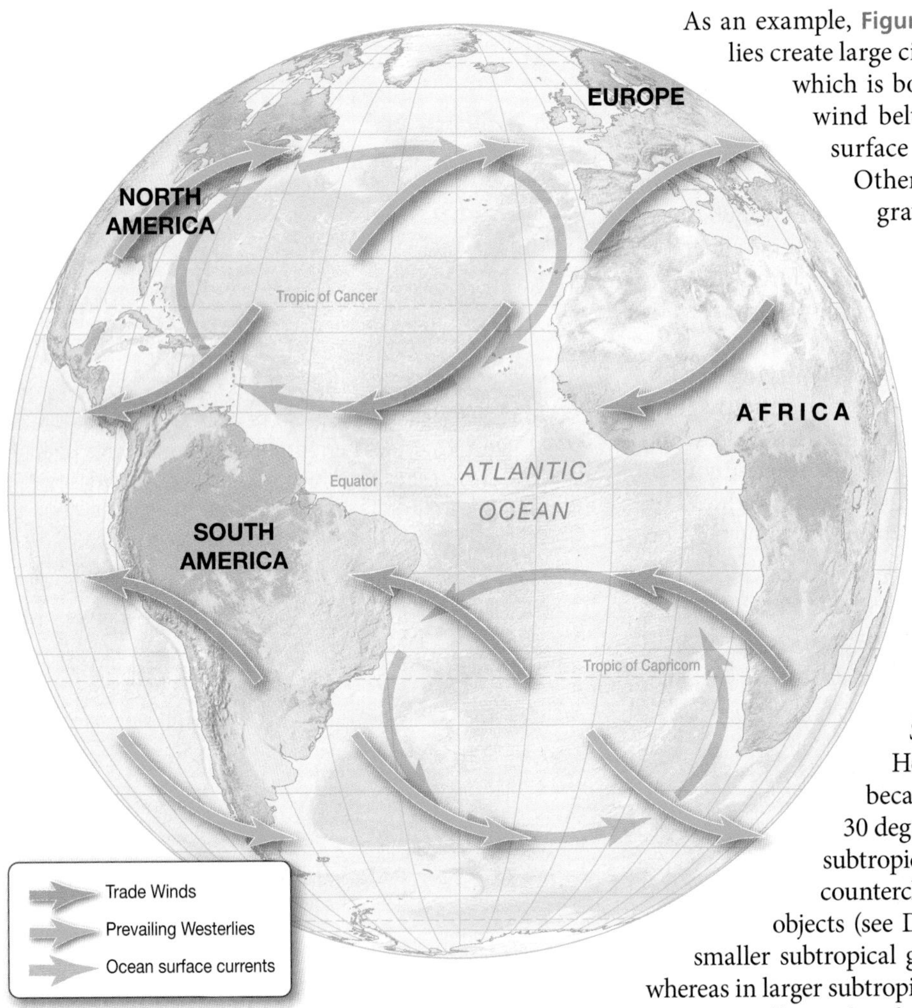

As an example, **Figure 7.4** shows how the trade winds and prevailing wester-lies create large circular-moving loops of water in the Atlantic Ocean basin, which is bounded by the irregular shape of continents. These same wind belts affect the other ocean basins, so a similar pattern of surface current flow also exists in the Pacific and Indian Oceans. Other factors that influence surface current patterns include gravity, friction, and the Coriolis effect.

Main Components Of Ocean Surface Circulation

Although ocean water continuously flows from one current into another, ocean currents can be organized into discrete patterns within each ocean basin.

SUBTROPICAL GYRES   The large, circular-moving loops of water shown in Figure 7.4 that are driven by the major wind belts of the world are called **gyres** (*gyros* = a circle). **Figure 7.5** shows the world's five **subtropical gyres**: (1) the *North Atlantic Subtropical Gyre*, (2) the *South Atlantic Subtropical Gyre*, (3) the *North Pacific Subtropical Gyre*, (4) the *South Pacific Subtropical Gyre*, and (5) the *Indian Ocean Subtropical Gyre* (which is mostly within the Southern Hemisphere). The reason they are called subtropical gyres is because the center of each gyre coincides with the subtropics at 30 degrees north or south latitude. As shown in Figures 7.4 and 7.5, subtropical gyres rotate clockwise in the Northern Hemisphere and counterclockwise in the Southern Hemisphere. Studies of floating objects (see Diving Deeper 7.1) indicate that the average drift time in a smaller subtropical gyre such as the North Atlantic Gyre is about three years, whereas in larger subtropical gyres such as the North Pacific Gyre it is about six years.

Generally, *each subtropical gyre is composed of four main currents* that flow progressively into one another (**Table 7.1**). The North Atlantic Gyre, for instance, is composed of the North Equatorial Current, the Gulf Stream, the North Atlantic Current, and the Canary Current (Figure 7.5). Let's examine each of the four main currents that comprise subtropical gyres.

Equatorial Currents   The trade winds, which blow from the southeast in the Southern Hemisphere and from the northeast in the Northern Hemisphere, set in motion the water masses between the tropics. The resulting currents are called **equatorial currents**, which travel westward along the equator and form the equatorial boundary current of subtropical gyres (Figure 7.5). They are called *north equatorial currents* or *south equatorial currents*, depending on their position relative to the equator.

Western Boundary Currents   When equatorial currents reach the western portion of an ocean basin, they must turn because they cannot cross land. The Coriolis effect deflects these currents away from the equator as **western boundary currents**, which comprise the western boundaries of subtropical gyres. Western boundary currents are so named because they travel along the western boundaries of their respective ocean basins.[3] For example, the Gulf Stream and the Brazil Current, which are shown in Figure 7.5, are western boundary currents. They come from equatorial regions, where water temperatures are warm, so they carry warm water to higher latitudes. Note that Figure 7.5 shows warm currents as red arrows.

Figure 7.4 Atlantic Ocean surface circulation pattern. The trade winds (*purple arrows*) in conjunction with the prevailing westerlies (*green arrows*) create circular-moving loops of water (*underlying blue arrows*) at the surface in both parts of the Atlantic Ocean basin. If there were no continents, the ocean's surface circulation pattern would closely match the major wind belts of the world.

Web Animation
Ocean Circulation

[3]Notice that western boundary currents are off the *eastern* coasts of adjoining continents. It's easy to be confused about this because we have a land-based perspective. From an *oceanic perspective*, however, the western side of the ocean basin is where western boundary currents reside.

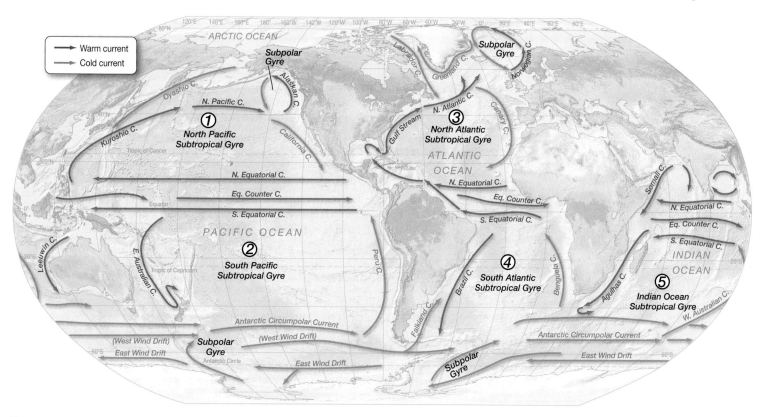

Figure 7.5 Wind-driven surface currents. Major wind-driven surface currents of the world's oceans during February–March. The five major subtropical gyres are ① the North Pacific Subtropical Gyre, ② the South Pacific Subtropical Gyre, ③ the North Atlantic Subtropical Gyre, ④ the South Atlantic Subtropical Gyre, and ⑤ the Indian Ocean Subtropical Gyre. Smaller subpolar gyres rotate in the reverse direction of their adjacent subtropical gyres.

TABLE 7.1   SUBTROPICAL GYRES AND SURFACE CURRENTS

Pacific Ocean	Atlantic Ocean	Indian Ocean
North Pacific Gyre	**North Atlantic Gyre**	**Indian Ocean Gyre**
North Pacific Current	North Atlantic Current	South Equatorial Current
California Current[a]	Canary Current[a]	Agulhas Current[b]
North Equatorial Current	North Equatorial Current	West Wind Drift
Kuroshio (Japan) Current[b]	Gulf Stream[b]	West Australian Current[a]
South Pacific Gyre	**South Atlantic Gyre**	**Other Major Currents**
South Equatorial Current	South Equatorial Current	Equatorial Countercurrent
East Australian Current[b]	Brazil Current[b]	North Equatorial Current
West Wind Drift	West Wind Drift	Leeuwin Current
Peru (Humboldt) Current[a]	Benguela Current[a]	Somali Current
Other Major Currents	**Other Major Currents**	
Equatorial Countercurrent	Equatorial Countercurrent	
Alaskan Current	Florida Current	
Oyashio Current	East Greenland Current	
Labrador Current	Falkland Current	

[a]Denotes an eastern boundary current of a gyre, which is relatively *slow*, *wide*, and *shallow* (and is also a *cold-water* current).
[b]Denotes a western boundary current of a gyre, which is relatively *fast*, *narrow*, and *deep* (and is also a *warm-water* current).

Northern or Southern Boundary Currents Between 30 and 60 degrees latitude, the prevailing westerlies blow from the northwest in the Southern Hemisphere and from the southwest in the Northern Hemisphere. These winds direct ocean surface water in an easterly direction across an ocean basin (see the North Atlantic Current and the Antarctic Circumpolar Current [West Wind Drift] in Figure 7.5). In the Northern Hemisphere, these currents comprise the northern parts of subtropical gyres and are called **northern boundary currents**; in the Southern Hemisphere, they comprise the southern parts of subtropical gyres and are called **southern boundary currents**.

Eastern Boundary Currents When currents flow back across the ocean basin, the Coriolis effect and continental barriers turn them toward the equator, creating **eastern boundary currents** of subtropical gyres along the eastern boundary of the ocean basins. Examples of eastern boundary currents include the Canary Current and the Benguela Current,[4] which are shown in Figure 7.5. They come from high-latitude regions where water temperatures are cool, so they carry cool water to lower latitudes. Note that Figure 7.5 shows cold currents as blue arrows.

EQUATORIAL COUNTERCURRENTS A large volume of water is driven westward by the north and south equatorial currents. Because of Ekman transport and the changing Coriolis effect on either side of the equator, there is a surface divergence of water near the equator that creates a flow of water back toward the east. This current, called the **equatorial countercurrent**, is a narrow, easterly flow of water that occurs counter to and between the adjoining equatorial currents.

Figure 7.5 shows that an equatorial countercurrent is particularly apparent in the Pacific Ocean. This is because of the large equatorial region that exists in the Pacific Ocean and because of a dome of equatorial water that becomes trapped in the island-filled embayment between Australia and Asia. Continual influx of water from equatorial currents builds the dome and creates an eastward countercurrent that stretches across the Pacific toward South America. The equatorial countercurrent in the Atlantic Ocean, on the other hand, is not nearly as well defined because of the shapes of the adjoining continents, which limit the equatorial area that exists in the Atlantic Ocean. The presence of an equatorial countercurrent in the Indian Ocean is strongly influenced by the monsoons, which will be discussed later in this chapter.

SUBPOLAR GYRES Northern or southern boundary currents that flow eastward as a result of the prevailing westerlies eventually move into subpolar latitudes (about 60 degrees north or south latitude). Here, they are driven in a westerly direction by the polar easterlies, producing **subpolar gyres** that rotate opposite the adjacent subtropical gyres. Subpolar gyres are smaller and fewer than subtropical gyres. Two examples include the subpolar gyre in the Atlantic Ocean between Greenland and Europe and in the Weddell Sea off Antarctica (Figure 7.5).

Other Factors Affecting Ocean Surface Circulation

Several other factors influence circulation patterns in subtropical gyres, including Ekman spiral and Ekman transport, geostrophic currents, and western intensification of subtropical gyres.

Web Video
Perpetual Ocean by NASA

ESSENTIAL CONCEPT 7.2A

The principal ocean surface current pattern consists of subtropical and subpolar gyres that are large, circular-moving loops of water powered by the major wind belts of the world.

---

[4]Currents are often named for a prominent geographic location near where they pass. For instance, the Canary Current passes the Canary Islands, and the Benguela Current is named for the Benguela Province in Angola, Africa.

EKMAN SPIRAL AND EKMAN TRANSPORT During the voyage of the *Fram* (see Web Diving Deeper 7.1), Norwegian explorer Fridtjof Nansen (1861–1930) observed that Arctic Ocean ice moved 20 to 40 degrees to the right of the wind blowing across its surface (Figure 7.6). Not only ice but surface waters in the Northern Hemisphere were observed to move to the right of the wind direction; in the Southern Hemisphere, surface waters move to the *left* of the wind direction. Why does surface water move in a direction different than the wind? **V. Walfrid Ekman** (1874–1954), a Swedish physicist, developed a circulation model in 1905 called the **Ekman spiral** (Figure 7.7) that explains Nansen's observations as a balance between frictional effects in the ocean and the Coriolis effect.

The Ekman spiral describes the speed and direction of flow of surface waters at various depths. Ekman's model assumes that a uniform column of water is set in motion by wind blowing across its surface. Because of the Coriolis effect, the immediate surface water moves in a direction 45 degrees to the right of the wind (in the Northern Hemisphere). The surface water moves as a thin "layer" on top of deeper layers of water. As the surface layer moves, other layers beneath it are set in motion, thus passing the energy of the wind down through the water column just like how a deck of cards can be fanned out by pressing on and rotating only the top card.

Current speed decreases with increasing depth, however, and the Coriolis effect increases curvature to the right (like a spiral). Thus, each successive layer of water is set in motion at a progressively slower velocity and in a direction progressively to the right of the one above it. At some depth, a layer of water may move in a direction *exactly opposite from the wind direction that initiated it!* If the water is deep enough, friction will consume the energy imparted by the wind, and no motion will occur below that depth. Although it depends on wind speed and latitude, this stillness normally occurs at a depth of about 100 meters (330 feet).

Figure 7.7 shows the spiral nature of this movement with increasing depth from the ocean's surface. The length of each arrow in Figure 7.7 is proportional to the velocity of the individual layer, and the direction of each arrow indicates the direction

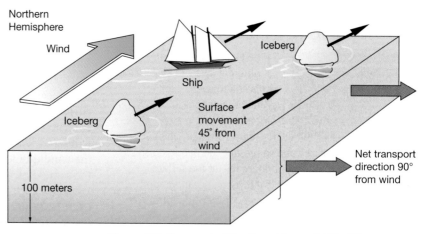

Figure 7.6 Transport of floating objects. Fridtjof Nansen first noticed that floating objects, such as icebergs and ships, were carried to the right of the wind direction in the Northern Hemisphere.

Web Animation
Ekman Spiral and Coastal Upwelling/Downwelling

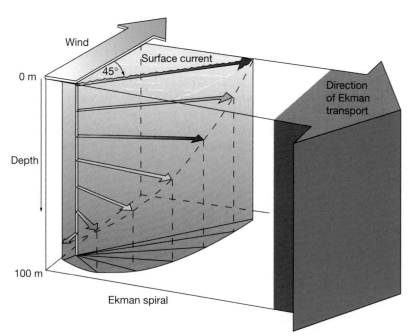

(a) Wind drives surface water in a direction 45 degrees to the right of the wind in the Northern Hemisphere. Deeper water continues to deflect to the right and moves at a slower speed with increased depth, causing the Ekman spiral.

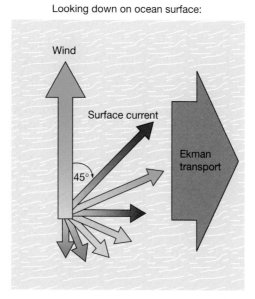

(b) Ekman transport, which is the average water movement for the entire column, is at a right angle (90 degrees) to the wind direction.

Figure 7.7 Ekman spiral. Perspective view **(a)** and top view **(b)** of same area showing how the Ekman spiral produces Ekman transport.

it moves.[5] Under ideal conditions, therefore, the surface layer should move at an angle of 45 degrees from the direction of the wind. All the layers combine, however, to create a net water movement that is 90 degrees from the direction of the wind. This average movement, called **Ekman transport**, is 90 degrees to the *right* in the Northern Hemisphere and 90 degrees to the *left* in the Southern Hemisphere.

"Ideal" conditions rarely exist in the ocean, so the actual movement of surface currents deviates slightly from the angles shown in Figure 7.7. Generally, surface currents move at an angle somewhat less than 45 degrees from the direction of the wind and Ekman transport in the open ocean is typically about 70 degrees from the wind direction. In shallow coastal waters, Ekman transport may be very nearly the same direction as the wind.

GEOSTROPHIC CURRENTS   Ekman transport deflects surface water to the right in the Northern Hemisphere, so a clockwise rotation develops within an ocean basin and produces the **Subtropical Convergence** of water in the middle of the gyre, causing water literally to pile up in the center of the subtropical gyre. Thus, there is a hill of water within all subtropical gyres that is as much as 2 meters (6.6 feet) high.

Surface water in the Subtropical Convergence tends to flow downhill in response to gravity. The Coriolis effect opposes gravity, however, deflecting the water to the right in a curved path (**Figure 7.8a**) into the hill again. When these two factors balance, the net effect is a **geostrophic** (*geo* = earth, *strophio* = turn) **current** that moves in a circular path around the hill and is shown in Figure 7.8a as the *path of ideal geostrophic flow*.[6] Friction between water molecules, however, causes the water to move gradually down the slope of the hill as it flows around it. This is the path of *actual geostrophic flow* labeled in Figure 7.8a.

If you reexamine the satellite image of sea surface elevation in Figure 7.2, you'll see that the hills of water within the subtropical gyres of the Atlantic Ocean are clearly visible. The hill in the North Pacific is visible as well, but the elevation of the equatorial Pacific is not as low as expected because the map shows conditions during a moderate El Niño event,[7] so there is a well-developed region of warm water across the equatorial Pacific that has an anomalously high sea surface height. Figure 7.2 also shows very little distinction between the North and South Pacific Gyres. Moreover, the South Pacific Gyre hill is less pronounced than in other gyres because (1) it covers such a large area, (2) it lacks confinement by continental barriers along its western margin, and (3) it is interfered with by numerous islands (really the tops of tall sea floor mountains). The South Indian Ocean hill is rather well developed in the figure, although its northeastern boundary stands high because of the influx of warm Pacific Ocean water through the East Indies islands.

WESTERN INTENSIFICATION OF SUBTROPICAL GYRES   Figure 7.8a shows that the apex (top) of the hill formed within a rotating gyre is closer to the western boundary than the center of the gyre. As a result, the western boundary currents of the subtropical gyres are faster, narrower, and deeper than their eastern boundary current counterparts. For example, the Kuroshio Current (a western boundary current) of the North Pacific Gyre is up to 15 times faster, 20 times narrower, and 5 times deeper than the California Current (an eastern boundary current). This phenomenon is called **western intensification**, and currents affected by this phenomenon are said to be *western intensified*. Note that the western boundary currents of *all* subtropical gyres are western intensified, *even in the Southern Hemisphere*.

A number of factors cause western intensification, including the Coriolis effect. The Coriolis effect increases toward the poles, so eastward-flowing high-latitude water turns toward the equator more strongly than westward-flowing equatorial

---

[5]The name Ekman *spiral* refers to the spiral observed by connecting the tips of the arrows shown in Figure 7.7.

[6] The term *geostrophic* for these currents is appropriate, since the currents behave as they do because of Earth's rotation.

[7]El Niño events are discussed later in this chapter, under "Pacific Ocean Circulation."

Northern Hemisphere Subtropical Gyre

Figure 7.8 **Geostrophic current and western intensification.**

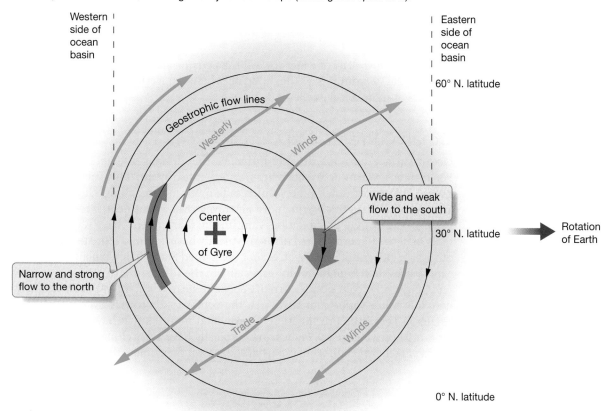

(a) Perspective view of a subtropical gyre showing how water literally piles up in the center, forming a hill up to 2 meters (6.6 feet) high. Gravity and the Coriolis effect balance to create an ideal geostropic current that flows in equilibrium around the hill. However, friction makes the current gradually run downslope (*actual geostrophic flow*).

(b) Corresponding map view of the same subtropical gyre, showing that the flow pattern is restricted (lines are closer together) on the western side of the gyre, resulting in western intensification.

water turns toward higher latitudes. This causes a wide, slow, and shallow flow of water toward the equator across most of each subtropical gyre, leaving only a narrow band through which the poleward flow can occur along the western margin of the ocean basin. If a constant volume of water rotates around the apex of the hill in Figure 7.8b, then the velocity of the water along the western margin will be much greater than the velocity around the eastern side.[8] In Figure 7.8b, the lines are close together along the western margin, indicating the faster flow. The end result is a high-speed western boundary current that flows along the hill's steeper westward

---

[8]A good analogy for this phenomenon is a funnel: In the narrow end of a funnel, the flow rates are speeded up (such as in western intensified currents); in the wide end, the flow rates are sluggish (such as in eastern boundary currents).

TABLE 7.2   CHARACTERISTICS OF WESTERN AND EASTERN BOUNDARY CURRENTS OF SUBTROPICAL GYRES

Current type	Examples	Width	Depth	Speed	Transport volume (millions of cubic meters per second[a])	Comments
Western boundary current	Gulf Stream, Brazil Current, Kuroshio Current	*Narrow:* usually less than 100 kilometers (60 miles)	*Deep:* to depths of 2 kilometers (1.2 miles)	*Fast:* hundreds of kilometers per day	*Large:* as much as 100 Sv[a]	Waters derived from low latitudes and are warm; little or no upwelling
Eastern boundary current	Canary Current, Benguela Current, California Current	*Wide:* up to 1000 kilometers (600 miles)	*Shallow:* to depths of 0.5 kilometer (0.3 mile)	*Slow:* tens of kilometers per day	*Small:* typically 10 to 15 Sv[a]	Waters derived from middle latitudes and are cool; coastal upwelling common

[a]One million cubic meters (35.3 million cubic feet) per second is a flow rate equal to one Sverdrup (Sv).

slope and a slow drift of water toward the equator along the more gradual eastern slope. Table 7.2 summarizes the differences between western and eastern boundary currents of subtropical gyres.

Ocean Currents and Climate

Ocean surface currents directly influence the climate of adjoining landmasses. For instance, warm ocean currents warm the nearby air. This warm air can hold a large amount of water vapor, which puts more moisture (high humidity) in the atmosphere. When this warm, moist air travels over a continent, it releases its water vapor in the form of precipitation. Continental margins that have warm ocean currents offshore (**Figure 7.9**, *red arrows*) typically have a humid climate. The presence of a warm current off the East Coast of the United States helps explain why the area experiences such high humidity, especially in the summer.

Conversely, cold ocean currents cool the nearby air, which is more likely to have low water vapor content. When the cool, dry air travels over a continent, it results in very little precipitation. Continental margins that have cool ocean currents offshore (Figure 7.9, *blue arrows*) typically have a dry climate. The presence of a cold current off California is part of the reason the climate there is so arid.

CONCEPT CHECK 7.2

1. How many subtropical gyres exist worldwide? How many main currents exist within each subtropical gyre?

2. On a base map of the world, plot and label the major currents involved in the surface circulation gyres of the oceans. Use colors to represent warm versus cool currents and indicate which currents are western intensified. On an overlay, superimpose the major wind belts of the world on the gyres and describe the relationship between wind belts and currents.

3. Explain why the subtropical gyres in the Northern Hemisphere move in a clockwise fashion while the subpolar gyres rotate in a counter-clockwise pattern.

4. Diagram and discuss how Ekman transport produces the "hill" of water within subtropical gyres that causes geostrophic current flow. As a starting place on the diagram, use the wind belts (the trade winds and the prevailing westerlies). What causes the apex of the geostrophic "hills" to be offset to the west of the center of the ocean gyre systems?

5. Describe western intensification, including the characteristics of western and eastern boundary currents of subtropical gyres.

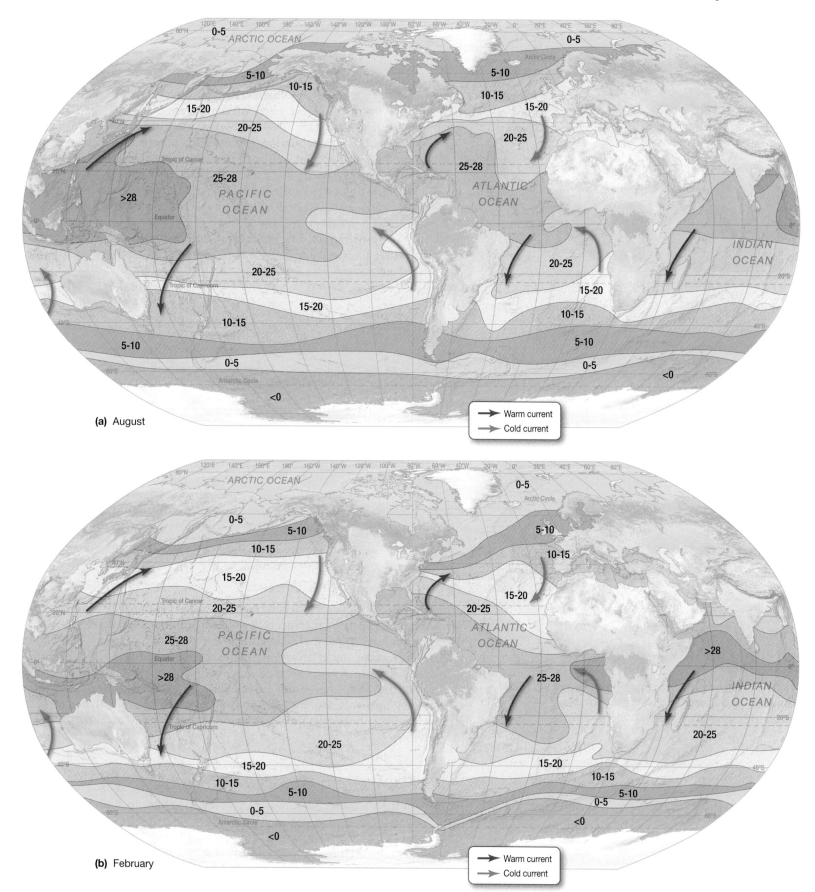

Figure 7.9 Sea surface temperature of the world ocean. Maps showing average sea surface temperature distribution in degrees centigrade for August **(a)** and for February **(b)**. Note that temperatures migrate north–south with the seasons. Red arrows indicate warm surface currents; blue arrows indicate cool surface currents.

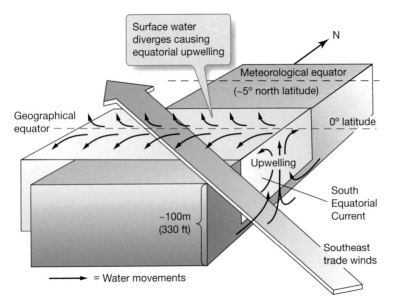

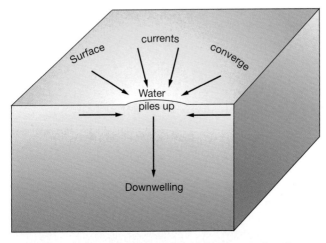

Figure 7.10 Equatorial upwelling. As the southeast trade winds pass over the geographical equator to the meteorological equator, they cause water within the South Equatorial Current north of the equator to veer to the right (northward) and water south of the equator to veer to the left (southward). Thus, surface water diverges, which causes equatorial upwelling.

Figure 7.11 Downwelling caused by convergence of surface currents. Where surface currents converge, water piles up and slowly sinks downward, creating downwelling.

Web Animation
Ekman Spiral and Coastal
Upwelling/Downwelling

7.3 What Causes Upwelling and Downwelling?

Upwelling is the vertical movement of cold, deep, nutrient-rich water to the surface; **downwelling** is the vertical movement of surface water to deeper parts of the ocean. Upwelling hoists chilled water to the surface. This cold water, rich in nutrients, creates high **productivity** (an abundance of microscopic algae), which establishes the base of the food web and, in turn, supports incredible numbers of larger marine life like fish and whales. Downwelling, on the other hand, is associated with much lower amounts of surface productivity but carries necessary dissolved oxygen to those organisms living on the deep-sea floor.

Upwelling and downwelling provide important mixing mechanisms between surface and deep waters and are accomplished by a variety of methods.

Diverging Surface Water

Current divergence occurs when surface waters move *away from* an area on the ocean's surface, such as along the equator. As shown in **Figure 7.10**, the South Equatorial Current occupies the area along the *geographical equator* (most notably in the Pacific Ocean; see Figure 7.5), while the *meteorological equator* (where the doldrums exist) typically occurs a few degrees of latitude to the north. As the southeast trade winds blow across this region, Ekman transport causes surface water north of the equator to veer to the right (northward) and water south of the equator to veer to the left (southward). The net result is a divergence of surface currents along the geographical equator, which causes upwelling of cold, nutrient-rich water. Because this type of upwelling is common along the equator—especially in the Pacific—it is called **equatorial upwelling**, and it creates areas of high productivity that are some of the most prolific fishing grounds in the world.

Converging Surface Water

Current convergence occurs when surface waters move *toward* each other. In the North Atlantic Ocean, for instance, the Gulf Stream, the Labrador Current, and the East Greenland Current all come together in the same vicinity. When currents converge, water stacks up and has no place to go but downward. The surface water slowly sinks in a process called *downwelling* (**Figure 7.11**). Unlike upwelling, areas of downwelling are not associated with prolific marine life because the necessary nutrients are not continuously replenished from the cold, deep, nutrient-rich water below the surface. Consequently, downwelling areas have low productivity.

Coastal Upwelling and Downwelling

Coastal winds can cause upwelling or downwelling due to Ekman transport. **Figure 7.12** shows a coastal region along the west coast of a continent in the Northern Hemisphere with winds moving parallel to the coast. If the winds are from the north (Figure 7.12a), Ekman transport moves the coastal water to the right of the wind direction, causing the water to flow *away from* the shoreline. Water rises from below to replace the water moving away from shore in a process called **coastal upwelling**. Areas where coastal upwelling occurs, such as the West Coast of the United States, are characterized by high concentrations of nutrients, resulting

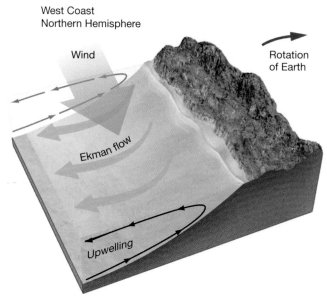

West Coast
Northern Hemisphere

Wind

Rotation
of Earth

Ekman flow

Upwelling

(a) Ekman transport carries surface water to the *right* of the wind
direction and away from a continent. Upwelling of deeper water
replaces the surface water that has moved away from the coast.

West Coast
Northern Hemisphere

Rotation
of Earth

Ekman flow

Downwelling

Wind

(b) A reversal of the wind direction still causes Ekman transport to
the right but results in water piling up against the shore,
producing downwelling.

Figure 7.12 Coastal upwelling and downwelling. Coastal upwelling and downwelling can be produced by winds that blow parallel to the coast. Similar situations exist for coastal winds and upwelling/downwelling in the Southern Hemisphere, except that Ekman transport is to the *left* of the wind direction.

in high productivity and rich marine life. This coastal up-welling also creates low water temperatures in areas such as San Francisco, providing a natural form of air conditioning (and much cool weather and fog) in the summer.

If the winds are from the south, Figure 7.12b shows that Ekman transport still moves the coastal water to the right of the wind direction but, in this case, the water flows *toward* the shoreline. The water stacks up along the shoreline and has nowhere to go but down, in a process called **coastal downwelling**. Areas where coastal downwelling occurs have low productivity and a lack of marine life. If the winds reverse, areas that are typically associated with coastal downwelling can experience upwelling.

A similar situation exists for coastal winds and upwelling/downwelling in the Southern Hemisphere, except that Ekman transport is to the *left* of the wind direction.

Other Causes Of Upwelling

Figure 7.13 shows how upwelling can be created by offshore winds, sea floor obstructions, or a sharp bend in a coastline. Upwelling also occurs in high-latitude regions, where there is no pycnocline (a layer of rapidly changing density). The absence of a pycnocline allows significant vertical mixing between high-density cold surface water and high-density cold deep water below. Thus, both upwelling and downwelling are common in high latitudes.

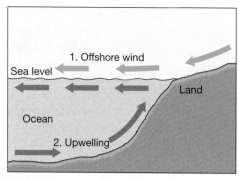

1. Offshore wind
Sea level
Land
Ocean
2. Upwelling

(a) Upwelling caused by offshore winds

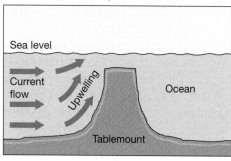

Sea level
Current flow
Upwelling
Ocean
Tablemount

(b) Upwelling caused by a tablemount

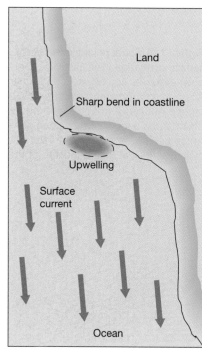

Land
Sharp bend in coastline
Upwelling
Surface current
Ocean

(c) Upwelling caused by a sharp bend in coastal geometry

Figure 7.13 Other types of upwelling. A variety of factors can cause upwelling, including **(a)** offshore winds, **(b)** a sea floor obstruction such as a tablemount, and **(c)** a sharp bend in a coastline.

CONCEPT CHECK 7.3

❶ Draw and describe several different oceanographic conditions that produce upwelling.

❷ Explain why upwelling areas are associated with an abundance of marine life.

ESSENTIAL CONCEPT 7.3

Upwelling and downwelling cause vertical mixing between surface and deep water. Upwelling brings cold, deep, nutrient-rich water to the surface, which results in high productivity.

7.4 What Are The Main Surface Circulation Patterns in Each Ocean Basin?

The pattern of surface currents varies from ocean to ocean, depending on the geometry of the ocean basin, the pattern of major wind belts, seasonal factors, and other periodic changes.

Antarctic Circulation

Antarctic circulation is dominated by the movement of water masses in the southern Atlantic, Indian, and Pacific Oceans south of about 50 degrees south latitude.

ANTARCTIC CIRCUMPOLAR CURRENT The main current in Antarctic waters is the **Antarctic Circumpolar Current**, which is also called the **West Wind Drift**. This current encircles Antarctica and flows from west to east at approximately 50 degrees south latitude but varies between 40 and 65 degrees south latitude. At about 40 degrees south latitude is the *Subtropical Convergence* (**Figure 7.14**), which forms the northernmost boundary of the Antarctic Circumpolar Current. The Antarctic Circumpolar Current is driven by the powerful prevailing westerly wind belt, which creates winds so strong that these Southern Hemisphere latitudes have been called the "Roaring Forties," "Furious Fifties," and "Screaming Sixties."

The Antarctic Circumpolar Current is the only current that completely circumscribes Earth, and it is allowed to do so because of the lack of land at high southern latitudes. It meets its greatest restriction as it passes through the *Drake Passage* (named for the famed English sea captain and ocean explorer Sir Francis Drake [1540–1596]) between the Antarctic Peninsula and the southern islands of South America, which is about 1000 kilometers (600 miles) wide. Although the current is not speedy (its maximum surface velocity is about 2.75 kilometers [1.65 miles] per hour), it transports on average about 130 Sverdrups[9] (130 million cubic meters [4.6 billion cubic feet] per second), which is more than any other surface current.

ANTARCTIC CONVERGENCE AND DIVERGENCE The **Antarctic Convergence** (Figure 7.14), or *Antarctic Polar Front*, is where colder, denser Antarctic waters converge with (and sink sharply below) warmer, less-dense sub-Antarctic waters at about 50 degrees south latitude. The Antarctic Convergence marks the northernmost boundary of the Southern, or Antarctic, Ocean.

The **East Wind Drift**, a surface current propelled by the polar easterlies, moves from an easterly direction around the margin of the Antarctic continent. The East Wind Drift is most extensively developed to the east of the Antarctic Peninsula in the Weddell Sea region and in the area of the Ross Sea (Figure 7.14). As the East Wind Drift and the Antarctic Circumpolar Current flow around Antarctica in opposite directions, they create a surface divergence. Recall that the Coriolis effect deflects moving masses to the left in the Southern Hemisphere, so the East Wind Drift is deflected toward the continent, and the Antarctic Circumpolar Current is deflected away from it. This creates a divergence of currents along a boundary called the **Antarctic Divergence**. The Antarctic Divergence has abundant marine life in the Southern Hemisphere summer because of the mixing of these two currents, which supplies nutrient-rich water to the surface through upwelling.

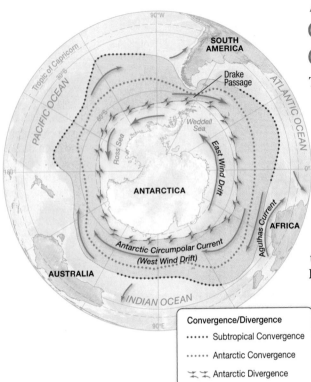

Figure 7.14 South polar view of Earth showing Antarctic surface circulation. The East Wind Drift is driven by the polar easterlies and flows around Antarctica from the east. The Antarctic Circumpolar Current (West Wind Drift) flows around Antarctica from the west but is further from the continent and is a result of the strong prevailing westerlies. The Antarctic Convergence and Antarctic Divergence are caused by interactions at the boundaries of these two currents.

---

[9]One million cubic meters (35.3 million cubic feet) per second is a useful flow rate for describing ocean currents, so it has become a standard unit, named the **Sverdrup (Sv)**, after Norwegian meteorologist and physical oceanographer Harald Sverdrup (1888–1957).

Atlantic Ocean Circulation

Figure 7.15 shows Atlantic Ocean surface circulation, which consists of two large subtropical gyres: the North Atlantic Gyre and the South Atlantic Gyre.

THE NORTH AND SOUTH ATLANTIC SUBTROPICAL GYRES   The **North Atlantic Subtropical Gyre** rotates clockwise, and the **South Atlantic Subtropical Gyre** rotates counterclockwise, due to the combined effects of the trade winds, the prevailing westerlies, and the Coriolis effect. Figure 7.15 shows that each gyre consists of a poleward-moving warm current (*red*) and an equatorward-moving cold "return" current (*blue*). The two gyres are partially offset by the shapes of the surrounding continents, and the **Atlantic Equatorial Countercurrent** moves in between them.

In the South Atlantic Gyre, the **South Equatorial Current** reaches its greatest strength just below the equator, where it encounters the coast of Brazil and splits in two. Part of the South Equatorial Current moves off along the northeastern coast of South America toward the Caribbean Sea and the North Atlantic. The rest is turned southward as the **Brazil Current**, which ultimately merges with the Antarctic Circumpolar Current (West Wind Drift) and moves eastward across the South Atlantic. The Brazil Current is much smaller than its Northern Hemisphere counterpart, the Gulf Stream, due to the splitting of the South Equatorial Current. The **Benguela Current**, slow moving and cold, flows toward the equator along Africa's western coast, completing the gyre.

Outside the gyre, the *Falkland Current* (Figure 7.15), which is also called the *Malvinas Current*, moves a significant amount of cold water along the coast of Argentina as far north as 25 to 30 degrees south latitude, wedging its way between the continent and the southbound Brazil Current.

THE GULF STREAM   The **Gulf Stream** is the world's best studied ocean current. It moves northward along the East Coast of the United States, warming coastal states and moderating winters in these and northern European regions.

Figure 7.16 shows the network of currents in the North Atlantic Ocean that contribute to the flow of the Gulf Stream. The **North Equatorial Current** moves parallel to the equator in the Northern Hemisphere, where it is joined by the portion of the South Equatorial Current that turns northward along the South American coast. This flow then splits into the **Antilles Current**, which passes along the Atlantic side of the West Indies, and the **Caribbean Current**, which passes through the Yucatán Channel into the Gulf of Mexico. These masses reconverge as the **Florida Current**.

The Florida Current flows close to shore over the continental shelf at a rate that at times exceeds 35 Sverdrups. As it moves off North Carolina's Cape Hatteras and flows across the deep ocean in a northeasterly direction, it is called the *Gulf Stream*. The Gulf Stream is a western boundary current, so it is subject to western intensification. Thus, it is only 50 to 75 kilometers (31 to 47 miles) wide, but it reaches depths of 1.5 kilometers (1 mile) and speeds from 3 to 10 kilometers (2 to 6 miles) per hour, making it the fastest current in the world ocean.

The western boundary of the Gulf Stream is usually abrupt, but it periodically migrates closer to and farther away from the shore. Its eastern boundary is very difficult to identify because it is usually masked by meandering water masses that change their position continuously.

The Sargasso Sea   The Gulf Stream gradually merges eastward with the water of the **Sargasso Sea**. The Sargasso Sea is the water that circulates around the rotation center of the North Atlantic gyre. The Sargasso Sea can be thought of as the stagnant eddy of the North Atlantic Gyre. Its name is derived from a type of floating marine alga called *Sargassum* (*sargassum* = grapes) that abounds on its surface.

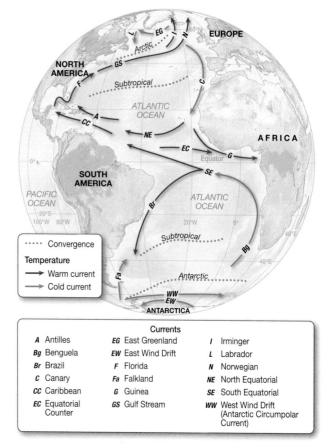

Currents		
A Antilles	**EG** East Greenland	**I** Irminger
Bg Benguela	**EW** East Wind Drift	**L** Labrador
Br Brazil	**F** Florida	**N** Norwegian
C Canary	**Fa** Falkland	**NE** North Equatorial
CC Caribbean	**G** Guinea	**SE** South Equatorial
EC Equatorial Counter	**GS** Gulf Stream	**WW** West Wind Drift (Antarctic Circumpolar Current)

Figure 7.15 Atlantic Ocean surface currents. Atlantic Ocean surface circulation is composed primarily of two subtropical gyres.

STUDENTS SOMETIMES ASK . . .

Is the Gulf Stream rich in life?

The Gulf Stream *itself* isn't rich in life, but its *boundaries* often are. The oceanic areas that have abundant marine life are typically associated with cool water—either in high-latitude regions, or in any region where upwelling occurs. These areas are constantly resupplied with oxygen- and nutrient-rich water, which results in high productivity. Warm-water areas develop a prominent thermocline that isolates the surface water from colder, nutrient-rich water below. Nutrients used up in warm waters tend not to be resupplied. The Gulf Stream, which is a western intensified, warm-water current, is therefore associated with low productivity and an absence of marine life. New England fishers knew about the Gulf Stream (see **Web Diving Deeper 7.3**) because they sought their catch along the sides of the current, where mixing and upwelling occur.

Actually, all western intensified currents are warm and are associated with low productivity. The Kuroshio Current in the North Pacific Ocean, for example, is named for its conspicuous absence of marine life. In Japanese, *Kuroshio* means "black current," in reference to its clear, lifeless waters.

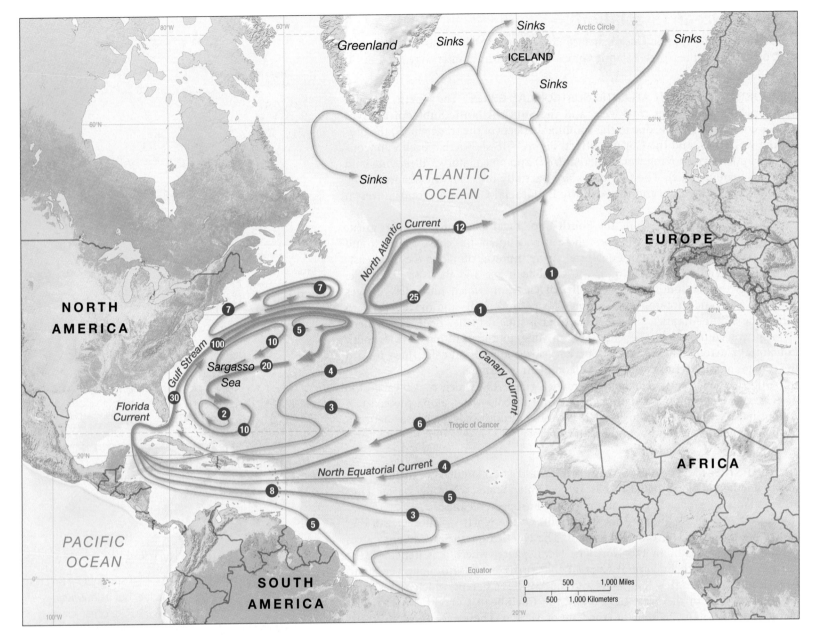

Figure 7.16 North Atlantic Ocean circulation. The North Atlantic Gyre, showing average flow rates in *Sverdrups* (1 Sverdrup = 1 million cubic meters [35.3 million cubic feet] per second). The four major currents include the western intensified Gulf Stream, the North Atlantic Current, the Canary Current, and the North Equatorial Current. Some water splits off in the North Atlantic, where it becomes cold and dense, so it sinks. The Sargasso Sea occupies the stagnant eddy in the middle of the subtropical gyre.

Web Animation
The Gulf Stream, Meanders, and the Formation of Cold-Core and Warm-Core Eddies

The transport rate of the Gulf Stream off Chesapeake Bay is about 100 Sverdrups,[10] which suggests that a large volume of water from the Sargasso Sea has combined with the Florida Current to produce the Gulf Stream. By the time the Gulf Stream nears Newfoundland, however, the transport rate is only 40 Sverdrups, which suggests that a large volume of water has returned to the diffuse flow of the Sargasso Sea.

Warm- and Cold-Core Rings The mechanisms that produce the dramatic loss of water as the Gulf Stream moves northward are yet to be determined. Meanders, however, may cause much of it. **Meanders** (*Menderes* = a river in Turkey that has a very sinuous course) are snakelike bends in the current that often disconnect from the Gulf Stream and form large rotating masses of water called *vortexes* (*vortex* = to

[10]The Gulf Stream's flow of 100 Sverdrups equates to a volume of about 100 major league sport stadiums passing by the southeast U.S. coast *each second* and is more than 100 times greater than the combined flow of *all* the world's rivers!

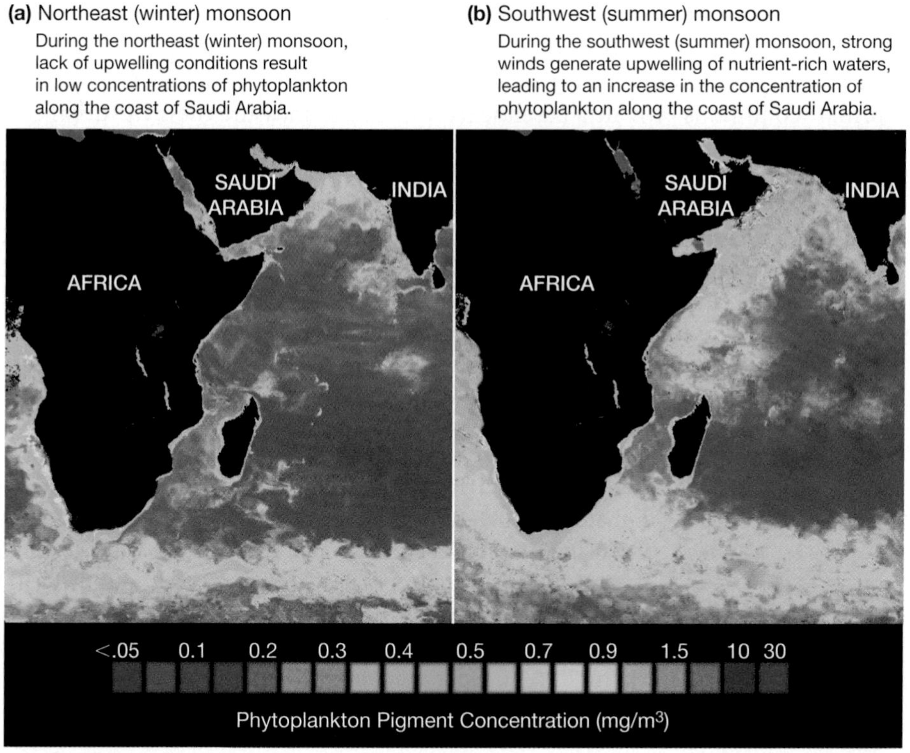

(a) Northeast (winter) monsoon

During the northeast (winter) monsoon, lack of upwelling conditions result in low concentrations of phytoplankton along the coast of Saudi Arabia.

(b) Southwest (summer) monsoon

During the southwest (summer) monsoon, strong winds generate upwelling of nutrient-rich waters, leading to an increase in the concentration of phytoplankton along the coast of Saudi Arabia.

<.05   0.1   0.2   0.3   0.4   0.5   0.7   0.9   1.5   10 30

Phytoplankton Pigment Concentration (mg/m³)

Figure 7.20 Seasonal variations in phytoplankton concentration in the Indian Ocean. Paired satellite images for the two monsoons showing oceanic phytoplankton pigment, where orange and red colors indicate higher amounts of phytoplankton and thus higher productivity.

recent years due to stronger winds caused by warming of the Eurasian landmass, resulting in higher-than-normal summer productivity in the Arabian Sea.

INDIAN OCEAN SUBTROPICAL GYRE   Surface circulation in the southern Indian Ocean (the **Indian Ocean Subtropical Gyre**) is similar to subtropical gyres observed in other southern oceans. When the northeast trade winds blow, the South Equatorial Current provides water for the Equatorial Countercurrent and the **Agulhas Current**,[11] which flows southward along Africa's east coast and joins the Antarctic Circumpolar Current (West Wind Drift). The *Agulhas Retroflection* is created when the Agulhas Current makes an abrupt turn as it meets the strong Antarctic Circumpolar Current (see the chapter-opening satellite image). Turning northward out of the Antarctic Circumpolar Current is the **West Australian Current**, an eastern boundary current that merges with the South Equatorial Current, completing the gyre.

LEEUWIN CURRENT   Eastern boundary currents in other subtropical gyres are cold drifts toward the equator that produce arid coastal climates (that is, they receive less than 25 centimeters [10 inches] of rain per year). In the southern Indian Ocean, however, the **Leeuwin Current** displaces the West Australian Current offshore. The Leeuwin Current is driven southward along the Australian coast from the warm-water dome piled up in the East Indies by the Pacific equatorial currents.

The Leeuwin Current produces a mild climate in southwestern Australia, which receives about 125 centimeters (50 inches) of rain per year. During El Niño events, however, the Leeuwin Current weakens, so the cold Western Australian Current brings drought instead.

Pacific Ocean Circulation

Two large subtropical gyres dominate the circulation pattern in the Pacific Ocean, resulting in surface water movement and climatic effects similar to those found in the Atlantic. However, the Equatorial Countercurrent is much better developed in the Pacific Ocean than in the Atlantic (**Figure 7.21**), largely because the Pacific Ocean basin is larger and more unobstructed than the Atlantic Ocean.

---
[11]The Agulhas Current is named for Cape Agulhas, which is the southernmost tip of Africa.

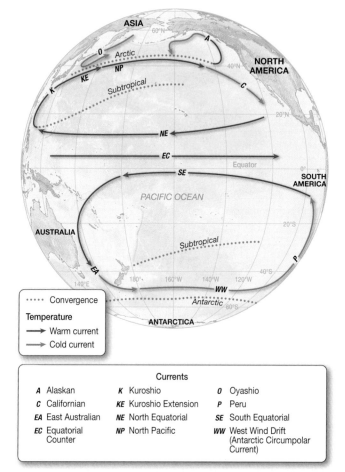

···· Convergence

Temperature
→ Warm current
→ Cold current

Currents					
A	Alaskan	*K*	Kuroshio	*O*	Oyashio
C	Californian	*KE*	Kuroshio Extension	*P*	Peru
EA	East Australian	*NE*	North Equatorial	*SE*	South Equatorial
EC	Equatorial Counter	*NP*	North Pacific	*WW*	West Wind Drift (Antarctic Circumpolar Current)

Figure 7.21 Pacific Ocean surface currents. Circulation in the Pacific Ocean consists of two large subtropical gyres similar to those in the Atlantic Ocean. However, the equatorial countercurrent is more strongly developed in the Pacific than in smaller ocean basins.

Web Animation
El Nino and La Nina

NORMAL CONDITIONS "Normal" conditions in the Pacific Ocean are a bit of a misnomer because they are experienced so infrequently. As we'll see, various atmospheric and oceanic disturbances dominate conditions in the Pacific. Still, "normal" conditions provide a baseline from which to measure these disturbances.

North Pacific Subtropical Gyre Figure 7.21 shows that the **North Pacific Subtropical Gyre** includes the North Equatorial Current, which flows westward into the western intensified **Kuroshio Current**[12] near Asia. The warm waters of the Kuroshio Current make Japan's climate warmer than would be expected for its latitude. This current flows into the **North Pacific Current**, which connects to the cool-water **California Current**. The California Current flows south along the coast of California to complete the loop. Some North Pacific Current water also flows to the north and merges into the **Alaskan Current** in the Gulf of Alaska.

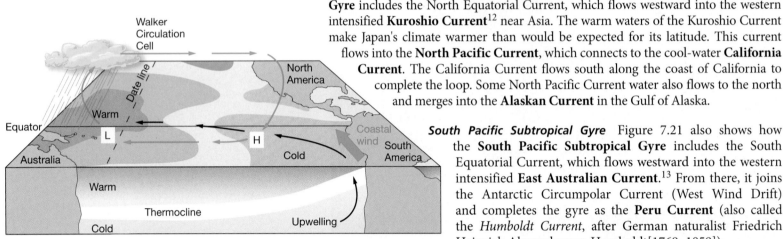

South Pacific Subtropical Gyre Figure 7.21 also shows how the **South Pacific Subtropical Gyre** includes the South Equatorial Current, which flows westward into the western intensified **East Australian Current**.[13] From there, it joins the Antarctic Circumpolar Current (West Wind Drift) and completes the gyre as the **Peru Current** (also called the *Humboldt Current*, after German naturalist Friedrich Heinrich Alexander von Humboldt[1769–1859]).

(a) Normal conditions

FISHERIES AND THE PERU CURRENT The cool waters of the Peru Current have historically been one of Earth's richest fishing grounds. What conditions produce such an abundance of fish? Figure 7.22a shows that along the west coast of South America, coastal winds create Ekman transport that moves water away from shore, causing upwelling of cool, nutrient-rich water. This upwelling increases productivity and results in an abundance of marine life, including small silver-colored fish called *anchovetas* (anchovies) that become particularly plentiful near Peru and Ecuador. Anchovies provide a food source for many larger marine organisms and also supply Peru's commercial fishing industry, which was established in the 1950s. Anchovies had been so abundant in the waters off South America that by 1970 Peru was the largest producer of fish from the sea in the world, with a peak production of 12.3 million metric tons (13.5 million short tons), accounting for about one-quarter of *all* fish from the sea worldwide.

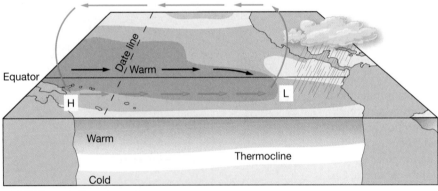

(b) El Niño conditions (strong)

WALKER CIRCULATION Figure 7.22a shows that high pressure and sinking air dominate the coastal region of South America, resulting in clear, fair, and dry weather. On the other side of the Pacific, a low-pressure region and rising air create cloudy conditions with plentiful precipitation in Indonesia, New Guinea, and northern Australia. This pressure difference causes the strong southeast trade winds to blow across the equatorial South Pacific. The resulting atmospheric circulation cell in the equatorial South Pacific Ocean is named the **Walker Circulation Cell** (*green arrows*) after Sir Gilbert T. Walker (1868–1958), the British meteorologist who first described the effect in the 1920s.

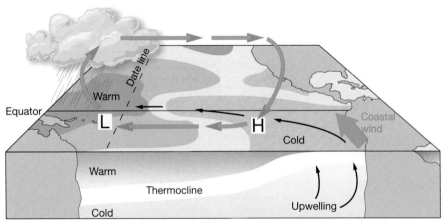

(c) La Niña conditions

Figure 7.22 Normal, El Niño, and La Niña conditions. Perspective views of oceanic and atmospheric conditions in the equatorial Pacific Ocean. **(a)** Normal conditions. **(b)** El Niño (ENSO warm phase) conditions. **(c)** La Niña (ENSO cool phase) conditions.

---

[12]Kuroshio is pronounced "kuhr-ROH-shee-oh." Because of its proximity to Japan, the Kuroshio Current is also called the *Japan Current.*

[13]Note that the western intensified East Australian Current was named because it lies off the *east coast* of Australia, even though it occupies a position along the *western* margin of the Pacific Ocean basin.

PACIFIC WARM POOL   The southeast trade winds set ocean water in motion across the Pacific from east to west. The water warms as it flows in the equatorial region and creates a wedge of warm water on the western side of the Pacific Ocean, called the **Pacific Warm Pool** (see Figure 7.9). Due to the movement of equatorial currents to the west, the Pacific Warm Pool is thicker along the western side of the Pacific than along the eastern side. The *thermocline* beneath the Warm Pool in the western equatorial Pacific occurs below 100 meters (330 feet) depth. In the eastern Pacific, however, the thermocline is within 30 meters (100 feet) of the surface. The difference in depth of the thermocline can be seen by the sloping boundary between the warm surface water and the cold deep water in Figure 7.22a.

EL NIÑO–SOUTHERN OSCILLATION (ENSO) CONDITIONS   Peru's residents have known for generations that every few years, a current of warm water reduces the population of anchovies in coastal waters. The decrease in anchovies causes a dramatic decline not only in the fishing industry, but also in marine life such as sea birds, sea lions, and seals that depend on anchovies for food. The warm current also brings about changes in the weather—usually intense rainfall—and even brings such interesting items as floating coconuts from tropical islands near the equator. At first, these events were called *años de abundancia* (years of abundance) because the additional rainfall dramatically increased plant growth on the normally arid land. What was once thought of as a joyous event, however, soon became associated with the ecological and economic disaster that is now a well-known consequence of the phenomenon.

This warm-water current usually occurred around Christmas and thus was given the name **El Niño**, Spanish for "the child," in reference to baby Jesus. In the 1920s, Walker was the first to recognize that an east–west atmospheric pressure seesaw accompanied the warm current and he called the phenomenon the **Southern Oscillation**. Today, the combined oceanic and atmospheric effects are called **El Niño–Southern Oscillation (ENSO)**, which periodically alternates between warm and cold phases and causes dramatic environmental changes.

ENSO Warm Phase (El Niño)   Figure 7.22b shows the atmospheric and oceanic conditions during an ENSO warm phase, which is known as *El Niño*. The high pressure along the coast of South America weakens, reducing the difference between the high- and low-pressure regions of the Walker Circulation Cell. This, in turn, causes the southeast trade winds to diminish. In very strong El Niño events, the trade winds actually blow in the *reverse* direction.

Without the trade winds, the Pacific Warm Pool that has built up on the western side of the Pacific begins to flow back across the ocean toward South America, creating a band of warm water that stretches across the equatorial Pacific Ocean (Figure 7.23a). The warm water usually begins to move in September of an El Niño year and reaches South America by December or January. During strong to very strong El Niños, the water temperature off Peru can be up to 10°C (18°F) higher than normal. In addition, the average sea level can increase as much as 20 centimeters (8 inches), simply due to thermal expansion of the warm water along the coast.

As the warm water increases sea surface temperatures across the equatorial Pacific, temperature-sensitive corals are decimated in Tahiti, the Galápagos, and other tropical Pacific islands. In addition, many other organisms are affected by the warm water (see Web Diving Deeper 7.2). Once the warm water reaches South America, it moves north and south along the west coast of the Americas, increasing average sea level and the number of tropical hurricanes formed in the eastern Pacific.

The flow of warm water across the Pacific also causes the sloped thermocline boundary between warm surface waters and the cooler waters below to flatten out and become more horizontal (Figure 7.22b). Near Peru, upwelling brings warmer, nutrient-depleted water to the surface instead of cold, nutrient-rich water. In fact, *downwelling* can sometimes occur as the warm water stacks up along coastal South America. Productivity diminishes and most types of marine life in the area are dramatically reduced.

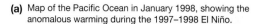
(a) Map of the Pacific Ocean in January 1998, showing the anomalous warming during the 1997–1998 El Niño.

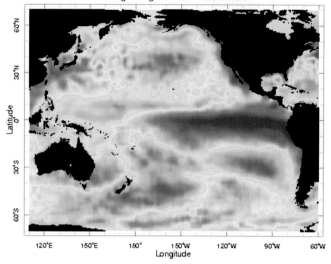

Sea Surface Temperature Anomaly

(b) Map of the same area in January 2000, showing cooling in the equatorial Pacific related to La Niña.

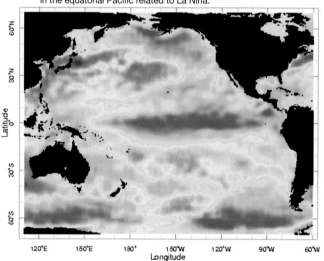

Figure 7.23   Sea surface temperature anomaly maps during El Niño and La Niña. Maps showing sea surface temperature anomalies, which represent departures from normal conditions. Red colors indicate water warmer than normal and blue colors represent water cooler than normal.

Web Video
Weekly Pacific Sea Surface
Temperature Anomalies

STUDENTS SOMETIMES ASK . . .

The amount of anchovies produced by Peru is impressive! Besides a topping for pizza, what are some other uses of anchovies?

Anchovies are an ingredient in certain dishes, hors d'oeuvres, sauces, and salad dressing, and fishers also use them as bait. Historically, however, most of the *anchoveta* caught in Peruvian waters were exported and used as fishmeal (consisting of ground anchovies). The fishmeal, in turn, was used largely in pet food and as a high-protein chicken feed. As unbelievable as it may seem, El Niños affected the price of eggs! Prior to the collapse of the Peruvian *anchoveta* fishing industry in 1972–1973, El Niño events significantly reduced the availability of *anchovetas*. This drastically cut the export of anchovies from Peru, causing U.S. farmers to pursue more expensive options for chicken feed. Thus, egg prices typically increased.

The collapse of the *anchoveta* fishing industry in Peru was triggered by the 1972–1973 El Niño event but was caused by chronic overfishing in prior years (see Web Diving Deeper 13.2). Interestingly, the shortage of fishmeal after 1972–1973 led to an increased demand for soyameal, an alternative source of high-quality protein. Increased demand for soyameal increased the price of soy commodities, thereby encouraging U.S. farmers to plant soybeans instead of wheat. Reduced production of wheat, in turn, caused a major global food crisis—and all this was triggered by an El Niño event.

As the warm water moves to the east across the Pacific, the low-pressure zone also migrates. In a strong to very strong El Niño event, the low pressure can move across the entire Pacific and remain over South America. The low pressure substantially increases precipitation along coastal South America. Conversely, high pressure replaces the Indonesian low, bringing dry conditions or, in strong to very strong El Niño events, drought conditions to Indonesia and northern Australia.

ENSO Cool Phase (La Niña) In some instances, conditions opposite of El Niño prevail in the equatorial South Pacific; these events are known as *ENSO cool phase* or **La Niña** (Spanish for "the female child"). Figure 7.22c shows La Niña conditions, which are similar to normal conditions but more intensified because there is a larger pressure difference across the Pacific Ocean. This larger pressure difference creates stronger Walker Circulation and stronger trade winds, which in turn cause more upwelling, a shallower thermocline in the eastern Pacific, and a band of cooler than normal water that stretches across the equatorial South Pacific (Figure 7.23b).

La Niña conditions commonly occur following an El Niño. For instance, the 1997–1998 El Niño was followed by several years of persistent La Niña conditions. The alternating pattern of El Niño–La Niña conditions since 1950 is shown by the multivariate **ENSO index** (**Figure 7.24**), which is calculated using a weighted average of atmospheric and oceanic factors, including atmospheric pressure, winds, and sea surface temperatures. Positive ENSO index numbers indicate El Niño conditions, whereas negative numbers reflect La Niña conditions. Normal conditions are indicated by a value near zero, and the greater the index value differs from zero (either negative or positive), the stronger the respective condition.

For a summary and comparison of atmospheric and oceanic conditions during normal, El Niño, and La Niña conditions, see **Web Tables 7.2** and **7.3**.

How Often Do El Niño Events Occur? Records of sea surface temperatures over the past 100 years reveal that throughout the 20th century, El Niño conditions occur on average about every 2 to 10 years, but in a highly irregular pattern. In some decades, for instance, there has been an El Niño event every few years, while in others there may have been only one. Figure 7.24 shows the pattern since 1950, revealing that the equatorial Pacific fluctuates between El Niño and La Niña conditions, with only a few years that could be considered "normal" conditions (represented by an ENSO index value close to zero). Typically, El Niño events last for 12 to 18 months and are followed by La Niña conditions that usually exist for a similar length of time. However, some El Niño or La Niña conditions can last for several years.

Recently recovered sediments from a South American lake provide a continuous 10,000-year record of the frequency of El Niño events. The sediments indicate

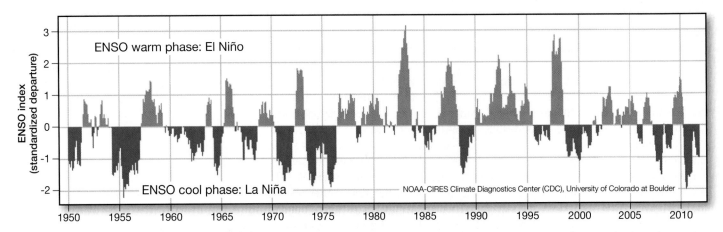

Figure 7.24 Multivariate ENSO index, 1950–present. The multivariate ENSO index is calculated using a variety of atmospheric and oceanic factors. ENSO index values greater than zero (*red areas*) indicate El Niño conditions while ENSO index values less than zero (*blue areas*) indicate La Niña conditions. The greater the value is from zero, the stronger the corresponding El Niño or La Niña. Updates are available at **www.cdc.noaa.gov/people/klaus.wolter/MEI/**.

that between 10,000 and 7000 years ago, no more than five strong El Niños occurred each century. The frequency of El Niños then increased, peaking at about 1200 years ago (and coinciding with the early Middle Ages in Europe), when they occurred every three years or so. If the pattern observed in the lake sediments continues, researchers predict that there should be an increase in El Niños in the early part of the 22nd century.

El Niño events—especially severe ones—may occur more frequently as a result of increased global warming. For instance, the two most severe El Niño events in the 20th century occurred in 1982–1983 and 1997–1998. Presumably, increased ocean temperatures could trigger more frequent and more severe El Niños. However, this pattern could also be a part of a long-term natural climate cycle. Recently, oceanographers have recognized a phenomenon called the **Pacific Decadal Oscillation (PDO)**, which lasts 20 to 30 years and appears to influence Pacific sea surface temperatures. Analysis of satellite data suggests that the Pacific Ocean was in the warm phase of the PDO from 1977 to 1999 and that it is now in its cool phase, which may suppress the initiation of El Niño events during the next few decades.

Effects of El Niños and La Niñas   Mild El Niño events influence only the equatorial South Pacific Ocean, while strong to very strong El Niño events can influence worldwide weather patterns. Typically, stronger El Niños alter the atmospheric jet stream and produce unusual weather in most parts of the globe. Sometimes the weather is drier than normal; at other times, it is wetter. The weather may also be warmer or cooler than normal. It is still difficult to predict exactly how a particular El Niño will affect any region's weather.

Figure 7.25 shows how very strong El Niño events can result in flooding, erosion, droughts, fires, tropical storms, and effects on marine life worldwide. These

Figure 7.25 Effects of severe El Niños. Flooding, erosion, droughts, fires, tropical storms, and effects on marine life are all associated with severe El Niño events.

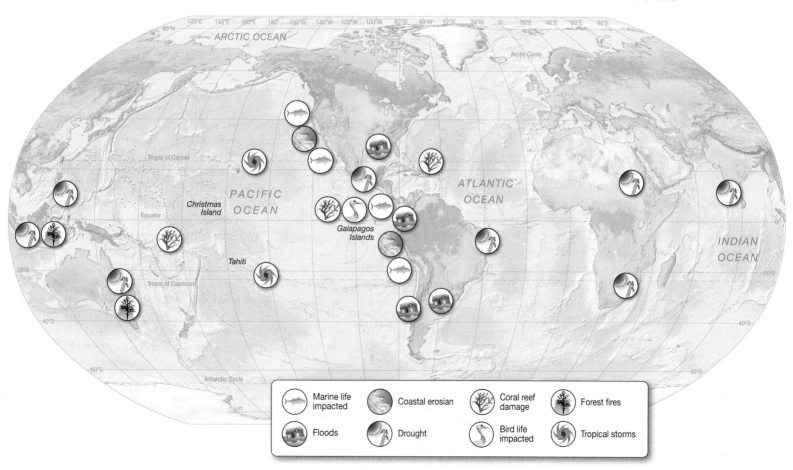

(a) Sea surface temperature anomaly map for January 1998, an El Niño year.

(b) Sea surface temperature anomaly map of the same region a year later (January 1999), during a La Niña event.

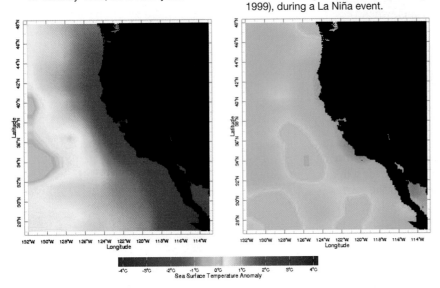

Figure 7.26 Sea surface temperatures off western North America during El Niño and La Niña. Satellite-derived sea surface temperature anomaly maps (in °C) along the west coast of North America. Red color represents water that is warmer than normal; blue color represents water that is colder than normal.

weather perturbations also affect the production of corn, cotton, and coffee. More locally, the satellite images in Figure 7.26 show that sea surface temperatures off western North America are significantly higher during an El Niño year.

Even though severe El Niños are typically associated with vast amounts of destruction, they can be beneficial in some areas. Tropical hurricane formation, for instance, is generally suppressed in the Atlantic Ocean, some desert regions receive much-needed rain, and organisms adapted to warm-water conditions thrive in the Pacific.

La Niña events are associated with sea surface temperatures and weather phenomena opposite those of El Niño. Indian Ocean monsoons, for instance, are typically drier than usual in El Niño years but wetter than usual in La Niña years.

Examples from Recent El Niños Recent El Niños provide an indication of the variability of the effects of El Niño events. For instance, in the winter of 1976, a moderate El Niño event coincided with northern California's worst drought of the 20th century, showing that El Niño events don't always bring torrential rains to the western United States. During that same winter, the eastern United States experienced record cold temperatures.

THE 1982–1983 EL NIÑO The 1982–1983 El Niño was the strongest ever recorded, causing far-ranging effects around the globe. Not only was there anomalous warming in the tropical Pacific, but the warm water also spread along the west coast of North America, influencing sea surface temperatures as far north as Alaska. Sea level was higher than normal (due to thermal expansion of the water), which, when high surf was experienced, caused damage to coastal structures and increased coastal erosion. In addition, the jet stream swung much farther south than normal across the United States, bringing a series of powerful storms that resulted in three times normal rainfall across the southwestern United States. The increased rainfall caused severe flooding and landslides as well as higher-than-normal snowfall in the Rocky Mountains. Alaska and western Canada had a relatively warm winter, and the eastern United States had its mildest winter in 25 years.

The full strength of El Niño was experienced in western South America. Normally arid Peru was drenched with more than 3 meters (10 feet) of rain, causing extreme flooding and landslides. Sea surface temperatures were so high for so long that temperature-sensitive coral reefs across the equatorial Pacific were decimated. Marine mammals and sea birds, which depend on the food normally available in the highly productive waters along the west coast of South America, went elsewhere or died. In the Galápagos Islands, for example, over half of the island's fur seals and sea lions died of starvation during the 1982–1983 El Niño.

French Polynesia had not experienced a hurricane in 75 years; in 1983, it endured six. The Hawaiian Island of Kauai also experienced a rare hurricane. Meanwhile, in Europe, severe cold weather prevailed. Elsewhere, droughts occurred in Australia, Indonesia, China, India, Africa, and Central America. Worldwide, more than 2000 deaths and at least $10 billion in property damage ($2.5 billion in the United States) were attributed to the 1982–1983 El Niño event.

THE 1997–1998 EL NIÑO The 1997–1998 El Niño event began several months earlier than normal and peaked in January 1998. The amount of Southern Oscillation and sea surface warming in the equatorial Pacific was initially as strong as in the 1982–1983 El Niño, which caused a great deal of concern. However, the 1997–1998

STUDENTS SOMETIMES ASK . . .

Do El Niño events occur in other ocean basins?

Yes, the Atlantic and Indian Oceans both experience events similar to the Pacific's El Niño. These events are not nearly as strong, however, nor do they influence worldwide weather phenomena to the same extent as those that occur in the equatorial Pacific Ocean. The great width of the Pacific Ocean in equatorial latitudes is the main reason that El Niño events occur more strongly in the Pacific.

In the Atlantic Ocean, this phenomenon is related to the North Atlantic Oscillation (NAO), which is a periodic change in atmospheric pressure between Iceland and the Azores Islands. This pressure difference determines the strength of the prevailing westerlies in the North Atlantic, which in turn affects ocean surface currents there. The Atlantic Ocean periodically experiences NAO events, which sometimes cause intense cold in the northeast United States, unusual weather in Europe, and heavy rainfall along the normally arid coast of southwest Africa.

El Niño weakened in the last few months of 1997 before reintensifying in early 1998. The impact of the 1997–1998 El Niño was felt mostly in the tropical Pacific, where surface water temperatures in the eastern Pacific averaged more than 4°C (7°F) warmer than normal, and, in some locations, reached up to 9°C (16°F) above normal (see Figure 7.23a). High pressure in the western Pacific brought drought conditions that caused wildfires to burn out of control in Indonesia. Also, the warmer than normal water along the west coast of Central and North America increased the number of hurricanes off Mexico.

In the United States, the 1997–1998 El Niño caused killer tornadoes in the Southeast, massive blizzards in the upper Midwest, and flooding in the Ohio River Valley. Most of California received twice the normal rainfall, which caused flooding and landslides in many parts of the state. The lower Midwest, the Pacific Northwest, and the eastern seaboard, on the other hand, had relatively mild weather. In all, the 1997–1998 El Niño caused 2100 deaths and $33 billion in property damage worldwide.

Predicting El Niño Events   The 1982–1983 El Niño event was not predicted, nor was it recognized until it was near its peak. Because it affected weather worldwide and caused such extensive damage, the **Tropical Ocean–Global Atmosphere (TOGA)** program was initiated in 1985 to study how El Niño events develop. The goal of the TOGA program was to monitor the equatorial South Pacific Ocean during El Niño events to enable scientists to model and predict future El Niño events. The 10-year program studied the ocean from research vessels, analyzed surface and subsurface data from radio-transmitting sensor buoys, monitored oceanic phenomena by satellite, and developed computer models.

These models have made it possible to predict El Niño events since 1987 as much as one year in advance. After the completion of TOGA, the **Tropical Atmosphere and Ocean (TAO)** project (sponsored by the United States, Canada, Australia, and Japan) has continued to monitor the equatorial Pacific Ocean with a series of 70 moored buoys, providing real-time information about the conditions of the tropical Pacific that is available on the Internet. Although monitoring has improved, the trigger mechanisms of El Niño events are still not fully understood.

ESSENTIAL CONCEPT 7.4

El Niño is a combined oceanic–atmospheric phenomenon that occurs periodically in the tropical Pacific Ocean, bringing warm water to the east. La Niña describes conditions opposite those of El Niño.

CONCEPT CHECK 7.4

1 Explain why Gulf Stream eddies that develop northeast of the Gulf Stream rotate clockwise and have warm-water cores, whereas those that develop to the southwest rotate counterclockwise and have cold-water cores.

2 Describe changes in atmospheric pressure, precipitation, winds, and ocean surface currents during the two monsoon seasons of the Indian Ocean.

3 Describe changes in atmospheric and oceanographic phenomena that occur during El Niño/La Niña events, including changes in atmospheric pressure, winds, Walker Circulation, weather, equatorial surface currents, coastal upwelling/downwelling and the abundance of marine life, sea surface temperature and the Pacific Warm Pool, sea surface elevation, and position of the thermocline.

4 How often do El Niño events occur? Using Figure 7.24, determine how many years since 1950 have been El Niño years. Has the pattern of El Niño events occurred at regular intervals?

5 How is La Niña different from El Niño? Describe the pattern of La Niña events in relation to El Niños since 1950 (see Figure 7.24).

6 Describe the global effects of severe El Niños.

7.5 What Deep-Ocean Currents Exist?

Deep currents occur in the deep zone below the pycnocline, so they influence about 90% of all ocean water. Density differences create deep currents. Although these density differences are usually small, they are large enough to cause denser waters to sink. Deep-water currents move larger volumes of water and are much slower than surface currents. Typical speeds of deep currents range from 10 to 20 kilometers (6 to 12 miles) per year. Thus, it takes a deep current an *entire year* to travel the same distance that a western intensified surface current can move in *one hour*.

Because the density variations that cause deep ocean circulation are caused by differences in temperature and salinity, deep-ocean circulation is also referred to as **thermohaline** (*thermo* = heat, *haline* = salt) **circulation**.

Origin Of Thermohaline Circulation

Recall from Chapter 5 that an increase in seawater density can be caused by a *decrease* in temperature or an *increase* in salinity. Temperature, though, has the greater influence on density. Density changes due to salinity are important only in very high latitudes, where water temperature remains low and relatively constant.

Most water involved in deep-ocean currents (thermohaline circulation) originates in high latitudes *at the surface*. In these regions, surface water becomes cold and its salinity increases as sea ice forms. When this surface water becomes dense enough, it sinks, initiating deep-ocean currents. Once this water sinks, it is removed from the physical processes that increased its density in the first place, so its temperature and salinity remain largely unchanged for the duration it spends in the deep ocean. Thus, a **temperature–salinity (T–S) diagram** can be used to identify deep-water masses based on their characteristic temperature, salinity, and resulting density. **Figure 7.27** shows a T–S diagram for the North Atlantic Ocean.

As these surface water masses become dense and are sinking (downwelling) in high-latitude areas, deep-water masses are also rising to the surface (upwelling). Because the water temperature in high-latitude regions is the same at the surface as it is down below, the water column is isothermal, there is no thermocline or associated pycnocline (see Chapter 5), and upwelling and downwelling can easily occur.

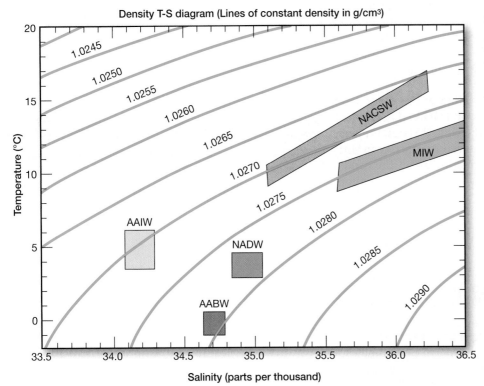

North Atlantic Water Masses:

- (AAIW) Antarctic Intermediate Water
- (AABW) Antarctic Bottom Water
- (NADW) North Atlantic Deep Water
- (NACSW) North Atlantic Central Surface Water
- (MIW) Mediterranean Intermediate Water

Figure 7.27 Temperature–salinity (T–S) diagram. A density T–S diagram for the North Atlantic Ocean. Lines of constant density are in grams/cm³. After various deep-water masses sink below the surface, they can be identified based on their characteristic temperature, salinity, and resulting density.

Sources Of Deep Water

In southern subpolar latitudes, huge masses of deep water form beneath sea ice along the margins of the Antarctic continent. Here, rapid winter freezing produces very cold, high-density water that sinks down the continental slope of Antarctica and becomes **Antarctic Bottom Water**, the densest water in the open ocean (**Figure 7.28**). Antarctic Bottom Water slowly sinks beneath the surface and spreads into all the world's ocean basins, eventually returning to the surface perhaps 1000 years later.

In the northern subpolar latitudes, large masses of deep water form in the Norwegian Sea. From there, the deep water flows as a subsurface current into the North Atlantic, where it becomes part of the **North Atlantic Deep Water**. North Atlantic Deep Water also comes from the margins of the Irminger Sea off southeastern Greenland, the Labrador Sea, and the dense, salty Mediterranean Sea. Like Antarctic Bottom Water, North Atlantic Deep Water spreads throughout the ocean basins. It is less dense, however, so it layers on top of the Antarctic Bottom Water (Figure 7.28).

Figure 7.28 Atlantic Ocean subsurface water masses. Schematic diagram of the various water masses in the Atlantic Ocean. Similar but less distinct layering based on density occurs in the Pacific and Indian Oceans as well. Upwelling and downwelling occur in the North Atlantic and near Antarctica (*inset*), creating deep-water masses.

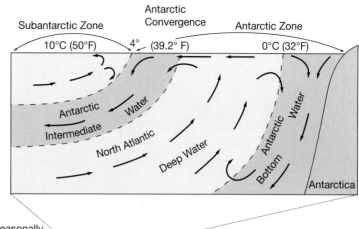

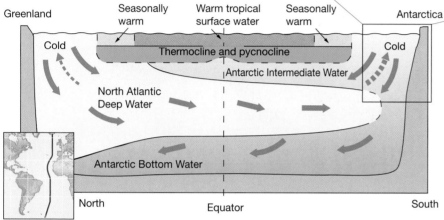

Surface water masses converge within the subtropical gyres and in the Arctic and Antarctic. Subtropical Convergences do not produce deep water, however, because the density of warm surface waters is too low for them to sink. Major sinking does occur, however, along the **Arctic Convergence** and Antarctic Convergence (Figure 7.28, *inset*). The deep-water mass formed from sinking at the Antarctic Convergence is called the **Antarctic Intermediate Water** mass (Figure 7.28), which remains one of world's most poorly studied water masses.

Figure 7.28 also shows that the highest-density water is found along the ocean bottom, with less-dense water above. In low-latitude regions, the boundary between the warm surface water and the deeper cold water is marked by a prominent thermocline and corresponding pycnocline that prevent vertical mixing. There is no pycnocline in high-latitude regions, so substantial vertical mixing (upwelling and downwelling) occurs.

This same general pattern of layering based on density occurs in the Pacific and Indian Oceans as well. They have no source of Northern Hemisphere deep water, however, so they lack a deep-water mass. In the northern Pacific Ocean, the low salinity of surface waters prevents them from sinking into the deep ocean. In the northern Indian Ocean, surface waters are too warm to sink. **Oceanic Common Water**, which is created when Antarctic Bottom Water and North Atlantic Deep Water mix, lines the bottoms of these basins.

Worldwide Deep-Water Circulation

For every liter of water that sinks from the surface into the deep ocean, a liter of deep water must return to the surface somewhere else. However, it is difficult to identify specifically *where* this vertical flow to the surface is occurring. It is generally believed that it occurs as a gradual, uniform upwelling throughout the ocean

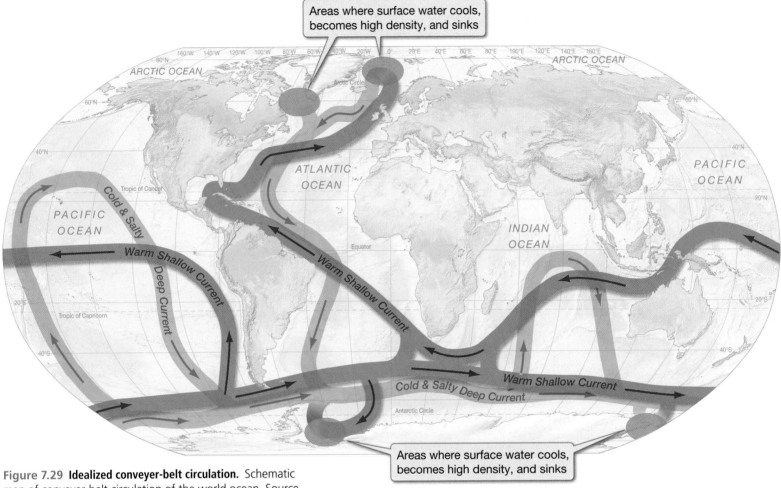

Areas where surface water cools, becomes high density, and sinks

Areas where surface water cools, becomes high density, and sinks

Figure 7.29 Idealized conveyer-belt circulation. Schematic map of conveyer-belt circulation of the world ocean. Source areas (*purple ovals*) exist in high-latitude regions where surface water cools and becomes high density. These source areas feed the flow of deep, high density waters (*blue bands*), which flow into all oceans. This water slowly ascends throughout the oceans and completes the conveyer by returning to the source areas as warm surface currents (*red bands*).

basins and that it may be somewhat greater in low-latitude regions, where surface temperatures are higher. Alternatively, recent research on turbulent mixing rates between deep-ocean and surface waters in the Southern Ocean suggests that deep water traveling across rugged bottom topography is a major factor in producing the upwelling that returns deep water toward the surface.

CONVEYER-BELT CIRCULATION An integrated model combining deep thermohaline circulation and surface currents is shown in **Figure 7.29**. Because the overall circulation pattern resembles a large conveyer belt, the model is called **conveyer-belt circulation**. Beginning in the North Atlantic, surface water carries heat to high latitudes via the Gulf Stream. During the cold winter months, this heat is transferred to the overlying atmosphere, warming northern Europe.

Cooling in the North Atlantic increases the density of this surface water to the point where it sinks to the bottom and flows southward, initiating the lower limb of the "conveyor." Here, seawater flows downward at a rate equal to 100 Amazon Rivers and begins its long journey into the deep basins of all the world's oceans. This limb extends all the way to the southern tip of Africa, where it joins the deep water that encircles Antarctica. The deep water that encircles Antarctica includes deep water that descends along the margins of the Antarctic continent. This mixture of deep waters flows northward into the deep Pacific and Indian Ocean basins, where it eventually surfaces and completes the conveyer belt by flowing west and then north again into the North Atlantic Ocean.

Does this simple conveyer-belt model of ocean circulation adequately reflect the movement of both surface and deep-ocean currents? Satellite and deep sampling of the oceans confirms the basic conveyer-belt movement of warm waters poleward at the surface and cold waters equatorward at depth. However, the conveyer-belt model ignores some crucial complex components of the ocean's circulation system—including small-scale eddies and oceanic fronts—that affect climate change and are now being studied at ever-finer scales.

DISSOLVED OXYGEN IN DEEP WATER   Cold water can dissolve more oxygen than warm water. Thus, deep-water circulation brings dense, cold, oxygen-enriched water from the surface to the deep ocean. During its time in the deep ocean, deep water becomes enriched in nutrients as well, due to decomposition of dead organisms and the lack of organisms using nutrients there.

At various times in the geologic past, warmer water probably constituted a larger proportion of deep oceanic waters. As a result, the oceans had a lower oxygen concentration than today because warm water cannot hold as much oxygen. Moreover, the oxygen content of the oceans has probably fluctuated widely throughout time.

If high-latitude surface waters did not sink and eventually return from the deep sea to the surface, the distribution of life in the sea would be considerably different. There would be very little life in the deep ocean, for instance, because there would be no oxygen for organisms to breathe. In addition, life in surface waters might be significantly reduced without the circulation of deep water that brings nutrients to the surface.

CONCEPT CHECK 7.5

1 Discuss the origin of thermohaline vertical circulation. Why do deep currents form only in high-latitude regions?

2 What are the two major deep-water masses? Where do they form at the ocean's surface?

3 Explain why no matter where you are in the ocean, if you go deep enough, you will encounter Oceanic Common Water.

4 Describe how the distribution of life in the ocean would be different if there were very little dissolved oxygen in deep-water currents.

7.6 Can Power from Currents Be Harnessed as A Source Of Energy?

The movement of ocean currents has often been considered to be capable of providing a source of renewable, clean energy similar to that of wind farms (see Chapter 6)—but underwater. Even though currents move much more slowly than winds, currents carry much more energy because water has about 800 times the density of air. As a result, currents have the potential to generate even more power than wind farms do.

One location that has received much consideration as a site for harnessing power from ocean currents is the Florida–Gulf Stream Current System, which is a fast, western intensified surface current that runs along the East Coast of the United States. In fact, researchers have determined that at least 2000 megawatts[14]

---

[14]Each megawatt of electricity is enough to serve the energy needs of about 800 average U.S. homes.

Figure 7.30 Power from ocean currents. A prototype ocean current power system, which uses a field of underwater turbines anchored to the ocean floor. As currents flow past the turbines, they turn the rotors and generate electricity. The turbines can swivel on their anchors to face oncoming currents, so this system is also useful in locations where strong tidal flows occur.

of electricity could be recovered from this ocean current system along the southeastern coast of Florida alone.

Various devices have been proposed to extract energy from ocean currents. All of them involve some sort of mechanism for converting the movement of water into electrical energy. **Figure 7.30** shows one solution. Here, a series of underwater turbines similar to windmills, are placed within a current and anchored to the ocean floor. The turbines can swivel on their anchors to face oncoming currents, so this system is also useful in locations where strong reversing tidal flows occur. Such a system of six turbines has been successfully tested in the East River near New York City. Once this system is expanded to its capacity of 300 turbines, it will have the ability to generate about 10 megawatts of electricity.

Systems that generate power from currents have some significant hurdles to overcome. For example, current systems are expensive, difficult to maintain, potentially hazardous to ship traffic, and they have unknown effects on marine life. Also, underwater deployment of complex machines with exposed moving parts for long periods of time presents challenges associated with *biofouling*—the accumulation of algae and other sea life on machinery—and corrosion. Further, the variability of current flow causes irregular power generation, which is problematic. Still, similar turbine systems powered by currents are in use in Strangford Lough, Ireland, and are planned for offshore South Korea.

ESSENTIAL CONCEPT 7.6

In spite of problems with placing mechanical devices in the ocean, there is vast potential for developing power from currents as a renewable source of energy.

CONCEPT CHECK 7.6

❶ Why do ocean currents have the potential to generate even more power than wind farms?

❷ Discuss the advantages and disadvantages of building an offshore current power system.

Essential Concepts Review

7.1 How are ocean currents measured?

▶ *Ocean currents are masses of water that flow* from one place to another and can be divided into *surface currents that are wind driven and deep currents that are density driven.* Currents can be measured directly or indirectly by various methods.

Study Resources
Online Study Guide Quizzes, Web Table 7.1

Critical Thinking Question

Describe the main differences between surface and deep currents including how they are both measured.

7.2 How are ocean surface currents organized?

▶ *Surface currents occur within and above the pycnocline.* They consist of circular-moving loops of water called gyres, set in motion by the major wind belts of the world. They are modified by the positions of the continents, the Coriolis effect, and other factors. There are *five major subtropical gyres in the world*; they rotate *clockwise in the Northern Hemisphere* and *counterclockwise in the Southern Hemisphere.* Water is pushed toward the center of the gyres, forming low "hills" of water.

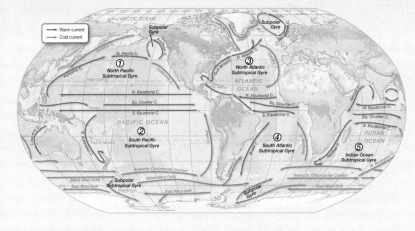

▶ The *Ekman spiral* influences shallow surface water and is *caused by winds and the Coriolis effect.* The average net flow of water affected by the Ekman spiral causes the water to move at *90-degree angles to the wind direction.* At the center of a gyre, the Coriolis effect deflects the water so that it tends to move into the hill, whereas gravity moves the water down the hill. When gravity and the Coriolis effect balance, a *geostrophic current* flowing parallel to the contours of the hill is established.

▶ *The apex (top) of the hill is located to the west of the geographical center of the gyre* due to Earth's rotation. A phenomenon called *western intensification* occurs in which *western boundary currents of subtropical gyres are faster, narrower, and deeper* than their eastern boundary counterparts.

Study Resources
Online Study Guide Quizzes, Web Diving Deeper 7.1, Web Animations, Web Videos

Critical Thinking Question

Describe what the pattern of ocean surface currents would look like if there were no continents on Earth.

7.3 What causes upwelling and downwelling?

▶ *Upwelling and downwelling help vertically mix deep and surface waters.* Upwelling—the movement of *cold, deep, nutrient-rich water to the surface*—stimulates productivity and creates a large amount of marine life. Upwelling and downwelling can occur in a variety of ways.

Study Resources
Online Study Guide Quizzes, Web Animation

Critical Thinking Question

From a practical standpoint, give several reasons why upwelling is a much more intensely studied process than downwelling.

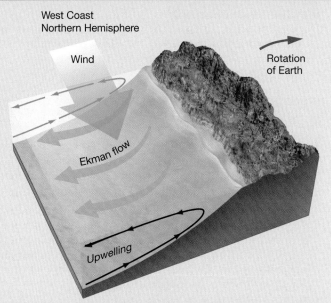

(a) Ekman transport carries surface water to the *right* of the wind direction and away from a continent. Upwelling of deeper water replaces the surface water that has moved away from the coast.

Essential Concepts Review (cont.)

7.4 What are the main surface circulation patterns in each ocean basin?

▶ *Antarctic circulation is dominated by a single large current, the Antarctic Circumpolar Current* (West Wind Drift), which flows in a clockwise direction around Antarctica and is driven by the Southern Hemisphere's prevailing westerly winds. Between the Antarctic Circumpolar Current and the Antarctic continent is a current called the *East Wind Drift*, which is powered by the polar easterly winds. The two currents flow in opposite directions, and the Coriolis effect deflects them away from each other, creating the *Antarctic Divergence*, an area of abundant marine life due to upwelling and current mixing.

(b) El Niño conditions (strong)

▶ *The North Atlantic Gyre and the South Atlantic Gyre dominate circulation in the Atlantic Ocean.* A poorly developed equatorial countercurrent separates these two subtropical gyres. The highest-velocity and best-studied ocean current is the *Gulf Stream*, which carries warm water along the southeastern U.S. Atlantic coast. Meanders of the Gulf Stream produce *warm- and cold-core rings*. The warming effects of the Gulf Stream extend along its route and reach as far away as northern Europe.

▶ *The Indian Ocean consists of one gyre, the Indian Ocean Gyre*, which exists mostly in the Southern Hemisphere. The *monsoon wind system*, which changes direction with the seasons, dominates circulation in the Indian Ocean. The monsoons blow from the northeast in the winter and from the southwest in the summer.

▶ *Circulation in the Pacific Ocean consists of two subtropical gyres: the North Pacific Gyre and the South Pacific Gyre*, which are separated by a well-developed equatorial countercurrent.

▶ *A periodic disruption of normal sea surface and atmospheric circulation patterns in the Pacific Ocean is called El Niño–Southern Oscillation (ENSO).* The *warm phase of ENSO (El Niño)* is associated with the eastward movement of the Pacific Warm Pool, halting or reversal of the trade winds, a rise in sea level along the equator, a decrease in productivity along the west coast of South America, and, in very strong El Niños, worldwide changes in weather. El Niños fluctuate with the *cool phase of ENSO (La Niña conditions)*, which are associated with cooler than normal water in the eastern tropical Pacific.

Study Resources
Online Study Guide Quizzes, Web Diving Deeper 7.2 and 7.3, Web Tables 7.2 and 7.3, Web Animations, Web Videos

Critical Thinking Question
During flood stage, the largest river in the world—the mighty Amazon River—dumps 200,000 cubic meters of water into the Atlantic Ocean each second. Compare its flow rate with the volume of water transported by the West Wind Drift and the Gulf Stream. How many times larger than the Amazon is each of these two ocean currents?

7.5 What deep-ocean currents exist?

▶ *Deep currents occur below the pycnocline.* They affect much larger amounts of ocean water and move much more slowly than surface currents. Changes in temperature and/or salinity at the surface create slight increases in density, which set deep currents in motion. Deep currents, therefore, are called *thermohaline circulation*.

▶ *The deep ocean is layered based on density. Antarctic Bottom Water*, the densest deep-water mass in the oceans, forms near Antarctica and sinks along the continental shelf into the South Atlantic Ocean. Farther north, at the Antarctic Convergence, the low-salinity *Antarctic Intermediate Water* sinks to an intermediate depth dictated by its density. Sandwiched between these two masses is the *North Atlantic Deep Water*, which has high nutrient levels after remaining below the surface for hundreds of years. Layering in the Pacific and Indian Oceans is similar, except there is no source of Northern Hemisphere deep water.

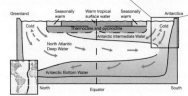

▶ *Worldwide circulation models that include both surface and deep currents resemble a conveyer belt.* Deep currents carry oxygen into the deep ocean, which is extremely important for life on the planet.

Study Resources
Online Study Guide Quizzes

Critical Thinking Question
Antarctic Intermediate Water can be identified throughout much of the South Atlantic based on its temperature, salinity, and dissolved oxygen content. Why is it colder and less salty—and why does it contain more oxygen—than the surface water mass above it and the North Atlantic Deep Water below it?

7.6 Can power from currents be harnessed as a source of energy?

▶ *Ocean currents can be harnessed as a source of power.* Although there is vast potential for developing this clean, renewable resource, significant problems must be overcome to make this a practical source of energy.

Study Resources
Online Study Guide Quizzes

Critical Thinking Question

What are some environmental factors that could inhibit the development of offshore ocean current power systems?

MasteringOceanography™

www.masteringoceanography.com

Looking for additional review and test prep materials? Visit the Study Area in MasteringOceanography to enhance your understanding of this chapter's content by accessing a variety of resources, including Self-Study Quizzes, Geoscience Animations, RSS feeds, flashcards, Web links, and an optional Pearson eText.

Surfer riding a giant wave at Maverick's. Located in Half Moon Bay off central California, Maverick's is rated as the world's top big wave surf spot. A unique combination of oceanographic factors produces waves so large here that only the most accomplished surfers even attempt to wade into the surf zone.

WAVES AND WATER DYNAMICS

ESSENTIAL CONCEPTS

At the end of this chapter, you should be able to:

8.1 Demonstrate an understanding of how waves are generated and how they move

8.2 Describe the characteristics of waves

8.3 Discuss how wind-generated waves develop

8.4 Explain how waves change in the surf zone

8.5 Describe the origin and characteristics of tsunami

8.6 Discuss advantages and disadvantages of harnessing waves as a source of energy

"Can ye fathom the ocean, dark and deep, where the mighty waves and the grandeur sweep?"
—Poet Fanny Crosby (1820–1915)

What combination of oceanographic factors causes waves to reach extreme heights at places such as Maverick's? This site, which is rated as the world's premier big wave surf spot, is located 0.5 kilometer (0.3 mile) offshore of Pillar Point in Half Moon Bay along the central California coast. One factor is that Maverick's is located offshore of a prominent point of land, which, as will be explained in this chapter, tends to concentrate wave energy due to *wave refraction*. Another factor is that the point juts directly into the North Pacific Ocean, which is known for its wintertime storms and giant waves. Still another factor is that the shoreline abruptly rises from deep depths to a shallowly submerged rock reef, which causes waves to build up to extreme heights in a very short distance. These factors combined with cold ocean temperatures, unforgiving boulders just below the surface, and the presence of large sharks make the site challenging to even the most skilled surfers. Still, surfing competitions are held every year at Maverick's for those brave enough to catch some of the world's most extreme waves.

Most waves are driven by the wind and are relatively small, so they release their energy gently, although ocean storms can build up waves to extreme heights. When these waves come ashore, they often produce devastating effects—or, in the case at Maverick's, a wild ride. Waves are *moving energy* traveling along the interface between ocean and atmosphere, often transferring energy from a storm far out at sea over distances of several thousand kilometers. That's why, even on calm days, the ocean is in continual motion as waves travel across its surface.

8.1 How Are Waves Generated, and How Do They Move?

All waves begin as *disturbances*; the energy that causes ocean waves to form is called a **disturbing force**. A rock thrown into a still pond, for example, creates waves that radiate out in all directions. Releases of energy, similar to the rock hitting the water, are the cause of all waves.

Disturbances Generate Ocean Waves

Wind blowing across the surface of the ocean generates most ocean waves. The waves radiate out in all directions, just as when the rock is thrown into the pond, but on a much larger scale.

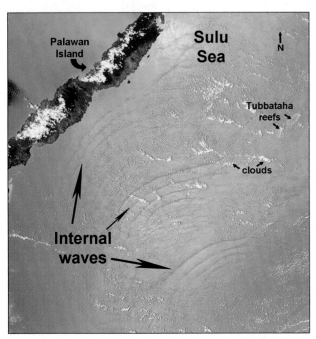

(a) Surface waves versus internal waves.

(b) Internal waves in the Sulu Sea between the Philippines and Malaysia.

Figure 8.1 Internal waves. (a) An internal wave moving along the density interface (pycnocline) below where surface waves occur. **(b)** Aerial view of internal waves. Image taken April 8, 2003, by the Moderate Resolution Imaging Spectroradiometer (MODIS) instrument aboard the Aqua satellite.

STUDENTS SOMETIMES ASK . . .

Can internal waves break?

Internal waves do not break in the way that surface waves break in the surf zone because the density difference across an interface at depth is much smaller than that between the atmosphere and the surface. When internal waves approach the edges of continents, however, they undergo physical changes similar to those of waves in the surf zone. This causes the waves to build up and expend their energy with much turbulent motion—in essence "breaking" against the continent.

The movement of fluids with different densities can also create waves. These waves travel along the interface (boundary) between two different fluids. Both the air and the ocean are fluids, so waves can be created along interfaces *between* and *within* these fluids as follows:

- Along an *air–water interface*, the movement of air across the ocean surface creates **ocean waves** (simply called *waves*).
- Along an *air–air interface*, the movement of different air masses creates **atmospheric waves**, which are often represented by ripplelike clouds in the sky. Atmospheric waves are especially common when cold fronts (high-density air) move into an area.
- Along a *water–water interface*, the movement of water of different densities creates **internal waves**, as shown in **Figure 8.1a**. Because these waves travel along the boundary between waters of different density, they are associated with a *pycnocline*.[1] Internal waves can be much larger than surface waves, with heights exceeding 100 meters (330 feet). Tidal movement, turbidity currents, wind stress, and even passing ships at the surface create internal waves, which can sometimes be observed from space (**Figure 8.1b**).

Internal waves can even be a hazard for submarines: If submarines are caught in an internal wave while testing their depth limits, the submarines can inadvertently be carried to depths exceeding their designed pressure strength. At the surface, parallel slicks caused by a film of surface debris may indicate the presence of internal waves below. On a smaller scale, internal waves are prominently featured sloshing back and forth in "desktop oceans," which contain two fluids that do not mix.

Mass movement into the ocean, such as coastal landslides or large icebergs that fall from coastal glaciers, also creates waves. These waves are commonly called *splash waves* (see **Web Diving Deeper 8.1** for a description of a large splash wave).

Another way in which large waves are created involves the uplift or downdropping of large areas of the sea floor, which can release large amounts of energy to the entire water column (compared to wind-driven waves, which affect only surface water). Examples include underwater avalanches (turbidity currents), volcanic eruptions, and fault slippage. The resulting waves are called *seismic sea waves* or *tsunami*, which are discussed later in this chapter. Fortunately, tsunami occur infrequently. When they do, however, they can flood coastal areas and cause large amounts of destruction.

The tides are also a type of wave, in this case caused mainly by the gravitational pull of the Moon and, to a lesser extent, the Sun. The tides are actually a wide-ranging and very predictable type of wave that will be discussed in Chapter 9, "Tides."

Human activities also cause ocean waves. When ships travel across the ocean, they leave behind a *wake*, which is a wave. In fact, smaller boats are often carried along in the wake of larger ships, and marine mammals sometimes play there.

In all cases, though, some type of energy release creates waves. **Figure 8.2** shows the distribution of energy in waves, indicating that most ocean waves are wind generated.

Wave Movement

A good way to think about waves is that waves are simply *energy in motion*. All types of waves transmit energy—or propagate—by means of cyclic movement through matter. The medium itself (solid, liquid, or gas) does not actually travel in the direction of the energy that is passing through it. The particles in the medium simply oscillate—or cycle—back and forth, up and down, or around and around,

---

[1]As discussed in Chapter 5, a *pycnocline* is a layer of rapidly changing density.

transmitting energy from one particle to another. If you thump your fist on a table, for example, the energy travels through the table as waves that someone sitting at the other end can feel, but the table itself does not move. Another useful analogy is a field of grain: If you have ever seen wind pass across it, you know that the wave form moves as the grains sway, but the individual grains stay in the same location.

Different types of waves move in a variety of ways. Simple *progressive waves* (Figure 8.3) are waves that oscillate uniformly and progress or travel without breaking. Progressive waves are divided into *longitudinal*, *transverse*, or a combination of the two motions, called *orbital*.

In **longitudinal waves** (also known as push–pull waves), the particles that vibrate "push and pull" in the same direction that the energy is traveling, like a spring whose coils are alternately compressed and expanded. The shape of the wave (called a *waveform*) moves through the medium by compressing and decompressing as it goes. Sound, for instance, travels as longitudinal waves. Clapping your hands initiates a percussion that compresses and decompresses the air as the sound moves through a room. Energy can be transmitted through all states of matter—gaseous, liquid, or solid—by this longitudinal movement of particles.

In **transverse waves** (also known as side-to-side waves), energy travels at right angles to the direction of the vibrating particles. If one end of a rope is tied to a doorknob while the other end is moved up and down (or side to side) by hand, for example, a waveform progresses along the rope and energy is transmitted from the motion of the hand to the doorknob. The waveform moves up and down (or side to side) with the hand, but the motion is at right angles to the direction in which energy is transmitted (from the hand to the doorknob). Generally, transverse waves transmit energy only through solids, because the particles in solids are bound to one another strongly enough to transmit this kind of motion.

Longitudinal and transverse waves are called *body waves* because they transfer energy through a body of matter. Ocean waves are body waves, too, because they transmit energy through the upper part of the ocean near the interface between the atmosphere and the ocean. The movement of particles in ocean waves involves components of *both* longitudinal and transverse waves, so particles move in circular orbits. Thus, waves at the ocean surface are **orbital waves** (also called *interface waves*).

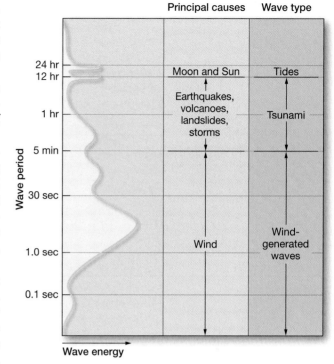

Figure 8.2 Distribution of energy in ocean waves. Most of the energy possessed by ocean waves exists as wind-generated waves, while other peaks of wave energy represent tsunami and ocean tides.

CONCEPT CHECK 8.1

1. Discuss several different ways in which waves form. How are most ocean waves generated?

2. Why is the development of internal waves likely within the pycnocline?

3. Describe the three types of motion in which progressive waves move. Which type includes waves at the ocean surface?

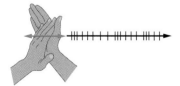

Figure 8.3 Types of progressive waves. Diagrammatic view of the three types of progressive waves. Examples include ① longitudinal waves, ② transverse waves, and ③ orbital waves.

① LONGITUDINAL WAVE
Hands clapping or thumping a table. Particles (*blue color*) move back and forth in the direction of energy transmission. These waves transmit energy through all states of matter.

② TRANSVERSE WAVE
A rope attached to a wall. Particles (*blue color*) move back and forth at right angles to the direction of energy transmission. These waves transmit energy only through solids.

③ ORBITAL WAVE
The movement of water waves. Particles (*blue color*) move in a circular path. These waves transmit energy along interface between two fluids of different density (liquids and/or gases).

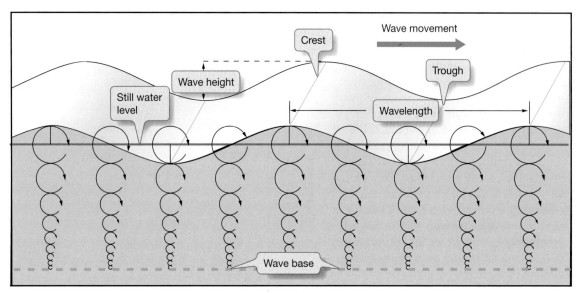

(a) Wave characteristics

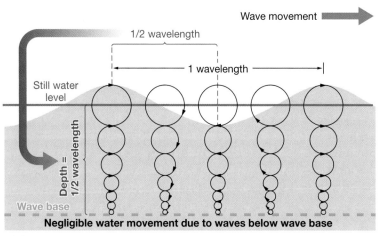

(b) Calculation of wave base

Figure 8.4 Characteristics of progressive waves. (a) A diagrammatic view of an idealized ocean wave, showing its characteristics. **(b)** Enlargement of the orbital motion of water particles in waves, showing that wave base (the depth at which orbital motion ceases) is at a depth of one-half the wavelength, measured from still water level.

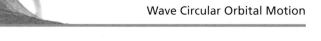

Web Animation
Wave Circular Orbital Motion

8.2 What Characteristics Do Waves Possess?

Figure 8.4a shows the characteristics of an idealized ocean wave. The simple, uniform, moving waveform transmits energy from a single source and travels along the ocean–atmosphere interface. These waves are also called *sine waves* because their uniform shape resembles the oscillating pattern expressed by a sine curve. Even though idealized waveforms do not exist in nature, they help us understand wave characteristics.

Wave Terminology

As an idealized wave passes a permanent marker (such as a pier piling), a succession of high parts of the waves, called **crests**, alternate with low parts, called **troughs**. Halfway between the crests and the troughs is the **still water level**, or *zero energy level*. This is the level of the water if there were no waves. The **wave height**, designated by the symbol H, is the vertical distance between a crest and a trough.

The horizontal distance between any two corresponding points on successive waveforms, such as from crest to crest or from trough to trough, is the **wavelength**, L. **Wave steepness** is the ratio of wave height to wavelength:

$$\text{Wave steepness} = \frac{\text{Wave height}(H)}{\text{Wavelength}(L)} \quad (8.1)$$

If the wave steepness exceeds $1/7$, the wave *breaks* (spills forward) because the wave is too steep to support itself. A wave can break anytime the 1:7 ratio is exceeded, either along the shoreline or out at sea. This ratio also dictates the maximum height of a wave. For example, a wave 7 meters long can only be 1 meter high; if the wave is any higher than that, it will break.

The time it takes one full wave—one wavelength—to pass a fixed position (such as a pier piling) is the **wave period**, T. Typical wave periods range between 6 and 16 seconds. The **frequency** (f) is defined as the number of wave crests passing a fixed location per unit of time and is the inverse of the period:

$$\text{Frequency}(f) = \frac{1}{\text{Period}(T)} \quad (8.2)$$

For example, consider waves with a period of 12 seconds. These waves have a frequency of $1/12$, or 0.083 waves per second, which converts to 5 waves per minute.

Circular Orbital Motion

Waves can travel great distances across ocean basins. In one study, waves generated near Antarctica were tracked as they traveled through the Pacific Ocean basin. After more than 10,000 kilometers (over 6000 miles), the waves finally expended their energy a week later along the shoreline of the Aleutian Islands of Alaska. The water itself doesn't travel the entire distance, but the waveform does. As the wave travels, the water passes the energy along by moving in a circle. This movement is called **circular orbital motion**.

Observation of an object floating in the waves reveals that it moves not only up and down but also slightly forward and backward with each successive wave. **Figure 8.5** shows that a floating object moves up and backward as the crest approaches, up and forward as the crest passes, down and forward after the crest, down and backward as the trough approaches, and up and backward again as the next crest advances. When the movement of the rubber ducky shown in Figure 8.5 is traced as a wave passes, it can be seen that the ducky moves in a circle and returns close to its original position.[2] This motion allows a waveform (the wave's shape) to move forward through the water while the individual water particles that transmit the wave move around in a circle and return to essentially the same place. Wind moving across a field of wheat causes a similar phenomenon: The wheat itself doesn't travel across the field, but the waves do.

The circular orbits of an object floating at the surface have a diameter equal to the wave height (Figure 8.4a). Figure 8.4b shows that circular orbital motion dies out quickly below the surface. At some depth below the surface, the circular orbits become so small that movement is negligible. This depth is called the **wave base**, and it is equal to one-half the wavelength (*L/2*) *measured from still water level*. Only wavelength controls the depth of the wave base, so the longer the wave, the deeper the wave base.

The decrease of orbital motion with depth has many practical applications. For instance, submarines can avoid large ocean waves simply by submerging below the wave base. Even the largest storm waves will go unnoticed if a submarine submerges to only 150 meters (500 feet). Floating bridges and floating oil rigs are constructed so that most of their mass is below wave base, so they will be unaffected by wave motion. In fact, offshore floating airport runways have been designed using similar principles (**Figure 8.6**). In addition, seasick scuba divers find relief when they submerge into the calm, motionless water below wave base. Finally, as you walk from the beach into the ocean, you reach a point where it is easier to dive under an incoming wave than to jump over it. It is easier to swim through the smaller orbital motion below the surface than to fight the large waves at the surface.

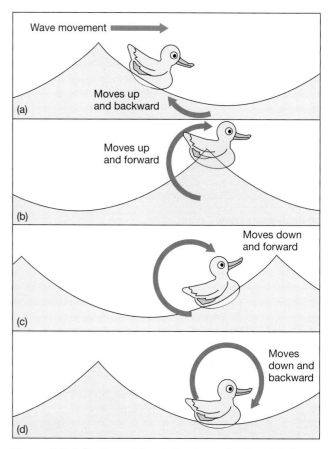

Figure 8.5 A floating rubber ducky shows circular orbital motion. As a wave passes to the right, the motion of a floating rubber ducky resembles that of a circle, which is known as circular orbital motion.

STUDENTS SOMETIMES ASK . . .

Since the water is actually going around in a circle and returns to just about the same place, what actually is being transmitted as waves move?

In a word: energy. The energy of waves has been known to wreak havoc on coastal structures, lift submerged obstacles that weigh many tons, and even change the shape of entire beaches. If you have ever watched really big waves breaking at a beach, you'll witness the impressive amount of energy waves can transmit.

Figure 8.6 Floating runway, Japan. This floating runway, called "Mega-Float," was built offshore Yokosuka in Tokyo Bay, Japan. It is the world's largest floating runway, with a length of 1000 meters (3300 feet). Floating runways and bridges use submerged pontoons (not shown) to make them stable at the surface; note that the majority of the structure's mass is below wave base.

---

[2]Actually, the circular orbit does not quite return the floating object to its original position because the half of the orbit accomplished in the trough is slower than the crest half of the orbit. This results in a slight forward movement (net mass transport), which is called *wave drift*.

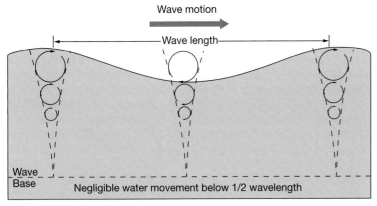

(a) **Deep–water wave:** Circular orbits diminish in size with increasing depth. Depth is greater than $1/2$ wavelength.

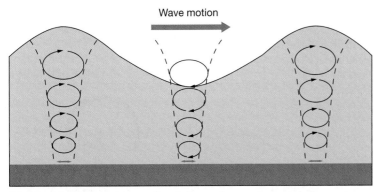

(b) **Transitional wave:** Intermediate between deep-water and shallow-water waves. Depth is greater than $1/20$ wavelength, but less than $1/2$ wavelength.

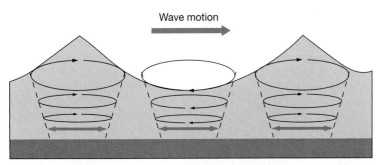

(c) **Shallow–water wave:** The ocean floor interferes with circular orbital motion, causing the orbits to become progressively flattened. Depth is less than $1/20$ wavelength.

Figure 8.7 Characteristics of deep-water, shallow-water, and transitional waves. Diagrammatic views showing the characteristics of **(a)** deep-water waves, **(b)** transitional waves, and **(c)** shallow-water waves. Diagrams are not to scale.

Deep-Water Waves

If the water depth (d) is greater than the wave base ($L/2$), the waves are called **deep-water waves** (Figure 8.7a). Deep-water waves have no interference with the ocean bottom, so they include all wind-generated waves in the open ocean, where water depths far exceed wave base.

Wave speed (S) is the rate at which a wave travels. Numerically, it is the distance traveled divided by the travel time; for a wave, it can be calculated as:

$$\text{Wave speed } (S) = \frac{\text{Wavelength } (L)}{\text{Period } (T)} \tag{8.3}$$

Wave speed is more correctly known as *celerity* (C), which is different from the traditional concept of speed. Celerity is used only in relation to waves where no mass is in motion, just the waveform.

According to the equations that govern the movement of progressive waves, the speed of deep-water waves is dependent upon (1) wavelength and (2) several other variables (such as gravitational attraction) that remain constant on Earth. So, by filling in the constants with numbers, the equation for wave speed of deep-water waves varies only with wavelength and becomes (in meters per second):

$$S \text{ (in meters per second)} = 1.25 \sqrt{L \text{ (in meters)}} \tag{8.4}$$

or, in feet per second:

$$S \text{ (in feet per second)} = 2.26 \sqrt{L \text{ (in feet)}} \tag{8.5}$$

We can also determine wave speed if we know only the period (T) because wave speed (S) is defined in Equation 8.3 as L/T. Filling in the known variables with numbers gives (in meters per second):

$$S \text{ (in meters per second)} = 1.56 \times T \tag{8.6}$$

or, in feet per second:

$$S \text{ (in feet per second)} = 5.12 \times T \tag{8.7}$$

The graph in **Figure 8.8** uses the above equations to relate the wavelength, period, and speed of deep-water waves. Of the three variables, the wave period is usually easiest to measure. Since all three variables are related, the other two can be determined by using Figure 8.8. For example, the vertical red line in Figure 8.8 shows that a wave with a period of 8 seconds has a wavelength of 100 meters. Thus, the speed of the wave is shown by the horizontal red line in Figure 8.8, which is:

$$\text{Speed } (S) = \frac{L}{T} = \frac{100 \text{ meters}}{8 \text{ seconds}} = 12.5 \text{ meters per second} \tag{8.8}$$

In summary, the general relationship shown by Equations 8.3 through 8.8 (and shown in Figure 8.8) for deep-water waves is *the longer the wavelength, the faster the wave travels*. A fast wave does not necessarily have a large wave height, however, because wave speed depends *only* on wavelength.

Shallow-Water Waves

Waves in which depth (*d*) is less than ¹⁄₂₀ of the wavelength (*L*/20) are called **shallow-water waves**, or *long waves* (**Figure 8.7c**). Shallow-water waves are said to *touch bottom* or *feel bottom* because they touch the ocean floor, which interferes with the wave's orbital motion.

The speed of shallow-water waves is influenced only by gravitational acceleration (*g*) and water depth (*d*). Since gravitational acceleration remains constant on Earth, the equation for wave speed becomes (in meters per second):

$$S \text{ (in meters per second)} = 3.13\sqrt{d} \text{ (in meters)} \quad (8.9)$$

or, in feet per second:

$$S \text{ (in meters per second)} = 5.67\sqrt{d} \text{ (in meters)} \quad (8.10)$$

Equations 8.9 and 8.10 show that wave speed in shallow-water waves is determined only by water depth, where *the deeper the water, the faster the wave travels.*

Examples of shallow-water waves include wind-generated waves that have moved into shallow nearshore areas; *tsunami* (seismic sea waves), generated by earthquakes in the ocean floor; and the *tides*, which are a type of wave generated by the gravitational attraction of the Moon and the Sun. Tsunami and the tides are very long-wavelength waves, which far exceed even the deepest ocean water depths.

Particle motion in shallow-water waves is in a very flat elliptical orbit that approaches horizontal (back-and-forth) oscillation. The vertical component of particle motion decreases with increasing depth below sea level, causing the orbits to become even more flattened.

Transitional Waves

Waves that have some characteristics of shallow-water waves and some of deep-water waves are called **transitional waves**. The wavelengths of transitional waves are between 2 times and 20 times the water depth (**Figure 8.7b**). The wave speed of shallow-water waves is a function of water depth; for deep-water waves, wave speed is a function of wavelength. Thus, the speed of transitional waves depends partially on water depth and partially on wavelength.

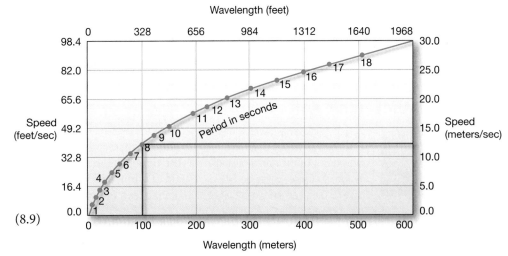

Figure 8.8 Speed of deep-water waves. Ideal relations among wavelength, period (*blue line*), and wave speed for deep-water waves. Red lines show a sample wave with a wavelength of 100 meters, a period of 8 seconds, and a corresponding speed of 12.5 meters per second.

ESSENTIAL CONCEPT 8.2B

Deep-water waves exist in water that is deeper than wave base and move at speeds controlled by wavelength; shallow-water waves exist in water shallower than ¹⁄₂₀ their wavelength and move at speeds controlled by water depth; transitional waves are intermediate between the two.

CONCEPT CHECK 8.2

1. Can a wave with a wavelength of 14 meters ever be more than 2 meters high? Why or why not?

2. What physical feature of a wave is related to the depth of the wave base? What is the difference between the *wave base* and *still water level*?

3. Calculate the speed (*S*) in meters per second for deep-water waves with the following characteristics:
   a. *L* = 351 meters, *T* = 15 seconds
   b. *T* = 12 seconds
   c. *f* = 0.125 wave per second

4. Explain why each of the following statements for deep-water waves is either true or false:
   a. The longer the wave, the deeper the wave base.
   b. The greater the wave height, the deeper the wave base.
   c. The longer the wave, the faster the wave travels.
   d. The greater the wave height, the faster the wave travels.
   e. The faster the wave, the greater the wave height.

8.3 How Do Wind-Generated Waves Develop?

Most ocean waves are generated by the wind and so are termed *wind-generated waves.*

Wave Development

The life history of a wind-generated wave includes its origin in a windy region of the ocean, its movement across great expanses of open water without subsequent aid of wind, and its termination when it breaks and releases its energy, either in the open ocean or against the shore.

CAPILLARY WAVES, GRAVITY WAVES, AND THE "SEA"   As the wind blows over the ocean surface, it creates pressure and stress. These factors deform the ocean surface into small, rounded waves with V-shaped troughs and wavelengths less than 1.74 centimeters (0.7 inch). Commonly called *ripples,* these waves are called **capillary waves** by oceanographers (**Figure 8.9**, *left*). The name comes from *capillarity,* a property that results from the surface tension of water.[3]

As capillary wave development increases, the sea surface takes on a rougher appearance. The water "catches" more of the wind, allowing the wind and ocean surface to interact more efficiently. As more energy is transferred to the ocean, **gravity waves** develop. These are symmetric waves that have wavelengths exceeding 1.74 centimeters (0.7 inch) (Figure 8.9, *middle*).

The length of gravity waves is generally 15 to 35 times their height. As additional energy is gained, wave height increases more rapidly than wavelength. The crests become pointed and the troughs are rounded, resulting in a *trochoidal* (*trokhos* = wheel) waveform (Figure 8.9, *right*).

Energy imparted by the wind increases the height, length, and speed of the wave. When wave speed equals wind speed, neither wave height nor wavelength can change because there is no net energy exchange and the wave has reached its maximum size.

The area where wind-driven waves are generated is called the **sea**, or *sea area*. It is characterized by choppiness and waves moving in many directions. The waves have a variety of periods and wavelengths (most of them short) due to frequently changing wind speed and direction.

FACTORS AFFECTING WAVE ENERGY **Figure 8.10** shows the factors that determine the amount of energy in

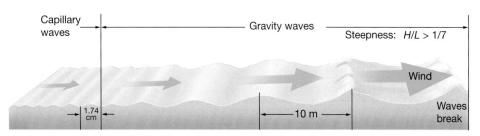

Figure 8.9 Wind creates capillary and gravity waves. As wind increases (*left to right*), the height and wavelength of waves increase; they begin as capillary waves and progress to gravity waves. When the wave steepness (*H/L*) exceeds a 1:7 ratio, the waves become unstable and break. Diagram is not to scale.

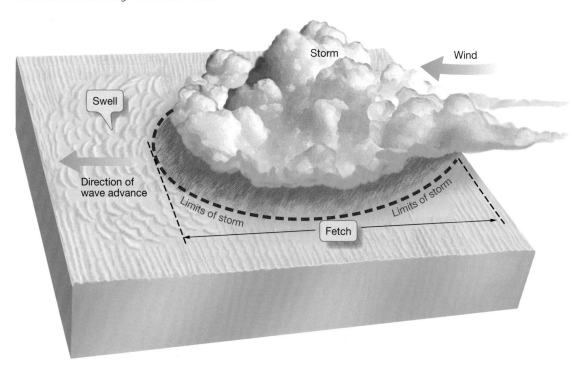

Figure 8.10 The "sea" and swell. As wind blows across the "sea" (*red dashed line*), wave size increases with increasing wind speed, duration, and fetch. As waves advance beyond their area of origination, they advance across the ocean surface and become sorted into uniform, symmetric swell.

---

[3]See Chapter 5 for a description of water's surface tension.

TABLE 8.1 BEAUFORT WIND SCALE AND THE STATE OF THE SEA

Beaufort number	Descriptive term	Wind speed km/hr	Wind speed mi/hr	Appearance of the sea
0	Calm	<1	<1	Like a mirror
1	Light air	1–5	1–3	Ripples with the appearance of scales, no foam crests
2	Light breeze	6–11	4–7	Small wavelets; crests of glassy appearance, no breaking
3	Gentle breeze	12–19	8–12	Large wavelets; crests begin to break, scattered whitecaps
4	Moderate breeze	20–28	13–18	Small waves, becoming longer; numerous whitecaps
5	Fresh breeze	29–38	19–24	Moderate waves, taking longer form; many whitecaps, some spray
6	Strong breeze	39–49	25–31	Large waves begin to form, whitecaps everywhere, more spray
7	Near gale	50–61	32–38	Sea heaps up and white foam from breaking waves begins to be blown in streaks
8	Gale	62–74	39–46	Moderately high waves of greater length, edges of crests begin to break into spindrift, foam is blown in well-marked streaks
9	Strong gale	75–88	47–54	High waves, dense streaks of foam and sea begins to roll, spray may affect visibility
10	Storm	89–102	55–63	Very high waves with overhanging crests; foam is blown in dense white streaks, causing the sea to appear white; the rolling of the sea becomes heavy; visibility reduced
11	Violent storm	103–117	64–72	Exceptionally high waves (small and medium-sized ships might for a time be lost from view behind the waves), the sea is covered with white patches of foam, everywhere the edges of the wave crests are blown into froth, visibility further reduced
12	Hurricane	118+	73+	The air is filled with foam and spray, sea completely white with driving spray, visibility greatly reduced

waves. These factors are (1) the *wind speed*, (2) the *duration*—the length of time during which the wind blows in one direction, and (3) the *fetch*—the distance over which the wind blows in one direction.

Wave height is directly related to the energy in a wave. Wave heights in a sea area are usually less than 2 meters (6.6 feet), but waves with heights of 10 meters (33 feet) and periods of 12 seconds are not uncommon. As "sea" waves gain energy, their steepness increases. When steepness reaches a critical value of $\frac{1}{7}$, open ocean breakers—called *whitecaps*—form. The **Beaufort Wind Scale** and the state of the sea (Table 8.1), which was initially devised by Admiral Sir Francis Beaufort (1774–1857) of the British Navy, describes the appearance of the sea surface from dead calm conditions to hurricane-force winds.

Figure 8.11 is a map based on satellite data of average wave heights during October 3–12, 1992. The waves in the Southern Hemisphere are particularly large because the prevailing westerlies between 40 and 60 degrees south latitude reach the highest average wind speeds on Earth, creating the latitudes called the "Roaring Forties," "Furious Fifties," and "Screaming Sixties."

HOW HIGH CAN WAVES BE? According to a U.S. Navy Hydrographic Office bulletin published in the early 1900s, the theoretical maximum height of wind-generated waves should be no higher than 18.3 meters (60 feet); this became known as the "60-foot rule." Although there were some isolated eyewitness accounts of larger waves, the U.S. Navy considered any sightings of waves over 60 feet to be exaggerations. Certainly, embellishment of reported wave height under conditions of extremely rough seas would be understandable. For many years, the "60-foot rule" was accepted as fact.

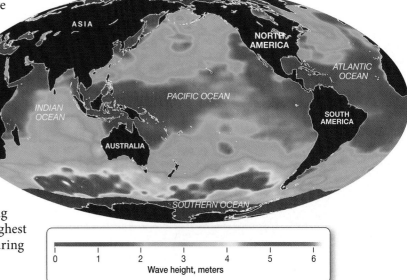

Figure 8.11 TOPEX/Poseidon wave height, October 3–12, 1992. The TOPEX/Poseidon satellite receives a return of stronger signals from calm seas and weaker signals from seas with large waves. Based on these data, a map of average wave height can be produced. Much of the ocean's tropical regions have small wave heights (*blue areas*) while the largest waves (*red areas*) are in the prevailing westerly wind belt in the Southern Hemisphere. Scale is in meters.

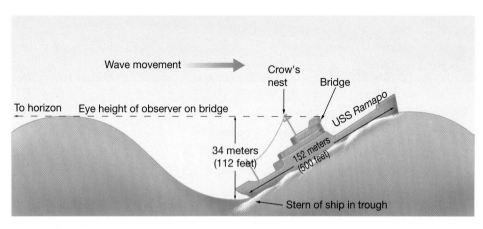

Figure 8.12 USS *Ramapo* in heavy seas. Bridge officers aboard the USS *Ramapo* measured the largest authentically recorded wave by sighting from the bridge across the crow's nest to the horizon while the ship's stern was directly in the trough of a large wave. A wave height of 34 meters (112 feet) was calculated based on geometric relationships of the vessel and the waves.

Careful observations made aboard the 152-meter-long (500-foot-long) U.S. Navy tanker USS *Ramapo* in 1933 proved the "60-foot rule" incorrect. The ship was caught in a typhoon in the western Pacific Ocean and encountered 108-kilometer-per-hour (67-mile-per-hour) winds en route from the Philippines to San Diego. The resulting waves were symmetrical, uniform, and had a period of 14.8 seconds. Because the *Ramapo* was traveling with the waves, the vessel's officers were able to measure the waves accurately. The officers used the dimensions of the ship, including the *eye height* of an observer on the ship's bridge (**Figure 8.12**). Geometric relationships revealed that the waves were 34 meters (112 feet) high, which is taller than an 11-story building! These waves proved to be a record that still stands today for the largest authentically recorded wind-generated waves, shattering the "60-foot rule." Although the *Ramapo* was largely undamaged, other ships traveling in rough seas aren't always so lucky (**Figure 8.13**). In fact, several large ships disappear at sea every year simply due to enormous waves.

FULLY DEVELOPED SEA For a given wind speed, **Table 8.2** lists the minimum fetch and duration of wind beyond which the waves cannot grow. Waves cannot grow because an equilibrium condition, called a **fully developed sea**, has been achieved. Waves can grow no further in a fully developed sea because they lose as much energy breaking as whitecaps under the force of gravity as they receive from the wind. Table 8.2 also lists the average characteristics of waves resulting from a fully developed sea, including the height of the highest 10% of the waves.

Figure 8.13 Wave damage on the aircraft carrier *Bennington*. The *Bennington* returns from heavy seas encountered in a typhoon off Okinawa in 1945, with part of its reinforced steel flight deck bent down over its bow. Damage to the flight deck, which is 16.5 meters (54 feet) above still water level, was caused by large waves.

SWELL As waves generated in a sea area move toward its margins, wind speeds diminish and the waves eventually move faster than the wind. When this occurs, wave steepness decreases and waves become long-crested waves called **swells** (*swellan* = swollen), which are uniform, symmetrical waves that have traveled out of their area of origination. Swells move with little loss of energy over large stretches of the ocean surface, transporting energy away from one sea area and depositing it in another. The movement of swells to distant areas is the reason why there can be waves at a shoreline even though there is no wind.

Waves with longer wavelengths travel faster and thus leave the sea area first. They are followed by slower, shorter **wave trains**, or groups of waves. The progression from long, fast waves to short, slow waves illustrates the principle of **wave dispersion** (*dis* = apart, *spargere* = to scatter)—the sorting of waves by their wavelength. Waves of many wavelengths are present in the generating area. Wave speed depends on wavelength in deep water (see Figure 8.8), however, so the longer waves "outrun" the shorter ones. The distance over which waves change from a choppy "sea" to uniform swell is called the **decay distance**, which can be up to several hundred kilometers.

As a group of waves leaves a sea area and becomes a *swell wave train*, the leading wave keeps disappearing. However, the same number of waves always remains in the group because as the leading wave disappears, a new wave replaces it at the

TABLE 8.2 CONDITIONS NECESSARY TO PRODUCE A FULLY DEVELOPED SEA AT VARIOUS WIND SPEEDS AND THE CHARACTERISTICS OF THE RESULTING WAVES

These conditions . . .			. . . produce these waves			
Wind speed in km/hr (mi/hr)	Fetch in km (mi)	Duration in hours	Average height in m (ft)	Average wave-length in m (ft)	Average period in seconds	Highest 10% of waves in m (ft)
20 (12)	24 (15)	2.8	0.3 (1.0)	10.6 (34.8)	3.2	0.8 (2.5)
30 (19)	77 (48)	7.0	0.9 (2.9)	22.2 (72.8)	4.6	2.1 (6.9)
40 (25)	176 (109)	11.5	1.8 (5.9)	39.7 (130.2)	6.2	3.9 (12.8)
50 (31)	380 (236)	18.5	3.2 (10.5)	61.8 (202.7)	7.7	6.8 (22.3)
60 (37)	660 (409)	27.5	5.1 (16.7)	89.2 (292.6)	9.1	10.5 (34.4)
70 (43)	1093 (678)	37.5	7.4 (24.3)	121.4 (398.2)	10.8	15.3 (50.2)
80 (50)	1682 (1043)	50.0	10.3 (33.8)	158.6 (520.2)	12.4	21.4 (70.2)
90 (56)	2446 (1517)	65.2	13.9 (45.6)	201.6 (661.2)	13.9	28.4 (93.2)

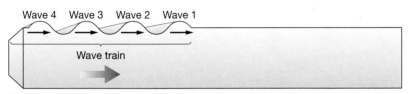

(a) Energy in the leading waves (*Waves 1 and 2*) is transferred into circular orbital motion.

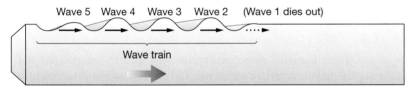

(b) Wave 1 dies out and is replaced by Wave 2; note new Wave 5 behind.

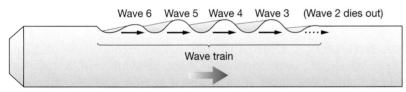

(c) Wave 2 dies out and is replaced by Wave 3; note new Wave 6 behind.

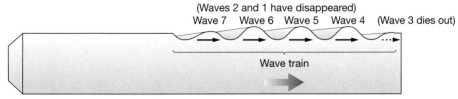

(d) Wave 3 dies out and is replaced by Wave 4; note new Wave 7 behind. Even though new waves take up the lead, the length of the wave train and the total number of waves remain the same. This causes the group speed to be one-half that of the individual wave.

Figure 8.14 Movement of a wave train. Diagrammatic sequence showing the movement of a wave train. Note how waves die out in the front but are replaced in the back, causing the length of the wave train and the total number of waves to remain the same.

STUDENTS SOMETIMES ASK . . .

What is the difference between "groundswell" and "wind swell"?

The term *groundswell* was originally a sailor's word (and is now a surfer's term) for deep-ocean swell, such as might be generated by a distant storm or earthquake. In its original sense, groundswell referred to waves so huge that their troughs bared the "ground" of the sea bottom. Groundswell is essentially the same thing as *wind swell*, although groundswell often refers to very large waves from a distant origin, whereas wind swell refers to smaller, locally produced waves.

back of the group (**Figure 8.14**). For example, if four waves are generated, the lead wave keeps dying out as the wave train travels, but one is created in the back, so the wave train stays four waves. Because of the progressive dying out and creation of new waves, the group moves across the ocean surface at only *half* the velocity of an individual wave in the group.

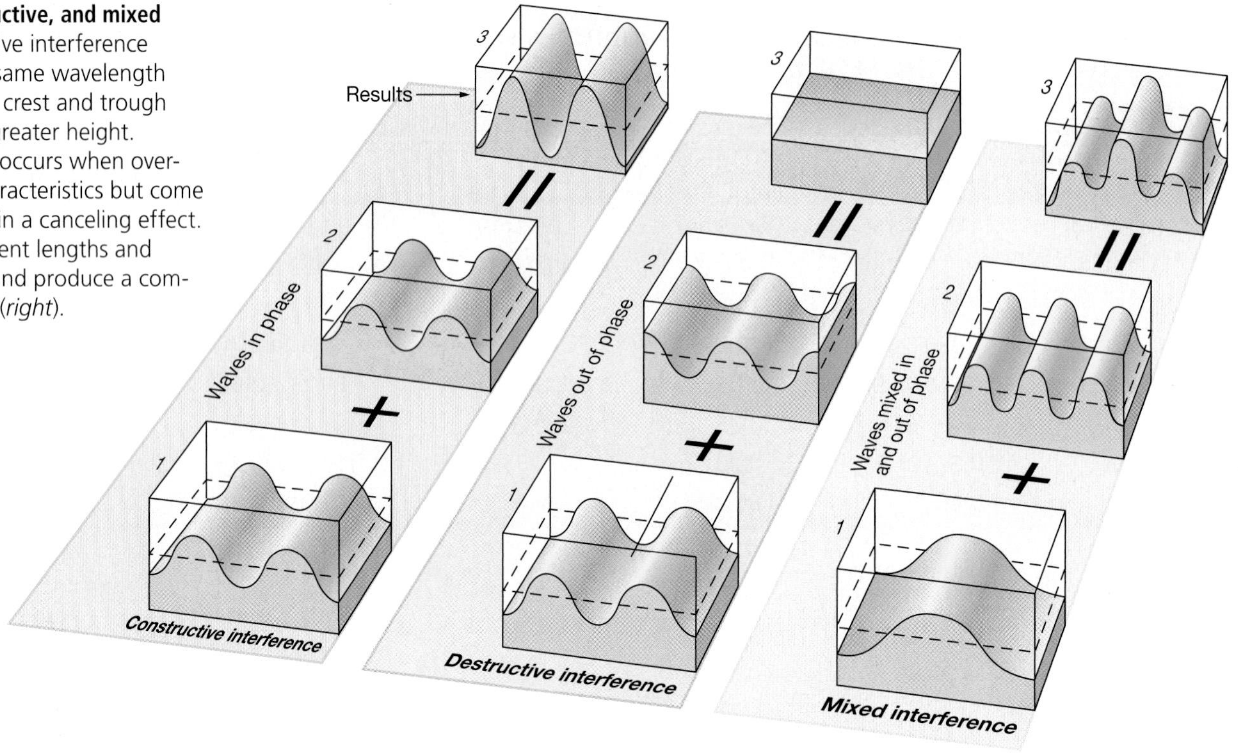

Figure 8.15 Constructive, destructive, and mixed interference patterns. Constructive interference (*left*) occurs when waves of the same wavelength come together in phase (crest to crest and trough to trough), producing waves of greater height. Destructive interference (*center*) occurs when overlapping waves have identical characteristics but come together out of phase, resulting in a canceling effect. More commonly, waves of different lengths and heights encounter one another and produce a complex, mixed interference pattern (*right*).

Web Animation
Interference Patterns in Waves

Interference Patterns

When swells from different storms run together, the waves clash, or *interfere* with one another, giving rise to **interference patterns**. An interference pattern produced when two or more wave systems collide is the sum of the disturbance that each wave would have produced individually. **Figure 8.15** shows that the result may be a larger or smaller trough or crest, depending on conditions.

CONSTRUCTIVE INTERFERENCE **Constructive interference** occurs when wave trains having the same wavelength come together *in phase*, meaning crest to crest and trough to trough. If the displacements from each wave are added together, the interference pattern results in a wave with the same wavelength as the two overlapping wave systems but with a wave height equal to the sum of the individual wave heights (Figure 8.15, *left*).

DESTRUCTIVE INTERFERENCE **Destructive interference** occurs when wave trains having the same wavelength come together *out of phase*, meaning the crest from one wave coincides with the trough from a second wave. If the waves have identical heights, the sum of the crest of one and the trough of another is zero, so the energies of these waves cancel each other (Figure 8.15, *center*).

MIXED INTERFERENCE In most ocean areas, it is likely that two or more swells of different heights and lengths will come together and produce a mixture of constructive and destructive interference. In this scenario, a more complex **mixed interference** pattern develops (Figure 8.15, *right*). Mixed interference patterns explain the varied sequence of higher and lower waves called **surf beat** that most people notice at the beach as well as other irregular wave patterns that occur when two or more swells approach the shore. In the open ocean, several swell systems often interact, creating complex wave patterns (**Figure 8.16**) and, sometimes, large waves that can be hazardous to ships.

Rogue Waves

Rogue waves are massive, solitary, spontaneous ocean waves that can reach enormous height and often occur at times when normal ocean waves are not unusually large. In a sea of 2-meter (6.5-foot) waves, for example, a 20-meter (65-foot) rogue wave may suddenly appear. *Rogue* means "unusual" and, in this case, the waves are unusually large. Rogue waves—sometimes called *superwaves*, *monster waves*, *sleeper waves*, or *freak waves*—are defined as individual waves of exceptional height or abnormal shape that are more than twice the average of the highest one-third of all the wave heights in a given wave record. Mariners often describe rogue wave crests as "mountains of water" and their troughs as "holes in the sea," creating a roller-coaster ride that can be disastrous.

Because of their size and destructive power, rogue waves have been popularized in literature and movies such as *The Perfect Storm* and the disaster film remake *Poseidon*. The impressive size of some rogue waves makes them quite hazardous to oil-drilling platforms and vessels at sea. In 1966, for example, the Italian luxury liner *Michelangelo* received extensive damage from a rogue wave it encountered during a storm in the North Atlantic (**Figure 8.17**). In a nearby area in 1995, the giant luxury liner *Queen Elizabeth II*, which carries 1500 passengers, plowed through a 29-meter (95-foot) wave generated by Hurricane Luis.

In the open ocean, 1 wave in 23 will be over twice the height of the wave average, 1 in 1175 will be three times as high, and 1 in 300,000 will be four times as high. The chances of a truly monstrous wave, therefore, are only 1 in several billion. Nevertheless, rogue waves do occur, and satellite observations over a three-week period in 2001 confirmed that rogue waves occur more frequently than was previously thought: The study documented more than 10 individual giant waves around the globe of over 25 meters (82 feet) in height. Even with satellite monitoring of ocean waves (see Web Table 3.1), it remains difficult to forecast specifically when or

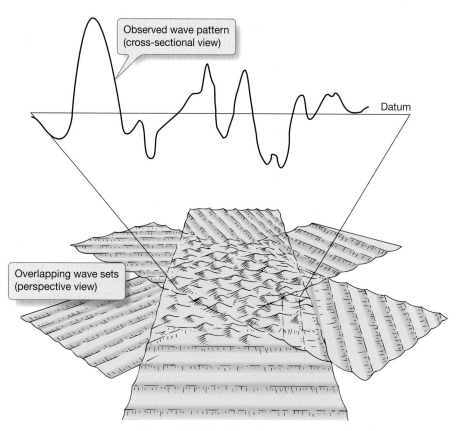

Observed wave pattern (cross-sectional view)

Datum

Overlapping wave sets (perspective view)

Figure 8.16 Mixed interference pattern. The observed wave pattern in the ocean (*above*) is often the result of mixed interference of many different overlapping wave sets (*below*).

Figure 8.17 Damage to the *Michelangelo* from a rogue wave. In April 1966, while making a crossing of the North Atlantic Ocean, the Italian luxury liner *Michelangelo* encountered a rogue wave amid a storm. As the wave crashed aboard the ship, it ripped off part of the ship's bow, tore a hole in its superstructure, smashed bridge windows 24 meters (80 feet) above the waterline, killed three people, and injured dozens more.

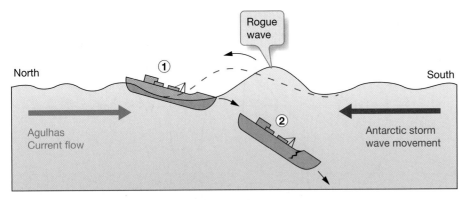

Figure 8.18 Rogue waves along Africa's "Wild Coast." When strong ocean currents (such as the Agulhas Current, *left*) flow in the opposite direction of large waves (such as those generated in Antarctic waters, *right*), rogue waves can be created. These rogue waves can overwhelm and sink even large, well-constructed ships.

where rogue waves will arise. For instance, the 17-meter (56-foot) NOAA research vessel R/V *Ballena* was flipped and sunk in 2000 by a 4.6-meter (15-foot) rogue wave off the California coast while conducting a survey in shallow, calm water. Fortunately, the three people on board survived the incident.

Worldwide each year, about 10 large ships such as supertankers or containerships are reported missing without a trace; as many as 1000 vessels of all sizes are lost each year, and rogue waves are the suspected cause of many of these sinkings. Still, scientists lack detailed shipboard measurements of rogue waves because they tend to appear without warning, and bobbing ships make poor observation platforms.

The main cause of rogue waves is theorized to be an extraordinary case of constructive wave interference where multiple waves overlap in-phase to produce an extremely large wave. Rogue waves also tend to occur more frequently near weather fronts and downwind from islands or shoals. Recent modeling of the wave conditions in the Pacific that caused a Japanese fishing vessel to capsize in 2008 suggests that rogue waves can be created when the low- and high-frequency components of ordinary ocean waves interact and channel their energy into a narrow frequency band.

Another way in which rogue waves can be created is when strong ocean currents focus and amplify opposing swells. Such conditions exist along the "Wild Coast" off the southeast coast of Africa, where the Agulhas Current flows directly against large Antarctic storm waves, creating rogue waves that can crash onto a ship's bow, overcome its structural capacity, and sink even large, well-constructed ships (**Figure 8.18**).

CONCEPT CHECK 8.3

❶ Define *swell*. Does swell necessarily imply a particular wave size? Why or why not?

❷ Waves from separate sea areas move away as swell and produce an interference pattern when they come together. If Sea A has wave heights of 1.5 meters (5 feet) and Sea B has wave heights of 3.5 meters (11.5 feet), what would be the height of waves resulting from constructive interference and destructive interference? Illustrate your answer (see Figure 8.15).

8.4 How Do Waves Change in the Surf Zone?

Most waves generated in the sea area by storm winds move across the ocean as swell. These waves then release their energy along the margins of continents in the **surf zone**, which is the zone of breaking waves. Breaking waves exemplify power and persistence, sometimes moving objects weighing several tons. In doing so, energy from a distant storm can travel thousands of kilometers until it is finally expended along a distant shoreline in a few wild moments.

Physical Changes as Waves Approach Shore

As deep-water waves of swell move toward continental margins over gradually **shoaling** (*shold* = shallow) water, they eventually encounter water depths that are less than one-half of their wavelength (**Figure 8.19**) and become transitional waves. Actually, any shallowly submerged obstacle (such as a coral reef, sunken wreck, or sand bar) will cause waves to release some energy. Navigators have long known that breaking waves indicate dangerously shallow water.

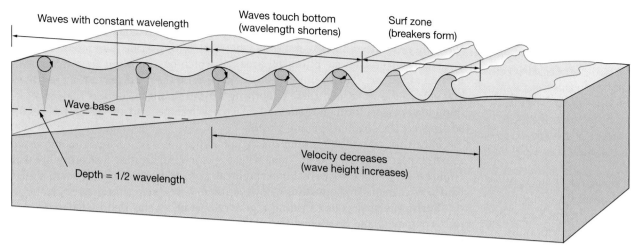

Waves with constant wavelength

Waves touch bottom (wavelength shortens)

Surf zone (breakers form)

Wave base

Depth = 1/2 wavelength

Velocity decreases (wave height increases)

Figure 8.19 Physical changes of a wave in the surf zone. As waves approach the shore and encounter water depths of less than one-half wavelength, the waves "feel bottom." The *wave speed decreases,* and waves stack up against the shore, causing the *wavelength to decrease.* This results in an *increase in wave height* to the point where the wave *steepness is increased* beyond the 1:7 ratio, causing the wave to pitch forward and break in the surf zone.

Many physical changes occur to a wave as it encounters shallow water, becomes a shallow-water wave, and breaks. The shoaling depths interfere with water particle movement at the base of the wave, so the *wave speed decreases.* As one wave slows, the following waveform, which is still moving at its original speed, moves closer to the wave that is being slowed, causing a *decrease in wavelength.* Although some wave energy is lost due to friction, the wave energy that remains must go somewhere, so *wave height increases.* This increase in wave height combined with the decrease in wavelength causes an *increase in wave steepness* (H/L). When the wave steepness reaches the 1:7 ratio, the waves break as surf (Figure 8.19).

If the surf is swell that has traveled from distant storms, breakers will develop relatively near shore, in shallow water. The horizontal motion characteristic of shallow-water waves moves water alternately toward and away from the shore as an oscillation. The surf will be characterized by parallel lines of relatively uniform breakers.

If the surf consists of waves generated by local winds, the waves may not have been sorted into swell. The surf may be mostly unstable, deep-water, high-energy waves with steepness already near the 1:7 ratio. In this case, the waves will break shortly after feeling bottom some distance from shore, and the surf will be rough, choppy, and irregular.

When the water depth is about one and one-third times the wave height, the crest of the wave breaks, producing surf.[4] When the water depth becomes less than $1/20$ the wavelength, waves in the surf zone begin to behave like shallow-water waves (see Figure 8.7). Particle motion is greatly impeded by the bottom, and a significant transport of water toward the shoreline occurs (Figure 8.19).

Waves break in the surf zone because particle motion near the bottom of the wave is severely restricted, slowing the waveform. At the surface, however, individual orbiting water particles have not yet been slowed because they have no contact with the bottom. In addition, the wave height increases in shallow water. The difference in speed between the top and bottom parts of the wave causes the top part of the wave to overrun the lower part, which results in the wave toppling over and breaking. Breaking waves are analogous to a person who leans too far forward. If you don't catch yourself, you may also "break" something when you fall.

Breakers and Surfing

There are three main types of breakers. **Figure 8.20a** shows a **spilling breaker**, which is a turbulent mass of air and water that runs down the front slope of the wave as it breaks. Spilling breakers result from a gently sloped ocean bottom, which

Web Animation
Wave Motion and Wave Refraction

STUDENTS SOMETIMES ASK . . .

How much more energy does a 3-foot wave pack compared to a 1-foot wave?

Interestingly, the energy that a wave produces is proportional to the height of the wave *squared.* Therefore, a 3-foot wave is nine—not three—times more powerful than a 1-foot wave. When onshore breaking waves reach 5 feet, the surf is generally too dangerous for swimming, and anyone who has been tumbled around by waves can attest to their incredible power. Although experienced surfers relish big waves, good surfing beaches are generally not safe for swimming.

ESSENTIAL CONCEPT 8.4A

As waves come into shallow water and feel bottom, their speed and wavelength decrease, while their wave height and wave steepness increase; this action causes the waves to break.

---

[4]Using this relationship provides a handy way of estimating water depth in the surf zone: The depth of the water where waves are breaking is one and one-third times the breaker height.

gradually extracts energy from the wave over an extended distance and produces breakers with low overall energy. As a result, spilling breakers have a longer life span and give surfers a long—but somewhat less exciting—ride than other breakers.

Figure 8.20b shows a **plunging breaker**, which has a curling crest that moves over an air pocket. The curling crest occurs because the particles in the crest literally outrun the wave, and there is nothing beneath them to support their motion. Plunging breakers form on moderately steep beach slopes and are the best waves for surfing (see the chapter-opening photo).

When the ocean bottom has an abrupt slope, the wave energy is compressed into a shorter distance, and the wave will surge forward, creating a **surging breaker** (**Figure 8.20c**). These waves build up and break right at the shoreline, so board surfers tend to avoid them. For body surfers, however, these waves present the greatest challenge.

Surfing is analogous to riding a gravity-operated water sled by balancing the forces of gravity and buoyancy. The particle motion of ocean waves (see Figure 8.4) shows that water particles move up into the front of the crest. This force, along with the buoyancy of the surfboard, helps maintain a surfer's position in front of a breaking wave. The trick is to perfectly balance the force of gravity (directed downward) with the buoyant force (directed perpendicular to the wave face) to enable a surfer to be propelled forward by the wave's energy. A skillful surfer, by positioning the board properly on the wave front, can regulate the degree to which the propelling gravitational forces exceed the buoyancy forces, and the surfer can obtain speeds up to 40 kilometers (25 miles) per hour while moving along the face of a breaking wave.

When the wave passes over water that is too shallow to allow the upward movement of water particles to continue, the wave has expended its energy, and the ride is over.

Wave Refraction

Waves seldom approach a shore at a perfect right angle (90 degrees). Instead, some segment of the wave will "feel bottom" first and will slow before the rest of the wave. This results in the **refraction** (*refringere* = to break up), or *bending*, of each *wave crest* (also called a *wave front*) as waves approach the shore.

Figure 8.21a shows how waves coming toward a straight shoreline are refracted and tend to align themselves *nearly* parallel to the shore. This explains why all waves come almost straight in toward a beach, no matter what their original orientation was. **Figure 8.21b** shows how waves coming toward an irregular shoreline refract so that they, too, nearly align with the shore. **Figure 8.21c** shows a classic example of wave refraction around Rincon Point in California.

(a) Spilling breaker, resulting from a gradual beach slope.

(b) Plunging breaker, resulting from a steep beach slope; these are the best waves for surfing.

(c) Surging breaker, resulting from an abrupt beach slope.

Figure 8.20 Types of breakers. Photos of the three types of breakers, which are a result of different beach slopes.

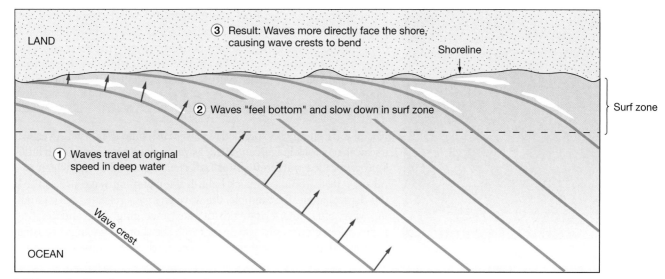

(a) Wave refraction along a straight shoreline.

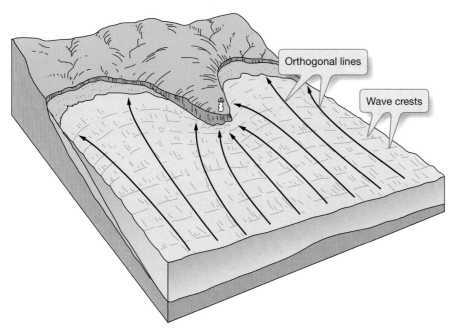

(b) Wave refraction along an irregular shoreline.

(c) Wave refraction at Rincon Point, California, looking west.

Figure 8.21 Wave refraction. Wave refraction is the bending of waves, as shown here along **(a)** a straight shoreline and **(b** and **c)** an irregular shoreline.

Web Animation
Wave Motion and
Wave Refraction

   The refraction of waves along an irregular shoreline distributes wave energy unevenly along the shore. To help illustrate how this works, notice the long black arrows in Figure 8.21b, which are called **orthogonal** (*ortho* = straight, *gonia* = angle) **lines**, or *wave rays*. Orthogonal lines are always oriented perpendicular to the wave crests, so they indicate the direction that waves travel. More importantly, orthogonals are spaced so that the energy between lines is equal at all times and can thus be used to show variations in wave energy. Far from shore, for example, notice how the orthogonals in Figure 8.21b are evenly spaced, which indicates that all parts of the wave have the same amount of energy. As the waves approach the shore and refract, however, notice how the orthogonals *converge* on headlands that jut into the ocean and *diverge* in bays. This means that wave energy is focused against the headlands but dispersed in bays. As a result, large waves occur at headlands, which are areas of good surfing[5] and sites of erosion. Conversely, smaller waves occur in bays, which

---

[5]Sailors have long known that "the points draw the waves." Surfers also know how wave refraction causes good "point breaks."

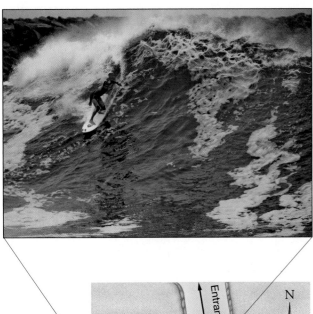

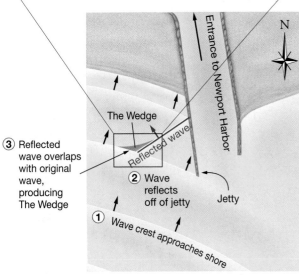

Figure 8.22 **Wave reflection and constructive interference at The Wedge, Newport Harbor, California.** As waves approach the shore ①, some of the wave energy is reflected off the long jetty at the entrance to the harbor ②. The reflected wave overlaps and constructively interferes with the original wave ③, resulting in a wedge-shaped wave (*dark blue triangle*) that may reach heights exceeding 8 meters (26 feet). Photo (*inset*) shows a surfer riding The Wedge; note the jetty in the background.

Web Video
Wave Dispersion, Wave Reflection, and Wave Interference

often provide areas for boat anchorages and are also associated with sediment deposition. Interestingly, the same waves break against both headlands and in nearby bays, but their energy is different because wave refraction causes the spacing of orthogonals (and thus energy) to change. In addition, waves approaching shore are also influenced by sea floor features such as shallow banks or submarine canyons.

Wave Reflection

Not all wave energy is expended as waves rush onto the shore. A vertical barrier, such as a seawall or a rock ledge, can reflect waves back into the ocean with little loss of energy—a process called **wave reflection** (*reflecten* = to bend back), which is similar to how a mirror reflects (bounces) back light. If the incoming wave strikes the barrier at a right (90-degree) angle, for example, the wave energy is reflected back parallel to the incoming wave, often interfering with the next incoming wave and creating unusual waveforms. More commonly, waves approach the shore at an angle, causing wave energy to be reflected at an angle equal to the angle at which the wave approached the barrier.

THE WEDGE: A CASE STUDY OF WAVE REFLECTION AND CONSTRUCTIVE INTERFERENCE An outstanding example of wave reflection and constructive interference occurs in an area called *The Wedge*, which develops west of the jetty that protects the harbor entrance at Newport Harbor, California (**Figure 8.22**). The jetty is a solid human-made object that extends into the ocean 400 meters (1300 feet) and has a near-vertical side facing the waves. As incoming waves strike the vertical side of the jetty at an angle, they are reflected at an equivalent angle. Because the original waves and the reflected waves have the same wavelength, a constructive interference pattern develops, creating plunging breakers that may exceed 8 meters (26 feet) in height (Figure 8.22, *inset*). Too dangerous for board surfers, these waves present a fierce challenge to the most experienced body surfers. The Wedge has crippled and even killed many who have come to challenge it.

STANDING WAVES, NODES, AND ANTINODES **Standing waves** (also called *stationary waves*) can be produced when waves are reflected at right angles to a barrier. Standing waves are the sum of two waves with the same wavelength moving in opposite directions, resulting in no net movement. Although the water particles continue to move vertically and horizontally, there is none of the circular motion that is characteristic of a progressive wave.

Figure 8.23 shows the movement of water during the wave cycle of a standing wave. Lines along which there is no vertical movement are called *nodes* (*nodus* = knot), or *nodal lines*. *Antinodes*, which are crests that alternately become troughs, are the points of greatest vertical movement within a standing wave.

When water sloshes back and forth in a basin, the maximum vertical displacements are at the antinodes. When water moves from crest to trough at an antinode,

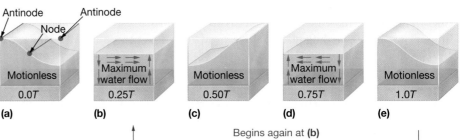

Figure 8.23 **Sequence of motion in a standing wave.** In a standing wave, water is motionless when antinodes reach maximum displacement (**a**, **c**, and **e**; **a** and **e** are identical). Water movement is at a maximum (*blue arrows*) when the water is horizontal (**b** and **d**). Movement is vertical beneath the antinodes, and maximum horizontal movement occurs beneath the node. After (**e**), the cycle begins again at (**b**).

the displaced volume of water has to move horizontally to raise the water level of the adjacent antinode from trough to crest. As a result, there is no vertical motion at the node; instead, there is only horizontal motion. At the antinodes, however, the movement of water particles is entirely vertical.

We'll consider standing waves further when tidal phenomena are discussed in Chapter 9, "Tides." Under certain conditions, the development of standing waves significantly affects the tidal character in coastal regions.

CONCEPT CHECK 8.4

1. Describe the physical changes that occur to a wave's wave speed (S), wavelength (L), height (H), and wave steepness (H/L) as the wave moves across shoaling water to break on the shore.

2. Describe the three different types of breakers and indicate the slope of the beach that produces each of the three types. How is the energy of the wave distributed differently within the surf zone by the three types of breakers?

3. Using examples, explain how wave refraction is different from wave reflection.

4. Using orthogonal lines, illustrate how wave energy is distributed along a shoreline with headlands and bays. Identify areas of high- and low-energy release.

8.5 How Are Tsunami Created?

The Japanese term for the large, sometimes destructive waves that occasionally roll into their harbors is **tsunami** (*tsu* = harbor, *nami* = wave[s]) Tsunami originate from sudden changes in the topography of the sea floor caused by such events as slippage along underwater faults, underwater avalanches such as turbidity currents or the collapse of large oceanic volcanoes, and underwater volcanic eruptions. Many people mistakenly call them "tidal waves," but tsunami are unrelated to the tides. The mechanisms that trigger tsunami are typically seismic events, so tsunami are *seismic sea waves*.

The majority of tsunami are caused by *fault movement*. Underwater fault movement displaces Earth's crust, generates earthquakes, and, if it ruptures the sea floor, produces a sudden change in water level at the ocean surface (Figure 8.24a). Faults that produce *vertical* displacements (the uplift or downdropping of ocean floor) change the volume of the ocean basin, which affects the entire water column and generates tsunami. Conversely, faults that produce *horizontal* displacements (such as the lateral movement associated with transform faulting) do not generally generate tsunami because the side-to-side movement of these faults does not change the volume of the ocean basin. Much less common events, such as underwater avalanches triggered by shaking or underwater volcanic eruptions—which create the largest waves—also produce tsunami. In addition, large objects that splash into the ocean, such as above-water coastal landslides (see Web Diving Deeper 8.1) or meteorite impacts, produce **splash waves**, which are a type of tsunami.

During the 20th century, 498 measurable tsunami occurred worldwide, 66 of them resulting in fatalities. The source events were earthquakes (86%), volcanic activity (5%), landslides (4%), and combinations of those processes (5%). Tsunami generated by meteorite or asteroid impacts occur much less frequently.

The wavelength of a typical tsunami exceeds 200 kilometers (125 miles), so it is a shallow-water wave everywhere in the ocean.[6] Because tsunami are shallow-water

Web Animation
Tsunami

---

[6]Recall that the depth of the wave base is equal to one-half a wave's wavelength. Thus, tsunami can typically be felt to depths of 100 kilometers (62 miles), which is deeper than even the deepest ocean trenches.

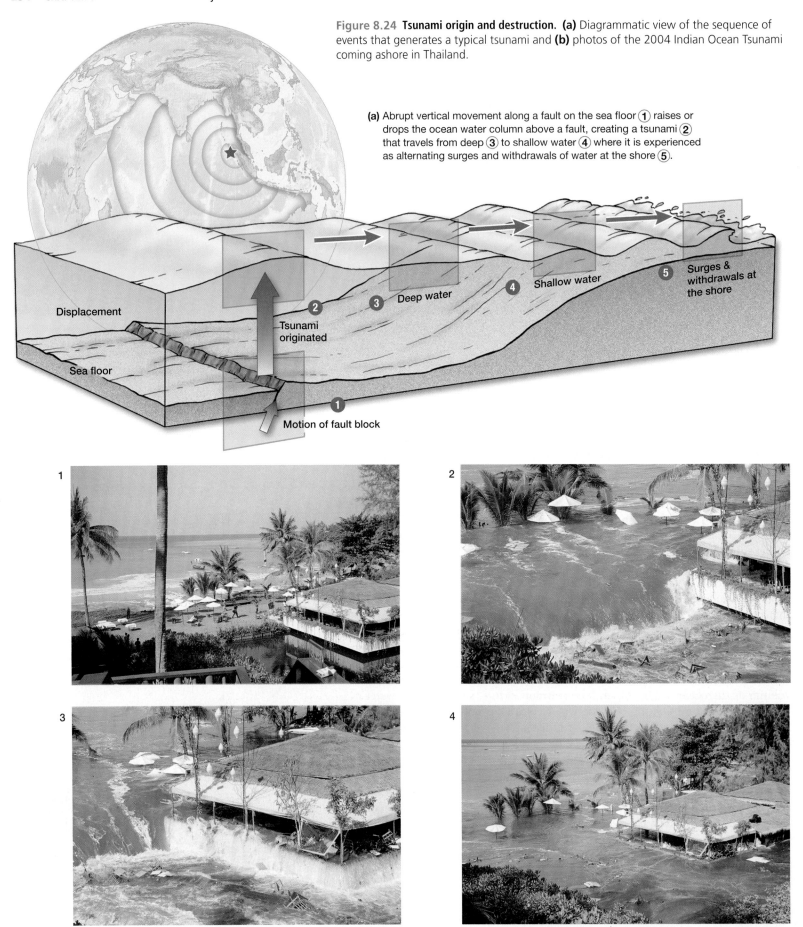

Figure 8.24 **Tsunami origin and destruction. (a)** Diagrammatic view of the sequence of events that generates a typical tsunami and **(b)** photos of the 2004 Indian Ocean Tsunami coming ashore in Thailand.

(a) Abrupt vertical movement along a fault on the sea floor ① raises or drops the ocean water column above a fault, creating a tsunami ② that travels from deep ③ to shallow water ④ where it is experienced as alternating surges and withdrawals of water at the shore ⑤.

Displacement

Tsunami originated

Sea floor

② ③ Deep water ④ Shallow water ⑤ Surges & withdrawals at the shore

① Motion of fault block

(b) Sequence of photos of the 2004 Indian Ocean Tsunami inundating the Chedi Resort in Phuket, Thailand, on December 26, 2004.

waves, their speed is determined only by water depth. In the open ocean, tsunami move at well over 700 kilometers (435 miles) per hour—they can easily keep pace with a jet airplane—and have heights of only about 1 meter (3.3 feet) or less. Even though they are fast, tsunami are small in the open ocean and pass unnoticed in deep water until they reach shore, where they slow in the shallow water and undergo physical changes (just as wind-generated waves do) that cause tsunami to increase in wave height.

Coastal Effects

Contrary to popular belief, a tsunami does not form a huge breaking wave at the shoreline. Instead, it is a strong flood or surge of water that causes the ocean to advance (or, in certain cases, retreat) dramatically. In fact, a tsunami resembles a sudden, *extremely* high tide, which is why they are misnamed "tidal waves." It may take many minutes for the tsunami to express itself fully, during which time sea level can rise up to 40 meters (131 feet) above normal, with normal waves superimposed on top of the higher sea level. The strong surge of water can rush into low-lying areas with destructive results (**Figure 8.24b**).

As the trough of the tsunami arrives at the shore, the water rapidly drains off the land. In coastal areas, it looks like a sudden and *extremely* low tide, where sea level is many meters lower than even the lowest low tide. Because tsunami are typically a series of waves, there is often an alternating series of dramatic surges and withdrawals of water that are widely separated in time. The first surge is only rarely the largest; instead, the third, fourth—or even seventh—surge may be the largest and can occur several hours later.

Depending on the geometry of sea floor motion that creates a tsunami, the trough can sometimes arrive at a coast first. For example, on the side of a fault where the sea floor drops down, the first part of the wave to propagate outward will be the trough, followed by the crest. Conversely, on the side of the fault that moves upward, the crest leads the trough. At a coast where the trough arrives first, the water drains out and exposes parts of the lowermost shoreline that are rarely seen. For people at the shoreline, the temptation is to explore these newly exposed areas and catch stranded organisms. Within a few minutes, however, a strong surge of water (the crest of the tsunami) is due to arrive.

The alternating surges and retreats of water by tsunami can severely damage coastal structures and can injure and kill people as well. The speed of the advance—up to 4 meters (13 feet) per second—is faster than any human can run. Those who are caught in a tsunami are often drowned or crushed by floating debris (**Figure 8.25**).

Some Examples Of Historic and Recent Tsunami

Although many small tsunami are created each year, most go unnoticed. On average, about 50 noticeable tsunami occur every decade, with a large tsunami occurring somewhere in the world every 2 to 3 years and an extremely large and damaging one occurring every 15 to 20 years. Remarkably, two of the largest and most deadly tsunami have occurred within the past decade.

Where do most tsunami occur? About 86% of all great waves are generated in the Pacific Ocean because large-magnitude earthquakes occur along the series of trenches that ring its ocean basin where oceanic plates are subducted along convergent plate boundaries. Volcanic activity is also common along the *Pacific Ring of Fire*, and the large earthquakes that occur along its margin are capable of producing extremely large tsunami. **Figure 8.26** shows significant tsunami that have struck since 1990—with the majority occurring along the Pacific Ring of Fire.

THE ERUPTION OF KRAKATAU (1883) One of the most destructive historical tsunami ever generated came from the eruption of the volcanic island of Krakatau[7] on August 27, 1883. Approximately the size of a small Hawaiian Island in what is now

---

[7]The island Krakatau (which is *west* of Java) is also called Krakatoa.

Figure 8.25 Tsunami damage in Hilo, Hawaii. Flattened parking meters in Hilo, Hawaii, caused by the 1946 tsunami that resulted in more than $25 million in damage and 159 deaths.

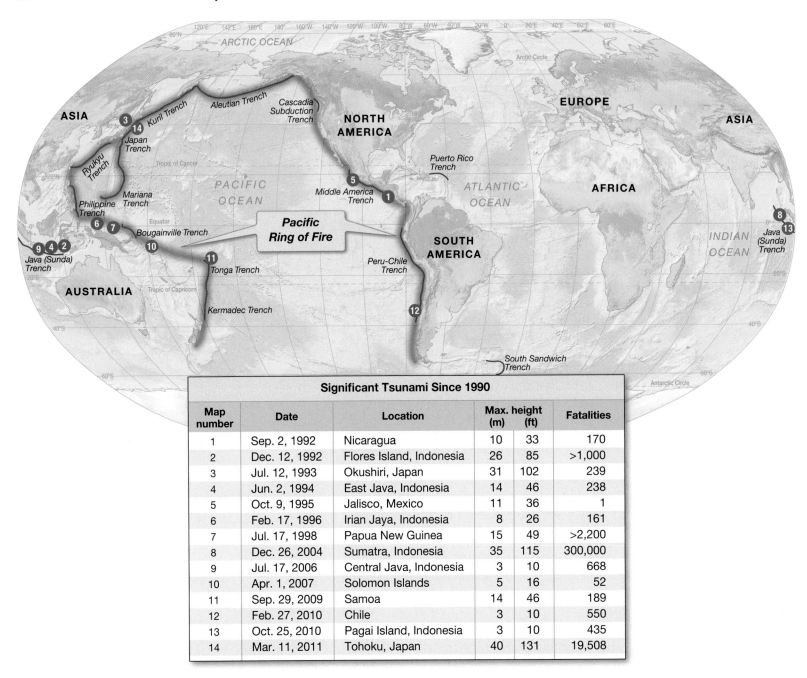

Map number	Date	Location	Max. height (m)	(ft)	Fatalities
1	Sep. 2, 1992	Nicaragua	10	33	170
2	Dec. 12, 1992	Flores Island, Indonesia	26	85	>1,000
3	Jul. 12, 1993	Okushiri, Japan	31	102	239
4	Jun. 2, 1994	East Java, Indonesia	14	46	238
5	Oct. 9, 1995	Jalisco, Mexico	11	36	1
6	Feb. 17, 1996	Irian Jaya, Indonesia	8	26	161
7	Jul. 17, 1998	Papua New Guinea	15	49	>2,200
8	Dec. 26, 2004	Sumatra, Indonesia	35	115	300,000
9	Jul. 17, 2006	Central Java, Indonesia	3	10	668
10	Apr. 1, 2007	Solomon Islands	5	16	52
11	Sep. 29, 2009	Samoa	14	46	189
12	Feb. 27, 2010	Chile	3	10	550
13	Oct. 25, 2010	Pagai Island, Indonesia	3	10	435
14	Mar. 11, 2011	Tohoku, Japan	40	131	19,508

Significant Tsunami Since 1990

Figure 8.26 Significant tsunami since 1990. Global locations and corresponding table showing significant tsunami (those with large tsunami height and/or responsible for a large number of fatalities) since 1990. Tsunami have claimed more than 300,000 lives worldwide since 1990. These killer waves are most often generated by earthquakes along colliding tectonic plates of the Pacific Rim, although the most deadly tsunami in history was the 2004 Indian Ocean Tsunami (*number 8*). Locations of ocean trenches are shown by dark red lines; Pacific Ring of Fire is shown by red shading.

Indonesia, Krakatau exploded with the greatest release of energy from Earth's interior ever recorded in historic times. The island, which stood 450 meters (1500 feet) above sea level, was nearly obliterated. The sound of the explosion was heard throughout the Indian Ocean, up to 4800 kilometers (3000 miles) away, and remains the loudest noise on human record. Dust from the explosion ascended into the atmosphere and circled Earth on high-altitude winds, producing brilliant red sunsets worldwide for nearly a year.

Not many people were killed by the outright explosion of the volcano because the island was uninhabited. However, the displacement of water from the energy released during the explosion was enormous, creating a tsunami that exceeded 35 meters (116 feet)—as high as a 12-story building. It devastated the coastal region of the Sunda Strait between the nearby islands of Sumatra and Java, drowning more than 1000 villages and taking more than 36,000 lives. The energy carried by this wave reached every ocean basin and was even detected by tide-recording stations as far away as London and San Francisco.

Like most of the other approximately 130 active volcanoes in Indonesia, Krakatau was formed along the Sunda Arc, a 3000-kilometer (1900-mile) curving chain of volcanoes associated with the subduction of the Australian Plate beneath the Eurasian Plate. Where these two sections of Earth's crust meet, earthquakes and volcanic eruptions are common.

THE SCOTCH CAP, ALASKA/HILO, HAWAII TSUNAMI (1946)   A strong and damaging tsunami hit the Hawaiian Islands on April 1, 1946, with north-facing shores including the port city of Hilo receiving the majority of damage. The tsunami was from a magnitude $M_w = 7.3$ earthquake in the Aleutian Trench off the island of Unimak, Alaska, more than 3000 kilometers (1850 miles) away. From that direction, the offshore bathymetry in Hilo Bay focused the tsunami's energy directly toward town, so the city of Hilo experienced tsunami surges of tremendous heights. In this case, the tsunami expressed itself as a strong recession followed by a surge of water nearly 17 meters (55 feet) above normal high tide, causing more than $25 million in damage and killing 159 people. Remarkably, it stands as Hawaii's worst natural disaster (Figure 8.25).

Closer to the source of the earthquake, the tsunami was considerably larger. The tsunami struck Scotch Cap, Alaska, on Unimak Island, where a two-story reinforced concrete lighthouse stood 14 meters (46 feet) above sea level at its base. The lighthouse was destroyed by a wave that is estimated to have reached 36 meters (118 feet), killing all five people inside the lighthouse at the time. Vehicles on a nearby mesa 31 meters (103 feet) above water level were also moved by the onrush of water.

PAPUA NEW GUINEA (1998)   In July 1998, off the north coast of Papua New Guinea in the western part of the Pacific Ring of Fire, a magnitude $M_w = 7.1$ earthquake was followed shortly thereafter by a 15-meter (49-foot) tsunami, which was up to five times larger than expected for a tsunami created by an earthquake that size. The tsunami completely overtopped a heavily populated low-lying sand bar, destroying three entire villages and resulting in at least 2200 deaths. Researchers who mapped the sea floor after the tsunami discovered the remains of a huge underwater landslide, which was apparently triggered by the shaking and caused the unusually large tsunami.

INDIAN OCEAN (2004)   Although most tsunami occur in the Pacific Ocean, they do sometimes occur in other ocean basins. On December 26, 2004, an enormous earthquake struck in a subduction zone off the west coast of Sumatra in Indonesia on the shores of the Indian Ocean. Known as the Sumatra–Andaman Earthquake, it was the second biggest earthquake recorded during the past century and the largest to be recorded by modern seismograph equipment. The earthquake was so large that it changed Earth's gravity field, triggered small earthquakes as far away as Alaska, and even altered Earth's rotation. Initially, it was deemed a magnitude $M_w = 9.0$, but it was subsequently upgraded to magnitude $M_w = 9.2$. The earthquake occurred about 30 kilometers (19 miles) beneath the sea floor, near the Sunda Trench, where the Indian Plate is being subducted beneath the Eurasian Plate. Moreover, about 1200 kilometers (750 miles) of sea floor was ruptured along the interface of these two tectonic plates, thrusting sea floor upward and generating about 10 meters (33 feet) of vertical displacement. This abrupt vertical movement of the sea floor is what generated the deadliest tsunami in history.

Once generated, the tsunami spread out at jetliner speeds across the Indian Ocean. Only 15 minutes after the earthquake, the tsunami hit the shores of Sumatra with a series of alternating rapid withdrawals and strong surges up to 35 meters (115 feet) high. Many coastal villages were completely washed away (Figure 8.27), causing several hundred thousand deaths. Sites farther from the quake zone experienced smaller but nevertheless deadly waves. Particularly affected were Thailand (see Figure 8.24b), which was struck by the tsunami about 75 minutes after the quake

(a) Before: January 10, 2003

(b) After: December 29, 2004

Figure 8.27 Satellite views of tsunami destruction in Indonesia. High-resolution satellite images of Lhoknga, in the province of Aceh, west coast of Sumatra, Indonesia. **(a)** Before the tsunami, on January 10, 2003. **(b)** The same area on December 29, 2004, three days after the tsunami that inundated the city with a 15-meter (50-foot) surge of water. In both images, note the mosque (*black arrow*), which was one of the few buildings that remained standing.

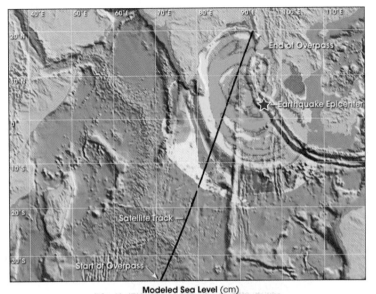

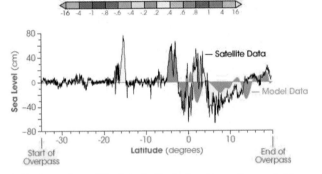

Figure 8.28 Jason-1 satellite detects the Indian Ocean Tsunami. The Indian Ocean Tsunami was initiated by a large earthquake offshore Sumatra on December 26, 2004 (*red star*). By a fortuitous circumstance, the Jason-1 satellite passed over the Indian Ocean (*black line*) two hours after the tsunami was generated. Its radar altimeter detected the crests and troughs of the tsunami (*colors*), which showed a wave height of about 1 meter (3.3 feet) in the open ocean. The graph of the satellite's overpass (*below*) shows the difference between the measured sea level from satellite data (*black line*) and the modeled wave height (*blue curve*).

STUDENTS SOMETIMES ASK . . .

Where and when will the next major tsunami hit?

Although most tsunami occur in the Pacific Ocean, a tsunami can affect any coastal region worldwide. Tsunami prediction is still in its infancy, yet seismologists who examine the history of large earthquakes (*paleoseismologists*) can produce forecasts of future tsunami based on evidence such as sediment records that indicate the size and frequency of past tsunami. Using this technique, paleoseismologists have issued forecasts for large, damaging tsunami in two regions: (1) In the Indian Ocean, where in 2004 the deadliest tsunami in history killed between 230,000 and 300,000 people, a similar tsunami has been forecast to occur sometime within the next 30 years; and (2) off the U.S. Pacific Northwest coast, where there is a subduction zone just offshore and a large tsunami generated by faulting in the offshore trench occurred in 1700; another tsunami is expected in the next 250 to 500 years. For both these locations, it's only a matter of time.

occurred, and Sri Lanka and India, which were pounded by devastating waves about three hours after the earthquake. After seven hours and at a distance of more than 5000 kilometers (3000 miles), the tsunami hit the east coast of Africa, where it still had enough power to kill more than a dozen people. Although much smaller, the tsunami was also detected in the Atlantic, Pacific, and Arctic Oceans.

In a remarkable coincidence, the Jason-1 satellite happened to be passing over the Indian Ocean two hours after the tsunami originated (**Figure 8.28**). The satellite's radar altimeter, which is designed to accurately measure the elevation of the ocean surface (see Diving Deeper 3.1), was able to detect the crests and troughs of the tsunami as it radiated out across the Indian Ocean with a wavelength of about 500 kilometers (300 miles). Although this sighting occurred about an hour before the first waves struck Sri Lanka and India, the satellite data couldn't have been used to warn tsunami victims because scientists needed several hours to analyze the information. However, these satellite data are particularly valuable because they will allow scientists to check the accuracy of open-ocean tsunami travel models, which are based on seismic data, bathymetric information, and coastal tide gauges.

In all, between 230,000 and 300,000 people in 11 countries were killed by the tsunami, ranking this tragedy as the most deadly tsunami in recorded history. The tsunami also caused billions of dollars of damage throughout the region and left millions homeless. Why was there such a large loss of life during the tsunami? One factor was the lack of a warning system in the Indian Ocean, like the network of buoys and deep-sea instruments that monitors earthquake and wave activity in the Pacific Ocean, where most tsunami have historically occurred. Another is the lack of an emergency response system to warn high-population-density coastal communities and beachgoers. Still another was the lack of good public awareness in recognizing the signs of an impending tsunami, such as when a rapid withdrawal of water is observed at the shore, which is caused by the trough of the tsunami arriving first and indicates that an equally strong surge of water will soon follow.

Studies of the effect of the tsunami on different coastlines suggest that those areas lacking protective coral reefs or mangroves received stronger surges. In many cases, the shape of the coastline as well as the geometry of the offshore sea floor affected the height of the tsunami and the amount of coastal destruction. Sediment cores reveal that several other large tsunami have occurred in the region during the past 1000 years.

On July 17, 2006, another strong earthquake—this time an $M_w = 7.7$ earthquake associated with faulting in the Java Trench off the southern coast of Java, Indonesia—triggered a 3-meter (10-foot) tsunami that killed 668 people, injured at least 600, and destroyed many coastal structures. This destruction near the area that experienced even greater damage from the 2004 tsunami emphasized the fact that the region still lacked a comprehensive tsunami warning system. In 2010, however, the Indian Ocean Tsunami Warning and Mitigation System became fully operational, with its network of deep-ocean pressure sensors, buoys, land seismographs, tidal gauges, data centers, and communications upgrades. The system will continue to be upgraded with more detection buoys in the coming years. With this new warning system in place and enhanced public education about what to do in the event of a tsunami, all countries bordering the Indian Ocean will be much better prepared for the destructive power of future tsunami.

JAPAN (2011) On March 11, 2011, an $M_w = 9.0$ earthquake—the fourth largest ever to be recorded—occurred in the Japan Trench offshore of northeastern Japan. Known as the Tohoku Earthquake, it uplifted the sea floor by several meters and created a large, devastating tsunami that was felt throughout the Pacific Ocean (**Diving Deeper 8.1**). In all, damages have been estimated at $235 billion, making it the most expensive natural disaster in world history.

WAVES OF DESTRUCTION: THE 2011 JAPANESE TSUNAMI

On March 11, 2011, the nation of Japan and seismologists around the world received a terrible surprise: A huge earthquake, significantly stronger than people had anticipated or prepared for in the region, struck off the northeastern shore of Japan. The great 2011 $M_w = 9.0$ Tohoku Earthquake—named for the region it struck—occurred along the Japan Trench where the Pacific Plate is being subducted underneath Japan (**Figure 8A**). It shifted the coast of Japan up to 8 meters (26 feet) eastward and also uplifted the sea floor by as much as 5 meters (16.4 feet) over an area comparable in size to the state of Connecticut; this sudden vertical movement of the sea floor imparted energy into the Pacific Ocean and generated one of the largest and best-studied tsunami in history.

Although Japan prides itself in disaster preparedness, many residents did not get accurate tsunami warnings, and many chose to stay in dangerous locations in part because they misunderstood the risks. Based on the initial estimates of the earthquake, the Japan Meteorological Agency (JMA) issued a tsunami warning within three minutes from the start of the shaking. The agency's warning predicted a tsunami surge of only 3 to 6 meters (10 to 20 feet), and unfortunately, many people did not take immediate action out of a belief that Japan's extensive network of tsunami walls would protect them. Within the next few minutes, JMA reissued its warning for a 10-meter (33-foot) tsunami, but many people did not receive the updated warning because of the destruction of the regional power supply system.

About 20 minutes after the earthquake began, the tsunami swept onto the coastline of northeastern Japan, where most fishing villages and towns were built at the ends of long, narrow bays because these bays offer protection from wind waves. However, these long narrow bays amplified the height of the tsunami. The initial surge of the tsunami reached heights of about 15 meters (49 feet) and easily overtopped harbor-protecting tsunami walls and coastal margins (**Figure 8B**), penetrating as far as 10 kilometers (6 miles) inland. In one location, the tsunami reached a record height of 40 meters (131 feet) because of amplification by offshore topography.

The Pacific Tsunami Warning Center, with its network of buoys that had been greatly expanded since the 2004 Indian Ocean tsunami, sent out tsunami warnings across the Pacific and accurately predicted its arrival, allowing coastal communities to prepare. The tsunami struck Hawaii seven hours after the earthquake—and the coast of California three hours later still—with a surge up to 2 meters (7 feet) that damaged harbors, marinas, and coastal resorts. Minor damage occurred in Peru and Chile when the wave struck their coastlines 20 to 22 hours after the event. The Philippines, Indonesia, and New Guinea were also affected by the tsunami.

In Japan, shaking from the earthquake caused few fatalities and did minimal damage due largely to Japan's strict building codes, thorough earthquake preparation, and advance warning systems. However, the tsunami destroyed many small towns and villages, killed 19,508 people, and initially displaced nearly half a million people; nearly two years later, thousands were still living in temporary shelters such as high school gymnasiums. The tsunami also disrupted the power supply needed to maintain water circulation for cooling the reactors at the Fukushima Daiichi nuclear power plant. Three reactors exploded, releasing radioactivity that continues to plague central Japan. Radioactivity was also released into the ocean, and studies are now under way to determine its effect on marine life.

The destructive power of the tsunami also generated a huge amount of floating debris, including boats, cars, building parts, household items, and even entire houses. Japanese authorities estimate that the tsunami released 1.4 million metric tons (1.5 million short tons) of floating debris. The tsunami debris is being monitored as it slowly spreads out and moves across the Pacific Ocean, transported by currents in the North Pacific Subtropical Gyre. Although some of the debris will sink, disperse, or biodegrade with time, the wreckage may impact Hawaiian coral reefs. Along the U.S. West Coast, a derelict Japanese fishing vessel and other tsunami debris has already arrived, and more is expected during 2013.

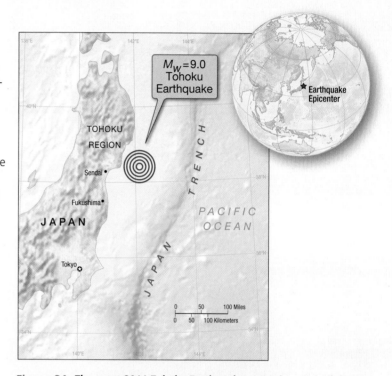

Figure 8A The great 2011 Tohoku Earthquake. Map location of the $M_w = 9.0$ Tohoku Earthquake, which was the most powerful earthquake ever to have hit Japan. It occurred along the Japan Trench and uplifted the sea floor, thereby creating a large, devastating tsunami.

Figure 8B The 2011 tsunami surges over a protective seawall in Miyako, northeastern Japan.

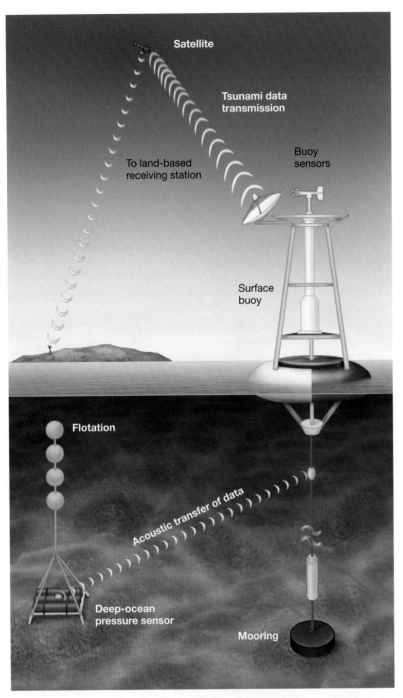

Figure 8.29 Deep-ocean Assessment and Reporting of Tsunamis (DART). The DART system consists of a deep-ocean pressure sensor that can detect a tsunami passing above. The pressure sensor relays information to a buoy at the surface that transmits the data via satellite, allowing oceanographers to detect the passage of a tsunami in the open ocean.

Tsunami Warning System

In response to the tsunami that struck Hawaii in 1946, a tsunami warning system was established throughout the Pacific Ocean. It led to what is now the **Pacific Tsunami Warning Center (PTWC)**, which coordinates information from 25 Pacific Rim countries and is headquartered in Ewa Beach (near Honolulu), Hawaii. The tsunami warning system uses seismic waves—some of which travel through Earth at speeds 15 times faster than tsunami—to forecast destructive tsunami. In addition, oceanographers have recently established a network of sensitive pressure sensors on the deep-ocean floor of the Pacific. The program, called **Deep-ocean Assessment and Reporting of Tsunamis (DART)**, utilizes sea floor sensors that are capable of picking up the small yet distinctive pressure pulse from a tsunami passing above. The pressure sensors relay information to a buoy at the surface that transmits the data via satellite, allowing oceanographers to detect the passage of a tsunami in the open ocean (**Figure 8.29**). DART buoys, which are essential components of tsunami warning systems, have now been deployed in all oceans.

TSUNAMI WATCHES AND TSUNAMI WARNINGS When a seismic disturbance occurs beneath the ocean surface that is large enough to be tsunamigenic (capable of producing a tsunami), a *tsunami watch* is issued. At this point, a tsunami may or may not have been generated, but the potential for one exists.

The PTWC is linked to a series of sea floor pressure sensors, ocean buoys, and tide-measuring stations throughout the Pacific, so the recording stations nearest the earthquake are closely monitored for any indication of unusual wave activity. If unusual wave activity is verified, the tsunami watch is upgraded to a *tsunami warning*. Generally, earthquakes smaller than magnitude $M_w = 6.5$ are not typically tsunamigenic because they lack the duration of ground shaking necessary to initiate a tsunami. In addition, transform faults do not usually produce tsunami because lateral movement does not offset the ocean floor and impart energy to the water column in the same way that vertical fault movements do.

When a tsunami is detected, warnings are sent to all the coastal regions that might encounter the destructive wave, along with its estimated time of arrival. This warning, usually just a few hours in advance of the tsunami, makes it possible to evacuate people from low-lying areas and remove ships from harbors before the waves arrive. If the disturbance is nearby, however, there is not enough time to issue a warning because a tsunami travels so rapidly. Unlike hurricanes, whose high winds and waves threaten ships at sea and send them to the protection of a coastal harbor, a tsunami washes ships from their coastal moorings into the open ocean or onto shore. The best strategy during a tsunami warning is to move ships out of coastal harbors and into deep water, where tsunami are not easily felt.

EFFECTIVENESS OF TSUNAMI WARNINGS Since the PTWC was established in 1948, it has effectively prevented loss of life due to tsunami when people have heeded the evacuation warnings. Property damage, however, has increased as more buildings have been constructed close to shore. To combat the damage caused by tsunami, countries that are especially prone to tsunami, such as Japan, have invested in shoreline barriers, seawalls, and other coastal fortifications.

Perhaps one of the best strategies to limit tsunami damage and loss of life is to restrict construction projects in low-lying coastal regions where tsunami have frequently struck in the past. However, the long time interval between large tsunami—typically 200 years or more—often causes people to let down their guard about remembering past disasters.

8.6 Can Power from Waves Be Harnessed as a Source Of Energy?

Moving water has a huge amount of energy, which is why there are so many hydro-electric power plants on rivers. Even greater energy exists in ocean waves, but significant problems must be overcome for the power to be harnessed efficiently. For example, a serious obstacle to the use of any device to harness wave energy is the monumental engineering problem of preventing the devices from being destroyed by the wave force they are built to harness.

Another key disadvantage of wave energy is that the system produces significant power only when large storm waves break against it, so the system could serve only as a power supplement. In addition, a series of 100 or more of these structures along the shore would be required. Structures of this type could have a significant impact on the environment, with negative effects on marine organisms that rely on wave energy for dispersal, transporting food supplies, or removing wastes. Also, harnessing wave energy might alter the transport of sand along the coast, causing erosion in areas deprived of sediment.

Still, the immense power contained in waves could be used for generating electricity. Offshore wave-generating plants would be able to tap into the higher wave energy found offshore, but they are more likely to be damaged in large waves and more difficult to maintain. The most promising locations for coastal power generation from waves are where waves refract (bend) and converge, such as at headlands, which tend to focus wave energy (see Figure 8.21b). Using this advantage, an array of wave power plants might extract up to 10 megawatts of power[8] per 1 kilometer (0.6 mile) of shoreline.

Internal waves are a potential source of energy, too. Along shores that have favorable sea floor shape for focusing wave energy, internal waves may be effectively concentrated by refraction and thus could power an energy-conversion device that generates electricity.

Wave Power Plants and Wave Farms

In 2000, the world's first commercial-scale wave power plant began generating electricity. Built by Voith Hydro Wavegen and called **LIMPET 500** (*L*and *I*nstalled *M*arine *P*owered *E*nergy *T*ransformer), the plant is located on Islay, a small island off the west coast of Scotland that experiences high wave energy potential.

[8]Each megawatt of electricity is enough to serve the energy needs of about 800 average U.S. homes.

STUDENTS SOMETIMES ASK . . .

If a tsunami warning is issued, what is the best thing to do?

The *smartest* thing to do is to stay out of coastal areas, but people often want to observe the tsunami firsthand. For instance, after the February 2010 magnitude $M_{ww} = 8.8$ Chilean earthquake, tsunami warnings were issued in 53 countries. Across the Pacific Ocean in Australia, widespread media broadcasts warned people of the impending danger, yet live television coverage showed hundreds of people standing on low-lying beaches waiting for the tsunami to arrive. Worse, some of them were deliberately swimming into the incoming tsunami, despite the efforts of volunteer lifeguards to prevent this.

If you *must* go to the beach to observe a tsunami, expect crowds, road closings, and general mayhem. It would be a good idea to stay at least 30 meters (100 feet) above sea level. If you happen to be at a remote beach and notice the water suddenly withdrawing, evacuate immediately to higher ground (**Figure 8.30**). And, if you happen to be at a beach when an earthquake occurs and shakes the ground so hard that you can't stand up, then *RUN*—don't walk—for high ground as soon as you *can* stand up!

After the first surge of the tsunami, stay out of low-lying coastal areas for several hours because several more surges (and withdrawals) can be expected. There are many documented cases of curious people being killed when they were caught in the third or fourth (. . . or ninth) surge of a tsunami.

Figure 8.30 Tsunami warning sign. This tsunami warning sign in coastal Oregon advises residents to evacuate low-lying areas during a tsunami.

(a) LIMPET 500, the world's first commercial wave power plant.

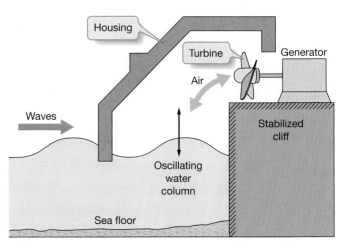

(b) As a wave advances into the power plant, it causes water to surge into the structure, forcing air out through a turbine. As the wave recedes, air is sucked back through the turbine. Electricity is generated when air moves both ways through the turbine as waves advance and recede.

Figure 8.31 How a wave power plant works. Photo **(a)** and a schematic cross-sectional view **(b)** of Limpet 500, the world's first commercial-scale wave power plant.

The plant consists of a partially submerged chamber facing the sea (**Figure 8.31a**). As a wave approaches, the water level inside the structure rises, compressing air in the top of the chamber that is expelled through a turbine, which rotates to generate electricity (**Figure 8.31b**). As the wave recedes, the water level falls in the chamber and air is drawn back into the structure through the turbine. Even though there are two directions of air flow through the turbine, it is designed to continually produce power throughout the wave cycle. LIMPET 500 was constructed as a research and test facility that was originally configured with a turbo-generator of only 500 kilowatts (capable of supplying the energy needs of about 400 average U.S. homes) but has since been upgraded with a more efficient turbine that can generate more power and is continually being improved.

Wavegen's technology has recently been deployed in the 16-unit Mutriku breakwater power plant. Located in the Basque Country in northern Spain and developed by the Basque Energy Board EVE, this power plant is the world's only commercially-operated wave energy plant. Another wave power plant using Wavegen technology will also be installed in Siadar on the Isle of Lewis off the west coast of Scotland. This installation will have a peak operating capacity of 30 megawatts and will be one of the world's first large-scale developments of wave energy.

In 2008, Pelamis Wave Power completed the world's first wave farm off the coast of northern Portugal. This project uses three 150-meter-long (500-foot-long) devices that resemble giant segmented snakes and float half-submerged in the ocean (**Figure 8.32**). As each segment surges up or down with the crest of an oncoming wave, its hydraulic power plant pumps a biodegradable hydraulic fluid through a turbine, thus generating electricity. These wave-energy devices are already supplying up to 2.25 megawatts of power to Portugal's electrical grid, and up to 26 more devices are planned for the future.

Currently, about 50 wave-energy projects are in development at various sites around the world. These projects use different methods to harness wave power, including floats or submerged pistons that move up and down with each passing wave, tethered paddles that oscillate back and forth, and collection of water from breaking waves that overtops coastal structures and then use of that water to turn turbines as it returns to the ocean.

Figure 8.32 Harnessing the power of ocean waves. This wave energy device resembles a large segmented floating snake and is designed to flex as waves pass, thus generating electricity. A wave farm of three of these devices is currently generating electricity off northern Portugal, and up to 26 more are planned.

Global Coastal Wave Energy Resources

Leading estimates suggest that the global resource for wave energy lies between 1 and 10 terawatts; the world currently produces about 12 terawatts from all sources. Where are the best places to develop additional wave power plants and wave farms? **Figure 8.33** shows the average wave heights experienced along coastal regions and indicates the sites most favorable for wave-energy generation (*red areas*). The map shows that west-to-east movement of storm systems in the middle latitudes between 30 and 60 degrees north or south latitude causes the western coasts of continents to be struck by larger waves than eastern coasts. Thus, more wave energy is generally available along western than eastern shores. Furthermore, some of the largest waves (and greatest potential for wave power) are associated with the prevailing westerly wind belt in the Southern Hemisphere middle latitudes.

ESSENTIAL CONCEPT 8.6

Ocean waves produce large amounts of energy. Although significant problems exist in harnessing wave energy effectively, several types of devices are extracting wave energy today.

CONCEPT CHECK 8.6

1 Discuss some problems that might result from developing facilities for conversion of wave energy to electrical energy.

2 Describe the locations and power-generating capability of existing offshore wave power plants and wave farms.

3 Worldwide, where are the best locations for new wave farms? Explain the oceanographic conditions that make these locations favorable.

Figure 8.33 Global coastal wave energy resources. This map of coastal wave energy shows that some of the highest wave energy is available along windward shorelines facing stormy seas, particularly those along the western shores of landmasses in the Southern Hemisphere. kW/m is kilowatts per meter (for example, every meter of "red" shoreline has the potential of generating over 60 kilowatts of electricity); corresponding average wave height is in meters.

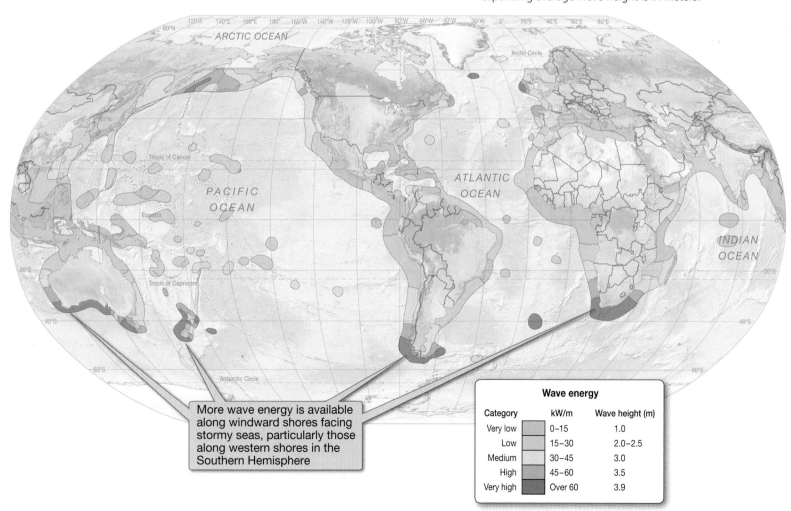

More wave energy is available along windward shores facing stormy seas, particularly those along western shores in the Southern Hemisphere

Wave energy		
Category	kW/m	Wave height (m)
Very low	0–15	1.0
Low	15–30	2.0–2.5
Medium	30–45	3.0
High	45–60	3.5
Very high	Over 60	3.9

Essential Concepts Review

8.1 How are waves generated, and how do they move?

▶ *All ocean waves begin as disturbances caused by releases of energy.* The releases of energy include wind, the movement of fluids of different densities (which create internal waves), mass movement into the ocean, underwater sea floor movements, the gravitational pull of the Moon and the Sun on Earth, and human activities in the ocean.

▶ Once initiated, *waves transmit energy through matter by setting up patterns of oscillatory motion* in the particles that make up the matter. Progressive waves are longitudinal, transverse, or orbital, depending on the pattern of particle oscillation. Particles in ocean waves move primarily in orbital paths.

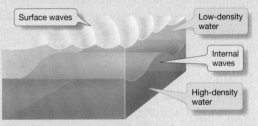

(a) Surface waves versus internal waves.

Study Resources
Online Study Guide Quizzes

Critical Thinking Question

Discuss longitudinal, transverse, and orbital wave phenomena, including the states of matter in which each can transmit energy.

8.2 What characteristics do waves possess?

▶ *Waves are described according to their wavelength* (L), *wave height* (H), *wave steepness* (H/L), *wave period* (T), *frequency* (f), *and wave speed* (S). As a wave travels, the water passes the energy along by moving in a circle, called *circular orbital motion*. This motion advances the waveform, not the water particles themselves. *Circular orbital motion decreases with depth*, ceasing entirely at wave base, which is equal to one-half the wavelength measured from still water level.

▶ If water depth is greater than one-half the wavelength, a progressive wave travels as a *deep-water wave with a speed that is directly proportional to wavelength*. If water depth is less than $1/20$ wavelength (L/20), the wave moves as a *shallow-water wave with a speed that is directly proportional to water depth*. *Transitional waves* have wavelengths between deep- and shallow-water waves, with speeds that depend on both wavelength and water depth.

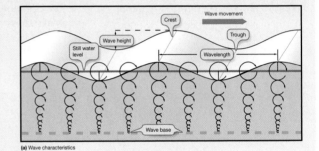

(a) Wave characteristics

Study Resources
Online Study Guide Quizzes, Web Animation

Critical Thinking Question

To help reinforce your knowledge of wave terminology, draw a diagram of a simple progressive wave from memory. Include wave orbitals and label the crest, trough, wavelength, wave height, wave base, and still water level.

8.3 How do wind-generated waves develop?

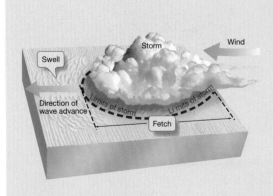

▶ As wind-generated waves form in a sea area, *capillary waves with rounded crests and wavelengths less than 1.74 centimeters (0.7 inch) form first. As the energy of the waves increases, gravity waves form*, with increased wave speed, wavelength, and wave height. Factors that influence the size of wind-generated waves include wind *speed*, *duration* (time), and *fetch* (distance). An equilibrium condition called a *fully developed sea* is reached when the maximum wave height is achieved for a particular wind speed, duration, and fetch.

▶ *Energy is transmitted from the sea area across the ocean by uniform, symmetrical waves called swell.* Different wave trains of swell can create either *constructive*, *destructive*, or *mixed interference* patterns. Constructive interference produces unusually large waves called rogue waves or superwaves.

Study Resources
Online Study Guide Quizzes, Web Animation, Web Video

Critical Thinking Question

Using the information about the giant waves experienced by the USS *Ramapo* in 1933, determine the waves' wavelength and speed.

8.4 How do waves change in the surf zone?

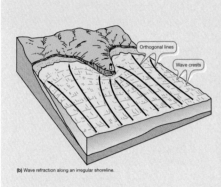

(b) Wave refraction along an irregular shoreline.

▶ *As waves approach shoaling water near shore, they undergo many physical changes.* Waves release their energy in the surf zone when their steepness exceeds a 1:7 ratio and break. If waves break on a relatively flat surface, they produce *spilling breakers.* The curling crests of *plunging breakers*, which are the best for surfing, form on steep slopes, and abrupt beach slopes create *surging breakers.*

▶ When swell approaches the shore, *segments of the waves that first encounter shallow water are slowed* whereas other segments of the wave in deeper water move at their original speed, *causing each wave to refract, or bend.* Refraction concentrates wave energy on headlands, while low-energy breakers are characteristic of bays.

▶ *Reflection of waves off seawalls or other barriers* can cause an interference pattern called a *standing wave.* The crests of standing waves do not move laterally as in progressive waves but alternate with troughs at antinodes. Between the antinodes are nodes, where there is no vertical movement of the water.

Study Resources
Online Study Guide Quizzes, Web Animations, Web Video

Critical Thinking Question
Explain the basic physics of surfing.

8.5 How are tsunami created?

▶ *Sudden changes in the elevation of the sea floor, such as from fault movement or volcanic eruptions, generate tsunami, or seismic sea waves.* These waves often have lengths exceeding 200 kilometers (125 miles) and travel across the open ocean with undetectable heights of about 0.5 meter (1.6 feet) at speeds in excess of 700 kilometers (435 miles) per hour. Upon approaching shore, a tsunami produces a *series of rapid withdrawals and surges*, some of which may increase the height of sea level by 40 meters (131 feet) or more.

▶ *Most tsunami occur in the Pacific Ocean*, where they have caused billions of dollars of coastal damage and taken tens of thousands of lives. A recent example is the *2011 Japanese Tohoku Earthquake and resulting tsunami.* However, the *2004 Indian Ocean tsunami killed as many as 300,000*, making it the most deadly tsunami in history. The *Pacific Tsunami Warning Center (PTWC)* has dramatically reduced fatalities by successfully predicting tsunami using real-time seismic information and a network of deep-ocean pressure sensors. A new tsunami warning system has become operational in the Indian Ocean.

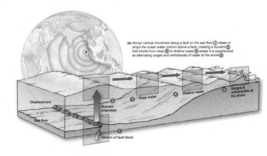

Study Resources
Online Study Guide Quizzes, Web Diving Deeper 8.1, Web Animation, Web Videos

Critical Thinking Question
What ocean depth would be required for a tsunami with a wavelength of 220 kilometers (136 miles) to travel as a deep-water wave? Is it possible that such a wave could become a deep-water wave anyplace in the world ocean? Explain.

8.6 Can power from waves be harnessed as a source of energy?

▶ *Ocean waves can be harnessed to produce hydroelectric power,* but significant problems must be overcome to make this a practical source of energy.

Study Resources
Online Study Guide Quizzes

Critical Thinking Question
What are some negative environmental factors that could inhibit the development of ocean wave power systems?

MasteringOceanography™

www.masteringoceanography.com

Looking for additional review and test prep materials? Visit the Study Area in MasteringOceanography to enhance your understanding of this chapter's content by accessing a variety of resources, including Self-Study Quizzes, Geoscience Animations, RSS feeds, flashcards, Web links, and an optional Pearson eText.

Extreme tidal variation. High and low tides in a small harbor near Blomidon Provincial Park, Nova Scotia, Canada, demonstrate the dramatic change in sea level experienced daily in the Bay of Fundy. This region experiences the largest tidal range in the world, up to 17 meters (56 feet).

Before you begin reading this chapter, use the glossary at the end of this book to discover the meanings of any of these words you don't already know:

Tidal range Aphelion
Tidal bore Ebb current Mixed tidal pattern
Perihelion High slack water
Neap tide Diurnal tidal pattern Grunion Apogee
Centripetal force
Spring tide Solar bulge Semidiurnal tidal pattern
Gravitational force Proxigean Perigee
Low slack water Tidal period Lunar bulge
Syzygy Flood current
Tide Lunar day

TIDES

ESSENTIAL CONCEPTS

At the end of this chapter, you should be able to:

9.1 Demonstrate an understanding of the forces that cause ocean tides

9.2 Explain how tides vary during a monthly tidal cycle

9.3 Discuss what tides look like in the ocean

9.4 Compare the characteristics and locations of the three types of tidal patterns

9.5 Describe tidal phenomena that occur in coastal regions

9.6 Discuss advantages and disadvantages of harnessing tides as a source of energy

"I derive from the celestial phenomena the forces of gravity with which bodies tend to the sun and several planets. Then from these forces, by other propositions which are also mathematical, I deduce the motions of the planets, the comets, the moon, and the sea."

—Sir Isaac Newton,
Philosophiae Naturalis Principia Mathematica
(Philosophy of Natural Mathematical Principles)
(1686)

Tides are the periodic raising and lowering of sea level that occurs daily throughout the ocean. As sea level rises and falls, the edge of the sea slowly shifts landward and seaward each day; as it rises, it often destroys sand castles that were built during low tide. Knowledge of tides is important in many coastal activities, including tide pooling, shell collecting, surfing, fishing, navigation, and preparing for storms. Tides are so important that accurate records have been kept at nearly every port for several centuries, and there are many examples of the term *tide* in everyday vocabulary (for instance, "to tide someone over," "to go against the tide," or to wish someone "good tidings").

Although the original inhabitants of coastal regions most certainly noticed the tides, the first written record of tides doesn't appear until about 450 B.C. Even the earliest sailors knew the Moon had some connection with the tides because both followed a similar pattern. For example, high tides were associated with either a full or new moon, and lower tides were associated with quarter moons. However, it wasn't until **Isaac Newton** (1642–1727) developed the universal law of gravitation that the tides could adequately be explained.

Although the study of tides can be complex, tides are fundamentally very long and regular shallow-water waves. As we shall see, their wavelengths are measured in thousands of kilometers and their heights range to more than 15 meters (50 feet).

9.1 What Causes Ocean Tides?

Simplistically, the gravitational attraction of the Sun and Moon on Earth creates ocean tides. In a more complete analysis, tides are generated by forces imposed on Earth that are caused by a combination of *gravity* and *motion* among Earth, the Moon, and the Sun.

Tide-Generating Forces

Newton's work on quantifying the forces involved in the Earth–Moon–Sun system led to the first understanding of the underlying forces that keep bodies in orbit around each other. It is well known that gravity is the force that interconnects the Sun, its planets, and their moons and keeps them in relatively fixed orbits. For example, most of us are taught that "the Moon orbits Earth," but it is not quite this simple. The two bodies actually rotate around a common center of mass called the **barycenter** (*barus* = heavy, *center* = center), which is the *balance point* of the system, located 1600 kilometers

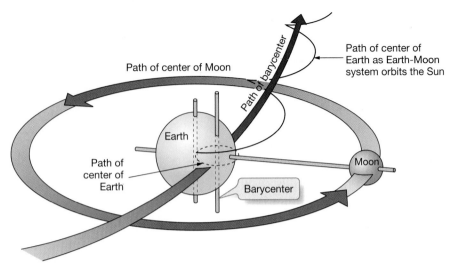

(a) The center of mass (*barycenter*) of the Earth-Moon system moves in a nearly circular orbit around the Sun.

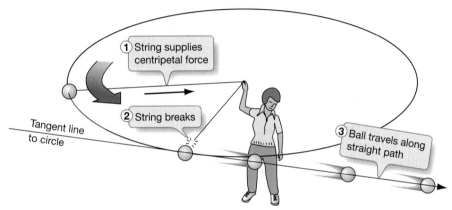

(b) If a ball with a string attached is swung overhead, it stays in a circular orbit because the string exerts a centripetal (center-seeking) force on the ball ①. If the string breaks ②, the ball will fly off along a straight path along a tangent to the circle ③.

Figure 9.1 Earth–Moon system rotation. Schematic diagrams showing the movement of the barycenter of the Earth-Moon system around the Sun **(a)** and the existence of the centripetal force **(b)**.

(1000 miles) beneath Earth's surface (**Figure 9.1a**). Why isn't the barycenter halfway between the two bodies? It's because Earth's mass is so much greater than that of the Moon. This can be visualized by imagining Earth and its Moon as ends of an object that is much heavier on one end than the other, such as a sledgehammer, which has a lighter handle and a much heavier head. If you were to balance the sledgehammer on its side using one finger, you'd find that its balance point is within the head of the hammer. Now imagine that the sledgehammer is flung into space, tumbling slowly end over end about its balance point. This is exactly the situation that describes the movement of the Earth–Moon system. The purple arrow in Figure 9.1a shows the smooth, nearly circular path of the Earth–Moon barycenter around the Sun.

If the Moon and Earth are attracted to one another, why don't the two collide? The Earth–Moon system is involved in a mutual orbit held together by gravity and motion, which prevents the Moon and Earth from colliding. This is how orbits that keep objects at more or less fixed distances are established.

Newton's work also allowed an understanding of why the tides behave as they do. Just as gravity and motion serve to keep bodies in mutual orbits, they also exert an influence on every particle of water on Earth, thus creating the tides.

GRAVITATIONAL AND CENTRIPETAL FORCES IN THE EARTH–MOON SYSTEM To understand how *tide-generating forces* influence the oceans, let's examine how *gravitational forces* and *centripetal forces* affect objects on Earth within the Earth–Moon system. (We'll ignore the influence of the Sun for the moment.)

The **gravitational force** is derived from **Newton's law of universal gravitation**, which states that *every object that has mass in the universe is attracted to every other object.* An object can be as small as an individual atomic particle or as large as a sun. The basic equation for this relationship is:

$$F_g = \frac{Gm_1m_2}{r^2} \tag{9.1}$$

This equation states that the gravitational force (F_g) is directly proportional to the product of the masses of the two bodies (m_1, m_2) and is inversely proportional to the square of the distance between the two masses (r^2). Note that G is the gravitational constant, so it does not change.

Let's simplify Newton's law of universal gravitation and examine the effect of both mass and distance on the gravitational force, which can be expressed with arrows (up arrow = increase, down arrow = decrease):

If mass increases (↑), then gravitational force increases (↑).

A practical example of this can be seen in an object with a large mass (such as the Sun), which produces a large gravitational attraction (**Figure 9.2a**).

Looking at how distance influences gravitational force, the relationship is:

If distance increases (↑), then gravitational force greatly decreases (↓ ↓).

Equation 9.1 shows that the gravitational attraction varies with the *square* of distance, so even a small *increase* in the distance between two objects significantly *decreases* the gravitational force between them—hence the double arrows in the distance relationship illustrated above. This means that when an object is twice as far away, the gravitational attraction is only one-quarter as strong. As a practical example, this is why astronauts experience weightlessness in space when they get far enough from Earth's gravitational pull (Figure 9.2b). In summary, then, the *greater* the mass of the objects and (especially) the *closer* they are together, the greater their gravitational attraction.

Figure 9.3 shows how gravitational forces for points on Earth (caused by the Moon) vary depending on their distances from the Moon. The greatest gravitational attraction (the longest arrow) is at Z, the **zenith** (*zenith* = a path over the head), which is the point closest to the Moon. The gravitational attraction is weakest at N, the **nadir** (*nadir* = opposite the zenith), which is the point farthest from the Moon. The direction of the gravitational attraction between most particles and the center of the Moon is at an angle relative to a line connecting the center of Earth and the Moon (Figure 9.3). This angle causes the force of gravitational attraction between each particle and the Moon to be slightly different.

The **centripetal** (*centri* = the center, *pet* = seeking) **force**[1] required to keep planets in their orbits is provided by the gravitational attraction between each of them and the Sun. Centripetal force connects an orbiting body to its parent, pulling the object *inward* toward the parent, "seeking the center" of its orbit. For example, if you tie a string to a ball and swing the ball around your head (Figure 9.1b), the string pulls the ball toward your hand. The string exerts a *centripetal force* on the ball, forcing the ball to *seek the center* of its orbit. If the string breaks, the force is gone, and the ball can no longer maintain its circular orbit. The ball flies off in a *straight* line,[2] *tangent* (*tangent* = touching) to the circle (Figure 9.1b).

Earth and the Moon are interconnected, too, but by gravity rather than by a string. Gravity provides the centripetal force that holds the Moon in its orbit around Earth. If all gravity in the solar system could be shut off, centripetal force would vanish, and the momentum of the celestial bodies would send them flying off into space along straight-line paths, tangent to their orbits.

RESULTANT FORCES Particles of identical mass rotate in identical-sized paths due to the Earth–Moon rotation system (Figure 9.4). Each particle requires

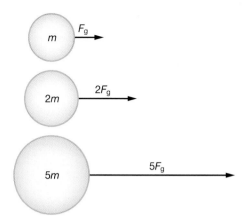

(a) The effect of mass on gravitational attraction

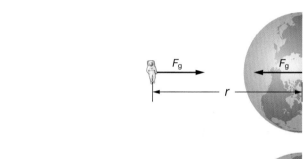

(b) The effect of distance on gravitational attraction

Figure 9.2 The relationship of gravitational force to mass and distance. (a) Gravitational force (F_g) is proportional to a body's mass; as mass increases, so does the gravitational force. **(b)** Gravitational forces between two bodies decrease rapidly as distance (r) increases.

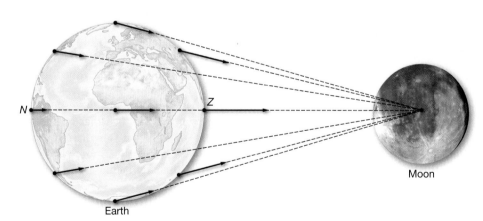

Figure 9.3 Gravitational forces on Earth due to the Moon. The gravitational forces on objects located at different places on Earth due to the Moon are shown by arrows. The length and orientation of the arrows indicate the strength and direction of the gravitational force. Notice the length and angular differences of the arrows for different points on Earth. The letter Z represents the zenith; N represents the nadir. Distance between Earth and Moon not shown to scale.

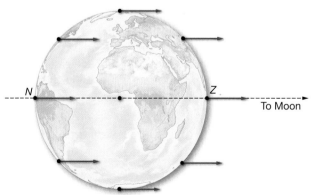

Figure 9.4 Required centripetal (center-seeking) forces. Centripetal forces required to keep identical-sized particles in identical-sized orbits as a result of the rotation of the Earth–Moon system around its barycenter. Notice that the arrows are all the same length and are oriented in the same direction for all points on Earth. Z = zenith; N = nadir.

---

[1]This is not to be confused with the so-called *centrifugal* (*centri* = the center, *fug* = flee) *force*, an apparent or fictitious force that is oriented outward.

[2]At the moment that the string breaks, the ball will continue along a straight-line path, obeying Newton's first law of motion (the *law of inertia*), which states that moving objects follow straight-line paths until they are compelled to change that path by other forces.

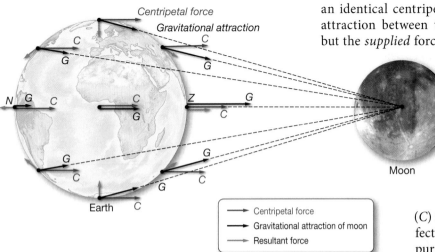

Figure 9.5 Resultant forces. Red arrows indicate centripetal forces (*C*), which are not equal to the black arrows that indicate gravitational attraction (*G*). The small blue arrows show resultant forces, which are established by constructing an arrow from the tip of the centripetal (*red*) arrow to the tip of the gravity (*black*) arrow and located where the red and black arrows begin. *Z* = zenith; *N* = nadir. Distance between Earth and Moon not shown to scale.

Web Animation
Tidal Forces

STUDENTS SOMETIMES ASK . . .

Are there also tides in other objects, such as lakes and swimming pools?

The Moon and the Sun act on all objects that have the ability to flow, so there are tides in lakes, wells, and swimming pools. In fact, there are even extremely tiny tidal bulges in a glass of water! However, the tides in the atmosphere and the "solid" Earth have greater significance. Tides in the atmosphere—called *atmospheric tides*—can be miles high and are also affected by solar heating. The tides inside Earth's interior—called *solid-body tides*, or *Earth tides*—cause a slight but measurable stretching of Earth's crust, typically only a few centimeters high, that has recently been identified as a trigger mechanism for tremors along certain weak faults.

The tides are caused by an imbalance between the required centripetal and the provided gravitational forces acting on Earth. This difference produces residual forces, the horizontal component of which pushes ocean water into two equal tidal bulges on opposite sides of Earth.

an identical centripetal force to maintain it in its circular path. Gravitational attraction between the particle and the Moon supplies the centripetal force, but the *supplied* force is different than the *required* force (because gravitational attraction varies with distance from the Moon) except at the center of Earth. This difference creates tiny **resultant forces**, which are the mathematical difference between the two sets of arrows shown in Figures 9.3 and 9.4.

Figure 9.5 combines Figures 9.3 and Figure 9.4 to show that resultant forces are produced by the difference between the required centripetal (*C*) and supplied gravitational (*G*) forces. However, do not think that both of these forces are being applied to the points because (*C*) is a force that would be required to keep the particles in a perfectly circular path, while (*G*) is the force actually provided for this purpose by gravitational attraction between the particles and the Moon. The resultant forces (*blue arrows*) are established by constructing an arrow from the tip of the centripetal (*red*) arrow to the tip of the gravity (*black*) arrow and located where the red and black arrows begin.

TIDE-GENERATING FORCES Resultant forces are small, averaging about one-millionth the magnitude of Earth's gravity. If the resultant force is vertical to Earth's surface, as it is at the zenith and nadir (oriented upward) and along an "equator" connecting all points halfway between the zenith and nadir (oriented downward), it has no tide-generating effect (**Figure 9.6**). However, if the resultant force has a significant *horizontal component*—that is, tangential to Earth's surface—it produces tidal bulges on Earth, creating what are known as the **tide-generating forces**. These tide-generating forces are quite small but reach their maximum value at points on Earth's surface at a "latitude" of 45 degrees relative to the "equator" between the zenith and nadir (Figure 9.6).

As previously discussed, gravitational attraction is inversely proportional to the *square* of the distance between two masses. The tide-generating force, however, is inversely proportional to the *cube* of the distance between each point on Earth and the *center* of the tide-generating body (Moon or Sun). Although the tide-generating force is derived from the gravitational force, it is not linearly proportional to it. As a result, distance is a more highly weighted variable for tide-generating forces.

The tide-generating forces push water into two simultaneous bulges: one on the side of Earth directed *toward* the Moon (the zenith) and the other on the side directed *away from* the Moon (the nadir) (**Figure 9.7**). On the side directly facing the Moon, the bulge is created because the provided gravitational force is greater than the required centripetal force. Conversely, on the side facing away from the Moon, the bulge is created because the required centripetal force is greater than the provided gravitational force. Although the forces are oriented in opposite directions on the two sides of Earth, the resultant forces are equal in magnitude, so the bulges are equal, too.

Tidal Bulges: The Moon's Effect

It is easier to understand how tides on Earth are created if we consider an ideal Earth and an ideal ocean. The ideal Earth has two tidal bulges, one toward the Moon and one away from the Moon (called the **lunar bulges**), as shown in Figure 9.7. The ideal ocean has a uniform depth, with no friction between the seawater and the sea floor. Newton made these same simplifications when he first explained Earth's tides.

If the Moon is stationary and aligned with the ideal Earth's equator, the maximum bulge will occur on the equator on opposite sides of Earth. If you were standing on the equator, you would experience two high tides each day. The time between high tides, which is the **tidal period**, would be 12 hours. If you moved to any latitude north or south of the equator, you would experience the same tidal period, but the high tides would be less high because you would be at a lower point on the bulge.

In most places on Earth, however, high tides occur every 12 hours 25 minutes because tides depend on the lunar day, not the solar day. The **lunar day** (also called a *tidal day*) is measured from the time the Moon is on the meridian of an observer—that is, directly overhead—to the next time the Moon is on that meridian and is 24 hours 50 minutes.[3] The **solar day** is measured from the time the Sun is on the meridian of an observer to the next time the Sun is on that meridian and is 24 hours. Why is the lunar day 50 minutes longer than the solar day? During the 24 hours it takes Earth to make a full rotation, the Moon has continued moving another 12.2 degrees to the east in its orbit around Earth (**Figure 9.8**). Thus, Earth must rotate an additional 50 minutes to "catch up" to the Moon so that the Moon is again on the meridian (directly overhead) of our observer.

The difference between a solar day and a lunar day can be seen in some of the natural phenomena related to the tides. For example, alternating high tides are normally 50 minutes *later* each successive day, and the Moon rises 50 minutes *later* each successive night.

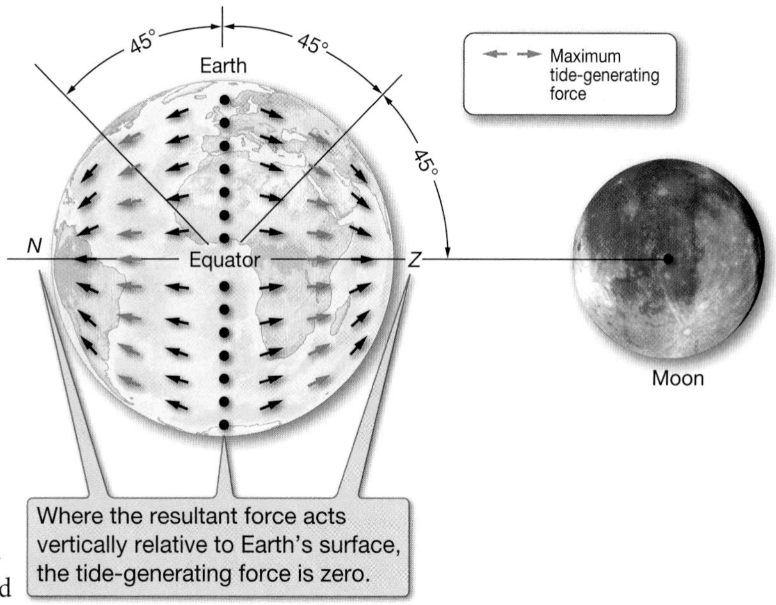

Tidal Bulges: The Sun's Effect

The Sun affects the tides, too. Like the Moon, the Sun produces tidal bulges on opposite sides of Earth, one oriented *toward* the Sun and one oriented *away from* the Sun. These **solar bulges**, however, are much smaller than the lunar bulges. Although the Sun is 27 million times more massive than the Moon, its tide-generating force is not 27 million times greater than the Moon's. This is because the Sun is 390 times farther from Earth than the Moon (**Figure 9.9**). Moreover, tide-generating forces vary inversely as the *cube* of the distance between objects. Thus, the tide-generating force is reduced by the cube of 390, or about 59 million times compared with that of the Moon. These conditions result in the Sun's tide-generating force being $^{27}/_{59}$ that of the Moon, or 46% (about one-half). Consequently, the solar bulges are only 46% the size of the lunar bulges and, as a result, the Moon exerts over two times the gravitational pull of the Sun on the tides.

Even though the Moon exerts over two times the gravitational pull of the Sun on Earth's tides, note that the Sun does not exert a smaller gravitational force on Earth as compared to the Moon. In fact, the Sun's total "pull" on all points on Earth is much greater than that of the Moon, but the *difference* across Earth is small because the diameter of Earth is very small in relation to the distance from the Sun. In contrast, the diameter of Earth is quite large in relation to the distance to the center of the Moon. In summary, the reason the Moon controls tides far more than the Sun is that the Moon is much closer to Earth, although it is much smaller in size and mass as compared to the Sun.

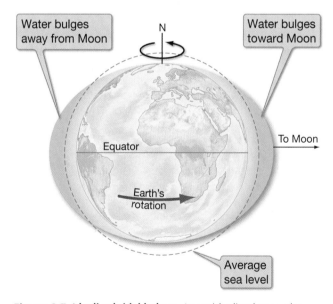

Figure 9.6 Tide-generating forces. Schematic diagram showing the tide-generating forces on Earth (*arrows*) that are caused by the Moon. The maximum tide-generating forces (*blue arrows*) occur where the resultant forces have a significant horizontal component. See the text for description. Z = zenith; N = nadir. Distance between Earth and Moon not shown to scale.

Figure 9.7 Idealized tidal bulges. In an idealized case, the Moon creates two bulges in the ocean surface: one that extends *toward* the Moon and the other *away from* the Moon. As Earth rotates, it carries various locations into and out of the two tidal bulges so that all points on its surface (except the poles) experience two high tides daily.

ESSENTIAL CONCEPT 9.1B

A solar day (24 hours) is shorter than a lunar day (24 hours and 50 minutes). The extra 50 minutes is caused by the Moon's movement in its orbit around Earth.

---

[3]A lunar day is exactly 24 hours, 50 minutes, 28 seconds long.

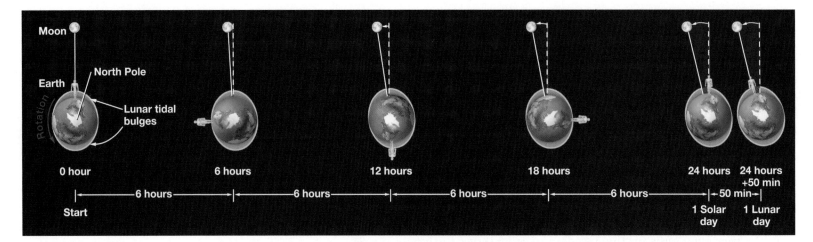

Figure 9.8 The lunar day. A lunar day is the time that elapses between when the Moon is directly overhead and the next time the Moon is directly overhead. During one complete rotation of Earth (the 24-hour solar day), the Moon moves eastward 12.2 degrees, and Earth must rotate an additional 50 minutes for the Moon to be in the exact same position overhead. Thus, a lunar day is 24 hours 50 minutes long.

Web Animation
The Lunar Day

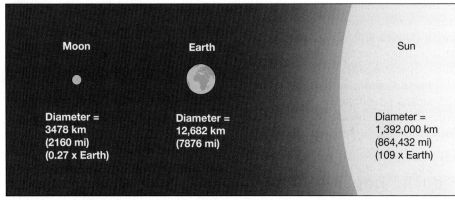

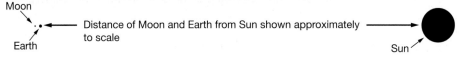

Figure 9.9 Relative sizes and distances of the Moon, Earth, and Sun. *Top*: The relative sizes of the Moon, Earth, and Sun, showing that the diameter of the Moon is roughly one-fourth that of Earth, while the diameter of the Sun is 109 times the diameter of Earth. *Bottom*: The relative distances of the Moon, Earth, and Sun are shown to scale.

ESSENTIAL CONCEPT 9.1C

The lunar bulges are about twice the size of the solar bulges. In an idealized case, the rise and fall of the tides are caused by Earth's rotation carrying various locations into and out of the tidal bulges.

Earth's Rotation and the Tides

The tides appear to move water in toward shore (the **flood tide**) and to move water away from shore (the **ebb tide**). However, according to the nature of the idealized tides presented so far, *Earth's rotation carries various locations into and out of the tidal bulges*, which are in fixed positions relative to the Moon and the Sun. In essence, alternating high and low tides are created as Earth constantly rotates inside fluid bulges that are supported by the Moon and the Sun.

CONCEPT CHECK 9.1

❶ Why are there tidal bulges on both sides of Earth (for example, not just the side of Earth that faces the Moon or the Sun)?

❷ Explain why the Sun's influence on Earth's tides is only 46% that of the Moon, even though the Sun is so much more massive than the Moon.

❸ Why is a lunar day 24 hours 50 minutes long, while a solar day is 24 hours long?

❹ Which is more technically correct: The tide comes in and goes out or Earth rotates into and out of the tidal bulges? Explain your reasoning.

❺ If Earth did not have the Moon orbiting it, would there still be tides? Why or why not?

9.2 How Do Tides Vary During a Monthly Tidal Cycle?

The monthly tidal cycle is 29½ days because that's how long it takes the Moon to complete an orbit around Earth.[4] During its orbit around Earth, the Moon's changing position influences tidal conditions on Earth.

The Monthly Tidal Cycle

During the monthly tidal cycle, the phase of the Moon changes dramatically. When the Moon is between Earth and the Sun, it cannot be seen at night; this phase is called **new moon**. When the Moon is on the side of Earth opposite the Sun, its entire disk is brightly visible; this phase is called **full moon**. A **quarter moon**—a moon that is half lit and half dark as viewed from Earth—occurs when the Moon is at right angles to the Sun relative to Earth.

Figure 9.10 shows the positions of Earth, the Moon, and the Sun at various points during the 29½-day lunar cycle. When the Sun and Moon are aligned, either with the Moon between Earth and the Sun (new moon; Moon in *conjunction*) or with the Moon on the side opposite the Sun (full moon; Moon in *opposition*), the tide-generating

Web Animation
Monthly Tidal Cycle

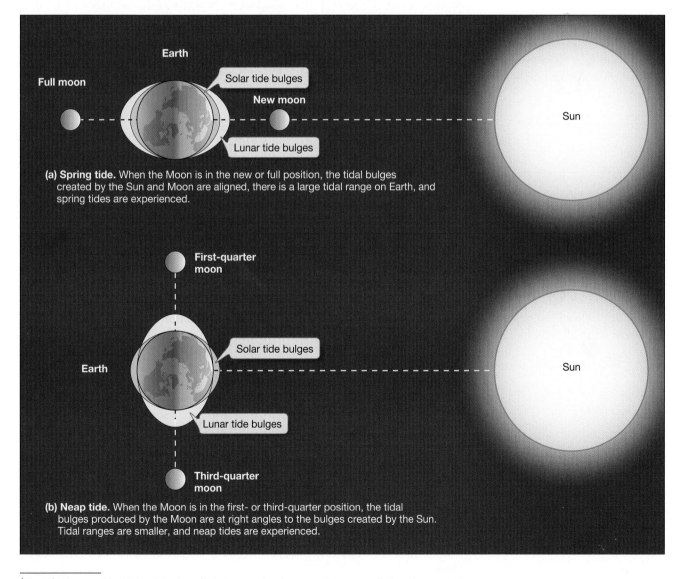

(a) Spring tide. When the Moon is in the new or full position, the tidal bulges created by the Sun and Moon are aligned, there is a large tidal range on Earth, and spring tides are experienced.

(b) Neap tide. When the Moon is in the first- or third-quarter position, the tidal bulges produced by the Moon are at right angles to the bulges created by the Sun. Tidal ranges are smaller, and neap tides are experienced.

Figure 9.10 Earth–Moon–Sun positions and the tides. Schematic diagram showing the positions of Earth, the Moon, and the Sun during spring tide (when the lunar and solar tidal bulges are aligned and a large tidal range exists on Earth) **(a)** and during neap tide (when the lunar and solar tidal bulges are at right angles to each other and a small tidal range exists on Earth **(b)**. Distances not to scale. Note that there is only one moon in orbit around Earth.

---

[4]The 29½-day monthly tidal cycle is also called a *lunar cycle*, a *lunar month*, or a *synodic* (*synod* = meeting) *month*.

Spring tides occur during the full moon and new moon phases, when the lunar and solar tidal bulges constructively interfere, producing a large tidal range. Neap tides occur during the quarter moon phases, when the lunar and solar tidal bulges destructively interfere, producing a small tidal range.

STUDENTS SOMETIMES ASK . . .

What would Earth be like if the Moon didn't exist?

For starters, Earth would spin faster, and days would be much shorter because the tidal forces that act as slow brakes on Earth's rotation wouldn't exist. In fact, geologists have evidence that an Earth day was originally five or six hours long in the distant geologic past; it might be just a little longer than that today if the Moon didn't exist. In the ocean, the tidal range would be much smaller because only the Sun would produce relatively small tidal bulges. Spring tides would not exist, and coastal erosion would be markedly reduced. There would be no moonlight, and nighttime would be much darker, which would affect nearly all life on Earth. There is even some speculation that life would not exist at all on Earth without the stabilizing effect of the Moon.

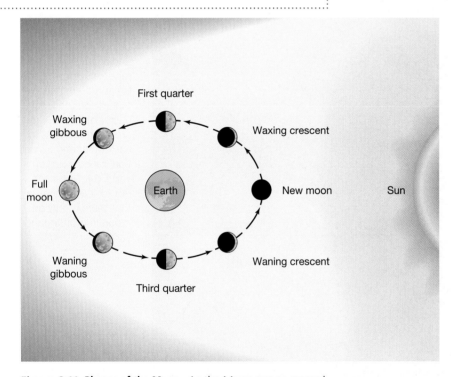

Figure 9.11 Phases of the Moon. As the Moon moves around Earth during its 29½-day lunar cycle, its phase changes depending on its position relative to the Sun and Earth. During a new moon, the dark side of the Moon faces Earth, while during a full moon, the lit side of the Moon faces Earth. Moon phases are shown diagrammatically as seen from Earth.

forces of the Sun and Moon combine (Figure 9.10, *top*). At this time, the **tidal range** (the vertical difference between high and low tides) is large (very *high* high tides and quite *low* low tides) because there is *constructive interference*[5] between the lunar and solar tidal bulges. The maximum tidal range is called a **spring** (*springen* = to rise up) **tide**[6] because the tide is extremely large, or "springs forth." When the Earth–Moon–Sun system is aligned, the Moon is said to be in **syzygy** (*syzygia* = union).

When the Moon is in either the first- or third-quarter[7] phase (Figure 9.10, *bottom*), the tide-generating force of the Sun is working at right angles to the tide-generating force of the Moon. The tidal range is small (*lower* high tides and *higher* low tides) because there is *destructive interference*[8] between the lunar and solar tidal bulges. This is called a **neap** (*nep* = scarcely or barely touching) **tide**,[9] and the Moon is said to be in **quadrature** (*quadra* = four).

The time between successive spring tides (full moon and new moon) or neap tides (first quarter and third quarter) is one-half the monthly lunar cycle, which is about two weeks. The time between a spring tide and a successive neap tide is one-quarter the monthly lunar cycle, which is about one week.

Figure 9.11 shows the pattern that the Moon experiences as it moves through its monthly cycle. As the Moon progresses from new moon to first-quarter phases, the Moon is a **waxing crescent** (*waxen* = to increase; *crescere* = to grow). In between the first-quarter and full moon phases, the Moon is a **waxing gibbous** (*gibbus* = hump). Between the Moon's full and third-quarter phases, it is a **waning gibbous** (*wanen* = to decrease). And, in between the third-quarter and new moon phases, the Moon is a **waning crescent**. The Moon has identical periods of rotation on its axis and revolution around Earth (a property called *synchronous rotation*). As a result, the same side of the Moon always faces Earth.

Complicating Factors

Besides Earth's rotation and the relative positions of the Moon and the Sun, there are many other factors that influence tides on Earth. Two of the most prominent of these factors are the declination of the Moon and Sun and the elliptical shapes of Earth's and the Moon's orbits. Let's examine both of these factors.

DECLINATION OF THE MOON AND SUN Up to this point, we have assumed that the Moon and Sun have remained directly overhead at the equator, but this is not usually the case. Most of the year, in fact, they are either north or south of the equator. The angular distance of the Sun or Moon above or below Earth's equatorial plane is called **declination** (*declinare* = to turn away).

Earth revolves around the Sun along an invisible ellipse in space. The imaginary plane that contains this ellipse is called the **ecliptic** (*ekleipein* = to fail to appear). Recall from Chapter 6 that Earth's axis of rotation is tilted 23.5 degrees with respect to the ecliptic and that this tilt causes Earth's seasons. It also means the maximum declination of the Sun relative to Earth's equator is 23.5 degrees.

To complicate matters further, the plane of the Moon's orbit is tilted 5 degrees with respect to the ecliptic. Thus, the

[5]As mentioned in Chapter 8, *constructive interference* occurs when two waves (or, in this case, two tidal bulges) overlap crest to crest and trough to trough.

[6]Spring tides have no connection with the spring season; they occur twice a month during the time when the Earth-Moon-Sun system is aligned.

[7]The third-quarter moon is often called the last-quarter moon, which is not to be confused with certain sports that have a fourth quarter.

[8]*Destructive interference* occurs when two waves (or, in this case, two tidal bulges) match up crest to trough and trough to crest.

[9]To help remember a *neap tide*, think of it as one that has been "*nipped in the bud*," indicating a small tidal range.

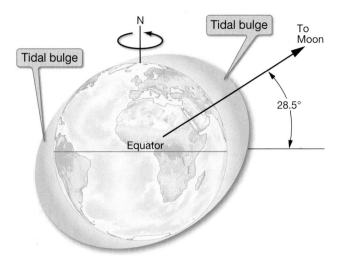

Figure 9.12 Maximum declination of tidal bulges from the equator. The center of the tidal bulges may lie at any latitude from the equator to a maximum of 28.5 degrees on either side of the equator, depending on the season of the year (solar angle) and the Moon's position.

maximum declination of the Moon's orbit relative to Earth's equator is 28.5 degrees (5 degrees plus the 23.5 degrees of Earth's tilt). The declination changes from 28.5 degrees south to 28.5 degrees north and back to 28.5 degrees south of the equator during the multiple lunar cycles within one year. As a result, tidal bulges are rarely aligned with the equator. Instead, they occur mostly north and south of the equator. The Moon affects Earth's tides more than the Sun, so tidal bulges follow the Moon, ranging from a maximum of 28.5 degrees north to a maximum of 28.5 degrees south of the equator (**Figure 9.12**).

EFFECTS OF ELLIPTICAL ORBITS Earth revolves around the Sun in an elliptical orbit (**Figure 9.13**) such that Earth is 148.5 million kilometers (92.2 million miles) from the Sun during the Northern Hemisphere winter and 152.2 million kilometers (94.5 million miles) from the Sun during summer. Thus, the distance between Earth and the Sun varies by 2.5% over the course of a year. Tidal ranges are largest when Earth is near its closest point, called **perihelion** (*peri* = near, *helios* = Sun) and smallest near its most distant point, called **aphelion** (*apo* = away from, *helios* = Sun). Thus, the greatest tidal ranges typically occur in January each year.

The Moon revolves around Earth in an elliptical orbit, too. The Earth–Moon distance varies by 8% (between 375,000 kilometers [233,000 miles] and 405,800 kilometers [252,000 miles]). Tidal ranges are largest when the Moon is closest to Earth, called **perigee** (*peri* = near, *geo* = Earth), and smallest when most distant, called **apogee** (*apo* = away from, *geo* = Earth) (Figure 9.13, *top*). The Moon cycles between perigee, apogee, and back to perigee every 27½ days[10]. Spring tides happen to coincide with perigee about every one-and-a-half years, producing **proxigean** (*proximus* = nearest, *geo* = Earth), or "closest of the close moon," tides. During this time, the tidal range is especially large and often results in the flooding of low-lying coastal areas; if a storm occurs simultaneously, damage can be extreme. In 1962, for example, a winter storm that occurred at the same time as a proxigean tide caused widespread damage along the entire U.S. East coast.

The elliptical orbits of Earth around the Sun and the Moon around Earth change the distances between Earth, the Moon, and the Sun, thus affecting Earth's tides. The net result is that spring tides have greater ranges during the Northern

---

[10]Note that this 27½-day perigee-apogee-perigee cycle is not to be confused with the 29½-day monthly tidal cycle; both cycles occur simultaneously and are ongoing but are independent of one another.

STUDENTS SOMETIMES ASK . . .

Do the phases of the Moon influence human behavior?

All kidding about werewolves aside, tidal and gravitational effects have been used to explain the apparent increase in people's bizarre behavior when the Moon is full. The belief is that since the phases of the Moon affect ocean tides, they may have similar effects on the bodily fluids of human beings, whose composition is about 65% water. A number of scientists have tried to determine whether the activity of the Moon is in fact related to human behavior by calculating the precise number of births, crimes, and unusual behaviors that purportedly occur during various phases of the Moon. A few such investigators have indeed found an association between a full moon and unintentional poisonings, absenteeism, and crime. More often, however, researchers have found no relationship between lunar cycles and human biology or behavior. And, it would seem that since tidal forces are so similar during full and new moon phases, there would be reports of similar unusual behavior during both full and new moons. It may simply be that certain people expect to be influenced by a full moon, so they may be more attentive and responsive to their internal sensations or desires at that time.

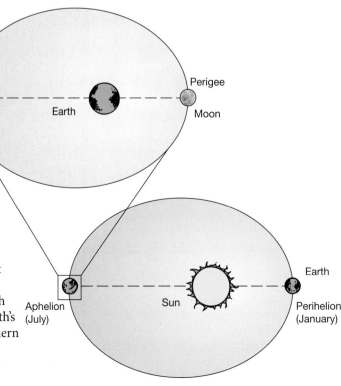

Figure 9.13 Effects of elliptical orbits. *Top*: The Moon moves from its most distant point (*apogee*) to its closest point to Earth (*perigee*), which causes greater tidal ranges every 27½ days. *Bottom*: The Earth also moves from its most distant point (*aphelion*) to its closest point (*perihelion*), which causes greater tidal ranges every year in January. Diagram is not to scale (the elliptical orbits are highly exaggerated).

STUDENTS SOMETIMES ASK . . .

I've heard of a blue moon. Is the Moon really blue then?

No. "Once in a blue moon" is a phrase that has gained popularity and is synonymous with a rather infrequent occurrence. A blue moon is the second full moon of any calendar month, which occurs when the 29½-day lunar cycle falls entirely within a 30- or 31-day month. Because the divisions between our calendar months were determined arbitrarily, a blue moon has no special significance aside from the fact that it occurs only once every 2.72 years (about 33 months). At that rate, it's certainly less common than a month of Sundays!

The origin of the term *blue moon* is not exactly known, but it probably has nothing to do with color—although large forest fires or volcanic eruptions can put enough soot and ash particles in the atmosphere to cause the Moon to appear blue. One likely explanation involves the Old English word *belewe*, meaning "to betray." Thus, the Moon is belewe because it betrays the usual perception of one full moon per month. Another explanation links the term to a 1946 article in *Sky and Telescope* that tried to correct a misinterpretation of the term *blue moon*, but the article itself was misinterpreted to mean the second full moon in a given month. Apparently, the erroneous interpretation was repeated so often that it eventually stuck.

Hemisphere winter than in the summer, and spring tides have greater ranges when they coincide with perigee.

Idealized Tide Prediction

The declination of the Moon determines the position of the tidal bulges. The example illustrated in **Figure 9.14** shows that the Moon is directly overhead at 28 degrees north latitude when its declination is 28 degrees north of the equator. Imagine standing at 28 degrees north latitude and experiencing tidal conditions during a day, which is the sequence shown in Figure 9.14a–d:

- With the Moon directly overhead, the tidal condition experienced will be high tide (Figure 9.14a).
- Low tide occurs 6 lunar hours later (6 hours 12½ minutes solar time) (Figure 9.14b).
- Another high tide, but one much lower than the first, occurs 6 lunar hours later (Figure 9.14c).
- Another low tide occurs 6 lunar hours later (Figure 9.14d).
- Six lunar hours later, at the end of a 24-lunar-hour period (24 hours 50 minutes solar time), you will have passed through a complete lunar-day cycle of two high tides and two low tides (returns to Figure 9.14a).

The graphs in Figure 9.14e show the heights of the tides observed during the same lunar day at 28 degrees north latitude, the equator, and 28 degrees south latitude when the declination of the Moon is 28 degrees north of the equator. Tide curves for 28 degrees north and 28 degrees south latitude have identically timed highs and lows, but the *higher* high tides and *lower* low tides occur 12 hours later. The reason they occur out of phase by 12 hours is because the bulges in the two hemispheres are on opposite sides of Earth in relation to the Moon. Web Table 9.1 summarizes the characteristics of the tides on the idealized Earth.

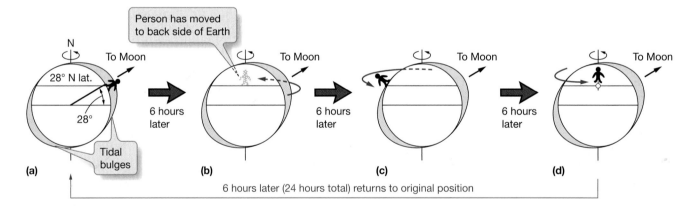

Figure 9.14 Predicted idealized tides. (a)–(d) Sequence showing the tide experienced every 6 lunar hours at 28 degrees north latitude when the declination of the Moon is 28 degrees north. **(e)** Tide curves for 28 degrees north, 0 degrees, and 28 degrees south latitudes during the lunar day shown in the sequence above. The tide curves for 28 degrees north and 28 degrees south latitude show that the higher high tides occur 12 hours later.

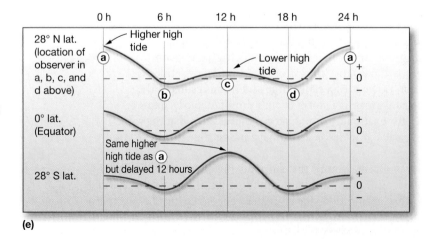

❶ To help reinforce your knowledge of tides, draw the positions of the Earth–Moon–Sun system during a complete monthly tidal cycle from memory. Indicate the tide conditions experienced on Earth, the phases of the Moon, the time between those phases, and syzygy and quadrature.

❷ Explain why the maximum tidal range (spring tide) occurs during new and full moon phases and the minimum tidal range (neap tide) at first-quarter and third-quarter moons.

❸ What is declination? Discuss the degree of declination of the Moon and Sun relative to Earth's equator. What are the effects of declination of the Moon and Sun on the tides?

❹ Diagram the Earth–Moon system's orbit about the Sun. Label the positions on the orbit at which the Moon and Sun are closest to and farthest from Earth, stating the terms used to identify them. Discuss the effects of the Moon's and Earth's positions on Earth's tides.

❺ Describe differences between the 27½-day perigee–apogee–perigee cycle and the 29½-day monthly tidal cycle.

STUDENTS SOMETIMES ASK . . .

What are tropical tides?

Differences between successive high tides and successive low tides occur each lunar day (see, for example, Figure 9.14e). Because these differences occur within a period of one day, they are called *diurnal* (*daily*) *inequalities*. These inequalities are at their greatest when the Moon is at its maximum declination, and such tides are called *tropical tides* because the Moon is over one of Earth's tropics. When the Moon is over the equator (*equatorial tides*), the difference between successive high tides and low tides is minimal.

9.3 What Do Tides Look Like in the Ocean?

If tidal bulges are wave crests separated by a distance of one-half Earth's circumference—about 20,000 kilometers (12,420 miles)—one would expect the bulges to move across Earth at about 1600 kilometers (1000 miles) per hour. Tides, however, are an extreme example of shallow-water waves, so their speed is proportional to the water depth. For a tide wave to travel at 1600 kilometers (1000 miles) per hour, the ocean would have to be 22 kilometers (13.7 miles) deep! However, the average depth of the ocean is only 3.7 kilometers (2.3 miles), so tidal bulges move as *shallow-water waves*, meaning that their speed is controlled by the shallowness of the ocean (see Chapter 8 for a description of the speed of shallow-water waves).

Based on the average depth of the ocean, the average speed at which tide waves can travel across the open ocean is only about 700 kilometers (435 miles) per hour. Thus, the idealized bulges that are oriented toward and away from a tide-generating body cannot exist because they cannot keep up with the rotational speed of Earth. Instead, ocean tides break up into distinct large circulation units called *cells*.

Amphidromic Points and Cotidal Lines

In the open ocean, the crests and troughs of the tide wave rotate around an **amphidromic** (*amphi* = around, *dromus* = running) **point** near the center of each cell. There is essentially no tidal range at amphidromic points, but radiating from each point are **cotidal** (*co* = with, *tidal* = tide) **lines**, which connect all nearby locations where high tide occurs simultaneously. The labels on the cotidal lines in **Figure 9.15** indicate the time of high tide in hours as they rotate around the cell.

The times in Figure 9.15 indicate that the tide wave rotates counterclockwise in the Northern Hemisphere and clockwise in the Southern Hemisphere. The wave must complete one rotation during the tidal period (usually 12 lunar hours), so this limits the size of the cells.

Low tide occurs 6 hours after high tide in an amphidromic cell. If high tide is occurring along the cotidal line labeled "10," for example, then low tide is occurring along the cotidal line labeled "4."

Effect Of the Continents

The continents affect tides, too, because they interrupt the free movement of the tidal bulges across the ocean surface. Tides are expressed in each ocean basin as freestanding waves that are affected by the position and shape of the continents that ring the ocean basin. In fact, two of the most important factors that influence tidal conditions along a coast are coastline shape and offshore depth.

Figure 9.15 Cotidal map of the world. Cotidal lines indicate times of the main lunar daily high tide in lunar hours after the Moon has crossed the Greenwich Meridian (0 degrees longitude). Tidal ranges generally increase with increasing distance along cotidal lines away from the amphidromic points (center of the cell). Where cotidal lines terminate at both ends in amphidromic points, maximum tidal range will be near the midpoints of the lines.

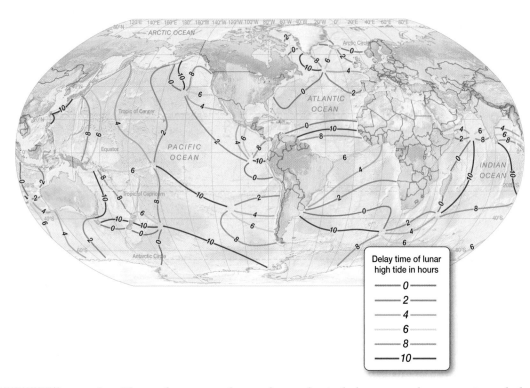

Delay time of lunar
high tide in hours

—— 0 ——
—— 2 ——
—— 4 ——
—— 6 ——
—— 8 ——
—— 10 ——

STUDENTS SOMETIMES ASK . . .

How often are conditions right to produce the maximum tide-generating force?

Maximum tides occur when Earth is closest to the Sun (at perihelion), the Moon is closest to Earth (at perigee), and the Earth–Moon–Sun system is aligned (at syzygy) with both the Sun and Moon at zero declination. This rare condition—which creates an absolute *maximum* spring tidal range—occurs once every 1600 years. Fortunately, the next occurrence is predicted for the year 3300.

However, there are other times when conditions produce large tide-generating forces. During early 1983, for example, large, slow-moving low-pressure cells developed in the North Pacific Ocean and caused strong northwest winds. In late January, the winds produced a near fully developed 3-meter (10-foot) swell that affected the West coast from Oregon to Baja California. The large waves would have been trouble enough under normal conditions, but there were also unusually high spring tides of 2.25 meters (7.4 feet) because Earth was near perihelion at the same time that the Moon was at perigee. In addition, a strong El Niño had raised sea level by as much as 20 centimeters (8 inches). When the waves hit the coast during these unusual conditions, they caused more than $100 million in damage, including the destruction of 25 homes, damage to 3500 others, the collapse of several commercial and municipal piers, and at least a dozen deaths.

Just like surface waves that undergo physical changes as they move into shallow water (such as slowing down and increasing in height; see Chapter 8), tides experience similar physical changes as they enter the shallow water of continental shelves. These changes tend to amplify the tidal range as compared to the deep ocean, where the maximum tidal range is only about 45 centimeters (18 inches).

In addition, increased turbulent mixing rates in deep water over areas of rough bottom topography (as discussed in Chapter 7) are associated with internal waves created by tides breaking on this rough topography and against continental slopes. These tide-generated internal waves have recently been observed along the chain of Hawaiian Islands, have heights of up to 300 meters (1000 feet), and contribute to increased turbulence and mixing, which strongly affect the tides.

Other Considerations

A detailed analysis of all the variables that affect the tides at any particular coast reveals that nearly 400 factors are involved—far more than can adequately be addressed here. The combination of all these factors creates some conditions that are unexpected based on a simple tidal model. For example, high tide rarely occurs when the Moon is at its highest point in the sky. Instead, the time between the Moon crossing the meridian and a corresponding high tide varies from place to place.

Because of the complexity of the tides, a completely mathematical model of the tides is beyond the limits of marine science. Instead, a combination of mathematical analysis and observation is required to adequately model the tides. Moreover, successful models must take into account at least 37 independent factors related to tides (the two most important are the Moon and the Sun) and are usually quite successful in predicting future tides.

CONCEPT CHECK 9.3

❶ Are tides considered deep-water waves anywhere in the ocean? Why or why not?

❷ What are amphidromic points and cotidal lines?

❸ Discuss reasons why the tides don't follow a simple tidal bulge model.

9.4 What Types Of Tidal Patterns Exist?

In theory, most areas on Earth should experience two high tides and two low tides of unequal heights during a lunar day. In practice, however, the various depths, sizes, and shapes of ocean basins modify tides so they exhibit three different patterns in different parts of the world. The three tidal patterns, which are illustrated in **Figure 9.16**, are *diurnal* (*diurnal* = daily), *semidiurnal* (*semi* = half, *diurnal* = daily), and *mixed*.[11]

Diurnal Tidal Pattern

A **diurnal tidal pattern** has one high tide and one low tide each lunar day. These tides are common in shallow inland seas such as the Gulf of Mexico and along the coast of Southeast Asia. Diurnal tides have a tidal period of 24 hours 50 minutes.

Web Animation
Tidal Patterns

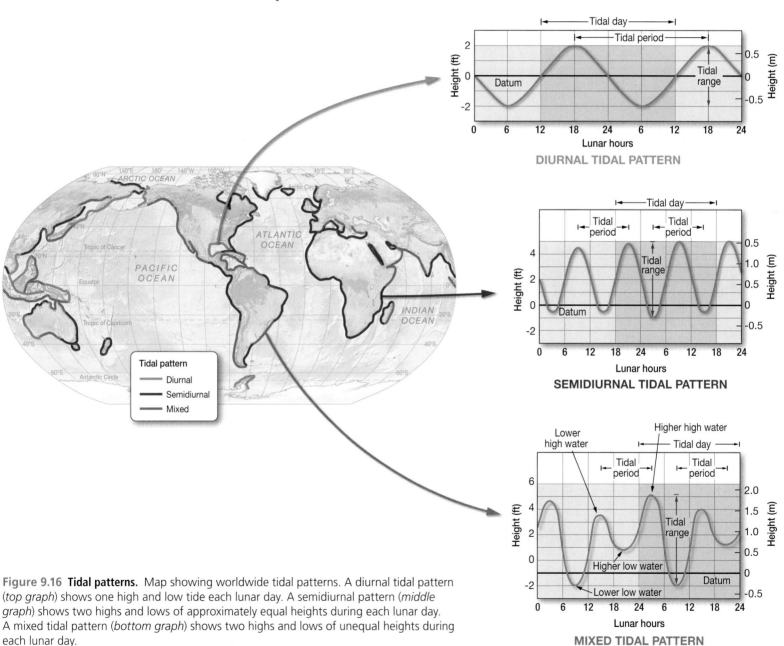

Figure 9.16 Tidal patterns. Map showing worldwide tidal patterns. A diurnal tidal pattern (*top graph*) shows one high and low tide each lunar day. A semidiurnal pattern (*middle graph*) shows two highs and lows of approximately equal heights during each lunar day. A mixed tidal pattern (*bottom graph*) shows two highs and lows of unequal heights during each lunar day.

---

[11]Sometimes a *mixed* tidal pattern is referred to as *mixed semidiurnal*.

STUDENTS SOMETIMES ASK . . .

I noticed that Figure 9.16 shows negative tides. How can there ever be a negative tide?

Negative tides occur because the *datum*—the starting point or reference point from which tides are measured—is an average of the tides over many years. Along the West coast of the United States, for instance, the datum is mean lower low water (MLLW), which is the average of the *lower* of the two low tides that occur daily in a mixed tidal pattern. Because the datum is an average, there will be some days when the tide is less than the average (similar to the distribution of exam scores, some of which will be below the average). These lower-than-average tides are given negative values, occur only during spring tides, and are often the best times to visit local tide pool areas.

ESSENTIAL CONCEPT 9.4

A diurnal tidal pattern exhibits one high and one low tide each lunar day; a semidiurnal tidal pattern exhibits two high and two low tides daily of about the same height; and a mixed tidal pattern usually has two high and two low tides of different heights daily but may also exhibit diurnal characteristics.

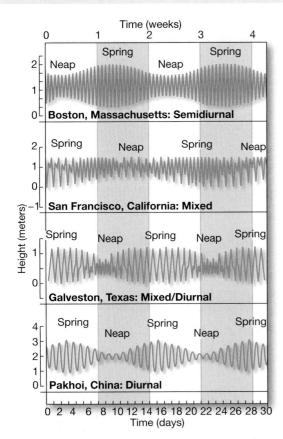

Figure 9.17 Monthly tidal curves. *Top*: Boston, Massachusetts, showing a semidiurnal tidal pattern. *Upper middle*: San Francisco, California, showing a mixed tidal pattern. *Lower middle*: Galveston, Texas, showing a mixed tidal pattern with strong diurnal tendencies. *Bottom*: Pakhoi, China, showing a diurnal tidal pattern.

Semidiurnal Tidal Pattern

A **semidiurnal tidal pattern** has two high tides and two low tides each lunar day. The heights of successive high tides and successive low tides are approximately the same.[12] Semidiurnal tides are common along the Atlantic coast of the United States. The tidal period is 12 hours 25 minutes.

Mixed Tidal Pattern

A **mixed tidal pattern** may have characteristics of both diurnal and semidiurnal tides. Successive high tides and/or low tides will have significantly different heights, a condition called *diurnal inequality*. Mixed tides commonly have a tidal period of 12 hours 25 minutes, but they may also exhibit diurnal periods. Mixed tides are the most common type in the world, including along the Pacific coast of North America.

Figure 9.17 shows examples of monthly tidal curves for various coastal locations. Even though a tide at any particular location follows a single tidal pattern, it still may pass through stages of one or both of the other tidal patterns. Typically, however, the tidal pattern for a location remains the same throughout the year. Also, the tidal curves in Figure 9.17 clearly show the weekly switching of the spring tide–neap tide cycle.

CONCEPT CHECK 9.4

❶ What do the terms *diurnal*, *semidiurnal*, and *mixed* mean, as related to tidal patterns?

❷ Describe the number of high and low tides in a lunar day, the period, and any inequality of the following tidal patterns: diurnal, semidiurnal, and mixed.

❸ Using the Internet, find a coastal tidal prediction for your birthday this year (or, if your birthday has already happened, find next year's tide prediction). What tidal pattern is displayed?

❹ Of the three tidal patterns, which one is most common along the U.S. East coast? The U.S. West coast? Worldwide?

9.5 What Tidal Phenomena Occur in Coastal Regions?

Remember that tides are fundamentally very long-wavelength waves. When tide waves enter coastal waters, they are subject to reflection and amplification similar to what wind-generated waves experience. In certain locations, reflected wave energy causes water to slosh around in a bay, producing *standing waves*.[13] As a result, interesting tidal phenomena are sometimes experienced in coastal waters.

Large lakes and coastal rivers experience tidal phenomena, too. In some low-lying rivers, for instance, an incoming high tide produces a *tidal bore* (**Diving Deeper 9.1**). And, as we'll see later in this section, the tides profoundly affect the spawning behavior of a certain type of marine fish.

---

[12]Because tides are always growing higher or lower at any location due to the spring tide-neap tide sequence, successive high tides and successive low tides can never be *exactly* the same at any location.

[13]See Chapter 8 for a discussion of standing waves, including the terms *node* and *antinode*.

Diving Deeper 9.1 Oceans and People

TIDAL BORES: BORING WAVES THESE ARE NOT!

A **tidal bore** (*bore* = crest, or wave) is a wall of water that moves up certain low-lying rivers due to an incoming tide. Because it is a wave created by the tides, it is a *true* tidal wave. When an incoming tide rushes up a river, it develops a steep forward slope because the flow of the river resists the advance of the tide (**Figure 9A**). This creates a tidal bore, which may reach heights of 5 meters (16.4 feet) or more and move at speeds up to 24 kilometers (15 miles) per hour.

Tidal bores develop only in coastal regions where the following conditions are present: (1) a large spring tidal range of at least 6 meters (20 feet); (2) a tidal cycle that has a very abrupt rise of the flood tide phase and an elongated ebb tide phase; (3) a low-lying river with a persistent seaward current during the time when an incoming high tide begins; (4) a progressive shallowing of the sea floor as the basin progresses inland; and (5) a progressive narrowing of the basin toward its upper reaches. Because of these unique circumstances, only about 60 places on Earth experience tidal bores.

Although tidal bores do not commonly attain the size of waves in the surf zone, tidal bores have successfully been rafted, kayaked, and even surfed (**Figure 9B**). They can give a surfer a very long ride because the bore travels many kilometers upriver. If you miss the bore, though, you have to wait about half a day before the next one comes along because the incoming high tide occurs only twice a day.

The Amazon River is probably the longest estuary affected by oceanic tides: Tides can be measured as far as 800 kilometers (500 miles) from the river's mouth, although the effects are quite small at this distance. Tidal bores near the mouth of the Amazon River can reach heights up to 5 meters (16.4 feet) and are locally called *pororocas*, meaning "mighty noise." Other rivers that have notable tidal bores include the Qiantang River in China (which has the largest tidal bores in the world, often reaching 8 meters [26 feet] high); the Petitcodiac River in New Brunswick, Canada; the River Seine in France; the Trent and Severn Rivers in England; and Cook Inlet

near Anchorage, Alaska (where the largest tidal bore in the United States can be found). Although the Bay of Fundy has the world's largest tidal range, its tidal bore rarely exceeds 1 meter (3.3 feet), mostly because the bay is so wide.

Figure 9A How a tidal bore forms (*figure*) and a tidal bore moving quickly upriver near Chignecto Bay, New Brunswick, Canada (*photo*).

Figure 9B Brazilian surf star Alex "Picuruta" Salazar tidal bore surfing on the Amazon River.

Why don't all areas of the world experience the same type of tidal pattern?

If Earth were a perfect sphere without large continents, all areas on the planet would experience two equally proportioned high and low tides every lunar day (a semidiurnal tidal pattern). The large continents on the planet, however, block the westward passage of the tidal bulges as Earth rotates. Unable to move freely around the globe, the tides instead establish complex patterns within each ocean basin that often differ greatly from tidal patterns of adjacent ocean basins or even other regions of the same ocean basin.

An Example Of Tidal Extremes: The Bay Of Fundy

The largest tidal range in the world is found in Nova Scotia's **Bay of Fundy**. With a length of 258 kilometers (160 miles), the Bay of Fundy has a wide opening into the Atlantic Ocean. At its northern end, however, it splits into two narrow basins, Chignecto Bay and Minas Basin (**Figure 9.18**). The period of free oscillation in the bay—the oscillation that occurs when a body is displaced and then released—is very nearly that of the tidal period. The resulting constructive interference—along with the narrowing and shoaling of the bay to the north—causes a buildup of tidal energy in the northern end of the bay. In addition, the bay curves to the right, so the Coriolis effect in the Northern Hemisphere adds to the extreme tidal range.

During maximum spring tide conditions, the tidal range at the mouth of the bay (where it opens to the ocean) is only about 2 meters (6.6 feet). However, the tidal range increases progressively from the mouth of the bay northward. In the northern end of Minas Basin, the maximum spring tidal range is 17 meters (56 feet), which leaves boats high and dry during low tide (Figure 9.18, *insets*).

Coastal Tidal Currents

The current that accompanies the slowly turning tide crest in a Northern Hemisphere basin rotates counterclockwise, producing a **rotary current** in the open portion of the basin. Friction increases in nearshore shoaling waters, so the rotary current changes to an alternating current, or **reversing current**, that moves into and out of restricted passages along a coast.

Figure 9.18 The Bay of Fundy, site of the world's largest tidal range. Even though the maximum spring tidal range at the mouth of the Bay of Fundy is only 2 meters (6.6 feet), amplification of tidal energy causes a maximum tidal range at the northern end of Minas Basin of 17 meters (56 feet), often stranding ships (*insets*).

The velocity of rotary currents in the open ocean is usually well below 1 kilometer (0.6 mile) per hour. Reversing currents, however, can reach velocities up to 44 kilometers (28 miles) per hour in restricted channels such as between islands of coastal waters.

Reversing currents also exist in the mouths of bays (and some rivers) due to the daily flow of tides. **Figure 9.19** shows that a **flood current** is produced when water rushes into a bay (or river) with an incoming high tide. Conversely, an **ebb current** is produced when water drains out of a bay (or river) because a low tide is approaching. No currents occur for several minutes during either **high slack water** (which occurs at the peak of each high tide) or **low slack water** (at the peak of each low tide).

Reversing currents in bays can sometimes reach speeds of 40 kilometers (25 miles) per hour, creating a navigation hazard for ships. On the other hand, the daily flow of these currents often keeps sediment from closing off the bay and replenishes the bay with seawater and ocean nutrients.

Tidal currents can be significant even in deep-ocean waters. For example, tidal currents were encountered shortly after the discovery of the remains of the *Titanic* at a depth of 3795 meters (12,448 feet) on the continental slope south of Newfoundland's Grand Banks in 1985. These tidal currents were so strong that they forced researchers to abandon the use of the camera-equipped, tethered, remotely operated vehicle *Jason Jr.*

Whirlpools: Fact or Fiction?

A **whirlpool**—a rapidly spinning body of water, which is also termed a *vortex* (*vertere* = to turn)—can be created in some restricted coastal passages due to reversing tidal currents. Whirlpools most commonly occur in shallow passages connecting two large bodies of water that have different tidal cycles. The different tidal heights of the two bodies cause water to move vigorously through the passage. As water rushes through the passage, it is affected by the shape of the shallow sea floor, causing turbulence, which, along with spin due to opposing tidal currents, creates whirlpools. The larger the tidal difference between the two bodies of water and the smaller the passage, the greater the vortex caused by the tidal currents. Because whirlpools can have high flow rates of up to 16 kilometers (10 miles) per hour, they can cause ships to spin out of control for a short time.

One of the world's most famous whirlpools is the *Maelstrom* (*malen* = to grind in a circle, *strom* = stream), which occurs in a passage off the west coast of Arctic Norway (**Figure 9.20**). This and another famous whirlpool in the Strait of Messina, which separates mainland Italy from Sicily, are probably the source of ancient legends of huge churning funnels of water that destroy ships and carry mariners to their deaths, although they are not nearly as deadly as legends suggest. Other notable whirlpools occur off the west coast of Scotland, in the Bay of Fundy at the border between Maine and the Canadian province of New Brunswick, and off Japan's Shikoku Island.

Grunion: Doing What Comes Naturally on the Beach

From March through September, shortly after the maximum spring tide has occurred, **grunion** (*Leuresthes tenuis*) come ashore along sandy beaches of Southern California and Baja California to bury their eggs. Grunion—slender, silvery fish up

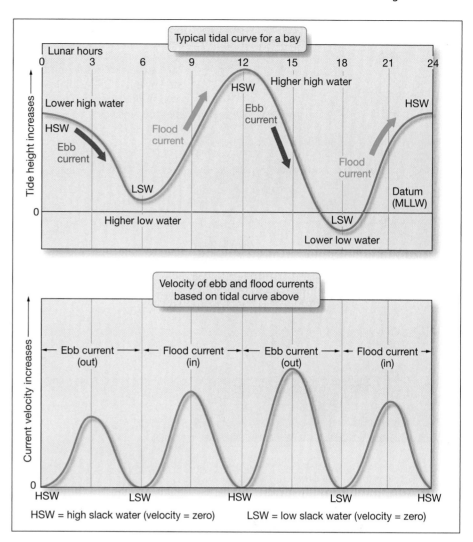

Figure 9.19 Reversing tidal currents in a bay. *Top*: Typical tidal curve for a bay that experiences a mixed tidal pattern, showing alternating ebb currents (approaching low tide) and flood currents (approaching high tide). No currents occur during either high slack water (*HSW*) or low slack water (*LSW*). The datum *MLLW* stands for *mean lower low water*, which is the average of the lower of the two low tides that occur daily in a mixed tidal pattern. *Bottom*: Corresponding chart showing the velocity of ebb and flood currents based on the tidal curve above.

Figure 9.20 **The Maelstrom.** The Maelstrom, located off the west coast of Norway, is one of the strongest whirlpools in the world and can cause ships to spin out of control. It is created by tidal currents that pass through a narrow, shallow passage between Vest Fjord and the Norwegian Sea.

to 15 centimeters (6 inches) long—are the only marine fish in the world that come completely out of water to spawn. The name *grunion* comes from the Spanish *gruñón*, which means "grunter" and refers to the faint noise they make during spawning.

A mixed tidal pattern occurs along Southern California and Baja California beaches. On most lunar days (24 hours and 50 minutes), there are two high tides and two low tides. There is usually a significant difference in the heights of the two high tides that occur each day. During the summer months, the higher high tide occurs at night. The night high tide becomes higher each night as the maximum spring tide range is approached, causing sand to be eroded from the beach (Figure 9.21, *graph*). After the maximum spring tide has occurred, the night high tide diminishes each night. As neap tide is approached, sand is deposited on the beach.

Grunion spawn only after each night's higher high tide has peaked on the three or four nights following the night of the highest spring high tide. This ensures that their eggs will be covered deeply in sand deposited by the receding higher high tides each succeeding night. The fertilized eggs buried in the sand are ready to hatch about ten days after spawning. By this time, another spring tide is approaching, so the night high tide is getting progressively higher each night again. The beach sand is eroding again, too, which exposes the eggs to the waves that break ever higher on the beach. The eggs hatch about three minutes after being freed in the water. Tests done in laboratories have shown that the grunion eggs will not hatch until agitated in a manner that simulates that of the eroding waves.

The spawning begins as the grunion come ashore immediately following an appropriate high tide, and it may last from one to three hours. Spawning usually peaks about an hour after it starts and may last an additional 30 minutes to an hour. Thousands of fish may be on the beach at this time. During a run, the females, which are larger than the males, move high on the beach. If no males are near, a female may return to the water without

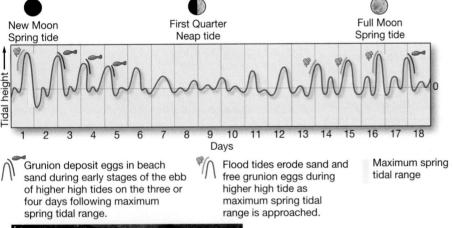

Figure 9.21 **The tidal cycle and spawning grunion.** During summer months and for 3 or 4 days after the highest spring tides (*graph*), grunion deposit their eggs on sandy beaches (*photo*). The successively lower high tides during the approaching neap tide conditions won't wash the eggs from the sand until they are ready to hatch about 10 days later. As the next spring tide is approached, successively higher high tides wash the eggs free and allow them to hatch. The spawning cycle begins a few days later, after the peak of spring tide conditions, with the next cycle of successively lower high tides.

depositing her eggs. In the presence of males, she drills her tail into the semifluid sand until only her head is visible. The female continues to twist, depositing her eggs 5 to 7 centimeters (2 to 3 inches) below the surface.

The male curls around the female's body and deposits his milt against it (Figure 9.21, *photo*). The milt runs down the body of the female to fertilize the eggs. When the spawning is completed, both fish return to the water with the next wave.

Larger females are capable of producing up to 3000 eggs for each series of spawning runs, which are separated by the two-week period between spring tides. As soon as the eggs are deposited, another group of eggs begins to form within the female. These eggs will be deposited during the next spring tide run. Early in the season, only older fish spawn. By May, however, even the one-year-old females are in spawning condition.

Young grunion grow rapidly and are about 12 centimeters (5 inches) long when they are a year old and ready for their first spawning. They usually live two or three years, but four-year-olds have been recovered. The age of a grunion can be determined by its scales. After growing rapidly during the first year, they grow very slowly. In fact, there is no growth at all during the six-month spawning season, which causes marks to form on each scale that can be used to identify a grunion's age.

It is not known exactly how grunion are able to time their spawning behavior so precisely with the tides. Research suggests that grunion are somehow able to sense very small changes in hydrostatic pressure caused by rising and falling sea level due to changing tides. Certainly, a very dependable detection mechanism keeps the grunion accurately informed of the tidal conditions because their survival depends on a spawning behavior precisely tuned to the tides.

> ### ESSENTIAL CONCEPT 9.5
>
> Coastal tidal phenomena include large tidal ranges (the largest of which—17 meters [56 feet]—occurs in the Bay of Fundy), tidal currents, rapidly spinning vortices called whirlpools, and grunion, which time their spawning cycles with the tides.

> ### CONCEPT CHECK 9.5
>
> ❶ Discuss factors that help produce the world's largest tidal range in the Bay of Fundy.
>
> ❷ Discuss the difference between rotary and reversing tidal currents.
>
> ❸ Of flood current, ebb current, high slack water, and low slack water, when is the best time to enter a bay by boat? When is the best time to navigate in a shallow, rocky harbor? Explain.
>
> ❹ Describe the spawning cycle of grunion, indicating the relationship among tidal phenomena, where grunion lay their eggs, and the movement of sand on the beach.

9.6 Can Tidal Power Be Harnessed as a Source Of Energy?

Throughout history, ocean tides have been used as a source of power. During high tide, water can be trapped in a basin and then harnessed to do work as it flows back to the sea. In the 12th century, for example, water wheels driven by the tides were used to power gristmills and sawmills. During the 17th and 18th centuries, much of Boston's flour was produced at a tidal mill.

Today, tidal power is considered a clean, renewable resource with vast potential. The initial cost of building a tidal power electricity-generating plant may be higher than the cost of building a conventional thermal power plant, but the operating costs are lower because it does not use fossil fuels or radioactive substances to generate electricity.

One disadvantage of tidal power, however, is the periodicity of the tides, which allows power to be generated only during a portion of a 24-hour day. People operate on a solar period, but tides operate on a lunar period, so the energy available from the tides would coincide with need only part of the time. Power would have to be distributed to the point of need at the moment it was generated, which could be a great distance away, resulting in an expensive transmission problem.

The power could be stored, but even this alternative presents a large and expensive technical problem.

To generate electricity effectively, electrical turbines (generators) need to run at a constant speed, which is difficult to maintain when generated by the variable flow of tidal currents in two directions (flood tide and ebb tide). Specially designed turbines that allow both advancing and receding water to spin their blades are necessary to solve the problem of generating electricity from the tides.

Another disadvantage of tidal power is harm to wildlife and change in habitat. For example, tidal power plants alter the normal flow of tidal currents, affecting those marine organisms that depend on these currents for bringing food or for aiding migrations. Another example is the harm done to marine organisms entrapped or injured by tidal power devices. In addition, a tidal power plant would likely interfere with many traditional uses of coastal waters, such as transportation and fishing.

Tidal Power Plants

Tidal power can be harnessed in one of two ways: (1) Tidal water trapped behind coastal barriers in bays and estuaries can be used to turn turbines and generate electrical energy, and (2) tidal currents that pass through narrow channels can be used to

Figure 9.22 La Rance tidal power plant at St. Malo, France. Electricity is generated at the La Rance tidal power plant at St. Malo, France, when water from a rising tide ① flows into the estuary and turns turbines; electricity is also generated when water from a falling tide ② exits the estuary and turns turbines in the other direction.

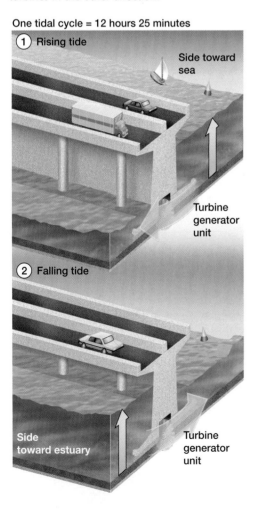

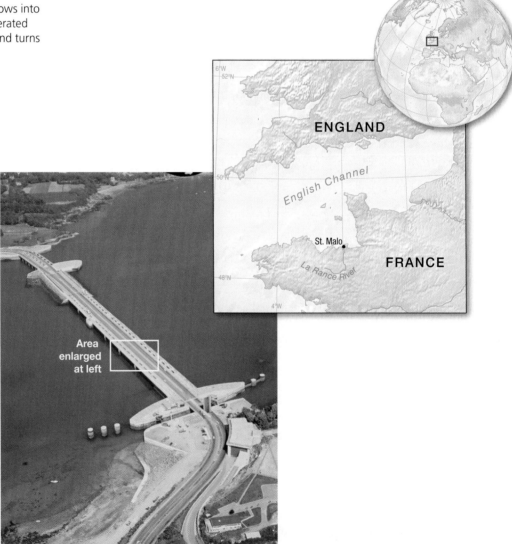

turn underwater pivoting turbines, which produce energy (see Chapter 7). Although the first type is much more commonly employed, Norway, the United Kingdom, and the United States have recently installed offshore turbines that harness swift coastal tidal currents and plan to expand these devices into tidal energy farms.

Worldwide, only a few small tidal power plants use water trapped behind coastal barriers. One successful tidal power plant has been operating in the estuary of La Rance River in northern France (**Figure 9.22**) since 1967. The estuary has a surface area of approximately 23 square kilometers (9 square miles), and the tidal range is 13.4 meters (44 feet). Usable tidal energy increases as the area of the basin increases and as the tidal range increases.

The power-generating barrier was built across the estuary a little over 3 kilometers (2 miles) upstream to protect it from storm waves. The barrier is 760 meters (2500 feet) wide and supports a two-lane road (Figure 9.22). Water passing through the barrier powers 24 electricity-generating units that operate beneath the power plant. At peak operating capacity, each unit can generate 10 megawatts of electricity, for a total of 240 megawatts.[14]

To generate electricity, the La Rance plant needs a sufficient water height between the estuary and the ocean—which occurs only about half of the time. Annual power production of about 540 million kilowatt-hours can be increased to 670 million kilowatt-hours by using the turbine generators as pumps to move water into the estuary at proper times.

Within the Bay of Fundy, which has the largest tidal range in the world, the Canadian province of Nova Scotia constructed a small tidal power plant in 1984 that can generate 20 megawatts of electricity. The plant is built on the Annapolis River estuary, an arm of the Bay of Fundy (see Figure 9.18), where maximum tidal range is 8.7 meters (26 feet).

In 2006, the first Asian tidal power plant came online in Daishan County of eastern China's Zhejiang province. This small power station has the capability to produce 40 kilowatts of electricity, and China has proposed building another larger plant.

Larger power plants that avoid some of the shortcomings of smaller plants have often been considered. For example, a tidal power plant could be made to generate electricity continually if it were located on the Passamaquoddy Bay near the U.S.–Canadian border near the entrance to the Bay of Fundy. Although a tidal power plant across the Bay of Fundy has often been proposed, it has never been built. The usable tidal energy is potentially large compared to the La Rance plant because the flow volume is over 100 times greater.

Recognizing the benefits of tidal power, the United Kingdom has proposed building a tidal power barrage across the Severn Estuary that separates England and Wales. The Severn River has the second-largest tidal range in the world and is a prime target for producing tidal power. If completed, it would be the world's largest tidal power plant, with a 12-kilometer- (7.5-mile-) long dam that could produce 8.6 gigawatts of energy, or about 5% of the electricity currently used in the United Kingdom.

CONCEPT CHECK 9.6

1 Discuss at least two positive and two negative factors related to tidal power generation.

2 Explain how a tidal power plant works, using as an example an estuary that has a mixed tidal pattern. Why does potential for usable tidal energy increase with an increase in the tidal range?

[14]Each megawatt of electricity is enough to serve the energy needs of about 800 average U.S. homes.

Essential Concepts Review

9.1 What causes ocean tides?

▶ *Gravitational attraction of the Moon and Sun create ocean tides*, which are fundamentally very long and regular shallow-water waves. According to a *simplified model of tides*, which assumes an ocean of uniform depth and ignores the effects of friction, small horizontal forces (the tide-generating forces) tend to push water into *two bulges on opposite sides of Earth*. One bulge is directly facing the tide-generating body (the Moon and the Sun), and the other is directly opposite.

▶ Despite its vastly smaller size, *the Moon has about twice the tide-generating effect of the Sun* because the Moon is so much closer to Earth. The tidal bulges due to the Moon's gravity (the lunar bulges) dominate, so lunar motions dominate the periods of Earth's tides. However, the changing position of the solar bulges relative to the lunar bulges modifies tides. According to the simplified idealized tide theory, *Earth's rotation carries locations on Earth into and out of the various tidal bulges.*

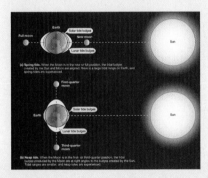

Study Resources
Online Study Guide Quizzes, Web Animations

Critical Thinking Question

Discuss the forces involved in creating ocean tides.

9.2 How do tides vary during a monthly tidal cycle?

▶ Tides would be easy to predict if Earth were a uniform sphere covered with an ocean of uniform depth. For most places on Earth, *the time between successive high tides would be 12 hours 25 minutes (half a lunar day)*. The 29½-day monthly tidal cycle would consist of tides with maximum tidal range (spring tides) and minimum tidal range (neap tides). *Spring tides would occur each new moon and full moon, and neap tides would occur each first- and third-quarter phases of the Moon.*

▶ The *declination of the Moon* varies between 28.5 degrees north or south of the equator during the lunar month, and the *declination of the Sun* varies between 23.5 degrees north or south of the equator during the year, so *the location of tidal bulges usually creates two high tides and two low tides of unequal height per lunar day*. Tidal ranges are greatest when Earth is nearest the Sun and Moon.

Study Resources
Online Study Guide Quizzes, Web Table 9.1, Web Animation

Critical Thinking Question

Observe the Moon from a reference location every night at about the same time for two weeks. Keep track of your observations about the shape (phase) of the Moon and its position in the sky. Then compare your observations to the reported tides in your area (or a coastal area you visit frequently) and report your findings.

9.3 What do tides look like in the ocean?

▶ *When friction and the true shape of ocean basins are considered, the dynamics of tides become more complicated.* Moreover, the two bulges on opposite sides of Earth cannot exist because they cannot keep up with the rotational speed of Earth. Instead, the bulges are broken up into *several tidal cells that rotate around an amphidromic point—a point of zero tidal range*. Rotation is counterclockwise in the Northern Hemisphere and clockwise in the Southern Hemisphere. *Many other factors influence tides on Earth*, too, such as the positions of the continents, the varying depth of the ocean, and coastline shape.

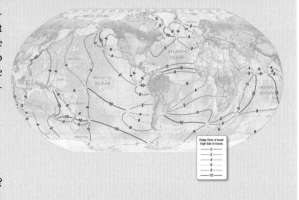

Study Resources
Online Study Guide Quizzes

Critical Thinking Question

Describe amphidromic points and cotidal lines. How do they influence tides in the open ocean?

9.4 What types of tidal patterns exist?

▶ The *three types of tidal patterns* observed on Earth are *diurnal* (a single high and low tide each lunar day), *semidiurnal* (two high and two low tides each lunar day), and *mixed* (characteristics of both). Mixed tidal patterns usually consist of semidiurnal periods with significant diurnal inequality. Mixed tidal patterns are the most common type in the world.

Study Resources
Online Study Guide Quizzes, Web Animation

Critical Thinking Question

Assume that there are two moons in orbit around Earth that are on the same orbital plane but always on opposite sides of Earth and that each moon is the same size and mass of our Moon. How would this affect the tidal range during spring and neap tide conditions? Also, how would this affect the tidal pattern observed?

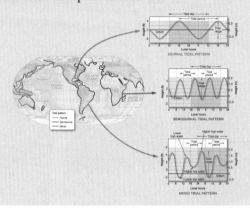

9.6 Can tidal power be harnessed as a source of energy?

▶ *Tides can be used to generate power* without need for fossil or nuclear fuel. There are some *significant drawbacks*, however, to creating successful tidal power plants. Still, many sites worldwide have the *potential for tidal power generation*.

Study Resources
Online Study Guide Quizzes

Critical Thinking Question

What are some negative environmental factors that could inhibit the development of ocean tidal power systems?

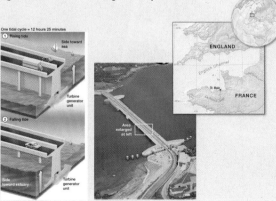

9.5 What tidal phenomena occur in coastal regions?

▶ There are many types of *observable tidal phenomena in coastal areas. Tidal bores are true tidal waves* (a wave produced by the tides) that occur in certain rivers and bays due to an incoming high tide. The effects of constructive interference together with the shoaling and narrowing of coastal bays creates the *largest tidal range in the world—17 meters (56 feet)—at the northern end of Nova Scotia's Bay of Fundy. Tidal currents follow a rotary pattern* in open-ocean basins but are converted to *reversing currents* along continental margins. The maximum velocity of reversing currents occurs during flood and ebb currents, when the water is halfway between high slack water and low slack water. *Whirlpools* can be created in some restricted coastal passages due to reversing tidal currents. *The tides are also important to many marine organisms.* For instance, *grunion*—small silvery fish that inhabit waters along the West coast of North America—time their spawning cycle to match the pattern of the tides.

Study Resources
Online Study Guide Quizzes, Web Videos

Critical Thinking Question

Examine a current monthly tide calendar for a coastal location. When is the best time for a field trip to the tide pools? If this were a grunion spawning month, when would the conditions be right for the next grunion spawning run?

MasteringOceanography™

Looking for additional review and test prep materials? Visit the Study Area in MasteringOceanography to enhance your understanding of this chapter's content by accessing a variety of resources, including Self-Study Quizzes, Geoscience Animations, RSS feeds, flashcards, Web links, and an optional Pearson eText.

Beaches of the world. Beaches are composed of whatever material is locally available and differ depending on a variety of oceanographic factors. Beach locations shown include California, Oregon, Hawaii, the Galápagos Islands, Ecuador, and Baja California, Mexico.

Before you begin reading this chapter, use the glossary at the end of this book to discover the meanings of any of these words you don't already know:

Beach face Rip current
Shore Bay barrier
Spit Sea arch Tombolo Barrier island
Longshore current Sea cave Longshore bar Jetty
Rip-rap
Beach replenishment Longshore drift Hard stabilization
Sea stack Coast Delta Breakwater
Groin field Beach Berm Seawall
Drowned river valley

THE COAST: BEACHES AND SHORELINE PROCESSES

"The waves which dash upon the shore are, one by one, broken, but the ocean conquers nevertheless. It overwhelms the Armada, it wears out the rock."

—Lord Byron (1821)

Humans have always been attracted to the coastal regions of the world for their moderate climate, seafood, transportation, recreational opportunities, and commercial benefits. In the United States, for example, 80% of the population now lives within easy access of the Atlantic, Pacific, and Gulf coasts, increasing the stress on these important national resources.

The coastal region is constantly changing because waves crash along most shorelines about 10,000 times a day, releasing their energy from distant storms. Waves cause erosion in some areas and deposition in others, resulting in changes that occur hourly, daily, weekly, monthly, seasonally, and yearly.

In this chapter, we'll examine the major features of the seacoast and shore and the processes that modify them. We'll also discuss ways people interfere with these processes, creating hazards to themselves and to the environment.

10.1 How Are Coastal Regions Defined?

The **shore** is a zone that lies between the lowest tide level (low tide) and the highest elevation on land that is affected by storm waves. The **coast** extends inland from the shore as far as ocean-related features can be found (**Figure 10.1**). The width of the shore varies between a few meters and hundreds of meters. The width of the coast may vary from less than 1 kilometer (0.6 mile) to many tens of kilometers. The **coastline** marks the boundary between the shore and the coast. It is the landward limit of the effect of the highest storm waves on the shore.

Beach Terminology

The beach profile in Figure 10.1 shows features characteristic of a cliffed shoreline. The shore is divided into the **backshore** and the **foreshore**.[1] The backshore is above the high tide shoreline and is covered with water only during storms. The foreshore is the portion exposed at low tide and submerged at high tide. The **shoreline** migrates back and forth with the tide and is the water's edge. The **nearshore** extends seaward from the low tide shoreline to the low tide breaker line. It is never exposed to the atmosphere, but it is affected by waves that touch bottom. Beyond the low-tide breakers is the **offshore** zone, which is deep enough that waves rarely affect the bottom.

---

[1]The foreshore is often referred to as the *intertidal zone*, or *littoral (litoralis = the shore) zone*.

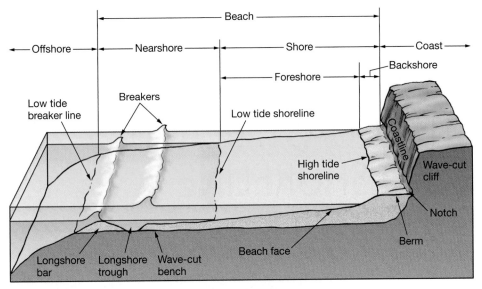

Figure 10.1 Diagrammatic view of a cliffed coastal region, showing beach terminology and landforms. The beach is defined as the entire active area of a coastline that is affected by waves; it extends from the low tide breaker line offshore (*left*) to the far end of the berm (*right*). Although most of these beach features can be found at any beach, not every beach has coastal cliffs.

A **beach** is a deposit of the shore area. It consists of wave-worked sediment that moves along the **wave-cut bench** (a flat, wave-eroded surface). A beach may continue from the coastline across the nearshore region to the line of breakers. Thus, the beach is the entire active area of a coast that experiences changes due to breaking waves. The area of the beach above the shoreline is often called the *recreational beach*.

The **berm** is the dry, gently sloping, slightly elevated margin of the beach that can be found at the foot of coastal cliffs or sand dunes. Because the berm is normally composed of dry sand, it is a favorite place of beachgoers for activities such as lying in the Sun, beach volleyball, barbecues, and bonfires. Moving offshore, the **beach face** is the wet, sloping surface that extends from the berm to the shoreline. It is more fully exposed during low tide and is also known as the *low tide terrace*. The beach face is a favorite place for runners because the sand is wet and hard packed. Offshore beyond the beach face is one or more **longshore bars**—sand bars that parallel the coast. A longshore bar may not always be present throughout the year, but when one is, it may be exposed during extremely low tides. Longshore bars can "trip" waves as they approach shore and cause them to begin breaking. Separating the longshore bar from the beach face is a **longshore trough**.

Beach Composition

Beaches are composed of whatever material is locally available (see the chapter-opening photos). When this material—sediment—comes from the erosion of beach cliffs or nearby coastal mountains, beaches are composed of mineral particles from these rocks and may be relatively coarse in texture. When the sediment comes primarily from rivers that drain lowland areas, beaches are finer in texture. Often, mud flats develop along the shore because only tiny clay-sized and silt-sized particles are emptied into the ocean. Such is the case for muddy coastlines such as along the coast of Suriname in South America and the Kerala coast of southwest India.

Other beaches have a significant biological component. For example, in low-relief, low-latitude areas such as southern Florida, there are no mountains or other sources of rock-forming minerals nearby. As a result, beaches in these areas are generally composed of shell fragments, broken coral, and the remains of organisms that live in coastal waters. Many beaches on volcanic islands in the open ocean are composed of black or green fragments of the basaltic lava that comprises the islands or of coarse debris from coral reefs that develop around islands in low latitudes.

Regardless of the composition, though, the material that comprises the beach does not stay in one place. Instead, the waves that crash along the shoreline are constantly moving it. Thus, beaches can be thought of as *material in transit along the shoreline*.

Web Animation
Summertime/Wintertime
Beach Conditions

10.2 How Does Sand Move on the Beach?

The movement of sand on the beach occurs both perpendicular to the shoreline (*toward* shore and *away from* shore) and parallel to the shoreline (often referred to as *upcoast* and *downcoast*).

Movement Perpendicular to the Shoreline

Sand on the beach moves perpendicular to the shoreline as a result of breaking waves.

MECHANISM   As each wave breaks, water rushes up the beach face toward the berm. Some of this **swash** soaks into the beach and eventually returns to the ocean. However, most of the water drains away from shore as **backwash**, though usually not before the next wave breaks and sends its swash over the top of the previous wave's backwash.

While standing in ankle-deep water at the shoreline, you can see that swash and backwash transport sediment up and down the beach face perpendicular to the shoreline. Whether swash or backwash dominates determines whether sand is deposited or eroded from the berm.

LIGHT VERSUS HEAVY WAVE ACTIVITY   During *light wave activity* (characterized by less energetic waves), much of the swash soaks into the beach, so backwash is reduced. The swash dominates the transport system, therefore causing a net movement of the sand up the beach face toward the berm, which creates a wide, well-developed berm.

During *heavy wave activity* (characterized by high-energy waves), the beach is saturated with water from previous waves, so very little of the swash soaks into the beach. Backwash dominates the transport system, therefore causing a net movement of sand down the beach face, which erodes the berm. When a wave breaks, moreover, the incoming swash comes *on top of* the previous wave's backwash, effectively protecting the beach from the swash and adding to the eroding effect of the backwash.

During heavy wave activity, where does the sand from the berm go? The orbital motion in waves is too shallow to move the sand very far offshore. Thus, the sand accumulates just beyond where the waves break and forms one or more offshore sand bars (the longshore bars).

SUMMERTIME AND WINTERTIME BEACHES   Light and heavy wave activity alternate seasonally at most beaches, so the characteristics of the beaches they produce change, too (**Table 10.1**). For example, light wave activity produces a wide sandy berm and an overall steep beach face—a **summertime beach**—at the expense of the longshore bar (**Figure 10.2a**). Conversely, heavy wave activity produces a narrow rocky berm and an overall flattened beach face—a **wintertime beach**—and builds prominent longshore bars (**Figure 10.2b**). A wide berm that takes several months to build can be destroyed in just a few hours by high-energy wintertime storm waves.

Movement Parallel to the Shoreline

At the same time that movement occurs perpendicular to shore, movement parallel to the shoreline also occurs.

MECHANISM   Recall from Chapter 8 that within the surf zone, waves *refract* (bend) and line up *nearly* parallel to shore. With each breaking wave, the swash moves up onto the exposed beach at a slight angle; then gravity pulls the backwash down

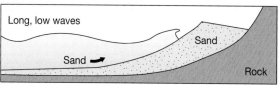

(a) Summertime beach (fair weather)

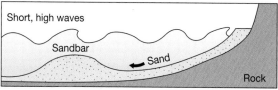

(b) Wintertime beach (storm)

Figure 10.2 Summertime and wintertime beach conditions. Dramatic differences occur between **(a)** summertime and **(b)** wintertime beach conditions at Boomer Beach in La Jolla, California.

ESSENTIAL CONCEPT 10.2A

Smaller, low-energy waves move sand up the beach face toward the berm and create a summertime beach; larger, high-energy waves scour sand from the berm and create a wintertime beach.

TABLE 10.1 CHARACTERISTICS OF BEACHES AFFECTED BY LIGHT AND HEAVY WAVE ACTIVITY

	Light wave activity	Heavy wave activity
Berm/longshore bars	Berm is built at the expense of the longshore bars	Longshore bars are built at the expense of the berm
Wave energy	Low wave energy (nonstorm conditions)	High wave energy (storm conditions)
Time span	Long time span (weeks or months)	Short time span (hours or days)
Characteristics	Creates summertime beach: sandy, wide berm, steep beach face	Creates wintertime beach: rocky, narrow berm, flattened beach face

(a) Waves approaching the beach at a slight angle near Oceanside, California, producing a longshore current moving toward the right of the photo.

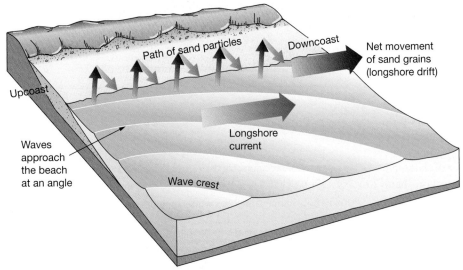

(b) A longshore current, caused by refracting waves, moves water in a zigzag fashion along the shoreline. This causes a net movement of sand grains (longshore drift) from upcoast to downcoast ends of a beach.

Figure 10.3 Longshore current and longshore drift. A longshore current is produced by waves approaching the beach at a slight angle. The longshore current produces longshore drift, which moves sand downcoast in a zigzag fashion.

the beach face at a slight downcoast angle. As a result, water moves in a zigzag fashion along the shore.

LONGSHORE CURRENT AND LONGSHORE DRIFT (LONGSHORE TRANSPORT) The zigzag movement of water along the shore is called a **longshore current** (Figure 10.3). Longshore currents have speeds up to 4 kilometers (2.5 miles) per hour. Speeds increase as beach slope increases, as the angle at which breakers arrive at the beach increases, as wave height increases, and as wave frequency increases.

Swimmers can inadvertently be carried by longshore currents and find themselves carried far from where they initially entered the water. This demonstrates that longshore currents are strong enough to move people as well as a vast amount of sand in a zigzag fashion along the shore.

Longshore drift (also called **longshore transport**, *beach drift*, or *littoral drift*) is the movement of *sediment* in a zigzag fashion caused by the longshore current (Figure 10.3b). Both longshore currents and longshore transport occur only within the surf zone and not farther offshore because the water is too deep there. Recall from Chapter 8 that the depth of a wave's *wave base* is one-half its wavelength, measured from still water level. Below this depth, waves don't touch bottom, and they don't refract; as a result, longshore currents can't form.

THE BEACH: A RIVER OF SAND By various processes, both rivers and coastal zones move water and sediment from one area (*upcoast* or *upstream*) to another (*downcoast* or *downstream*). As a result, the beach has often been referred to as a "river of sand." There are, however, differences between how beaches and rivers transport sediment. For example, a longshore current moves in a zigzag fashion, while rivers flow mostly in a turbulent, swirling fashion. In addition, the direction of flow of longshore currents along a shoreline can change, whereas rivers always flow in the same basic direction (downhill). The longshore current can change direction because the waves that approach the beach typically come from different directions in different seasons. Nevertheless, the longshore current generally flows *southward along both the Atlantic and Pacific coasts of the United States.*

10.3 What Features Exist Along Erosional and Depositional Shores?

Sediment eroded from the beach is transported along the shore and deposited in areas where wave energy is low. Even though all shores experience some degree of both erosion and deposition, shores can often be identified primarily as one type or the other. **Erosional shores** typically have well-developed cliffs and are in areas where tectonic uplift of the coast occurs, such as along the U.S. Pacific coast.

The U.S. southeastern Atlantic coast and the Gulf coast, on the other hand, are primarily **depositional shores**. Sand deposits and offshore barrier islands are common there because the shore is gradually subsiding. Erosion can still be a major problem on depositional shores, especially when human-made coastal structures interfere with natural coastal processes (as discussed later in this chapter).

STUDENTS SOMETIMES ASK . . .

How much sand is moved along coasts by longshore drift?

Very impressive amounts! For example, longshore drift rates are typically in the range of 75,000 to 230,000 cubic meters (100,000 to 300,000 cubic yards) per year. To help you visualize how much sand this is, think of a typical dump truck, which has a volume of about 45 cubic meters (60 cubic yards). In essence, longshore drift carries the equivalent of thousands of full dump trucks along coastal regions each year. And a few coastal regions have longshore drift rates as high as 765,000 cubic meters (1,000,000 cubic yards) per year.

Features Of Erosional Shores

Because of wave refraction, wave energy is concentrated on any **headlands** that jut out from the continent, while the amount of energy reaching the shore in bays is reduced. Headlands, therefore, are eroded and the shoreline retreats. Some of these erosional features are shown in **Figure 10.4**.

Web Animation
Longshore Current and Longshore (Beach) Drift

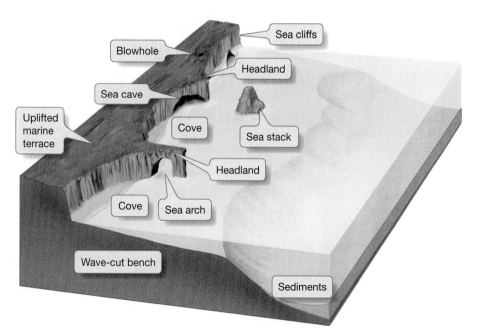

Figure 10.4 Features of erosional coasts. Diagrammatic view of features characteristic of erosional coasts.

Figure 10.5 Sea arches and sea stack at Praia da Marinha Beach, near Armacao de Pera, Algarve, Portugal. When the roof of a sea arch (*behind*) collapses, a sea stack (*middle*) is formed.

STUDENTS SOMETIMES ASK . . .

What is the difference between a rip current and a rip tide? Are they the same thing as an undertow?

Like tidal waves (tsunami), rip tides are a misnomer and have nothing to do with the tides. Rip tides are more correctly called *rip currents*. Perhaps rip currents have incorrectly been called rip tides because they occur suddenly (like an incoming tide). The origin of rip currents and their associated dangers are discussed in **Diving Deeper 10.1**.

An *undertow*, similar to a rip current, is a flow of water away from shore. An undertow is much wider, however, and is usually more concentrated along the ocean floor. An undertow is really a continuation of backwash that flows down the beach face and is strongest during heavy wave activity. Undertows can be strong enough to knock people off their feet, but they are confined to the immediate floor of the ocean and only within the surf zone.

Figure 10.6 Wave-cut bench and marine terrace. A wave-cut bench is exposed at low tide along the California coast at Bolinas Point near San Francisco. An elevated wave-cut bench, called a marine terrace, is shown at right.

Waves pound relentlessly away at the base of headlands, undermining the upper portions, which eventually collapse to form **wave-cut cliffs**. The waves may form **sea caves** at the base of the cliffs.

As waves continue to pound the headlands, the caves may eventually erode through to the other side, forming openings called **sea arches** (**Figure 10.5**). Some sea arches are large enough to allow a boat to maneuver safely through them. With continued erosion, the tops of sea arches eventually crumble to produce **sea stacks** (Figure 10.5). Waves also erode the bedrock of the bench. Uplift of the wave-cut bench creates a gently sloping **marine terrace** above sea level (**Figure 10.6**).

Rates of coastal erosion are influenced by the degree of exposure to waves, the amount of tidal range, and the composition of the coastal bedrock. Regardless of

WARNING: RIP CURRENTS . . . DO YOU KNOW WHAT TO DO?

The backwash from breaking waves usually returns to the open ocean as a flow of water across the ocean bottom, so it is commonly referred to as "sheet flow." Some of this water, however, can return to the ocean in strong, narrow surface currents called **rip currents** that flow away from shore and are generally oriented perpendicular to the beach.

Rip currents are between 15 and 45 meters (15 and 150 feet) wide and can attain velocities of 7 to 8 kilometers (4 to 5 miles) per hour—faster than most people can swim for any length of time. In fact, it is useless to swim for long against a current stronger than about 2 kilometers

(1.2 miles) per hour. Rip currents can travel hundreds of meters from shore before they break up. If a light-to-moderate swell is breaking, numerous rip currents that are moderate in size and velocity may develop. A heavy swell usually produces fewer, more concentrated, and stronger rips. They can often be recognized by the way they interfere with incoming waves, by their characteristic brown color caused by suspended sediment, or by their foamy and choppy appearance (**Figure 10A**).

The rip currents that occur during heavy swell are a significant hazard to coastal swimmers. In fact, rip currents cause an estimated 70 to 100 drownings annually in the United States, and 80% of rescues at beaches by lifeguards involve people who are trapped in rip currents. What is the best thing to do if you are caught in a rip current? Swimmers can escape rip currents by swimming parallel to the shore for a short distance (simply swimming out of the narrow rip current) and then riding the waves in toward the beach. However, even excellent swimmers who panic or try to fight the current by swimming directly into it are eventually overcome by exhaustion and may drown. Even though most beaches have warnings posted and are frequently patrolled by lifeguards, many people tragically lose their lives each year because of rip currents.

Figure 10A **Rip current and warning sign.** A rip current (*red arrow*), which extends outward from shore and interferes with incoming waves, and a warning sign (*inset*).

the erosion rate, all coastal regions follow the same developmental path. As long as there is no change in the elevation of the landmass relative to the ocean surface, the cliffs will continue to erode and retreat until the beaches widen sufficiently to prevent waves from reaching them. The eroded material is carried from high-energy areas and deposited in low-energy areas.

Web Video
Rip Currents

Features Of Depositional Shores

Coastal erosion of sea cliffs produces large amounts of sediment. Additional sediment, which is carried to the shore by rivers, comes from the erosion of inland rocks. Waves then distribute all this sediment along the continental margin.

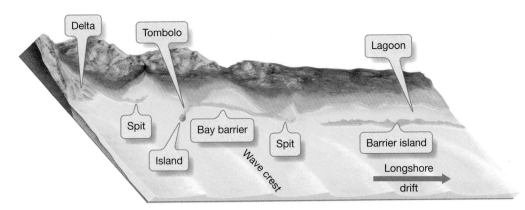

Figure 10.7 **Features of depositional coasts.** Diagrammatic view of features characteristic of depositional coasts.

Figure 10.7 shows some of the features of depositional coasts. These features are primarily deposits of sand moved by longshore drift but are also modified by other coastal processes. Some are partially or wholly separated from the shore.

A **spit** (*spit* = spine) is a linear ridge of sediment that extends in the direction of longshore drift from land into the deeper water near the mouth of a bay. The end of the spit normally curves into the bay due to the movement of currents.

Tidal currents or currents from river runoff are usually strong enough to keep the mouth of the bay open. If not, the spit may eventually extend across the bay and connect to the mainland, forming a **bay barrier**, or **bay-mouth bar** (**Figure 10.8a**), which cuts off the bay from the open ocean. Although a bay barrier is a buildup of sand usually less than 1 meter (3.3 feet) above average sea level, permanent buildings are often constructed on them.

A **tombolo** (*tombolo* = mound) is a sand ridge that connects an island or a sea stack to the mainland (**Figure 10.8b**). Tombolos can also connect two adjacent islands. Formed in the wave-energy shadow of an island, a tombolo is usually oriented perpendicular to the average direction of incoming waves.

BARRIER ISLANDS Extremely long offshore deposits of sand that are parallel to the coast are called **barrier islands** (**Figure 10.9**). They form a first line of defense against rising sea level and high-energy storm waves, which would otherwise exert their full force directly against the shore. The origin of barrier islands is complex, but many appear to have developed during the worldwide rise in sea level that is associated with the melting of glaciers at the end of the most recent ice age, about 18,000 years ago.

According to a recent study of global satellite images, 2149 barrier islands have been identified worldwide in every climate and every tide–wave combination.

Figure 10.8 **Coastal depositional features.** Photos showing various coastal depositional features, including a bay barrier, spit, and tombolo.

(a) Barrier coast, spit, and bay barrier along the coast of Martha's Vineyard, Massachusetts.

(b) Tombolo at Goat Rock Beach, California.

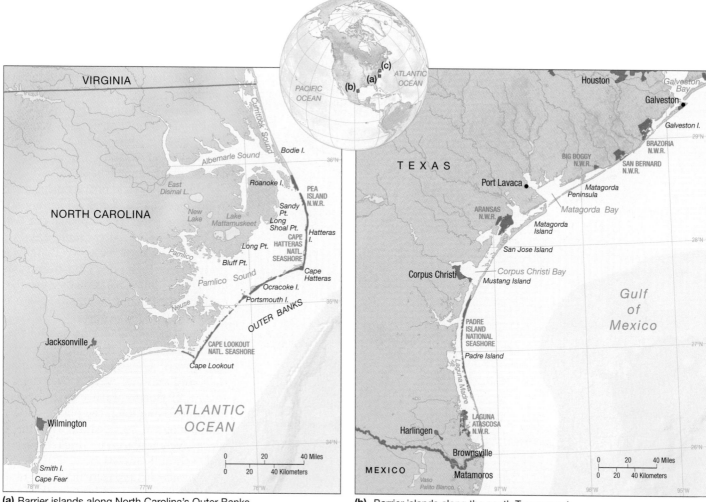

(a) Barrier islands along North Carolina's Outer Banks.

(b) Barrier islands along the south Texas coast.

Figure 10.9 Examples of barrier islands. Maps and aerial photo of barrier islands in **(a)** North Carolina, **(b)** Texas, and **(c)** New Jersey.

Nearly 300 barrier islands ring the Atlantic and Gulf coasts of the United States (**Figure 10.10**). They exist from Massachusetts to eastern Florida in a nearly continuous line; they also occur discontinuously in the Gulf of Mexico from western Florida into Mexico. Barrier islands may exceed 100 kilometers (60 miles) in length, have widths of several kilometers, and are separated from the mainland by a lagoon. Notable barrier islands include Fire Island off the New York coast, North Carolina's Outer Banks, and Padre Island off the coast of Texas.

One human-related environmental issue of barrier islands is their attractiveness as building sites because of their proximity to the ocean. For example, the population densities of barrier islands are three times higher than those of adjoining coastal states. In addition, the overall population on barrier islands increased 14% from 1990 to 2000 and continues to increase. Although it seems unwise to build a coastal structure on a narrow, low-lying, shifting strip of sand, many large buildings have been constructed on barrier islands (see Figure 10.9c). Some of these structures have either fallen into the ocean or have needed to be moved (see **Web Diving Deeper 10.1**).

(c) A portion of a heavily developed barrier island near Tom's River, New Jersey.

Figure 10.10 Barrier island locations along the U.S. Atlantic and Gulf coasts. Barrier islands occur from Maine to eastern Florida along the Atlantic coast and from western Florida to Mexico along the Gulf coast. Barrier islands do not occur along the Pacific coast.

Features of Barrier Islands A typical barrier island has the physiographic features shown in **Figure 10.11a**. From the ocean landward, they are (1) ocean beach, (2) dunes, (3) barrier flat, (4) high salt marsh, (5) low salt marsh, and (6) lagoon between the barrier island and the mainland.

During the summer, gentle waves carry sand to the *ocean beach*, so it widens and becomes steeper. During the winter, higher-energy waves carry sand offshore and produce a narrow, gently sloping beach.

Winds blow sand inland during dry periods to produce coastal *dunes*, which are stabilized by dune grasses. These plants can withstand salt spray and burial by sand. Dunes protect the lagoon against excessive flooding during storm-driven high tides. Numerous passes exist through the dunes, particularly along the southeastern Atlantic coast, where dunes are less well developed than to the north.

The *barrier flat* forms behind the dunes from sand driven through the passes during storms. Grasses quickly colonize these flats, and seawater washes over them during storms. If storms wash over the barrier flat infrequently enough, the plants undergo natural biological succession, with the grasses successively replaced by thickets, woodlands, and eventually forests.

Salt marshes typically lie inland of the barrier flat. They are divided into the *low marsh*, which extends from about mean sea level to the high neap-tide line, and the *high marsh*, which extends to the highest spring-tide line. The low marsh is by far the most biologically productive part of the salt marsh.

New marshland is formed as overwash carries sediment into the lagoon, filling portions so they become intermittently exposed by the tides. Marshes may be poorly developed on parts of the island that are far from floodtide inlets. Their development is greatly restricted on barrier islands, where people perform artificial dune enhancement and fill inlets in an attempt to prevent overwashing and flooding.

The gradual sea level rise experienced along the eastern North American coast is causing barrier islands to migrate landward. The movement of the barrier island is similar to a slowly moving tractor tread, with the entire island rolling over itself, impacting structures built on these islands. *Peat deposits*, which are formed by the accumulation of organic matter in marsh environments, provide further evidence of barrier island migration (**Figure 10.11b**). As the island slowly rolls over itself and migrates toward land, it buries ancient peat deposits. These peat deposits can be found beneath the island and may even be exposed on the ocean beach when the barrier island has moved far enough.

DELTAS Some rivers carry more sediment to the ocean than longshore currents can distribute. These rivers develop a **delta** (*delta* = triangular) deposit at their mouths.

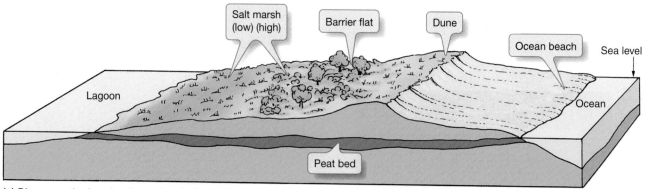

(a) Diagrammatic view showing major physiographic zones of a barrier island. The peat bed represents ancient marsh environments.

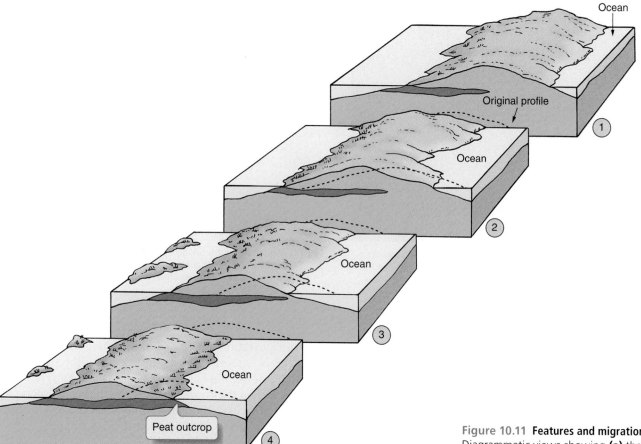

(b) Sequence (1–4) showing how a barrier island migrates toward the mainland in response to rising sea level and exposes peat deposits that have been covered by the island.

Figure 10.11 Features and migration of barrier islands. Diagrammatic views showing **(a)** the features of a typical barrier island and **(b)** the migration of a barrier island in response to rising sea level.

The Mississippi River, which empties into the Gulf of Mexico (**Figure 10.12a**), forms one of the largest deltas on Earth. Deltas are fertile, flat, low-lying areas that are subject to periodic flooding.

Delta formation begins when a river has filled its mouth with sediment. The delta then grows through the formation of *distributaries*, which are branching channels that deposit sediment as they radiate out over the delta in finger-like extensions (Figure 10.12a). When the fingers get too long, they become choked with sediment. At this point, a flood may easily shift the distributary's course and provide sediment to low-lying areas between the fingers. When depositional processes exceed coastal erosion and transportation processes, a branching "bird's foot" delta similar to the Mississippi River Delta results.

Web Animation
Movement of a Barrier Island in
Response to Rising Sea Level

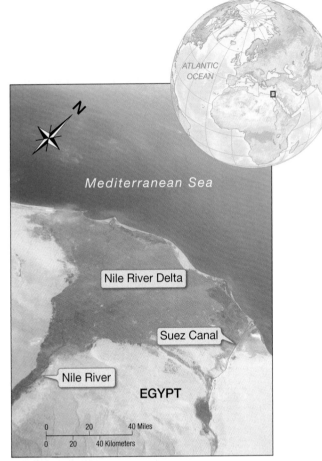

(a) Satellite image of the branching "bird's foot" structure of the Mississippi River Delta, which flows into the Gulf of Mexico and shows suspended sediment in the water.

(b) Photograph from the Space Shuttle of Egypt's Nile River Delta, which has a smooth, curved shoreline as it extends into the Mediterranean Sea.

Figure 10.12 Examples of deltas. Aerial photos showing **(a)** the Mississippi River Delta and **(b)** the Nile River Delta.

When erosion and transportation processes exceed deposition, on the other hand, a delta shoreline is smoothed to a gentle curve, like that of the Nile River Delta in Egypt (**Figure 10.12b**). The Nile Delta is presently eroding because sediment is trapped behind the Aswan High Dam. Before the completion of the dam in 1964, the Nile carried huge volumes of sediment into the Mediterranean Sea.

BEACH COMPARTMENTS   A **beach compartment** consists of three main components: (1) a series of rivers that supply sand to a beach; (2) the beach itself, where sand is moving due to longshore transport; and (3) offshore submarine canyons, where sand is drained away from the beach. The map in **Figure 10.13** shows that the coast of Southern California contains four separate beach compartments.

Within an individual beach compartment, sand is supplied primarily by rivers (Figure 10.13, *inset*), but in areas that have coastal bluffs, a substantial proportion of sand may also be supplied by sea cliff erosion. The sand moves south with the longshore current, so beaches are wider near the southern (*down-coast*) end of each beach compartment. Although some sand is washed offshore along the way or blows further inland to produce coastal sand dunes, most sand eventually moves near the head of a submarine canyon. Surprisingly, many submarine canyons come very close to shore. This allows sand to be drained off away from the beach and onto the ocean floor, lost from the beach forever. To

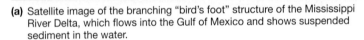

STUDENTS SOMETIMES ASK . . .

Can submarine canyons fill with sediment?

Yes. In many beach compartments, the submarine canyons that drain sand from the beach empty into deep basins offshore. However, given several million years and tons of sediments per year sliding down submarine canyons, offshore basins begin to fill up and can eventually be exposed above sea level. In fact, the Los Angeles basin in California was filled in by sediment derived from local mountains in this manner during the geologic past.

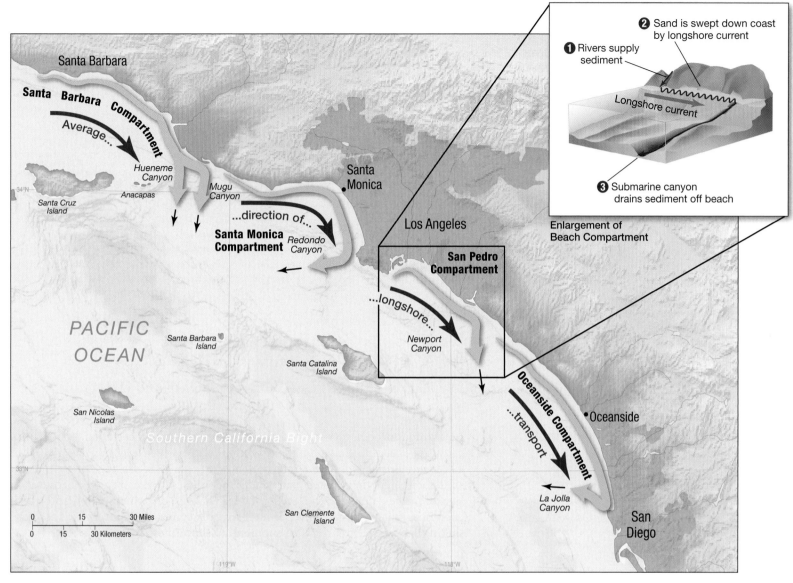

Figure 10.13 Beach compartments. Southern California has several beach compartments, which include rivers that bring sediment to the beach, the beach that experiences longshore transport, and the submarine canyons that remove sand from the beaches. Average longshore transport is toward the south.

the south of this beach compartment, the beaches are typically thin and rocky, without much sand. The process begins all over again at the upcoast end of the next beach compartment, where rivers add their sediment. Farther downcoast, the beach widens and has an abundance of sand until that sand is also diverted down a submarine canyon.

Web Animation
Movement of Sand in a Beach Compartment

Beach Starvation Human activities have altered the natural system of beach compartments. When a dam is built along one of the rivers that feed into a beach compartment, it deprives the beach of sand. Lining rivers with concrete for flood control further reduces the sediment load delivered to coastal regions. Longshore transport continues to sweep the shoreline's sand into the submarine canyons, so the beaches become narrower and experience **beach starvation**. If all the rivers are blocked, the beaches may nearly disappear.

What can be done to prevent beach starvation in beach compartments? One obvious solution is to eliminate the dams, which would allow rivers to supply sand to the beach and return beach compartments to a natural balance. However, most

ESSENTIAL CONCEPT 10.3

Erosional shores are characterized by erosional features such as cliffs, sea arches, sea stacks, and marine terraces. Depositional shores are characterized by depositional features such as spits, tombolos, barrier islands, deltas, and beach compartments.

dams are built for flood protection, water storage, and the generation of hydropower, so it is unlikely that many will be removed. Another option, called *beach nourishment*, is discussed later in this chapter.

CONCEPT CHECK 10.3

① Discuss the formation of such erosional features as wave-cut cliffs, sea caves, sea arches, sea stacks, and marine terraces.

② Describe the origin of these depositional features: spit, bay barrier, tombolo, and barrier island.

③ Describe the response of a barrier island to a rise in sea level. Why do some barrier islands develop peat deposits running through them from the ocean beach to the salt marsh?

④ Discuss why some rivers have deltas and others do not. What are the factors that determine whether a "bird's-foot" delta (like the Mississippi Delta) or a smoothly curved delta (like the Nile Delta) will form?

⑤ Describe the three parts of a beach compartment. What happens when dams are built across all the rivers that supply sand to the beach?

10.4 How Do Changes in Sea Level Produce Emerging and Submerging Shorelines?

In addition to being described as primarily erosional or depositional, shorelines can also be classified based on their position relative to sea level. *Sea level, however, has changed throughout time*, intermittently exposing large regions of continental shelf and then plunging them back under the sea. Sea level can change because the level of the land changes, the level of the sea changes, or a combination of the two. Shorelines that are rising above sea level are called **emerging shorelines**, and those sinking below sea level are called **submerging shorelines**.

Features Of Emerging Shorelines

Marine terraces (see Figures 10.6, 10.14, and 10.19) are one feature characteristic of emerging shorelines. Marine terraces are flat platforms backed by cliffs, which form when a wave-cut bench is exposed above sea level. **Stranded beach deposits** and other evidence of marine processes may exist many meters above the present shoreline, indicating that the former shoreline has risen above sea level.

Features Of Submerging Shorelines

Features characteristic of submerging shorelines include **drowned beaches** (**Figure 10.14**), **submerged dune topography**, and **drowned river valleys** along the present shoreline.

Changes in Sea Level

What causes the changes in sea level that produce submerging and emerging shorelines? One mechanism is the raising or lowering of the land surface relative to sea level through the movement of Earth's crust. Another mechanism is the alteration of the level of the sea itself through worldwide changes in sea level.

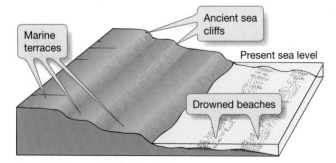

Figure 10.14 Evidence of ancient shorelines. Marine terraces result from exposure of ancient sea cliffs and wave-cut benches above present sea level. Below sea level, drowned beaches indicate the sea level has risen relative to the land.

MOVEMENT OF EARTH'S CRUST   The elevation of Earth's crust relative to sea level can be affected by tectonic movements and by isostatic adjustment.[2] These are called *changes in relative sea level*, because it's the land that has changed, not the sea.

Tectonic Movements   The most dramatic changes in sea level during the past 3000 years have been caused by *tectonic movements*, which affect the elevation of the land. These changes include uplift or subsidence of major portions of continents or ocean basins, as well as localized folding, faulting, or tilting of the continental crust.

Most of the U.S. Pacific coast, for example, is an emerging shoreline because continental margins where plate collisions occur are tectonically active, producing earthquakes, volcanoes, and mountain chains paralleling the coast. Most of the U.S. Atlantic coast, on the other hand, is a submerging shoreline. When a continent moves away from a spreading center (such as the Mid-Atlantic Ridge), its trailing edge subsides because of cooling and the additional weight of accumulating sediment. Passive margins experience only a low level of tectonic deformation, earthquakes, and volcanism, making the Atlantic coast far more quiet and stable than the Pacific coast.

Isostatic Adjustment   Earth's crust also undergoes *isostatic adjustment*: It sinks under the accumulation of heavy loads of ice, vast piles of sediment, or outpourings of lava, and it rises when heavy loads are removed (**Figure 10.15**).

For example, at least four major accumulations of glacial ice—and dozens of smaller ones—have occurred in high-latitude regions over the past 3 million years. Although Antarctica is still covered by a very large, thick ice cap, much of the ice that once covered northern Asia, Europe, and North America has melted.

The weight of ice sheets as much as 3 kilometers (2 miles) thick caused the crust beneath to sink (Figure 10.15). Today, these areas are still slowly rebounding, 18,000 years after the ice began to melt. The floor of Hudson Bay, for example, which is now about 150 meters (500 feet) deep, will be close to or above sea level by the time it stops isostatically rebounding. Another example is the Gulf of Bothnia (between Sweden and Finland), which has isostatically rebounded 275 meters (900 feet) during the past 18,000 years.

Generally, tectonic and isostatic changes in sea level are confined to a segment of a continent's shoreline. For a *worldwide* change in sea level, there must be a change in seawater volume or ocean basin capacity.

WORLDWIDE (EUSTATIC) CHANGES IN SEA LEVEL   Changes in sea level that are experienced worldwide due to changes in seawater volume or ocean basin capacity are called **eustatic** (*eu* = **good**, *stasis* = **standing**)[3] **sea level changes**. The formation or destruction of large inland lakes, for example, causes small eustatic changes in sea level. When lakes form, they trap water that would otherwise run off the land into the ocean, so sea level is lowered worldwide. When lakes are drained and release their water back to the ocean, sea level rises.

Another example of a eustatic change in sea level is through changes in sea floor spreading rates, which can change the capacity of the ocean basin and affect sea level worldwide. Fast spreading, for instance, produces larger rises, such as the East

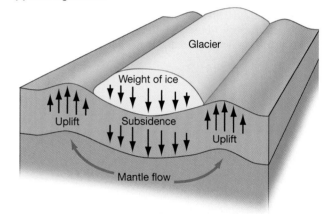

(a) Before glaciation

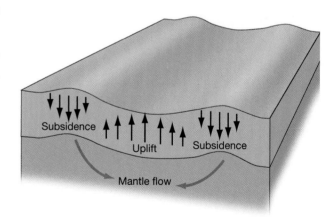

(b) During glaciation

(c) After glaciation

Figure 10.15 **Isostatic adjustment caused by glacial ice.**
(a) Block diagram view of an area of Earth's crust before an ice age. **(b)** The weight of a thick mass of glacial ice causes the crust to subside by isostatic adjustment; note the flow in the mantle below. **(c)** When the ice melts, the weight is removed and the crust rises up by the process of isostatic adjustment, which takes thousands of years.

---

[2]Recall that isostatic adjustment of Earth's crust is also discussed in Chapter 1.

[3]The term *eustatic* refers to a highly idealized situation in which all of the continents remain static (in *good standing*), while only the sea rises or falls.

Web Animation
Glacial Isostacy

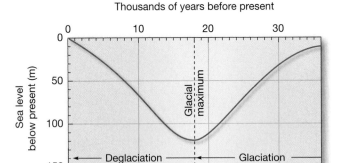

Figure 10.16 Sea level change during the most recent advance and retreat of Pleistocene glaciers. Sea level dropped worldwide by about 120 meters (400 feet) as the last glacial advance removed water from the oceans and transferred it to continental glaciers. About 18,000 years ago, sea level began to rise as the glaciers melted and water was returned to the oceans.

Pacific Rise, which displace more water than slow-spreading ridges such as the Mid-Atlantic Ridge. Thus, fast spreading raises sea level, whereas slower spreading lowers sea level worldwide. Significant changes in sea level due to changes in spreading rate typically take hundreds of thousands to millions of years and may have changed sea level by 1000 meters (3300 feet) or more in the geologic past.

Changes to Sea Level during Ice Ages Ice ages cause eustatic sea level changes, too. As glaciers form, they tie up vast volumes of water on land, eustatically lowering sea level. An analogy to this effect is a sink of water representing an ocean basin. To simulate an ice age, some of the water from the sink is removed and frozen, causing the water level of the sink to be lower. In a similar fashion, worldwide sea level is lower during an ice age. During interglacial stages (such as the one we are in at present), the glaciers melt and release great volumes of water that drain to the sea, eustatically raising sea level. This would be analogous to putting a frozen chunk of ice on the counter near the sink and letting the ice melt, causing the water to drain into the sink and raise "sink level."

During the *Pleistocene Epoch*,[4] glaciers advanced and retreated many times on land in middle- to high-latitude regions, causing sea level to fluctuate considerably. The thermal contraction and expansion of the ocean as its temperature decreased and increased, respectively, affected sea level, too. The thermal contraction and expansion of seawater work much like a mercury thermometer: As the mercury inside the thermometer warms, it expands and rises up the thermometer; as it cools, it contracts. Similarly, cooler seawater contracts and occupies less volume, thereby eustatically *lowering* sea level. Warmer seawater expands, eustatically *raising* sea level.

Although it is difficult to state definitely the range of shoreline fluctuation during the Pleistocene, evidence suggests that it was at least 120 meters (400 feet) below the present shoreline (**Figure 10.16**). It is also estimated that if *all* the remaining glacial ice on Earth were to melt, sea level would rise another 70 meters (230 feet). Thus, the maximum sea level change during the Pleistocene would have been on the order of 190 meters (630 feet), most of which was due to the capture and release of Earth's water by land-based glaciers and polar ice sheets.

The combination of tectonic and eustatic changes in sea level is very complex, so it is difficult to classify coastal regions as purely emergent or submergent. In fact, most coastal areas show evidence of *both* submergence and emergence in the recent past. Evidence suggests, however, that until recently, sea level has experienced only minor changes as a result of melting glacial ice during the past 3000 years.

More recently, there has been a documented sea level rise as a result of human-induced climate change. This topic is discussed in Chapter 16, "The Oceans and Climate Change."

STUDENTS SOMETIMES ASK . . .

Because of plate motions, I know that the continents have not always remained in the same geographic positions. Has the movement of the continents ever affected sea level?

Remarkably, yes. When plate motion moves large continental masses into polar regions, thick continental glaciation can occur (such as in Antarctica today). Glacial ice forms from water vapor in the atmosphere (in the form of snow), which is ultimately derived from the evaporation of seawater. Thus, water is removed from the oceans when continents assume positions close to the poles that provide a platform for large land-based ice accumulation, thereby lowering sea level worldwide.

Sea level is affected by the movement of land and changes in seawater volume or ocean basin capacity. Sea level has changed dramatically in the past because of changes in Earth's climate.

CONCEPT CHECK 10.4

1 Compare the causes and effects of tectonic versus eustatic changes in sea level.

2 List the two basic processes by which coasts advance seaward and list their counterparts that lead to coastal retreat.

3 Describe how an ice age affects sea level.

---

[4]The Pleistocene Epoch of geologic time, which is also called the "Ice Age," occurred 2.6 million to 10,000 years ago (see the Geologic Time Scale, Figure 1.26).

10.5 What Characteristics Do U.S. Coasts Exhibit?

Whether the dominant process along a coast is erosion or deposition depends on the combined effect of many variables, such as composition of coastal bedrock, the degree of exposure to ocean waves, tidal range, tectonic subsidence or emergence, isostatic subsidence or emergence, and eustatic sea level change.

Although many factors contribute to shoreline retreat, sea level rise is the main factor driving worldwide coastal land loss. In fact, more than 70% of the world's sandy beaches are currently eroding, and the percentage increases to nearly 90% for well-studied U.S. sandy coasts. Studies supported by the U.S. Geological Survey produced the rates of shoreline change presented in **Figure 10.17**, where erosion rates are shown as *negative* values and deposition rates are shown as *positive* values.

The Atlantic Coast

Figure 10.17 shows that the U.S. Atlantic coast has a variety of complex coastal conditions:

- Most of the Atlantic coast is exposed to storm waves from the open ocean. Barrier islands from Massachusetts southward, however, protect the mainland from large storm waves.
- Tidal ranges generally increase from less than 1 meter (3.3 feet) along the Florida coast to more than 2 meters (6.5 feet) in Maine.

Figure 10.17 Rates of deposition/erosion along U.S. coasts. Map showing the average rate of coastal deposition/erosion for each coastal state (*colors*) and an accompanying bar graph that shows the average rate of deposition/erosion for each coast, including Alaska and Hawaii. The map also shows the average direction of longshore drift (*purple arrows*), mean spring tide range (*light blue lines*), and main offshore surface current (*blue arrows*).

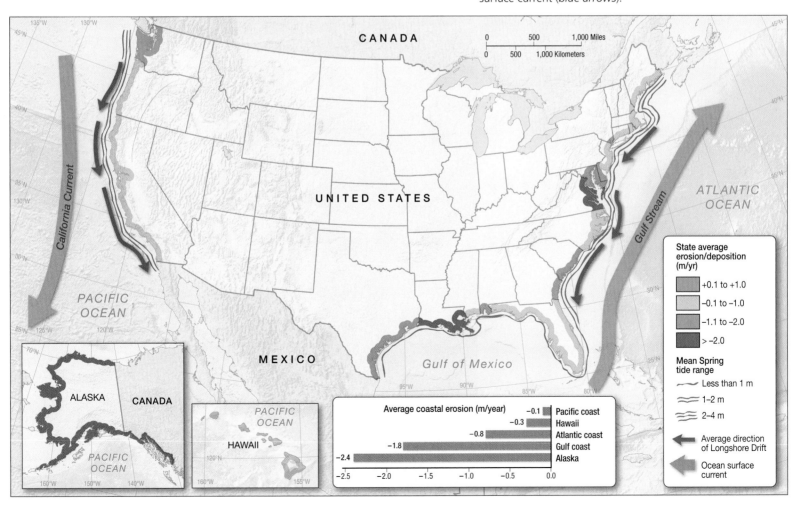

STUDENTS SOMETIMES ASK . . .

Along the East Coast of the United States, how can the longshore current move to the south when the strong Gulf Stream current is moving to the north?

Longshore currents and major ocean surface currents are different things and completely independent of one another. For one thing, longshore currents occur only within the surf zone, while ocean surface currents are much wider and occur farther from shore. In addition, longshore currents are caused by waves coming into shore at an angle (and so can reverse), while ocean surface currents are caused by the major wind belts of the world and modified by the Coriolis effect (and so rarely reverse). And remember that waves (which cause longshore currents) can move in the opposite direction from ocean surface currents.

Along the East Coast of the United States, the reason that the longshore current goes to the south is because the major storm centers that create waves occur in the stormy northern part of the North Atlantic Ocean. As waves radiate southward from these storm centers, a southward-moving longshore current is produced along the East Coast. A similar situation occurs in the North Pacific Ocean, creating a southward-moving longshore current along the West Coast, which just happens to move in the same direction as the California Current (see Figure 10.17).

• Bedrock for most of Florida is a resistant type of sedimentary rock called *limestone*. Most of the bedrock northward through New Jersey, however, consists of nonresistant sedimentary rocks formed in the recent geologic past. As these rocks rapidly erode, they supply sand to barrier islands and other depositional features common along the coast. The bedrock north of New York consists of very resistant rock types.

• From New York northward, continental glaciers affected the coastal region directly. Many coastal features, including Long Island and Cape Cod, are glacial deposits (called *moraines*) left behind when the glaciers melted.

North of Cape Hatteras in North Carolina, the coast is subject to very high-energy waves during fall and winter, when powerful storms called *nor'easters* (northeasters) blow in from the North Atlantic. The energy of these storms generates waves up to 6 meters (20 feet) high, with a 1-meter (3.3-foot) rise in sea level that follows the low pressure as it moves northward. Such high-energy conditions seriously erode coastlines that are predominantly depositional.

Sea level along most of the Atlantic coast appears to be rising at a rate of about 0.3 meter (1 foot) per century. Drowned river valleys, for instance, are common along the coast and form large bays (**Figure 10.18**). In northern Maine, however, sea level

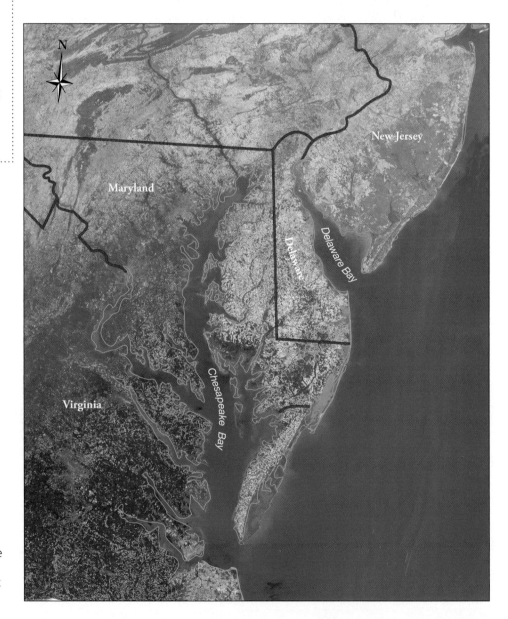

Figure 10.18 Drowned river valleys. Satellite false-color image of drowned river valleys along the East Coast of the United States, including the Chesapeake and Delaware Bays, which were formed by a relative rise in sea level that followed the end of the Pleistocene Ice Age.

may be dropping as the continent rebounds isostatically from the melting of the Pleistocene ice sheet.

The Atlantic coast has an average annual rate of erosion of 0.8 meter (2.6 feet), which means the sea is migrating landward each year by a distance approximately equal to the length of your legs! In Virginia, the loss is over five times that rate, at 4.2 meters (13.7 feet) per year but, this loss is largely confined to barrier islands.

Erosion rates for Chesapeake Bay are about average for the Atlantic coast, but rates for Delaware Bay (1.6 meters [5.2 feet] per year) are about twice the average. Of the observations made along the Atlantic coast, 79% showed some degree of erosion. Delaware, Georgia, and New York have depositional coasts despite serious erosion problems in these states as well.

The Gulf Coast

The Mississippi River Delta, which is deposited in an area with a tidal range of less than 1 meter (3.3 feet), dominates the Louisiana–Texas portion of the Gulf coast. Except during hurricane season (June to November), wave energy is generally low. Tectonic subsidence is common throughout the Gulf coast, and the average rate of sea level rise is similar to that of the southeast Atlantic coast: about 0.3 meter (1 foot) per century. Some areas of coastal Louisiana have experienced a 1-meter (3.3-foot) rise during the past century due to the compaction of Mississippi River sediments by overlying weight.

The average rate of erosion is 1.8 meters (6 feet) per year in the Gulf coast. The Mississippi River Delta experiences the greatest rate, averaging 4.2 meters (13.7 feet) per year. Erosion is made worse by barge channels dredged through marshlands, and Louisiana has lost more than 1 million acres of delta since 1900. Louisiana is now losing marshland at a rate exceeding 130 square kilometers (50 square miles) per year.

Although all Gulf states show a net loss of land, and the Gulf coast has a greater erosion rate than the Atlantic coast, only 63% of the shore is receding because of erosion. The high average rate of erosion reflects the heavy losses in the Mississippi River Delta.

The Pacific Coast

The Pacific coast is generally experiencing less erosion than the Atlantic and Gulf coasts. Along the Pacific coast, relatively young and easily eroded sedimentary rocks dominate the bedrock, with local outcrops of more resistant rock types. Tectonically, the coast is rising, as shown by marine (wave-cut) terraces (**Figure 10.19**). Sea level still shows at least small rates of rise, except for segments along the coast of Oregon and Alaska. The tidal range is mostly between 2.0 and 3.6 meters (6.6 and 12 feet).

The Pacific coast is fully exposed to large storm waves and as a result is said to have *open exposure*. High-energy waves may strike the coast in winter, with typical wave heights of 1 meter (3.3 feet). Frequently, the wave height increases to 2 meters (6.6 feet), and a few times

Figure 10.19 Marine (wave-cut) terraces. Each marine terrace on San Clemente Island offshore of Southern California was created by wave activity at sea level. Subsequently, each terrace has been exposed by tectonic uplift. The highest (oldest) terraces near the top of the photo are now about 400 meters (1320 feet) above sea level.

per year 6-meter (20-foot) waves hammer the shore. These high-energy waves erode sand from many beaches. The exposed beaches, which are composed primarily of pebbles and boulders during the winter months, regain their sand during the summer, when smaller waves occur.

Many Pacific coast rivers have been dammed for flood control and hydro-electric power generation. The amount of sediment supplied by rivers to the shoreline for longshore transport is reduced, resulting in beach starvation in some areas.

With an average erosion rate of only 0.005 meter (0.016 foot)[5] per year and only 30% of the coast showing erosion loss, the Pacific coast is eroded much less than the Atlantic and Gulf coasts. Nevertheless, high wave energy and relatively soft rocks result in high erosion rates of 0.3 meters (1 foot) per year or more in some parts of the Pacific coast. In some parts of Alaska, for example, the average rate of erosion is 2.4 meters (7.9 feet) per year.

Of the Pacific states, only Washington shows a net sediment deposition. The long, protected Washington shoreline within Puget Sound helps skew the Pacific coast values (see Figure 10.17). Although the average erosion rate for California is only 0.1 meter (4 inches) per year, over 80% of the California coast is experiencing erosion, with rates as high as 0.6 meter (2 feet) per year.

ESSENTIAL CONCEPT 10.5

U.S. coastal regions are affected by many variables, including composition of the coastal bedrock, degree of exposure to waves, and tidal range. Most U.S. coastal regions are experiencing erosion.

CONCEPT CHECK 10.5

1 List and discuss four factors that influence the classification of a coast as either erosional or depositional.

2 Describe the tectonic and/or depositional processes that affect the Atlantic coast.

3 Describe the tectonic and/or depositional processes that affect the Gulf coast.

4 Describe the tectonic and/or depositional processes that affect the Pacific coast.

STUDENTS SOMETIMES ASK . . .

I have the opportunity to live in a house at the edge of a coastal cliff where there is an incredible view along the entire coast. Is it safe from coastal erosion?

Based on what you've described, most certainly not! Geologists have long known that cliffs are naturally unstable. Even if cliffs appear to be stable (or have been stable for a number of years), they can be severely damaged during just one significant storm.

The most common cause of coastal erosion is direct wave attack, which undermines the support and causes the cliff to fail. You might want to check the base of the cliff and examine the local bedrock to determine for yourself if you think it will withstand the pounding of powerful storm waves that can move rocks weighing several tons. Other dangers include drainage runoff, weaknesses in the bedrock, slumps and landslides, seepage of water through the cliff, and even burrowing animals. Although all states enforce a setback from the edge of the cliff for all new buildings, sometimes that isn't enough because large sections of "stable" cliffs can fail all at once. For instance, several city blocks of real estate have been eroded from the edge of cliffs during the past 100 years in some areas of Southern California. Even though the view sounds outstanding, you may find out the hard way that the house is built a little *too* close to the edge of a cliff!

10.6 How Does Hard Stabilization Affect Coastlines?

Coastal residents continually modify coastal sediment erosion/deposition in attempts to improve or preserve their property. Structures built to protect a coast from erosion or to prevent the movement of sand along a beach are known as **hard stabilization**, or *armoring of the shore*. Hard stabilization can take many forms and often results in predictable yet unwanted outcomes.

Groins and Groin Fields

One type of hard stabilization is a **groin** (*groin* = ground). Groins are built perpendicular to a coastline and are specifically designed to trap sand moving along the coast in longshore transport (**Figure 10.20**). They are constructed of many types of material, but the most common is large blocks of rocky material called **rip-rap**. Sometimes groins are even constructed of sturdy wood pilings (similar to a fence built out into the ocean).

Although a groin traps sand on its *upcoast side*, erosion occurs immediately downcoast of the groin because the sand that is normally found just downcoast of the groin is trapped on the groin's upcoast side. To lessen the erosion, another

---

[5]0.005 meter is equal to 5 millimeters (0.2 inch).

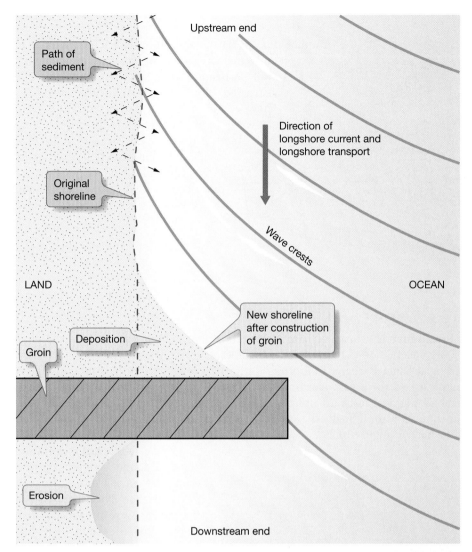

Figure 10.20 Interference of sand movement. Hard stabilization like the groin shown here interferes with the movement of sand along the beach, causing deposition of sand upcoast of the groin and erosion immediately downcoast, modifying the shape of the beach.

groin can be constructed downcoast, which in turn also creates erosion downcoast from it. More groins are needed to alleviate the beach erosion, and soon a **groin field** is created (Figure 10.21).

Does a groin (or a groin field) actually retain more sand on the beach? Sand eventually migrates around the end of the groin, so there is no additional sand on the beach; it is only *distributed differently*. With proper engineering and by taking into account the regional sand transport budget and seasonal wave activity, an equilibrium may be reached that allows sufficient sand to move along the coast before excessive erosion occurs downcoast from the last groin. However, some serious erosional problems have developed in many areas because of attempts to stabilize sand on the beach through the excessive use of groins.

Figure 10.21 Groin field. A series of groins has been built along the shoreline north of Ship Bottom, New Jersey, in an attempt to trap sand, altering the distribution of sand on the beach. The view is toward the north, and the primary direction of longshore current is toward the bottom of the photo (toward the south).

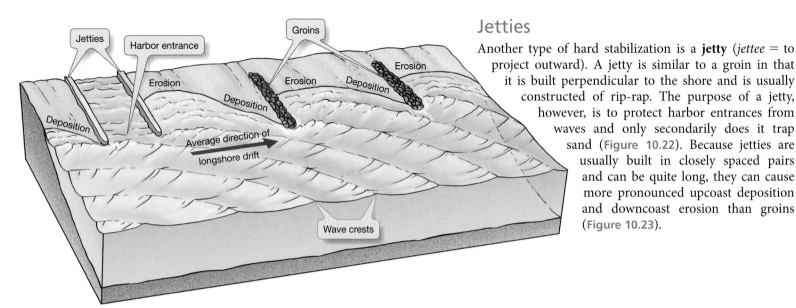

Figure 10.22 Effect of jetties and groins. Jetties protect a harbor entrance and usually occur in pairs. Groins are built specifically to trap sand moving in the longshore transport system and occur individually or as a groin field. Both structures cause deposits of sand on their upcoast sides and an equal amount of erosion downcoast.

Jetties

Another type of hard stabilization is a **jetty** (*jettee* = to project outward). A jetty is similar to a groin in that it is built perpendicular to the shore and is usually constructed of rip-rap. The purpose of a jetty, however, is to protect harbor entrances from waves and only secondarily does it trap sand (**Figure 10.22**). Because jetties are usually built in closely spaced pairs and can be quite long, they can cause more pronounced upcoast deposition and downcoast erosion than groins (**Figure 10.23**).

Breakwaters

A **breakwater** is a form of hard stabilization that is built parallel to a shoreline (**Figure 10.24**). **Figure 10.25** shows a breakwater that was constructed to create the harbor at Santa Barbara, California. California's longshore drift is predominantly southward, so the breakwater on the western side of the harbor accumulated sand that had migrated eastward along the coast. The beach to the west of the harbor continued to grow until finally the sand moved around the breakwater and began to fill in the harbor (Figure 10.25).

While abnormal deposition occurred to the west, erosion proceeded at an alarming rate east of the harbor. The waves east of the harbor were no greater than before, but the sand that had formerly moved down the coast was now trapped behind the breakwater.

A similar situation occurred in Santa Monica, California, where a breakwater was built to provide a boat anchorage. A bulge in the beach soon formed behind (inshore of) the breakwater, and severe erosion occurred downcoast (**Figure 10.26**). The breakwater interfered with the natural transport of sand by blocking the waves that used to keep the sand moving. If something was not done to put energy back into the system, the breakwater would soon be attached by a tombolo of sand, and further erosion downcoast might destroy coastal structures.

In Santa Barbara and Santa Monica, dredging was used to compensate for erosion downcoast from the breakwater and to

Figure 10.23 Jetties at Santa Cruz Harbor, California. These jetties protect the inlet to Santa Cruz Harbor and interrupt the flow of sand, which is toward the right (southward). Notice the buildup of sand to the left (upcoast) of the jetties and the corresponding erosion to the right (downcoast).

keep the harbor or anchorage from filling with sand. Sand dredged from behind the breakwater is pumped down the coast so it can reenter the longshore drift and replenish the eroded beach.

The dredging operation has stabilized the situation in Santa Barbara, but at considerable (and ongoing) expense. In Santa Monica, dredging was conducted until the breakwater was largely destroyed during winter storms in 1982–1983. Shortly thereafter, wave energy was able to move sand along the coast again, and the system was restored to normal conditions. When people interfere with natural processes in the coastal region, they must provide the energy needed to replace what they have misdirected through modification of the shore environment.

Seawalls

One of the most destructive types of hard stabilization is a **seawall** (Figure 10.27), which is built parallel to the shore, along the landward side of the berm. The purpose of a seawall is to armor the coastline and protect landward developments from ocean waves.

Once waves begin breaking against a seawall, however, turbulence generated by the abrupt release of wave energy quickly erodes the sediment on its seaward side, which can eventually cause it to collapse into the surf (Figure 10.27). In many cases where seawalls have been used to protect property on barrier islands, the seaward slope of the island beach has steepened and the rate of erosion has increased, causing the destruction of the recreational beach.

A well-designed seawall may last for many decades, but the constant pounding of waves eventually takes its toll (Figure 10.28). In the long run, the cost of repairing or replacing seawalls will be more than the property is worth, and the sea will claim more of the coast through the natural processes of erosion. It's just a matter of time for homeowners who live too close to the coast, many of whom are gambling that their houses won't be destroyed in their lifetimes.

Figure 10.24 Breakwater at Nea Fokea, northern Greece. Breakwaters are built parallel to the shore and are composed of rocky, blocky material that is piled up a meter or so (several feet) above sea level. They are designed to reduce wave energy, thus creating a protected area of quiet water inshore of (behind) the breakwater.

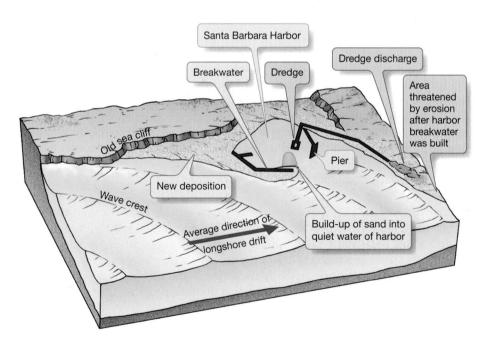

Figure 10.25 Breakwater at Santa Barbara Harbor, California. Construction of a shore-connected breakwater at Santa Barbara Harbor interfered with the longshore drift, creating a broad beach. As the beach extended around the breakwater into the harbor, the harbor was in danger of being closed off by accumulating sand. As a result, dredging operations were initiated to move sand from the harbor downcoast, where it helped reduce coastal erosion.

Web Animation
Coastal Stabilization Structures

Figure 10.26
Breakwater at Santa Monica, California. Paired aerial photos of the Santa Monica pier and shoreline **(a)** before a breakwater was built and **(b)** after the breakwater, showing how the breakwater caused a bulge of sand on the beach. After the breakwater was destroyed by waves in 1983, the bulge disappeared and the shoreline returned to a straight shoreline. North is to the right in both images.

(a) The shoreline and pier at Santa Monica as it appeared in September 1931, before the breakwater was constructed in 1933. Note that the pier is on stilts and thus does not affect longshore transport.

(b) The same area in 1949, showing that the construction of the breakwater to create a boat anchorage disrupted the longshore transport of sand and caused a bulge of sand on the beach in the wave shadow behind the breakwater.

Figure 10.27 Seawalls and beaches.
Diagrammatic sequence of the negative consequences that can occur when a seawall is built along a barrier island beach to protect beachfront property.

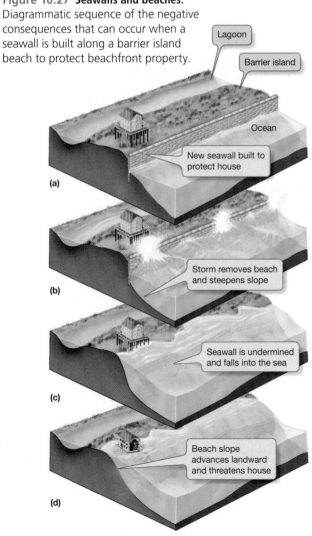

Alternatives to Hard Stabilization

Is it a good idea to preserve the houses of a few people who have built too close to the shore by armoring the coast with hard stabilization even though it destroys the recreational beach? If you own coastal property, your response would probably be different from the response of the general beachgoing public. Because hard stabilization has been shown to have negative environmental consequences, alternatives have been sought.

CONSTRUCTION RESTRICTIONS   One of the simplest alternatives to the use of hard stabilization is to restrict construction in areas prone to coastal erosion. Unfortunately, this is becoming less and less an option as coastal regions experience population increases and governments increase the risk of damage and injuries because of programs like the *National Flood Insurance Program* (*NFIP*). Since its inception in 1968, NFIP has paid out billions of dollars in federal subsidies to repair or replace high-risk coastal structures. As a result, NFIP has actually *encouraged* construction in exactly the unsafe locations it was designed to prevent![6] Further, many homeowners spend large amounts of money rebuilding structures and fortifying their property.

BEACH REPLENISHMENT   Another alternative to hard stabilization is **beach replenishment** (also called **beach nourishment**), in which sand is added to the beach to replace lost sediment (**Figure 10.29**). Although rivers naturally supply sand to most beaches, dams on rivers restrict the sand supply that would normally arrive at beaches. When inland dams are built, their effects on beaches far downstream are rarely considered. It's not until beaches begin disappearing that the rivers are seen as parts of much larger systems that operate along the coast.

Beach replenishment is expensive, however, because huge volumes of sand must be continually supplied to the beach. The cost of beach replenishment depends on the type and quantity of material placed on the beach, how far the material must be transported, and how it is to be distributed on the beach. Most sand used for replenishment comes from offshore areas, but sand that is dredged from nearby rivers, drained dams, harbors, and lagoons is also used.

---

[6]Recent changes in regulations of the Federal Emergency Management Agency (FEMA), which oversees NFIP, are intended to curb this practice.

Figure 10.28 Seawall damage. A seawall in Solana Beach, California, that has been damaged by waves and needs repair. Although seawalls appear to be sturdy, they can be destroyed by the continual pounding of high-energy storm waves.

The average cost of sand used to replenish beaches is between $5 and $10 per 0.76 cubic meter (1 cubic yard). In comparison, a typical top-loading trash dumpster holds about 2.3 cubic meters (3 cubic yards) of material, and a typical dump truck holds a volume of about 45 cubic meters (60 cubic yards). The drawbacks of beach replenishment projects are that a huge volume of sand must be moved and that new sand must be supplied on a regular basis. These problems often cause replenishment projects to exceed the monetary limits of what can be reasonably accomplished. For example, a small beach replenishment project of several hundred cubic meters can cost around $10,000 per year. Larger projects—requiring several thousand cubic meters of sand—cost several million dollars per year.

RELOCATION   U.S. coastal policy has recently shifted from defending coastal property in high-hazard areas to removing structures and letting nature reclaim the beach. This approach, called **relocation**, involves moving structures to safer locations as they become threatened by erosion. One example of the successful use of this technique is the relocation of the Cape Hatteras Lighthouse in North Carolina (see **Web Diving Deeper 10.1**). Relocation, if done wisely, can allow humans to live in balance with the natural processes that continually modify beaches.

Dredged sand exiting pipe

Figure 10.29 Beach replenishment. Beach replenishment projects, such as this one in Carlsbad, California, are used to widen beaches. Beach replenishment involves dredging sand from offshore or coastal locations, pumping it through a pipe (*lower right*), and spreading it across the beach.

CONCEPT CHECK 10.6

1. List the types of hard stabilization and describe what each is intended to do.

2. Overall, does a groin add any additional sand to the beach? Explain.

3. Why do groins often multiply to form a groin field?

4. When a breakwater was built in Santa Monica, what unexpected problem occurred? What was done to alleviate the problem (before the breakwater was destroyed by waves)?

5. Describe alternatives to hard stabilization, including potential drawbacks of each.

ESSENTIAL CONCEPT 10.6

Hard stabilization includes groins, jetties, breakwaters, and seawalls, all of which alter the coastal environment and result in changes in the shape of the beach. Alternatives to hard stabilization include construction restrictions, beach replenishment, and relocation.

Essential Concepts Review

10.1 How are coastal regions defined?

▶ *The coastal region changes continuously.* The *shore* is the region of contact between the oceans and the continents, lying between the lowest low tides and the highest elevation on the continents affected by storm waves. The *coast* extends inland from the shore as far as marine-related features can be found. The *coastline* marks the boundary between the shore and the coast. The shore is divided into the *foreshore*, extending from low tide to high tide, and the *backshore*, extending beyond the high tide line to the coastline. Seaward of the low tide shoreline are the *nearshore* zone, extending to the breaker line, and the *offshore* zone beyond.

▶ *A beach is a deposit of the shore area*, consisting of wave-worked sediment that moves along a wave-cut bench. It includes the *recreational beach*, *berm*, *beach face*, *low tide terrace*, *longshore trough*, and one or more *longshore bars*. Beaches are composed of whatever material is locally available.

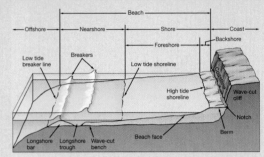

Study Resources
Online Study Guide Quizzes

Critical Thinking Question

To help reinforce your knowledge of beach terminology, construct and label your own diagram similar to Figure 10.1 from memory.

10.2 How does sand move on the beach?

▶ *Waves that break at the shore move sand perpendicular to shore* (toward and away from shore). In *light wave activity, swash dominates the transport system*, and sand is moved up the beach face toward the berm. In *heavy wave activity, backwash dominates the transport system*, and sand is moved down the beach face, away from the berm and toward longshore bars. In a natural system, there is a *balance between light and heavy wave activity*, alternating between sand piled on the berm (*summertime beach*) and sand stripped from the berm (*wintertime beach*), respectively.

▶ *Sand is moved parallel to the shore, too.* Waves breaking at an angle to the shore create a *longshore current* that results in a zigzag movement of sediment called *longshore drift* (*longshore transport*). Each year, millions of tons of sediment are moved from upcoast to downcoast ends of beaches. Most of the year, longshore drift moves southward along both the Pacific and Atlantic coasts of the United States.

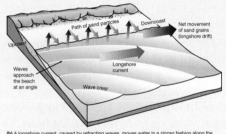

(b) A longshore current, caused by refracting waves, moves water in a zigzag fashion along the shoreline. This causes a net movement of sand grains (longshore drift) from upcoast to downcoast ends of a beach.

Study Resources
Online Study Guide Quizzes, Web Animations, Web Videos

Critical Thinking Question

How is the flow of water in a stream similar to a longshore current? How are the two different?

10.3 What features exist along erosional and depositional shores?

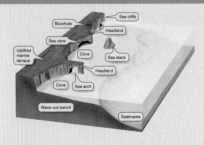

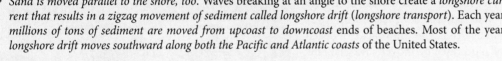

▶ *Erosional shores are characterized by headlands, wave-cut cliffs, sea caves, sea arches, sea stacks, and marine terraces* (caused by uplift of a wave-cut bench). Wave erosion increases as more of the shore is exposed to the open ocean, tidal range decreases, and bedrock weakens.

▶ *Depositional shores are characterized by beaches, spits, bay barriers, tombolos, barrier islands, deltas, and beach compartments.* Viewed from ocean side to lagoon side, barrier islands commonly have an ocean beach, dunes, barrier flat, and salt marsh. Deltas form at the mouths of rivers that carry more sediment to the ocean than the longshore current can carry away. *Beach starvation* occurs in beach compartments and other areas where the sand supply is interrupted.

Study Resources
Online Study Guide Quizzes, Web Diving Deeper 10.1, Web Animations

Critical Thinking Question

Describe the characteristics and coastal features that differentiate erosional and depositional shores.

10.4 How do changes in sea level produce emerging and submerging shorelines?

▶ *Shorelines are also classified as emerging or submerging, based on their position relative to sea level.* Ancient wave-cut cliffs and stranded beaches well above the present shoreline may indicate a drop in sea level relative to land. Old drowned beaches, submerged dunes, wave-cut cliffs, or drowned river valleys may indicate a rise in sea level relative to land. *Changes in sea level may result from tectonic processes causing local movement of the landmass or from eustatic processes changing the amount of water in the oceans or the capacity of ocean basins.* Melting of continental ice caps and glaciers during the past 18,000 years has caused a *eustatic rise in sea level of about 120 meters (400 feet).*

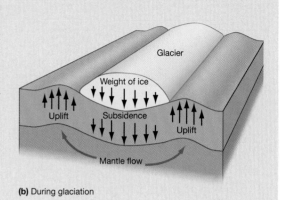

(b) During glaciation

Study Resources
Online Study Guide Quizzes, Web Animation

Critical Thinking Question

Describe the characteristics and coastal features that differentiate emerging and submerging shorelines.

10.5 What characteristics do U.S. coasts exhibit?

▶ *Sea level is rising along the Atlantic coast* about 0.3 meter (1 foot) per century, and the average erosion rate is 0.8 meter (2.6 feet) per year. *Along the Gulf coast,* sea level is rising 0.3 meter (1 foot) per century, and the average rate of erosion is 1.8 meters (6 feet) per year. The Mississippi River Delta is eroding at 4.2 meters (13.7 feet) per year, resulting in a large loss of wetlands every year. *Along the Pacific coast,* the average erosion rate is only 0.005 meter (0.016 foot) per year. Different shorelines erode at different rates, depending on wave exposure, amount of uplift, and type of bedrock.

Study Resources
Online Study Guide Quizzes

Critical Thinking Question

Compare the Atlantic coast, Gulf coast, and Pacific coast by describing the conditions and features of emergence/submergence and erosion/deposition that are characteristic of each.

MasteringOceanography™

www.masteringoceanography.com

Looking for additional review and test prep materials? Visit the Study Area in MasteringOceanography to enhance your understanding of this chapter's content by accessing a variety of resources, including Self-Study Quizzes, Geoscience Animations, RSS feeds, flashcards, Web links, and an optional Pearson eText.

10.6 How does hard stabilization affect coastlines?

▶ *Hard stabilization—such as groins, jetties, breakwaters, and seawalls— is often used in an attempt to stabilize a shoreline.* Groins (built to trap sand) and *jetties* (built to protect harbor entrances) widen the beach by trapping sediment on their upcoast side, but erosion usually becomes a problem downcoast. Similarly, *breakwaters* (built parallel to a shore) trap sand behind the structure but cause unwanted erosion downcoast. *Seawalls* (built to armor a coast) often cause loss of the recreational beach. Eventually, the constant pounding of waves destroys all types of hard stabilization.

▶ *Alternatives to hard stabilization* include *construction restrictions* in areas prone to coastal erosion, *beach replenishment* (*beach nourishment*), which is an expensive and temporary method of reducing beach starvation, and *relocation,* which is a technique that has been successfully used to protect coastal structures.

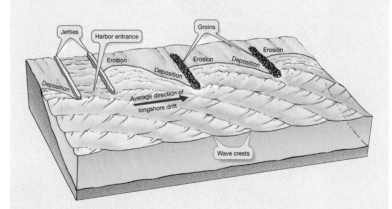

Study Resources
Online Study Guide Quizzes, Web Diving Deeper 10.1, Web Animation

Critical Thinking Question

Draw an aerial view of a shoreline to show the effect on erosion and deposition caused by constructing a groin, a jetty, a breakwater, and a seawall within the coastal environment.

The *Deepwater Horizon* oil drilling platform on fire in the Gulf of Mexico. In 2010, the *Deepwater Horizon* oil drilling platform exploded and caught fire in the Gulf of Mexico, producing the world's largest accidental oil spill.

Before you begin reading this chapter, use the glossary at the end of this book to discover the meanings of any of these words you don't already know:

Halocline Nonpoint source pollution
Wetland Eastern Pacific Garbage Patch Fjord
Salt marsh DDT Anoxic Bioamplification Mangrove swamp Bioassay
Minamata disease Estuarine circulation pattern Thermocline Law of the Sea
Toxic compound Coastal plain estuary Lagoon Sewage sludge
PCBs Nurdle
Bar-built estuary Pollution Estuary
Biomagnification

THE COASTAL OCEAN

ESSENTIAL CONCEPTS

At the end of this chapter, you should be able to:

11.1 Demonstrate an understanding of the laws that govern ocean ownership

11.2 Describe the characteristics that coastal waters exhibit

11.3 Compare the various types of coastal waters

11.4 Discuss the issues that face coastal wetlands

11.5 Explain how pollution is defined

11.6 Describe the main types of marine pollution

"Most people think of oceans as so immense and bountiful that it's difficult to imagine any significant impact from human activity. Now we've begun to recognize how much of an impact we do have."

—Jane Lubchenco, marine ecologist (2002)

The coastal ocean is a very busy place, filled with life, commerce, recreation, fisheries, and waste. Of the world fishery,[1] about 95% is obtained within 320 kilometers (200 miles) of shore. Coastal waters also support about 95% of the total mass of life in the oceans. Further, coastal estuary and wetland environments are among the most biologically productive ecosystems on Earth and serve as nursery grounds for many species of marine organisms that inhabit the open ocean. In addition, these waters are the focal point of most shipping routes, oil and gas production, and recreational activities.

Coastal waters are also the conduits through which land-derived compounds must pass to reach the open ocean. Numerous chemical, physical, and biological processes occur in these environments that tend to protect the quality of the water in the open ocean. The ocean has a tremendous ability to assimilate waste materials, yet negative results are beginning to be felt worldwide. Recently, the effects of cumulative stresses on the oceans have become large enough for humans to finally acknowledge the finiteness and fragility of the world environment. In the United States, for example, comprehensive reports such as those from the U.S. Commission on Ocean Policy and the Pew Oceans Commission have identified an emerging national crisis regarding damage being done to ocean and coastal resources, calling for a plan of action to restore coastal environments.

Earth's rapidly expanding human population has put ever-increasing stress on the marine environment. Human activities are increasingly altering coastal environments in two main ways: (1) the destruction of coastal ecosystems through development and exploitation and (2) the addition of land-based waste products into coastal waters. Pollution in coastal waters comes from many sources, such as accidental spills of petroleum, the accumulation of sewage, and certain chemicals (such as DDT, PCBs, and mercury). These pollutants—either alone or in combination with each other—often have severe deleterious effects on marine organisms.

In this chapter, we'll first examine the legal framework of who owns the ocean. Then we'll explore the properties of coastal waters before discussing how they're being affected and what can be done to reduce or eliminate pollution from coastal waters.

---

[1]The term *fishery* refers to fish caught by commercial fishers. Marine fisheries are discussed in Chapter 13, "Biological Productivity and Energy Transfer."

11.1 What Laws Govern Ocean Ownership?

Who owns the ocean? Who owns the sea floor? If a company wanted to drill for oil offshore between two different countries, would it have to obtain permission from either country? The current extraction of minerals and petroleum on the coastal sea floor necessitates laws that unambiguously answer these questions. Exploration and development of marine resources is also occurring further offshore, far from the jurisdiction of any country. Furthermore, overfishing and pollution are worsening. Are these kinds of problems covered by long-established laws? The answer is yes . . . and no.

Mare Liberum and the Territorial Sea

In 1609, Hugo Grotius, a Dutch jurist and scholar whose writings eventually helped formulate international law, urged freedom of the seas to all nations in his treatise *Mare Liberum* (*mare* = sea, *liberum* = free), which was premised on the assumption that the sea's major known resource—fish—exists in inexhaustible supply. Nevertheless, controversy continued over whether nations could control a *portion* of an ocean, such as the ocean adjacent to a nation's coastline.

Dutch jurist Cornelius van Bynkershoek attempted to solve this problem in *De Dominio Maris* (*de* = of, *dominio* = domain, *maris* = sea), published in 1702. It provided for national domain over the sea out to the distance that could be protected by cannons from the shore, an area called the **territorial sea**. Just how far from shore did the territorial sea extend? The British had determined in 1672 that cannon range extended 1 league (3 nautical miles) from shore. Thus, every country with a coastline maintained ownership over a *3-mile territorial limit* from shore.

Law Of the Sea

In response to new technology that facilitated mining the ocean floor, the first **United Nations Conference on the Law of the Sea**, held in 1958 in Geneva, Switzerland, established that prospecting and mining of minerals on the continental shelf was under the control of the country that owned the nearest land. Because the continental shelf is the portion of the sea floor extending from the coastline to where the slope markedly increases, the seaward limit of the shelf is subject to interpretation. Unfortunately, the continental shelf was not well defined in the treaty, which led to disputes. In 1960, the second United Nations Conference on the Law of the Sea was also held in Geneva, but it made little progress toward an unambiguous and fair treaty concerning ownership of the coastal ocean.

Meetings of the third Law of the Sea Conference were held during 1973–1982. A new Law of the Sea treaty was adopted by a vote of 130 to 4, with 17 abstentions. Most developing nations that could benefit significantly from the treaty voted to adopt it. The United States, Turkey, Israel, and Venezuela opposed the new treaty because of concerns that it made sea floor mining unprofitable. The abstaining countries included the Soviet Union, Great Britain, Belgium, the Netherlands, Italy, and West Germany, all of which were interested in sea floor mining, too. Nevertheless, the treaty was ratified by the required 60th nation in 1993, establishing it as international law.

Meanwhile, in the early 1990s, the United States and other nations negotiated an agreement to the Law of the Sea that fundamentally overhauled the deep sea floor mining provisions of the 1982 convention. The agreement, adopted in 1994, modified the convention to include the following provisions: (1) Eliminate production controls on sea floor mining, (2) scale back the structure of the organization that administers deep seabed mining, (3) ensure that the United States has

Figure 11.1 Territorial sea. The territorial sea, which extends 3 nautical miles from the nearest shore, was originally determined by the reach of cannons fired from land. Cannons shown are in Korcula, Croatia.

a permanent seat and political say on any amendment to deep sea floor mining provisions, (4) eliminate the objectionable provisions on mandatory technology transfer, and (5) provide assured access for any future qualified miners. Even with the successful negotiation of these provisions, the United States did not ratify the modified treaty because of lingering concerns about seabed mining.

Despite the fact that 162 coastal nations—including all the countries in the European Community—have ratified the treaty and have made it the definitive word on coastal law, the United States has still not officially ratified the Law of the Sea treaty. Moreover, this means that the United States doesn't have the legal right to extend its maritime claims or hold a seat on the commission that reviews the plans of seabed mining.

Recently, the United States has shown renewed interest in ratifying the Law of the Sea treaty. In 2004 and 2007, for example, U.S. Senate committees voted for ratifying the treaty, only to have a vote by the full Senate blocked both times by senators who were concerned about U.S. access to the seabed, potential loss of U.S. sovereignty, and other factors.

In 2012, the U.S. Senate re-examined joining the Law of the Sea Convention and made it a top priority for the United States, stating that the convention sets forth a comprehensive legal framework governing uses of the oceans and protects and advances a broad range of U.S. interests, including U.S. national security and economic interests. In fact, current and past presidential administrations (both Republican and Democratic, which demonstrates broad bipartisan support), the U.S. military, and relevant industry and other groups all strongly support joining the convention. Experts agree that only by joining the convention can the United States maximize legal certainty and have a defined path to securing international recognition of its continental shelf.

The Law of the Sea treaty specifies how coastal nations watch over their natural resources, settle maritime boundary disputes, and—especially in the Arctic—extend their rights to any riches on or beneath the adjacent sea floor. The four main components of the treaty are:

1. **Coastal nations jurisdiction.** The treaty established a uniform 19-kilometer (12-mile) territorial sea and a 370-kilometer (200-nautical-mile) **exclusive economic zone (EEZ)** from all land (including islands) within a nation. Each of the 151 coastal nations has jurisdiction over mineral resources, fishing, and pollution regulation within its EEZ. If the continental shelf (defined geologically) exceeds the 370-kilometer EEZ, the EEZ is extended to 648 kilometers (350 nautical miles) from shore. Although the United States is not yet a party to the treaty, it is actively exploring and mapping its extended continental shelf.

2. **Ship passage.** The right of free passage for all vessels on the high seas is preserved. The right of free passage is also provided within territorial seas and through straits used for international navigation.

3. **Deep-ocean mineral resources.** Private exploitation of sea floor resources may proceed under the regulation of the International Seabed Authority (ISA), within which a mining company will be strictly controlled by the United Nations. This provision required mining companies to fund two mining operations—their own and one operated by the regulatory United Nations. As discussed above, some industrialized nations still oppose this portion of the law even though it was modified by an agreement to the convention in 1994 that eliminated some of the restrictive regulatory components.

4. **Arbitration of disputes.** A United Nations Law of the Sea tribunal will arbitrate any disputes in the treaty or disputes concerning ownership rights.

The Law of the Sea treaty puts 42% of the world's oceans under the control of coastal nations. The EEZ of the United States consists of about 11.5 million square

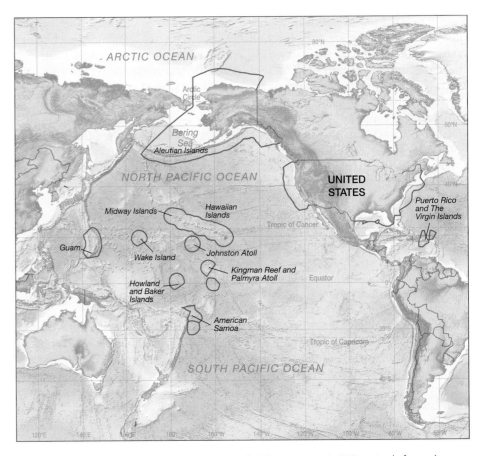

Figure 11.2 The U.S. exclusive economic zone (EEZ). A country's EEZ extends from shore to a distance of 370 kilometers (200 nautical miles) from the continent or islands. If the continental shelf (defined geologically) exceeds the 370-kilometer EEZ, the EEZ is extended to 648 kilometers 350 nautical miles) from shore.

ESSENTIAL CONCEPT 11.1

Ownership of the ocean and sea floor is regulated by the internationally ratified Law of the Sea treaty, which gives nations control of waters immediately adjacent to their coasts.

kilometers (4.2 million square miles) (Figure 11.2), which is about 30% more than the entire land area of the United States and its territories. This huge offshore area is widely believed to have tremendous economic potential.

CONCEPT CHECK 11.1

1 What was the basis for establishing the 3-mile territorial limit of ownership from shore?

2 Create a timeline outlining the development of the Law of the Sea treaty. What major country has not yet adopted the treaty?

3 Describe the four main components of the Law of the Sea treaty.

11.2 What Characteristics Do Coastal Waters Exhibit?

Coastal waters are the relatively shallow-water areas that adjoin continents or islands. If the continental shelf is broad and shallow, coastal waters can extend several hundred kilometers from land. If the continental shelf has significant relief or drops rapidly onto the deep-ocean basin, on the other hand, coastal waters will occupy a relatively thin band near the margin of the land. Beyond coastal waters lies the *open ocean*.

Because of their proximity to land, coastal waters are directly influenced by processes that occur on or near land. River runoff and tidal currents, for example, have a far more significant effect on coastal waters than on the open ocean.

Salinity

Freshwater is less dense than seawater, so river runoff does not mix well with seawater along the coast. Instead, the freshwater forms a wedge at the surface, which creates a well-developed **halocline**[2] (**Figure 11.3a**). When water is shallow enough, however, tidal mixing causes freshwater to mix with seawater, thus reducing the salinity of the water column (**Figure 11.3c**). There is no halocline here; instead, the water column is **isohaline** (*iso* = same, *halo* = salt).

Freshwater runoff from the continents generally lowers the salinity of coastal regions compared to the open ocean. Where precipitation on land is mostly rain, river runoff peaks in the rainy season. Where runoff is due mainly to melting snow and ice, on the other hand, runoff always peaks in summer.

Prevailing offshore winds can increase the salinity in some coastal regions. As winds travel over a continent, they usually lose most of their moisture. When these dry winds reach the ocean, they typically evaporate considerable amounts of water as they move across the surface of the coastal waters. The increased evaporation rate increases surface salinity, creating a halocline (**Figure 11.3b**). The gradient of the halocline, however, is reversed compared to the one developed from the input of freshwater (Figure 11.3a).

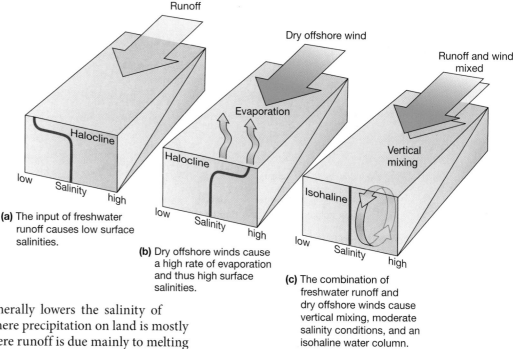

(a) The input of freshwater runoff causes low surface salinities.

(b) Dry offshore winds cause a high rate of evaporation and thus high surface salinities.

(c) The combination of freshwater runoff and dry offshore winds cause vertical mixing, moderate salinity conditions, and an isohaline water column.

Figure 11.3 Salinity variation in the coastal ocean. Diagrammatic views showing various factors that affect the salinity of coastal oceans. Red curves represent vertical salinity profiles.

Temperature

In low-latitude coastal regions, where circulation with the open ocean is restricted, surface waters are prevented from mixing thoroughly, so sea surface temperatures may approach 45°C (113°F) (**Figure 11.4a**). Alternatively, sea ice forms in many high-latitude coastal areas where water temperatures are uniformly cold—generally lower than −2°C (28.4°F) (**Figure 11.4b**). In both low- and high-latitude coastal waters, **isothermal** (*iso* = same, *thermo* = heat) conditions prevail.

Surface temperatures in middle-latitude coastal regions are coolest in winter and warmest in late summer. A strong **thermocline**[3] may develop from surface water being warmed during the summer (**Figure 11.4c**) and cooled during the winter (**Figure 11.4d**). In summer, very high-temperature surface water may form a relatively thin layer. Vertical mixing reduces the surface temperature by distributing the heat through a greater volume of water, thus pushing the thermocline deeper and making it less pronounced. In winter, cooling increases the density of surface water, which causes it to sink.

Prevailing offshore winds can significantly affect surface water temperatures. These winds are relatively warm during the summer, so they increase the ocean surface temperature and seawater evaporation. During winter, they are much cooler than the ocean surface, so they absorb heat and cool surface water near

---

[2]Recall that a halocline (*halo* = salt, *cline* = slope) is a layer of rapidly changing salinity, as discussed in Chapter 5.

[3]Recall that a *thermocline* (*thermo* = heat, *cline* = slope) is a layer of rapidly changing temperature, as discussed in Chapter 5.

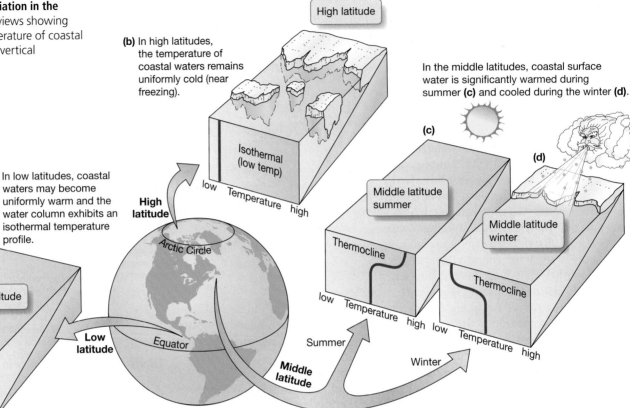

Figure 11.4 Temperature variation in the coastal ocean. Diagrammatic views showing how latitude affects the temperature of coastal oceans. Red curves represent vertical temperature profiles.

(a) In low latitudes, coastal waters may become uniformly warm and the water column exhibits an isothermal temperature profile.

(b) In high latitudes, the temperature of coastal waters remains uniformly cold (near freezing).

In the middle latitudes, coastal surface water is significantly warmed during summer **(c)** and cooled during the winter **(d)**.

shore. Mixing from strong winds may drive the thermoclines in Figures 11.4c and 11.4d deeper and even mix the entire water column, producing isothermal conditions. Tidal currents can also cause considerable vertical mixing in shallow coastal waters.

Coastal Geostrophic Currents

Recall from Chapter 7 that *geostrophic* (*geo* = earth, *strophio* = turn) *currents* move in a circular path around the middle of a current gyre. Wind and runoff create geostrophic currents in coastal waters, too, where they are called **coastal geostrophic currents**.

Where winds blow in a certain direction parallel to a coastline, they transport water toward the coast, where it piles up along the shore. Gravity eventually pulls this water back toward the open ocean. As it runs downslope away from the shore, the Coriolis effect causes it to curve to the right in the Northern Hemisphere and to the left in the Southern Hemisphere. Thus, in the Northern Hemisphere, the coastal geostrophic current curves *northward* on the western coast and *southward* on the eastern coast of continents. These currents are reversed in the Southern Hemisphere.

A high-volume runoff of freshwater produces a surface wedge of freshwater that slopes away from the shore (**Figure 11.5**). This causes a surface flow of low-salinity water toward the open ocean, which the Coriolis effect curves to the right in the Northern Hemisphere and to the left in the Southern Hemisphere.

Coastal geostrophic currents are variable because they depend on the wind and the amount of runoff for their strength. If the wind is strong and the volume of runoff is high, then the currents are relatively strong. They are bounded on the ocean side by the steadier eastern or western boundary currents of subtropical gyres.

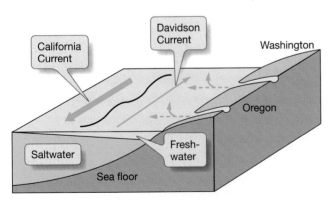

Figure 11.5 Davidson coastal geostrophic current. The Davidson Current is a coastal geostrophic current that flows north along the coast of Washington and Oregon. During the winter, runoff produces a freshwater wedge (*light blue*) that thins away from shore. This causes a surface flow of low-salinity water toward the open ocean, which is acted upon by the Coriolis effect and curves to the right.

An example of a coastal geostrophic current is the **Davidson Current** that develops along the coast of Washington and Oregon during winter (Figure 11.5). Heavy precipitation (which produces high volumes of runoff) combines with strong southwesterly winds to produce a relatively strong northward-flowing current. It flows between the shore and the southward-flowing California Current.

ESSENTIAL CONCEPT 11.2

The shallow coastal ocean adjoins land and experiences changes in salinity and temperature that are more dramatic than the open ocean. Coastal geostrophic currents can also develop.

CONCEPT CHECK 11.2

1 For coastal oceans where deep mixing does not occur, discuss the effect that offshore winds and freshwater runoff will have on salinity distribution. How will the winter and summer seasons affect the temperature distribution in the water column?

2 For the low, high, and middle latitudes, describe the temperature variation in the coastal ocean. In each location, explain why a thermocline exists or why the water column is isothermal.

3 Describe how coastal runoff of low-salinity water produces a coastal geostrophic current and give a specific location where a coastal geostrophic current can be found.

11.3 What Types Of Coastal Waters Exist?

The most important types of coastal waters include estuaries, lagoons, and marginal seas.

Estuaries

An **estuary** (*aestus* = tide) is a partially enclosed coastal body of water in which freshwater runoff from a river dilutes the input of salty ocean water. Estuaries are marine environments whose pH, salinity, temperature, and water levels vary, depending on the mixing between the river that feeds the estuary and the ocean from which it derives its salinity. The most common example of an estuary is a river mouth, where a river empties into the sea. Other coastal bodies of water such as bays, inlets, gulfs, and sounds are considered estuaries, too.

The mouths of large rivers form the most economically significant estuaries because many are seaports, centers of ocean commerce, and important commercial fisheries. Examples include Baltimore, New York, San Francisco, Buenos Aires, London, Tokyo, and many others.

ORIGIN OF ESTUARIES   The estuaries of today exist because sea level has risen approximately 120 meters (400 feet) since major continental glaciers began melting about 18,000 years ago. As described in Chapter 10, these glaciers covered portions of North America, Europe, and Asia during the Pleistocene Epoch, which is also referred to as the *Ice Age*.

Four major types of estuaries can be identified based on their geologic origin (Figure 11.6):

1. A **coastal plain estuary** forms as sea level rises and floods existing river valleys. These estuaries, such as the Chesapeake Bay in Maryland and Virginia, are called *drowned river valleys* (Figure 11.6a).

2. A **fjord**[4] forms as sea level rises and floods a glaciated valley. Water-carved valleys have V-shaped profiles, but fjords are U-shaped valleys with steep

---

[4]The Norwegian term *fjord* is pronounced "FEE-yord" and means a long, narrow sea inlet bordered by steep cliffs.

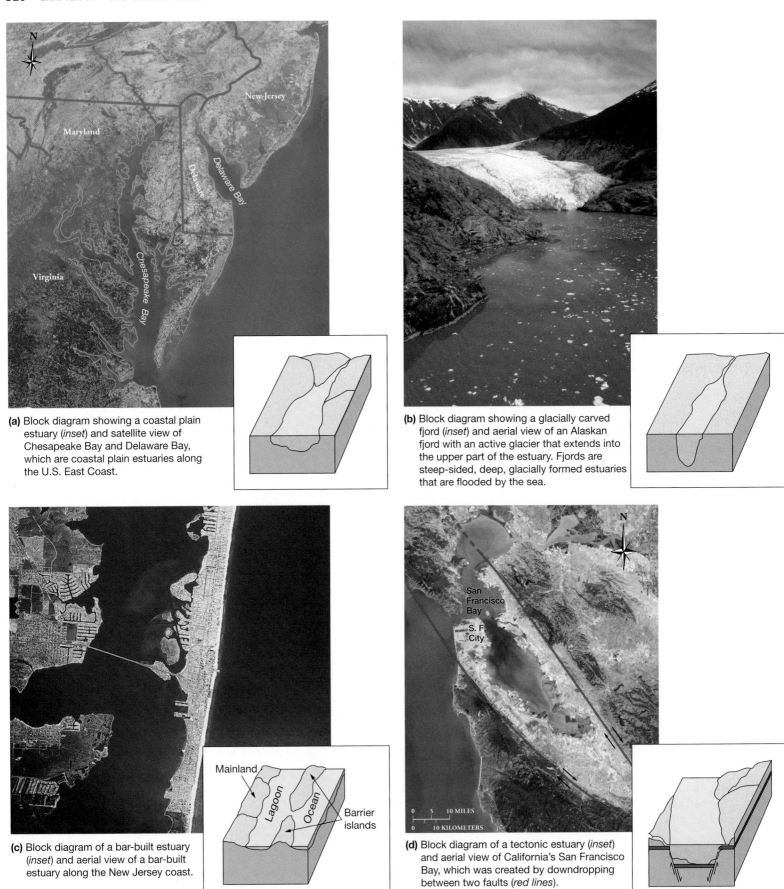

(a) Block diagram showing a coastal plain estuary (*inset*) and satellite view of Chesapeake Bay and Delaware Bay, which are coastal plain estuaries along the U.S. East Coast.

(b) Block diagram showing a glacially carved fjord (*inset*) and aerial view of an Alaskan fjord with an active glacier that extends into the upper part of the estuary. Fjords are steep-sided, deep, glacially formed estuaries that are flooded by the sea.

(c) Block diagram of a bar-built estuary (*inset*) and aerial view of a bar-built estuary along the New Jersey coast.

(d) Block diagram of a tectonic estuary (*inset*) and aerial view of California's San Francisco Bay, which was created by downdropping between two faults (*red lines*).

Figure 11.6 Classifying estuaries by geologic setting. Block diagrams and corresponding photos of the four types of estuaries, based on geologic setting.

walls. Commonly, a shallowly submerged glacial deposit of debris (called a *moraine*) is located near the ocean entrance, marking the farthest extent of the glacier. Fjords are common along the coasts of Alaska, Canada, New Zealand, Chile, and Norway (Figure 11.6b).

3. A **bar-built estuary** is shallow and is separated from the open ocean by sand bars that are deposited parallel to the coast by wave action. Lagoons that separate *barrier islands* from the mainland are bar-built estuaries. They are very common along the U.S. Gulf Coast and East Coast (see Figure 10.10). Examples include Laguna Madre in Texas and Pamlico Sound in North Carolina (Figure 11.6c).

4. A **tectonic estuary** forms when faulting or folding of rocks creates a restricted downdropped area into which the sea has flooded. California's San Francisco Bay is in part a tectonic estuary (Figure 11.6d), formed by movement along faults, including the San Andreas Fault.

WATER MIXING IN ESTUARIES   Because freshwater from a river is less dense than seawater, the basic flow pattern in an estuary is a surface flow of less dense freshwater toward the ocean and an opposite flow below the surface of salty seawater into the estuary. Mixing takes place where these two water masses are in contact with one another.

Based on the physical characteristics of the estuary and the resulting mixing of freshwater and seawater, estuaries are classified into one of four main types, as shown in **Figure 11.7**:

1. A **vertically mixed estuary** is a shallow, low-volume estuary where the net flow always proceeds from the head of the estuary toward its mouth. Salinity at any point in the estuary is uniform from surface to bottom because river water mixes evenly with ocean water at all depths. Salinity simply increases from the head to the mouth of the estuary, as shown in Figure 11.7a. Salinity lines curve at the edge of the estuary because the Coriolis effect influences the inflow of seawater.

2. A **slightly stratified estuary** is a somewhat deeper estuary in which salinity increases from the head to the mouth at any depth, as in a vertically mixed estuary. However, two water layers can be identified. One is the less-saline, less-dense upper water from the river, and the other is the more-saline, more-dense deeper water from the ocean. These two layers are separated by a zone of mixing. The circulation that develops in slightly stratified estuaries is a net surface flow of low-salinity water toward the ocean and a net subsurface flow of seawater toward the head of the estuary (Figure 11.7b), which is called an **estuarine circulation pattern**.

Figure 11.7 Classifying estuaries by mixing. Block diagrams of the four types of estuaries, based on mixing. Numbers represent salinity in ‰; arrows indicate flow directions.

3. A **highly stratified estuary** is a deep estuary in which upper-layer salinity increases from the head to the mouth, reaching a value close to that of open-ocean water. The deep-water layer has a rather uniform open-ocean salinity at any depth throughout the length of the estuary. An estuarine circulation pattern is well developed in this type of estuary (Figure 11.7c). Mixing at the interface of the upper water and the lower water creates a net movement from the deep-water mass into the upper water. Less-saline surface water simply moves from the head toward the mouth of the estuary, growing more saline as water from the deep mass mixes with it. Relatively strong haloclines develop at the contact between the upper and lower water masses.

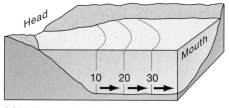

(a) Vertically mixed estuary

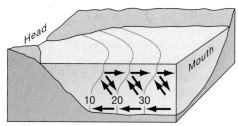

(b) Slightly stratified estuary

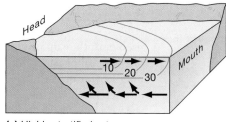

(c) Highly stratified estuary

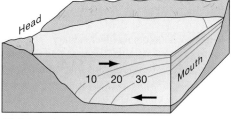

(d) Salt wedge estuary

4. A **salt wedge estuary** is an estuary in which a wedge of salty water intrudes from the ocean beneath the river water. This kind of estuary is typical of the mouths of deep, high-volume rivers. No horizontal salinity gradient exists at the surface because surface water is essentially fresh throughout the length of—and even beyond—the estuary (Figure 11.7d). There is, however, a *horizontal* salinity gradient at depth and a very pronounced vertical salinity gradient (a halocline) at any location throughout the length of the estuary. This halocline is shallower and more highly developed near the mouth of the estuary.

Within all estuaries, the predominant mixing pattern may vary with location, season, or tidal conditions. In addition, mixing patterns in real estuaries are rarely as simple as the models presented here.

ESTUARIES AND HUMAN ACTIVITIES   Estuaries are important breeding grounds and protective nurseries for many marine animals, so the ecological well-being of estuaries is vital to fisheries and coastal environments worldwide. Nevertheless, estuaries support shipping, logging, manufacturing, waste disposal, and other activities that can potentially damage the environment.

Estuaries are most threatened where human population is large and expanding, but they can be severely damaged where populations are still modest, too. Development in the Columbia River estuary, for example, demonstrates how a relatively small population can damage an estuary.

Columbia River Estuary   The Columbia River, which forms most of the border between Washington and Oregon, has a long salt-wedge estuary at its entrance to the Pacific Ocean (**Figure 11.8**). The strong flow of the river and tides drive a salt wedge as far as 42 kilometers (26 miles) upstream and raise the river's water level more than 3.5 meters (12 feet). When the tide falls, the huge flow of freshwater (up to 28,000 cubic meters [1,000,000 cubic feet] per second) creates a freshwater wedge that can extend hundreds of kilometers into the Pacific Ocean.

Most rivers create floodplains along their lower courses, which have rich soil that can be used for growing crops. In the late 19th century, farmers moved onto the floodplains to establish agriculture along the Columbia River. Eventually, protective dikes were built to prevent agricultural damage done by annual flooding. Flooding brings new nutrients, however, so the dikes deprived the floodplain of the nutrients necessary to sustain agriculture.

The river has been the principal conduit for the logging industry, which dominated the region's economy through most of its modern history. Fortunately, the river's ecosystem has largely survived the additional sediment caused by clear cutting by the logging industry. The construction of more than 250 dams along the river and its tributaries, on the other hand, has permanently altered the river's ecosystem. Many of these dams, for example, do not have salmon ladders, which help fish "climb" in short vertical steps around the dams to reach their spawning grounds at the headwaters of their home streams.

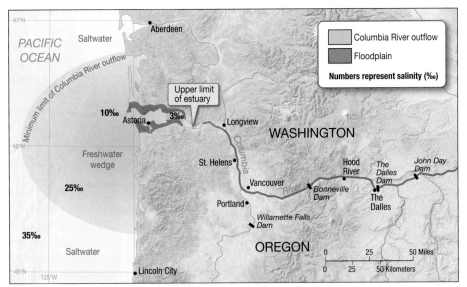

Figure 11.8 Columbia River estuary. The long estuary at the mouth of the Columbia River has been severely affected by interference with floodplains that have been diked, by logging activities, and—most significantly—by the construction of hydroelectric dams. The tremendous outflow of the Columbia River creates a large wedge of low-density freshwater that remains traceable far out at sea.

Even though the dams have caused a multitude of problems, they do provide flood control, electrical power, and a dependable source of water, all of which have become necessary to the region's economy. To aid shipping operations, the river receives periodic dredging of sediment, which brings an increased risk for pollution. If these kinds of problems have developed in such sparsely populated areas as the Columbia River estuary, then larger environmental effects must exist in more highly populated estuaries, such as the Chesapeake Bay.

ESSENTIAL CONCEPT 11.3A

Estuaries were formed by the rise in sea level after the last ice age. They can be classified based on geologic origin as coastal plain, fjord, bar-built, or tectonic estuaries. Estuaries can also be classified based on mixing as vertically mixed, slightly stratified, highly stratified, or salt wedge.

Chesapeake Bay Estuary Chesapeake Bay is about 320 kilometers (200 miles) long and 56 kilometers (35 miles) wide at its widest point, making it the largest (and best studied) estuary in the United States (Figure 11.9). It drains a watershed of about 166,000 square kilometers (64,000 square miles) spread over six states that includes a population of over 15 million people. The length of the bay's shoreline is an astonishing 17,700 kilometers (11,000 miles) because of all the inlets created by the 19 major rivers and 400 creeks and tributaries that flow into it. The bay formed when the lower parts of the Susquehanna River were drowned by rising sea level after the most recent ice age.

Chesapeake Bay is a slightly stratified estuary that experiences large seasonal changes in salinity, temperature, and dissolved oxygen. Figure 11.9a shows the estuary's average surface salinity, which increases oceanward. The salinity lines are oriented virtually north–south in the middle of the bay because of the Coriolis effect. Recall that the Coriolis effect causes flowing water to curve to the right in the Northern Hemisphere, so seawater entering the bay tends to hug the bay's *eastern* side, and freshwater flowing through the bay toward the ocean tends to hug its *western* side.

With maximum river flow in the spring, a strong halocline (and *pycnocline*[5]) develops, preventing the fresh surface water and saltier deep water from mixing. Beneath the pycnocline, which can be as shallow as 5 meters (16 feet), waters may become **anoxic** (*a* = without, *oxic* = oxygen) from May through August, as dead organic matter decays in the deep water (Figure 11.9b). Major kills of commercially important blue crab, oysters, and other bottom-dwelling organisms occur during this time.

The degree of stratification and extent of mortality of bottom-dwelling animals have increased since the early 1950s. Increased nutrients from sewage and agricultural fertilizers have been added to the bay during this time, too, which has increased the productivity of microscopic algae (algal blooms). When these organisms die, their remains accumulate as organic matter at the bottom of the bay and promote the development of anoxic conditions. In drier years with less river runoff, however, anoxic conditions aren't as widespread or severe in bottom waters (Figure 11.9c) because fewer nutrients are supplied.

Lagoons

Landward of barrier islands lie protected, shallow bodies of water called **lagoons** (see Figure 11.6c). Lagoons form in a bar-built type of estuary. Because of restricted circulation between lagoons and the ocean, three distinct zones can usually be identified within lagoons (Figure 11.10): (1) A *freshwater zone* that lies near the head of the lagoon where rivers enter, (2) a *transitional zone* of brackish[6] water that occurs near the middle of the lagoon, and (3) a *saltwater zone* that lies close to the lagoon's mouth.

Salinity within a lagoon is highest near the entrance and lowest near the head (Figure 11.10b). In latitudes that have seasonal variations in temperature and precipitation, ocean water flows through the entrance during a warm, dry summer to compensate for the volume of water lost through evaporation, thus increasing the salinity in the lagoon. Lagoons actually may become hypersaline[7] in arid regions, where evaporation rates are extremely high. Even though water flows into the lagoon from the open ocean to replace water lost by evaporation, the dissolved components do not evaporate and sometimes accumulate to extremely high levels. During the rainy season, the lagoon becomes much less saline as freshwater runoff increases.

---

[5]Recall that a *pycnocline* (*pycno* = density, *cline* = slope) is a layer of rapidly changing density, as discussed in Chapter 5. A pycnocline is caused by a change in temperature and/or salinity with depth.

[6]Brackish water is water with salinity between that of freshwater and seawater.

[7]Hypersaline conditions are created when water becomes excessively salty.

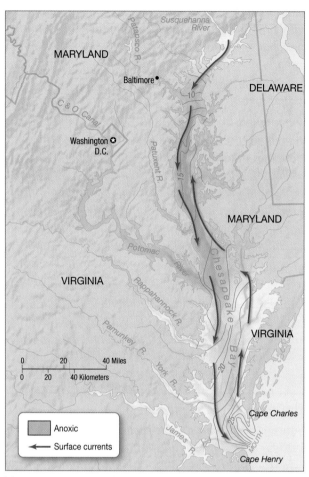

(a) Map of Chesapeake Bay, showing average surface salinity (*blue lines*) in ‰. The purple area in the middle of the bay represents anoxic (oxygen-poor) waters.

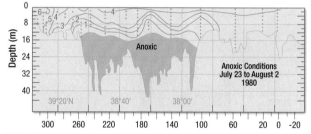

(b) Profile along length of Chesapeake Bay showing dissolved oxygen concentration (in ppm) during July-August 1980, indicating deep anoxic waters (*dark blue*).

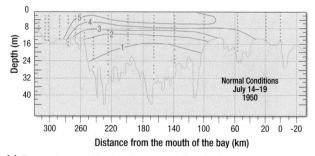

(c) Comparison profile showing normal dissolved oxygen concentration (in ppm) during July 1950.

Figure 11.9 Chesapeake Bay salinity and dissolved oxygen.

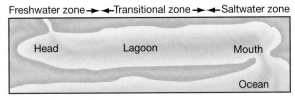

(a) Typical geometry

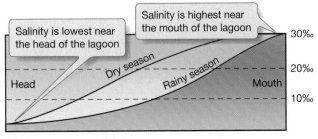

(b) Corresponding salinity

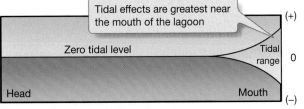

(c) Corresponding tidal effects

Figure 11.10 Lagoons. Diagrammatic representations of the general features of lagoons.

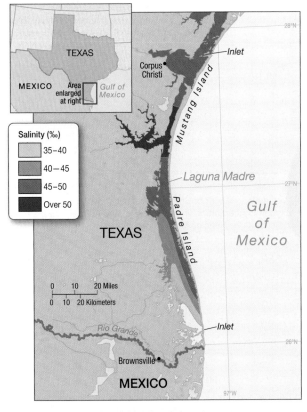

Figure 11.11 Laguna Madre summer surface salinity. Map showing geometry of Laguna Madre, Texas, and typical summer surface salinity (in ‰).

Tidal effects are greatest near the entrance to the lagoon (Figure 11.10c) and diminish inland from the saltwater zone until they are nearly undetectable in the freshwater zone.

LAGUNA MADRE   Laguna Madre is located along the Texas coast between Corpus Christi and the mouth of the Rio Grande (**Figure 11.11**). This long, narrow body of water is protected from the open ocean by Padre Island, a barrier island 160 kilometers (100 miles) long. The lagoon probably formed about 6000 years ago, as sea level approached its present height.

The tidal range of the Gulf of Mexico in this area is about 0.5 meter (1.6 feet). The inlets at each end of Padre Island are quite narrow (Figure 11.11), so there is very little tidal interchange between the lagoon and the open sea.

Laguna Madre is a hypersaline lagoon, and much of it is less than 1 meter (3.3 feet) deep. As a result, there are large seasonal changes in temperature and salinity. Water temperatures reach 32°C (90°F) in the summer and can dip below 5°C (41°F) in winter. Salinities range from 2‰ when infrequent local storms provide large volumes of freshwater to over 100‰ during dry periods. High evaporation generally keeps salinity well above 50‰.[8]

Because even salt-tolerant marsh grasses cannot withstand such high salinities, the marsh has been replaced by an open sand beach on Padre Island. At the inlets, ocean water flows in as a surface wedge *over* the denser water of the lagoon and water from the lagoon flows out as a *subsurface* flow, which is the exact opposite of a typical estuarine circulation pattern.

Marginal Seas

At the margins of the ocean are relatively large semi-isolated bodies of water called **marginal seas**. Most of these seas result from tectonic events that have isolated low-lying pieces of ocean crust between continents, such as the Mediterranean Sea, or are created behind volcanic island arcs, such as the Caribbean Sea. These waters are shallower than and have varying degrees of exchange with the open ocean, depending on climate and geography; as a result, salinities and temperatures are substantially different from those of typical open ocean seawater.

A CASE STUDY: THE MEDITERRANEAN SEA   The **Mediterranean** (*medi* = middle, *terra* = land) **Sea** is actually a number of small seas connected by narrow necks of water into one larger sea. It is the remnant of the ancient Tethys Sea that existed when all the continents were combined about 200 million years ago. It is more than 4300 meters (14,100 feet) deep and is one of the few inland seas in the world underlain by oceanic crust. Thick salt deposits and other evidence on the floor of the Mediterranean suggest that it nearly dried up about 6 million years ago, only to refill with a large saltwater waterfall (see Diving Deeper 4.1).

The Mediterranean is bounded by Europe and Asia Minor on the north and east and Africa on the south (**Figure 11.12a**). It is surrounded by land except for very shallow and narrow connections to the Atlantic Ocean through the Strait of Gibraltar (about 14 kilometers [9 miles] wide), and to the Black Sea through the Bosporus (roughly 1.6 kilometers [1 mile] wide). In addition, the Mediterranean Sea has a human-made passage to the Red Sea via the Suez Canal, a waterway 160 kilometers (100 miles) long that was completed in 1869. The Mediterranean Sea has a very irregular coastline, which divides it into subseas such as the Aegean Sea and Adriatic Sea, each of which has a separate circulation pattern.

An underwater ridge called a **sill**, which extends from Sicily to the coast of Tunisia at a depth of 400 meters (1300 feet), separates the Mediterranean into two major basins. This sill restricts the flow between the two basins, resulting in strong

---

[8]Recall that normal salinity in the open ocean averages 35‰.

Figure 11.12 Mediterranean Sea bathymetry and circulation.

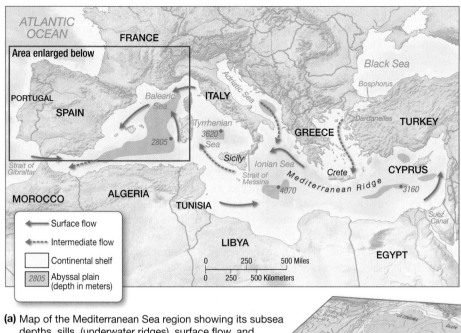

(a) Map of the Mediterranean Sea region showing its subsea depths, sills, (underwater ridges), surface flow, and intermediate flow.

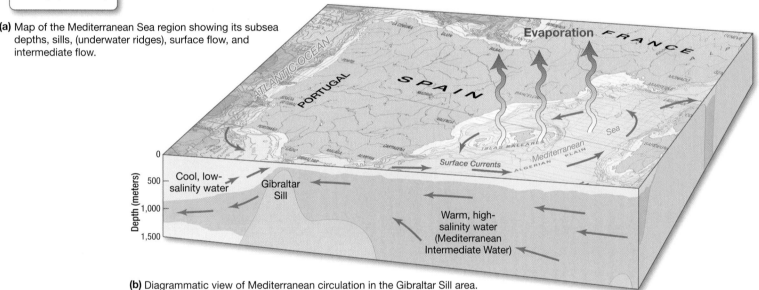

(b) Diagrammatic view of Mediterranean circulation in the Gibraltar Sill area.

currents that run between Sicily and the Italian mainland through the Strait of Messina (Figure 11.12a).

Mediterranean Circulation   The Mediterranean Sea has a very unusual and unique circulation pattern. This circulation is caused by the dry, intense heat of the Middle East, where a huge volume of water evaporates from the eastern Mediterranean and causes a tremendous surface inflow of Atlantic Ocean water through the Strait of Gibraltar to replace the evaporated water. In fact, the water level in the eastern Mediterranean is generally 15 centimeters (6 inches) lower than at the Strait of Gibraltar. The surface flow follows the northern coast of Africa throughout the length of the Mediterranean and spreads northward across the sea (Figure 11.12a).

The remaining Atlantic Ocean water continues eastward to Cyprus. During winter, it sinks to form what is called the *Mediterranean Intermediate Water*, which has a temperature of 15°C (59°F) and a salinity of 39.1‰. This water flows westward at a depth of 200 to 600 meters (660 to 2000 feet) and returns to the North Atlantic as a *subsurface* flow through the Strait of Gibraltar (Figure 11.12b).

By the time the Mediterranean Intermediate Water passes through Gibraltar, its temperature has dropped to 13°C (55°F) and its salinity to 37.3‰. It is still denser

STUDENTS SOMETIMES ASK . . .

How can Mediterranean Intermediate Water sink if it's so warm?

While it is true that warm water has low density, remember that *both* salinity and temperature affect sea-water density. In the case of the Mediterranean Intermediate Water, it has high enough salinity to increase its density despite being warm. Once its density increases enough, it sinks beneath the surface and retains its temperature and salinity characteristics as it flows out through the Strait of Gibraltar into the North Atlantic.

than even Antarctic Bottom Water and much denser than water at this depth in the Atlantic Ocean, so it moves down the continental slope. While descending, it mixes with Atlantic Ocean water and becomes less dense. At a depth of about 1000 meters (3300 feet), its density equals that of the surrounding Atlantic Ocean, so it spreads in all directions (Figure 11.12b), sometimes forming deep-ocean eddies that last for more than two years and can be detected by satellite as far north as Iceland.

Circulation between the Mediterranean Sea and the Atlantic Ocean is typical of closed, restricted basins where evaporation exceeds precipitation. Low-latitude restricted basins such as this always rapidly lose water to evaporation, so surface flow from the open ocean must replace it. Evaporation of inflowing water from the open ocean increases the sea's salinity to very high values. This denser water eventually sinks and returns to the open ocean as a subsurface flow.

This circulation pattern, which is called **Mediterranean circulation**, is opposite that of most estuaries, which experience estuarine circulation where freshwater flows at the surface into the open ocean and salty water flows below the surface into the estuary. In estuaries, however, freshwater input exceeds water loss to evaporation, whereas evaporation exceeds input in the Mediterranean.

> **ESSENTIAL CONCEPT 11.3B**
>
> High evaporation rates in the Mediterranean Sea cause it to have a shallow inflow of surface seawater and a subsurface high-salinity outflow—a circulation pattern opposite that of most estuaries.

CONCEPT CHECK 11.3

1. Describe the four main types of estuaries, based on geologic origin.
2. Describe the difference between vertically mixed and salt wedge estuaries in terms of salinity distribution, depth, and volume of river flow. Which displays the more classical estuarine circulation pattern?
3. Discuss factors that cause the surface salinity of Chesapeake Bay to be greater along its east side. Also, why are periods of summer anoxia in Chesapeake Bay's deep water becoming increasingly worse?
4. What factors lead to a wide seasonal range of salinity in Laguna Madre?
5. Describe the circulation between the Atlantic Ocean and the Mediterranean Sea, and explain how and why it differs from typical estuarine circulation.

11.4 What Issues Face Coastal Wetlands?

Wetlands are ecosystems in which the water table is close to the surface, so they are typically saturated most of the time. Wetlands can border either freshwater or coastal environments. Coastal wetlands occur along the margins of coastal waters such as estuaries, lagoons, and marginal seas; they include swamps, tidal flats, coastal marshes, and bayous.

Types Of Coastal Wetlands

The two most important types of coastal wetlands are **salt marshes** and **mangrove swamps**. Both experience intermittent submergence by ocean water and contain salt-adapted plants, oxygen-depleted muds, and accumulations of organic matter called *peat deposits*.

Salt marshes generally occur between about 30 and 65 degrees latitude (**Figures 11.13a** and **11.13b**) and support a variety of salt-tolerant grasses and other low-lying plants that are termed *halophytic* (*halo* = salt, *phyto* = plant). Examples of halophytic grasses include cordgrass and salt-meadow cordgrass, both of which belong to the genus *Spartina* and have the ability to get rid of excess salt by producing exterior salt crystals. Other plants that live in this habitat, like pickleweed (*Salicornia*), accumulate salts in their tissues and dispose of excess salts by breaking off the tissues once they become highly salty. Well-developed salt marsh habitats are found

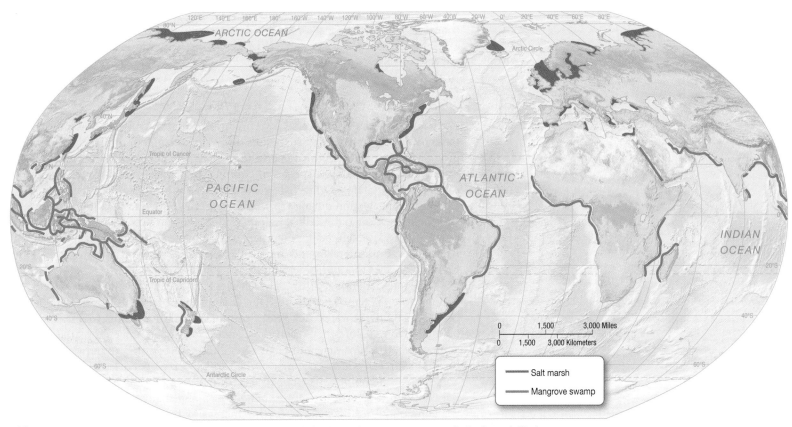

(a) Map showing the distribution of salt marshes in the higher latitudes and mangrove swamps in the lower latitudes.

(c) A dense mangrove swamp bordering a seaway in the Florida Keys.

(b) A typical salt marsh along the coast of Germany that borders the North Sea.

Figure 11.13 **Salt marshes and mangrove swamps.** Map showing the distribution of salt marshes and mangrove swamps, with accompanying photos.

along most coasts of the continental United States and also along the coasts of Europe, Japan, and eastern South America.

Mangrove swamps are restricted to tropical regions (below 30 degrees latitude; Figures 11.13a and 11.13c) and support various species of salt-tolerant mangrove trees, shrubs, and palms. To live in these salty conditions, some mangroves produce tall tripod-like root systems to stay above the salty water; others crystalize excess salt on their leaves. Mangrove swamps occur throughout the Caribbean and Florida; the most extensive mangroves in the world are found throughout Southeast Asia.

Characteristics Of Coastal Wetlands

Wetlands are home to a diverse assortment of plants and animals and are some of the most highly productive ecosystems on Earth. When left undisturbed, wetlands provide enormous economic benefits. Salt marshes, for example, serve as nurseries for more than half the species of commercially important fishes in the southeastern United States. Other fish, such as flounder and bluefish, use marshes for feeding and protection during the winter. Fisheries of oysters, scallops, clams, eels, and smelt are located directly in marshes, too. Mangrove ecosystems are important nursery areas and habitats for commercially valuable shrimp, prawn, shellfish, and fish species. Both marshes and mangroves also serve as important stopover points for many species of waterfowl and migrating birds.

Wetlands are also amazingly efficient at cleansing polluted water. Just 0.4 hectare (1 acre) of wetlands, for example, can filter up to 2,760,000 liters (730,000 gallons) of water each year, cleaning agricultural runoff, toxins, and other pollutants long before they reach the ocean. Wetlands remove inorganic nitrogen compounds (from sewage and fertilizers) and metals (from groundwater polluted by land sources), which become attached to clay-sized particles in wetland mud. Some nitrogen compounds trapped in sediment are decomposed by bacteria that release the nitrogen to the atmosphere as gas, and many of the remaining nitrogen compounds fertilize plants, further increasing the productivity of wetlands. As marsh plants die, their remains either accumulate as peat deposits or are broken up to become food for bacteria, fungi, and fish.

In addition, wetlands protect shorelines from erosion and serve as a first line of defense against hurricanes and tsunami by dissipating wave energy and absorbing excess water. The 2004 Indian Ocean tsunami, for example, devastated some coastal regions, yet others with protective offshore coral reefs or coastal mangroves experienced much less damage. And the loss of protective coastal wetlands in the Mississippi River Delta contributed to the extensive flooding associated with the storm surge caused by Hurricane Katrina in 2005 (see Chapter 6).

Serious Loss Of Valuable Wetlands

Despite all the benefits wetlands provide, more than half of the nation's wetlands have vanished. Of the original 87 million hectares (215 million acres) of wetlands that once existed in the conterminous United States, only about 43 million hectares (106 million acres) remain (**Figure 11.14**). Wetlands have been filled in and developed for housing, industry, and agriculture because people want to live near the oceans and because they often view wetlands as unproductive, useless land that harbors diseases. In many places, wetlands loss is compounded by the lack of fresh sediment from regular river floods. Instead, flooding rivers and their sediment are channeled away from wetland areas.

Louisiana's coastal wetlands, for example, are among those that are steadily disappearing. Over time, the soil in wetlands naturally compresses under its own weight, in a process called *subsidence*. Normally, the growth of plants and the infusion of fresh sediment from river floodwaters offset subsidence. With these factors reduced or eliminated, many wetlands are sinking into the ocean faster than they are building up. For example, scientists estimate that subsidence of the Mississippi River Delta—along with rising sea level—will cause about 10% of Louisiana to sink beneath the ocean surface by the end of the century.

Other countries have experienced similar losses of wetlands, too. In fact, scientists estimate that 50% of wetlands worldwide have been destroyed in the past

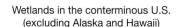

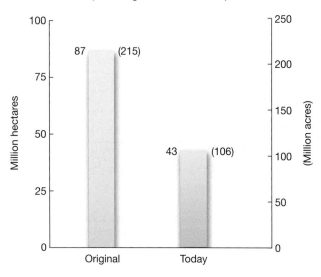

Figure 11.14 Wetland loss in the conterminous United States. Bar graph illustrating the amount of wetland loss in the conterminous United States (excluding Alaska and Hawaii), showing a reduction by more than half. Wetlands have been drained and converted to other uses (primarily development, industry, and agriculture).

century. Mangroves, for example, are already critically endangered or approaching extinction in 26 out of the 120 countries that have mangroves. Indonesia has lost over 50% of its mangroves in the past three decades, and the Philippines has reported losing 70% of its original mangrove cover. Of the mangrove swamps that remain, many are in critical condition or seriously damaged. At the current rate of mangrove loss, there is increasing concern that all mangrove ecosystems worldwide will be destroyed within the next 100 years.

To help prevent the loss of remaining wetlands, the U.S. Environmental Protection Agency (EPA) established an Office of Wetlands Protection (OWP) in 1986. At that time, wetlands were being lost to development at a rate of 121,000 hectares (300,000 acres) per year! In 1997, the rate of coastal wetland loss had slowed to about 8100 hectares (20,000 acres) per year. The goal of the OWP is to reduce the loss of wetlands in the United States to zero by actively enforcing regulations against wetlands pollution and identifying the most valuable wetlands to be protected or restored.

Recent trends suggest an overall increase in U.S. wetlands. In fact, a study from 1998 to 2004 revealed that the conterminous United States gained an estimated 13,000 hectares (32,000 acres) of wetlands each year. This small gain was primarily due to the large increase in freshwater wetlands; coastal wetlands were still decreasing, but at a slower rate than previously reported. The fact that coastal regions were losing wetlands despite the national trend of a net gain in wetlands points to the need for more research on the natural and human forces behind these trends and to an expanded effort on conservation of wetlands, particularly in coastal areas.

Future sea level rise is predicted to exacerbate the loss of wetlands. Even using a conservative estimate of sea level rise over the next 100 years of 50 centimeters (20 inches), it is estimated that as much as 61% of existing U.S. coastal wetlands will be lost. Human-caused global warming could cause additional sea level rise and thus even more wetland loss. Some of this wetland loss, however, would be partially offset by new wetland formation on former upland areas, although even under ideal circumstances, not all lost wetlands would be replaced.

ESSENTIAL CONCEPT 11.4

Coastal wetlands such as salt marshes and mangrove swamps are highly productive areas that serve as important nurseries for many marine organisms and act as filters for polluted runoff.

CONCEPT CHECK 11.4

1 Name the two types of coastal wetland environments and the latitude ranges where each will likely develop.

2 How do wetlands contribute to the biology of the oceans and the cleansing of polluted river water?

3 How large an area of wetlands has been lost in the conterminous United States? What percentage of original wetlands still remains? What is the United States doing to reverse this trend?

11.5 What Is Pollution?

As the use of coastal areas has increased for residences, recreation, and commerce, pollution of coastal waters has increased as well.

Marine Pollution: A Definition

Pollution can broadly be defined as *any harmful substance*, but how do scientists determine which substances are harmful? For example, a substance may be esthetically unappealing to people yet not be harmful to the environment. Conversely, certain types of pollution cannot easily be detected by humans, yet they can do harm to the environment. A substance may not immediately be harmful, but it may cause harm years, decades, or even centuries later. Also, to whom must this harm be done? For instance, some marine species thrive when exposed to a particular compound that is quite toxic to other species. Interestingly, natural conditions in coastal waters, such as dead seaweed on the beach, may be considered "pollution" by some people. It should be remembered, however, that although nature may produce conditions we dislike, it does not pollute. The amount of a pollutant is also important: If a substance that causes pollution is present in extremely tiny amounts, can it still be characterized as a pollutant? All of these questions are difficult to answer.

The World Health Organization defines pollution of the marine environment as follows:

> *The introduction by man, directly or indirectly, of substances or energy into the marine environment, including estuaries, which results or is likely to result in such deleterious effects as harm to living resources and marine life, hazards to human health, hindrance to marine activities, including fishing and other legitimate uses of the sea, impairment of quality for use of sea water and reduction of amenities.*

It is often difficult to determine the degree to which pollution affects the marine environment. Most areas were not studied sufficiently before they were polluted, so scientists do not have an adequate baseline from which to determine how pollutants have altered the marine environment. The marine environment is affected by decade- to century-long cycles, too, so it is difficult to determine whether a change is due to a natural biological cycle or any number of introduced pollutants, many of which combine to produce new compounds.

STUDENTS SOMETIMES ASK . . .

Is dilution the solution to ocean pollution?

That's certainly a catchy phrase, but the implications are controversial. It suggests that the oceans can be used as a repository for society's wastes, as long as the wastes are diluted to the point that they no longer threaten marine organisms (which is often difficult to determine). Because the oceans are vast and consist of a good solvent (water), they appear ideally suited to this disposal strategy. In addition, the oceans have good mixing mechanisms (currents, waves, and tides), which dilute many forms of pollution.

Air pollution was once viewed in a similar manner. Disposal of pollutants into the atmosphere was thought to be acceptable, as long as they were dispersed widely and high enough—so tall smokestacks were constructed. Over time, however, pollutants such as nitric acid and acid sulfates increased in the atmosphere to the point that acid rain is now a problem. The ocean, like the atmosphere, has a finite *holding capacity* for pollutants although experts disagree on exactly how much that is.

As disposal sites on land begin to fill, the ocean is increasingly evaluated as an area for disposal of society's wastes. One thing that we can all do is to limit the amount of waste we generate, alleviating some of the problem of where to put waste. It is likely, however, that the ocean will continue to be used as a dumping ground in the foreseeable future. Despite many new disposal techniques, a long-term "solution" to ocean pollution hasn't yet appeared.

Environmental Bioassay

One of the most widely used techniques for determining the concentration of pollutants that negatively affect the living resources of the ocean is to conduct a carefully controlled experiment to assess how a particular pollutant impacts marine organisms. Such an experiment is called an *environmental biological assay*, or **environmental bioassay** (*bio* = biologic, *essaier* = to weigh out). For example, regulatory agencies such as the U.S. EPA use an environmental bioassay to determine the concentration of a pollutant that causes 50% mortality among a specific group of test organisms within a prescribed period of time. If a pollutant exceeds a 50% mortality rate, then concentration limits are established for the discharge of the pollutant into coastal waters.

There are several drawbacks to using a specific environmental bioassay to draw general conclusions about a pollutant. One shortcoming is that it does not predict the long-term effect of pollution on marine organisms. Another is that it does not take into account how pollutants may combine with other substances to create new types of pollutants. In addition, environmental bioassays are often time-consuming, laborious, and organism specific and so may result in data that isn't applicable to other species.

The Issue Of Waste Disposal in the Ocean

Waste disposal facilities on land (such as landfills) have limited capacities that are already being exceeded in many cases. Should additional waste be discarded in the open ocean? Unlike coastal areas, the open ocean has mixing mechanisms (waves, tides, and currents) that distribute pollutants over a wide area—including an entire ocean basin. Diluting pollutants often renders them less harmful. On the other hand, do we really want to distribute a pollutant across an entire ocean without knowing what its long-term effects might be?

Some experts believe that we should not dump *anything* in the ocean, while others believe that the ocean can be a repository for many of society's wastes, as long as proper monitoring is conducted. Unfortunately, there are no easy answers and the issues are complex. What is clear is that more research is needed to assess the impact of pollutants in the ocean.

CONCEPT CHECK 11.5

1 Examine the World Health Organization's definition of pollution. Why does it need to be so specific?

2 What is an environmental bioassay? What are some drawbacks to using an environmental bioassay to determine whether a substance should be classified as a pollutant?

3 Why do some experts believe that the ocean can be a repository for many of society's wastes? If we do use the ocean as a disposal site, what conditions should be required?

11.6 What Are the Main Types Of Marine Pollution?

Marine pollution comes from substances such as petroleum, sewage, and various chemical compounds, all of which can have severe deleterious effects on marine organisms, particularly in coastal environments. Generally, coastal waters are more polluted than the open ocean for two main reasons: (1) Most pollution from land is dumped into nearby coastal waters, and (2) coastal waters are shallower and not as well circulated as the open ocean.

Petroleum

Major oil (**petroleum**) spills into the ocean are a fact of our modern oil-powered economy. Some oil spills result from oil extraction, such as the 2010 Gulf of Mexico *Deepwater Horizon* oil-drilling platform blowout. Other spills result from loading/unloading accidents or tanker collisions. Still others are caused by oil tankers running aground, as in the case of the 1989 oil spill from the *Exxon Valdez* tanker in Prince William Sound, Alaska.

THE 1989 *EXXON VALDEZ* OIL SPILL A large percentage of oil enters the oceans from spills by tankers and transportation operations. One of the most publicized oil spills was from the supertanker *Exxon Valdez*, which occurred in Prince William Sound, Alaska.

Crude oil produced from the North Slope of Alaska is carried by pipeline to the southern port of Valdez, Alaska, where it is loaded onto supertankers such as the *Exxon Valdez*, which are capable of holding almost 200 million liters (53 million gallons) when full. On March 24, 1989, the tanker left Valdez with a full load of crude oil and was headed toward refineries in California. She was only 40 kilometers (25 miles) out of Valdez when the ship's officers noted icebergs from nearby Columbia Glacier within the shipping channel. While maneuvering around the icebergs, the ship ran aground on a shallowly submerged rocky outcrop known as Bligh Reef (**Figure 11.15**), rupturing 8 of the ship's 11 cargo tanks. About 22% of her cargo—almost 44 million liters (about 11.6 million gallons) of oil—spilled into the pristine waters of Prince William Sound, where it subsequently spread into the Gulf of Alaska and fouled over 1775 kilometers (1100 miles) of shoreline.

Estimates suggest that at least 1000 sea otters and between 100,000 and 700,000 seabirds were killed outright by the spill, based on an extrapolation of the number of oiled carcasses that were collected on beaches and from the water. The area's

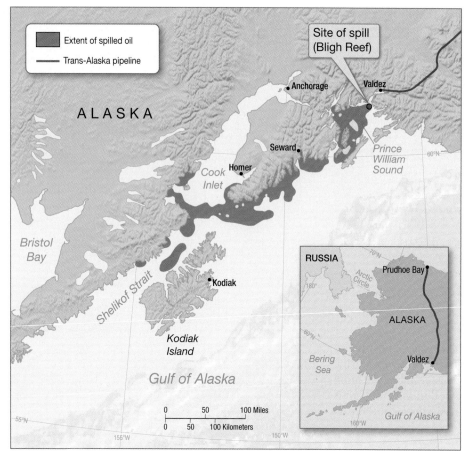

(a) Map showing the location of the *Exxon Valdez* oil spill.

(b) Aerial view of the supertanker *Exxon Valdez* grounded on Bligh Reef.

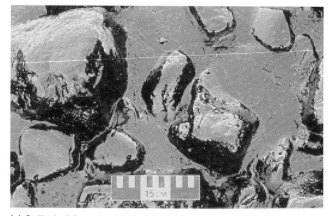

(c) Spilled oil from the *Exxon Valdez* coating an Alaskan beach.

Figure 11.15 The 1989 *Exxon Valdez* oil spill, Prince William Sound, Alaska.

remote location and the size of the affected area limit an accurate accounting of marine animal fatalities.

Immediately after the spill, Exxon spent more than $2 billion in cleanup efforts and another $900 million in subsequent years for restoration. Absorbent materials and skimming devices were used to remove oil from the water; to clean oil from rocky beaches, super-hot water (60°C [140°F]) was sprayed through high-pressure hoses. The hot water removed the oil but also killed most shoreline organisms. Analysis of the cleanup effort, when compared to areas that were left to biodegrade naturally, reveals that the beaches that were left alone recovered more quickly and more completely than the cleaned beaches.

Oil from the *Exxon Valdez* spill that settled on active shorelines—where wave action or microbes could break it down—disappeared fairly quickly. But surface puddles of crude oil from the spill are still found more than 20 years later, and pockets of undegraded oil rest just below the surface of some beaches and in coastal wetlands. Experts state that marine animals are no longer at risk of exposure to toxic oil and that natural processes will continue to biodegrade the remaining oil.

As large and damaging as the *Exxon Valdez* spill was, it ranks as the 2nd-largest oil spill in U.S. waters but only the *54th* largest oil spill worldwide (**Figure 11.16**; see also Web Table 11.1). The world's largest oil spill occurred because of intentional dumping by the Iraqi army during their invasion of Kuwait during the 1991 Persian Gulf War. By the time the Iraqis were driven out of Kuwait and the leaking oil wells and sabotaged production facilities were brought under control, more than 908 million liters (240 million gallons) of oil had spilled into the Persian Gulf (**Figure 11.17**)—more than 20 times the amount spilled by the *Exxon Valdez*.

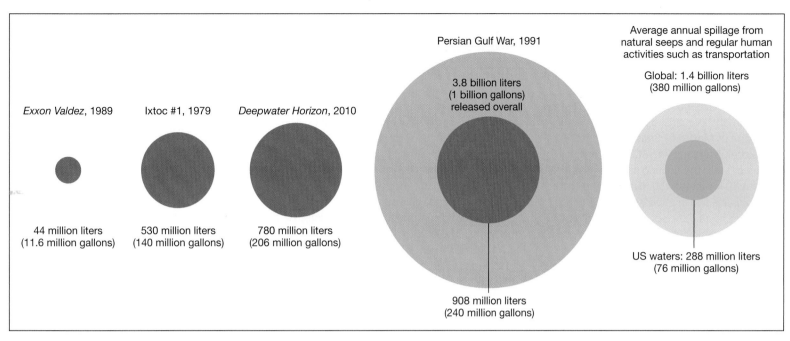

Figure 11.16 Comparison of selected oil spills.

In 2010, the *Deepwater Horizon* oil-drilling platform in the Gulf of Mexico exploded, causing the deep well it was drilling to flow unabated for about three months before it was brought under control (**Diving Deeper 11.1**). During that time, the well released the equivalent of an *Exxon Valdez*–sized spill about every four days. Remarkably, the *Deepwater Horizon* spill ranks as the world's largest accidental spill into the ocean, surpassed only by the intentional release of oil during the 1991 Persian Gulf War (see Figure 11.16).

OTHER OIL SPILLS   Although large spills such as from the *Exxon Valdez* occur infrequently, smaller-scale spills are more common. In fact, nearly 100 documented spills of at least 3800 liters (1000 gallons) of oil and more than 10,000 smaller spills have been reported annually in U.S. waters. Some oil spills are intentionally created, such as the oil that was spilled during the Persian Gulf War in 1991 (see Figure 11.17).

Other types of oil spills are caused by the blowout of undersea oil wells during drilling or pumping, such as the 2011 *Deepwater Horizon* oil-drilling platform blowout (see Diving Deeper 11.1) and the 1979 Gulf of Mexico oil spill. In 1979, the Petróleos Mexicanos (PEMEX) oil-drilling platform Ixtoc #1 in the Bay of Campeche off the Yucatán Peninsula, Mexico, blew out and caught fire, creating what was then the world's largest oil spill. (The spill has subsequently been topped by the 1991 Persian Gulf War oil spill and the 2010 *Deepwater Horizon* oil well blowout; see Figure 11.16.) Before the well was capped nearly 10 months later, it spewed 530 million liters (140 million gallons)

(b) A Saudi Arabian government official examining some of the oil spill damage along a Persian Gulf beach.

(a) Map showing the location of the Persian Gulf oil spill, with offshore point locations (oil platforms) shown as red dots. Most of the oil was confined to the northwest coast of the Perisan Gulf by currents and winds out of the southeast.

Figure 11.17 Oil pollution from the 1991 Persian Gulf War.

Diving Deeper 11.1 Focus on the Environment

THE 2010 GULF OF MEXICO *DEEPWATER HORIZON* OIL SPILL

On April 20, 2010, the *Deepwater Horizon*, an offshore drilling platform operating in the Gulf of Mexico by British Petroleum (BP), was completing the final stages of drilling a deep oil well. The floating platform was located about 80 kilometers (50 miles) off the coast of Louisiana and in 1500 meters (5000 feet) of water beyond the edge of the continental shelf. Unexpectedly, it received a "kick" from a large bubble of natural gas that was under high pressure at the bottom of the well 4 kilometers (2.5 miles) beneath the sea floor. The blowout preventer on the sea floor failed, and the natural gas bubble rushed to the drilling platform, which exploded and caught fire (see the chapter-opening photo), killing 11 crew members. Two days after the explosion, the *Deepwater Horizon* sank, leaving the well gushing crude oil at the seabed at a rate of about 9 million liters (2.4 million gallons) per day (**Figure 11A**). Three months later, an underwater robotic vehicle was finally able to cap the well, but the amount of oil released became the world's second-largest oil spill (the world's largest accidental spill into the ocean) and by far the largest oil spill in U.S. waters.

In all, 780 million liters (206 million gallons) of oil were released from the well. A deluge of chemical dispersants—detergent-like solvents designed to break up the oil but considered more toxic to marine life than the oil itself—were applied both at the surface and in deep water. Scientists estimate that BP removed about a quarter of the oil, most of it recovered directly from the well, burned at sea, or skimmed by boats. Another quarter of the oil evaporated or dissolved into scattered molecules.

And a third quarter was either naturally dispersed in the water as small droplets (which might still be toxic to some organism) or chemically dispersed. But the last quarter—around five times the amount released by the *Exxon Valdez*—formed slicks or sheens on the water (Figure 11A), washed onto local beaches and marshes, and accumulated as tar balls on the sea floor. Studies indicate that some oil never made it to the surface; instead, the oil formed diffuse plumes more than 1000 meters (3300 feet) below the surface, the effects of which are still not clear.

Fortunately, most of the oil remained offshore, where wave energy combined with oxygen, sunlight, and the Gulf's abundant oil-eating bacteria were able to naturally biodegrade it. Some of the oil, however, washed onto local beaches and salt marshes, fouling the shore and killing marine animals. The oil that sank to the bottom and became entrained in low-oxygen sediments like those on the sea floor—or a marsh—can hang around for decades, degrading the environment. The cost of the cleanup, which will be funded by BP and its partners over the next 20 years, is estimated to exceed $40 billion.

Birds, sea turtles, marine mammals, fish, and shellfish were some of the most affected marine organisms. Local fisheries were shut down after the spill but have been reopened since. It is unknown how this vast release of oil and chemical dispersants will affect marine organisms and the Gulf ecosystem, especially over time. Long-term studies are under way that will reveal how much damage the spill has done to the Gulf's marine life.

Aerial view of a ship passing though an oil slick.

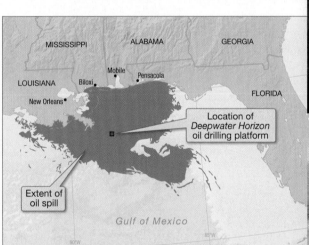

Map showing the extent of the 2010 Gulf of Mexico *Deepwater Horizon* oil spill.

Figure 11A **The 2010 Gulf of Mexico oil spill.**

Oil washing ashore at Fourchon Beach, Louisiana.

of crude oil into the Gulf of Mexico, and some of the oil washed up along the coast of Texas (**Figure 11.18**).

HOW DAMAGING IS OIL POLLUTANT IN THE OCEAN? Oil is a mixture of various **hydrocarbons**, which means it is composed of the elements *hydrogen* and *carbon*. Hydrocarbons are organic substances, so they can be broken down, or *biodegraded*,

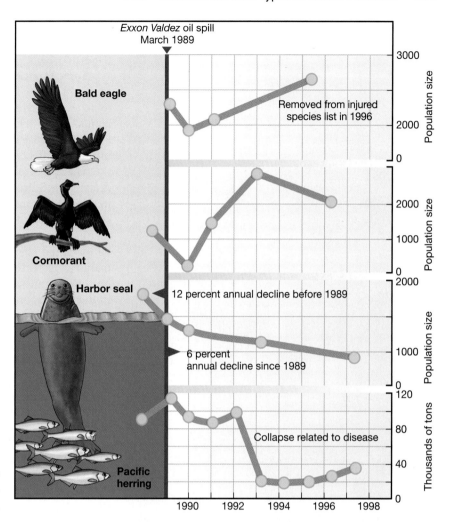

Figure 11.18 **Blowout from the Ixtoc #1 oil well, Gulf of Mexico.** Location of the 1979 Bay of Campeche blowout and oil slick that affected the Texas coast. The well blew out, caught fire, and flowed for 10 months, spilling 530 million liters (140 million gallons) of oil into the Gulf of Mexico.

Figure 11.19 **Recovery of organisms affected by the *Exxon Valdez* oil spill.** The populations of several key organisms in the Prince William Sound area of Alaska have rebounded since the 1989 *Exxon Valdez* oil spill. The bald eagle is so numerous that it was removed from the endangered species list in 1996. The collapse of the Pacific herring's population is thought to be caused by other factors not related to the effects of the oil spill.

by microorganisms. Because hydrocarbons are largely biodegradable, many marine pollution experts consider oil to be among the *least* damaging pollutants introduced into the ocean! Certainly oil spills appear ghastly as the oil coats the water, the shore, and helpless marine organisms, including seabirds, and it can cause grievous short-term damage. But oil dissipates and breaks down, becoming food for various microbes. As an example, vast quantities of oil were spilled in the Pacific and, especially, the Atlantic during World War II. In fact, some U.S. East Coast beaches were coated by oil several centimeters deep, yet not a trace of those spills remains today. In a broader view, some natural undersea oil seeps have occurred for millions of years, and the ocean ecosystem seems unaffected—or even *enhanced*—by them (because oil is a source of energy). As we'll see, other types of pollutants last far longer and can do much more damage than an oil spill.

Data from the *Exxon Valdez* oil spill are a case in point. The oil spill released almost 44 million liters (11.6 million gallons) of oil into a pristine wilderness area in Alaska. The affected waters were expected to have a long, slow recovery, but the fisheries that closed in 1989 bounced back with record takes in 1990. A study conducted 10 years after the spill revealed that several key species had rebounded to the point where their numbers were greater than before the spill (**Figure 11.19**). Recent research on the long-term effects of the spill, however, reveals that a significant amount of poorly weathered oil has seeped into intertidal sediments and remains below the surface.

OTHER CONCERNS ABOUT OIL IN THE OCEAN Oil is a complex mixture of various hydrocarbons and other substances, including the elements oxygen, nitrogen, sulfur, and various trace metals. When this complex chemical mixture combines with seawater—another complex chemical mixture, which also contains organisms—the results are usually devastating for marine organisms. Studies have revealed, for example, that crude oil in seawater at concentrations of only 0.7 parts per billion

Figure 11.20 A bird covered by oil from the Gulf of Mexico *Deepwater Horizon* oil spill. When marine organisms are covered by oil from an oil spill, their feathers or fur lose their insulation properties, resulting in high fatality rates. Some marine organisms such as this pelican were rescued and cleaned of oil.

kills or damages certain fish eggs. In addition, many organisms are killed outright when they are coated by oil, rendering their insulating feathers or fur useless (Figure 11.20).

Toxic (*toxicum* = poison) **compounds**[9] in crude oil vary, but the most worrisome are polycyclic aromatic hydrocarbons (PAHs), such as napthalenes, benzene, toluene, and xylenes. Even in small doses, they can sicken humans, animals, and plants. These hydrocarbons are particularly dangerous if inhaled or ingested by animals because they can be transformed into even more toxic products, which can ultimately affect an organism's genetic materials.

There is also concern about the long-term effects of oil spills, such as chronic, delayed, or indirect impacts, many of which are often difficult to document and link to the spill because of the considerable time lag. For example, exposure to even tiny concentrations of the chemicals present in oil can cause harmful biological effects that usually go unnoticed. A recent study found that fish exposed to PAHs in crude oil exhibited changes in gene expression that are linked to developmental abnormalities, decreased embryo survival, and lower reproductive success that wouldn't be noticed until years later. Another example is from the oil tanker that ran aground in the Galápagos Islands in 2001, spilling roughly 3 million liters (800,000 gallons) of diesel and bunker oil. The oil spread westward and was dispersed by strong currents, so only a few marine animals were immediately killed. Marine iguanas on a nearby island, however, suffered a massive 62% mortality in the year after the accident, due to a small amount of residual oil contamination in the sea.

In addition, spills from oil tankers may release a wide variety of petroleum products (not just crude oil), each of which has a different level of toxicity and behaves differently in the environment. For example, refined oil such as fuel oil is rich in compounds that are much more toxic to the environment than crude oil.

Although spills from oil tankers receive much media attention, they are not the primary source of oil to the oceans. Figure 11.21 shows that 47% of worldwide oil to the oceans is caused by underwater *natural oil seeps* (many of which occur in U.S. waters); the remaining 53% comes from *human sources*. The figure also shows that of human-caused oil to the oceans, 72% comes from *petroleum consumption*, which includes nontank vessels, runoff from increasingly paved urban areas, and individual car, boat, and watercraft owners; 22% comes from *petroleum transportation*, including refining and distribution activities; and only 6% comes from *petroleum extraction*, which is associated with oil and gas exploration and production. Remarkably, the overwhelming majority of the petroleum that enters the oceans due to human activity is a result of small but frequent and widespread releases of oil related to activities that consume petroleum.

CLEANING OIL SPILLS When oil enters the ocean, it initially floats because oil is less dense than water and forms a slick at the surface, where it starts to break down through natural processes (Figure 11.22). The volatile, lighter components of crude oil evaporate over the first few days, leaving behind a more viscous substance that aggregates into tar balls and eventually sinks. The tarry oil also coats suspended particles, which settle to the sea floor, too.

If the floating oil hasn't dispersed, it can be collected with specially designed skimmers or absorbent materials. The collected oil (or oiled materials), however, must still be disposed of elsewhere. Waves, winds, and currents serve to further disperse an oil slick and mix the remaining oil with water to make a frothy emulsion called *mousse*. In addition, bacteria and photo-oxidation act to break down the oil into compounds that dissolve in water.

Microorganisms such as bacteria and fungi naturally biodegrade oil, so they can be used to help clean oil spills—a method called **bioremediation** (*bio* = biologic,

---

[9]A *toxic compound* is a poisonous substance that has the capability of causing injury or death, especially by chemical means.

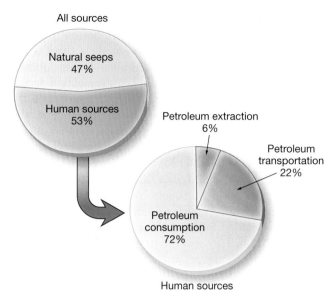

Figure 11.21 **Sources of oil to the oceans.** Of worldwide oil to the oceans, 47% comes from natural seeps, while 53% comes from human sources. Of all human sources, 72% comes from petroleum consumption activities such as individual car and boat owners, nontank vessels, and runoff from increasingly paved urban areas. Surprisingly, combined petroleum transportation and extraction account for only 28% of all human-caused oil to the oceans.

remedium = to heal again). Virtually all marine ecosystems harbor naturally occurring bacteria that degrade hydrocarbons. Although certain types of bacteria and fungi can break down particular kinds of hydrocarbons, none is effective against all forms. In 1980, however, microbiologists discovered a microorganism capable of breaking down nearly two-thirds of the hydrocarbons in most crude oil spills.

Releasing bacteria directly into the marine environment is one form of bioremediation. For example, a strain of oil-degrading bacteria was released into the Gulf of Mexico to test its effectiveness in cleaning up about 15 million liters (4 million gallons) of crude oil spilled after an explosion disabled the tanker *Mega Borg* in 1990. Preliminary results indicate that the bacteria reduced the amount of oil and had no negative effects on the area's ecology.

Providing conditions that stimulate the growth of naturally occurring oil-degrading bacteria is another form of bioremediation. Exxon, for example, spent $10 million to spread fertilizers rich in phosphorus and nitrogen on Alaskan shorelines to boost the growth of indigenous oil-eating bacteria after the *Exxon Valdez* spill. The resulting cleanup rate was more than twice the rate that occurs under natural conditions.

PREVENTING OIL SPILLS One of the best ways to protect areas from oil spills is to prevent spills from occurring in the first place. Because our society relies on petroleum products, however, oil spills are a likely occurrence in the future (**Figure 11.23**), especially as petroleum reserves beneath the continental shelves of the world are increasingly exploited.

After the *Exxon Valdez* oil spill in 1989, Congress enacted the Oil Pollution Act of 1990, which defined responsibility for fiscal damage and cleanup. The act also

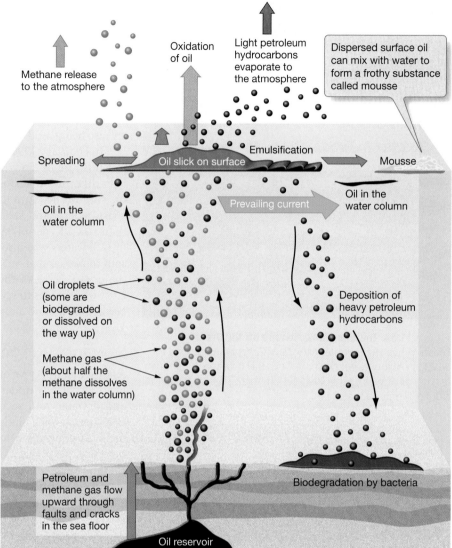

Figure 11.22 **Processes acting on oil spills.** When oil enters the ocean, it is acted upon by various natural processes that break up the oil. The lighter components evaporate, while the heavier components form tar balls or coat suspended particles and sink. The remaining dispersed oil photo-oxidizes or can mix with water, creating a frothy substance called "mousse."

Figure 11.23 **Who is at fault?**

Figure 11.24 The *New Carissa* on fire off the Oregon coast. When the freighter M/V *New Carissa* ran aground in 1999 in shallow water offshore Coos Bay, Oregon, and began leaking oil, it was intentionally set on fire to prevent further oil from spilling into the ocean.

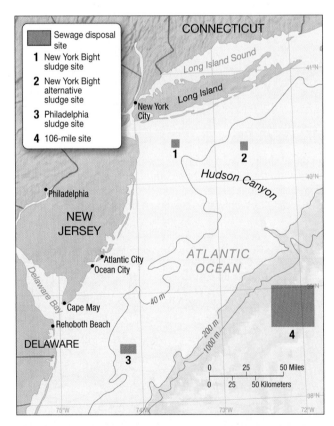

Figure 11.25 Atlantic sewage sludge disposal sites. More than 8 million metric tons (8.8 million short tons) of sewage sludge was dumped by barge annually at the New York Bight sludge sites (*1* and *2*) and the Philadelphia sludge site (*3*). After 1986, the new dump site is the larger and deeper-water 170-kilometer (106-mile) site (*4*).

phased out single-hulled oil tankers traveling in U.S. waters and mandated double-hulled construction by 2015. Currently, single-hulled tankers are barred from U.S. ports, and European countries such as France and Spain do not allow them within 320 kilometers (200 miles) of their coasts. A double hull houses two layers: An inner hull can prevent oil spillage if damage occurs to the outer hull. Studies of hull designs during groundings and collisions indicate that double-hull designs are more effective overall at reducing oil spills. However, analysis of the *Exxon Valdez* spill suggests that even a double-hulled tanker would not have prevented the disaster. Tanker designs are also being modified to limit the amount of oil spilled in the event of a hull rupture.

In February 1999, the Japanese-owned freighter M/V *New Carissa* ran aground just offshore of Coos Bay, Oregon, with nearly 1,500,000 liters (400,000 gallons) of tarlike fuel oil aboard that began leaking through cracks in its hull. When the ship washed into the surf zone and an approaching storm threatened to tear it apart, federal and state authorities decided to ignite the vessel and its fuel rather than risk a larger oil spill (**Figure 11.24**). This was the first time that oil on a ship in U.S. waters was intentionally burned to prevent an oil spill. Eventually, the ship split in two, and about half of its oil burned, limiting the amount of oil spilled into the ocean. Most of the remaining oil was sunk with the wrecked ship a month later, when it was towed offshore and sunk in water 3 kilometers (1.9 miles) deep by Naval gunfire and a torpedo.

Sewage Sludge

Another main type of marine pollution is **sewage sludge**. Sewage treated at a facility typically undergoes **primary treatment**, where solids are allowed to settle and dewater, and **secondary treatment**, where it is exposed to bacteria-killing chlorine. Sewage sludge is the semisolid material that remains after such treatment. It contains a toxic brew of human waste, oil, zinc, copper, lead, silver, mercury, pesticides, and other chemicals. Since the 1960s, at least 500,000 metric tons (550,000 short tons) of sewage sludge have been dumped into the coastal waters of Southern California, and more than 8 million metric tons (8.8 million short tons) have been dumped in the New York Bight between Long Island and the New Jersey shore.

Although the Clean Water Act of 1972 prohibited the dumping of sewage into the ocean after 1981, the high cost of treating and disposing of sewage sludge on land resulted in extension waivers being granted to many municipalities. In the summer of 1988, however, nonbiodegradable debris including medical waste—probably carried by heavy rains into the ocean through storm drains—washed up on Atlantic coast beaches and adversely affected the tourist business. Although this event was completely unrelated to sewage disposal at sea, it focused public awareness on ocean pollution and helped pass new legislation to terminate sewage disposal at sea.

NEW YORK'S SEWAGE SLUDGE DISPOSAL AT SEA Sewage sludge from New York and Philadelphia has traditionally been transported offshore by barge and dumped in the ocean at sites totaling 150 square kilometers (58 square miles) within the New York Bight sludge site and the Philadelphia sludge site (**Figure 11.25**).

The water depth is about 29 meters (95 feet) at the New York Bight sludge site and about 40 meters (130 feet) at the Philadelphia sludge site. The water column in such shallow water is relatively uniform, so even the smallest sludge particles reach the bottom without undergoing much horizontal transport, and the ecology of the dump site can be severely affected. At the very least, such a concentration of organic and inorganic matter seriously disrupts the chemical cycling of nutrients. Greatly reduced species diversity results, and in some locations, the environment becomes devoid of oxygen (*anoxic*).

In 1986, the shallow-water sites were abandoned, and sewage was subsequently transported to a deep-water site 171 kilometers (106 miles) out to sea (Figure 11.25). The deep-water site is beyond the continental shelf break, so there is usually a well-developed density gradient that separates low-density, warmer surface water from high-density, colder deep water. Internal waves moving along this density gradient can horizontally transport particles at rates 100 times greater than they sink.

Local fishermen reported adverse effects on their fisheries soon after deep-water dumping began. Also, concern was expressed that the sewage could be transported great distances in eddies of the Gulf Stream (see Chapter 7), even as far as the coast of the United Kingdom. This program was terminated in 1993, and municipalities must now dispose of their sewage on land.

BOSTON HARBOR SEWAGE PROJECT   Some 48 different communities that comprise the greater Boston area have, until recently, used an antiquated sewage system to dump sludge and partially treated sewage at the entrance to Boston Harbor. Tidal currents often swept the sewage back into the bay, and at other times, the system became overloaded and dumped raw sewage directly into the bay, making Boston Harbor one of the most polluted bays in the country.

A court-ordered cleanup of Boston Harbor in the 1980s resulted in the construction of a new waste treatment facility at Deer Island that came online in 1998. It treats all sewage with bacteria-killing chlorine and carries it through a tunnel 15.3 kilometers (9.5 miles) long into deeper waters offshore (**Figure 11.26a**), which prevents it from returning to the bay. Since the cleanup of Boston Harbor, beaches have reopened, clammers are digging again, and marine life—including harbor seals, porpoises, and even whales—has returned. To pay for the $3.8 billion sewage system, however, the average annual sewage bill for a Boston-area household is now about $1200, more than five times what it used to be.

Some fear that the project will degrade the environment in Cape Cod Bay and Stellwagen Bank (**Figure 11.26b**), which is an important whale habitat. The area's recent designation as a National Marine Sanctuary may affect the feasibility of dumping there.

DDT and PCBs

The pesticide **DDT** (dichloro-diphenyl-trichloroethane) and the industrial chemicals called **PCBs** (polychlorinated biphenyls) are now found throughout the marine environment. They are persistent, biologically active chemicals that have been

Figure 11.26 Boston Harbor sewage project. Diagrammatic view of the Boston Harbor sewage project tunnels **(a)** and a map of the sewage outfall area **(b)**.

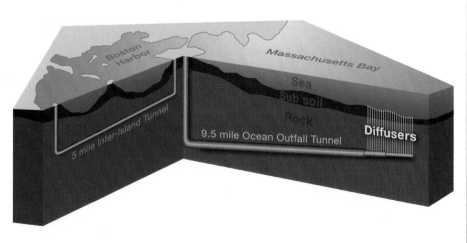

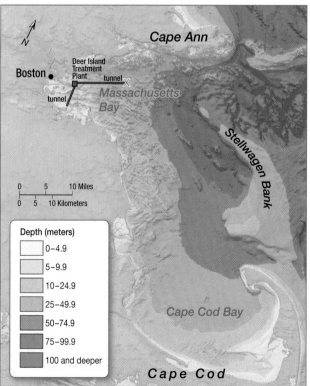

(a) The Boston Harbor sewage project includes tunnels that transport sewage to an outfall 15 kilometers (9.5 miles) offshore at a depth of 76 meters (250 feet) beneath the ocean floor.

(b) Bathymetric map of the coastal ocean in the Boston-Cape Cod area, showing the proximity of the new sewage outfall (*red x*) to Stellwagen Bank, which is a National Marine Sanctuary.

STUDENTS SOMETIMES ASK . . .

I've heard that some organizations want to lift the ban on DDT. Why would they want to do that?

Since production of DDT was banned in 1971, outbreaks of malaria have dramatically increased because DDT was the most effective and readily available pesticide used to kill mosquitoes that transmit malaria. According to the World Health Organization, malaria infects up to 500 million people a year—mostly in tropical regions—and kills as many as 2.7 million, including at least 1 child every 30 seconds. In addition, drug-resistant strains of malaria have begun to show up worldwide. This resurgence of malaria has caused many health organizations to call for an exception to the ban on DDT—despite its well-documented perseverance and negative effects on the environment—so that it can be used selectively to spray houses in malaria-prone areas such as tropical Africa and Indonesia. The ambitious Global Malaria Action Plan seeks to eliminate all malaria deaths by 2015 through research into vaccines and the use of bed nets, drugs, and spraying.

Figure 11.27 Survival of brown pelicans threatened by DDT. Brown pelicans (*Pelecanus occidentalis*) that breed on Anacapa Island offshore Southern California were found to have high levels of DDT, which decreased the thickness of their eggshells. Since DDT has been banned, healthy pelicans have returned to these waters, such as this one in breeding plumage (*photo*).

introduced into the oceans entirely as a result of human activities. Because of their toxicity, persistence, and propensity for accumulating in food chains, these and other chemicals have been classified as *persistent organic pollutants (POPs)* capable of causing cancer, birth defects, and other grave harm.

DDT was widely used in agriculture during the 1950s and improved crop production throughout developing countries for several decades. However, its extreme effectiveness as an insecticide and persistence as a toxin in the environment eventually resulted in a host of environmental problems, including devastating effects on marine food chains. In 1962, biologist Rachel Carson published her seminal book *Silent Spring* about the dangers of DDT and other chemicals in the environment. Her book played an important role in the environmental movement and DDT was subsequently banned in the United States in 1972.

PCBs are industrial chemicals that were once widely used as liquid coolants and insulation in industrial equipment such as power transformers, where they were released into the environment. PCBs were also widely used in wiring, paints, caulks, hydraulic oils, carbonless copy paper, and a host of other products. PCBs have been shown to cause liver cancer and harmful genetic mutations in animals. PCBs can also affect animal reproduction: They have been indicated as causes of spontaneous abortions in sea lions and the death of shrimp in Escambia Bay, Florida.

DDT AND EGGSHELLS Since 1972, the EPA has banned the use of DDT in the United States. Worldwide, the pesticide is banned from agricultural use, but it continues to be used in limited quantities for public health purposes. Ironically, U.S. companies continue to produce DDT and supply it to other countries.

The danger of excessive use of DDT and similar pesticides first became apparent in the marine environment when it affected marine bird populations. During the 1960s, there was a serious decline in the brown pelican population of Anacapa Island off Southern California (Figure 11.27). High concentrations of DDT in the fish eaten by the birds had caused them to produce eggs with excessively thin shells because DDT made it more difficult for calcium to be incorporated into the egg shells.

The osprey is a common bird of prey in coastal waters that is similar to a large hawk. The osprey population of Long Island Sound declined in the late 1950s and 1960s because DDT contamination caused them to produce eggs with thin shells, too. Since the ban on DDT, the osprey, brown pelican, and many other species affected by the chemical are making remarkable comebacks.

DDT AND PCBS LINGER IN THE ENVIRONMENT DDT and PCBs (which were banned in 1977) generally enter the ocean through the atmosphere and river runoff. They are initially concentrated in the thin slick of organic chemicals at the ocean surface, and then they gradually sink to the bottom, attached to sinking particles. A study off the coast of Scotland indicated that open-ocean concentrations of DDT and PCBs are 10 and 12 times less, respectively, than in coastal waters. Long-term studies have shown that DDT residue in mollusks along the U.S. coasts peaked in 1968.

Although most countries have banned their use, DDT and PCBs are so pervasive in the marine environment that even Antarctic marine organisms contain measurable quantities of them. There has been no agriculture or industry in Antarctica to introduce them directly, so they must have been transported from distant sources by winds and ocean currents.

Mercury and Minamata Disease

The elemental metal **mercury**, which is a silvery liquid at room temperature, has many industrial uses. When it enters the ecosystem, however, mercury is consumed by bacteria and converted to an organic form of mercury (*methylmercury*) that is generally toxic to most living things and has serious implications for human health.

Figure 11.28 A victim of Minamata disease. Ingestion or contact with the toxic metal mercury can cause Minamata disease, which affects the human nervous system and can result in brain damage, birth defects, paralysis, and even death.

MINAMATA DISEASE A chemical plant built in 1938 on Minamata Bay, Japan, produced acetaldehyde, which requires mercury in its manufacture. Industrial wastewater that contained methylmercury was discharged into Minamata Bay, where the chemical was later ingested and concentrated in the tissues of marine organisms such as fish and shellfish. The first ecological changes in Minamata Bay were reported in 1950, and human effects were noted as early as 1953. The mercury poisoning that is now known as **Minamata disease** became epidemic in 1956, when the plant was only 18 years old.

Minamata disease is a degenerative neurological disorder that affects the human nervous system and causes sensory disturbances, including blindness and tremor, brain damage, birth defects, paralysis, and even death. This mercury poisoning was the first major human disaster resulting from ocean pollution. However, the Japanese government did not declare mercury as the cause of the disease until 1968. The plant was immediately shut down, but more than 100 people were known to acutely suffer from Minamata disease by 1969 (**Figure 11.28**), and almost half of them died as a result of the disease. A second acetaldehyde plant was closed in 1965 in Niigata, Japan, because it, too, was discharging methylmercury that poisoned people. As of 2001, 2265 Japanese victims of Minamata disease had been officially recognized (1784 of whom had died). Today the concentration of methylmercury in Minamata Bay is no longer unusually high, suggesting that there has been enough time for the mercury to be widely dispersed within the marine environment.

BIOACCUMULATION AND BIOMAGNIFICATION During the 1960s and 1970s, methylmercury contamination in seafood received considerable attention. Certain marine organisms concentrate within their tissues many substances found in minute concentrations in seawater in a process called **bioaccumulation**. When animals eat other animals, some of these substances (including toxic chemicals) move up food chains and become concentrated in the tissues of larger animals, in a process called **biomagnification** (**Figure 11.29**). Because the amount of mercury in the ocean has been increasing, some seafood such as tuna and swordfish contains unusually high amounts of mercury.

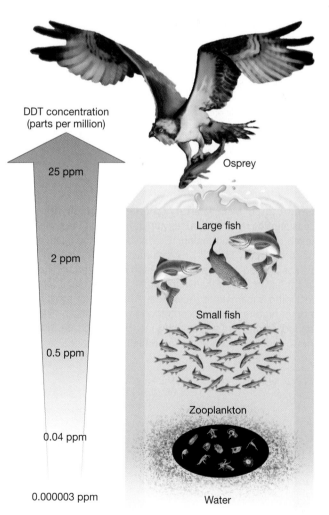

DDT concentration (parts per million)

25 ppm

2 ppm

0.5 ppm

0.04 ppm

0.000003 ppm

Osprey

Large fish

Small fish

Zooplankton

Water

Figure 11.29 Biomagnification. Certain toxins can accumulate in the tissues of organisms and thus can be concentrated (magnified) in higher levels of marine food chains.

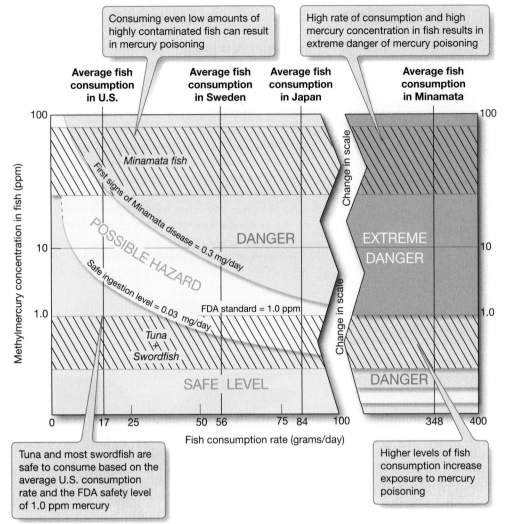

Figure 11.30 Methylmercury concentration in fish, fish consumption rates for various populations, and the danger levels of mercury poisoning. Graph showing the relative risk of contracting Minamata disease based on the amount of fish consumed and the concentration of toxic methylmercury in fish for people in the United States, Sweden, and Japan—including Minamata. It shows that a high rate of consumption of high mercury concentration in fish results in extreme danger of mercury poisoning (*upper right*) and that tuna and most swordfish are safe to consume at average U.S. consumption rates (*lower left*).

HOW MUCH MERCURY-RICH SEAFOOD IS SAFE TO EAT? Studies conducted on the amount of seafood consumed by various human populations have helped establish safe levels of methylmercury in marketed fish. (Note that most animals can get rid of elemental mercury quickly, and so it does not pose the same threat as methylmercury.) To establish these levels, three variables were considered:

1. The rate at which each group of people consumed fish

2. The methylmercury concentration in the fish consumed by that population

3. The minimum ingestion rate of methylmercury that induces disease symptoms

These three variables help establish a maximum allowable methylmercury concentration that will safeguard people from mercury poisoning, as long as they don't exceed the recommended intake of fish.

Figure 11.30 shows the relative risk of contracting Minamata disease for people in the United States, Sweden, and Japan, including those from the Minamata fishing community. The graph shows that the risk increases with increased consumption of fish and that the higher the methylmercury concentration of the fish, the greater the risk.

Figure 11.30 is a graph that shows the methylmercury concentration in fish, fish consumption rates for various populations, and the danger levels of mercury poisoning. Scientists have determined that the minimum level of methylmercury consumption that causes poisoning symptoms is 0.3 milligrams (0.00001 ounce) per day. Using a safety factor of 10 times, the safe ingestion level is set at 0.03 milligrams (0.000001 ounce) of methylmercury per day. The graph shows that the higher the consumption of fish—especially fish contaminated with mercury—the greater the chance of mercury poising symptoms. For example, the graph shows that a high rate of consumption of high mercury concentration fish (such as Minamata fish) results in extreme danger of mercury poisoning. The graph also shows that people in the United States consume, on average, 17 grams (0.6 ounce) of fish per day. At this rate, the chance of mercury poisoning symptoms first occurs when methylmercury concentrations in fish exceed 20 parts per million (ppm), and the safe ingestion level is 2.0 ppm methylmercury in fish.

Using the established safe U.S. ingestion level of 2.0 ppm methylmercury in fish, the U.S. Food and Drug Administration (FDA) doubled the safety factor and established a limit of methylmercury concentration for fish at 1.0 ppm. Based on consumption rates, this limit has adequately protected the health of U.S. citizens because essentially all tuna and most swordfish fall below this concentration. Still, the FDA issued an advisory in 2001 stating that pregnant women, women of child-bearing age, nursing mothers, and young children should avoid eating certain kinds of fish that may contain high levels of methylmercury, such as swordfish, sharks, king mackerel, and tilefish.

Figure 11.30 shows that for people in Sweden and Japan, the methylmercury concentration of fish deemed to be at a safe level is lower because these populations eat more fish. The graph also shows the extreme danger to which residents of Minamata were inadvertently subjected when they ate so much of the highly contaminated fish from Minamata Bay.

Nonpoint Source Pollution and Trash

Nonpoint source pollution—also called *poison runoff*—is any type of pollution entering the ocean from multiple sources rather than from a single discrete source, point, or location. In most urban areas, nonpoint source pollution arrives at the ocean via runoff from storm drains, many of which now have labels indicating that they lead to the ocean (Figure 11.31). The U.S. National Academy of Sciences estimates that 5.8 million metric tons (6.4 million short tons) of litter enters the world's oceans each year.

Because nonpoint source pollution comes from many different locations, it is difficult to pinpoint where it originates, although the *cause* of the pollution may readily be apparent. One such example is trash that is washed down storm drains to the ocean and then washes up on beaches (Figure 11.32). Others include pesticides and fertilizers from agriculture and oil from automobiles that are washed to the ocean whenever it rains. In fact, the amount of road oil and improperly disposed oil regularly discharged each year into U.S. waters as nonpoint source pollution is as much as one-and-a-half times the amount of the *Deepwater Horizon* oil spill!

Trash enters the ocean as a result of ocean dumping, too. According to existing laws (Figure 11.33), certain types of trash such as glass, metal, rags, and food can legally be dumped in the ocean—as long as they are dumped far enough away from shore or ground up small enough. Mostly, this material sinks or biodegrades and does not accumulate at the surface. The exception is plastics.

PLASTICS AS MARINE DEBRIS Globally, **plastics** (*plasticus* = molded) constitute the vast majority of marine debris. About 80% of marine debris comes from land-based sources, and the majority of that is plastics. When plastics enter the ocean, they float and are not readily biodegradable (Figure 11.34). As a result, plastics can remain in the marine environment almost indefinitely, affecting marine organisms

NO DUMPING
I LIVE DOWNSTREAM

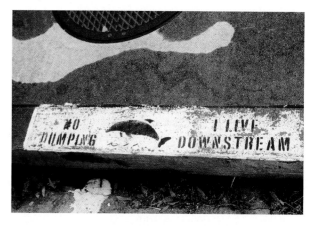

Figure 11.31 A labeled storm drain that leads to the ocean. Although many people believe that storm drain runoff is processed by sewage treatment plants, any material that enters storm drains goes directly into streams or the ocean.

Cigarettes/cigarette filters 22%
Other 22%
Food containers, wrappers 9%
Bottle caps, lids 9%
Cups, plates, plastic utensils, straws, stirrers 12%
Bottles and cans 13%
Plastic bags 13%

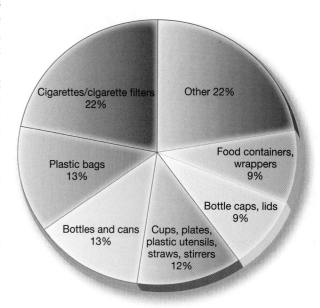

Figure 11.32 Most commonly found items collected during beach cleanups. Most of the trash collected during beach cleanups consists of single-use items of which people dispose every day.

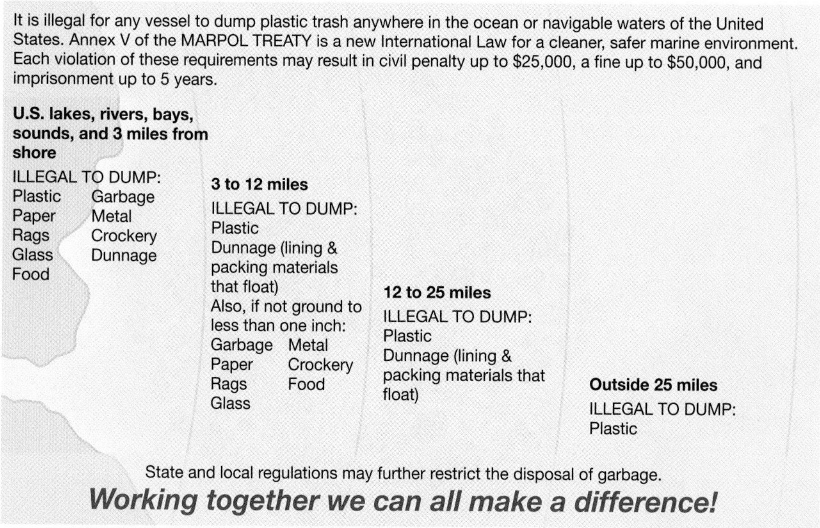

It is illegal for any vessel to dump plastic trash anywhere in the ocean or navigable waters of the United States. Annex V of the MARPOL TREATY is a new International Law for a cleaner, safer marine environment. Each violation of these requirements may result in civil penalty up to $25,000, a fine up to $50,000, and imprisonment up to 5 years.
U.S. lakes, rivers, bays, sounds, and 3 miles from shore
ILLEGAL TO DUMP:
Plastic Garbage
Paper Metal
Rags Crockery
Glass Dunnage
Food
3 to 12 miles
ILLEGAL TO DUMP:
Plastic
Dunnage (lining & packing materials that float)
Also, if not ground to less than one inch:
Garbage Metal
Paper Crockery
Rags Food
Glass
12 to 25 miles
ILLEGAL TO DUMP:
Plastic
Dunnage (lining & packing materials that float)
Outside 25 miles
ILLEGAL TO DUMP:
Plastic
State and local regulations may further restrict the disposal of garbage.
Working together we can all make a difference!

Figure 11.33 Current law regulating ocean dumping. Many types of trash can legally be dumped in the ocean, as long as it is ground fine enough and not composed of plastic. In fact, plastic is the only substance that cannot be dumped anywhere in the ocean.

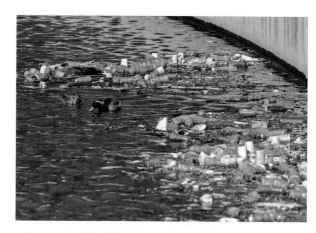

Figure 11.34 Floating plastic trash. When plastic trash enters the ocean, it floats and is not readily biodegradable.

STUDENTS SOMETIMES ASK . . .

Don't storm drains receive treatment before emptying into the ocean?

Contrary to popular belief, water (and any other material) that goes down a storm drain does not receive any treatment before being emptied into a river or directly into the ocean. Sewage treatment plants receive enough waste to process without the additional run-off from storms, so it is important to monitor carefully what is disposed into storm drains. For instance, some people discharge used motor oil into storm drains, thinking that it will be processed by a sewage plant. A good rule of thumb is this: *Don't put anything down a storm drain that you wouldn't put directly into the ocean itself.*

through entanglement and ingestion. In fact, there are many documented cases of plastic waste such as six-pack rings or packing straps strangling fish, marine mammals, and birds that have been entangled in plastic trash (**Figures 11.35a** and **11.35b**). Marine birds have also ingested so much floating plastic trash that it fills their stomachs and they die from starvation (**Figure 11.35c**). In probably the most well-known case of lethal ingestion of plastic trash, marine turtles have been killed when they eat floating plastic bags, evidently mistaking them for jellyfish or other transparent plankton on which they typically feed.

Entanglement and ingestion, however, are not the worst problems caused by the ubiquitous plastic pollution in the ocean. Researchers have recently discovered that floating plastic pieces have a high affinity for non-water-soluble toxic compounds, most notably DDT, PCBs, and other oily pollutants. As a result, some plastic pieces accumulate poisons to levels as high as a million times their concentrations in seawater. When marine organisms ingest the toxic plastic pieces, they accumulate vast quantities of the toxic materials.

(a) A female northern elephant seal (*Mirounga angustirostris*) with a plastic packing strap tight around her neck.

(b) A herring gull (*Larus argentatus*) entangled in a plastic six-pack ring.

Web Video
Plastic Pollution: Our Synthetic Sea

Figure 11.35 Examples of how floating plastic trash endangers marine life.

(c) Photo of a young laysan albatross (*Phoebastria immutabilis*) from Midway Island ot the northwestern end of the Hawwaiian chain that died from a stomach full of pieces of floating plastic trash (*right*) including lighters, bottle caps, electrical connectors, and other materials unintentionally fed to it by its parents.

Despite the fact that plastics are one of the few substances that are illegal to dump anywhere in the ocean (see Figure 11.33), their properties and abundant usage in society contribute to their increasing profusion in the marine environment.

A BRIEF HISTORY OF PLASTICS   Even though human-produced plastics made their debut in 1862 at the Great International Exhibition in London, their commercial development didn't occur until World War II, when shortages of rubber and other materials created great demand for alternative products. Plastic products are *lightweight*, *strong*, *durable*, and *inexpensive*, so they have many advantages over products made from other materials. By the 1970s, plastic products found their way into virtually every aspect of human lives, from airplane parts to zippers (**Figure 11.36**). Humans, for example, wear plastics, cook with plastics, drive in plastics, and even have internal artificial parts made of plastics. The convenience of plastic items intended for one-time use has also contributed to their popularity.

What was once thought of as a miracle substance, however, has been found to have several disadvantages. Disposing of plastics has already strained the capacity of land-based solid-waste disposal systems. Plastic waste is now an increasingly abundant component of oceanic flotsam (floating refuse). Unfortunately, the very same properties that make plastics so advantageous make them unusually persistent and damaging when released into the marine environment:

- They are *lightweight*, so they float and concentrate at the surface.
- They are *strong*, so they entangle marine organisms.
- They are *durable*, so they don't biodegrade easily, causing them to last almost indefinitely.
- They are *inexpensive*, so they are mass-produced and used in almost everything.

PLASTIC NURDLES IN THE MARINE ENVIRONMENT   Today, nearly all plastic products are produced from small pre-production plastic pellets called **nurdles** that range in size from a BB to a pea (**Figure 11.37**). Nurdles are transported in bulk aboard commercial vessels and are found throughout the oceans, probably due to spillage at loading terminals.

Because of ocean currents that wash plastics ashore, plastic nurdles and other trash can be found on all beaches—even in remote areas. In one of the best-documented studies of trash on beaches, for example, researchers painstakingly counted all debris larger than a pinhead along selected transits in Orange County, California. Based on what was collected during the six-week sampling period, researchers project that each year Orange County beaches receive an astonishing 106.9 million pieces of debris, 98% (105.2 million) of which are nurdles. Other studies have shown that some Bermuda beaches contain up to 10,000 nurdles per square meter (925 nurdles per square foot) and some beaches on Martha's Vineyard in Massachusetts contain 16,000 nurdles per square meter (1480 nurdles per square foot).

THE ISSUE OF PLASTICS IN THE OCEAN   Oceanographers have recently documented the accumulation of floating plastic trash in unprecedented amounts in the open ocean such as the centers of the major subtropical gyres, which have a sluggish circular motion and tend to accumulate floating debris. Indeed, equipment deployed from research and commercial fishing vessels often returns entangled in various types of plastic trash. Floating plastic trash is now so abundant that it is being used as artificial habitat in the open ocean by marine creatures from microbes to fish. As such, it is beginning to affect the composition of marine ecosystems.

What makes matters worse is that over time, floating plastics *photodegrade*, a process in which sunlight breaks them into progressively smaller pieces, which only serves to facilitate the ingestion of plastics by all types of marine organisms, particularly microscopic organisms that constitute lower levels of marine food webs.

Figure 11.36 Examples of plastic products. A variety of plastic products are used in everyday life.

Figure 11.37 Plastic pellets (*nurdles*) found at a beach. These pre-production plastic nurdles were found at a Southern California beach. Nurdles have been found floating in the ocean and on all beaches, even in remote areas.

Web Video
Seas of Plastic

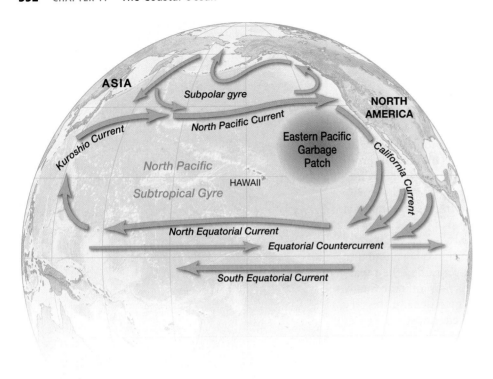

Figure 11.38 The Eastern Pacific Garbage Patch. Map showing the approximate location of the convergence of major ocean surface currents in the North Pacific Gyre that concentrates floating trash and produces the Eastern Pacific Garbage Patch. Note that other subtropical gyres contain similar garbage patches.

Although some plastics wash up on beaches, the vast majority of plastics don't ever leave the ocean—they just break down into smaller pieces and enter marine food webs at progressively lower levels.

Studies have revealed that marine plastic particles increased tenfold every 10 years in the 1970s and 1980s—and then tenfold again in just 3 years in the 1990s. Where does all this floating plastic trash go? Some of it washes up on various beaches, and some is eaten by marine animals, but most of it remains at sea, constantly breaking down into smaller pieces. Microscopic examination of these bits of floating plastic reveals that they are host to an abundance of marine bacteria, which further break the plastic down into smaller pieces.

In addition, oceanographers have discovered large floating regions of trash in all oceans of the world where plastics outnumber marine organisms several times over. These regions are located near the center of the world's gyres (see Chapter 7), where trash becomes trapped in large and relatively stable vortices caused by the convergence of major ocean surface currents. One of the best studied of these regions is the **Eastern Pacific Garbage Patch** (Figure 11.38), which is reportedly about twice the size of Texas and is estimated to contain a staggering 91 million metric tons (100 million short tons) of debris.

STUDENTS SOMETIMES ASK . . .

What does the garbage patch look like? Can it be seen from space? How can it be cleaned up?

One of the problems with determining the size of the garbage patch is that it lacks definitive edges, and so there are no accurate estimates of its size. Contrary to what many people have heard about it, the patch is not a floating island of garbage. Since most of the plastic pieces are about the size of a fingernail or smaller, the garbage patch cannot easily be seen from a boat, let alone from space. (Oceanographers use fine mesh trawl nets to determine its existence.) Imagine a very thin soup with some tiny scattered pieces of plastic.

Removing the floating garbage is not as easy as it sounds. For one thing, the trash is unevenly distributed. For another, it involves gathering small pieces that are spread across a huge area, which would require an armada of ships working continuously for many years to collect all the trash. Even if large amounts of floating trash were collected, what would be done with all of it once back on shore, especially the parts that can't be recycled? Certainly a bigger issue is that skimming the ocean's surface waters for floating trash would also remove plankton that are crucial to marine food webs. At the moment, the best approach appears to be preventing the trash from getting to the ocean in the first place.

REDUCING THE AMOUNT OF PLASTICS IN THE OCEAN What can be done to limit the amount of plastics in the marine environment? For starters, people can limit their use of single-use plastic, recycle plastic material, and dispose of their plastic trash properly, including not dumping any plastics at sea. At least 50 California municipalities have recently banned single-use plastic bags and polystyrene takeout containers, with mounting pressure to enforce statewide prohibition on these items. Other cities such as Seattle (Washington) and Toronto (Ontario, Canada) have recently passed laws to ban the sale of plastic bags. Even entire countries such as China, Australia, Italy, South Africa, Tanzania, and Bangladesh have banned plastic grocery bags. In addition, beach cleanups are responsible for removing an amazing amount of trash from beaches (see Figure 11.32). All of these actions help reduce the amount of plastics in the marine environment.

In the 1980s, increasing concern about floating trash—in particular syringes and other medical waste that closed long stretches of beaches in the New York area—prompted nations of the world to come together to try to correct the problem. In 1988, the International Convention for the Prevention of Pollution from Ships, which is commonly known as **MARPOL** (short for "Marine Pollution"), proposed a treaty banning the disposal of all plastics and regulating the dumping of most other garbage at sea (see Figure 11.33). By 2005, 122 nations had ratified MARPOL. Some studies suggest that MARPOL has reduced both marine debris and entanglements by discarded fishing nets in some places, most notably in waters off Alaska and California. Other studies, however, show no improvement in areas such as the Southern Ocean, South Atlantic, and Hawaiian Islands. As with many other environmental protections and international treaties, enforcement lags far behind stated intentions. Although an annex to MARPOL requires that nations that have signed the treaty provide shore-based facilities where ships can easily dispose of their garbage, many developing countries have not been able to provide these facilities. As a result, even well-meaning captains and shippers who want to comply with existing international marine pollution laws may find it difficult to do so.

Biological Pollution: Non-native Species

Non-native species (also called *exotic, alien,* or *invasive species*) are species that originate in a particular area but are introduced into new environments either by the deliberate or accidental actions of humans and so are classified as biological pollutants. Because non-native species inhabit new areas where they lack predators or other natural controls, they can wreak ecological havoc by outcompeting and dominating native populations. Non-native species can also introduce new parasites and/or diseases. In some cases, non-native species completely transform ecosystems. In the United States alone, more than 7000 introduced species (not counting microorganisms) have been documented, of which about 15% cause ecological and economic damage. In fact, invasive species cause an estimated $137 billion in loss and damages in the United States each year.

The seaweed *Caulerpa taxifolia*, which is native in tropical waters, is one example of an invasive, non-native marine species. It is ideal as a decorative alga in saltwater aquariums because it is hardy, fast growing, and not edible by most fish. However, when it is introduced into suitable new habitats (probably as a result of the dumping of household saltwater aquariums), it becomes a dominant and persistent species that displaces native seaweeds and other marine life. In 1984, a cold-tolerant clone of *Caulerpa taxifolia* produced for the aquarium industry was first introduced into the Mediterranean Sea, where it has overwhelmed aquatic ecosystems and continues to spread. This clone has also been found in the lagoons of Southern California (**Figure 11.39**), where it was first reported in 2000, but eradication efforts appear to have been successful. However, because *Caulerpa taxifolia* can regenerate from very small fragments, repeated surveys are being conducted in these lagoons to eliminate all remaining occurrences of the seaweed. Recently, a new infestation of *Caulerpa taxifolia* has been reported near Sydney, Australia.

Another example of a non-native aquatic species is the **zebra mussel** (*Dreissena polymorpha*), which has invaded the Great Lakes region. After entering North America more than a decade ago (probably in the ballast water[10] of a freighter from Europe), these mussels have proliferated rapidly in waters of eastern Canada and the United States. In doing so, they've driven out native mussels, altered the ecology of freshwater lakes and streams, and blocked the water-carrying pipes of power plants and many other industrial facilities. Although zebra mussels are exceedingly hardy organisms, researchers are working to identify predators, parasites, and infectious microbes that can kill zebra mussels but leave native populations unharmed.

Other notable examples of harmful non-native aquatic species include the Atlantic comb jelly *Mnemiopsis leidyi*, which was transported in ballast water to the Black Sea

STUDENTS SOMETIMES ASK . . .

All this pollution is distressing. Is there anything I can do about it?

Yes, there are many things you can do to help protect the ocean, some of which are listed in this book's Afterword. They all involve making intelligent choices and minimizing your impact on the environment. Nonpoint source pollution, for example, is something that the general public is directly responsible for, so one of the best methods of prevention may be educating people. Once people understand the impact their choices have on the environment, they often realize that the solution is up to *all* of us.

Figure 11.39 Invasion of *Caulerpa taxifolia*. The seaweed *Caulerpa taxifolia* is thought to have been illegally dumped from an aquarium into Southern California coastal lagoons. The alga was eradiated because of good public awareness and quick action by local authorities.

[10]Ballast water is taken into the hold of a ship to enhance stability and then released in a port when it is no longer needed.

ESSENTIAL CONCEPT 11.6

Examples of marine pollution include petroleum from oil spills; sewage sludge; chemicals such as DDT, PCBs, and mercury; nonpoint source pollution such as road oil and trash (including plastics); and biological pollution such as non-native species.

and has done extensive damage; the Atlantic cordgrass *Spartina alterniflora*, which has invaded soft-bottom coasts of California and Washington; the water hyacinth *Eichhornia crassipes*, which infests tropical estuaries and other water bodies; and the European green crab *Carcinus maenas*, which has invaded the Pacific Coast and is altering coastal food webs.

CONCEPT CHECK 11.6

❶ Why would many marine pollution experts consider oil among the *least* damaging pollutants in the ocean?

❷ Discuss techniques used to clean oil spills. Why is it important to begin the cleanup immediately?

❸ Describe the world's largest accidental and intentionally released oil spills. How many times larger was each than the *Exxon Valdez* oil spill?

❹ How would dumping sewage in deeper water off the East Coast help reduce negative impacts on the ocean floor?

❺ Discuss the animal populations that clearly suffered from the effects of DDT and the way in which this negative effect was manifested.

❻ What causes Minamata disease? What are the symptoms of the disease in humans?

❼ What is nonpoint source pollution, and how does it get to the ocean? What other ways does trash get into the ocean?

❽ What properties contributed to plastics being considered a miracle substance? How do those same properties cause them to be unusually persistent and damaging in the marine environment?

❾ What are non-native species? Why can they be so damaging to ecosystems?

Essential Concepts Review

11.1 What laws govern ocean ownership?

▶ *Coastal waters support about 95% of the total mass of life in the oceans*, and they are important areas for commerce, recreation, fisheries, and the disposal of waste. *Coastal law is specified in the United Nations Convention on the Law of the Sea*, which has been ratified by 153 coastal nations and in force since 1994. The treaty delineates *national obligations, rights, and jurisdiction in coastal waters.* In spite of its near-universal acceptance, *the United States has still not ratified the treaty.*

Study Resources
Online Study Guide Quizzes

Critical Thinking Question

Discuss reasons less-developed nations believe that the open ocean is the common heritage of all, whereas the more developed nations believe that the open ocean's resources belong to those who recover them.

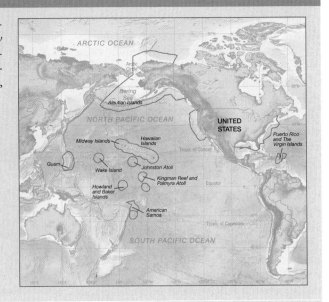

11.2 What characteristics do coastal waters exhibit?

▶ *The temperature and salinity of the coastal ocean vary over a greater range than the open ocean* because the coastal ocean is shallow and experiences river runoff, tidal currents, and seasonal changes in solar radiation. *Coastal geostrophic currents are produced from freshwater runoff and coastal winds.*

Study Resources
Online Study Guide Quizzes

Critical Thinking Question

Describe several differences between the coastal ocean and the open ocean.

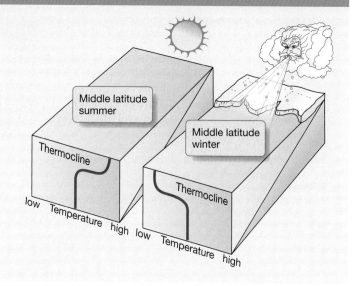

11.3 What types of coastal waters exist?

▶ *Estuaries are partially enclosed bodies of water where freshwater runoff from the land mixes with ocean water.* Estuaries are *classified by their geologic origin* as coastal plain, fjord, bar built, or tectonic. Estuaries are also *classified by their mixing patterns* of freshwater and saltwater as vertically mixed, slightly stratified, highly stratified, and salt wedge. *Typical circulation in an estuary consists of a surface flow of low-salinity water toward its mouth and a subsurface flow of marine water toward its head.*

▶ *Estuaries provide important breeding and nursery areas for many marine organisms but often suffer from human population pressures.* The *Columbia River Estuary*, for example, has degraded from agriculture, logging, and the construction of dams upstream. In the *Chesapeake Bay*, an anoxic zone occurs during the summer that kills many commercially important species.

▶ *Long offshore deposits called barrier islands protect marshes and lagoons.* Some lagoons have restricted circulation with the ocean, so water temperatures and salinity may vary widely with the seasons.

▶ *Circulation in the Mediterranean Sea is characteristic of restricted bodies of water in areas where evaporation greatly exceeds precipitation.* Called *Mediterranean circulation*, it is the reverse of estuarine circulation.

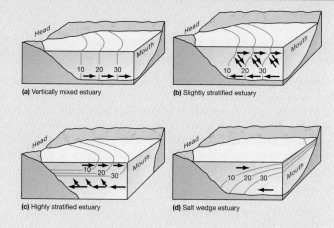

(a) Vertically mixed estuary

(b) Slightly stratified estuary

(c) Highly stratified estuary

(d) Salt wedge estuary

Study Resources
Online Study Guide Quizzes

Critical Thinking Question

Based on their geologic origin, draw and describe the four major types of estuaries and give an example of where each one occurs.

Essential Concepts Review (cont.)

11.4 What issues face coastal wetlands?

▶ *Wetlands are some of the most biologically productive regions on Earth. Salt marshes* and *mangrove swamps* are important *examples of coastal wetlands. Wetlands are ecologically important* because they *remove land-derived pollutants from water before it reaches the ocean* and they *provide critical habitat* for many seagoing species. Nevertheless, *human activities continue to destroy wetlands.*

Study Resources
Online Study Guide Quizzes

Critical Thinking Question

Discuss reasons wetlands are being destroyed worldwide in spite of their many benefits.

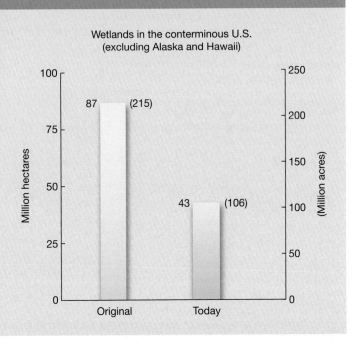

Wetlands in the conterminous U.S. (excluding Alaska and Hawaii)

11.5 What is pollution?

▶ *Although marine pollution seems easily defined, an all-encompassing definition is quite detailed.* It is often difficult to establish the degree to which pollution affects ocean areas. The most widely used technique for determining the effects of pollution is the *environmental bioassay*, which determines the concentration of a pollutant that causes a *50% mortality rate* among test organisms. The debate continues about whether society's wastes should be dumped in the ocean.

Study Resources
Online Study Guide Quizzes

Critical Thinking Question

From memory, define *pollution*. Then consider these items and determine whether each one is a pollutant based on your definition (and refine your definition as necessary):

a. Dead seaweed on a beach

b. Natural oil seeps

c. A small amount of sewage

d. Warm water from a power plant dumped into the ocean

e. Sound from boats in the ocean

11.6 What are the main types of marine pollution?

▶ *Oil is a complex mixture of hydrocarbons and other substances,* most of which are naturally *biodegradable.* Thus, *many marine pollution experts consider oil to be among the least damaging of all substances* introduced into the marine environment. In areas that have experienced oil spills, *recovery can be as rapid as a few years.* Still, *oil spills can cover large areas and kill many animals.* After the much-publicized *Exxon Valdez* oil spill in Alaska, *many novel approaches were used to clean the spilled oil,* including releasing oil-eating bacteria (*bioremediation*).

▶ *Millions of tons of sewage sludge have been dumped offshore in coastal waters.* Although 1972 legislation required an *end to dumping of sewage* in the coastal ocean by 1981, *exceptions continue to be made.* Increased public concern resulted in new legislation to prohibit sewage dumping in the ocean.

▶ *DDT and PCBs are persistent, biologically hazardous chemicals that have been introduced into the ocean by human activities.* DDT pollution produced a decline in the Long Island *osprey population* in the 1950s and the *brown pelican population* of the California coast in the 1960s. *Virtual cessation of DDT use* in the Northern Hemisphere in 1972 allowed the *recovery of both populations.* The DDT thinned the eggshells and reduced the number of successful hatchings. PCBs have been implicated in causing health problems in sea lions and shrimp.

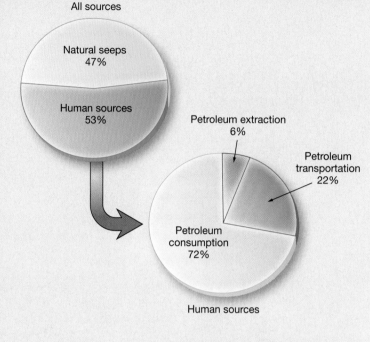

All sources

Natural seeps 47%

Human sources 53%

Petroleum extraction 6%

Petroleum transportation 22%

Petroleum consumption 72%

Human sources

▶ *Mercury poisoning was the first major human disaster resulting from ocean pollution.* It is now called *Minamata disease,* after the bay in Japan where it first occurred in 1953. The toxic form of mercury is *methylmercury, which bioaccumulates in the tissues of many large fish,* most notably tuna and swordfish, and *works its way up the food web* in a process called *biomagnification.* To prevent mercury poisonings in the United States, *stringent methylmercury contamination levels in fish have been established by the FDA.*

▶ *Nonpoint source pollution includes road oil and trash. Plastics* have gained popularity in modern culture because they *are lightweight, strong, durable, and inexpensive.* Unfortunately, these same properties make them a relentless *source of floating trash* in the ocean, especially because they *break down into progressively smaller pieces* over time. *The amount of plastics accumulating in the oceans has increased dramatically.* Certain types of plastics are known to be *lethal to marine mammals, birds, and turtles.* International legislation such as *MARPOL regulates the disposal of trash in the ocean* but lacks enforcement.

▶ The introduction of *biological pollution* such as *non-native species* (including *Caulerpa taxifolia* and zebra mussels) into new environments can cause *severe ecological and economic damage.*

Study Resources
Online Study Guide Quizzes, Web Table 11.1, Web Videos

Critical Thinking Question

List and describe the three main problems that floating plastic trash presents to marine organisms. Of the three, which one is potentially the most serious threat? Explain.

MasteringOceanography™

www.masteringoceanography.com

Looking for additional review and test prep materials? Visit the Study Area in MasteringOceanography to enhance your understanding of this chapter's content by accessing a variety of resources, including Self-Study Quizzes, Geoscience Animations, RSS feeds, flashcards, Web links, and an optional Pearson eText.

Unique adaptations of marine life. The goosefish (*Lophius americanus*; also called monkfish, or angler) has unique adaptations, including feathery appendages as camouflage and a modified dorsal fin that serves as a lure to attract prey. It has been described as mostly mouth with a tail attached, and it can eat prey almost as big as itself. The bottom-dwelling goosefish lives in water as deep as 800 meters (2600 feet) and ranges along the Atlantic coast from Grand Banks, Newfoundland, to Cape Hatteras, North Carolina.

Before you begin reading this chapter, use the glossary at the end of this book to discover the meanings of any of these words you don't already know:

Species · Crepuscular · Streamlining · Zooplankton · Nekton · Phytoplankton · Deep scattering layer (DSL) · Bioluminescence · Hypotonic · Euphotic zone · Detritus · Biomass · Benthic environment · Archaea · Abyssal · Oxygen minimum layer (OML) · Viscosity · Hypertonic · Countershading · Osmosis · Epifauna · Pelagic environment · Taxonomy · Plankton · Disruptive coloration

MARINE LIFE AND THE MARINE ENVIRONMENT

ESSENTIAL CONCEPTS

At the end of this chapter, you should be able to:

12.1 Discuss the characteristics of life and how living things are classified

12.2 Describe how marine organisms are classified

12.3 Demonstrate an understanding of the number of marine species that exist

12.4 Explain how marine organisms are adapted for the physical conditions of the ocean

12.5 Compare the main divisions of the marine environment

"A species is a masterpiece of evolution, a million-year-old entity encoded by five billion genetic letters, exquisitely adapted to the niche it inhabits."

—E. O. Wilson, biologist and global conservation advocate (2001)

A wide variety of organisms inhabits the marine environment. These organisms range in size from microscopic bacteria and algae to the blue whale, which is as long as three buses lined up end to end. Marine biologists have identified more than 250,000 marine species; this number is constantly increasing as new organisms are discovered.

Most marine organisms live within the sunlit surface waters of the ocean. Strong sunlight supports photosynthesis by marine algae, which either directly or indirectly provides food for the vast majority of marine organisms. All marine algae must live near the surface because they need the sunlight; most marine animals live near the surface because this is where food can be obtained. In shallow-water areas close to land, sunlight reaches all the way to the ocean floor, resulting in an abundance of marine life.

There are advantages and disadvantages to living in the marine environment. One advantage is that there is an abundance of water available, which is necessary for maintaining all types of life. One disadvantage is that maneuvering in water can be difficult because the high density of water impedes movement. The individual success of species depends on their ability to find food, avoid predators, and cope with the many physical barriers to their movement.

12.1 What Are Living Things, and How Are They Classified?

Living things are classified based on their physical characteristics, with organisms that are closely related to one another sharing a host of common traits. Recently, genome sequencing of the DNA of many organisms (including human, mouse, dog, cat, cow, elephant, honeybee, platypus, diatom, red algae, sea urchin, dolphin, and hundreds of bacteria and viruses, among others) has allowed genetic comparison, sometimes confirming many of the classification groupings based on structure and other times indicating some unexpected relationships. First, however, let's examine what characteristics qualify an entity as being alive.

A Working Definition Of Life

It might seem easy to differentiate those that are living from the nonliving, but the unusual nature of some life-forms makes defining *life* a challenging task.

Also, both living and nonliving things are composed of the same basic building blocks: atoms, which move continuously in and out of living and nonliving systems. This free exchange of identical components between life and nonlife is one of the factors that complicates attempts at formally defining life.

A simple definition of life is that it consumes energy from its environment. Using this definition, a car engine could probably be classified as alive. An engine, however, cannot self-replicate or otherwise reproduce itself, which is another key component of life.

Several other qualities are crucial in defining life. Water probably needs to be a part of a living organism because living things need a solvent for biochemical reactions—though ammonia or sulfuric acid might also work. A living thing probably has to have some sort of a membrane to distinguish itself from its environment. In addition, most living things tend to respond to stimuli or adapt to their environment. Finally, life as we know it is carbon based, since carbon is so useful in making chemical compounds. Because NASA's definition of life must encompass the potential of extraterrestrial life, it uses the following simple working definition of life: "Life is a self-sustained chemical system capable of undergoing Darwinian evolution."[1] However, even this definition is problematic in that it would likely require observation of several successive generations over a considerable length of time to verify evolution in a life-form.

A good working definition of life, then, should incorporate most of these ideas: that living things can capture, store, and transmit energy; they are capable of reproduction; they can adapt to their environment; and they change through time.

The Three Domains of Life

All living things belong to one of three domains, or "superkingdoms," of life: Bacteria (or Eubacteria), Archaea, and Eukarya (Figure 12.1a). Domain **Bacteria** (*bakterion* = rod) includes simple life-forms with cells that usually lack a

Figure 12.1 The three domains of life and the five kingdoms of organisms. Starting with a common ancestral community of primitive cells, life on Earth can be visualized as branching out into **(a)** the three domains of life and **(b)** the five kingdoms of organisms, which shows representative organisms in each kingdom.

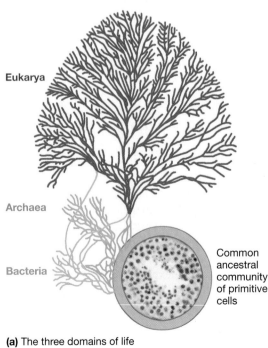

(a) The three domains of life

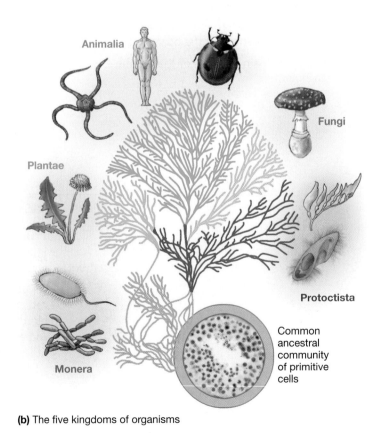

(b) The five kingdoms of organisms

---

[1]See Diving Deeper 1.2 for a description of Darwinian evolution.

nucleus, including purple bacteria, green nonsulfur bacteria, and cyanobacteria (blue-green algae). Domain **Archaea** (*achaeo* = ancient) is a group of simple, microscopic, bacteria-like creatures that includes methane producers and sulfur oxidizers that inhabit deep-sea vents and seeps,[2] as well as other forms—many of which prefer environments with extreme temperatures and/or pressures. Domain **Eukarya** (*eu* = good, *karuon* = nut) includes complex organisms: multicellular plants; multicellular animals; fungi; and protoctists, which are a diverse array of mostly microscopic organisms that don't fit into any other group. The main component of eukaryotes, DNA, is housed in a discrete nucleus, and their cells contain structures that supply energy to build the cell and maintain its functions.

What were the ancestors of these three domains of life? According to modern evolutionary theory (and first suggested by Charles Darwin), all organisms on Earth share a common genetic heritage, each being the genealogical descendant of a single primitive species from the distant past. This idea, called *universal common ancestry*, has recently stood up to rigorous statistical analyses using a set of commonly retained proteins found in a wide range of living organisms. This means that the ancestors of the three domains of life are thought to have consisted of a community of early primitive cells, some of which apparently acquired new genetic material by engulfing microbial neighbors, including their genetic codes. Through *symbiosis* (*sym* = together, *bios* = life),[3] these groups of organisms helped each other coexist for their mutual benefit, and as a result evolved into new organisms that contained merged genes. Today, in fact, it is well established that such adoption of foreign genetic material is common among single-celled bacteria and other microbes but also occurs in some fungi, plants, insects, worms, and other animals. For example, a recent study of venomous jellies has revealed that one of the genes necessary for jellies to sting is identical to a gene in bacteria, suggesting that the ancestors of jellies picked up the gene from microbes. This process, called *lateral gene transfer*, was likely quite important in the evolution of some of the first complex cells.

The Five Kingdoms of Organisms

Within the three domains of life, a system of five kingdoms was first proposed by ecologist and biologist Robert H. Whittaker in 1969. Although the five kingdoms have been modified somewhat since they were first proposed, the five kingdoms are Monera, Plantae, Animalia, Fungi, and Protoctista (or Protista) (**Figure 12.1b**). More recently, some biologists now recognize additional kingdoms within the Archaea domain.

Kingdom **Monera** (*monos* = single) includes some of the simplest organisms. These organisms are single celled but lack discrete nuclei and the internal organelles present in all other organisms. Included in this kingdom are cyanobacteria (blue-green algae), heterotrophic bacteria, and archaea. Recent discoveries have shown bacteria to be a much more important part of marine ecology than previously believed. These organisms are found throughout the breadth and depth of the oceans.

Kingdom **Plantae** (*planta* = plant) comprises the multicelled plants, all of which photosynthesize. Only a few species of true plants—such as surf grass (*Phyllospadix*) and eelgrass (*Zostera*)—inhabit shallow coastal environments. In the ocean, photosynthetic marine algae occupy the ecological niche of land plants. However, certain plants are vital parts of coastal ecosystems, including mangrove swamps and salt marshes.

STUDENTS SOMETIMES ASK . . .

What is coevolution? | The term *coevolution* refers to the way species adapt reciprocally to one another as they evolve. Organism interactions actually form a large web, but we tend to think of pairs of interacting species. The evolutionary dance between a predator and its prey is a good example. It is a kind of arms race, with each improvement in the prey's defense—such as better armor or camouflage—challenging the predator to come up with better adaptations that counter the advantage—such as stronger jaws or more sensitive eyes. And it also works vice versa. Because it's a mutual evolution of adaptations, it's hard to tell if the chicken or the egg came first (so to speak).

---

[2] A *seep* is an area where various fluids trickle out of the sea floor.
[3] For more on the various types of symbiosis, see Chapter 14, "Animals of the Pelagic Environment."

Figure 12.2 **Carolus Linnaeus.** An 1805 engraving of Swedish botanist and the father of taxonomic classification, Carolus Linnaeus, who is wearing traditional Lapland clothing.

Kingdom **Animalia** (*anima* = breath) comprises the multicelled animals. Organisms from kingdom Animalia range in complexity from the simple sponges to complex vertebrates (animals with backbones), which also includes humans.

Kingdom **Fungi** (*fungus* is probably from the Greek *sp(h)ongos* = sponge) includes 100,000 species of mold and lichen, though less than one-half of 1% of them are sea dwellers. Fungi exist throughout the marine environment, but they are much more common in the intertidal zone, where they live with cyanobacteria or green algae to form lichen. Other fungi remineralize organic matter and function primarily as decomposers in the marine ecosystem.

Kingdom **Protoctista** (*proto* = first, *ktistos* = to establish) includes a diverse collection of single-celled and multicelled organisms that have a nucleus. Examples of organisms in kingdom Protoctista include various types of marine *algae* (aquatic photosynthetic organisms that can be either microscopic single-celled or macroscopic multicelled) and single-celled organisms called **protozoa** (*proto* = first, *zoa* = animal).

Linnaeus and Taxonomic Classification

In an effort to determine the relationships of all living things on Earth, Swedish botanist Carl von Linné—who Latinized his name to **Carolus Linnaeus** (1707–1778) (**Figure 12.2**)—created a system in 1758 that is the basis of the modern scientific system of classification used today. Linnaeus developed a system similar to the social hierarchy of his day, with its kingdoms, countries, provinces, parishes, and villages. Linnaeus was a gifted observationist and tireless collector who spent much of his time naming species and organizing them into groups. In fact, Linnaeus's original names for the roughly 12,000 organisms he examined over the course of his life became the starting point for all biological classification. Linnaeus's organizational scheme can be visualized as a series of nested boxes (**Figure 12.3**).

Today, the systematic classification of organisms—called **taxonomy** (*taxix* = arrangement, *nomia* = a law)—involves using physical characteristics as well as genetic information to recognize organism similarities and then grouping them into the following increasingly specific categories:

- Kingdom   (Less specific grouping)
- Phylum[4]
- Class
- Order
- Family
- Genus
- Species   (More specific grouping)

All organisms that share a common category (for instance, a *family*—such as the cat or dolphin family) have certain characteristics and evolutionary similarities. In some cases, subdivisions of these categories are also used, such as subphylum (**Table 12.1**). The categories assigned to an individual species must be agreed upon by an international panel of experts.

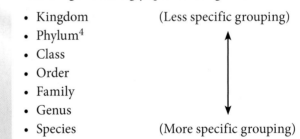

Genus • *Delphinus*
Species • *delphis*
(Common dolphin)

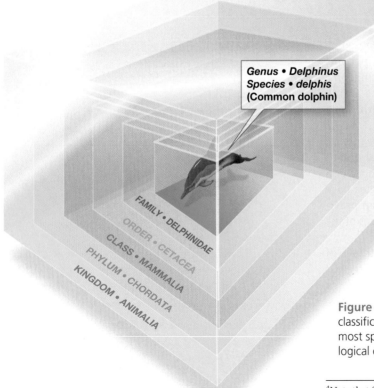

FAMILY • DELPHINIDAE
ORDER • CETACEA
CLASS • MAMMALIA
PHYLUM • CHORDATA
KINGDOM • ANIMALIA

Figure 12.3 **A visualization of Linnaeus's classification scheme.** The Linnaean system of classification can be visualized as a series of nested boxes, with genus and species the most specific groupings (smallest boxes). The box-in-box representation shows genealogical connections among organisms.

---

[4]Note that for plants, the term "division" is used instead of "phylum"; the current International Code of Botanical Nomenclature, however, allows the use of either term.

TABLE 12.1   TAXONOMIC CLASSIFICATION OF SELECTED ORGANISMS

Category	Human	Common dolphin	Killer whale	Bat star	Giant kelp
Kingdom	Animalia	Animalia	Animalia	Animalia	Protoctista
Phylum	Chordata	Chordata	Chordata	Echinodermata	Phaeophyta
Subphylum	Vertebrata	Vertebrata	Vertebrata		
Class	Mammalia	Mammalia	Mammalia	Asteroidea	Phaeophycae
Order	Primates	Cetacea	Cetacea	Valvatida	Laminariales
Family	Hominidae	Delphinidae	Delphinidae	Oreasteridae	Lessoniaceae
Genus	*Homo*	*Delphinus*	*Orcinus*	*Asterina*	*Macrocystis*
Species	*sapiens*	*delphis*	*orca*	*miniata*	*pyrifera*

The fundamental unit of taxonomic classification is the **species** (*species* = a kind). Species consist of populations of genetically similar, interbreeding (or potentially interbreeding) individuals that share a collection of inherited characteristics whose combination is unique.

As an offshoot of his classification scheme, Linnaeus also invented *binomial nomenclature*, where every living thing is known by just two Latin names (previously, organisms were known by a combination of as many as a dozen Latin names). In this way, every type of organism has a unique two-word scientific name composed of its genus and species, which is italicized with the first letter of the generic (genus) name capitalized—for example, *Delphinus delphis*. Most organisms also have one or more common names. *Delphinus delphis*, for instance, is the common dolphin, and *Orcinus orca* is the killer whale or orca. If the scientific name is referred to repeatedly within a document, it is often shortened by abbreviating the genus name to its first letter. Thus, *Delphinus delphis* becomes *D. delphis*.

The Linnaean system, even in its modern form, is far from perfect. New evidence routinely requires taxonomists to move species from one genus to another, or even to an entirely different order. Linnaeus, however, is credited with forging a single, flexible, universally applicable scientific language that is still in use over 250 years after he invented it.

CONCEPT CHECK 12.1

1 What characteristics should be included in a good working definition of life?

2 List the three major domains of life and the five kingdoms of organisms.

Describe the fundamental criteria used in assigning organisms to these divisions.

3 What are thought to be the ancestors of the three domains of life?

Web Animation
A Visualization of Linnaeus's Classification Scheme

ESSENTIAL CONCEPT 12.1

Living organisms use energy, reproduce, adapt, and change through time. Living things can be classified into one of three domains and five kingdoms, each of which is split into increasingly specific groupings of phylum, class, order, family, genus, and species.

12.2 How Are Marine Organisms Classified?

Marine organisms can be classified according to where they live (their habitat) and how they move (their mobility). Organisms that inhabit the water column can be classified as either *plankton* (drifters) or *nekton* (swimmers). All other organisms are *benthos* (bottom dwellers).

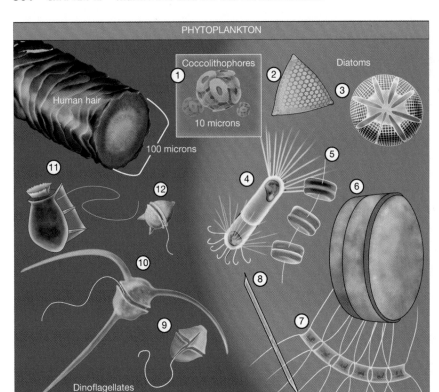

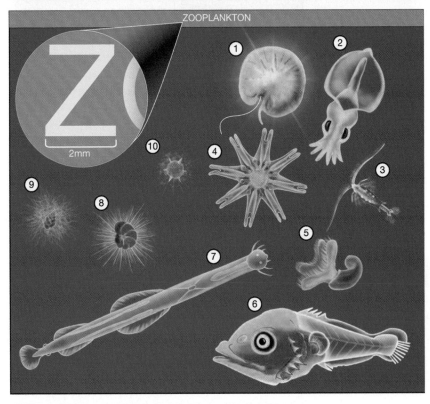

Figure 12.4 Phytoplankton and zooplankton (floaters).
Schematic drawings of various phytoplankton (*above*) and
zooplankton (*below*). **Phytoplankton: (1)** Coccolithophores;
(2–8) Diatoms; **(9–12)** Dinoflagellates. Width of human hair =
100 microns (0.004 in). **Zooplankton: (1)** Noctiluca, a predatory
dinoflagellate; **(2)** Squid larva; **(3)** Copepod; **(4)** Jelly
larva; **(5)** Snail larva; **(6)** Fish larva; **(7)** Arrowworm;
(8–9) Foraminifers; **(10)** Radiolarian. Scale = 2 mm (0.08 in).

Plankton (Drifters)

Plankton (*planktos* = wandering) include all organisms—
algae, animals, and bacteria—that drift with ocean currents.
An individual organism is called a **plankter**. Just because
plankters drift does not mean they are unable to swim. In fact,
many plankters can swim but either move only weakly or move
only vertically. As such, they cannot determine their horizontal
position within the ocean.

Plankton are unbelievably abundant and important within
the marine environment. In fact, *most of Earth's **biomass**—
the mass of living organisms—consists of plankton adrift in the
oceans.* Even though 98% of marine *species* are bottom dwelling,
the vast majority of the ocean's *biomass* is planktonic.

TYPES OF PLANKTON Plankton can be classified based on
their feeding styles. If an organism can photosynthesize and
therefore produce its own food, it is termed **autotrophic**
(*auto* = self, *tropho* = nourishment). Autotrophic plankton are
called **phytoplankton** (*phyto* = plant, *planktos* = wandering)
and can range in size from microscopic algae to larger species
of drifting kelp. If an organism cannot produce its own food
and relies instead on food produced by other organisms, it is
termed **heterotrophic** (*hetero* = different, *tropho* = nourishment). Heterotrophic plankton are called **zooplankton** (*zoo* =
animal, *planktos* = wandering), which includes drifting marine
animals. Representative members of each group are shown in
Figure 12.4.

Plankton also include bacteria. It has recently been discovered that free-living **bacterioplankton** are much more
abundant and far more widely distributed than previously
thought. Having an average diameter of only one-half of a micrometer[5] (0.00002 inch), they were missed in earlier studies
because they are so incredibly small. Recently, microbiologists
have begun to study oceanic bacterioplankton and have even
discovered an extremely small yet abundant bacterium that
has been estimated to constitute at least half of the ocean's
total photosynthetic biomass and is likely the most abundant
photosynthetic organism on Earth.

Plankton also include viruses, which are called
virioplankton. Virioplankton are an order of magnitude
smaller than bacterioplankton and are similarly little known.
The role of viruses in planktonic communities is not well understood, but they may limit the abundance of other types of
plankton through infection.

Although plankton can be classified as either phytoplankton, zooplankton, bacterioplankton, or virioplankton, they can
also be classified according to the portion of their life cycle spent
as plankton. Organisms that spend their entire lives as plankton are **holoplankton** (*holo* = whole, *planktos* = wandering).
Many organisms that spend their adult lives as nekton or
benthos spend their juvenile and/or larval stages as

---

[5]One micrometer (also known as a *micron*) is one-millionth of a meter and is designated by the symbol μm.

plankton (**Figure 12.5**). These organisms are called **meroplankton** (*mero* = a part, *planktos* = wandering).

Finally, plankton can also be classified based on size. For example, large floating animals and algae, such as jellies and Sargassum,[6] are called **macroplankton** (*macro* = large, *planktos* = wandering) and measure 2 to 20 centimeters (0.8 to 8 inches). Plankton also include *bacterioplankton*, which are so small that they can be removed from the water only with special microfilters. These very tiny drifters are called **picoplankton** (*pico* = small, *planktos* = wandering) and measure 0.2 to 2 microns (0.000008 to 0.00008 inch).

Nekton (Swimmers)

Nekton (*nektos* = swimming) include all animals capable of moving independently of the ocean currents by swimming or other means of propulsion. They are capable not only of determining their own positions within the ocean but also, in many cases, of long migrations. Nekton include most adult fish, marine mammals, marine reptiles, and some marine invertebrates such as squid (**Figure 12.6**). When you go ocean swimming, you become nekton, too.

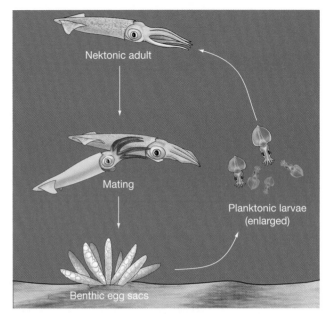

Figure 12.5 Typical life cycle of a squid. Squid are meroplankton because they are planktonic only during their larval stage. Adult squid are nekton, and their egg sacs are benthos.

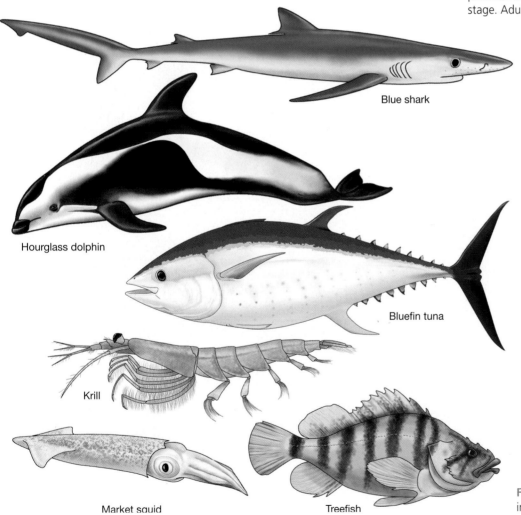

Blue shark

Hourglass dolphin

Bluefin tuna

Krill

Market squid

Treefish

Figure 12.6 Nekton (swimmers). Schematic drawing of various nektonic organisms.

---

[6]*Sargassum* is a floating type of brown macro marine algae commonly referred to as a "seaweed" that is particularly abundant in the Sargasso Sea.

Although nekton move freely, many are unable to move throughout the breadth of the ocean. Gradual changes in temperature, salinity, viscosity, and availability of nutrients effectively limit their lateral range. The deaths of large numbers of fish, for example, can be caused by temporary horizontal shifts of water masses in the ocean. Changes in water pressure normally limit the vertical range of nekton.

Fish may appear to exist everywhere in the oceans, but they are most abundant near continents and islands and in colder waters. Some fish, such as salmon, ascend freshwater rivers to spawn. Many eels do just the reverse, growing to maturity in freshwater and then descending the streams to breed in the great depths of the ocean.

Benthos (Bottom Dwellers)

The term **benthos** (*benthos* = bottom) describes organisms living on or in the ocean bottom. **Epifauna** (*epi* = upon, *fauna* = animal) live on the surface of the sea floor, either attached to rocks or moving along the bottom. **Infauna** (*in* = inside, *fauna* = animal) live buried in the sand, shells, or mud. Some benthos, called **nektobenthos** (*nektos* = swimming, *benthos* = bottom), live on the bottom yet also have the ability to swim or crawl through the water above the ocean floor (such as flat-fish, octopuses, crabs, and sea urchins). Examples of benthos are shown in **Figure 12.7**.

Figure 12.7 Benthos (bottom dwellers): Representative intertidal and shallow subtidal forms. Schematic drawing of various benthos organisms. Organisms include **(1)** sponges, **(2)** sand dollars, **(3)** crinoid, **(4)** sea anemones (open and closed), **(5)** barnacles, **(6)** mussels, **(7)** sea urchin, **(8)** sea cucumber, **(9)** sea hare, **(10)** shore crab, **(11)** sea star, **(12)** abalone, **(13)** ghost crab, **(14)** lug worm, **(15)** annelid worm, and **(16)** clam.

The shallow coastal ocean floor contains a wide variety of physical and nutritive conditions, which have allowed a great number of animal species to develop. Moving across the bottom from the shore into deeper water, the *number* of benthos species per square meter may remain relatively constant, but the *biomass* of benthos organisms decreases. Because these shallow sea floor areas receive sufficient sunlight, they can also support many species of large marine algae (often called "seaweeds") that are attached to the bottom.

Throughout most of the deeper parts of the sea floor, animals live in perpetual darkness, where photosynthetic production cannot occur. They must feed on each other or on whatever outside nutrients fall from the productive zone near the surface.

The deep-sea bottom is an environment of coldness, stillness, and darkness. Under these conditions, life progresses slowly, and organisms that live in the deep sea usually are widely distributed because physical conditions vary little on the deep-ocean floor, even over great distances.

HYDROTHERMAL VENT BIOCOMMUNITIES   Prior to the discovery of deep-sea hydrothermal vent biocommunities in 1977, marine scientists believed that only sparse and small life existed on the deep-ocean floor. Then the first biocommunity at a hydrothermal vent was discovered in waters 2500 meters (8200 feet) deep in the Galápagos Rift off South America, demonstrating for the first time that high concentrations of abundant and large deep-ocean benthos are possible. Because the primary factor that limits life on the deep-ocean floor is sparse food supply, the scientists wondered how the hydrothermal vent organisms were able to obtain enough food to exist.

It turns out that bacteria-like archaea, which produce food not by photosynthesis (for no sunlight is available) but instead thrive on sea floor chemicals, are the base of this marine food web. As a result, the size of individuals and the total biomass in hydrothermal communities far exceed those previously known for deep-ocean benthos. These biocommunities are discussed in Chapter 15, "Animals of the Benthic Environment."

CONCEPT CHECK 12.2

① Describe the lifestyles of plankton, nekton, and benthos. Why does plankton account for a much larger percentage of the ocean's biomass than benthos and nekton combined?

② List the subdivisions of plankton and benthos and the criteria used for assigning individual species to each.

③ For the following marine organisms, determine if they are plankton, nekton, or benthos: (a) sharks, (b) octopuses, (c) clams, (d) diatoms, (e) corals, (f) crabs, (g) giant kelps, (h) jellies, (i) dolphins.

STUDENTS SOMETIMES ASK . . .

What is the difference between kelp, seaweed, and marine algae?

In common usage, these terms all refer to large, branching, photosynthetic marine organisms that contain various pigments that give these organisms their color. However, differences among the terms do exist.

Early mariners probably coined the term *seaweed* because they thought these organisms were a nuisance. They clogged harbors, entangled vessels, and washed up on the beach in useless bunches after storms. Although they could be eaten, they could not be consumed in large quantities. Historically, all marine algae (except microscopic species) had similarities to weeds on land, so they have become known collectively as "seaweeds."

It turns out, however, that these organisms are vital to coastal ecosystems, and so marine biologists prefer to call them "marine algae." To differentiate them from the microscopic planktonic species of algae, large marine algae are often called "marine macro algae." The branching types of brown algae (phylum Phaeophyta) are called "kelp."

Today marine macro algae have many uses. They are used as a thickener and emulsifier in many products (such as toothpaste) and foods (such as ice cream). Cooking recipes sometimes call for small quantities of "sea vegetable," and nori, a type of red alga, is used to wrap sushi. Marine algae are used as fertilizer, and some species have recently been touted as health foods. Marine algae are discussed in Chapter 13, "Biological Productivity and Energy Transfer."

ESSENTIAL CONCEPT 12.2

Marine organisms can be classified according to their habitat and mobility as plankton (drifters), nekton (swimmers), or benthos (bottom dwellers).

12.3 How Many Marine Species Exist?

The total number of cataloged species on Earth in both marine and terrestrial environments is currently 1.8 million; the figure is constantly increasing as new species are discovered. Experts suggest that there must be many marine species (maybe millions) that have not yet been identified due in large part to the difficulty and expense of exploring the deep sea. Overall, as many as 2000 new marine and terrestrial species are discovered each year. Therefore, the total number of species on Earth, known and still undiscovered, has been estimated to be between 3 million and 100 million; the most likely total number of species on Earth is estimated to be between 6 million and 12 million.[7] Recently, a sophisticated analysis that takes

---

[7]Note that this figure does not include the millions of species throughout geologic time that were once living but are now extinct.

Figure 12.8 **Yeti crab.** In 2005, Census of Marine Life (CoML) researchers on a voyage to study the deep sea floor in the South Pacific discovered a white crab with long hairy arms, which they dubbed the yeti crab (*Kiwa hirsuta*). The crab uses its hairs to grow bacteria, which it eats or uses to detoxify poisonous minerals from the water emitted by hydrothermal vents where the crab lives. The length of the yeti crab is 15 centimeters (6 inches).

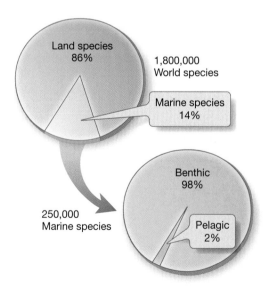

Figure 12.9 **Distribution of species on Earth.** Of the 1.8 million known species on Earth, 86% inhabit land environments and 14% inhabit the ocean. Of the 250,000 known marine species, 98% inhabit the benthic environment and live in or on the ocean floor, while only 2% inhabit the pelagic environment and live within the water column as either plankton or nekton.

advantage of a natural mathematical pattern in the biodiversity produced by evolution proposed that there must be 8.7 million eukaryotic species on Earth (give or take 1 million). Remarkably, newly discovered species are not all microbes or small invertebrates. In fact, new species of frogs, lizards, birds, fish, mammals, and even primates previously unknown to science have been recently discovered in remote areas of the world.

Some of the difficulties of describing the extent to which the marine environment is inhabited include the immense size of the marine habitat and its inaccessibility. In addition, many marine organisms have not been studied in any detail, relationships between organisms are poorly known, and some populations fluctuate greatly each season. To help address these shortcomings, an ambitious $650 million 10-year program called the *Census of Marine Life (CoML)* was completed in 2010. The CoML included ocean surveys conducted by thousands of researchers in more than 80 countries that assessed the diversity, distribution, and abundance of marine life. These organisms included not only fish, but sea birds, marine mammals, invertebrates, and microscopic sea life. The census produced thousands of scientific publications and led to the creation of the Ocean Biogeographic Information System database, which holds the millions of records generated by the surveys. The CoML discovered at least 1200 new marine species, such as the yeti crab (**Figure 12.8**), and will undoubtedly add more species as the surveys are fully analyzed. In all, there are currently 250,000 documented marine species, which represent only about 14% of all known species on Earth (**Figure 12.9**).

Why Are There So Few Marine Species?

If the ocean is such a prime habitat for life and if life originated there, why do so few of the world's species live in the ocean? The disparity is likely a result of the fact that the marine environment is more stable than the terrestrial environment. A major factor that leads to the creation of different species is the variability of the environment: The more variable the environment, the more species are generally present. The higher environmental variability on land presents many opportunities for natural selection to produce new species to inhabit varied new niches, which is one reason tropical rainforests have such a high diversity of species. Conversely, the relatively uniform conditions of the open ocean do not pressure organisms to adapt, so there are fewer species there. In addition, ocean temperatures are not only stable but also relatively low below the sunlit surface waters. The rate of chemical reaction is slowed, which may further reduce the tendency for speciation to occur.

Species in Pelagic and Benthic Environments

Figure 12.9 shows that only 2% or about 5000 of the 250,000 known marine species inhabit the **pelagic** (*pelagios* = of the sea) **environment** and live within the water column. The other 98% inhabit the **benthic** (*benthos* = bottom) **environment** and live either in or on the sea floor. These numbers are minimums, however, because recent discoveries indicate that many more species may inhabit the benthic environment than previously thought.

Why do most marine species inhabit the benthic environment? The ocean floor contains numerous benthic environments (such as rocky, sandy, muddy, flat, sloped, irregular, and mixed bottoms) that create different habitats to which organisms have adapted. On the other hand, most of the pelagic environment—especially that below the sunlit surface waters—is a watery world that is quite uniform from one region to the next and does not experience extreme environmental variability to which organisms need to be adapted in order to survive.

CONCEPT CHECK 12.3

❶ How many total species have been cataloged on Earth? How many marine species exist (both number and percentage of total)? Of marine species, how many are benthic versus pelagic?

❷ Why do scientists think there are many more undiscovered species on Earth?

❸ What factors account for the fact that most marine species inhabit the benthic environment?

12.4 How Are Marine Organisms Adapted for the Physical Conditions Of the Ocean?

For organisms to survive, they must be able to adapt to conditions of their environment. The ocean's physical conditions provide benefits but also present challenges to anything living within it.

For example, the marine environment—particularly its temperature—is far more stable than the terrestrial environment. As a result, ocean-dwelling organisms have not developed highly specialized regulatory systems to adjust to sudden changes that might occur within their environment. Marine organisms can, therefore, be affected adversely by very small changes in temperature, salinity, turbidity, pressure, or other environmental conditions.

Water constitutes more than 80% of the mass of **protoplasm** (*proto* = first, *plasm* = something molded), which is the substance of living matter. In fact, more than 65% of your body's weight—and 95% of a jelly's weight—is water (**Figure 12.10**). Water carries dissolved within it the gases and minerals organisms need to survive. Water is also a raw material in the photosynthesis of food by marine phytoplankton.

Land plants and animals have developed complex "plumbing systems" to retain water and to distribute it throughout their bodies. The inhabitants of the open ocean do not risk atmospheric *desiccation* (drying out), however, because they live in an environment of abundant water.

Need for Physical Support

One basic need of all plants and animals is for simple physical support. Land plants, for example, have vast root systems that anchor the plants securely to the ground. Land animals have skeletons and combinations of appendages—legs, arms, fingers, and toes—to support their entire weight.

In the ocean, water physically supports marine plants and animals. Organisms such as photosynthetic phytoplankton, which must live in the upper surface waters of the ocean, depend primarily upon buoyancy and frictional resistance to sinking in order to maintain their desired position. Still, maintaining position can be difficult, and some organisms have developed special adaptations to increase their efficiency. These adaptations are discussed in this and succeeding chapters.

ESSENTIAL CONCEPT 12.3

Marine species represent only 14% of the total number of known species on Earth. The benthic environment, which has large environmental variability, is home to 98% of the 250,000 known marine species.

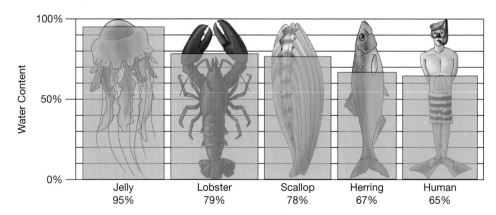

Figure 12.10 Percentage water content of selected organisms. Bar graph showing the water content of selected organisms. Jellies are 95% water, and humans are 65% water.

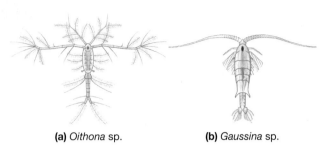

(a) *Oithona* sp.   **(b)** *Gaussina* sp.

Figure 12.11 Water temperature and appendages. Similar species display different adaptations based on water temperature. For example, the copepod *Oithona* displays the ornate plumage characteristic of warm-water varieties **(a)** and the copepod *Gaussina* displays the less ornate appendages found on temperate and cold-water forms **(b)**. Length of copepods = 2 millimeters (0.08 inch).

Water's Viscosity

Viscosity (*viscos* = sticky) is a substance's *internal resistance to flow*. Recall from Chapter 1 that a substance that has high resistance to flow (high viscosity)—such as toothpaste— does not flow easily. Conversely, a substance that has low viscosity—such as water— flows more readily. Viscosity is strongly affected by temperature. Tar, for example, must be heated to decrease its viscosity before it can be spread onto roofs or roads.

The viscosity of ocean water increases as salinity increases and temperature decreases. Thus, single-celled organisms that float in colder, higher-viscosity waters have less need for extensions to help them maintain their positions near the surface. **Figure 12.11** shows, for example, that a warm-water floating crustacean has ornate, featherlike appendages, whereas a cold-water variety does not.

THE IMPORTANCE OF ORGANISM SIZE   The basic requirements of phytoplankton are that they (1) stay in the upper portion of ocean water where solar radiation is available, (2) have available necessary nutrients, (3) efficiently take in these nutrients from surrounding waters, and (4) expel waste materials. Their ingenious size and shape help single-celled phytoplankton satisfy these requirements without needing specialized multiple cells.

Phytoplankton cannot propel themselves, so they use frictional resistance to maintain their general position near the surface of the water. Frictional resistance to sinking increases as an organism's ratio of surface area to volume (mass) increases. For example, **Figure 12.12** shows the surface area to volume ratio of three different cubes and illustrates that the ratio increases as an organism's size decreases. In the figure, Cube a has twice the surface area per unit of volume of Cube b and four times the surface area per unit of volume as Cube c. If the cubes were plankton, Cube a would have four times the resistance to sinking per unit of mass as Cube c, so Cube a would need to exert far less energy to stay afloat. Single-celled organisms, which make up the bulk of photosynthetic marine life, clearly benefit from being as small as possible. They are so small, in fact, that one needs a microscope to see them!

Photosynthetic cells take in nutrients from surrounding water and expel waste through their cell membranes. The efficiency of both functions increases with a higher surface area to volume ratio. Thus, if Cubes a and c in Figure 12.12 were planktonic algae, Cube a could take in nutrients and dispose of waste four times more efficiently than Cube c. This is why cells in all plants and animals are microscopic, regardless of the overall size of the organism.

Diatoms—one of the most important groups of phytoplankton—often have unusual appendages, needlelike extensions, or even rings (**Figure 12.13**) to increase their surface area, thus preventing them from sinking below sunlit surface waters. Other planktonic marine organisms—particularly warmwater species—use similar strategies to stay afloat.

Some small organisms produce a tiny droplet of oil, which lowers their overall density and increases buoyancy. Interestingly, accumulations of vast amounts of these organisms in sea floor sediment can produce offshore oil deposits. This happens when the droplets of oil combine, and, over millions of years, become buried deep below the sea floor, which exposes the oil to the natural heat and high pressure of Earth. Once the oil has undergone chemical changes, it can migrate and become trapped in an oil reservoir.

Despite adaptations to remain in the upper layers of the ocean, organisms still have a higher density than seawater, so they tend to sink, if ever so slowly. This is not a serious handicap, however, because wind causes considerable mixing and turbulence near the surface. Turbulence, in turn, keeps these organisms positioned to

Dimensions of Cube a
Side = 1
Surface = 6
Volume = 1^3 = 1
$\dfrac{S}{V} = \dfrac{6}{1} = 6$

 Crab larva

(a)

Dimensions of Cube b
Side = 2
Surface = $2^2 \times 6 = 24$
Volume = $2^3 = 8$
$\dfrac{S}{V} = \dfrac{24}{8} = 3$

 Amphipod

(b)

Box crab

Dimensions of Cube c
Side = 4
Surface = $4^2 \times 6 = 96$
Volume = $4^3 = 64$
$\dfrac{S}{V} = \dfrac{96}{64} = 1.5$

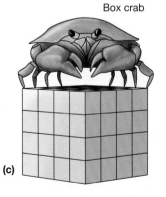

(c)

Oxygen diffusion across skin

95%

40%

5%

Figure 12.12 Surface area to volume ratio of cubes of different sizes. As the linear dimension of a cube increases, the ratio of surface area to volume decreases. Thus, smaller bodies have a higher surface area to volume ratio, which allows them to stay afloat more easily, exchange nutrients and wastes more efficiently, and diffuse oxygen across their skin more effectively.

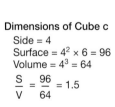

bask in the solar radiation needed to photosynthesize, producing the energy used by essentially all other members of the marine community.

VISCOSITY AND STREAMLINING   As organisms increase in size, viscosity ceases to enhance survival and instead becomes an obstacle. This is particularly true of large organisms that swim freely in the open ocean. They must pursue prey or flee predators, yet the faster they swim, the more the viscosity of water impedes their progress. Not only must water be displaced ahead of the swimmer, but water also must move in behind it to occupy the space that the animal has vacated.

Figure 12.14 shows the advantage of **streamlining**, which is having a shape that offers the least resistance to fluid flow. Streamlining allows marine organisms to overcome water's viscosity and move more easily through water. A streamlined shape usually consists of a flattened body, which presents a small cross section at the front end and a gradual tapering at the back end to reduce the wake created by eddies. It is exemplified in the shape of free-swimming fish (and in marine mammals such as whales and dolphins).

REPRODUCTION   Marine organisms also take advantage of water's high viscosity to enhance chances of reproduction and to populate new habitats. For example, many organisms use the reproductive strategy called **broadcast spawning**, which involves releasing eggs and sperm directly into seawater, like pollen adrift in the wind. In some instances, broadcast spawning occurs during large assemblages when organisms of the opposite sex are in close proximity to one another. In other cases, large quantities of reproductive materials are released into the water, assuming that at least a few will be used elsewhere. In addition, larvae are sometimes released in great numbers into ocean currents, which carry juvenile organisms into new habitats.

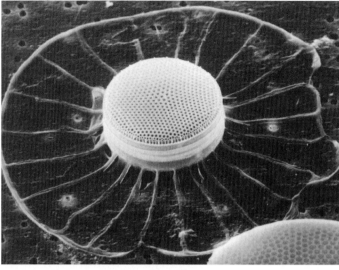

Figure 12.13 Warm-water diatom. Scanning electron micrograph of the warm-water diatom *Planktoniella sol*, which has a prominent marginal ring and spokes to increase its surface area, thus preventing it from sinking (diameter = 60 microns [0.0024 inch]).

Temperature

Figure 12.15 compares extremes in land and ocean surface temperatures and shows that ocean temperatures have a far narrower range than temperatures on land. The minimum surface temperature of the open ocean is seldom much below $-2°C$ (28.4°F), and the maximum surface temperature seldom exceeds 32°C (89.6°F), except in some shallow-water coastal regions, where the temperature may reach 40°C (104°F). On land, however, extremes in temperatures have ranged from $-88°C$ ($-127°F$) to 58°C (136°F), which represents a temperature range more than four times greater than that experienced by the ocean. This *continental effect* was discussed in Chapter 5.

Further, the ocean has a smaller daily, seasonal, and annual temperature range than that experienced on land, which provides a stable environment for marine organisms. The reasons for this are fourfold:

1. Recall from Chapter 5 that the heat capacity of water is much higher than that of land, which causes land to heat up by a greater amount and much more rapidly than the ocean.

2. The warming of the ocean is reduced substantially because of evaporation, a cooling process that stores excess heat as latent heat.

3. Radiation received at the surface of the ocean can penetrate several tens of meters deep and distribute its energy throughout a very large mass. In contrast, solar radiation absorbed by land heats only a very thin surface layer.

4. Unlike solid land surfaces, water has good mixing mechanisms, such as currents, waves, and tides that allow heat from one area to be transported to other areas.

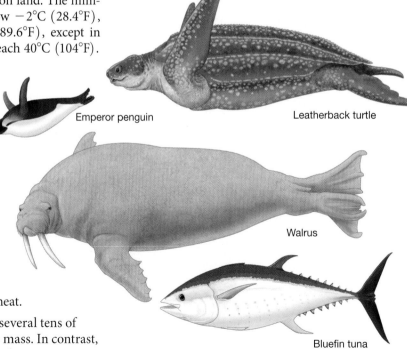

Emperor penguin

Leatherback turtle

Walrus

Bluefin tuna

Figure 12.14 Streamlining. Schematic drawing showing examples of marine organisms that have a streamlined body shape, which enables them to efficiently move through and displace water with minimum resistance (small wake).

Figure 12.15
Comparison of extremes in ocean and land surface temperatures. In the open ocean, the maximum temperature range is limited to only 34°C (61°F); it increases to 42°C (76°F) in the coastal ocean. On land, the maximum temperature range is 146°C (263°F), which is more than four times the temperature range in the open ocean.

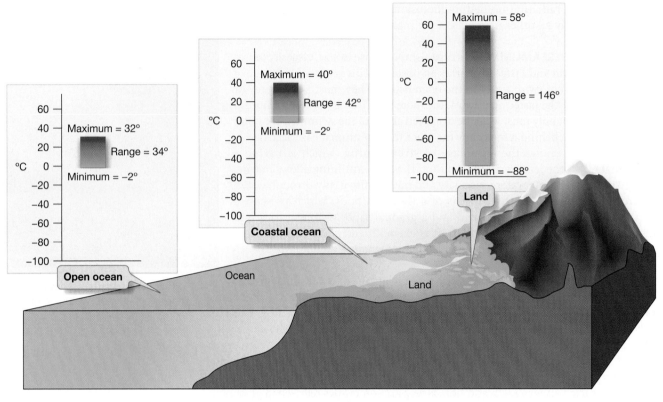

In addition, the small daily and seasonal temperature variations are confined to ocean surface waters and decrease with depth, becoming insignificant throughout the deeper parts of the ocean. At ocean depths that exceed 1.5 kilometers (0.9 mile), for example, temperatures hover around 3°C (37.4°F) year-round, regardless of latitude.

COMPARING COLD- AND WARM-WATER SPECIES Cold water is denser and has a higher viscosity than warm water. These factors, among others, profoundly influence marine life, resulting in the following differences between warm-water and cold-water species in the marine environment:

- Floating organisms are physically smaller in warm waters than in colder waters. Small organisms expose more surface area per unit of body mass, which helps them maintain their position in the lower viscosity and density of warm seawater more easily.

- Warm-water species often have ornate plumage to increase surface area, which is strikingly absent in the larger cold-water species (see Figures 12.11 and 12.13).

- Warmer temperatures increase the rate of biological activity, which more than doubles with an increase of 10°C (18°F). Tropical organisms apparently grow faster, have a shorter life expectancy, and reproduce earlier and more frequently than those in colder water.

- There are more *species* in warm waters, but the total *biomass* of plankton in colder, high-latitude waters greatly exceeds that of the warmer tropics. Note that the high biomass of plankton in high-latitude regions is not directly caused by temperature or viscosity but rather indirectly by allowing the upwelling of nutrients for phytoplankton.

Some animal species can live only in cooler waters, whereas others can live only in warmer waters. Many of these organisms can withstand only very small temperature changes and are called **stenothermal** (*steno* = narrow, *thermo* = temperature).

Stenothermal organisms are found predominantly in the open ocean, at depths where large temperature ranges do not occur.

Other species are little affected by different temperatures and can withstand large and even rapid changes in temperature. These organisms are called **eurythermal** (*eury* = wide, *thermo* = temperature) and are predominantly found in shallow coastal waters—where the largest temperature ranges are found—and in surface waters of the open ocean.

Salinity

The sensitivity of marine animals to changes in their environment varies from organism to organism. Those that inhabit estuaries, for example, such as oysters, must be able to withstand considerable fluctuations in salinity. The daily rise and fall of the tides forces salty ocean water into river mouths and draws it out again, changing the salinity considerably. During floods, the salinity in estuaries can reach extremely low levels. The coastal organisms that can tolerate large changes in salinity are known as **euryhaline** (*eury* = wide, *halo* = salt).

Marine organisms that inhabit the open ocean, on the other hand, are seldom exposed to a large variation in salinity. They have adapted to a constant salinity and can tolerate only very small changes. These organisms are called **stenohaline** (*steno* = narrow, *halo* = salt).

EXTRACTION OF SALINITY COMPONENTS   Some organisms extract minerals from ocean water—particularly *silica* (SiO_2) and *calcium carbonate* ($CaCO_3$)—to construct the hard parts of their bodies, which serve as protective coverings. In doing so, they reduce the amount of dissolved material in ocean water. For example, phytoplankton (including diatoms) and microscopic protozoans such as radiolarians and silicoflagellates extract silica from seawater. Coccolithophores, foraminifers, most mollusks, corals, and some algae that secrete a calcium carbonate skeletal structure extract calcium carbonate from seawater.

DIFFUSION   Molecules of soluble substances, such as nutrients, move through water from areas of *high* concentration to areas of *low* concentration until the distribution of the substance is uniform (**Figure 12.16a**). This process, called **diffusion** (*diffuse* = dispersed), is caused by random motion of molecules.

The outer membrane of a living cell is permeable to many molecules. Organisms may take in nutrients they need from the surrounding water by diffusion of the nutrients through their cell walls. Because nutrients are usually plentiful in seawater, they pass through the cell wall into the interior, where nutrients are less concentrated (**Figure 12.16b**). In addition to passive diffusion, organisms also import nutrients to cells by active transport.

After a cell uses the energy stored in nutrients, it must dispose of waste. Waste passes out of a cell by diffusion, too. As the concentration of waste materials becomes greater within the cell than in the water surrounding it, these materials pass from within the cell into the surrounding fluid. The waste products are then carried away by circulating fluid that services cells in higher animals or by the water that surrounds simple one-celled organisms.

OSMOSIS   When water solutions of unequal salinity are separated by a semipermeable membrane (such as skin or the membrane around a living cell), water molecules (but not dissolved ions) diffuse through the membrane. Water molecules always move from the *less* concentrated solution into the *more* concentrated solution in a process called **osmosis** (*osmos* = to push) (**Figure 12.17a**). **Osmotic pressure** is the pressure that must be applied to the more concentrated solution to prevent water molecules from passing into it.

Osmosis causes water to move through an organism's skin (its semipermeable membrane) and affects both marine and freshwater organisms. If the salinity of an

Diffusion

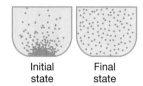

(a) A water-soluble substance that is added to the bottom of a container of water (*initial state*) will eventually become evenly distributed throughout the water by diffusion (*final state*).

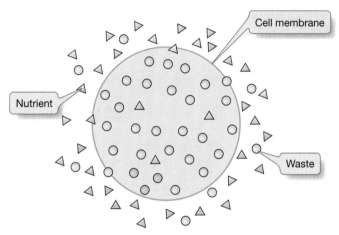

(b) Nutrients (*triangles*) are in high concentration outside the cell and diffuse into the cell through the cell membrane. Waste particles (*circles*) are in high concentration inside the cell and diffuse out of the cell through the cell membrane.

Figure 12.16 Diffusion. Diffusion is the process by which materials move through fluids from higher to lower concentrations and become more evenly distributed.

Osmosis

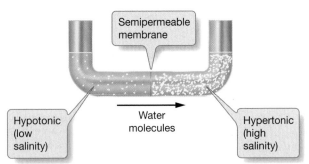

(a) A semipermeable membrane that separates two water solutions of different salinities allows water molecules (but not dissolved substances) to pass through it by the process of osmosis. Water molecules will diffuse through the membrane from the less concentrated (hypotonic) solution (*left*) into the more concentrated (hypertonic) solution (*right*).

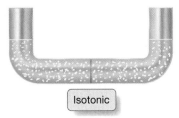

(b) If the salinity of the two solutions is the same (isotonic), there is no net movement of water molecules.

Figure 12.17 **Osmosis.** Osmosis is the process by which water molecules move through a semipermeable membrane from higher water molecule concentration (lower salinity) to lower water molecule concentration (higher salinity).

organism's body fluid equals that of the ocean, it is **isotonic** (*iso* = same, *tonos* = tension) and has equal osmotic pressure, and so no net transfer of water will occur through the membrane in either direction (**Figure 12.17b**).

If seawater has a lower salinity than the fluid within an organism's cells, seawater will pass through the cell walls into the cells (always toward the more concentrated solution). This organism is **hypertonic** (*hyper* = over, *tonos* = tension), which means it is saltier than the surrounding seawater.

If the salinity within an organism's cells is less than that of the surrounding seawater, water from the cells will pass through the cell membranes out into the seawater (toward the more concentrated solution). This organism is **hypotonic** (*hypo* = under, *tonos* = tension) relative to the water outside its body.

In essence, osmosis is diffusion that produces a net transfer of water molecules through a semipermeable membrane from the side with the *greatest concentration* of water molecules to the side with the *lesser concentration* of water molecules.

During osmosis, three things can occur simultaneously across the cell membrane:

1. Water molecules move through the semipermeable membrane toward the side with the lower concentration of water.

2. Nutrient molecules or ions move from where they are more concentrated into the cell, where they are used to maintain the cell.

3. Waste molecules move from within the cell to the surrounding seawater.

Molecules or ions of all the substances in the system are passing through the membrane in both directions. A net transport of molecules of a given substance always occurs from the side on which they are most highly concentrated to the side where the concentration is less, until equilibrium is attained.

The body fluids of marine invertebrates (those without backbones) such as worms, mussels, and octopuses, and the seawater in which they live, are nearly isotonic. As a result, these organisms have not had to evolve special mechanisms to maintain their body fluids at a proper concentration. This gives them an advantage over their freshwater relatives, whose body fluids are hypertonic.

AN EXAMPLE OF OSMOSIS: SALTWATER VERSUS FRESHWATER FISH Saltwater fish have body fluids that are only slightly more than one-third as saline as ocean water, possibly because they evolved in low-salinity coastal waters. They are, therefore, hypotonic (less salty) compared to the surrounding seawater.

This salinity difference means that saltwater fish, without some means of regulation, would lose water from their body fluids into the surrounding ocean and eventually dehydrate. This loss is counteracted, however, because saltwater fish drink ocean water and excrete the salts through special chloride-releasing cells located in their gills. These fish also help maintain their body water by discharging a very small amount of very highly concentrated urine (**Figure 12.18a**).

STUDENTS SOMETIMES ASK . . .

Why do my fingers get wrinkly when I stay in the water for a long time?

It's caused by water molecules passing through your skin (a semipermeable membrane) due to osmosis. The water molecules outside your body—contained in either freshwater or seawater—flow into your skin cells in an attempt to dilute the dissolved particles inside those cells. After a short time in water, your skin cells contain many more water molecules than before, which hydrates your skin and causes it to look wrinkly. The effect is most obvious on appendages such as fingers and toes because they have a high surface area per unit of volume. It's only a temporary condition, though, because your body absorbs and excretes the excess water molecules after you leave the water.

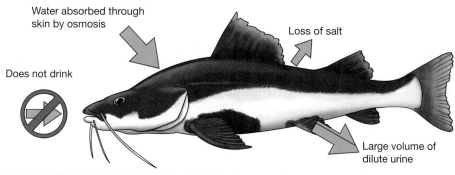

Water absorbed through skin by osmosis

Loss of salt

Does not drink

Large volume of dilute urine

Freshwater fish	Hypotonic = low osmotic pressure
Saltwater fish	Hypertonic = high osmotic pressure

Loss of water by osmosis

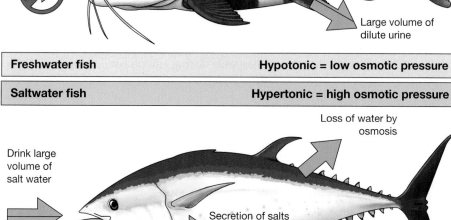

Drink large volume of salt water

Secretion of salts

Small volume of concentrated urine

Figure 12.18 Salinity adaptations of freshwater and saltwater fish. Osmotic processes cause freshwater and saltwater fish to have different adaptations to their environment. As a result, freshwater fish are hypotonic relative to their environment, and saltwater fish are hypertonic relative to theirs.

Freshwater fish are hypertonic (internally more saline) compared to the very dilute water in which they live. The osmotic pressure of the body fluids of such fish may be 20 to 30 times greater than that of the freshwater that surrounds them, so freshwater fish risk rupturing cell walls from taking in excessive quantities of water through osmosis.

To prevent this, freshwater fish do not drink water, and their cells have the capacity to absorb salt. They also excrete large volumes of very dilute urine to reduce the amount of water in their cells (**Figure 12.18b**).

ESSENTIAL CONCEPT 12.4A

Osmosis produces a net transfer of water molecules through a semipermeable membrane from the side with the greatest concentration of water molecules to the side with the lesser concentration.

Dissolved Gases

The amount of gases that dissolve in seawater increases as the temperature of seawater decreases, so cold water dissolves more gas than warm water. This helps vast phytoplankton communities develop in high latitudes during summer, when solar energy becomes available for photosynthesis. These cold waters contain an abundance of dissolved gases, specifically carbon dioxide (which phytoplankton need for photosynthesis) and oxygen (which all organisms need to metabolize their food). In addition, the cold, oxygen-rich water of high-latitude regions sinks and flows along the ocean bottom, supplying deep-sea organisms with an abundant supply of dissolved oxygen.

Most animals that live in the ocean—except air-breathing marine mammals and certain fishes—must extract dissolved oxygen from seawater. How do they do this? Most marine animals have specially designed fibrous respiratory organs called **gills** that exchange oxygen and carbon dioxide directly with seawater. Most fish, for instance, take water in through their mouths (which gives them the appearance of "breathing" underwater), pass it through their gills to extract oxygen, and then expel it through the gill slits on the sides of their bodies (**Figure 12.19**). Most fish need at least 4.0 parts per million (ppm by weight) of dissolved oxygen in seawater to survive for long periods, and they need even more for activity and rapid growth. That's why aquariums need a pump to continually resupply oxygen to the water in the tank.

During low-oxygen conditions, most marine animals with gills cannot simply breathe air at the surface. Their adaptations allow them to use only oxygen that is

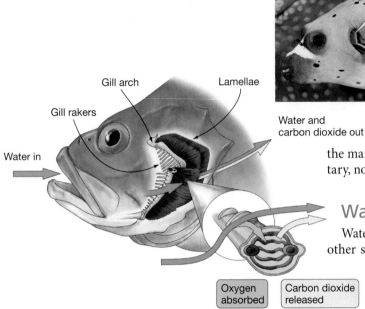

Figure 12.19 Gills on fish. Water is taken in through the mouth and passes through the gills, which extract dissolved oxygen. Afterward, water and carbon dioxide are expelled through the gill slits. Photo (*inset*) shows a blue-streaked cleaner wrasse (*Labroides dimidiatus*) cleaning the gills of a blackspotted puffer (*Arothron nigropunctatus*).

Figure 12.20 Moon jellies. Most jellies such as these moon jellies (*Aurelia aurita*) are nearly transparent, making them very difficult for predators to see. To enhance their appearance in this photograph, they are illuminated by a strong white light from above.

dissolved in water. If dissolved oxygen levels become low enough (such as after an algae bloom, when decomposition consumes dissolved oxygen), it may cause many marine organisms to suffocate and die unless they can move to other oxygen-rich areas.

Gill structure and location vary among animals of different groups. In fishes, gills are located at the rear of the mouth and contain capillaries. In higher aquatic invertebrates, they protrude from the body surface and contain extensions of the vascular system. In mollusks, they are inside the mantle cavity. In higher vertebrates (including humans), they occur as rudimentary, nonfunctional gill slits, which disappear during embryonic development.

Water's High Transparency

Water—including seawater—has relatively high transparency compared to many other substances, allowing sunlight to penetrate to a depth of about 1000 meters (3300 feet) in the open ocean. The actual depth depends on the amount of turbidity (suspended sediment) in the water, the amount of plankton in the water, latitude, time of day, and the season. Even though there are few places to hide in the open ocean, marine organisms employ many clever adaptations to prevent being seen.

TRANSPARENCY Because of water's high transparency, many marine organisms have developed large eyes, which help them locate and capture prey in dim light. To combat keen-eyed predators, many marine organisms such as jellies are themselves nearly transparent, which helps them blend into their environment (**Figure 12.20**). In fact, almost all open-ocean animals not otherwise protected by teeth, toxins, speed, or small size have some degree of invisibility. Only at depths where sunlight never penetrates is transparency uncommon. Another strategy used to increase an organism's transparency is to have silver sides that act like a mirror and reflect back weak light. Not only can transparency help organisms elude their predators, it can also help transparent organisms stalk their prey.

CAMOUFLAGE AND COUNTERSHADING Some marine organisms hide by using their coloration pattern as camouflage (**Figure 12.21a**). Still others use **countershading**, which means they are dark colored on top and light colored on bottom (**Figure 12.21b**), to blend into their environment. Many fish—especially flat fish—have countershading so they cannot easily be seen against the dark background of deep water or the ocean floor and they blend into the sunlight when viewed from below. Countershading also helps predators sneak up on prey.

DAILY VERTICAL MIGRATION: THE DEEP SCATTERING LAYER (DSL) Many marine organisms undertake a daily vertical migration to deeper, darker parts of the ocean to avoid becoming prey. These organisms create a curious feature called the **deep scattering layer (DSL)**, which was discovered when the U.S. Navy was testing sonar equipment to detect enemy submarines early in World War II. On many of the sonar recordings, a mysterious sound-reflecting surface appeared that was much too shallow to be the ocean floor. It was often referred to as a "false bottom" (see Figure 3.1). What was even more surprising was that the depth of the deep scattering layer changed with time. It was at a depth of about 100 to 200 meters (330 to 660 feet) during the night but was as deep at 900 meters (3000 feet) during the day.

With the help of marine biologists, sonar specialists were able to determine that sonar signals were reflecting off densely packed concentrations of marine organisms. Investigation with plankton nets, submersibles, and detailed sonar revealed that the DSL contains many different organisms, including copepods (which constitute a large proportion of planktonic animals), krill, and lantern fish (family Myctophidae) (**Figure 12.22**, *enlargement*).

(a) The head and eye of a well-camouflaged rock fish.

(b) Halibut on a dock in Alaska show countershading.

Figure 12.21 Examples of camouflage and countershading. (a) The head and eye of a well-camouflaged rock fish. **(b)** Halibut on a dock in Alaska showing countershading.

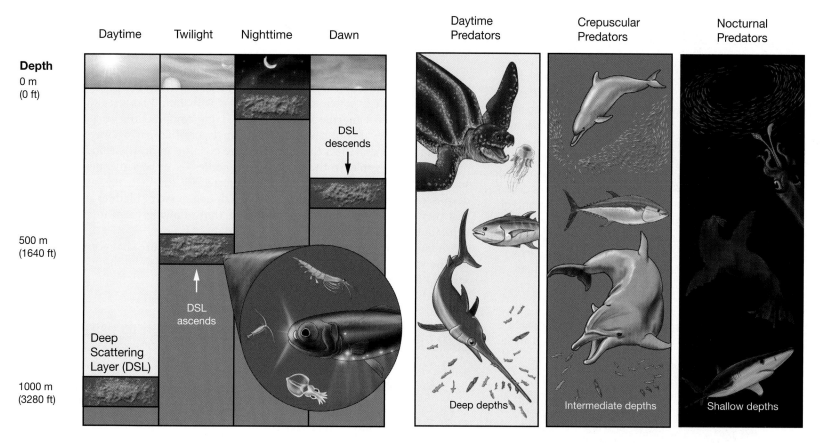

Figure 12.22 Daily movement and organisms of the deep scattering layer. Organisms in the deep scattering layer (DSL) migrate to deep depths during the day to hide from predators and rise at night to feed in surface waters. Organisms that feed in the DSL include deep-diving daytime, crepuscular (twilight), and nocturnal predators.

Web Animation
Daily Movement of the Deep Scattering Layer (DSL)

The daily movement of the deep scattering layer is caused by the vertical migration of marine organisms that feed in the highly productive surface waters but must protect themselves from being seen by predators. These predators include daytime, **crepuscular** (*crepusculum* = twilight), and nocturnal (*nocturnus* = night) feeders (Figure 12.22). Organisms within the DSL ascend to the surface only at night to feed under cover of darkness and then migrate to deeper (darker) water to hide during the day.

Research reveals that the daily migration of the DSL causes increased vertical mixing of ocean waters. A recent study, in fact, suggests that energy from small animals

Figure 12.23 Use of color by tropical fish. Many tropical fish such as this mandarinfish (*Synchiropus splendidus*) have bright colors and bold patterns, which allow them to blend into the environment by using disruptive coloration. Alternatively, they may also stand out so that they can advertise their identity, sex, or weaponry.

STUDENTS SOMETIMES ASK . . .

If marine animals that have large air pockets inside are affected by the extreme pressures at depth, how are sperm whales able to dive so deeply?

All marine mammals have lungs and breathe air, and certain ones have special adaptations that allow them to make extremely deep dives. Sperm whales, for example, can dive to more than 2800 meters (9200 feet) and stay submerged for more than two hours during their search for food! They are able to use small amounts of oxygen very efficiently and have a collapsible rib cage, which forces air out and collapses the lungs, thereby closing the air cavities inside their bodies. Marine mammals and their adaptations are discussed in Chapter 14, "Animals of the Pelagic Environment."

Figure 12.24 Swim bladder. Many marine fish possess a swim bladder that they use to regulate their position in the water column.

Swim bladder

ESSENTIAL CONCEPT 12.4B

The ocean's physical support, viscosity, temperature, salinity, sunlit surface waters, dissolved gases, high transparency, and pressure create conditions to which marine organisms are superbly adapted.

such as those found in the DSL is a major component of the total contributed to the ocean by all swimming creatures, which is comparable to that of winds and tides.

DISRUPTIVE COLORATION In contrast to species trying to blend into their environment, many species of tropical fish display bright colors (**Figure 12.23**). Why would they be brightly colored if it makes them easily seen by predators? The bright markings of tropical fish may be an example of **disruptive coloration,** where large, bold patterns of contrasting colors tend to make an object blend in when viewed against an equally variable, contrasting background such as a tropical reef. Zebras use this principle to evade predators, tigers use it to conceal themselves from their prey, and military uniforms are camouflaged in a similar way.

Even considering disruptive coloration, many tropical fish still don't seem to blend into their environment. Perhaps the bright colors and distinctive markings that make tropical fish more apparent make it easier to advertise their identity, to attract mates, or to display weaponry such as spines or poison. Scientists have yet to agree on why tropical fish have such vivid colors, but it must serve some biological advantage, or the fish wouldn't be so brightly colored.

Pressure

Water pressure increases about 1 kilogram per square centimeter (1 atmosphere, or 14.7 pounds per square inch) with every 10 meters (33 feet) of water depth. Humans are not well adapted to the high pressures that exist below the surface (**Diving Deeper 12.1**). Even when diving to the bottom of the deep end of a swimming pool, one can feel the dramatic increase in pressure in one's ears.

In the deep ocean, water pressure is on the order of several hundred kilograms per square centimeter (several hundred atmospheres, or several tons per square inch). How do deep-water marine organisms withstand pressures that can easily kill humans? Most marine organisms lack large compressible air pockets inside their bodies. They do not have lungs, ear canals, or other passageways as we do, so these organisms don't feel the high pressure pushing in on their bodies. Water is nearly incompressible, so their water-filled bodies have the same amount of pressure pushing outward, and they are unaffected by the high pressures found in deep-ocean environments. Instead, many fish have a **swim bladder** (**Figure 12.24**) that they can use to adjust their buoyancy, thus allowing them to regulate their depth in the water column.

A few species appear to be extremely tolerant of pressure changes. In fact, some marine species that are found in nearshore areas can also be found at depths of several kilometers.

CONCEPT CHECK 12.4

1. Discuss some adaptations other than size that organisms use to increase their resistance to sinking.

2. List differences between cold- and warm-water species in the marine environment.

3. Describe the process of osmosis. How is it different from diffusion? What three things can occur simultaneously across the cell membrane during osmosis?

4. What is the problem requiring osmotic regulation faced by *hypotonic* fish in the ocean? What adaptation do these animals have to overcome this problem?

5. How does water temperature affect the water's ability to hold gases? How do marine organisms extract dissolved oxygen from seawater?

6. How does the depth of the deep scattering layer vary over the course of a day? Why does it do this? Which organisms comprise the DSL?

DIVING INTO THE MARINE ENVIRONMENT

Throughout history, humans have submerged themselves in the marine environment to observe it directly for scientific exploration, profit, or adventure (**Figure 12A**). As early as 4500 B.C., brave and skillful divers reached depths of 30 meters (100 feet) on one breath of air to retrieve red coral and mother-of-pearl shells. Later, diving bells (bell-shaped structures full of trapped air) were lowered into the sea to provide passengers or underwater divers with an air supply. In 360 B.C., Aristotle, in his *Problematum*, recorded the use by Greek sponge divers of kettles full of air lowered into the sea. However, technology to move around freely while breathing underwater was not developed until 1943, when Jacques-Yves Cousteau and Émile Gagnan invented the fully automatic, compressed-air Aqualung. The equipment was later dubbed **scuba**, an acronym for *self-contained underwater breathing apparatus*, and it is used by millions of recreational divers today. By using scuba, divers can experience the ocean firsthand, leading to a fuller appreciation of the wonder and beauty of the marine environment.

Those who venture underwater must contend with many obstacles inherent in ocean diving, such as low temperatures, darkness, and the effects of greatly increased pressure. To combat low temperatures, specially designed clothing is worn. Waterproof, high-intensity diving lights are used to combat darkness. To combat the deleterious effects of pressure, depth and duration of dives must be limited. As a result, most scuba divers rarely venture below a depth of 30 meters (100 feet)—where the pressure is three times that at the surface—and they stay there less than 30 minutes.

It is relatively dangerous for humans to enter the marine environment because our bodies are adapted to living in the relatively low pressure of the atmosphere. In water, pressure increases rapidly with depth, to which anyone who has been to the bottom of the deep end of a swimming pool can attest. The increased pressure at depth in the ocean can cause problems for divers. For instance, higher pressure causes more nitrogen to be dissolved in a diver's body and may cause a disorienting condition known as *nitrogen narcosis*, or *rapture of the deep*. Further, if a diver surfaces too rapidly, expanding gases within the body can catastrophically rupture cell membranes (a condition called *barotrauma*).

In addition, when divers return to the surface, they may experience *decompression sickness*, which is also called *caisson disease*, or *the bends*. The bends affects divers who ascend to the lower pressure at the surface too rapidly, causing nitrogen bubbles to form in the bloodstream and other tissues (analogous to the bubbles that form in a carbonated beverage when the container is opened). Various symptoms can result, from nosebleed and joint pain (which causes divers to stoop over, hence the term *the bends*) to permanent neurological injury and even fatal paralysis. To avoid it, divers must ascend slowly, allowing time for excess dissolved nitrogen to be eliminated from the blood via the lungs.

Despite these risks, divers venture to greater and greater depths in the ocean. In 1962, Hannes Keller and Peter Small made an open-ocean dive from a diving bell to a then-record-breaking depth of 304 meters (1000 feet). Although they used a special gas mixture, Small died once they returned to the surface. Presently, the record ocean dive is 534 meters (1750 feet), but researchers who study the physiology of deep divers have simulated a dive to 701 meters (2300 feet) in a pressure chamber using a special mix of oxygen, hydrogen, and helium gases. Researchers believe that humans will eventually be able to stay underwater for extended periods of time at depths below 600 meters (1970 feet).

Figure 12A **Oceanographer and explorer Willard Bascom in early diving gear.**

12.5 What Are the Main Divisions Of the Marine Environment?

The oceans can be divided into two main environments. The *ocean water itself is the pelagic environment*, where drifters and swimmers play out their lives in a complex food web. The *ocean bottom is the benthic environment*, where marine algae and animals that do not float or swim (or at least not very well) spend their lives.

Figure 12.25 Oceanic biozones of the pelagic and benthic environments. Pelagic environments are shown in blue, and benthic environments are shown in brown. Pelagic and benthic environments are both based on depth, not necessarily on distance from shore. Sea floor features and sunlight zones are shown in black lettering.

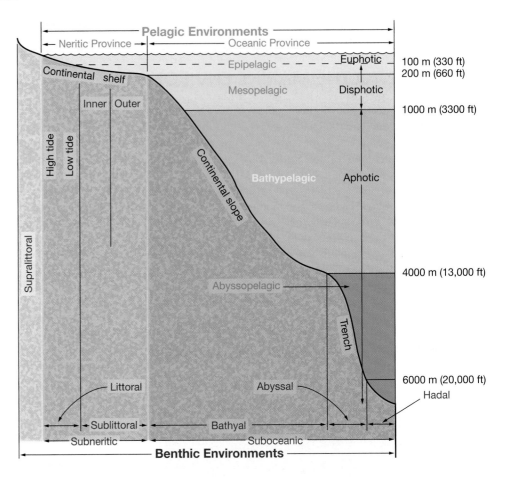

Pelagic (Open Sea) Environment

The pelagic environment can be divided into distinctive life zones called **biozones** that possess unique physical characteristics, as shown in **Figure 12.25**.

The pelagic environment is divided into neritic and oceanic provinces (Figure 12.25). The **neritic** (*neritos* = of the coast) **province** extends from the shore seaward and includes all water less than 200 meters (660 feet) deep. Seaward of the neritic province is the **oceanic province**, where depth increases beyond 200 meters (660 feet). The oceanic province is further subdivided into four biozones:

1. The **epipelagic** (*epi* = top, *pelagios* = of the sea) **zone**, from the surface to a depth of 200 meters (660 feet)

2. The **mesopelagic** (*meso* = middle, *pelagios* = of the sea) **zone**, from 200 to 1000 meters (660 to 3280 feet)

3. The **bathypelagic** (*bathos* = depth, *pelagios* = of the sea) **zone**, from 1000 to 4000 meters (3300 to 13,000 feet)

4. The **abyssopelagic** (*a* = without, *byssus* = bottom, *pelagios* = of the sea) **zone**, which includes all the deepest parts of the ocean below 4000 meters (13,000 feet)

The single most important factor that determines the distribution of life in the oceanic province is the availability of sunlight. Thus, in addition to the four biozones, the distribution of life in the ocean is also divided into three zones based on the availability of sunlight as follows:

- The **euphotic** (*eu* = good, *photos* = light) **zone** extends from the surface to a depth where enough light still exists to support photosynthesis, which is rarely deeper than 100 meters (330 feet). The euphotic zone—often called the thin sunlit surface layer—accounts for only about 2.5% of the marine environment, but as discussed in subsequent chapters, this is where most marine life exists.

- The **disphotic** (*dis* = apart from, *photos* = light) **zone** has small but measurable quantities of light. It extends from the euphotic zone to a depth where light no longer exists—usually about 1000 meters (3300 feet).
- The **aphotic** (*a* = without, *photos* = light) **zone** has no light, and it exists below about 1000 meters (3300 feet).

EPIPELAGIC ZONE The upper half of the epipelagic zone is the only place in the ocean where there is sufficient light to support photosynthesis. The boundary between the epipelagic and mesopelagic zones, at 200 meters (660 feet), is also where the level of dissolved oxygen begins to decrease significantly (**Figure 12.26**, *red curve*).

Oxygen decreases at this depth because no photosynthetic algae live below about 150 meters (500 feet), and dead organic tissue descending from the biologically productive upper waters is decomposing by bacterial oxidation. Nutrient content also increases abruptly below 200 meters (600 feet) (Figure 12.26, *green curve*) and it is the approximate bottom of the mixed layer, seasonal thermocline, and surface water mass.

MESOPELAGIC ZONE A dissolved **oxygen minimum layer (OML)** occurs at a depth of about 700 to 1000 meters (2300 to 3280 feet) (Figure 12.26). The intermediate-water masses that move horizontally in this depth range often possess the highest levels of nutrients in the ocean.

Organisms capable of **bioluminescence** (*bio* = life, *lumen* = light, *esc* = becoming), which have the ability to produce light biologically and "glow in the dark," are

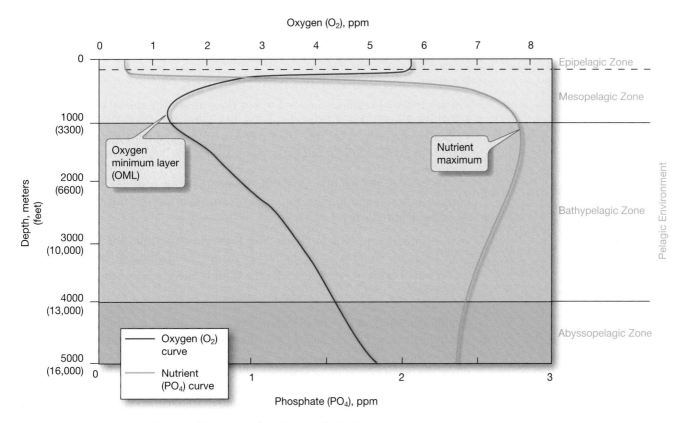

Figure 12.26 Abundance of dissolved oxygen and nutrients with depth. In surface water, oxygen is abundant due to mixing with the atmosphere and plant photosynthesis, and nutrient content (phosphate) is low due to uptake by algae. At deeper depths, oxygen decreases and produces an oxygen minimum layer (OML), which coincides with a nutrient maximum. Below that, nutrient levels remain high and oxygen increases as it is replenished with high-oxygen cold water from polar regions. Note that ppm = parts per million by weight.

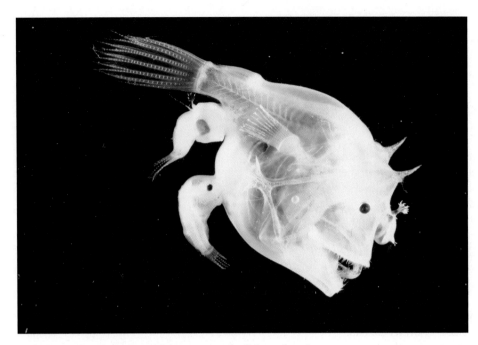

Figure 12.27 Adaptations of deep-sea anglerfish. A female deep-sea anglerfish (*Edriolychnus schmidti*) has a transparent body, small eyes, and sharp teeth. The feathery structure projecting from the front of its head is a bioluminescent lure, which is used to attract prey. Two much smaller parasitic males are attached to the bottom of the female's body. The length of the female is about 10 centimeters (4 inches).

common in the mesopelagic and deeper zones. In these areas below the sunlit surface waters, having the ability to produce light has many advantages, and so the vast majority of organisms are capable of producing light. Examples of bioluminescent organisms include certain species of shrimp, squid, and especially deep-sea fish (**Figure 12.27**). The mechanism involved in allowing organisms to bioluminesce is discussed more fully in Chapter 14, "Animals of the Pelagic Environment."

BATHYPELAGIC AND ABYSSOPELAGIC ZONES The aphotic (lightless) bathypelagic and abyssopelagic zones represent over 75% of the living space in the oceanic province. Many completely blind fish exist in this region of total darkness, and all are small, bizarre-looking, and predaceous.

Many species of shrimp that normally feed on **detritus**[8] become predators at these depths, where the food supply is greatly reduced compared to surface waters. Animals that live in these deep zones feed mostly upon one another. They have evolved impressive warning devices and unusual apparatuses to make them extremely efficient predators (Figure 12.27). Many also have sharp teeth and extremely large mouths relative to their body size.

Oxygen content increases with depth below the oxygen minimum layer because it is replenished by deep currents originating in polar regions as cold surface water high in oxygen. The abyssopelagic zone is the realm of the bottom-water masses, which commonly move in the direction opposite the deep-water masses in the bathypelagic zone.

Benthic (Sea Bottom) Environment

The transitional region from land to sea floor above the spring high tide line is called the **supralittoral** (*supra* = above, *littoralis* = the shore) **zone** (see Figure 12.25). Commonly called the spray zone, it is covered with water only during periods of extremely high tides and when tsunami or large storm waves break on the shore.

The rest of the benthic, or sea floor, environment is divided into two main units that correspond to the neritic and oceanic provinces of the pelagic environment (see Figure 12.25):

- The **subneritic province** extends from the spring high tide shoreline to a depth of 200 meters (660 feet), approximately encompassing the continental shelf.
- The **suboceanic province** includes the benthic environment below 200 meters (660 feet).

SUBNERITIC PROVINCE The subneritic province is subdivided into the littoral and sublittoral zones. The *intertidal zone* (the zone between high and low tides) coincides with the **littoral** (*littoralis* = the shore) **zone**. The **sublittoral** (*sub* = below, *littoralis* = the shore) **zone**, or *shallow subtidal zone*, extends from low tide shoreline out to a depth of 200 meters (660 feet).

The sublittoral zone consists of inner and outer regions. The **inner sublittoral zone** extends to the depth at which marine algae no longer grow attached to the ocean bottom (approximately 50 meters [160 feet]), so the seaward boundary varies.

---

[8]*Detritus* (*detritus* = to lessen) is a catchall term for dead and decaying organic matter, including waste products.

Figure 12.28 Benthic organisms produce tracks on the ocean floor. As benthic organisms—such as this sea urchin (*below*) and brittle star (*above*)—move across or burrow through the ocean bottom, they often leave tracks in the sediment on the ocean floor.

All photosynthesis seaward of the inner sublittoral zone is carried out by floating microscopic algae.

The **outer sublittoral zone** extends from the inner sublittoral zone out to a depth of 200 meters (660 feet) or the shelf break, which is the seaward edge of the continental shelf.

SUBOCEANIC PROVINCE   The suboceanic province is subdivided into bathyal, abyssal, and hadal zones. The **bathyal** (*bathus* = deep) **zone** extends from a depth of 200 to 4000 meters (660 to 13,000 feet) and corresponds generally to the continental slope.

The **abyssal** (*a* = without, *byssus* = bottom) **zone** extends from a depth of 4000 to 6000 meters (13,000 to 20,000 feet) and includes more than 80% of the benthic environment. The ocean floor of the abyssal zone is covered by soft oceanic sediment, primarily abyssal clay. Tracks and burrows of animals that live in this sediment can be seen in **Figure 12.28**.

The **hadal** (*hades* = hell)[9] **zone** extends below 6000 meters (20,000 feet), so it consists only of deep trenches along the margins of continents. animal communities that are found in these deep environments have been isolated from each other, often resulting in unique adaptations.

ESSENTIAL CONCEPT 12.5

The pelagic environment includes the water column and the benthic environment includes the sea bottom. Subdivisions of pelagic and benthic environments are based on depth, which influences the amount of sunlight.

CONCEPT CHECK 12.5

❶ Construct a table listing the subdivisions of the pelagic and benthic environments and the physical factors used in assigning their boundaries.

❷ Describe the three zones based on the availability of sunlight. Which one is where most marine life exists?

❸ Explain why the two curves in Figure 12.26 have the shape that they do (for example, why the oxygen minimum layer exists, and why it coincides with the nutrient maximum).

[9]The inhospitable high-pressure environment of the hadal zone is, in fact, aptly named.

Essential Concepts Review

12.1 What are living things, and how are they classified?

▶ *A wide variety of organisms lives in the ocean*, ranging in size from microscopic bacteria and algae to blue whales. *All living things belong to one of three major domains (branches) of life: Archaea*, simple microscopic bacteria-like creatures; *Bacteria*, simple life-forms consisting of cells that usually lack a nucleus; and *Eukarya*, complex organisms (including plants and animals) consisting of cells that have a nucleus.

▶ *Organisms are further divided into five kingdoms: Monera*, single-celled organisms without a nucleus; *Protoctista*, single-celled and multicelled organisms with a nucleus; *Fungi*, mold and lichen; *Plantae*, many-celled plants; and *Animalia*, many-celled animals. *Classification of organisms involves placing individuals within the kingdoms into increasingly specific groupings of phylum, class, order, family, genus, and species*, the last two of which denote an organism's scientific name. Many organisms also have one or more common names.

Study Resources
Online Study Guide Quizzes, Web Animation

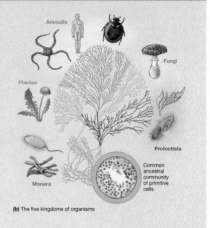

(b) The five kingdoms of organisms

Critical Thinking Question

Discuss why it is often difficult to differentiate between living and nonliving things.

12.2 How are marine organisms classified?

▶ *Marine organisms can be classified into one of three groups, based on habitat and mobility.* Plankton are free-floating forms with little power of locomotion, *nekton* are swimmers, and *benthos* are bottom dwellers. *Most of the ocean's biomass is planktonic.*

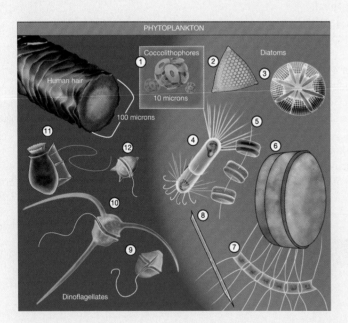

Study Resources
Online Study Guide Quizzes

Critical Thinking Question

Explain why most of the ocean's biomass is planktonic.

12.3 How many marine species exist?

▶ *Only about 14% of all known species inhabit the ocean, and more than 98% of marine organisms are benthic.* The *marine environment—especially the pelagic environment—is much more stable than the terrestrial environment*, so there is less pressure on marine organisms to diversify.

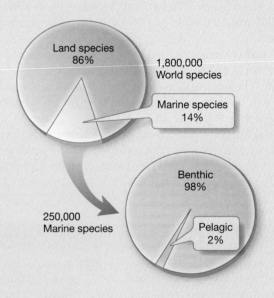

Study Resources
Online Study Guide Quizzes

Critical Thinking Question

Explain how environmental variability affects the number of species present.

12.4 How are marine organisms adapted for the physical conditions of the ocean?

▶ *Marine organisms are well adapted to life in the ocean.* Those organisms that have established themselves on land have had to develop complex systems for support and for acquiring and retaining water.

▶ Both *algae*, which must stay in surface waters to receive sunlight, and *small animals* that feed on them *lack an effective means of locomotion.* To keep from sinking below sunlit surface waters, they depend on their *small size* and other adaptations to *increase their ratio of surface area to body mass*, which gives them high frictional resistance to sinking. Their small size also allows them to efficiently absorb nutrients and dispose of wastes. Many nektonic organisms have developed *streamlined bodies so they can overcome the viscosity of seawater* and move through it more easily.

▶ *Surface temperature of the world ocean does not vary* on a daily, seasonally, or yearly basis *as much as on land. Organisms living in warm water tend to be individually smaller, have ornate plumage, comprise a greater number of species, and constitute a much smaller total biomass than organisms living in cold water.* Warm-water organisms also tend to live shorter lives and reproduce earlier and more frequently than cold-water organisms.

▶ *Osmosis is the passing of water molecules through a semipermeable membrane from a region of higher concentration to a region of lower concentration.* If the body fluids of an organism and ocean water are separated by a membrane that allows water molecules to pass through, the organism may become severely dehydrated from osmosis. Many marine invertebrates are essentially *isotonic*: The salinity of their body fluids is similar to that of ocean water. Most marine vertebrates are *hypotonic*: The salinity of their body fluids is lower than that of ocean water, so they tend to lose water through osmosis. Freshwater organisms are essentially all *hypertonic*: The salinity of their body fluids is greater than the water in which they live, so they tend to gain water through osmosis.

▶ *Most marine animals extract oxygen through their gills.* Many marine organisms have well-developed eyesight because water is so transparent. *To avoid being seen and consumed by predators*, many marine organisms are transparent, camouflaged, countershaded, or disruptively colored. Unlike humans, most *marine organisms are unaffected by the high pressure at depth* because they do not have large internal air pockets that can be compressed.

Study Resources
Online Study Guide Quizzes, Web Animation, Web Videos

Critical Thinking Question
Determine the surface to volume ratio of an organism whose average linear dimension is (a) 1 centimeter (0.4 inch), (b) 3 centimeters (1.1 inches), and (c) 5 centimeters (2 inches). Which one is better able to resist sinking, and why?

12.5 What are the main divisions of the marine environment?

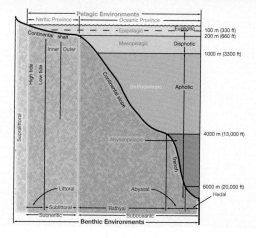

▶ *The marine environment is divided into pelagic (open sea) and benthic (sea bottom) environments.* These regions are further divided based on depth and have varying physical conditions to which marine life is superbly adapted. One of the most important layers of the pelagic environment is the *euphotic zone*, which includes the sunlit surface waters and *contains enough sunlight to support photosynthesis.*

Study Resources
Online Study Guide Quizzes, Web Diving Deeper 12.1

Critical Thinking Question
To help reinforce your knowledge of oceanic biozones, construct and label your own diagram similar to Figure 12.25 from memory.

MasteringOceanography™

www.masteringoceanography.com

Looking for additional review and test prep materials? Visit the Study Area in MasteringOceanography to enhance your understanding of this chapter's content by accessing a variety of resources, including Self-Study Quizzes, Geoscience Animations, RSS feeds, flashcards, Web links, and an optional Pearson eText.

Small fish and tomato grouper (*Cephalopholis sonnerati*) near a rocky reef. Sunlight powers nearly all marine food webs, which are generally based on the passing of nutrients from one organism to another through eating. Marine organisms usually feed on organisms smaller than they are, although some animals prove to be exceptions to this rule.

Before you begin reading this chapter, use the glossary at the end of this book to discover the meanings of any of these words you don't already know:

Bycatch Biomass pyramid
Harmful algal bloom (HAB) Primary productivity Secchi disk
Detritus Euphotic zone Overfishing
Producer Maximum sustainable yield (MSY) Driftnet
Food chain Trophic level Photosynthesis
Standing stock Eutrophication
Dead zone Decomposer Food web Upwelling
Consumer
Thermocline Phytoplankton Gill net
Red tide

BIOLOGICAL PRODUCTIVITY AND ENERGY TRANSFER

ESSENTIAL CONCEPTS

At the end of this chapter, you should be able to:

13.1 Demonstrate an understanding of primary productivity

13.2 Describe various kinds of photosynthetic marine organisms

13.3 Explain variations in regional primary productivity

13.4 Discuss how energy and nutrients are passed along in marine ecosystems

13.5 Describe several issues that affect marine fisheries

Producers are organisms that photosynthesize their own food from carbon dioxide, water, and sunlight. Through the process of photosynthesis, producers capture solar energy and create food sugars that sustain all other organisms in the marine biological community (except those near hydrothermal vents,[1] where *chemosynthesis* is the major source of "food" energy). As such, the ocean's producers are the foundation of the oceanic food web.

Photosynthetic producers in the ocean include plants, algae, and bacteria. Because there are few true marine plants and large species of marine algae play only a minor role, microscopic marine algae and bacteria comprise the majority of organisms responsible for the conversion of solar energy. These microscopic organisms—called *phytoplankton*—are mostly scattered throughout the ocean's sunlit surface waters and represent the largest community of organisms in the marine environment.

In this chapter, we'll examine primary productivity and the factors that cause it to vary, describe various types of photosynthetic marine organisms, discuss productivity in different regions of the ocean, examine feeding relationships such as food chains and food webs, and explore environmental issues related to marine fisheries.

13.1 What Is Primary Productivity?

Primary productivity is the rate at which organisms store energy through the formation of organic matter (carbon-based compounds), using energy derived from solar radiation during **photosynthesis** (*photo* = light, *syn* = with, *thesis* = an arranging) or from chemical reactions during **chemosynthesis** (*chemo* = chemistry, *syn* = with, *thesis* = an arranging).[2] Other organisms may then use this organic matter as food. Although chemosynthesis supports hydrothermal vent biocommunities along oceanic spreading centers, it is much less significant than photosynthesis in worldwide marine primary production. In fact, 99.9% of the ocean's **biomass**[3] relies either directly or indirectly on organic matter supplied by photosynthetic primary productivity as its source of food. Therefore, the discussion of primary productivity presented here focuses on photosynthetic productivity.

---

[1]Hydrothermal vent biocommunities are discussed in more detail in Chapter 15, "Animals of the Benthic Environment."

[2]More details about chemosynthesis can be found in Chapter 15, "Animals of the Benthic Environment."

[3]Recall that *biomass* is the mass of living organisms.

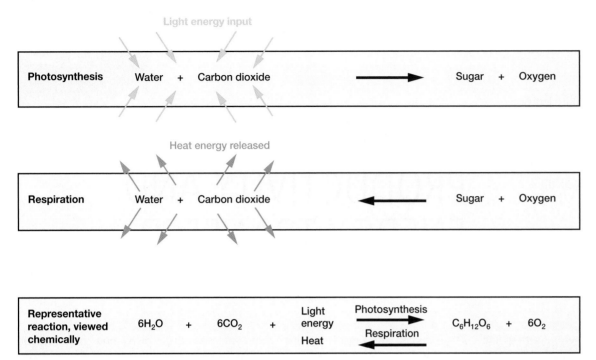

Figure 13.1 Photosynthesis (top), respiration (middle), and representative reactions viewed chemically (bottom). The process of photosynthesis, which is accomplished by plants, is represented in the upper panel. The second panel shows respiration, which is done by animals. Both processes are shown chemically in the third panel. (Note that this is the same image as Figure 1.22.)

Chemically, photosynthesis is a reaction in which energy from the Sun is stored in organic molecules. In photosynthesis (**Figure 13.1**, *top*), plant cells capture light energy and store it as sugars. As a chemical reaction, the photosynthetic equation is reversible; this process is called respiration (Figure 13.1, *middle*). The equations are also shown chemically (Figure 13.1, *bottom*). Note that this is the same figure from Chapter 1, where photosynthesis and respiration were previously discussed.

Measurement Of Primary Productivity

Various properties of the ocean can be measured to give an approximation of the amount of primary productivity. One of the most direct at-sea methods is to capture plankton in cone-shaped nylon **plankton nets** (**Figure 13.2**). These fine mesh nets—which resemble windsocks at airports—filter plankton from the ocean as they are towed at a specific depth by research vessels. Analysis of the amounts and types of organisms captured reveals much about the productivity of the area.

Other methods include lowering specially designed bottles into the ocean, analyzing the amount of radioactive carbon in seawater, and even monitoring ocean color using orbiting satellites. Photosynthetic organisms such as **phytoplankton** (*phyto* = plant, *planktos* = wandering) use the green pigment **chlorophyll** (*khloros* = green, *phylum* = leaf) to capture energy from the Sun and perform photosynthesis. The color of surface waters is strongly affected by the amount of chlorophyll, so ocean color can be used as an approximation of phytoplankton abundance and, in turn, productivity. One such instrument that collected ocean color measurements was the **SeaWiFS** (Sea-viewing Wide Field-of-view Sensor) instrument aboard the *SeaStar* satellite, which operated from 1997 to 2010. It replaced *Nimbus-7*'s Coastal Zone Color Scanner instrument, which operated between 1978 and 1986. During its operational lifespan, SeaWiFS measured the color of Earth's surface with a radiometer and provided global coverage of estimated ocean chlorophyll levels as well as the abundance of land vegetation. Today, ocean color is collected worldwide every two days by MODIS (Moderate Resolution Imaging Spectroradiometer) instruments aboard the *Terra* and *Aqua* satellites. MODIS measures 36 spectral frequencies of light, including ocean fluorescence, which provides a wealth of information about ocean phytoplankton productivity, health, and efficiency.

Factors Affecting Primary Productivity

In the ocean, the two main factors that limit the amount of photosynthetic primary productivity are the availability of nutrients and the availability of solar radiation. Sometimes other variables—such as the amount of carbon dioxide—can also limit primary productivity if they become scarce in seawater.

AVAILABILITY OF NUTRIENTS The distribution of life throughout the ocean's breadth and depth depends mainly on the availability of nutrients such as nitrogen, phosphorus, iron, and silica that phytoplankton need. Where the physical conditions supply large quantities of nutrients, marine populations reach their greatest concentration. To understand where these areas are found, the sources of nutrients must be considered.

Water in the form of runoff erodes the continents, carrying material to the oceans and depositing it as sediment on the continental margins. Runoff also dissolves and transports compounds such as nitrates and phosphates, which are the main nutrients for phytoplankton. Nitrates and phosphates are also the basic ingredients in all garden and farm fertilizers.

Through the process of photosynthesis, phytoplankton combine these nutrients with carbon dioxide and water to produce the carbohydrates, proteins, and fats that the vast majority of the ocean's biological community depends upon for food. The continents are the major sources of these nutrients, so the greatest concentrations of marine life are found along the continental margins. The concentration of marine life decreases as the distance from the continental margins into the open sea increases. The vast depth of the world's oceans and the great distance between the open ocean and the coastal regions where nutrients are concentrated account for these differences.

Often, the lack of certain nutrients, particularly nitrogen (as nitrates) and phosphorus, can limit productivity. As a result, these compounds are among the most studied in chemical oceanography. Comparatively, nitrogen compounds involved in photosynthesis may be 10 times the total nitrogen compound concentration that can be measured as a yearly average. This level implies that the soluble nitrogen compounds must be completely recycled up to 10 times per year. Available phosphates may be turned over up to 4 times per year.

Carbon is an important element in productivity, too, because carbon is the basic component of all organic compounds (including carbohydrates, proteins, and fats). In the ocean, however, various forms of carbon are quite abundant, so there is no scarcity of carbon for photosynthetic production. Thus, carbon does not limit productivity.

When nutrients are not limiting productivity, the ratio of carbon to nitrogen to phosphorus in the tissues of algae is in the proportion of 106:16:1 (C:N:P), which is called the *Redfield ratio*, after American oceanographer Alfred C. Redfield, who first described it in 1963. This ratio is also observed in zooplankton that feed on diatoms, and in most ocean water samples taken worldwide. Moreover, phytoplankton take up nutrients in the ratio in which they are available in the ocean water and pass them on to zooplankton in the same ratio. When these plankton and animals die, carbon, nitrogen, and phosphorus are restored to the water in this same ratio.

Recent studies in the waters near Antarctica and the Galápagos Islands have revealed that photosynthetic production is low even though the concentration of all nutrients—except iron—is high.[4] Production is high only in regions of shallow water downcurrent from islands or landmasses where a significant amount of iron from rocks and sediments is dissolved in water.

AVAILABILITY OF SOLAR RADIATION Photosynthesis cannot proceed unless light energy (solar radiation) is available. Despite the atmosphere's thickness of more than 80 kilometers (50 miles), its high transparency allows sunlight to penetrate it quite readily, so land-based plants almost always have an abundance of solar radiation to conduct photosynthesis.

In the clearest ocean water, solar energy may be detected to depths of only about 1 kilometer (0.6 mile) and, even then, the amount reaching these depths is inadequate for photosynthesis. Photosynthesis in the ocean, therefore, is restricted to the uppermost surface waters and those areas of the sea floor where the water is shallow enough to allow light to penetrate.

The depth at which net photosynthesis becomes zero is called the **compensation depth for photosynthesis**. The **euphotic** (*eu* = good, *photos* = light) **zone** extends

Figure 13.2 Plankton nets. These large, cone-shaped, fine-mesh plankton nets being washed are lowered into the water and towed behind a research vessel to collect plankton.

---

[4]The idea of fertilizing the ocean with iron to stimulate productivity and increase the amount of carbon dioxide gas absorbed by the ocean is discussed in Chapter 16, "The Oceans and Climate Change."

from the surface down to the compensation depth for photosynthesis, which is approximately 100 meters (330 feet) in the open ocean. Near the coast, the euphotic zone may extend to less than 20 meters (66 feet) because the water contains more suspended inorganic material (turbidity) or microscopic organisms that limit light penetration.

How do the two factors necessary for photosynthesis—the supply of nutrients and the presence of solar radiation—differ between coastal areas and the open ocean? In the open ocean (far from continental margins), solar energy extends deeper into the water column, but concentration of nutrients is low. In coastal regions, on the other hand, light penetration is much less, but the concentration of nutrients is much higher. Because the coastal zone is much more productive, nutrient availability must be the most important factor affecting the distribution of life in the oceans.

Light Transmission in Ocean Water

The graph in **Figure 13.3** shows that most solar energy falls in the range of wavelengths called **visible light**. This radiant energy from the Sun powerfully affects three major components of the oceans:

1. The major wind belts of the world, which produce ocean currents and wind-driven ocean waves, ultimately derive their energy from solar radiation. Wind belts and ocean currents strongly influence world climates.

2. A thin layer of warm water at the ocean surface, created by solar heating, overlies the great mass of cold water that fills most of the ocean basins. This is the "life layer" where most marine life exists.

3. Photosynthesis can occur only where sunlight penetrates the ocean water, so phytoplankton and most animals that eat them must live where the light is, in the relatively thin layer of sunlit surface water.

THE ELECTROMAGNETIC SPECTRUM   The Sun radiates a wide range of wavelengths of electromagnetic radiation. Together they comprise the **electromagnetic spectrum**, which is shown in the upper part of Figure 13.3. Only a very narrow portion of the electromagnetic spectrum is visible to humans as visible light. We call it "visible" light because our electromagnetic sensors—our eyes—are adapted to detect only the wavelengths in the visible region. In essence, our eyes "tune into" the visible light wavelengths, just as a radio "tunes into" specific radio waves.

Visible light can be further divided by wavelength into violet, blue, green, yellow, orange, and red energy levels. Together, these different wavelengths produce white light. The shorter wavelengths of energy to the left of visible light (for example, X-rays and gamma rays) damage tissue in high enough doses. The longer wavelengths of energy to the right of visible light (for example, infrared, microwaves, and radio waves) are used for heat transfer and communication.

THE COLOR OF OBJECTS   Light from the Sun includes all the visible colors. Most of the light we see is reflected from objects. All objects absorb and reflect different wavelengths of light, and each wavelength represents a color in the visible spectrum. Vegetation, for example, absorbs most wavelengths except green and yellow, which they reflect, so most plants look green. Similarly, a red jacket absorbs all wavelengths of color except red, which is reflected.

The lower part of Figure 13.3 shows how the ocean selectively absorbs the longer-wavelength colors (red, orange, and yellow) of visible light. The true colors of objects can be observed in natural light only in the surface waters because only there can all wavelengths of the visible spectrum be found. Red light is absorbed within the upper 10 meters (33 feet) of the ocean, and yellow is completely absorbed before a depth of 100 meters (330 feet). Thus, the shorter-wavelength portion of the visible spectrum is all that can be transmitted to greater depths (mostly blue light, with some violet and green wavelengths), and even then, their intensity is low. In the open ocean, sunlight strong enough to support photosynthesis occurs only

Figure 13.3 The electromagnetic spectrum and transmission of visible light in seawater. The electromagnetic spectrum (*top*) extends from extremely short cosmic rays (*left side*) to progressively increasing wavelengths (*right*). The narrow portion of the spectrum that we see as visible light is shown passing through seawater (*bottom*), which absorbs the longer wavelengths (red, orange, and yellow) above a depth of 100 meters (330 feet).

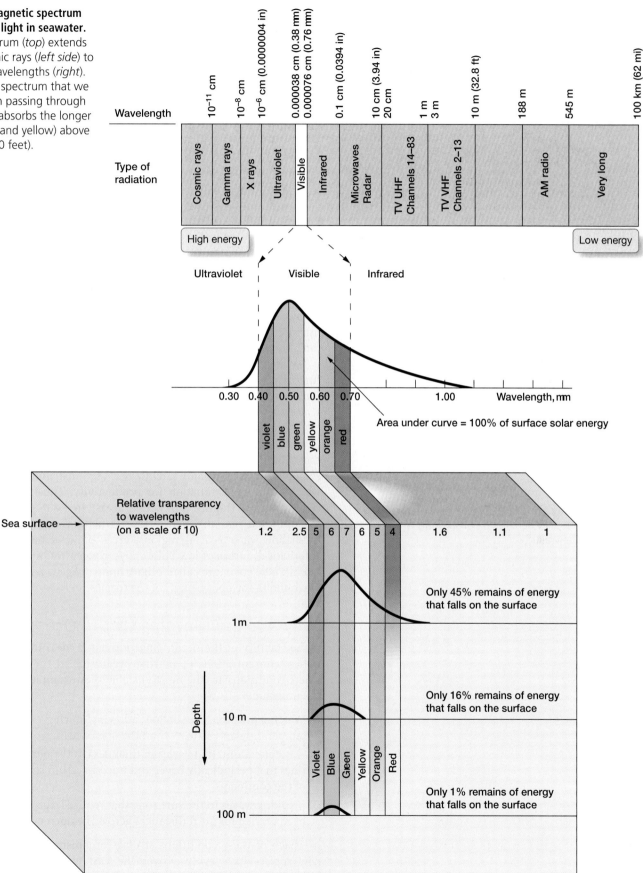

Figure 13.4 Using a Secchi disk. A Secchi disk is used to measure the depth of penetration of sunlight, which indicates the clarity of the water.

within the euphotic zone to a depth of 100 meters (330 feet), and no sunlight penetrates below a depth of about 1000 meters (3300 feet).

A **Secchi** (pronounced "SECK-ee") **disk**, such as the one shown in **Figure 13.4**, is used to measure water transparency, and based on that, the depth of light penetration can be estimated. The Secchi disk is named after its inventor, Angelo Secchi (1818–1878), an Italian astronomer who first used the device to measure water clarity of lakes in 1865. It consists of a disk 20 to 40 centimeters (8 to 16 inches) in diameter attached to a line that is marked off at regular intervals. As the disk is slowly lowered into the ocean, the depth at which it can last be seen indicates the water's clarity. Increased turbidity, which includes microorganisms and suspended sediment, increases the degree of light absorption, thus decreasing the depth to which visible light can penetrate into the ocean.

WATER COLOR AND LIFE IN THE OCEANS The color of the ocean ranges from deep indigo (blue) to yellow-green. Why are some areas of the ocean blue, whereas others appear green? Ocean color is influenced by (1) the amount of turbidity from runoff and (2) the amount of photosynthetic pigment, which increases with increasing biological production.

Coastal waters and upwelling areas are biologically very productive and almost always yellow-green in color because they contain large amounts of yellow-green microscopic marine algae and suspended particles. These materials disperse solar radiation so that the wavelengths for greenish or yellowish light are scattered most.

Water in the open ocean—particularly in the tropics—lacks productivity (and turbidity), so it is usually a clear, indigo-blue color. Water molecules disperse solar radiation so that the wavelength for blue light is scattered most. The atmosphere scatters blue light, too, which is why clear skies are blue.

Although photosynthetic marine algae and bacteria are microscopic, they occur in such large numbers that they can change the color of the ocean to such a degree that orbiting satellites are able to measure such changes from space. **Figure 13.5**, for example, shows a *SeaStar* satellite/SeaWiFS instrument view of ocean chlorophyll, which is an approximation for productivity. The figure shows high chlorophyll concentrations (highly productive areas) in light green color, which are called **eutrophic** (*eu* = good, *tropho* = nourishment). Generally, eutrophic waters are found in shallow-water coastal regions, areas of upwelling, and high-latitude regions. Alternatively, areas of low chlorophyll concentration (low productivity) are shown in dark blue color, are called **oligotrophic** (*oligo* = few, *tropho* = nourishment), and are found in the open oceans of the tropics.

Why Are the Margins Of the Oceans So Rich in Life?

If the stability of the ocean environment is ideal for sustaining life, why are the richest concentrations of marine organisms in the very margins of the oceans, where conditions are the most *un*stable? For example, characteristics of the coastal ocean include:

- Water depths that are shallow, allowing much greater seasonal variations in temperature and salinity than the open ocean.
- A water column that varies in thickness in the nearshore region in response to tides that periodically cover and uncover a thin strip of land along the margins of the continents.
- Breaking waves in the surf zone that release large amounts of energy, which has been carried for great distances across the open ocean.

Each of these conditions stresses organisms. In spite of hardships, however, new species have evolved over the vast expanse of geologic time that spans billions of years by the process of natural selection[5] to fit every imaginable

[5]See Diving Deeper 1.2 for a description of evolution and natural selection.

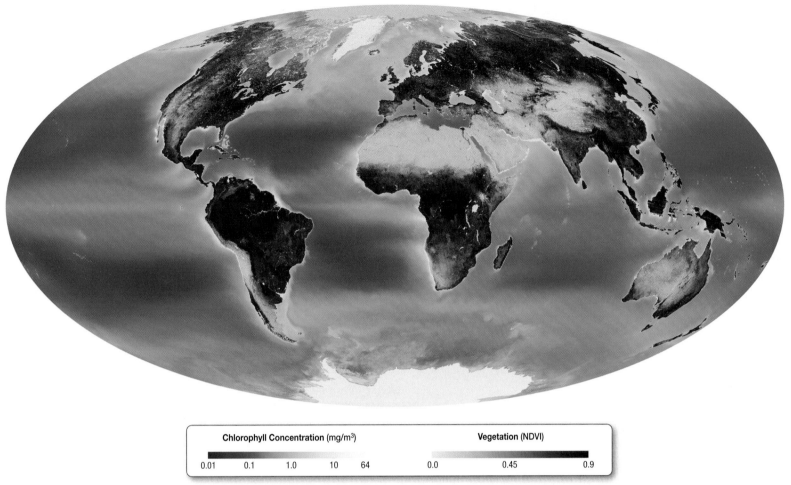

Chlorophyll Concentration (mg/m³)   Vegetation (NDVI)

0.01   0.1   1.0   10   64   0.0   0.45   0.9

Figure 13.5 Satellite image of ocean chlorophyll. Satellite data (1998–2010) showing average ocean chlorophyll concentration, which is an approximation for productivity. Data were gathered during the entire 13-year operation of the SeaWiFS instrument aboard the *SeaStar* satellite, which detected changes in seawater color caused by changing concentrations of chlorophyll that varies with photosynthetic productivity. Ocean chlorophyll concentrations are reported in milligrams per cubic meter (mg/m^3). On land, data are depicted using the Normalized Difference Vegetation Index (NDVI), which shows the density of green vegetation.

biological niche—even in environments that pose difficulties for organisms. In fact, many organisms have adapted to live under adverse conditions—such as coastal environments—as long as nutrients are available.

Along continental margins, some areas have more abundant life than others. What characteristics create such an uneven distribution of life? Again, only the basic requirements for the production of food need be considered. For example, areas that have the greatest biomass have the lowest water temperatures, too, because cold water contains higher amounts of nutrients than warm water. These nutrients stimulate phytoplankton growth, which profoundly affects the distribution of all other life in the oceans.

UPWELLING AND NUTRIENT SUPPLY As discussed in Chapter 7, **upwelling** is a flow of deep water toward the surface that brings water from depths below the euphotic zone. This deep water is rich in nutrients and dissolved gases because there are no phytoplankton at these depths to consume these compounds. When chilled water from below the surface rises, it hoists nutrients from the depths to the surface, where phytoplankton thrive and make food for larger organisms—copepods, fish, and on up to larger organisms such as whales.

Highly productive areas of *coastal upwelling* are found along the western margins of continents, where surface currents are moving toward the equator (**Figure 13.6**). Ekman transport (see Chapter 7) causes surface water to move away from these coasts,

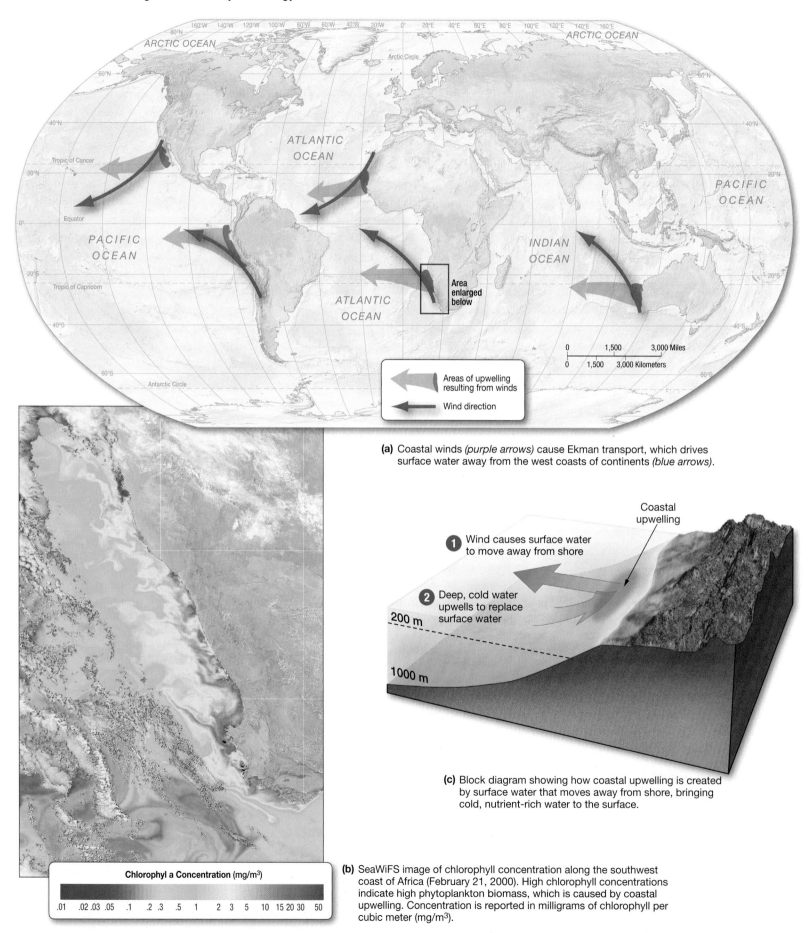

(a) Coastal winds *(purple arrows)* cause Ekman transport, which drives surface water away from the west coasts of continents *(blue arrows)*.

Coastal upwelling

❶ Wind causes surface water to move away from shore

❷ Deep, cold water upwells to replace surface water

200 m

1000 m

(c) Block diagram showing how coastal upwelling is created by surface water that moves away from shore, bringing cold, nutrient-rich water to the surface.

Chlorophyl a Concentration (mg/m³)

.01 .02 .03 .05   .1   .2 .3   .5   1   2 3   5   10 15 20 30 50

(b) SeaWiFS image of chlorophyll concentration along the southwest coast of Africa (February 21, 2000). High chlorophyll concentrations indicate high phytoplankton biomass, which is caused by coastal upwelling. Concentration is reported in milligrams of chlorophyll per cubic meter (mg/m³).

Figure 13.6 **Coastal upwelling.**

so nutrient-rich water from depths of 200 to 1000 meters (660 to 3300 feet) constantly rises to replace it.

CONCEPT CHECK 13.1

1 Discuss chemosynthesis as a method of primary productivity. How does it differ from photosynthesis?

2 An important variable in determining the distribution of life in the oceans is the availability of nutrients. How are the following variables related: proximity to the continents, availability of nutrients, and the concentration of life in the oceans?

3 Another important determinant of productivity is the availability of solar radiation. Why is biological productivity relatively low in the tropical open ocean, where the penetration of sunlight is greatest?

4 Discuss the characteristics of the coastal ocean where unusually high concentrations of marine life are found.

5 Explain why everything in the ocean at depths below the shallowest surface water takes on a blue-green appearance.

Web Animation
Ekman Spiral and Coastal
Upwelling/Downwelling

13.2 What Kinds Of Photosynthetic Marine Organisms Exist?

Many types of marine organisms photosynthesize. Mostly, they are represented by microscopic bacteria and algae but also include larger forms of algae and seed-bearing plants.

Seed-Bearing Plants (Anthophyta)

The only members of kingdom Plantae that exist in the marine environment belong to the highest group of plants, the seed-bearing members of phylum Anthophyta (*antho* = flower, *phytum* = plant), which occur exclusively in shallow coastal areas. Eelgrass (*Zostera*), for example, is a grasslike plant with true roots that exists primarily in the quiet waters of bays and estuaries from the low-tide zone to a depth of 6 meters (20 feet). Surf grass (*Phyllospadix*) (**Figure 13.7**), which is also a seed-bearing plant with true roots, is typically found in the high-energy environment of exposed rocky coasts from the intertidal zones down to a depth of 15 meters (50 feet).

Other seed-bearing plants are found in salt marshes and include grasses (mostly of the genus *Spartina*), whereas mangrove swamps contain primarily mangroves (genera *Rhizophora* and *Avicennia*). All these plants are important sources of food and protection for the marine animals that inhabit nearshore environments.

Macroscopic (Large) Algae

Various types of marine macro algae (the "seaweeds") are typically found in shallow waters along the ocean margins. These algae are usually attached to the bottom, but a few species float. Algae are classified in part on the color of the pigment they contain (**Figure 13.8**). Although modern classification of algae uses more than just its color, the division of algae into groups based on color is still a useful means of describing the different types of algae.

BROWN ALGAE   The brown algae of the phylum Phaeophyta (*phaeo* = dusky, *phytum* = plant) include the largest members of the attached (that is, not free-floating) species of marine algae. Their color ranges from very light brown to black. Brown algae occur primarily in middle latitude, cold-water areas.

The sizes of the brown algae range widely. One of the smallest is *Ralfsia*, which occurs as a dark brown encrusting patch in upper and middle intertidal zones. One of the largest is bull kelp (*Pelagophycus*), which may grow in water deeper than 30 meters (100 feet) and extend to the surface. Other types of brown algae include *Sargassum* (Figure 13.8a) and *Macrocystis* (Figure 13.8b).

Figure 13.7 Surf grass. Green surf grass (*Phyllospadix*) and various species of brown algae are exposed during an extremely low tide in this California tide pool. When the tide rises, the anchored surf grass floats and provides a protective hiding place for many tide pool organisms.

Figure 13.8 Examples of macroscopic algae.

(a) Brown algae *Sargassum*. This attached form is similar to the floating form that is the namesake of the Sargasso Sea.

(b) A small strand of brown algae *Macrocystis*, which is a major component of kelp beds.

(c) Green algae *Codium fragile*, also known as sponge weed or dead man's fingers.

(d) Red algae of two different types, *Bossiella californica* (*left, center*) and *Corallina* sp. (*right*). Both have tips that show portions of their internal calcareous skeleton (*white*).

STUDENTS SOMETIMES ASK . . .

What type of algae was involved in the massive cleanup effort prior to the 2008 Olympic sailing events in China?

In the summer of 2008, a bloom of the floating green algae *Enteromorpha* struck the coast of Qingdao, China, where the 2008 Olympic sailing events were to be held (**Figure 13.9**). The bloom lasted for more than one month, was linked to high levels of nitrates from runoff into coastal waters, and covered nearly the entire sailing venue. Threatened by having to cancel the sailing competition, the Chinese government initiated one of the world's largest algae-cleanup efforts that involved tens of thousands of citizens and defense personnel, thousands of fishing boats, and hundreds of dump trucks. The massive cleanup effort was able to collect enough of the floating algae that China was able to successfully host the Olympic sailing events. In all, about 682 million metric tons (750 million short tons) of algae was removed from coastal waters at a cost of $87.3 million. According to news reports, much of the algae was transported to farms as fertilizer or feed for livestock.

GREEN ALGAE Although green algae of phylum Chlorophyta (*khloros* = green, *phytum* = plant) are common in freshwater environments, they are not well represented in the ocean. Most marine species are intertidal or grow in shallow bay waters. They contain the pigment chlorophyll, which gives them their green color. They grow only to moderate size, seldom exceeding 30 centimeters (12 inches) in the largest dimension. Forms range from finely branched filaments to thin sheets.

Various species of sea lettuce (*Ulva*), a thin membranous sheet only two cell layers thick, are widely scattered throughout cold-water areas. Sponge weed (*Codium*), a two-branched form more common in warm waters, can exceed 6 meters (20 feet) in length (Figure 13.8c).

RED ALGAE Red algae of phylum Rhodophyta (*rhodos* = red, *phytum* = plant) are the most abundant and widespread of marine macroscopic algae. Over 4000 species occur from the very highest intertidal levels to the outer edge of the inner sublittoral zone. Many are attached to the bottom, either as branching forms (Figure 13.8d) or as forms that encrust surfaces. They are very rare in freshwater. Red algae range from being just barely visible to the unaided eye to being 3 meters (10 feet) long. Red algae are found in both warm and cold waters, but the warm-water varieties are relatively small.

The color of red algae varies considerably, depending on its depth in the intertidal or inner sublittoral zones. In upper, well-lighted areas, it may be green to black or purplish. In deeper water zones, where less light is available, it may be brown to pinkish red.

The vast majority of marine photosynthetic productivity occurs within the surface layer of the ocean to a depth of 100 meters (330 feet), which corresponds to the depth of the euphotic zone. At this depth, the amount of light is reduced to 1%

of that available at the surface. Remarkably, some deep-water species can survive on the very faint amount of sunlight that exists below the euphotic zone. For example, a species of red alga has been documented growing at a depth of 268 meters (880 feet) on a seamount near San Salvador in The Bahamas, where available light is only 0.0005% of that available at the surface.

Microscopic (Small) Algae

Microscopic algae are either directly or indirectly the source of food for more than 99% of marine animals. Most microscopic algae are phytoplankton—photosynthetic organisms that live in the upper surface waters and drift with currents—although some live on the bottom in the nearshore environment, where sunlight reaches the shallow ocean floor.

GOLDEN ALGAE The golden algae of phylum Chrysophyta (*chrysus* = golden, *phytum* = plant) contain the orange-yellow pigment *carotin*. They consist of diatoms and coccolithophores, both of which store food as carbohydrates and oils, as discussed in Chapter 4.

Figure 13.9 Algae cleanup in Qingdao, China, before the 2008 Olympic sailing events. The Chinese government undertook a massive cleanup of floating green algae in order to host the 2008 Olympic sailing competition. The cleanup effort involved tens of thousands of people that removed about 682 million metric tons (750 million short tons) of algae at a cost of $87.3 million.

Diatoms **Diatoms** (*diatoma* = cut in half) are a class of algae that are contained in a microscopic shell called a **test** (*testa* = shell). Diatom tests are composed of opaline silica ($SiO_2 \cdot nH_2O$) and are important geologically because they accumulate on the ocean bottom, producing **diatomaceous earth**. Some deposits of diatomaceous earth that have been elevated above sea level by tectonic forces are mined and used in filtering devices and numerous other applications (see Diving Deeper 4.1). Diatoms are the most productive group of marine algae.

Diatom tests have a variety of shapes, but all have a top half and a bottom half that fit together (**Figure 13.10a**). The single cell is contained within this test, and it exchanges nutrients and waste with the surrounding water through holes in its test.

Coccolithophores **Coccolithophores** (*coccus* = berry, *lithos* = stone, *phorid* = carrying) are covered with small calcareous plates called **coccoliths**, made of calcium carbonate ($CaCO_3$) (**Figure 13.10b**). Each individual plate is about the size of a bacterium, and the entire organism is too small to be captured in plankton nets. Coccolithophores live in temperate and warmer surface waters and contribute significantly to calcareous sea floor deposits.

DINOFLAGELLATES **Dinoflagellates** (*dino* = whirling, *flagellum* = whip) belong to the phylum Pyrrophyta (*pyrrhos* = fire, *phytum* = plant) (Figures 13.10c and d). They possess **flagella** (small, whiplike

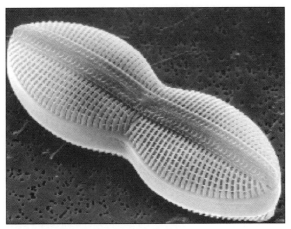

(a) Peanut-shaped diatom *Diploneis* (length = 50 microns [0.002 inch]).

(b) Coccolithophore *Emiliania huxleyi*, showing disk-shaped calcium carbonate ($CaCO_3$) plates—called coccoliths—that cover the organism (bar scale = 1 micron [0.00004 inch]).

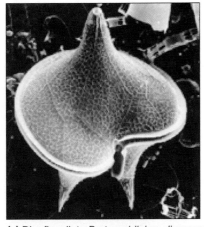

(c) Dinoflagellate *Protoperidinium divergens* (length = 70 microns [0.003 inch]).

(d) Leaflike tropical dinoflagellate *Heterodinium whittingae* (length = 100 microns [0.004 inch]).

Figure 13.10 Examples of microscopic algae.

Figure 13.11 Red tide in the ocean. When conditions are right, dinoflagellates reproduce and populate surface waters in staggering numbers, turning the water red. Although this phenomenon is called a red tide, it has nothing to do with the tides.

structures) for locomotion, giving them a slight capacity to move into areas that are more favorable for photosynthetic productivity. Dinoflagellates are rarely important geologically because their tests are made of cellulose, which is biodegradable and not preserved as deposits on the sea floor.

Red Tides Dinoflagellates sometimes exist in such great abundance that they color surface waters red, producing the phenomenon known as a **red tide** (Figure 13.11), which has nothing to do with tidal phenomena. When conditions are right, phytoplankton populations can grow explosively and create a phenomenon known as a *bloom*, which can be seen in satellite images. Red tides and associated algal blooms that do not color the water red but are detrimental to marine animals, humans, or the environment are more accurately called **harmful algal blooms (HABs)**. These toxic blooms can kill marine life and people who eat contaminated seafood. In addition to producing toxins, many of the 1100 species of dinoflagellates undergo bizarre structural changes in response to changes in their environment (see Web Diving Deeper 13.1).

What causes red tides? Natural oceanographic conditions sometimes stimulate the productivity of certain dinoflagellates. During these times, up to 2 million dinoflagellates may be found in 1 liter (about 1 quart) of water, giving the water a reddish color (Figure 13.11). In other instances, red tides appear to be associated with

STUDENTS SOMETIMES ASK . . .

Why does a red tide glow bluish-green at night?

Many of the species of dinoflagellates that produce red tides (most notably those of genus *Gonyaulax*) also have bioluminescent capabilities—that is, they can produce light organically. When the organisms are disturbed, they emit a faint bluish-green glow. When waves break during a red tide at night, the waves are often spectacularly illuminated by millions of bioluminescent dinoflagellates. During these times, one can easily observe marine animals moving through the water because their bodies are silhouetted by bioluminescent dinoflagellates that light up as they pass over the animals' bodies.

Figure 13.12 Routes through which dinoflagellate toxins spread to marine organisms and humans. During algal blooms, certain species of dinoflagellates and other phytoplankton produce powerful toxins, which can spread throughout marine food webs. These toxic blooms can kill marine life and even people who eat contaminated seafood.

Humans

Marine mammals   Sea birds

Nekton   Fish   Nekton

Zooplankton

Plankton   Toxins from red tide dinoflagellates   Plankton

Fish larvae Benthic larvae   Microbial food web

Bivalves (clams, mussels, scallops, etc.)

Benthos   Benthos

Crabs, whelks, drills, etc.

nutrient-laden runoff from land. Red tides are by no means a new phenomenon. In fact, the Old Testament and other ancient literature make reference to waters turning blood red, which most likely influenced the colorful names of certain seas such as the Red Sea and the Vermillion Sea.

Dinoflagellate Toxins   Although many red tides are harmless to marine animals and humans, they can still be responsible for mass die-offs of marine organisms.[6] When huge numbers of dinoflagellates die, oxygen is removed from seawater during decomposition, and many types of marine life literally suffocate to death. In other cases, dinoflagellates that are responsible for many red tides produce biotoxins that can spread to many different types of organisms—including humans (**Figure 13.12**). *Karenia* and *Gonyaulax*, for example, are two common genera of dinoflagellates in red tides that produce water-soluble toxins. Certain filter-feeding shellfish called bivalves—various clams, mussels, and oysters—then strain the dinoflagellates from the water for food. *Karenia* toxin kills fish and shellfish. *Gonyaulax* toxin is not poisonous to shellfish, but it concentrates in their tissues and is poisonous to humans who eat the shellfish, even after the shellfish are cooked. This malady is called *paralytic shellfish poisoning (PSP).*

The symptoms of PSP in humans are similar to those of drunkenness—incoherent speech, uncoordinated movement, dizziness, and nausea—and can

STUDENTS SOMETIMES ASK . . .

What other strange occurrences have been linked to the ingestion of algae toxins?

Worldwide, a number of strange occurrences and mysterious poisonings have implicated various species of toxic marine microorganisms that can spread throughout the marine food web and even to humans. For instance, domoic acid—a biotoxin produced by a diatom (*Pseudo-nitzschia*)—was first recognized in 1987 as the poison that infected more than 100 people who ate contaminated mussels from Prince Edward Island, Canada. When ingested, domoic acid passes through the blood and binds to receptors in the brain, causing symptoms such as confusion, disorientation, seizures, coma, and even death. In the Prince Edward Island incident, 4 of the victims died, and 10 suffered permanent short-term memory loss, which led researchers to call poisoning by domoic acid *amnesic shellfish poisoning.*

Interestingly, recent research has linked domoic acid poisoning with famously odd behavior by birds in California's Monterey Bay during the summer of 1961. Studies of the gut contents of archival samples gathered in 1961 ship surveys have found that toxin-making algae were present in 79% of zooplankton. Domoic acid was concentrated in plankton-eating fish, which are unaffected by the toxin. In turn, the fish were eaten by migratory sooty shearwaters (*Puffinus griseus*), which flew erratically, smashed into structures, rammed cars, and pecked eight people. The incident was reported to have inspired a local visitor, Alfred Hitchcock, to include similar events in his classic thriller *The Birds* (**Figure 13.13**), which was released two years later.

There have been several other incidents involving toxic diatoms in Monterey Bay. In 1991, for example, brown pelicans and Brandt's cormorants exhibited odd behavior— they acted drunk, swam in circles, and made loud squawking sounds. Research revealed that toxic diatoms (*Pseudo-nitzschia australis*) were to blame for their odd behavior and caused more than 100 birds to wash up dead at the shore. In 1998, another toxic diatom bloom caused the death of 400 California sea lions that ate fish contaminated with domoic acid.

Figure 13.13 **Actress Tippi Hedren fights off a menacing seagull in Hitchcock's 1963 classic suspense film *The Birds.*** The inspiration for Hitchcock's classic thriller was an actual event that occurred in Monterey Bay in 1961, when birds acted erratically after eating fish contaminated with domoic acid, a biotoxin produced by diatoms.

---

[6]Interestingly, this is what is thought to be the evolutionary advantage of toxin use by dinoflagellates: to subdue other types of algae as a food source (rather than to defend themselves from being eaten).

occur only 30 minutes after ingesting contaminated shellfish. There is no known antidote for the toxin, which attacks the human nervous system, but the critical period usually passes within 24 hours. At least 300 fatal and 1750 nonfatal cases of PSP have been documented worldwide.

April through September are particularly dangerous months for red tides in the Northern Hemisphere. In most areas, quarantines exist to prohibit harvesting shellfish that feed on toxic microscopic organisms.

Dinoflagellates are also associated with various types of seafood poisoning. One example is **ciguatera**, which is a type of seafood poisoning caused by ingestion of certain tropical reef fish (most notably barracuda, red snapper, and grouper) that have high levels of naturally occurring dinoflagellate toxins. Symptoms of ciguatera in humans usually involve a combination of gastrointestinal, neurological, and cardiovascular disorders. Worldwide, ciguatera causes more cases of human illness than any other form of seafood poisoning.

Ocean Eutrophication and Dead Zones

Ocean **eutrophication** (*eu* = good, *tropho* = nourishment, *ation* = action) is the artificial enrichment of waters by a previously scarce nutrient that can trigger an overabundance of algae such as an HAB. Human activities have been shown to contribute to ocean eutrophication when excess nutrients in the form of fertilizer, sewage, and animal waste make their way into coastal waters.

Large areas of ocean eutrophication are associated with extensive *hypoxic* (*hypo* = under, *oxic* = oxygen) **dead zones** of oxygen-poor water that often occur near the mouths of major rivers after large spring runoffs (**Figure 13.14**). When rivers deliver excess nutrients to the ocean, they feed profuse, widespread blooms of algae that later die and decompose, robbing the water of oxygen. Oxygen levels within these dead zones drop from above 5.0 parts per million (ppm by weight) to below 2.0 ppm, which is lower than most marine animals can tolerate. Some of the most mobile marine organisms that swim well can flee the area, but the dead zone suffocates and kills many bottom-dwelling organisms such as crabs, sea stars, and sea slugs. Low oxygen levels have also been shown to limit the growth and reproduction of marine life.

The number of dead zones has doubled every decade from the1960s into the twenty-first century, and scientists currently have documented the existence of more than 500 worldwide (**Figure 13.15**). In the future, the size and number of dead zones is expected to increase because of continuing human impacts such as polluted runoff and destruction of coastal wetland habitats. In fact, the number of dead zones is expected to double between 2010 and 2100.

The world's largest dead zone is in the Baltic Sea, where a combination of agricultural runoff, deposition of nitrogen from burning fossil fuels, and human waste discharge has over-fertilized the sea. The countries surrounding the Baltic Sea have recently established a commission to help protect the environmental health of the sea by implementing policies and measures to reduce its incoming load of nutrients.

The second-largest dead zone in the world is the one that forms each summer near the mouth of the Mississippi River in the Gulf of Mexico off Louisiana (**Figure 13.16**). In 2002, in fact, it reached a record size of 22,000 square kilometers (8500 square miles)—about the size of the state of New Jersey. Smaller dead zones have occurred each summer in the region for decades, but they dramatically increased in size after the record-breaking Midwest floods in 1993 and 2011. Over the past several years, the average size of the dead zone in the northern Gulf of Mexico has been about 17,000 square kilometers (6560 square miles), the size of Lake Ontario.

The Gulf's dead zone appears to be related to runoff of nutrients—especially nitrates—from agricultural activities that are washed down the Mississippi River

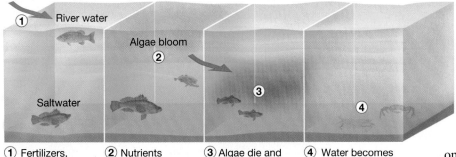

① River water

Algae bloom
②

Saltwater

③

④

① Fertilizers, sewage, and nutrients from farming flow down rivers.

② Nutrients stimulate massive algae blooms.

③ Algae die and sink to the bottom, where bacterial decomposition uses up dissolved oxygen.

④ Water becomes anoxic and kills marine life that cannot flee.

Figure 13.14 How dead zones form. Dead zones are low-oxygen regions related to excess nutrients from agricultural runoff that stimulates algal blooms. When the algae die, decomposition consumes vast amounts of oxygen, creating an anoxic dead zone that affects a multitude of marine life.

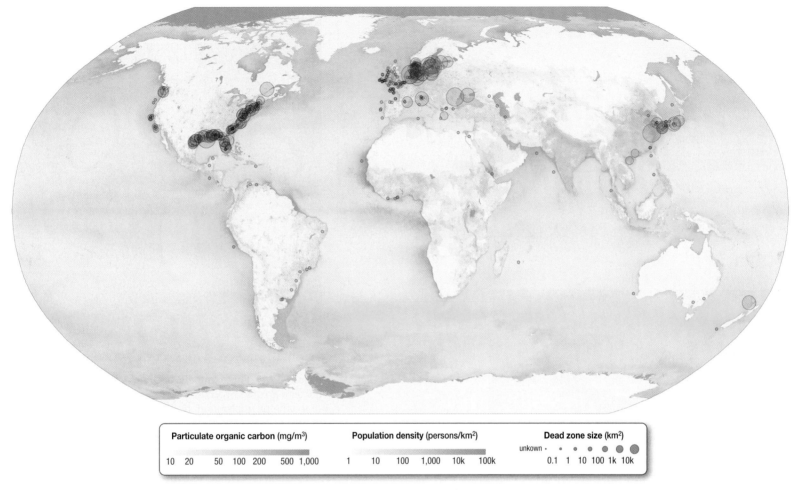

Figure 13.15 Worldwide dead zones. Map of oceanic particulate organic carbon (*blue*), population density (*brown*), and documented dead zones (*red circles*, based on size). Small black circles show where dead zones have been reported, but their size is unknown. Notice that many dead zones occur where large rivers enter the sea.

and eventually reach the Gulf, where they trigger algal blooms (see Figure 13.14). Once these algae die and rain down on the sea floor, bacteria feed on them and on fecal matter, depleting the water of oxygen along the bottom. Other factors—most notably stratification of the water column—can influence the formation of dead zones, which can persist for several months until there is strong mixing of ocean waters from a hurricane or other storm.

Proposals to combat the spread of the Gulf's dead include controlling nutrient runoff from agriculture, preserving and utilizing wetlands that filter runoff before it enters the Gulf, planting buffer strips of trees and grasses between farm fields and streams, altering the times when fertilizers are applied, improving crop rotation, and enforcing existing clean water regulations. In this way, scientists, land planners, and policymakers are working to institute an action plan to shrink the yearly Gulf of Mexico dead zone.

Photosynthetic Bacteria

Until recently, the role of marine bacteria in photosynthesis has been largely ignored. Because of the extremely small size of bacteria, earlier samplings of marine life completely overlooked them. Recent improvements in sampling methods for bacteria-sized organisms and genome sequencing studies have revealed bacteria's incredible abundance and importance in the oceans.

For example, one of the first types of marine photosynthetic bacteria to be identified was *Synechococcus*, which are extremely abundant in coastal and open-ocean environments, sometimes reaching densities greater than 100,000 cells per milliliter

STUDENTS SOMETIMES ASK . . .

Are red tides becoming a more common occurrence? If so, why?

In many places worldwide, such as the northern Gulf of Mexico, algal blooms known as red tides have been documented in some of the earliest written records. Yet it is unclear how large a role human-caused versus natural factors play in their development. For example, the occurrence of algal blooms in some locations appears to be entirely natural, while in others the blooms appear to be a result of increased nutrients such as nitrates and phosphates from land-based runoff. In addition to human-caused coastal water nutrient loading, the recently documented increase in seawater surface temperatures has also been implicated as a contributing factor in algal blooms. Currently, it is under debate in the scientific community whether the apparent increase in frequency and severity of algal blooms in various parts of the world is in fact a real increase or whether it is simply due to an enhanced observation and reporting network.

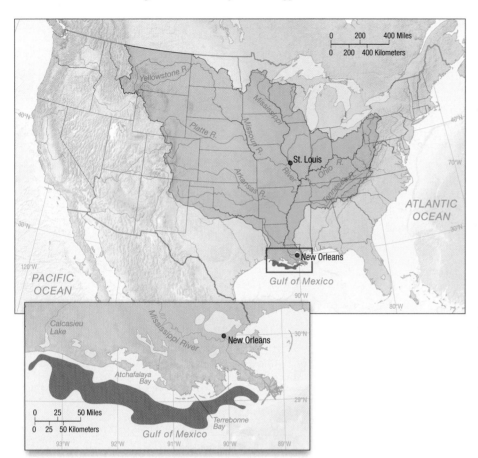

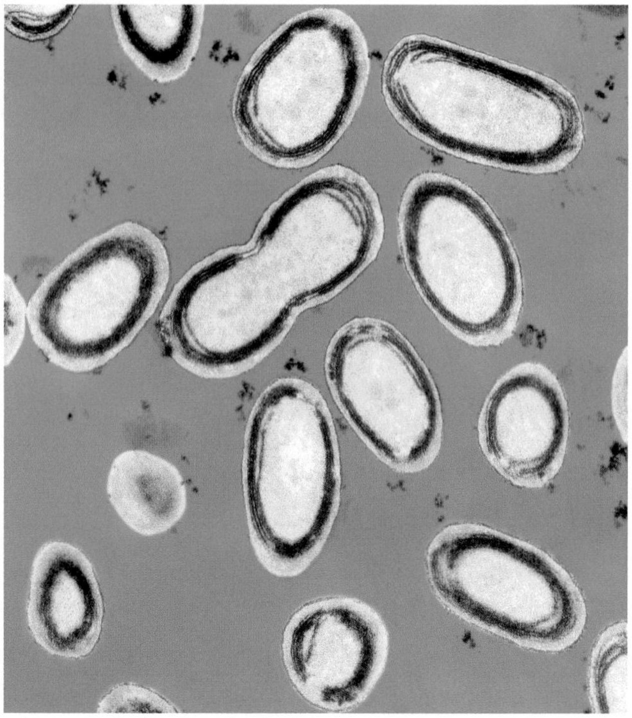

Figure 13.17 **Photosynthetic bacteria *Prochlorococcus*.** The bacteria *Prochlorococcus* is the most abundant and smallest of marine phytoplankton, reaching only about 0.6 microns (0.00002 inch) in diameter.

Figure 13.16 The Gulf of Mexico dead zone. Map (*above*) showing Mississippi River drainage basin (*gold*) and enlargement (*below*) showing the extent of the 1999 Gulf of Mexico dead zone (*red*). The northern Gulf of Mexico dead zone is the second largest in the world, after the Baltic Sea.

(0.03 ounce) of seawater. At certain times and places, these cells can be responsible for as much as half of the primary production of food in the ocean. More recently, microbiologists have discovered an extremely small but abundant bacteria named *Prochlorococcus* (**Figure 13.17**) that reaches concentrations many times that of *Synechococcus*. In fact, *Prochlorococcus* has been estimated to constitute at least half of the world ocean's total photosynthetic biomass, which means that it is probably the most abundant photosynthetic organism on Earth.

In addition, recent large-scale gene sequencing of microbes in the Sargasso Sea has revealed a host of new types of bacteria, suggesting substantial yet previously unrecognized oceanic microbial diversity. Clearly, microbes exert a critical influence on marine ecosystems and carry significant implications for sustainability, global climate change, ocean system cycles, and human health.

ESSENTIAL CONCEPT 13.2

Marine photosynthetic organisms include seed-bearing plants (such as surf grass), macroscopic algae (seaweeds), microscopic algae (diatoms, coccolithophores, and dinoflagellates), and bacteria. Red tides are caused by an overabundance of dinoflagellates.

CONCEPT CHECK 13.2

❶ Compare the macroscopic algae in terms of color, maximum depth in which they grow, common species, and size.

❷ The golden algae include two classes of important phytoplankton. Compare their composition and the structure of their tests and explain their importance in the geologic fossil record.

❸ What is a red tide? What conditions cause red tides?

❹ How does paralytic shellfish poisoning (PSP) differ from amnesic shellfish poisoning? What types of microorganisms create each?

❺ What conditions create ocean eutrophication (dead zones)? What can be done to limit their spread?

13.3 How Does Regional Primary Productivity Vary?

Primary photosynthetic production in the oceans varies dramatically from place to place (see Figure 13.5). Typical units of photosynthetic production are in weight of carbon (*grams of carbon*) per unit of area (*square meter*) per unit of time (*year*), which is abbreviated as $gC/m^2/yr$. Values range from as low as $1\ gC/m^2/yr$ in some areas of the open ocean to as much as $4000\ gC/m^2/yr$ in some highly productive coastal estuaries (Table 13.1). This variability is a result of the uneven distribution of nutrients throughout the photosynthetic zone and seasonal changes in the availability of solar energy.[7]

About 90% of the biomass generated in the euphotic (sunlit) zone of the open ocean is decomposed into inorganic nutrients before descending below this zone. The remaining 10% of this organic matter sinks into deeper water, where all but about 1% of it is decomposed. The 1% that reaches the deep-ocean floor accumulates there. The way in which material is removed from the euphotic zone to the sea floor is called a **biological pump** because it "pumps" carbon dioxide and nutrients from the upper ocean and concentrates them in deep-sea waters and sea floor sediments.

Throughout much of the subtropical gyres, a permanent **thermocline** (and resulting **pycnocline**[8]) develops. It forms a barrier to vertical mixing, preventing the resupply of nutrients to the sunlit surface layer. In the middle latitudes, a thermocline develops only during the summer season; in polar regions, a thermocline does not usually develop. The degree to which waters develop a thermocline profoundly affects the patterns of biological production observed at different latitudes.

TABLE 13.1   VALUES OF NET PRIMARY PRODUCTIVITY FOR VARIOUS ECOSYSTEMS

Ecosystem	Primary productivity	
	Range ($gC/m^2/yr$)	Average ($gC/m^2/yr$)
Oceanic		
Algae beds and coral reefs	1000–3000	2000
Estuaries	500–4000	1800
Upwelling zone	400–1000	500
Continental shelf	300–600	360
Open ocean	1–400	125
Land		
Freshwater swamp and marsh	800–4000	2500
Tropical rainforest	1000–5000	2000
Middle latitude forest	600–2500	1300
Cultivated land	100–4000	650

---

[7]For a review of Earth's seasons, see Chapter 6.

[8]Recall that a *thermocline* is a layer of rapidly changing temperature, and a *pycnocline* is a layer of rapidly changing density. Development of ocean thermoclines and pycnoclines is discussed in Chapter 5.

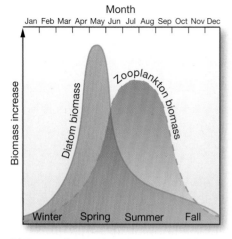

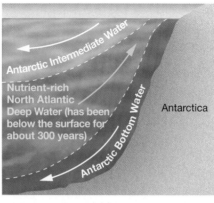

(a) Barents Sea productivity graph, which shows the dramatic springtime increase of diatom biomass that results in an increase in zooplankton abundance.

(b) The upwelling of cold, nutrient-rich North Atlantic Deep Water near Antarctica continually supplies Antarctic waters with nutrients.

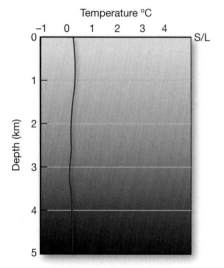

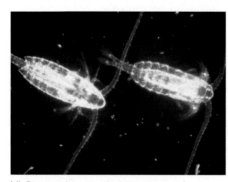

(c) Temperature graph of Antarctic waters showing a nearly uniform water temperature with depth (an isothermal water column).

(d) Copepods are an important type of zooplankton in polar oceans. These copepods are of the genus *Calanus*, each of which is about 8 millimeters (0.3 inch) in length.

Figure 13.18 Productivity in polar oceans.

Productivity in Polar (High Latitude) Oceans

Polar regions such as the Arctic Ocean's Barents Sea, which is off the northern coast of Europe, experience continuous darkness for about three months of winter and continuous illumination for about three months during summer. Diatom productivity peaks in the Barents Sea during May (**Figure 13.18a**), when the Sun rises high enough in the sky so that there is deep penetration of sunlight into the water. As soon as diatoms develop, zooplankton—mostly small crustaceans (**Figure 13.18d**)—begin feeding on them. The zooplankton biomass peaks in June and continues at a relatively high level until winter darkness begins in October.

In the Antarctic region—particularly at the southern end of the Atlantic Ocean—productivity is somewhat greater. This is caused by the upwelling of North Atlantic Deep Water, which forms on the opposite side of the ocean basin, where it sinks and moves southward below the surface. Hundreds of years later, it rises to the surface near Antarctica, carrying with it high concentrations of nutrients (**Figure 13.18b**). When the Sun provides sufficient solar radiation in summer, there is an explosion of biological productivity. A recent study of Antarctic waters, however, documented as much as a 12% decrease in phytoplankton productivity because of increased ultraviolet radiation as a result of the Antarctic ozone hole.

Blue whales—the largest of all whales (see Figure 14.20)—eat mostly zooplankton and time their migration through middle latitude and polar oceans to coincide with maximum zooplankton productivity. This enables the whales to develop and support calves that can exceed 7 meters (23 feet) in length at birth. The mother blue whale suckles the calf with rich, high-fat milk for six months. By the time the calf is weaned, it is over 16 meters (50 feet) long. In two years, it will be 23 meters (75 feet) long, and after about three years, it will weigh 55 metric tons (60 short tons)! This phenomenal growth rate gives some indication of the enormous biomass of small copepods and krill upon which these large mammals feed.[9]

Density and temperature change very little with depth in polar regions (**Figure 13.18c**), so these waters are **isothermal** (*iso* = same, *thermo* = temperature), and there is no barrier to mixing between surface waters and deeper, nutrient-rich waters. In the summer, however, melting ice creates a thin, low-salinity layer that does not readily mix with the deeper waters. This stratification is crucial to summer production because it helps prevent phytoplankton from being carried into deeper, darker waters. Instead, they are concentrated in the sunlit surface waters, where they reproduce continuously.

Nutrient concentrations (mostly nitrates and phosphates) are usually adequate in high-latitude surface waters, so the availability of solar energy limits photosynthetic productivity in these areas more than does the availability of nutrients.

---

[9]As a similar size analogy, consider how many ants you would have to eat as a child to grow to adult size!

Productivity in Tropical (Low-Latitude) Oceans

Perhaps surprisingly, productivity is low in tropical regions of the open ocean. Because the Sun is more directly overhead, light penetrates much more deeply into tropical oceans than in middle latitude and polar waters, and solar energy is available year-round. However, productivity is low in tropical regions of the open ocean because a permanent thermocline produces a stratification (layering) of water masses. This prevents mixing between surface waters and nutrient-rich deeper waters, effectively eliminating any supply of nutrients from deeper waters below (**Figure 13.19**).

At about 20 degrees north and south latitude, phosphate and nitrate concentrations are commonly less than $\frac{1}{100}$ of their concentrations in middle latitude oceans during winter. In fact, nutrient-rich waters in the tropics lie below 150 meters (500 feet), with the highest concentrations between 500 and 1000 meters (1640 and 3300 feet). So, productivity in tropical regions is limited by the lack of nutrients (unlike in polar regions, where productivity is limited by the lack of sunlight).

Generally, primary production in tropical oceans occurs at a steady but rather low rate. The total annual production of tropical oceans is only about half of that found in middle latitude oceans.

Exceptions to the general pattern of low productivity in tropical oceans include the following:

1. **Equatorial upwelling.** Where trade winds drive westerly equatorial currents on either side of the equator, Ekman transport causes surface water to diverge toward higher latitudes (see Figure 7.10). This surface water is replaced by nutrient-rich water from depths of up to 200 meters (660 feet). Equatorial upwelling is best developed in the eastern Pacific Ocean.

2. **Coastal upwelling.** Where the prevailing winds blow toward the equator and along western continental margins, surface waters are driven away from the coast. They are replaced by nutrient-rich waters from depths of 200 to 900 meters (660 to 2950 feet). This upwelling promotes high primary production along the west coasts of continents (see Figure 13.6), which can support large fisheries.

3. **Coral reefs.** Organisms that comprise and live among coral reefs are superbly adapted to low-nutrient conditions, similar to the way certain organisms are adapted to desert life on land. Symbiotic algae living within the tissues of coral and other species allow coral reefs to be highly productive ecosystems. Coral reefs also tend to retain and recycle what little nutrients exist. Coral reef ecosystems are discussed further in Chapter 15, "Animals of the Benthic Environment."

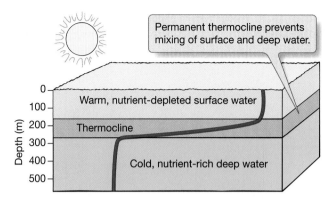

Figure 13.19 Productivity in tropical oceans. Although tropical regions receive adequate sunlight year-round, a permanent thermocline prevents the mixing of surface water and deep water. As phytoplankton consume nutrients in the surface layer, productivity is limited because the thermocline prevents replenishment of nutrients from deeper water. Thus, productivity remains at a steady, low level. The red curve indicates the temperature of the water column, with high temperature toward the right.

STUDENTS SOMETIMES ASK . . .

The number and variety of tropical species on land is astounding. I don't understand how the tropical oceans can have such low productivity.

Life on land does not necessarily correspond to life in the ocean! Tropical rainforests support an amazing diversity of species and an enormous biomass. In the tropical ocean, however, a strong, permanent thermocline limits the availability of nutrients that are necessary for the growth of phytoplankton. Without abundant phytoplankton, not much else can live in the ocean. In fact, these areas are often considered biological deserts. It is ironic that the clear blue water of the tropics so prominently displayed in tourist brochures indicates seawater that is biologically quite sterile!

Productivity in Middle Latitude (Temperate) Oceans

Productivity is limited by available sunlight in polar regions and by nutrient supply in the low-latitude tropics. In middle latitude (temperate) oceans, a combination of these two limiting factors controls productivity, as shown in **Figure 13.20a** (which shows the pattern for the Northern Hemisphere; in the Southern Hemisphere, the seasons are reversed).

WINTER   Productivity in middle latitude oceans is very low during winter, even though nutrient concentration is *highest* at this time (Figure 13.20a). The water column is isothermal, too, like in polar regions, so nutrients are well distributed throughout the water column. **Figure 13.20b** (*winter*) shows, however, that the Sun is at its lowest position above the horizon during winter, so a high percentage of the available solar energy is reflected, leaving only a small

Figure 13.20 Productivity in middle latitude (temperate) oceans in the Northern Hemisphere.

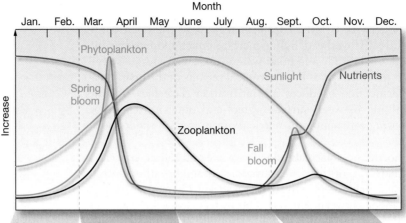

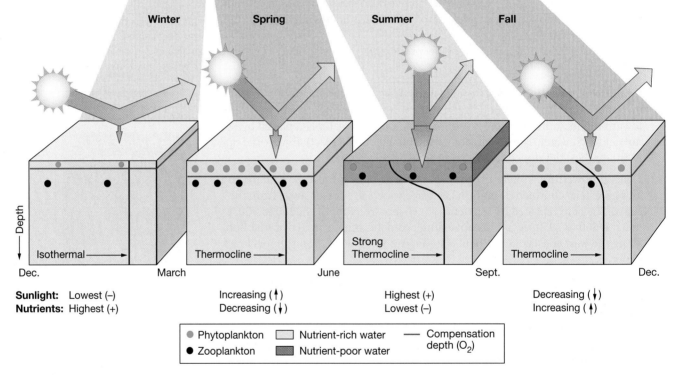

(a) Relationship among phytoplankton, zooplankton, amount of sunlight, and nutrient levels for surface waters in northern temperate latitudes.

Sunlight:	Lowest (–)	Increasing (↑)	Highest (+)	Decreasing (↓)
Nutrients:	Highest (+)	Decreasing (↓)	Lowest (–)	Increasing (↑)

Legend:
● Phytoplankton   ▢ Nutrient-rich water   — Compensation depth (O₂)
● Zooplankton   ▨ Nutrient-poor water

(b) The seasonal cycle of sunlight affects the presence and depth of the thermocline, which affects the availability of nutrients. This, in turn, affects the abundance of phytoplankton and other organisms such as zooplankton that rely on phytoplankton for food.

percentage to be absorbed into surface waters. As a result, the compensation depth for photosynthesis is so shallow that phytoplankton do not grow much. The absence of a thermocline, moreover, allows algal cells to be carried down beneath the euphotic zone for extended periods by turbulence associated with winter waves.

Web Animation
Oceanic Midlatitude Productivity

SPRING The Sun rises higher in the sky during spring than during winter (Figure 13.20b, *spring*), so the compensation depth for photosynthesis deepens. A **spring bloom** of phytoplankton occurs (Figure 13.20a) because solar energy and nutrients are available, and a seasonal thermocline develops (due to increased solar heating) that traps algae in the euphotic zone (Figure 13.20b). This creates a tremendous demand for nutrients in the euphotic zone, so the supply becomes limited, causing productivity to decrease sharply. Even though the days are lengthening and sunlight is increasing, productivity during the spring bloom is limited by the lack of nutrients. In most areas of the Northern Hemisphere, therefore, phytoplankton populations decrease in April due to insufficient nutrients and because their population is being consumed by zooplankton (grazers).

SUMMER The Sun rises even higher in the summer than in the spring (Figure 13.20b, *summer*), so surface waters in the middle latitudes continue to warm. A strong seasonal thermocline is created at a depth of about 15 meters (50 feet). The thermocline, in turn, prevents vertical mixing, so nutrients depleted from surface waters cannot be replaced by those from deeper waters. Throughout summer, the phytoplankton population remains relatively low (Figure 13.20a). Even though the compensation depth for photosynthesis is at its maximum, phytoplankton can actually become scarce in late summer.

FALL Solar radiation diminishes in the fall, as the Sun moves lower in the sky (Figure 13.20b, *fall*), so surface temperatures drop, and the summer thermocline breaks down. Nutrients return to the surface layer as increased wind strength mixes surface waters with deeper waters. These conditions create a **fall bloom** of phytoplankton, which is much less dramatic than the spring bloom (Figure 13.20a). The fall bloom is very short-lived because sunlight (not nutrient supply, as in the spring bloom) becomes the limiting factor as winter approaches to repeat the seasonal cycle.

Comparing Regional Productivity

Figure 13.21 compares the seasonal variation in phytoplankton biomass of tropical, north polar, and north middle latitude regions, where the total area under each curve represents photosynthetic productivity. The figure shows the dramatic peak in productivity in polar oceans during the summer; the steady, low rate of productivity year-round in the tropical ocean; and the seasonal pattern of productivity that occurs in middle latitude oceans. It also shows that the highest overall productivity occurs in middle latitude regions.

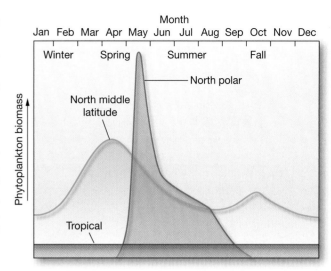

Figure 13.21 Comparison of productivity in tropical, middle latitude, and polar oceans (Northern Hemisphere). Seasonal variations in phytoplankton biomass, where the total area under each curve represents annual photosynthetic productivity.

CONCEPT CHECK 13.3

1. Describe how a biological pump works. What percentage of organic material from the euphotic zone accumulates on the sea floor?

2. Describe the yearly productivity pattern in polar oceans, including the main factor that limits productivity in polar oceans.

3. Why is the productivity in tropical oceans uniformly low year-round?

   What are three environments that are exceptions to this, and what factors contribute to their higher productivity?

4. Describe the yearly productivity pattern in middle latitude oceans. Describe the limiting factor(s) for the spring phytoplankton bloom and for the fall phytoplankton bloom.

13.4 How Are Energy and Nutrients Passed Along in Marine Ecosystems?

A **biotic community** is an assemblage of organisms that live together within some definable area. An **ecosystem** includes the biotic community plus the environment with which it exchanges energy and chemical substances. A kelp forest biotic

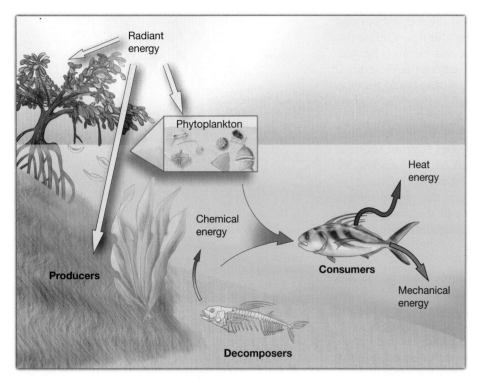

Figure 13.22 Energy flow through a photosynthetic marine ecosystem. Energy enters a marine ecosystem as radiant solar energy and is converted to chemical energy through photosynthesis by producers. Metabolism in the fish (a consumer) then releases the chemical energy for conversion to mechanical energy. Energy is also dissipated within the ecosystem as heat energy. Decomposers work to break down the remaining energy after an organism dies.

community, for instance, includes all organisms living within or near the kelp and receiving some benefit from it. A kelp forest ecosystem, on the other hand, includes all those organisms plus the surrounding seawater, the hard substrate onto which the kelp is attached, and the atmosphere where gases are exchanged.

Two of the most important commodities passed along in marine ecosystems are energy and nutrients.

Flow Of Energy in Marine Ecosystems

Energy flow in photosynthetic marine ecosystems is not a cycle but a *unidirectional flow* based on a continuous supply of solar energy. Consider an algae-supported biotic community (**Figure 13.22**), where energy enters the system and algae absorb solar radiation. Photosynthesis converts this solar energy into chemical energy, which is used for the algae's respiration. This chemical energy is also passed on to the animals that consume the algae for their growth and other life functions. The animals expend mechanical and heat energy, which are progressively less recoverable forms of energy, until this residual energy becomes dissipated within the ecosystem. In essence, the ecosystem relies on a constant input of energy in the form of sunlight.

PRODUCERS, CONSUMERS, AND DECOMPOSERS   Generally, three basic categories of organisms exist within an ecosystem: **producers**, **consumers**, and **decomposers** (Figure 13.22).

Producers can nourish themselves through either photosynthesis or chemosynthesis. Examples of producers include algae, plants, archaea, and photosynthetic bacteria, all of which are called **autotrophic** (*auto* = self, *tropho* = nourishment) organisms. Consumers and decomposers, on the other hand, are called **heterotrophic** (*hetero* = different, *tropho* = nourishment) organisms because they depend either directly or indirectly on the organic compounds produced by autotrophs for their food supply.

Consumers eat other organisms and are categorized as either **herbivores** (*herba* = grass, *vora* = eat), which feed directly on plants or algae; **carnivores** (*carni* = meat, *vora* = eat), which feed only on other animals; **omnivores** (*omni* = all, *vora* = eat), which feed on both; or **bacteriovores** (*bacterio* = bacteria, *vora* = eat), which feed only on bacteria.

Decomposers such as bacteria break down organic compounds that comprise **detritus** (*detritus* = to lessen)—dead and decaying remains and waste products of organisms—for their own energy requirements. In the decomposition process, compounds are released that are again available for use by autotrophs as nutrients.

Flow Of Nutrients in Marine Ecosystems

Unlike the noncyclic, unidirectional flow of energy through a biotic community, the flow of nutrients depends on **biogeochemical cycles**.[10] That is, matter does not dissipate (as energy does) but is *cycled* from one chemical form to another by the various members of the biotic community.

---

[10]Biogeochemical cycling is so named because it involves *bio*logical, *geo*logical (Earth processes), and *chemical* components.

BIOGEOCHEMICAL CYCLING   **Figure 13.23** shows the biogeochemical cycling of matter within the marine environment. The chemical components of organic matter enter the biological system through photosynthesis (or, less commonly, through chemosynthesis at hydrothermal vents). These chemical components are passed on to animal populations (consumers) through feeding. When organisms die, some of the material is used and reused within the euphotic zone, while some sinks as detritus. Some of this detritus feeds organisms living in deep water or on the sea floor, while some undergoes bacterial or other decomposition processes that convert organic remains into usable nutrients (nitrates and phosphates). When upwelling hoists these nutrients to the surface again, they can be used by algae and plants to begin the cycle anew.

Oceanic Feeding Relationships

As producers make food (organic matter) available to the consuming animals of the ocean, the food passes from one feeding population to the next. Only a small percentage of the energy taken in at any level is passed on to the next because energy is consumed and lost at each level. As a result, the producers' biomass in the ocean is many times greater than the mass of the top consumers, such as sharks or whales.

FEEDING STRATEGIES   For most marine animals, obtaining food occupies the majority of their time. Some animals are fast and agile and are able to obtain food through active predation. Other animals move more leisurely or not at all and instead obtain food by filtering it from seawater or locating it as deposits on the sea floor. **Figure 13.24** shows several modes of feeding along sediment-covered shores.

In **suspension feeding**, which is also called **filter feeding**, organisms use specially designed structures to filter plankton from seawater. For example, barnacles (**Figure 13.25**) attach to hard surfaces and use their legs to strain passing food particles from the water. Clams (Figure 13.24a) bury themselves in sediment and extend siphons up to the surface. They pump in overlying water and filter from it suspended plankton and other organic matter.

In **deposit feeding**, organisms feed on food items that occur as deposits. These deposits include detritus—dead and decaying organic matter and waste products—and the sediment itself, which is coated with organic matter. Some deposit feeders, such as the segmented worm *Arenicola* (Figure 13.24b), feed by ingesting sediment and extracting organic matter from it. Others, such as the amphipod *Orchestoidea* (Figure 13.24c), feed on more concentrated deposits of organic matter (detritus) on the sediment surface.

In **carnivorous feeding**, organisms directly capture and eat other animals. This predation can be either passive or active. Passive predation involves waiting for prey items and ensnaring them such as how a sea anemone feeds. Active predation involves seeking prey; examples include sharks and the sand star *Astropecten* (Figure 13.24d), which can burrow rapidly into sandy beaches and feeds voraciously on crustaceans, mollusks, worms, and other echinoderms.

TROPHIC LEVELS   Chemical energy stored in the mass of the ocean's algae (the "grass of the sea") is transferred to the animal community mostly through feeding. Most zooplankton are *herbivores*, like cows,[11] so they eat diatoms and other

---

[11]In fact, zooplankton could be considered the miniature drifting "cows of the sea."

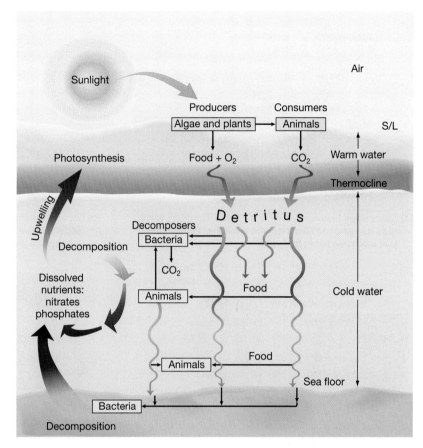

Figure 13.23 Biogeochemical cycling of matter. The chemical components of organic matter enter the biological system through photosynthesis and are passed on to consumers through feeding. Detritus sinks and feeds organisms living below the surface or undergoes decomposition, which returns nutrients to the water that can be hoisted to the surface by upwelling.

STUDENTS SOMETIMES ASK . . .

How likely am I to see a whale during a whale-watching trip?

It depends on the area and the time of the year, but generally your chances are quite low because large marine organisms comprise such a small proportion of marine life. In fact, it has been estimated that large nektonic organisms such as whales comprise only *one-tenth of 1%* of all biomass in the sea! With your knowledge of food pyramids, it should be no surprise that the majority of the ocean's biomass is comprised of phytoplankton. Perhaps commercial boat operators should conduct *plankton*-watching trips! Everyone would be guaranteed to see dozens of different species—and all that would be required would be a plankton net and a microscope.

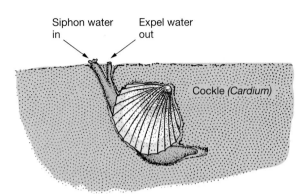

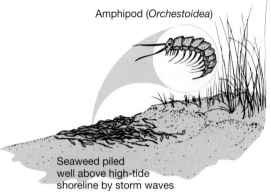

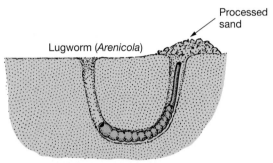

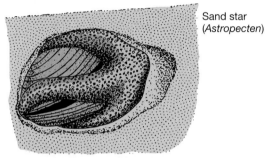

(a) Suspension feeding by the clam (cockle) *Cardium*, which uses its siphon to filter plankton and other organic matter that is suspended in the water.

(b) Deposit feeding by the amphipod *Orchestoidea*, which feeds on more concentrated deposits of organic matter (detritus) on the sediment surface.

(c) Deposit feeding by the segmented worm *Arenicola*, which feeds by ingesting sediment and extracting organic matter from it.

(d) Carnivorous feeding of a clam by the sand star *Astropecten*.

Figure 13.24 Modes of feeding along sediment-covered shores.

Figure 13.25 Barnacles filter feeding. Barnacles are shelled organisms that attach to hard surfaces and filter feed by using their feathery legs to strain microscopic food particles from water. Height of each barnacle = 2 centimeters (0.8 inch).

microscopic marine algae. Larger herbivores feed on the larger algae and marine plants that grow attached to the ocean bottom near the shore.

The herbivores are then eaten by larger animals, the *carnivores*. They, in turn, are eaten by another population of larger carnivores, and so on. Each of these feeding stages is called a **trophic** (*tropho* = nourishment) **level**.

Generally, individual members of a feeding population are larger—but not too much larger—than the organisms they eat. There are conspicuous exceptions, however, such as the blue whale. At 30 meters (100 feet) long, it is possibly the largest animal that has ever existed on Earth, yet it feeds on krill, which have a maximum length of only 6 centimeters (2.4 inches).

The transfer of energy from one population to another is a continuous *flow* of energy. Small-scale recycling and storage interrupt the flow; this in turn slows the conversion of potential (chemical) energy to kinetic energy, then to heat energy, and finally the energy is so dissipated it becomes useless.

TRANSFER EFFICIENCY The transfer of energy between trophic levels is very inefficient. The efficiencies of different algal species vary, but the average is only about 2%, which means that 2% of the light energy absorbed by algae is ultimately synthesized into food and made available to herbivores.

The **gross ecological efficiency** at any trophic level is the ratio of energy passed on to the next higher trophic level divided by the energy received from the trophic level below. The ecological efficiency of herbivorous anchovies, for example, would be the energy consumed by carnivorous tuna that feed on the anchovies divided by the energy contained in the phytoplankton that the anchovies consumed.

Figure 13.26 shows that some of the chemical energy taken in as food by herbivores is excreted as feces, and the rest is assimilated. Of the assimilated chemical energy, much is converted through respiration to kinetic energy for maintaining life, and what remains is available for growth and reproduction. Thus, only about 10% of the food mass consumed by herbivores is available to the next trophic level.

Figure 13.27 shows the passage of energy between trophic levels through an entire ecosystem, from the solar energy assimilated by phytoplankton through all trophic levels to the ultimate carnivore—humans. Because energy is lost at each trophic level, it takes thousands of smaller marine organisms to produce a *single* fish that is so easily consumed during a meal!

The efficiency of energy transfer between trophic levels depends on many variables. Young animals, for example, have a higher growth efficiency than older animals. In addition, when food is plentiful, animals expend more energy in digestion and assimilation than when food is scarce.

Most ecological efficiencies in natural ecosystems average about 10% but range between 6% and 15%. There is some evidence, however, that ecological efficiencies in populations important to present fisheries may run as high as 20%. The true value of this efficiency is of practical importance because it determines the size of the fish harvest that can safely be taken from the oceans without damaging the ecosystem.

FOOD CHAINS, FOOD WEBS, AND THE BIOMASS PYRAMID   The loss of energy between each feeding population limits the number of feeding populations in an ecosystem. If there were too many levels, there would not be enough energy to support the organisms in higher and higher trophic levels. In addition, each feeding population must necessarily have less mass than the population it eats. As a result, individual members of a feeding population are generally *larger in size* and *less numerous* than their prey.

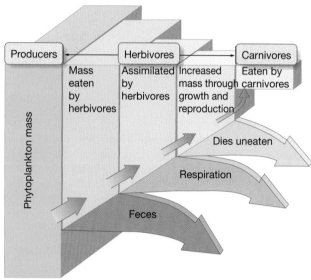

Figure 13.26 Passage of energy through trophic levels. As food mass initially produced by phytoplankton passes from herbivores to carnivores, a large percentage is excreted as feces, used during respiration, or dies uneaten. Thus, only about 10% of the food mass consumed by herbivores is available for consumption by carnivores.

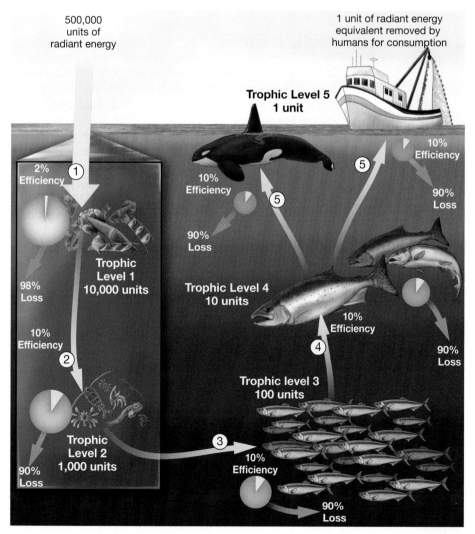

Figure 13.27 Ecosystem energy flow and efficiency.
For every 500,000 units of radiant energy input available to producers (phytoplankton) at trophic level 1, only 1 unit of equivalent mass is added to the fifth trophic level. For trophic level 1, the average transfer efficiency is 2% (98% loss); all other trophic levels average 10% efficiency (90% loss).

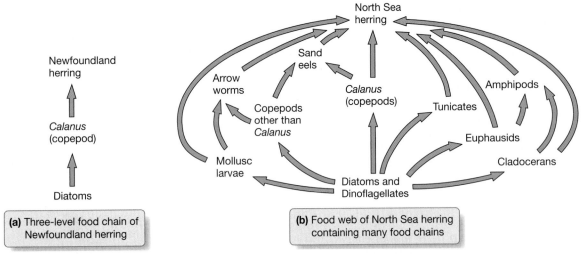

(a) Three-level food chain of Newfoundland herring

(b) Food web of North Sea herring containing many food chains

Figure 13.28 Comparison between a food chain and a food web. (a) An example of a food chain, showing the passage of energy along a single path, such as from diatoms to copepods to Newfoundland herring in three trophic levels. **(b)** An example of a food web for a similar species, showing the multiple paths for food sources of the North Sea herring, which may be at the third or fourth trophic level.

Food Chains A **food chain** is a sequence of organisms through which energy is transferred, starting with an organism that is the primary producer, then an herbivore, then one or more carnivores, and finally culminating with the "top carnivore," which is not usually preyed upon by any other organism.

Because energy transfer between trophic levels is inefficient, it is advantageous for fishers to choose a population that feeds as close to the primary producing population as possible. This increases the biomass available for food and the number of individuals available to be taken by the fishery. Newfoundland herring, for example, are an important fishery that usually represents the third trophic level in a food chain. They feed primarily on small crustaceans (copepods) that feed, in turn, upon diatoms (**Figure 13.28**).

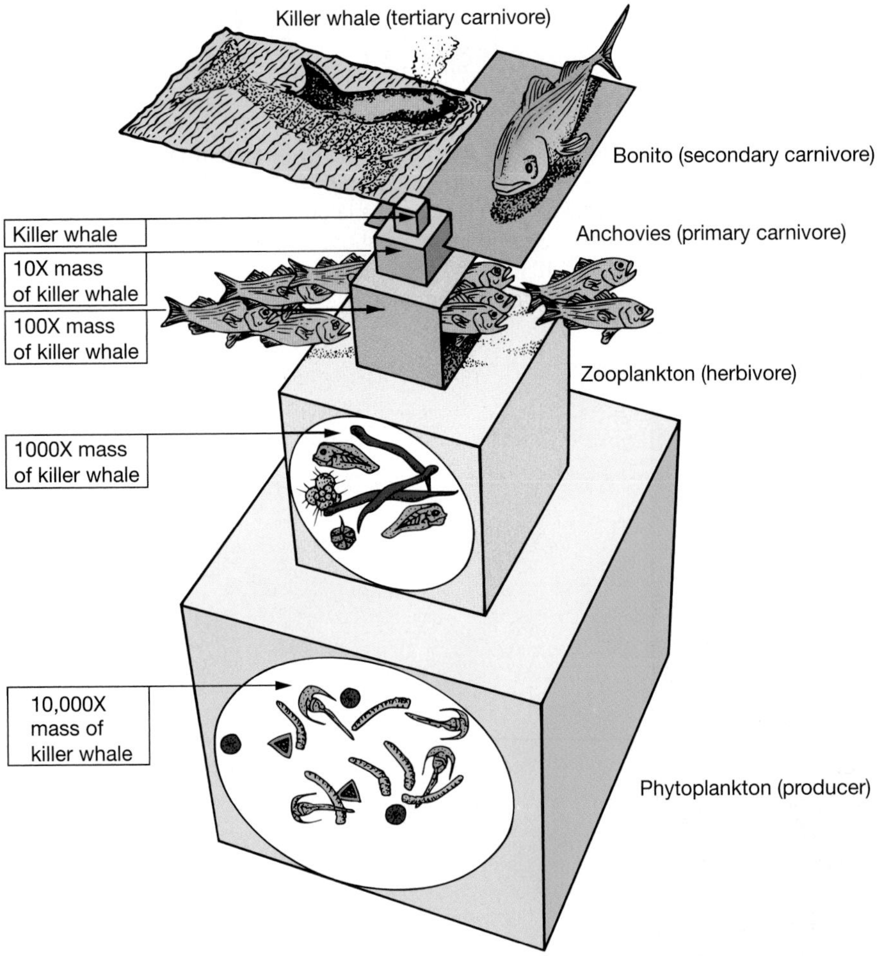

Food Webs Feeding relationships are rarely as simple as that of the Newfoundland herring. More often, top carnivores in a food chain feed on a number of different animals, each of which has its own simple or complex feeding relationships. This constitutes a **food web**, as shown in Figure 13.28b for North Sea herring.

Animals that feed through a food web rather than a food chain are more likely to survive because they have alternative foods to eat should one of their food sources diminish in quantity or even disappear. Newfoundland herring, which feed through a food chain, eat only copepods, so the disappearance of copepods would catastrophically affect their population. Conversely, Newfoundland herring are more likely to have a larger biomass to eat, because they are only two steps removed from the producers, whereas North Sea herring are three steps removed in some of the food chains within their web.

Biomass Pyramid The ultimate effect of energy transfer between trophic levels can be seen in the oceanic **biomass pyramid** shown in **Figure 13.29**. It shows that for the survival of each large marine organism, many levels of progressively larger populations of smaller-sized organisms must exist to support those higher on the pyramid. In essence, the *number of individuals* and *total biomass* decrease at successive trophic levels because the amount of available energy decreases. The figure also shows that organisms *increase in size* at successive trophic levels up the pyramid.

In some marine environments, "inverted" biomass pyramids can occur. An "inverted" biomass pyramid can exist, for example, where a smaller population of phytoplankton with a very high turnover rate supports a larger population of

Killer whale (tertiary carnivore)

Bonito (secondary carnivore)

Anchovies (primary carnivore)

Killer whale

10X mass of killer whale

100X mass of killer whale

Zooplankton (herbivore)

1000X mass of killer whale

10,000X mass of killer whale

Phytoplankton (producer)

Figure 13.29 Oceanic biomass pyramid. A huge mass of phytoplankton constitutes the base of the biomass pyramid. At each step up the pyramid, there are larger organisms but fewer individuals and a smaller total biomass because transfer efficiency between steps averages only 10%.

zooplankton. As a result, oceanic biomass pyramids can have a variety of shapes, depending on the turnover rates of different trophic levels.

13.5 What Issues Affect Marine Fisheries?

Since well before the beginning of recorded history, humans have used the sea as a source of food. Over the past several decades, **fisheries** (fish caught from the ocean by commercial fishers) have provided billions of people around the world nearly 20% of their protein intake, with less-developed nations relying on fish for 27% of their dietary protein.

Marine Ecosystems and Fisheries

Figure 13.30 shows that the world marine fishery is drawn from five ecosystems. In descending order, they are (1) nontropical shelves, (2) tropical shelves, (3) upwellings, (4) coastal and coral systems, and (5) open ocean. The largest proportion of the marine fishery is found in highly productive shallow shelf and coastal waters, whereas low-productivity open ocean areas comprise only 3.8% of the total. Nearly 21% of the world's total catch is from very highly productive upwelling areas, which represent only about 0.1% of the ocean surface area.

Overfishing

Fisheries harvest from the **standing stock** of a population, which is the mass present in an ecosystem at a given time. Successful fisheries leave enough individuals from the standing stock to repopulate the ecosystem after fisheries have made their harvest.

Overfishing occurs when harvesting of fish stocks takes place so rapidly that the majority of the population is sexually immature and is therefore unable to reproduce. Predictably, overfishing results in the decline of marine fish populations, the overall size of fish in a population, and the reduction of a fishery's **maximum sustainable yield (MSY)** (the maximum fishery biomass that can be removed yearly and still be sustained by the fishery ecosystem).

According to a report by the Food and Agriculture Organization (FAO) of the United Nations, 80% of the 523 world marine fish stocks for which assessment information is available are classified as fully exploited, overexploited, or depleted/recovering from depletion (**Figure 13.31**). In U.S. waters, the story is much the same: 80% of the 191 commercial fish stocks are fully exploited or overfished. In U.K. waters, stocks of palatable fish—such as cod—have been reduced by overfishing to less than 10% of what they were 100 years ago.

Web Video
The Secret Life of Plankton

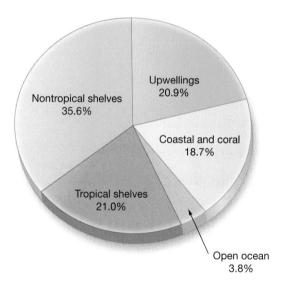

Figure 13.30 Marine fishery ecosystems. Pie diagram showing the relative contribution of various ecosystems to the total world marine fishery.

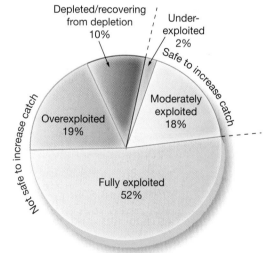

Figure 13.31 Exploitation status of world marine fisheries. Pie diagram showing the current status of world fishery stocks. The only fisheries safe to increase are those that are moderately exploited and underexploited.

ECOSYSTEM EFFECTS OF OVERFISHING LARGE PREDATORS Recent studies show that large predatory fish are a vital part of healthy marine ecosystems. For instance, they prevent smaller fish from overrunning and potentially destroying ocean ecosystems. They also increase ecosystem health by culling sick, lame, and old populations of marine herbivores. Yet fishing practices have removed 90% of large predatory fish species from the world ocean (**Diving Deeper 13.1**).

Removing large predators often has unintended consequences. One example is the recent abundance of the American (Maine) lobster (*Homarus americanus*) in the North Atlantic Ocean. Since the population of its main predator—cod—was overfished and collapsed, the American lobster population has significantly rebounded, which seems like a good thing but has unknown ecological effects.

With the reduction of large fish, the fishing industry has concentrated on smaller fish that occupy successively lower trophic levels. In fact, low-trophic level species now account for over 30% of global fisheries production. Removing so many individuals from lower trophic levels can cause an unwanted cascade of effects on other parts of marine ecosystems, particularly for other organisms that rely on these small prey as a food source.

Not only are big fish disappearing at an alarming rate, but the ocean's top predators are now only about one-half to one-fifth the size they once were, which also affects entire marine ecosystems. As a case in point, the removal of large predatory sharks along the East Coast by fishing practices during the past 35 years has allowed one of their prey items, the cownose ray (*Rhinoptera bonasus*), to increase 20-fold. As a result, shellfish populations such as the bay scallop—which are a major prey item of the rays—have been decimated. In fact, this overabundance of rays caused the closure of North Carolina's century-old bay scallop fishery in 2004.

Coral reef ecosystems are being affected by the removal of top predators, too. For example, the removal of large reef residents such as sharks, parrot fish, and surgeonfish can clear the way for smaller fish to proliferate because of decreased predation and competition. These smaller fish tend to increase the amount of algae on reefs, which competes with corals and suppresses their growth.

THE END OF FISH? Recent research shows that fishing can induce inherited changes that cut the average growth rate, maturing time, and size in fish from generation to generation, which has some severe implications for the health and longevity of fisheries. In fact, a recent study of the effect of pollution and fisheries on marine ecosystem biodiversity projects that fish stocks will be depleted by 2048. Not only will the loss of seafood threaten humans' food supplies, but the loss of fish will also seriously damage marine ecosystems.

RECREATIONAL FISHING Although it is widely assumed that recreational fishing has a minimal influence on most species of fish, recreational fishing has been shown to have an impact on fish populations, too. A recent rigorous analysis of multiple U.S. fisheries records shows that for some threatened sportfishing species, such as red drum, bocaccio rockfish, and red snapper, recreational fishing actually poses more of a threat than commercial fishing because of the size and number of fish taken by recreational fishers. Thus, new fishing regulations that target recreational fishers are needed to complement existing catch restrictions on the commercial sector.

WORLD FISH PRODUCTION Figure 13.32 shows the world total fish production in marine waters since 1950. After increasing steadily for 35 years, the world catch of ocean fish finally leveled off in the 1990s at about 85 million metric tons (93.5 million short tons). In the 2000s, the catch stagnated and has yet to improve. In fact, data from 2008 indicate that the world marine fish catch had shrunk to 79.5 million metric tons (87.5 million short tons).

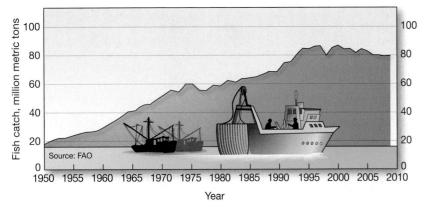

Figure 13.32 World total marine fish production since 1950. Graph showing the increasing amount of reported world total marine fish production, which begins to level off during the 1990s. The graph shows a quadrupling of fish catch since 1950.

Diving Deeper 13.1 Focus on the Environment

FISHING DOWN THE FOOD WEB: SEEING IS BELIEVING

Marine scientists can measure the health of an ecosystem by determining its original fish abundance, diversity, and size. But how can these variables be assessed effectively for periods before extensive scientific field surveys were conducted? Recently, scientists have been using novel ways to estimate these parameters. A new field of research called *historical marine ecology* uses items such as old photographs, newspaper accounts, ships' logs, cannery records, and even old restaurant menus to estimate the type and quantity of fish that used to live in the sea.

Striking evidence of past fish abundance and size comes from old photographs. Certainly, people have always enjoyed having their picture taken with the fish they catch. Loren McClenachan, a graduate student at Scripps Institute of Oceanography at the University of California, San Diego, found a historical archive of old photographs in Key West, Florida, in the Monroe County Public Library. The archive allowed her to examine fish caught by day-trippers at coral reef sites aboard a series of vessels called *Gulf Stream* and *Greyhound*, which have operated out of Key West for the past 50 years.

By comparing old photographs with newer ones (**Figure 13A**), it is apparent that there has been a pronounced decrease in both fish abundance and fish size through the years. Research suggests that chronic overfishing by both commercial and recreational fishers is the likely source of this decline. In fact, some species completely disappear over time, reflecting a loss of the largest fish from the coral reef environment. For example, in the 1950s, fishers caught huge grouper and sharks. In the 1970s, they landed a few grouper but more jack. Today's main catch lacks grouper and instead consists of small snapper, which once weren't even deemed worthy of a photograph; instead, they were just piled up below the hanging racks. Historical records from other regions throughout the world also reveal astonishing declines in most fish stocks. Scientists use the phrase "fishing down the food web" to describe how key large species in healthy functioning ecosystems have been replaced by progressively smaller, less valuable species that occupy lower levels of the food web.

Worldwide, fishers typically catch the biggest animals first, whether turtles, whales, cod, or grouper. Then they catch whatever is left—including animals so young that they haven't yet reproduced—until, in some cases, the animals are gone. To change human habits, it is important to gain a clearer picture of what has been lost. In the Florida Keys, historical photographs provide such a window into a more pristine coral reef ecosystem that existed a half-century ago.

Figure 13A Decreasing size of fish catch. Historical photos from chartered fishing vessels that concentrated on the fish of the coral reefs surrounding Key West, Florida, show decreasing catch size and abundance from 1958 (*above*), 1980s (*middle*), and 2007 (*below*).

How much fish can the oceans produce? The National Research Council estimated in 1999 that the maximum catch that might be expected from the oceans is about 100 million metric tons (110 million short tons). This amount might be exceeded, however, if new species—such as Antarctic krill, squid, and pelagic crabs—become significant new fisheries in the future. Based on current yields of traditional fisheries, the United Nations FAO estimates that the maximum catch that might be expected from the oceans is about 120 million metric tons (132 million short tons).

The reported worldwide catch, however, does not include additional organisms caught but deemed undesirable and discarded, so the actual catch is likely much higher than the reported value. Unfortunately, this suggests that the amount of biomass removed by fishing exceeds that produced in the ocean, which means that the oceans are being overfished.

Figure 13.33 Fishing bycatch. In commercial fishing operations, about one-fourth of the catch is discarded as unwanted bycatch, which includes birds, turtles, sharks, dolphins, and many species of noncommercial fish.

Incidental Catch

Incidental catch, or **bycatch**, includes any marine organisms that are caught incidentally by fishers seeking commercial species. On average, close to one-fourth of the catch is discarded, although for some fisheries, such as shrimp, the incidental catch may be up to eight times larger than the catch of the target species. Incidental catch includes birds, turtles, sharks, and dolphins, as well as many species of noncommercial fish (**Figure 13.33**). In most cases, these animals die before they are thrown back overboard, even though some of them are protected by U.S. and international law. Globally, an estimated 20 million metric tons (32 million short tons) of bycatch is produced each year by the fishing industry, accounting for nearly one-fourth of the world's total marine fish catch.

TUNA AND DOLPHINS Schools of yellowfin tuna are commonly found swimming beneath spotted and spinner dolphins in the eastern Pacific Ocean (**Figure 13.34**). Fishers commonly used these dolphins to locate tuna and set a *purse seine net* around the entire school (**Figure 13.35**). When an underwater line is drawn tight, the net traps the tuna underwater as well as the dolphins at the surface.

The problem of dolphin deaths caused by tuna fishing was graphically presented to the world in 1988 through video footage taken by biologist Samuel F. La Budde of dolphins struggling in tuna fishing nets. In 1990, under intense public outcry and a boycott on tuna, the U.S. tuna canning industry declared that it would not buy or sell tuna caught using methods that kill or injure dolphins. In 1992, a special addendum was added to the **Marine Mammals Protection Act**, further protecting dolphins. As a result of these measures, purse seine nets were modified so that dolphins could be released alive. In spite of reducing dolphin mortality as bycatch, dolphin populations have not rebounded accordingly. Research suggests that tuna fishing operations are still having a negative effect on dolphin populations by reducing dolphin survival and birth rates.

Figure 13.34 Spotted dolphin (*Stenella attnuata*), which are commonly associated with yellowfin tuna. Eastern Pacific spotted dolphins are often found swimming above yellowfin tuna. This relationship is thought to exist because the dolphins use tuna to help find schools of prey items (squid and small fish), which are eaten by both animals.

DRIFTNETS Another means of netting tuna and other species is by use of **driftnets**, or **gill nets**, which are made of monofilament fishing line that is virtually invisible and cannot be detected by most marine animals. Depending on the size of the holes in the net, it is highly effective at catching anything large enough to become entangled in it. As a result, driftnets often have high amounts of bycatch.

Up until 1993, Japan, Korea, and Taiwan had the largest driftnet fleets, deploying as many as 1500 fishing vessels into the North Pacific and setting over

48,000 kilometers (30,000 miles) of driftnets in one day. Although driftnetting was supposed to be restricted to specific fisheries, some fishers who claimed to be fishing for squid were involved in illegally taking large quantities of salmon and steelhead trout. Driftnetters were also targeting immature tuna in the South Pacific, which could result in the reduced abundance of South Pacific tuna. In addition, tens of thousands of birds, turtles, dolphins, and other species were killed annually in these nets.

A prohibition against the importation of fish caught in driftnets was included in the International Convention of Pacific Long Driftnet Fishing, signed in 1989. Although driftnets are now banned by international law, some fishers continue to use them illegally.

Fisheries Management

Fisheries management is the organized effort directed at regulating fishing activity with the goal of maintaining a long-term fishery. Fisheries management practices include assessing ecosystem health, determining fish stocks, analyzing fishing practices (including recommending gear modification), establishing closed areas, and setting and enforcing catch limits. Unfortunately, however, fisheries management has historically been more concerned with maintaining human employment than preserving a self-sustaining marine ecosystem. For example, fisheries such as anchovy, cod, flounder, haddock, herring, and sardine are suffering from overfishing in spite of being managed (see Web Diving Deeper 13.2).

One of the most pressing problems facing fisheries management results from the fact that some fisheries encompass the waters of many different countries, including a variety of ecosystems. Some species of commercial fish, for instance, are raised in coastal estuaries and migrate long distances across international waters to their preferred environment. Fishing limits are difficult to enforce internationally, and if human interference occurs at any location where the fish exist, these species can be severely reduced in number. Other problems include the degradation of many ecosystems that sustain fisheries, limited scientific analysis, a lack of regulations enforcement (particularly on illegal/private fishers), poaching, misreporting of catch and bycatch, political barriers, and inadequate guidance for minimally regulated new fisheries.

REGULATION OF FISHING VESSELS A major regulatory failure has been the absence of restrictions on the number of fishing vessels. In 2004, according to the FAO, the world's fishing fleet consisted of about 4 million vessels. Of these, about 1.3 million were engine-powered, mechanized-decked vessels engaged in commercial fishing, including 40,000 vessels over 100 gross tons in size (generally more than 24 meters [80 feet] in length). Figure 13.36 shows that the number of decked fishing vessels in the world more than doubled between 1970 and 1995, and it continues to increase. Many of these larger vessels use nets that can hold up to 27,000 kilograms (60,000 pounds) of fish in one haul. In addition, there are more than 2.1 million smaller nondecked fishing vessels that are owned and operated by artisanal fishers, mostly in Asia, Africa, and the Middle East.

The increase in fishing vessels has resulted in an increased fishing effort, which often leads to overfishing. In some locations, fish are becoming so scarce that the fishing effort costs more than what the catch is worth! In 1995, for instance, the world fishing fleet spent $124 billion to catch $70 billion worth of fish. To make up for the shortfall, many governments have given fishers subsidies in the form of cash payments totaling an estimated $30–34 billion yearly, which compounds the problem by maintaining (or even *increasing*) the number of fishing vessels.

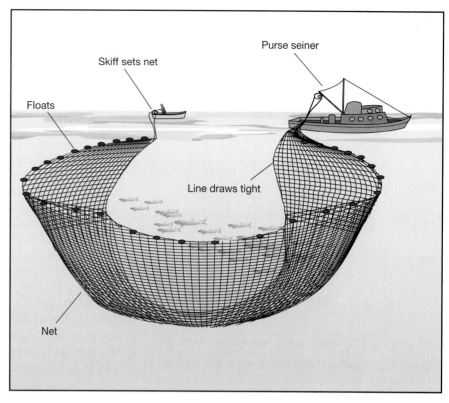

Figure 13.35 Purse seiner. Purse seine nets are set around schools of yellowfin tuna, often trapping dolphins at the surface as bycatch.

STUDENTS SOMETIMES ASK . . .

Is tuna labeled as "dolphin-safe" really safe to dolphins?

It depends on where the tuna is from, but the answer is probably yes. Tuna and tuna products harvested in the eastern tropical Pacific Ocean can only be labeled as "dolphin-safe" in the United States if no nets were intentionally set on dolphins during fishing and no dolphins were killed or seriously injured during the set in which the tuna were caught. In the United States, the National Marine Fisheries Service (NMFS) has developed an extensive monitoring, tracking, and verification program to ensure that only tuna and tuna products that meet the definition of dolphin-safe are indeed labeled as "dolphin-safe." This program, along with consumer awareness, has substantially reduced dolphin mortality, which numbered in the hundreds of thousands per year in the 1980s. Recently, however, changes to the definition of "dolphin-safe" have been accepted by the NMFS that address international trade concerns and allow more tuna to be imported from international sources, particularly Mexico. Tuna from international sources caught along with dolphins may be labeled "dolphin-safe" in the United States as long as observers aboard fishing vessels certify that no dolphins were killed or seriously injured during the catch—but this relies on the honesty of the observers.

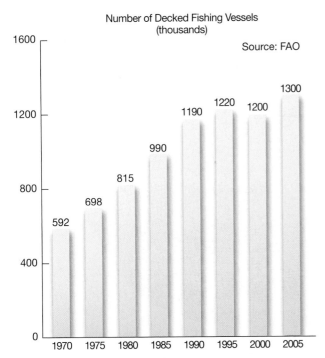

Number of Decked Fishing Vessels
(thousands)

Source: FAO

Figure 13.36 Number of decked fishing vessels in the world (in thousands). Bar graph showing that the worldwide number of engine-powered, mechanized-decked vessels engaged in commercial fishing has more than doubled since 1970. Even though the number of vessels has leveled off recently, the existence of so many well-equipped fishing vessels results in an increased fishing effort, which often leads to overfishing.

STUDENTS SOMETIMES ASK . . .

As the world's oceans are over-fished and degraded, are they being taken over by jellies?

There's an increasing concern that "jellies gone wild" is in fact happening now. As predators, jellies appear to be slow and passive. Unable to see or chase their prey, most drift along, creating tiny eddies that guide food particles toward their tendrils. Yet in many parts of the ocean, jellies are thriving as many of their competitors are eliminated by overfishing and other human impacts, creating environmental conditions that favor jellies. All around the world, jellies are creating havoc by reproducing in astonishing numbers and congregating where they've rarely been seen before. For example, jellies have shut down nuclear power plants that use seawater as a coolant, disabled a military aircraft carrier by blocking its seawater intake system, halted sea floor mining operations by clogging dredges, capsized fishing vessels by their sheer numbers caught in nets, endangered bathers, and even stung and killed hoards of farmed fish. Although some studies show that the apparent increase in the size and frequency of jelly blooms in coastal waters and estuaries around the world during the past few decades is part of a natural cycle, jellies' impact on marine food webs is likely to increase in the future, leading to speculation that "the rise of slime" may lead to a "gelatinous ocean."

FISHERIES IN THE NORTHWEST ATLANTIC The effects of inadequate fisheries management are illustrated by the history of fisheries in the northwest Atlantic. Under management by the International Commission for the Northwest Atlantic Fisheries, the fishing capacity of the international fleet increased 500% from 1966 to 1976; the total catch, however, rose by only 15%. This was a significant decrease in the catch per unit of effort, which is a good indication that the fishing stocks of the Newfoundland–Grand Banks area were being overexploited. The biologists who recommended the total allowable fishing quotas for the major species within this region complained that enforcement was lacking and that countries' quotas were being bartered in a game of international politics. As a result, fishers exceeded the total allowable catch set by the commission.

The difficulty of enforcing regulations by the international commission was largely responsible for Canada's unilateral decision to extend its right to control fish stocks for a distance of 200 nautical miles (370 kilometers) from its shores beginning January 1, 1977. The United States followed with a similar action only two months later.

Claiming coastal waters, however, proved to have limited effectiveness. After essentially all coastal nations assumed regulatory control over their coastal waters, the situation continued to deteriorate, and overfishing became even more of a problem. The Canadian government, in fact, had to shut down the Grand Banks fishery off Newfoundland in 1992, which resulted in the loss of about 40,000 jobs and subsequent government outlays of more than $3 billion in welfare. This cost far exceeds the value of the fishery, which generated no more than $125 million during one of its best years.

A similar situation has emerged on Georges Bank, which lies in Canadian and U.S. waters. On Georges Bank, however, tough international restrictions and proper enforcement are beginning to bring back haddock and yellowtail flounder fisheries.

In spite of protection, some fish stocks in the North Atlantic have not rebounded as anticipated. Atlantic cod stocks, for example, continue to decrease despite sharply curtailed annual catches. To help halt the decline of cod—one of the North Atlantic's most impacted fish species—the Canadian government in 2003 declared cod completely off-limits to fishing. A similar decline in cod stocks is also occurring in European waters, but so far officials have rejected a proposed cod-fishing ban.

DEEP-WATER FISHERIES One of the results of depleting and/or banning certain fish stocks is that the fishing industry then expands its efforts into the deep sea, where there are few regulations. As compared to surface species, most deep-water species have low metabolic and reproduction rates, which cause them to be severely impacted by fishing. For example, the collapse of Atlantic cod stocks resulted in fishers seeking the deep-water Greenland halibut as a substitute. Predictably, that species is now in danger of becoming overfished throughout the Atlantic.

The deep-water orange roughy fishery near Australia is another prime example. It was developed to satisfy the middle-American market for a bland white fish. Even the name was picked through careful supermarket research (its original moniker—the slimehead—sounded far less appealing). And when that fishery began its inevitable decline, the industry moved on to another deep-water species, the Patagonian toothfish. It was renamed the Chilean sea bass, even though it is neither a bass nor exclusively Chilean. The precipitous decline of these deep-water fish has qualified them for endangered status.

Deep-sea ecosystems are beginning to feel the effect of this increased fishing effort, too. For example, tall seamounts rise precipitously from the deep sea floor and provide a unique environment that harbors many species of fish and deep-water coral. To effectively catch fish here, fishers use large bottom-dragging trawl nets. These nets, however, do long-lasting damage to the sea floor and have proven to negatively affect these slow-growing yet important ecosystems.

ECOSYSTEM-BASED FISHERY MANAGEMENT The historic record of fishery management shows many examples of attempts to regulate a fish stocks by looking only at an individual species. If fisheries are to remain viable in the future, marine ecosystems must be studied to document the natural relationships among organisms in marine food webs, the critical environmental factors for the health of the fishery, and the effects of removing so many organisms as fisheries catch. In addition, critical fish habitats must be protected and fishing limits must be upheld despite political factors. Only after such obstacles are overcome can a fishery be successfully managed.

To ensure sustainable fisheries and a healthy marine environment in U.S. waters, there has been a recent push to institute *ecosystem-based fishery management*, which employs a more comprehensive approach to understanding fish stocks including analysis of such variables as fish habitat, migration routes, and predator–prey interactions. Ecosystem-based fishery management essentially reverses the order of management priorities, starting with an ecosystem rather than a target species. It also focuses on rebuilding fisheries on a global basis.

Research has shown that sustainable fisheries can be achieved if management strategies are adopted that remove subsidies and give individual fishers a right to a proportion of the total catch. By removing competition, fishers of all types would be encouraged to act to sustain the entire fishery because, as the common fish stock improves, each individual's quota will increase. In this way, fishers will have the incentive to manage the stock sustainably, thus avoiding overfishing and the eventual collapse of the fishery.

Recently, a study was conducted on the effectiveness of fisheries management practices. Scientists analyzed six key parameters, including the scientific quality of management recommendations, the transparency of converting recommendations into policy, the enforcement of policies, the influence of subsidies, fishing effort, and the extent of fishing by foreign entities. Results of the study (**Figure 13.37**) show that fisheries management, despite broad acceptance and commitments by governments to initiatives for improvements, remains largely ineffective.

STUDENTS SOMETIMES ASK . . .

Does fish farming relieve some of the demand for wild fish?

Even though global production of the nearly 250 species of farmed fish and shellfish has quadrupled in the past 20 years and now provides nearly half of all seafood directly consumed by humans, some types of aquaculture have actually *increased* the demand for wild fish. This is because the farming of carnivorous species such as salmon, tuna, cod, and seabass requires large inputs of wild fish for feed. For example, it takes 3 kilograms (6.6 pounds) of wild mackerel or anchovies to produce 1 kilogram (2.2 pounds) of farmed salmon or shrimp.

Some aquaculture systems further reduce wild fish supplies through habitat modification, the collection of wild fish for initially stocking aquaculture operations, and other negative ecological impacts, including waste disposal, exotic species introduction, and pathogen invasions, all of which have probably contributed to the collapse of fishery stocks worldwide. If the aquaculture industry is to sustain its contribution to world fish supplies, it must reduce wild fish inputs as food and adopt more ecologically sound management practices.

Figure 13.37 Effectiveness of world fisheries management.
Map showing an assessment of the effectiveness of the world's fisheries management programs. Due to a variety of problems, fisheries management remains largely ineffective.

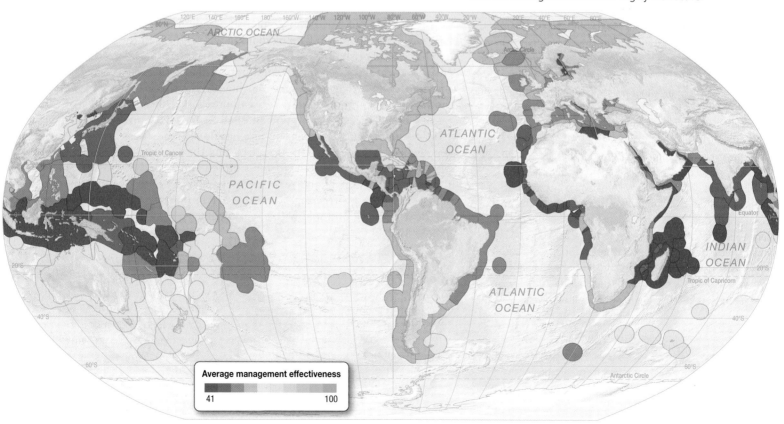

Average management effectiveness

41   100

Best Choices	Good Alternatives	Avoid	Support Ocean-friendly Seafood
Arctic Char (farmed) Barramundi (US farmed) Catfish (US farmed) Clams (farmed) Cod: Pacific (US non-trawled) Crab: Dungeness, Stone Halibut: Pacific (US) Lobster: California Spiny (US) Mussels (farmed) Oysters (farmed) Saberfish/Black Cod (Alaska & Canada) Salmon (Alaska wild) Sardines: Pacific (US) Scallops (farmed) Shrimp: Pink (OR) Striped Bass (farmed & wild*) Tilapia (US farmed) Trout: Rainbow (US farmed) Tuna: Albacore (Canada & US Pacific, troll/pole) Tuna: Skipjack, Yellowfin (US troll/pole)	Basa/Pangasius/Swai (farmed) Caviar, Sturgeon (US farmed) Clams (wild) Cod: Atlantic (imported) Cod: Pacific (US trawled) Crab: Blue*, King (US), Snow Flounders, Soles (Pacific) Flounder: Summer (US Atlantic) Grouper: Black, Red (US Gulf of Mexico) Herring: Atlantic Lobster: American/Maine Mahi mahi (US) Oysters (wild) Pollock: Alaska (US) Saberfish/Black Cod (CA, OR, WA) Salmon (CA, OR, WA*, wild) Scallops (wild) Shrimp (US, Canada) Squid Swai, Basa (farmer) Swordfish (US)* Tilapia (Central & South America farmed) Tuna: Bigeye, Tongoi, Yellowtail (US farmed)	Caviar, Sturgeon* (imported wild) Chilean Seabass/Toothfish* Cobia (imported farmed) Cod: Atlantic (Canada & US) Crab: King (imported) Flounders, Halibut, Soles (US Atlanitc except summer flounder) Groupers (US Atlantic)* Lobster: Spiny (Brazil) Mahi mahi (imported longline) Marlin: Blue, Striped (Pacific)* Monkfish Orange Roughy* Salmon (farmed, including Atlantic)* Sharks* & Skates Shrimp (imported) Snapper: Red (US Gulf of Mexico) Swordfish (imported)* Tilapia (Asia farmed) Tuna: Albacore*, Bigeye*, Skipjack, Tongol, Yellowfin* (except troll/pole) Tuna: Bluefin* Tuna: Canned (except troll/pole)	**Best Choices** are abundant, well-managed and caught or farmed in environmental friendly ways. **Good Alternatives** are an option, but there are concerns with how they're caught or farmed–or with the health of their habitat due to other human impacts. **Avoid** for now as these items are caught or farmed in ways that harm other marine life or the environment. **Key** CA = California OR = Oregeon WA = Washington *Limit consumption due to concerns about mercury or other contaminants. Visit www.edf.org/seafoodhealth Contaminant information provided by: ENVIRONMENTAL DEFENSE FUND Seafood may appear in more than one column

Figure 13.38 Recommended seafood choices. The Monterey Bay Aquarium Seafood Watch shows recommended seafood choices (*green*), seafood with some concerns (*yellow*), and seafood to avoid (*red*). Printable regional pocket guides and mobile apps are available at **www.seafoodwatch.org**.

STUDENTS SOMETIMES ASK . . .

I've heard of a movement to ban shark fin soup. What is shark fin soup?

Shark fin soup is a Chinese delicacy dating back to the Sung Dynasty (around 960 A.D.). The soup is traditionally made with chicken, ham broth, vegetables, and strands of cartilage from shark fin (and likely some MSG). Interestingly, none of the soup's flavor actually comes from shark fin, so including shark fin is a symbolic gesture. Like many other dishes in traditional cultures, shark fin soup is served as a symbol of class and wealth. The dish has become an ingrained tradition of status, "face," and respect; as such, it has come to be expected at big celebrations, especially weddings. Unfortunately, obtaining shark fins is a wasteful practice. Each year, up to 73 million sharks are killed primarily for their fins, threatening one-third of open-ocean sharks with extinction. In response to this environmental issue, grassroots organizations spearheaded by marine biologists have sprung up to protect sharks and get wedding couples to stop serving shark fin soup. Two such examples are Shark Truth (**www.sharktruth.com**) and Happy Hearts Love Sharks (**www.happyheartslovesharks.org**), which sponsors a fin-free wedding contest each year.

ESSENTIAL CONCEPT 13.5

The marine fishing industry suffers from overfishing, wasteful practices that produce a large amount of unwanted bycatch, and a lack of adequate fisheries management.

Seafood Choices

Consumer demand has driven some fish populations to the brink of disappearing. However, consumers can help by making wise choices in the fish they consume by purchasing only fish that are from healthy, thriving fisheries. Certainly, some types of seafood carry less environmental impact than others because of differences in abundance, how they're caught, and how well fishing is managed. **Figure 13.38** shows some recommendations for seafood (both fish and shellfish) in three categories: best choices (*green*), good alternatives (*yellow*), and seafood to avoid (*red*).

Consumers can also help by purchasing seafood that carries a "sustainable" label, which is issued by two organizations that certify fisheries as scientifically sustainable: the Marine Stewardship Council and Friend of the Sea. Such schemes aim to help consumers and retailers to support fisheries that are sustainable and not exploited by overfishing. Recent research, however, has indicated that about one-quarter of seafood sold with a "sustainable" label is not meeting the criteria of sustainability.

CONCEPT CHECK 13.5

1 Define *overfishing*. When a species is overfished, what changes are there in the standing stock, the size of individuals in the remaining fish population, and the maximum sustainable yield?

2 Describe three unintended consequences that have occurred in marine ecosystems when top predators are removed.

3 Of the following types of seafood on a menu, which is the best choice for supporting sustainable fisheries: (a) farmed salmon, (b) monkfish, (c) orange roughy, (d) Atlantic cod, or (e) Pacific halibut?

Essential Concepts Review

13.1 What is primary productivity?

▶ *Microscopic planktonic bacteria and algae that photosynthesize represent the largest biomass in the ocean.* They are the ocean's *primary producers*—the foundation of the ocean's food web. Organic biomass is also produced near deep-sea hydrothermal springs through *chemosynthesis*, in which bacteria-like organisms trap chemical energy by the oxidation of hydrogen sulfide.

▶ *The availability of nutrients and the amount of solar radiation limit the photosynthetic productivity* in the oceans. *Nutrients*—such as *nitrogen, phosphorus, iron,* and *silica*—are most abundant in coastal areas, due to runoff and upwelling. The depth at which net photosynthesis is zero is the *compensation depth for photosynthesis.* Generally, algae cannot live below this depth, which may be less than 20 meters (65 feet) in turbid coastal waters or as much as 100 meters (330 feet) in the open ocean.

▶ *Marine life is most abundant along continental margins,* where nutrients and sunlight are optimal. It decreases with distance from the continents and with increased depth. In addition, *cool water typically supports more abundant life than warm water* because cool water can dissolve more of the gases necessary for life (oxygen and carbon dioxide). *Areas of upwelling bring cold, nutrient-rich water to the surface and have some of the highest productivities.*

▶ *Ocean water selectively absorbs the colors of the visible spectrum. Red and yellow light are absorbed at relatively shallow depths,* whereas *blue and green light are the last to be removed.* Ocean water of *low biological productivity* scatters the short wavelengths of visible light, producing a *blue color.* Turbidity and photosynthetic algae in *more productive ocean water* scatter more green light wavelengths, which produces a *green color.*

Study Resources
Online Study Guide Quizzes, Web Animation

Critical Thinking Question

Describe the color difference between coastal waters and less-productive open-ocean water and explain why each has the color that it does.

Chlorophyl a Concentration (mg/m³)

.01   .02 .03 .05   .1   .2 .3   .5   1   2   3   5   10 15 20 30   50

(c) SeaWiFS image of chlorophyll concentration along the southwest coast of Africa (February 21, 2000). High chlorophyll concentrations indicate high phytoplankton biomass, which is caused by coastal upwelling. Concentration is reported in milligrams of chlorophyll per cubic meter (mg/m³).

13.2 What kinds of photosynthetic marine organisms exist?

▶ *There are many different types of photosynthetic marine organisms.* The seed-bearing Anthophyta are represented by a few genera of *nearshore plants* such as eelgrass (*Zostera*), surf grass (*Phyllospadix*), marsh grass (*Spartina*), and mangrove trees (*Rhizophora* and *Avicennia*). *Macroscopic algae* include *brown algae* (Phaetophyta), *green algae* (Chlorophyta), and *red algae* (Rhodophyta). *Microscopic algae* include *diatoms* and *coccolithophores* (Chrysophyta), and *dinoflagellates* (Pyrrophyta).

▶ *Dinoflagellates sometimes exist in such great abundance that they color surface waters red, producing a red tide,* which is more accurately called a *harmful algal bloom (HAB).* Dinoflagellates also *produce powerful biotoxins* that can lead to poisonings. *Ocean eutrophication is the artificial enrichment of waters* by a previously scarce nutrient—usually provided by river runoff—that can *trigger an overabundance of algae* and lead to the *creation of an oxygen-poor dead zone.*

Study Resources
Online Study Guide Quizzes, Web Diving Deeper 13.1, Web Video

Critical Thinking Question

Explain similarities and differences among the following: red tides, harmful algal blooms, ocean eutrophication, and dead zones.

Essential Concepts Review (cont.)

13.3 How does regional primary productivity vary?

▶ *A thermocline is generally absent in high-latitude (polar) areas, so upwelling can readily occur.* The *availability of solar radiation limits productivity* more than the availability of nutrients. In *low-latitude (tropical) regions*, however, a *strong thermocline may exist year-round*, so the *lack of nutrients limits productivity*, except in areas of upwelling or near coral reefs. In *middle latitude (temperate) oceans, productivity peaks in the spring and fall and is limited by lack of solar radiation in the winter and lack of nutrients in the summer.*

Study Resources
Online Study Guide Quizzes, Web Animation

Critical Thinking Question

Compare the biological productivity of polar, middle latitude, and tropical regions of the oceans. Consider seasonal changes, the development of a thermocline, the availability of nutrients, and solar radiation.

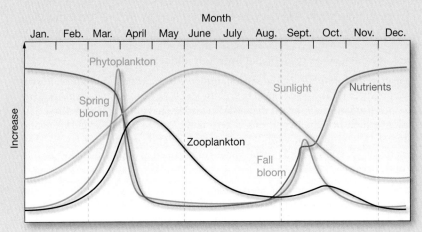

(a) Relationship among phytoplankton, zooplankton, amount of sunlight, and nutrient levels for surface waters in northern temperate latitudes.

13.4 How are energy and nutrients passed along in marine ecosystems?

▶ *Radiant energy captured by algae is converted to chemical energy* and passed through the different *trophic levels* of a *biotic community*. It is expended as *mechanical and heat energy* and ultimately becomes biologically useless. Upon death, organisms are *decomposed* to an inorganic form that algae can use again for nutrients.

▶ *Marine ecosystems are composed of populations of organisms called producers* (which photosynthesize or chemosynthesize), *consumers* (which eat producers), and *decomposers* (which break down detritus). Animals can be categorized as *herbivores* (eat plants), *carnivores* (eat animals), *omnivores* (eat both), or *bacteriovores* (eat bacteria). Through *biogeochemical cycles*, the organisms of a biotic community *cycle nutrients and other chemicals* from one form to another.

▶ *Feeding strategies* include *suspension* or *filter feeding* (filtering planktonic organisms from seawater), *deposit feeding* (ingesting sediment and detritus), and *carnivorous feeding* (preying directly upon other organisms). On average, *only about 10% of the mass taken in at one feeding level is passed on to the next.* As a result, the *size of individuals increases* but the *number of individuals decreases* with each *trophic level* of a *food chain* or *food web*. Overall, the *total biomass of populations decreases the higher they are in the biomass pyramid.*

Study Resources
Online Study Guide Quizzes

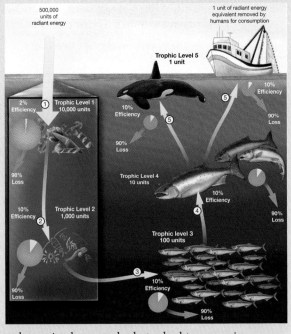

Critical Thinking Question

What is the average efficiency of energy transfer between trophic levels? Use this efficiency to determine how much phytoplankton mass is required to add just 1 gram (0.04 ounce) of new mass to a killer whale, which is a third-level carnivore. Include a diagram that shows the different trophic levels and the relative size and abundance of organisms at different levels. How would your answer change if the efficiency were half the average rate? Twice the average rate?

13.5 What issues affect marine fisheries?

▶ *Marine fisheries harvest standing stocks of populations from various ecosystems*, particularly shallow shelf and coastal waters and areas of upwelling. *Overfishing occurs when adult fish are harvested faster than they can reproduce* and results in the decline of fish populations as well as a reduction of a fishery's *maximum sustainable yield (MSY)*. Many fishing practices capture unwanted *bycatch*. Despite the *management of fisheries, many fish stocks worldwide are declining. Wise seafood choices can help* reverse the decline in fish populations.

MasteringOceanography™

Looking for additional review and test prep materials? Visit the Study Area in MasteringOceanography to enhance your understanding of this chapter's content by accessing a variety of resources, including Self-Study Quizzes, Geoscience Animations, RSS feeds, flashcards, Web links, and an optional Pearson eText.

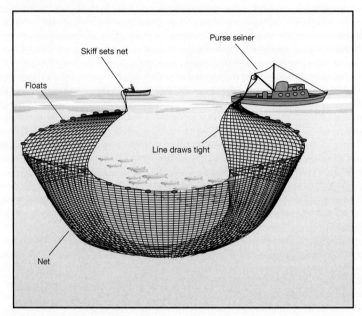

Study Resources
Online Study Guide Quizzes, Web Diving Deeper 13.2

Critical Thinking Question

Describe several significant problems with fisheries management.

A superbly streamlined shark glides through the ocean. Sharks like this silky shark (*Carcharhinus falciformis*) have unique adaptations that make them efficient ocean predators.

Before you begin reading this chapter, use the glossary at the end of this book to discover the meanings of any of these words you don't already know:

Cruiser Lunger Symbiosis Photophore Baleen Melon Krill Seal falcate Counterillumination Biomass Cetacea Sea lion School Sirenia Foraminifer Copepod Odontoceti Mysticeti Echolocation Nitrogen narcosis Bioluminescence Swim bladder Radiolarian Detritus

ANIMALS OF THE PELAGIC ENVIRONMENT

ESSENTIAL CONCEPTS

At the end of this chapter, you should be able to:

14.1 Describe various methods by which marine organisms are able to avoid sinking through the water column

14.2 Discuss adaptations that pelagic organisms possess for seeking prey

14.3 Discuss adaptations that pelagic organisms possess for avoiding being prey

14.4 Describe characteristics and examples of marine mammals

14.5 Demonstrate an understanding of why gray whales migrate

Pelagic organisms live suspended in seawater (not on the ocean floor) and comprise the vast majority of the ocean's **biomass**.[1] Phytoplankton and other photosynthesizing microbes live within the sunlit surface waters of the ocean and are the food source for nearly all other marine life. As a result, many marine animals live in surface waters so they can be close to their food supply. One of the most important challenges facing many marine organisms is to stay afloat and not sink below surface waters into the immense depth of the oceans.

Phytoplankton and other photosynthesizing microbes depend primarily on their small size to provide a high degree of frictional resistance to sinking. Most animals, however, are more dense than ocean water and have less surface area per unit of body mass. Therefore, they tend to sink more rapidly than phytoplankton.

To remain in surface waters where the food supply is greatest, pelagic marine animals must increase their buoyancy or swim continually. Animals apply one or both of these strategies in a wonderful variety of adaptations and lifestyles.

14.1 How Are Marine Organisms Able to Stay Above the Ocean Floor?

Some animals increase their buoyancy to remain in near-surface waters. They may have containers of gas, which significantly reduce their average density, or they may have soft bodies void of hard, high-density parts. Larger animals often have the ability to swim, but if their bodies are denser than seawater, they must exert more energy to propel themselves through the water.

Use Of Gas Containers

Air is approximately 1000 times less dense than water at sea level, so even a small amount of air inside an organism can dramatically increase its buoyancy. Generally, animals use either a rigid gas container or a swim bladder to achieve *neutral buoyancy*, which means the amount of air in their bodies regulates their density, so they can remain at a particular depth.

---

[1]Remember that *biomass* is the mass of living organisms.

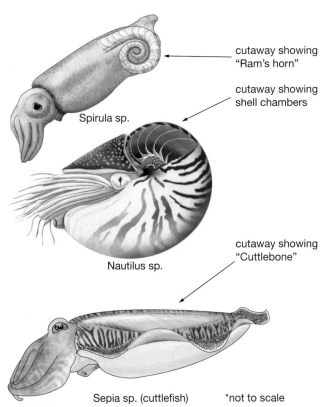

cutaway showing "Ram's horn"

Spirula sp.

cutaway showing shell chambers

Nautilus sp.

cutaway showing "Cuttlebone"

Sepia sp. (cuttlefish)   *not to scale

Figure 14.1 Gas containers in cephalopods. The *Nautilus* has an external chambered shell, while *Sepia* and *Spirula* have rigid internal chambered structures that can be filled with gas to provide buoyancy.

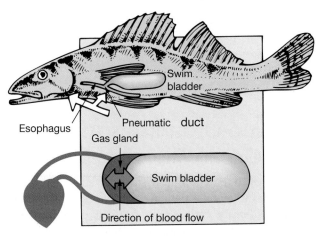

Swim bladder

Esophagus

Pneumatic duct

Gas gland

Swim bladder

Direction of blood flow

Figure 14.2 Swim bladder. Some bony fish have a swim bladder, which is connected to the esophagus by the pneumatic duct and allows air to be added or removed rapidly. In fish with no pneumatic duct, all gas must be added or removed through the blood, which requires more time.

ESSENTIAL CONCEPT 14.1A

Marine organisms use a variety of adaptations to stay within the sunlit surface waters, such as rigid gas containers, swim bladders, spines to increase their surface area, soft bodies, and the ability to swim.

RIGID GAS CONTAINERS   Some animals, such as the cephalopods (*cephalo* = head, *poda* = foot), have rigid gas containers in their bodies. For instance, the genus *Nautilus* has an external shell, whereas the cuttlefish *Sepia*[2] and deep-water squid *Spirula* have an internal chambered structure (**Figure 14.1**).

Because the pressure in their air chambers is always 1 kilogram per square centimeter (1 atmosphere, or 14.7 pounds per square inch), the *Nautilus* must stay above a depth of approximately 500 meters (1640 feet) to prevent collapse of its chambered shell. The *Nautilus*, therefore, rarely ventures below about 250 meters (800 feet).

SWIM BLADDER   Instead of using rigid gas containers, some slow-moving fish use an internal organ called a **swim bladder** (**Figure 14.2**) to achieve neutral buoyancy. Very active swimmers (such as tuna) or fish that live on the bottom, however, do not usually have a swim bladder because they don't have a problem maintaining their positions in the water column.

A change in depth either expands or contracts the swim bladder, so fish must remove or add gas to maintain a constant volume. In some fish, a pneumatic duct connects the swim bladder to the esophagus, so these fish can quickly add or remove gases through the duct. In other fish without the pneumatic duct, the gases of the swim bladder must be added or removed more slowly by an interchange with the blood, so they cannot withstand rapid changes in depth.

The composition of gases in the swim bladders of shallow-water fish is similar to that of the atmosphere. At the surface, the concentration of oxygen in the swim bladder is about 20%, and as depth increases, the oxygen concentration increases to more than 90%. Fish with swim bladders have been captured from as deep as 7000 meters (23,000 feet), where the pressure is 700 kilograms per square centimeter (700 atmospheres, or 10,300 pounds per square inch). Pressure this high compresses the gas to a density of 0.7 gram per cubic centimeter (0.025 pound per cubic inch).[3] This is approximately the same density as fat, and as a result, many deep-water fish have special organs for buoyancy that are filled with fat instead of compressed gas.

Ability to Float

Floating animals range in size from microscopic shrimplike organisms to relatively large species, such as the familiar jellies. These floating organisms—collectively called *zooplankton*—comprise the second largest biomass in the ocean after phytoplankton and other photosynthesizing microbes. Microscopic forms of zooplankton usually have a hard shell called a **test** (*testa* = shell). Many larger forms have soft, gelatinous bodies with little if any hard tissue, which reduces their density and allows them to stay afloat.

Microscopic zooplankton are incredibly abundant in the ocean. They are primary consumers because they eat microscopic phytoplankton, the primary producers. Thus, many zooplankton are herbivores. Others are omnivores because they eat other zooplankton in addition to phytoplankton. Most types of microscopic zooplankton have adaptations to increase the surface area of their bodies (or shells) so they can remain in the sunlit surface waters near their food source.[4]

In addition, some organisms produce low-density fats or oils to stay afloat. Many zooplankton, for example, produce tiny droplets of oil to help maintain neutral or nearly neutral buoyancy. As another example, sharks have a very large, oil-rich liver to help reduce their density and float more easily.

---

[2]Many species of cephalopods have an inking response. In fact, the ink of *Sepia* was used as writing ink (with the brand name Sepia) before alternatives were developed.

[3]For comparison, note that the density of water is 1.0 gram per cubic centimeter (0.04 pound per cubic inch).

[4]For a more complete discussion of how increased surface area affects an organism's ability to float, see Chapter 12.

Ability to Swim

Many larger pelagic animals such as fish and marine mammals can maintain their position in the water column by swimming and can also swim easily against currents. These organisms are called *nekton* (*nektos* = swimming). Because of their swimming ability, some of these organisms undertake long migrations.

The Diversity Of Planktonic Animals

Diversity in planktonic marine animals stems from a shifting balance between competition for food and avoidance of predators. As a result, a wide variety of planktonic animals inhabit the oceans.

EXAMPLES OF MICROSCOPIC ZOOPLANKTON   Three of the most important groups of microscopic zooplankton are the radiolarians, foraminifers (both of which are discussed in more detail in Chapter 4, "Marine Sediments"), and copepods.

Radiolarians (*radio* = spoke or ray) are single-celled, microscopic protozoans (*proto* = first, *zoa* = animal) that build their hard shells (*tests*) out of silica (**Figure 14.3**). Their tests have intricate ornamentation, including long projections. Although the spikes and spines appear to be a defense mechanism against predators, they increase the test's surface area so the organism won't sink through the water column.

Foraminifers (*foramen* = an opening) are microscopic to barely macroscopic single-celled protozoans. While the most abundant types of foraminifers are planktonic, the most diverse (in terms of number of species) are benthic. Foraminifers produce a hard test made of calcium carbonate (**Figure 14.4**) that is segmented or chambered, with a prominent opening in one end. The tests of both radiolarians and foraminifers are common components of deep-sea sediment.

Copepods (*kope* = oar, *poda* = foot) are microscopic shrimplike animals of the subphylum Crustacea, which also includes shrimps, crabs, and lobsters. Like other crustaceans, copepods have a hard exoskeleton (*exo* = outside) and a segmented body with jointed legs (**Figure 14.5**). Most copepods have forked tails and distinct and elaborate antennae.

More than 7500 species of copepods exist, with most possessing special adaptations for filtering tiny floating food particles from water. Some copepods are herbivores that eat algae, others are carnivores that eat other zooplankton, and still others are parasitic.

All copepods lay eggs, which are sometimes carried in egg sacs attached to the abdomen but generally released into seawater, where they hatch in about a day. Their rapid reproduction allows great numbers of copepods to occur wherever favorable conditions—generally an abundance of food, which is mostly algae—exist.

Although copepods are small (only a few species are larger than 1 millimeter [0.04 inch]), they are extremely numerous in the ocean. They are so abundant, in fact, that they are one of the most numerically dominant types of multicelled organisms on the planet. As such, copepods comprise the majority of the ocean's zooplankton biomass and are a vital link in many marine food webs between phytoplankton (producers) and larger species such as plankton-eating fish.

EXAMPLES OF MACROSCOPIC ZOOPLANKTON   Many types of zooplankton are large enough to be seen without the aid of a microscope yet don't swim well. Two of

(a) *Euphysetta elegans,* ×280

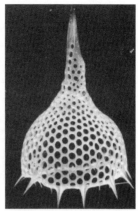
(b) *Anthocyrtidium ophirense,* ×230

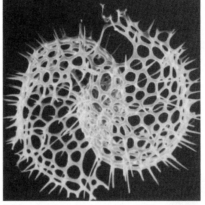
(c) *Larcospira quadrangula,* ×190

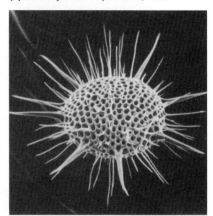
(d) *Heliodiscus asteriscus,* ×200

Figure 14.3 Radiolarians. Scanning electron micrographs of the tests of various radiolarians.

Figure 14.4 Foraminifers. Photomicrograph of the tests of various species of foraminifers, the largest of which is about 1 millimeter (0.04 inch) long. These pelagic foraminifers were collected from the Mediterranean Sea.

Figure 14.5 Copepods.
Line drawings of various copepods (shown many times
in their actual size) from
Wilhelm Giesbrecht's 1892
book on the flora and fauna
of the Gulf of Naples.

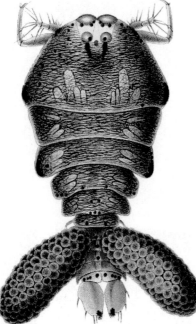

(a) The adult female *Sapphirina auranitens*
carries a pair of lobe-like egg sacs.

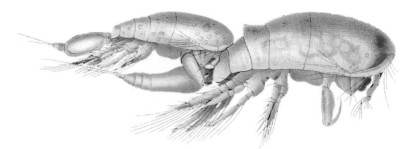

(b) Copulating pair of *Oncaea conifera*.

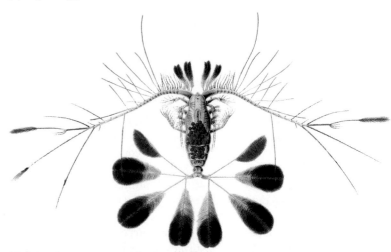

(c) *Calocalanus pavo* showing elaborate feathery appendages that are
characteristic of warm-water species.

(d) *Copilia vitrea* uses its appendages to cling to large particles in the water
column or to larger zooplankton.

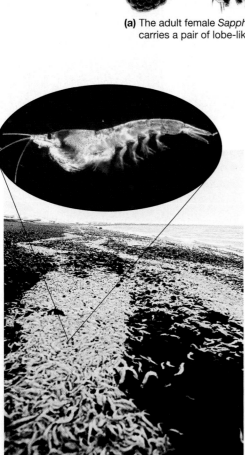

Figure 14.6 Krill. Krill that has washed up on a beach in
Antarctica and close-up view of *Meganyctiphanes norvegica*,
which is about 3.8 centimeters (1.5 inches) long (*inset*).

the most important groups of macroscopic zooplankton are krill and various types
of cnidarians.

Krill, which means "young fry of fish" in Norwegian, are actually in the subphylum Crustacea (genus *Euphausia*) and resemble minishrimp or large copepods
(**Figure 14.6**). There are more than 1500 species of krill, most of which achieve a
length no longer than 5 centimeters (2 inches). They are abundant near Antarctica
and form a critical link in the food web there, supplying food for many organisms
from sea birds to the largest whales in the world.

Cnidarians (*cnid* = nematocyst [*nemato* = thread, *cystis* = bladder]), which
were formerly known as *coelenterates* (*coel* = hollow, *enteron* = intestines), have soft
bodies that are more than 95% water and tentacles armed with stinging cells called
nematocysts. Macroscopic zooplankton that are cnidarians fall into one of two basic
groups: the *hydrozoans* and the *scyphozoans* (jellies).

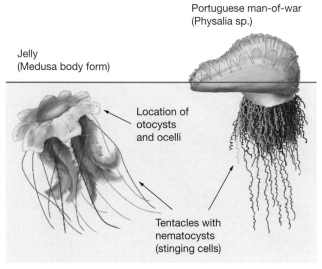
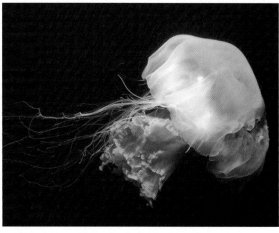

Portuguese man-of-war
(*Physalia* sp.)

Jelly
(Medusa body form)

Location of
otocysts
and ocelli

Tentacles with
nematocysts
(stinging cells)

(a) Line drawings of a typical medusa jelly (*left*) and a Portuguese man-of-war (*Physalia*) (*right*).

(b) Photo of medusa jelly (*Cyanea capillata*), which has a bell that is 50 centimeters (20 inches) wide.

Figure 14.7 Planktonic cnidarians. Line drawings and a photo of various species of planktonic cnidarians.

Hydrozoan (*hydro* = water, *zoa* = animal) *cnidarians* are represented in all oceans by the "Portuguese man-of-war" (genus *Physalia*) and the "by-the-wind sailor" (genus *Velella*). Their gas chambers, called *pneumatophores* (*pneumato* = breath, *phoros* = bearing), serve as floats and sails that allow the wind to push them across the ocean surface (**Figure 14.7a**, *left*). Sometimes the wind pushes large numbers of these organisms toward a beach, where they can wash ashore and die. In the living organism, a colony of tiny individuals is suspended beneath the float. Portuguese man-of-war tentacles may be many meters long and, because they are armed with nematocysts, they have been known to inflict a painful and occasionally dangerous neurotoxin poisoning in humans.

Jellies, or **scyphozoan** (*skyphos* = cup, *zoa* = animal) *cnidarians*, have a bell-shaped body with a fringe of tentacles and a mouth at the end of a clapperlike extension hanging beneath the bell-shaped float (**Figure 14.7**). Ranging in size from nearly microscopic to 2 meters (6.6 feet) in diameter, most jellies are less than 0.5 meter (1.6 feet) in diameter. The largest jellies can have tentacles as long as 60 meters (200 feet).

Jellies move by muscular contraction. Water enters the cavity under the bell and is forced out when muscles that circle the bell contract, jetting the animals ahead in short spurts. To allow the animal to orient itself generally in an upward direction, light-sensitive or gravity-sensitive organs exist around the outer edge of the bell. The ability to orient is important because jellies feed by swimming to the surface and sinking slowly through the rich surface waters.

Other types of macroscopic zooplankton include tunicates and salps (barrel-shaped, often colonial organisms), ctenophores (comb jellies or sea gooseberries), and chaetognaths (arrowworms).

EXAMPLES OF SWIMMING ORGANISMS   Swimming, or *nektonic*, organisms include invertebrate squids, fish, sea turtles, and marine mammals.

Swimming squid include the common squid (genus *Loligo*), flying squid (*Ommastrephes*), and giant squid (*Architeuthis*), all of which are active predators of fish. Most squid possess long, slender bodies with paired fins (**Figure 14.8**) and must remain active to stay afloat. However, a few species, such as *Sepia* and *Spirula* (see Figure 14.1), have a hollow gas-filled chamber in their bodies to help them remain afloat and so they can be less active.

Squid can swim about as fast as any fish their size and do so by drawing water into their body cavity and then expelling the water out through their siphon,

Web Video
Giant Squid Encounter

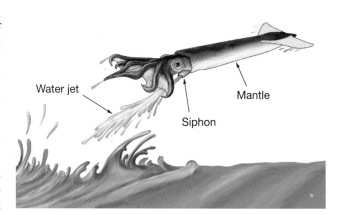

Water jet

Mantle

Siphon

Figure 14.8 Squid. Squid move by trapping water in their mantle cavity between their soft body and penlike shell. They then jettison the water through their siphon for rapid propulsion as this flying squid (*Ommastrephes*) does to become airborne.

Figure 14.9 Swimming motion and general features of fish. Fish send a wave of body curvature along their bodies to produce a forward thrust (*left*). General features of fish (*right*), with the major fins labeled.

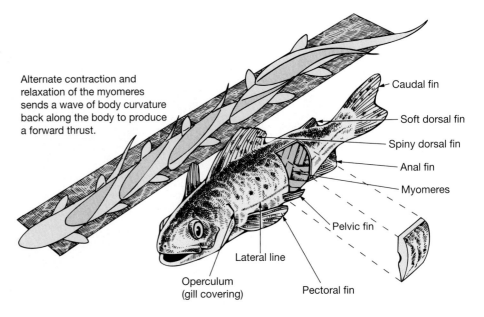

Alternate contraction and relaxation of the myomeres sends a wave of body curvature back along the body to produce a forward thrust.

Caudal fin

Soft dorsal fin

Spiny dorsal fin

Anal fin

Myomeres

Pelvic fin

Lateral line

Operculum (gill covering)

Pectoral fin

propelling themselves backward. To capture prey, squid use two long arms with pads containing suction cups at the ends (Figure 14.8). Eight shorter arms with suckers convey the prey to the mouth, where it is crushed by a mouthpiece that resembles a parrot's beak.

In fish, locomotion occurs when a wave of lateral body curvature passes from the front of the fish to the back. This is achieved by the alternate contraction and relaxation of muscle segments, called *myomeres* (*myo* = muscle, *merous* = parted), along the sides of the body. The backward pressure of the fish's body and fins produced by the movement of this wave provides the forward thrust (Figure 14.9).

FIN DESIGNS IN FISH   Most active swimming fish use two sets of paired fins—*pelvic fins* and *pectoral* (*pectoralis* = breast) *fins*—to turn, brake, and balance (Figure 14.9). When not in use, these fins can be folded against the body. Vertical fins, both *dorsal* (*dorsum* = back) and *anal*, serve primarily as stabilizers.

The fin that is most commonly used to propel high-speed fish is the tail fin, which is called the *caudal* (*cauda* = tail) *fin*. Caudal fins flare vertically to increase the surface area available to develop thrust against the water.[5] The increased surface area also increases frictional drag. The efficiency of the design of a caudal fin depends on its shape, which is one of the five basic forms illustrated in Figure 14.10 and keyed by letter to the following descriptions:

 a. *Rounded fins* are flexible and useful in accelerating and maneuvering at slow speeds.

 b. and c. *Truncate fins* (b) and *forked fins* (c) are found on faster fish; the fins are somewhat flexible for better propulsion but are also used for maneuvering.

 d. *Lunate fins* are found on fast-cruising fish such as tuna, marlin, and swordfish; the fins are very rigid and useless for maneuverability but very efficient for propulsion.

---

[5]This is equivalent to humans donning swimming fins on their feet to enable them to swim more efficiently.

(b) Truncate caudal fin on a gray angelfish (other examples: salmon, bass).

(c) Forked caudal fin on a yellowfin doctorfish (other examples: herring, goatfish).

(a) Rounded caudal fin on a blue-face angel (other examples: sculpin, flounder).

(d) Lunate caudal fin on a striped marlin (other examples: bluefish, tuna).

(e) Heterocercal caudal fin on a gray reef shark (other examples: many other types of sharks).

Figure 14.10 Caudal fin shapes. Photos of various fish displaying different caudal fin shapes, including **(a)** rounded, **(b)** truncate, **(c)** forked, **(d)** lunate, and **(e)** heterocercal.

e. *Heterocercal* (*hetero* = uneven, *cercal* = tail) *fins* are asymmetrical, with most of their mass and surface area in the upper lobe. Sharks have a heterocercal fin, which produces significant lift and is necessary because sharks have no swim bladder and tend to sink when they stop moving. In fact, the fundamental body design of sharks has many adaptations to compensate for the shark's negative buoyancy. The pectoral (chest) fins, for example, are large and flat and positioned so they function like airplane wings to lift the front of the shark's body, balancing the rear lift supplied by its heterocercal caudal fin. Although the shark gains tremendous lift from this adaptation of its pectoral fins, it sacrifices maneuverability. This is why sharks tend to swim in broad circles—like a circling airplane—and do not make sharp turns while swimming.

CONCEPT CHECK 14.1

❶ Discuss why the rigid gas chamber in cephalopods limits the depth to which they can descend. Why do fish with a swim bladder not have this limitation?

❷ Explain why hydrozoans and scyphozoans (jellies) are classified as macroscopic forms of zooplankton. Why are they not considered nekton?

❸ From memory, name and describe the different types of fins that fish exhibit. What are the five basic shapes of caudal fins, and what are their uses?

14.2 What Adaptations Do Pelagic Organisms Possess for Seeking Prey?

Pelagic organisms possess several adaptations that are designed to enhance their ability to seek and capture food. These adaptations include mobility (lunging versus cruising), swimming speed, body temperature, and specially designed circulatory systems. In addition, deep-water nekton have some unique adaptations to help them succeed in capturing prey in their world of darkness.

Mobility: Lungers versus Cruisers

Some fish wait patiently for prey and exert themselves only in short bursts as they lunge at the prey. Others cruise relentlessly through the water, seeking prey. A marked difference occurs in the musculature of fish that use these different styles of obtaining food.

For example, **lungers** (such as the grouper shown in Figure 14.11a) sit and wait for prey to come close by. Lungers have truncate caudal fins for speed and maneuverability, and almost all their muscle tissue is white.

On the other hand, **cruisers** (such as the tuna shown in Figure 14.11b) actively seek prey. Less than half of a cruiser's muscle tissue is white; most is red.

What is the significance of red versus white muscle tissue? Red muscle tissue contains fibers that are 25 to 50 microns in diameter (0.01 to 0.02 inch), whereas the fibers in white muscle tissue are 135 microns (0.05 inch) and contain much lower concentrations of **myoglobin** (*myo* = muscle, *globus* = sphere), a red pigment with an affinity for oxygen. Moreover, red muscle tissue supplies a much greater amount of oxygen and supports a much higher metabolic rate than white tissue. This is why cruisers have so much red muscle tissue: It allows them to have the endurance needed to support their active lifestyle.

Lungers, on the other hand, need very little red tissue because they do not move continually. Instead, they need white tissue, which fatigues much more rapidly than red tissue, for quick bursts of speed to capture prey. Cruisers use white tissue, too, for short periods of acceleration while on the attack.

Swimming Speed

Although rapid swimming consumes much energy, it can help organisms capture prey. Fish normally swim slowly when cruising, fast when hunting for prey, and fastest of all when trying to escape from predators.

Figure 14.11 Feeding styles of lungers and cruisers. Photos of **(a)** a typical lunger and **(b)** cruisers.

(a) Lungers, such as this tiger grouper, sit patiently on the bottom and capture prey with quick, short lunges.

(b) Cruisers, such as these yellowfin tuna, swim constantly in search of prey and capture it with short periods of high-speed swimming.

Diving Deeper 14.1 Oceans and People

SOME MYTHS (AND FACTS) ABOUT SHARKS

"Now we know that almost every attack on a human is an accident: the shark mistakes the human for its normal prey."

—*Peter Benchley (2000), author of* Jaws

Sharks (**Figure 14A**) are the fish humans most fear. Their strength, large size, sharp teeth, and unpredictable nature are enough to keep some people from *ever* entering the ocean. The media hype generated by occasional shark attacks on humans has led to many myths about sharks, including the following:

- *Myth #1: All sharks are dangerous.* Of the world's roughly 400 shark species, 80% are unable to hurt people or rarely encounter people. The shark species that are responsible for most attacks on humans are great white (*Carcharodon carcharias*), tiger (*Galeocerdo cuvier*), and bull sharks (*Carcharhinus leucas*). The world's largest shark is the whale shark (*Rhincodon typus*), which reaches lengths of up to 15 meters (50 feet) but is a filter-feeder that eats only plankton and so is not considered dangerous.
- *Myth #2: Sharks are voracious eaters that must eat continuously.* Like other large animals, sharks eat periodically, depending on their metabolism and the availability of food. Humans are not a primary food source of any shark, and many large sharks prefer the higher fat content of seals and sea lions.
- *Myth #3: Most people attacked by sharks are killed.* Of every 100 people attacked by sharks, 85 survive. Many large sharks commonly attack by biting their prey to immobilize it before trying to eat it. Consequently, many potential prey escape and survive. The most common targets of shark attacks are surfers (49%), swimmers/waders (29%), divers/snorkelers (15%), and kayakers (6%). Most shark encounters with humans are investigative, not predatory.
- *Myth #4: Many people are killed by sharks each year.* The chances of a person being killed by a shark are quite low (**Table 14A**). Over the past several decades, sharks have killed an average of only 5 to 15 people worldwide each year. Humans, on the other hand, kill as many as 100 million sharks a year (mostly as bycatch from fishing activities).

Compared to many other fish, sharks have low reproduction rates and grow slowly, which may result in many sharks being designated as endangered species in the future.

- *Myth #5: The great white shark is a common, abundant species found off most beaches.* Great white sharks are relatively uncommon predators that prefer cooler waters. At most beaches, great whites are rarely encountered.
- *Myth #6: Sharks are not found in freshwater.* A specialized osmoregulatory system enables some species (such as the bull shark) to cope with dramatic changes in salinity, from the high salinity of seawater to the low salinity of freshwater rivers and lakes.
- *Myth #7: All sharks need to swim constantly.* Some sharks can remain at rest for long periods on the bottom and obtain enough oxygen by opening and closing their mouths to pump water across their gills. Typically, sharks swim very slowly—cruising speeds are less than 9 kilometers (6 miles) per hour—but they can swim at bursts of over 37 kilometers (23 miles) per hour.
- *Myth #8: Sharks have poor vision.* The lens of a shark's eye is up to seven times more powerful than that of a human's. Sharks can even distinguish color.
- *Myth #9: Eating shark meat makes one aggressive.* There is no indication that eating shark meat will alter a person's temperament. The firm texture, white flesh, low fat content, and mild taste of shark meat have made it a favorite seafood in many countries.
- *Myth #10: No one would ever want to enter water filled with sharks.* Long regarded with fear and suspicion, sharks have more recently been viewed as skilled predators that are vitally important to the health of ocean ecosystems. Diving tours specializing in close encounters with sharks are becoming increasingly popular.

Figure 14A **Great white shark (*Carcharodon carcharias*).**

TABLE 14A TYPES AND NUMBERS OF OCCURRENCES IN THE UNITED STATES

Occurrence to people in the U.S.	Average number per year
Transportation fatalities	42,000
Death involving slipping, tripping, or stumbling	565
Struck by lightning	352
Killed by lightning	50
Bitten by a squirrel in New York City	88
Killed by a dog bite	26
Killed by a snake bite	12
Bitten by a shark	10
Killed by a shark	0.4

"Today I could not, for instance, portray the shark as a villain, especially not as a mindless omnivore that attacks boats and humans with reckless abandon. No, the shark in an updated Jaws could not be the villain; it would have to be written as the victim, for, world-wide, sharks are much more the oppressed than the oppressors."

—*Peter Benchley (2005), author of* Jaws

Generally, when comparing fish of similar shapes, the larger the fish, the faster it can swim. For tuna, well adapted for sustained cruising and short, high-speed bursts, cruising speed averages about 3 body lengths per second. They can, however, maintain a maximum speed of about 10 body lengths per second, but only for about 1 second. Remarkably, yellowfin tuna (*Thunnus albacares*) have been clocked at 74.6 kilometers (46 miles) per hour! Even though this speed is more than 20 body lengths per second, the tuna could maintain it for only a *fraction* of a second. Theoretically, a 4-meter (13-foot) bluefin tuna (*Thunnus thynnus*) can reach speeds up to about 144 kilometers (90 miles) per hour.[6]

Like fish, many of the toothed whales are fast swimmers, too. For instance, spotted dolphins of the genus *Stenella* have been clocked at 40 kilometers (25 miles) per hour, and killer whales may exceed 55 kilometers (34 miles) per hour during short bursts.

Cold-Blooded versus Warm-Blooded Organisms

Fish are mostly **cold-blooded**, or **poikilothermic** (*poikilos* = spotted, *theromos* = heat), so their body temperatures are nearly the same as their environment. Usually, these fish are not fast swimmers. The mackerel (*Scomber*), yellowtail (*Seriola*), and bonito (*Sarda*), on the other hand, are indeed fast swimmers. They also have body temperatures that are 1.3, 1.4, and 1.8°C (2.3, 2.5, and 3.2°F), respectively, higher than the surrounding seawater.

Mackerel sharks (genera *Lamna* and *Isurus*) and tuna (genera *Thunnus*) have body temperatures much higher than their environment. Bluefin tuna can maintain a body temperature of 30 to 32°C (86 to 90°F) regardless of the water temperature, which is characteristic of organisms that are **warm-blooded**, or **homeothermic** (*homeo* = alike, *thermos* = heat). Although these tuna are more commonly found in warmer water, where the temperature difference between fish and water is no more than 5°C (9°F), body temperatures of 30°C (86°F) have been measured in bluefin tuna swimming in 7°C (45°F) water.

Why do these fish exert so much energy to maintain their body temperatures at high levels, when other fish do quite well with ambient body temperatures? It may be that for cruisers, any adaptation (high temperature and high metabolic rate) that increases the power output of their muscle tissue helps them seek and capture prey.

Adaptations Of Deep-Water Nekton

Living below the surface water but still above the ocean floor are deep-water nektonic species—mostly various species of fish—that are specially adapted to the deep-water environment, where it is very still and completely dark. Their food source is either **detritus** (*detritus* = to lessen)—dead and decaying organic matter including waste products that slowly sink downward from surface waters—or each other. The lack of abundant food limits the *number* of organisms (total biomass) and the *size* of these organisms. As a result, small populations of these organisms exist, and most individuals are less than 30 centimeters (1 foot) long. Many have low metabolic rates to conserve energy, too.

These **deep-sea fish** (Figure 14.12) have special adaptations to efficiently find and collect food. They have good sensory devices, for example, such as long antennae or sensitive lateral lines that can detect movement of other organisms within the water column.

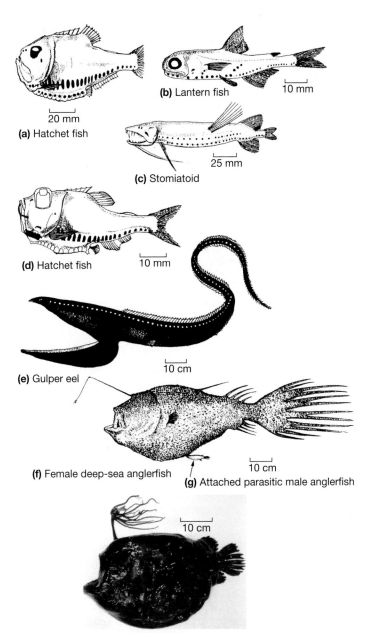

(a) Hatchet fish — 20 mm

(b) Lantern fish — 10 mm

(c) Stomiatoid — 25 mm

(d) Hatchet fish — 10 mm

(e) Gulper eel — 10 cm

(f) Female deep-sea anglerfish

(g) Attached parasitic male anglerfish — 10 cm

(h) Photo of female deep-sea anglerfish (*Himantolophus danae*) that washed ashore in Carlsbad, California, in 2001.

Figure 14.12 Deep-sea fish. Line drawings and a photo of various deep-sea fish.

---

[6]Imagine how difficult it would be to clock a bluefin tuna in the ocean at this speed!

Well over half of deep-sea fish and even some deep-water species of shrimp and squid can **bioluminesce** (*bios* = life, *lumen* = light, *esc* = becoming), which means they can produce light biologically and "glow in the dark." Fireflies and glowworms are well-known animals on land that have bioluminescent capability. Approximately 80% of bioluminescent organisms have light-producing organs called **photophores** (*photo* = light, *phoros* = bearing), which can be simple luminous spots or may be quite complex and equipped with lenses, shutters, color filters, and reflectors.

Bioluminescent light is produced from compounds during the digestion of prey, from specialized cells in the organism, or associated with symbiotic bacteria that is cultured and lives inside the organism. The light is produced when molecules of the biological pigment *luciferin* (*lucifer* = light bringing) are excited and emit photons of light in the presence of oxygen. Remarkably, only a 1% loss of energy is required to produce this illumination. New research suggests that some sharks control their bioluminescence by using hormones.

In a world of darkness, the ability to bioluminesce is useful for a variety of purposes, including:

- Searching for food items in the dark
- Attracting prey (for example, the deep-sea anglerfish in Figure 14.12f uses its specially modified dorsal fin as a bioluminescent lure)
- Staking out territory by constantly patrolling an area
- Communicating or seeking a mate by sending signals
- Escaping from predators by using a flash of light to temporarily blind or distract them
- Avoiding predators by use of a "burglar alarm" by attracting unwanted attention with brilliant displays of bioluminescence
- Camouflaging by using belly lights to match the color and intensity of dim filtered sunlight from above and obliterate a telltale shadow; this is known as **counterillumination**

To take advantage of bioluminescent light, many deep-sea fish species have large and sensitive eyes—perhaps 100 times more sensitive to light than human eyes—that enable them to see potential prey. To avoid being prey, most species are dark in color so that they blend into the environment. Still other species are blind and rely on senses such as smell to track down prey.

Other adaptations that various deep-sea fish species possess include large sharp teeth, expandable bodies to accommodate large food items, hinged jaws that can unlock to open widely, and mouths that are huge in proportion to their bodies (**Figure 14.13**). These adaptations allow deep-sea fish to ingest species that are larger than they are and to process food efficiently whenever it is captured.

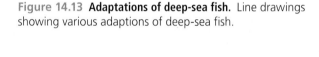

Web Video
Bioluminescent Organisms

Figure 14.13 **Adaptations of deep-sea fish.** Line drawings showing various adaptions of deep-sea fish.

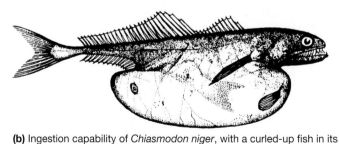

(b) Ingestion capability of *Chiasmodon niger*, with a curled-up fish in its stomach that is longer than it is.

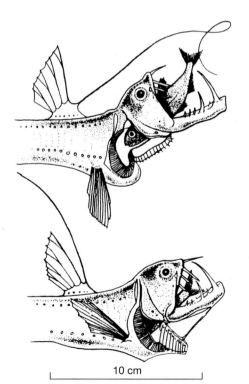

10 cm

(a) Large teeth, hinged jaw, and swallowing mechanism of the deep-sea viper fish *Chauliodus sloani*.

ESSENTIAL CONCEPT 14.2

Adaptations of pelagic organisms for seeking prey include mobility (lunging versus cruising), high swimming speed, and high body temperature. Deep-water nekton exhibit a host of unusual adaptations—including bioluminescence—that allow them to survive in deeper waters.

CONCEPT CHECK 14.2

1. Are most fast-swimming fish cold-blooded or warm-blooded? What advantage does this provide?

2. What are the two food sources of deep-water nekton? List several adaptations of deep-water nekton that allow them to survive in their environment.

3. Describe the mechanism by which bioluminescence is accomplished in deep-sea organisms. What is bioluminescence useful for in the marine environment?

14.3 What Adaptations Do Pelagic Organisms Possess to Avoid Being Prey?

Many animals have unique adaptations to avoid being captured and eaten. Examples of adaptations that organisms use to enhance their survival include schooling and symbiosis.

Schooling

The term **school** refers to large numbers of fish, squid, or shrimp that form well-defined social groupings. Although vast populations of phytoplankton and zooplankton may be highly concentrated in certain areas of the ocean, they are not usually referred to as schools.

The number of individuals in a school can vary from a few larger predaceous fish (such as bluefin tuna) to hundreds of thousands of small filter feeders (such as anchovies). Within the school, individuals move in the same direction and are evenly spaced. Spacing is probably maintained through visual contact and, in the case of fish, by use of the lateral line system (see Figure 14.9) that detects vibrations of swimming neighbors. The school can turn abruptly or reverse direction as individuals at the head or rear of the school assume leadership positions (**Figure 14.14**).

What are the advantages of schooling? One advantage is that during spawning, schooling ensures that there will be males to release sperm to fertilize the eggs released into the water or deposited on the bottom by females. Another advantage is that schools of smaller fish can invade the territory of larger aggressive species and feed there because the "owner" of the territory can never chase away the whole school. The most important function of schooling in small fish, however, is protection from predators.

It may seem illogical that schooling would be protective. For instance, schooling creates tighter groupings of organisms so that any predator lunging into a school would surely catch something, just as land predators run a herd of grazing animals until one weakens and becomes a meal. So, aren't the smaller fish making it easier for the predators by forming a large target? Scientists who study fish behavior suggest that schooling does indeed serve to protect a group of organisms based on strategies that give them "safety in numbers," much like flocks of birds. In many parts of the marine environment, such as the open ocean where there is no place to hide, schooling has the following advantages:

1. When members of a species form schools, they reduce the percentage of ocean volume in which a cruising predator might find one of their kind.

Figure 14.14 Schooling. A school of blue-lined snapper near a reef in the Maldives, Indian Ocean. Schooling increases chances of survival, and more than half of all fish species are known to join schools during at least a portion of their lives.

2. When a predator encounters a large school, it is less likely to consume the entire unit than if it encounters a small school or an individual.

3. The school may appear as a single large and dangerous opponent to a potential predator and prevent some attacks.

4. Predators may find the continually changing position and direction of movement of fish within the school confusing, making attack particularly difficult for predators, which can attack only one fish at a time.

In addition, the fact that more than half of all fish species join schools during at least a portion of their lives suggests that schooling enhances survival of species, especially for those with no other means of defense. Schooling may also help fish swim greater distances than individuals because each schooling fish gets a boost from the vortex created by the fish swimming in front of it.

Recently, a new ocean predator has developed a method to take advantage of the schooling behavior of many species of fish. Human fishers have developed nets large enough to encircle whole schools of fish. These nets are very efficient at catching fish, thereby leading to the decline of many fish stocks (see Chapter 13).

Symbiosis

Many marine organisms seek relationships with other organisms to help them survive. One such relationship is **symbiosis** (*sym* = together, *bios* = life), which occurs when two or more organisms associate in a way that benefits at least one of them. There are three main types of symbiotic relationships: commensalism, mutualism, and parasitism.

In **commensalism** (*commensal* = sharing a meal, *ism* = process), a smaller or less dominant participant benefits without harming its host, which affords subsistence or protection to the other. A remora, for example, attaches itself to a shark or another fish to obtain food and transportation, generally without harming its host (**Figure 14.15a**).

In **mutualism** (*mutuus* = borrowed, *ism* = process), both participants benefit. For example, the stinging tentacles of the sea anemone protect the clown fish (**Figure 14.15b**), and the clown fish, which is small but aggressive, chases away any fish that tries to feed on the anemone itself. In addition, the clown fish helps clean the anemone and may even supply scraps of food. Remarkably, clown fish are not stung by the anemone because clown fish have a protective agent in the mucus that coats their bodies.

In **parasitism** (*parasitos* = a person who eats at someone else's table, *ism* = process), one participant (the parasite) benefits at the expense of the other (the host). Many fish are hosts to isopods, which attach to the fish and derive their nutrition from the body fluids of the fish, thereby robbing the host of some of its energy supply (**Figure 14.15c**). Usually, the parasite does not rob enough energy to kill the host because if the host dies, so does the parasite.

Recently, symbiosis had been discovered to be an important component that drives evolution. For example, the sequencing of the genome of a species of diatom reveals that it apparently acquired new genes by engulfing microbial neighbors. The research suggests that early on in the evolution of diatoms, the most significant acquisition was an algal cell that provided the diatom with photosynthetic machinery.

Other Adaptations

Marine animals exhibit a variety of behaviors that serve as defensive mechanisms to help them ward off predators—or to be more successful predators themselves. These include using speed, secreting poisons, and mimicking other poisonous or

STUDENTS SOMETIMES ASK . . .

What are the world's largest and smallest fish?

The world's largest fish is the whale shark (*Rhincodon typus*), which reaches lengths of up to 15 meters (50 feet) and can weigh up to 13.6 metric tons (15 short tons). Its mouth is an enormous 1.5 meters (5 feet) wide. It is a slow-moving, wide-ranging, filter-feeding animal that exists almost entirely on plankton. Alternatively, the smallest known fish is *Paedocypris progenetica*, a relative of carp and minnows. Its adult length is only 7.9 millimeters (0.31 inch), which is about as thick as a pencil. It lives exclusively in Indonesian swamps and, remarkably, was only discovered in 2005. There is, however, one other fish that is smaller: the male deep-sea anglerfish *Photocorynus spiniceps*, which has an adult length of only 6.2 millimeters (0.24 inch). It is generally not considered the world's smallest fish because it is not a self-sustaining organism. Instead, the parasitic males bite into larger females and fuse for life (see Figure 14.12g).

(a) Commensalism occurs when an organism benefits without harming its host, such as these remoras attached to a lemon shark.

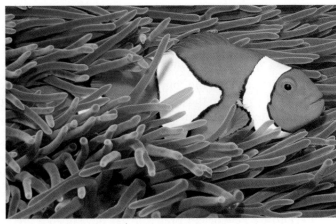

(b) Mutualism occurs when both participants benefit, such as this clown anemonefish and sea anemone.

(c) Parasitism occurs when one participant benefits at the expense of the other, such as this isopod that has attached itself to the head of a whitetip soldierfish.

Figure 14.15 Types of symbiosis. The three main types of symbiosis are **(a)** commensalism, **(b)** mutualism, and **(c)** parasitism.

ESSENTIAL CONCEPT 14.3

Many pelagic species (especially fish) school, engage in symbiosis, or have other adaptations to increase their chances of survival by avoiding predators.

distasteful species. Others use transparency, camouflage, or countershading, as discussed in Chapter 12.

CONCEPT CHECK 14.3

1 What are several benefits of schooling?

2 What are the three types of symbiosis, and how do they differ?

3 Besides schooling and symbiosis, what other adaptations do pelagic animals possess to avoid being prey?

14.4 What Characteristics Do Marine Mammals Possess?

Marine mammals include some of the largest, best known, and most charismatic animals in the sea, such as seals, sea lions, manatees, porpoises, dolphins, and whales.

Although all marine mammals have an aquatic existence, their ancestors were land animals. For example, a series of striking fossil discoveries of ancient whales in Pakistan, India, and Egypt provide strong evidence that whales evolved from mammals on land about 50 million years ago. Some whale ancestors had small, unusable hind legs, suggesting that the land mammal predecessor had no need for its hind legs when it developed a large paddle-shaped structure for a tail used to swim through water. Other fossils show a remarkable progression of skeletal adaptations for an increasingly aquatic existence, such as the migration of the blowhole (nostrils) toward the top of the head, upper vertebrae that become increasingly fused, shrinking hip and ankle structures, and jawbone and ear components designed for underwater hearing. Additional lines of evidence—including DNA analysis of modern whales and a host of anatomical similarities between land and marine mammals—confirm the evolution of whales from a hippopotamus-like land-dwelling ancestor.

The geologic record shows that life on land evolved from marine organisms millions of years ago. Why would a land mammal migrate back to the sea? One hypothesis suggests that they may have returned to the sea because of more abundant food sources. Another is that the extinction of many large marine predators that occurred at the same time as the demise of the dinosaurs allowed mammals to expand into a new environment: the sea. Interestingly, recent research suggests that the rise of diatoms as dominant marine primary producers and global temperature change were key factors that influenced the evolution of modern whales.

Mammalian Characteristics

All organisms in class Mammalia (including marine mammals) share the following characteristics:

- They are warm-blooded.
- They breathe air.
- They have hair (or fur) during at least some stage of their development.
- They bear live young.[7]
- The females of each species have mammary glands that produce milk for their young.

Marine mammals include at least 117 species within the orders Carnivora, Sirenia, and Cetacea. The major groups of marine mammals are shown in **Figure 14.16** and described below.

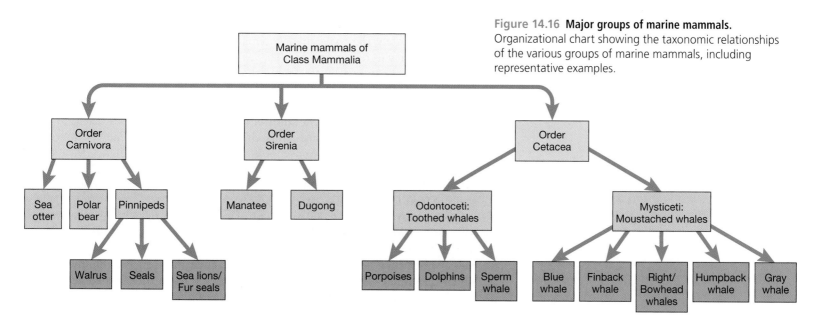

Figure 14.16 Major groups of marine mammals.
Organizational chart showing the taxonomic relationships of the various groups of marine mammals, including representative examples.

---

[7]This is true except for a few egg-laying monotreme mammals of Australia from the subclass Prototheria, which includes the duck-billed platypus and the spiny anteater (echidna).

(a) Sea otter (*Enhydra lutris*).

(b) Polar bear (*Ursus maritimus*).

(c) Walrus (*Odobenus rosmarus*).

(d) Harbor seals (*Phoca vitulina*).

(e) California sea lions (*Zalophus californianus*).

Figure 14.17 Marine mammals of order Carnivora. Photos of various types of marine mammals of order Carnivora, including **(a)** sea otter and **(b)** polar bear. Pinnipeds include **(c)** walrus, **(d)** harbor seals, and **(e)** California sea lions.

Order Carnivora

All animals within order **Carnivora** (*carni* = meat, *vora* = eat)—such as the familiar cat and dog families on land—have prominent canine teeth. Marine representatives of order Carnivora include sea otters, polar bears, and the **pinnipeds** (*pinna* = feather, *ped* = foot), which include walruses, seals, sea lions, and fur seals. The name *pinniped* describes these organisms' prominent skin-covered flippers, which are well adapted for propelling them through water.

Sea otters (Figure 14.17a) inhabit kelp beds in coastal waters of the eastern North Pacific Ocean. They are some of the smallest marine mammals, with adults reaching lengths up to 1.2 meters (4 feet). Sea otters lack an insulating layer of blubber but have extremely dense fur, which is extremely luxurious and was highly sought for pelts; as a result, they were hunted to the brink of extinction in the late 1800s. Fortunately, they have made a remarkable comeback and now inhabit most areas where they were formerly hunted. Sea otters seem particularly playful because of their habit of continually scratching themselves, which serves to clean their fur and adds an insulating layer of air. Because they lack the insulative benefit of blubber, they have high caloric requirements and are voracious eaters.

Sea otters eat more than 50 kinds of marine life, including sea urchins, crabs, lobsters, sea stars, abalone, clams, mussels, octopuses, and fish. They are one of the few types of animals known to use tools. During dives for prey items, they keep a tool—usually a rock—tucked under one arm while they use their dexterous hands to obtain food. When they return to the surface, they use the tool to break open the shells of their food while floating on their backs.

Polar bears (Figure 14.17b) are a type of marine mammal with massive webbed paws that make them excellent swimmers. The polar bear's fur is thick, and each hair is hollow for better insulation. Polar bears also have large teeth and sharp claws, which they use for prying and killing. Their diet consists mainly of seals, which they often capture at holes in the Arctic ice when the seals come up for a breath of air. The topic of shrinking Arctic sea ice and its effect on polar bear populations is discussed in Chapter 16, "The Oceans and Climate Change."

Walruses have large bodies, and adults—both male and female—have ivory tusks up to 1 meter (3 feet) long (Figure 14.17c). Their tusks are used for territorial fighting, for hauling themselves onto icebergs, and sometimes for stabbing their prey.

Skeleton of crabeater seal, *Lobodon* sp.   Skeleton of Steller sea lion, *Eumetopias* sp.

Figure 14.18 Skeletal and morphological differences between seals and sea lions. The most significant differences between seals and sea lions are that seals have smaller fore flippers with visible nails; the hip structure of sea lions allows them to bend their rear flippers underneath their bodies, which allows them good mobility on land; and sea lions have an external ear flap (not shown).

Seals—also called the *earless seals* or *true seals*—differ from the **sea lions** and **fur seals**—also called the *eared seals*—in the following ways:

- Seals lack prominent ear flaps that are specific to sea lions and fur seals. (Look closely and compare **Figures 14.17d** and **14.17e**.)
- Seals have smaller and less-prominent front flippers (called *fore flippers*) than sea lions and fur seals.
- Seals have prominent claws that extend from their fore flippers that sea lions and fur seals lack (**Figure 14.18**).
- Seals have a different hip structure than sea lions and fur seals. As a result, seals cannot move their rear flippers underneath their bodies in the way that sea lions and fur seals do (Figure 14.18).
- Seals, with their smaller front flippers and different hip structure, do not move around on land very well and can only slither along like caterpillars. Sea lions and fur seals, on the other hand, use their large front flippers and their rear flippers, which they can turn under their bodies, to walk easily on land and can even ascend steep slopes, climb stairs, and do other acrobatic tricks.
- Seals propel themselves through the water by using a back-and-forth motion of their rear flippers (similar to a wagging tail), whereas sea lions and fur seals flap their large front flippers.

Order Sirenia

Animals of order **Sirenia** (*siren* = a mythical mermaidlike creature with an enticing voice) include the *manatees* and *dugongs*, collectively known as "sea cows."[8] Manatees are concentrated in coastal areas of the tropical Atlantic Ocean, while dugongs populate the tropical regions of the Indian and Western Pacific Oceans.

Both manatees and dugongs have a paddlelike tail and rounded front flippers (**Figure 14.19**). Their bodies are covered with sparse hairs, which are concentrated around the mouth. They are large animals that can reach lengths of up to 4.3 meters (14 feet) and weigh more than 1360 kilograms (3000 pounds). The land-dwelling ancestors of sirenians were elephantlike; in fact, the front flippers of manatees have prominent nails that bear a striking resemblance to the nails on elephant feet.

(a) West Indian manatees, which have a rounded tail fin and nails on their front flippers.

(b) Indian Ocean dugong, which has a fluked tail fin similar to that of a whale and has no visible nails on its front flippers.

Figure 14.19 Marine mammals of order Sirenia. Photos of various marine mammals of order Sirenia, including **(a)** West Indian manatees and **(b)** Indian Ocean dugong.

---

[8]Sea cows also include the cold-water Steller's sea cow (*Hydrodamalis gigas*), which was driven to extinction in 1768 by early whalers only 27 years after its discovery.

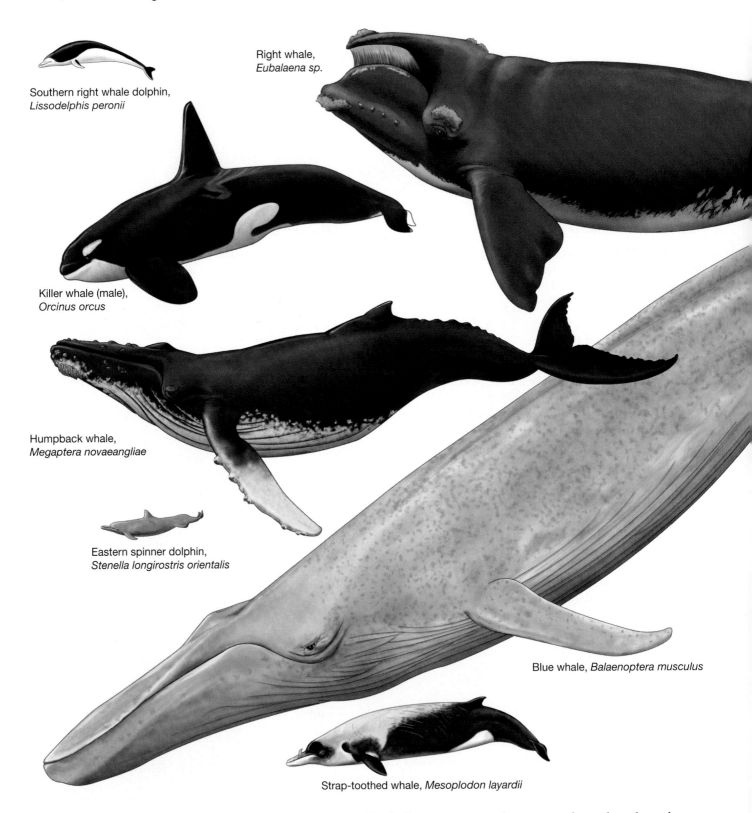

Southern right whale dolphin,
Lissodelphis peronii

Right whale,
Eubalaena sp.

Killer whale (male),
Orcinus orcus

Humpback whale,
Megaptera novaeangliae

Eastern spinner dolphin,
Stenella longirostris orientalis

Blue whale, *Balaenoptera musculus*

Strap-toothed whale, *Mesoplodon layardii*

Figure 14.20 Marine mammals of order Cetacea. A composite drawing of representatives of the two cetacean suborders, Odontoceti (toothed whales) and Mysticeti (baleen whales). All whales and dolphins are drawn to relative scale; also note the human diver in the upper right.

Sirenians eat only shallow-water coastal grasses and are thus the only vegetarian marine mammals. They spend most of their lives in coastal waters that are heavily used by humans for commerce, recreation, development, and waste disposal, and so a major concern for sirenian survival is habitat destruction. A recent worldwide study of biologically sensitive seagrass habitat—which is vital for sirenians—indicates that it is being destroyed so rapidly that it is one of the most threatened ecosystems on Earth. The rate of seagrass loss, in fact, is similar to that of other endangered ecosystems, such as mangroves, coral reefs, and rainforests.

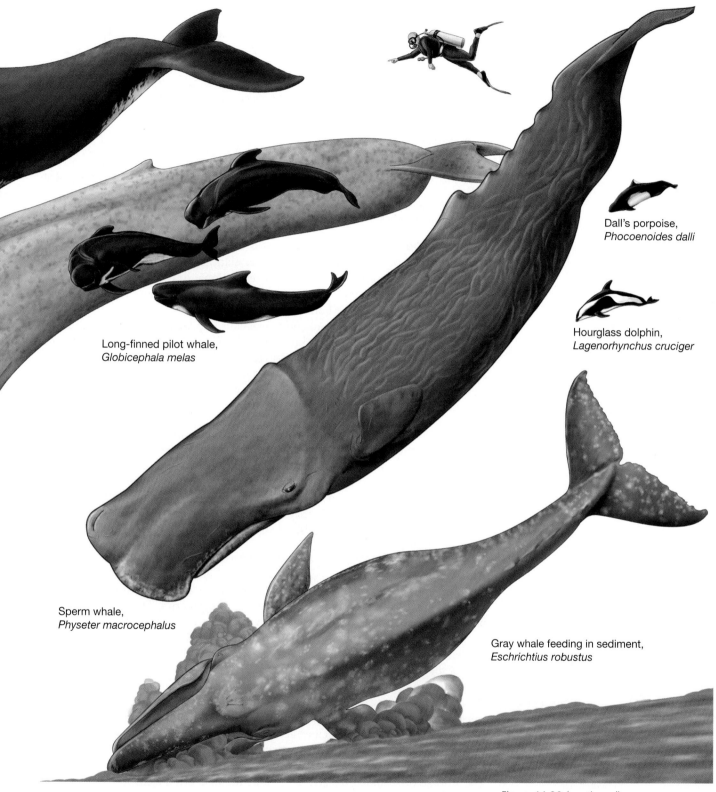

Dall's porpoise,
Phocoenoides dalli

Hourglass dolphin,
Lagenorhynchus cruciger

Long-finned pilot whale,
Globicephala melas

Sperm whale,
Physeter macrocephalus

Gray whale feeding in sediment,
Eschrichtius robustus

Figure 14.20 (continued)

In addition, there have been many accidents with motorboats that run over these slow-moving animals. In 2002, for example, the Florida Fish and Wildlife Commission reported the deaths of 305 Florida manatees, of which 31% were attributed to boat collisions. Scientists have recently developed a forward-looking sonar system called a "manatee finder" to help boaters locate and avoid the animals. The continuing decline in populations of manatees and dugongs, however, has led to their being classified as endangered species.

Web Video
Whale Evolution

Order Cetacea

The order **Cetacea** (*cetus* = whale) includes the whales, dolphins, and porpoises (**Figure 14.20**). The cetacean body is more or less cigar shaped and insulated with a thick layer of blubber. Cetacean forelimbs are modified into flippers that move only at the "shoulder" joint. The hind limbs are vestigial (rudimentary), not attached to the rest of the skeleton, and are usually not visible externally. All cetaceans share the following characteristics:

- An elongated (telescoped) skull
- Blowholes on top of the skull
- Very few hairs
- A horizontal tail fin called a *fluke* that is used for propulsion by vertical movements

These characteristics make cetaceans' bodies very streamlined, allowing them to be excellent swimmers.

MODIFICATIONS TO INCREASE SWIMMING SPEED Cetaceans' muscles are not a great deal more powerful than those of other mammals, so their ability to swim fast must result from modifications that reduce frictional drag. The muscles of a small dolphin, for example, would need to be five times stronger than they are to swim at 40 kilometers (25 miles) per hour in turbulent flow.

In addition to a streamlined body, cetaceans improve the flow of water around their bodies with a specialized skin structure. Their skin consists of a soft outer layer that is 80% water and has narrow canals filled with spongy material, and a stiffer inner layer composed mostly of tough connective tissue. The soft layer decreases the pressure differences at the skin–water interface by compressing when pressure is high and expanding when pressure is low, reducing turbulence and drag.

MODIFICATIONS TO ALLOW DEEP DIVING Humans can free-dive to a maximum depth of 130 meters (428 feet) and hold their breath in rare instances for up to six minutes. In contrast, sperm whales (*Physeter macrocephalus*) dive deeper than 2800 meters (9200 feet), and northern bottlenose whales (*Hyperoodon ampullatus*) can stay submerged for up to two hours. These remarkable feats require unique adaptations, such as special structures that allow them to use oxygen efficiently, muscular adaptations, and an ability to resist nitrogen narcosis.

Oxygen Usage **Figure 14.21** shows the internal structures that allow cetaceans to remain submerged for extended periods. Inhaled air finds its way to tiny terminal chambers, the *alveoli* (*alveus* = small hollow). Alveoli are lined with a thin alveolar membrane that is in contact with a dense bed of capillaries. The exchange of gases between the inhaled air and the blood (oxygen in, carbon dioxide out) occurs across the alveolar membrane. Some cetaceans have an exceptionally large concentration of capillaries surrounding the alveoli (**Figure 14.21b**), which have muscles that move air against the membrane by repeatedly contracting and expanding.

Cetaceans take from 1 to 3 breaths per minute while resting, compared with about 15 in humans. Because they hold the inhaled breath much longer, and because of the large capillary mass in contact with the alveolar membrane and the circulation of the air by muscular action, cetaceans can extract almost 90% of the oxygen in each breath, whereas terrestrial mammals extract only 4 to 20%.

To use oxygen efficiently during long dives, cetaceans store it and limit its use. The storage of so much oxygen is possible because prolonged divers have such a large blood volume per unit of body mass.

Some cetaceans have twice as many red blood cells per unit of blood volume and up to nine times as much myoglobin in their muscle tissue as terrestrial animals. As a result, large supplies of oxygen can be stored chemically in **hemoglobin** (*hemo* = blood, *globus* = sphere) within red blood cells and in myoglobin within muscles.

Muscular Adaptations Cetaceans' muscles are well-adapted for deep dives. One adaptation is that their muscle tissue is relatively insensitive to high levels of carbon dioxide, which builds up in the body through respiration, especially during deep

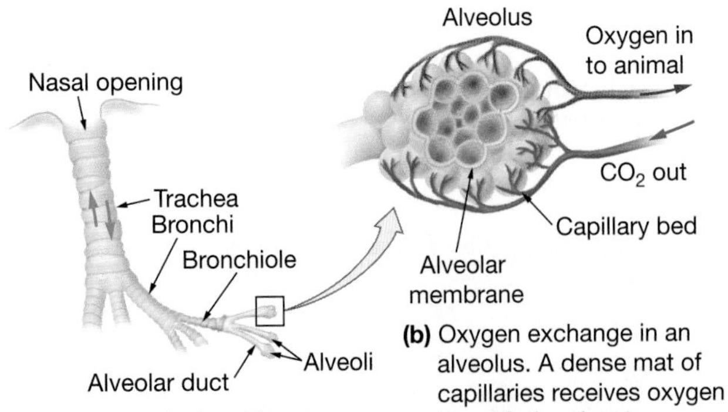

(b) Oxygen exchange in an alveolus. A dense mat of capillaries receives oxygen through the alveolar membrane, allowing whales to extract as much as 90% of oxygen from each breath.

(a) Basic lung design. Air enters the lung through the trachea, and oxygen is absorbed into the blood through the walls of the alveoli.

Figure 14.21 Internal adaptations of cetaceans that facilitate prolonged submergence. Diagrammatic figure showing internal adaptations of cetaceans that enable them to stay underwater for extended periods, including **(a)** basic lung design and **(b)** oxygen exchange in the alveolus.

dives. Another adaptation is that their muscles can continue to function through anaerobic respiration when oxygen becomes depleted.

In addition, research has shown that cetaceans' swimming muscles can still function during a dive, even in the absence of oxygen. This suggests that these muscles and other organs, such as the digestive tract and kidneys, may be sealed off from the circulatory system by constriction of key arteries. The circulatory system would then service only essential components, such as the heart and brain. Because of the decreased circulatory requirements, the heart rate can be reduced by 20 to 50% of normal. Other research has shown, however, that no such reduction in heart rate occurs during dives by the common dolphin (*Delphinus delphis*), the white whale (*Delphinapterus leucas*), or the bottlenose dolphin (*Tursiops truncatus*).

Do Cetaceans Suffer from the Effects of Deep Diving? One difficulty with deep and prolonged dives is the absorption of compressed gases into the blood. When humans dive using compressed air—which includes nitrogen and oxygen—the higher pressure at depth causes more nitrogen to be dissolved in a diver's body and can cause divers to experience **nitrogen narcosis**, which is also called *rapture of the deep*. The effect of nitrogen narcosis is similar to drunkenness and can occur when a diver either goes too deep or stays too long at depths greater than 30 meters (100 feet) (see Diving Deeper 12.1).

Another difficulty can occur when divers return to the surface too rapidly, in which case they may experience **decompression sickness**, which is also called *the bends*. During a rapid ascent, the lungs cannot remove excess gases from the bloodstream fast enough, and the reduced pressure causes small bubbles of nitrogen to form in a diver's blood and tissue. This process is analogous to the bubbles that form in a carbonated beverage when the container is opened. The bubbles interfere with blood circulation and can cause bone damage, excruciating pain, severe physical debilitation, or even death.

Until recently, it was believed that cetaceans and other marine mammals had adaptations that prevented them from suffering from the effects of deep diving. However, research involving a rigorous examination of sperm whale skeletons reveals that sperm whales acquire progressive bone damage caused by recurring decompression sickness. The researchers conclude that sperm whales are neither anatomically nor physiologically immune to the effects of deep diving.

Still, the debilitating effects of deep diving are minimized in cetaceans because they have collapsible rib cages. By the time a cetacean has reached a depth of 70 meters (230 feet), its rib cage has collapsed under the 8 kilograms per square centimeter (8 atmospheres, or 118 pounds per square inch) of pressure. The lungs within the rib cage also collapse, removing all air from the alveoli. This, in turn, prevents the blood from absorbing additional gases across the alveolar membrane, minimizing nitrogen narcosis.

In addition, cetaceans may be naturally resilient to the buildup of nitrogen gas in their bodies. In a study where enough nitrogen was put into the tissue of a dolphin to give a human a severe case of the bends, for instance, the dolphin appeared to suffer no ill effects. This suggests that dolphins (as well as other marine mammals) may have simply evolved an insensitivity to the buildup of excess nitrogen gas.

SUBORDER ODONTOCETI Members of order Cetacea can be divided into two suborders: Odontoceti (the *toothed whales*) and Mysticeti (the *baleen whales*). Suborder **Odontoceti** (*odonto* = tooth, *cetus* = whale) includes the dolphins, porpoises, killer whales, and sperm whales.[9]

Characteristics of Toothed Whales All toothed whales have prominent teeth that are used to hold and position fish and squid to facilitate swallowing them whole. Killer whales, however, are known to feed on a variety of larger animals, including other whales. The toothed whales form complex and long-lived social groups. A toothed whale has one external nasal opening (blowhole), while a baleen whale has

---

[9]Recent genetic analyses suggest that sperm whales are more closely related to baleen whales than to other toothed whales.

STUDENTS SOMETIMES ASK . . .

I've heard of the problem with military sonar causing whale strandings. How does sonar harm cetaceans?

There is some recent evidence linking mass strandings of cetaceans—predominately beaked whales, a group of deep-diving whales that have been very poorly studied—with the deployment of mid-frequency sonar used for detecting submarines during military exercises. Although the link is circumstantial, researchers have documented gas-bubble lesions in stranded cetaceans consistent with rapid decompression that is normally associated with decompression sickness. It is hypothesized that the sonar may be responsible for the formation of these bubbles by causing the cetaceans to ascend to the surface too rapidly or by frightening them into undertaking dangerously deep dives; it is also possible that the sonar itself has physical effects on cetaceans' nitrogen-saturated tissues. Recent research has shown that beaked whales stop their echolocations and flee the area when sonar is in use, suggesting that these whales may be more sensitive than other species to sound. Other strandings, however, have occurred during times when military sonar is not in use but stranded animals are still found to have similar bubbles in their tissues. So, the relationship between military sonar and strandings may be just an unfortunate coincidence. Clearly, more research is needed in this important area.

Documented whale strandings have occurred since as far back as the time of Aristotle, which implies that many strandings are a natural phenomenon. For example, strandings due to pneumonia and trauma after storms are common reasons. Other reasons for strandings are obvious, such as shark attacks or even assaults by members of the same species. In other cases, both human-derived pollutants and natural toxins, such as biotoxins from algae (see Chapter 13), are implicated in mass strandings. Parasites and pathologies—including bacterial and viral infections—are sometimes to blame, too. Even using diagnostic techniques common in advanced human medicine, such as CT and MRI scanning and molecular studies, there are still many instances of whale strandings with no single, clear answer. For example, of 55 cetacean strandings in the United States since 1991, scientists have classified 29 as "undetermined."

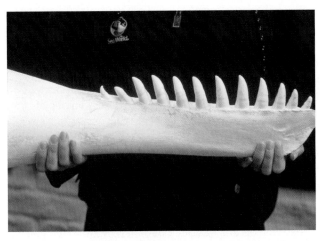

Figure 14.22 Jawbone of a killer whale. The lower jawbone of a killer whale (*Orcinus orca*), showing large teeth that end in points. Thus, killer whales are in the dolphin family.

Web Video
Killer Whales Feeding on Sea Lions in South America

Web Video
The Hunting Sperm Whale

two. Although both toothed and baleen whales can emit and receive sounds, the ability to use sound is best developed in toothed whales (particularly sperm whales, which are the most vocal cetaceans).

Differences between Dolphins and Porpoises Dolphins and porpoises are small toothed whales of suborder Odontoceti. They have similarities in appearance, behavior, and range and so are easily confused. For instance, both dolphins and porpoises (as well as seals, sea lions, and fur seals) can exhibit a behavior known as "porpoising," which is leaping out of the water while swimming. However, there are several morphological differences between dolphins and porpoises.

Porpoises are somewhat smaller and have a more stout (bulky and robust) body shape than dolphins, which are more elongated and streamlined. Generally, porpoises have a blunt snout (rostrum), while dolphins have a longer rostrum. Porpoises have a smaller and more triangular (or, on one species, no) dorsal fin, whereas a dolphin's dorsal fin is sickle-shaped, or **falcate** (*falcatus* = sickle), and appears hooked and curved backward in profile view.

Dolphins and porpoises also have differences in the shape of their teeth, although it is often difficult to get close enough to see them. The teeth of dolphins end in points, while the teeth of porpoises are blunt or flat (shovel shaped) and resemble our incisors (front teeth). Killer whales have teeth that end in points (Figure 14.22), confirming that they are indeed members of the dolphin family.

Echolocation and Hearing in Toothed Whales All marine mammals have good vision, but ocean conditions often limit its effectiveness. In coastal waters (where suspended sediment and dense plankton blooms make the water turbid) and in deeper waters (where light is limited or absent), the use of sound—which travels well in water—has many advantages over sight.

Despite their lack of vocal cords, toothed whales can produce a variety of sounds—some of which are within the range of human hearing. Speculations about the sounds' purpose range from **echolocation**—using sound to determine the direction and distance of objects—(clearly true) to a highly developed language (doubtful). In fact, what marine biologists know about cetaceans' use of sound is limited.

Sound generation is a complex process. In small toothed whales such as dolphins and porpoises (Figure 14.23a), sounds are emitted from the blowhole. Contractions of muscles produce sounds that reflect off the top of the skull, which is bowl shaped and resembles a radar dish. The sound is concentrated as it passes through an organ that sits atop the skull called the **melon**, which can form various lenslike shapes and sizes and act as an acoustical lens, focusing the sound before it leaves the body.

In sperm whales (Figure 14.23b), sounds are generated when the whale forces air through its right nasal passage to a special structure called the *museau du singe* ("monkey's muzzle"), which claps shut. The resulting *click!* travels through the **spermaceti** (*sperm* = seed, *cetus* = whale) **organ**[10] to the skull. From there, the sound is reflected forward through another organ called *the junk* and amplified out into the watery world. Research suggests that sperm whales are able to manipulate the shape of both the spermaceti organ and the junk, allowing them to focus the sound and perhaps even to aim their clicks.

Using lower-frequency clicks at great distance and higher frequency at closer range, the bottlenose dolphin (*Tursiops truncatus*) can detect a school of fish at distances exceeding 100 meters (330 feet). It can pick out an individual fish 13.5 centimeters (5.3 inches) long at a distance of 9 meters (30 feet). Sperm whales can detect their main prey—squid—from distances of up to 400 meters (1300 feet).

To locate an object and determine its distance, toothed whales send sound signals through the water, some of which are reflected from various objects and are returned to the animal and interpreted (Figure 14.24). Because sound penetrates

---

[10]The spermaceti organ is composed of a white, waxy substance; early whalers noticed its similarity to human sperm, hence the name. The spermaceti organ, however, has nothing to do with reproduction but instead is used for echolocation.

objects, echolocation can produce a three-dimensional image of the object's internal structure and density (which is more than eyesight alone can do). Recent research suggests that some toothed whales may also use a sharp burst of sound to stun their prey before they close in for the kill.

How do toothed whales hear underwater? Toothed whales have specialized fats associated with their jaws that efficiently convey sound waves from the ocean to their ears. These structures insulate the inner ear housing from the rest of the skull, thereby allowing them to differentiate the sounds they hear underwater. In many toothed whales, sounds are picked up by the thin, flaring jawbone and passed to the inner ear via the connecting, fat-filled body.[11] The signals are then sent to the brain, where the sounds are interpreted.

There is growing concern that noise pollution in the ocean is affecting cetaceans. This increased noise comes from the greater number of ships plying the world's oceans as well as the larger size, higher speeds, and enhanced propulsion power of individual ships. Over the past 50 years, the world's growing trade fleet has contributed to a 32-fold increase in low-frequency noise in some parts of the ocean, which represents a doubling of noise every decade. The impact of this increased underwater noise on cetacean hearing, behavior, and communication is unknown.

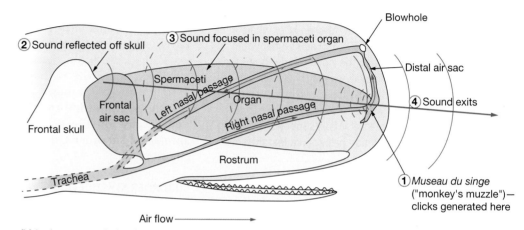

(a) In small toothed whales, clicks are generated within the blowhole, reflected off the frontal surface of the skull, and then focused by the melon, which can form lenses to focus the sound.

(b) In the sperm whale, air passes from the trachea through the right nasal passage across the *museau du singe* ("monkey's muzzle"), where clicks are generated. The air passes along the distal air sac, past the closed blowhole, and returns to the lungs along the left nasal passage. The sounds are reflected off the frontal surface of the skull and are focused by the spermaceti organ before leaving the whale.

Figure 14.23 Generation of echolocation clicks in small toothed and sperm whales. Diagrammatic figure showing how echolocation clicks are generated in **(a)** small toothed whales and **(b)** a sperm whale.

How Intelligent Are Toothed Whales?

The question of cetacean intelligence is a topic of much debate. Although there may not be a definitive answer, the following facts about toothed whales imply a certain level of intelligence:

- They can communicate with each other by using sound.
- They have large brains relative to their body size.[12]

Figure 14.24 Echolocation. Toothed whales generate sounds that bounce off objects in the ocean to determine their size, shape, distance, movement, density, and even internal structure.

---

[11]To simulate this, try pushing the end of a vibrating tuning fork into your chin. The sound is transmitted through your jaw directly to your ear.

[12]In fact, sperm whales have the largest brain of any animal on the planet—it has been reported to weigh up to 9 kilograms (20 pounds), which is over six times the weight of a typical human brain.

- Their brains are highly convoluted—a characteristic shared by many organisms that are considered to have highly developed intelligence (such as humans and other primates).
- Some wild dolphins have been reported to assist drowning humans in the ocean.
- Some dolphins have been trained to respond to hand signals and do tricks on command (such as retrieve objects).

Although toothed whales have remarkable abilities, this does not necessarily imply intelligence. Pigeons, for instance, which are not known for being highly intelligent, have also been trained to retrieve objects by using hand signals. Perhaps many of us would like to think that whales and dolphins are more intelligent than they really are because humans feel an attachment to these charismatic, seemingly ever-smiling, air-breathing creatures. It is interesting to note that even experts in the field of animal intelligence disagree on how to assess *human* intelligence accurately, let alone that of a marine mammal.

If the large brain that toothed whales possess is not an indication of intelligence, then why is their brain so large? Leading whale researchers don't exactly know, but it might be because toothed whales need a large brain to process the wealth of information they receive from the sound echoes they transmit. Because intelligence is difficult to measure, perhaps it is best to say that animals of suborder Odontoceti are tremendously well adapted to the marine environment.

SUBORDER MYSTICETI   Suborder **Mysticeti** (*mystic* = moustache, *cetus* = whale)—also known as the *baleen whales*—includes the world's largest whales (the blue whale, finback whale, and humpback whale) and the gray whale (a bottom feeder).

Baleen whales are generally much larger than toothed whales because of differences in food sources. Baleen whales eat lower on the food web (including zooplankton, such as krill and small nektonic organisms), which are relatively abundant in the marine environment. How are the largest whales in the world able to survive on eating such small prey, especially when these smaller organisms are widely dispersed in the marine environment?

Use of Baleen   To concentrate small prey items and separate them from seawater, baleen whales have parallel rows of **baleen** (*balaena* = whale) plates in their mouths instead of teeth (Figure 14.25a). These baleen plates hang from the whale's upper jaw and, when the whale opens its mouth, the baleen resembles a moustache (except that it is on the *inside* of their mouths), which is why these whales are sometimes called *moustached whales* (Figures 14.25c and 14.26). Baleen is made of flexible keratin—the same as human nails and hair—and can be up to 4.3 meters (14 feet) long (Figure 14.25d).[13]

To feed, the largest baleen whales fill their mouths with a body-weight of water and prey (Figure 14.25b), allowing their pleated lower jaw to balloon in size (Figure 14.26). The whales force the water out between the fibrous plates of baleen, trapping small fish, krill, and other plankton inside their mouths. Mostly, baleen whales feed at or near the surface, sometimes working together in large groups and surfacing in vertical lunges (Figure 14.27). The gray whale, however, has short baleen slats and feeds by straining benthic organisms such as amphipods and shellfish from bottom sediment.

Baleen Whale Families   Baleen whales are grouped into the following three families:

1. The **gray whale**, which has short, coarse baleen, no dorsal fin, and only two to five ventral grooves on its lower jaw. The gray whale is a bottom-feeder.

STUDENTS SOMETIMES ASK . . .

In a battle between a killer whale and a great white shark, which one would win?

Although many people who are fascinated with large and powerful wild animals have often wondered which of the two would win such a fight, there was little evidence to settle the dispute until recently. A remarkable video was taken in waters off northern California in 1997, documenting a battle between a 6-meter (20-foot) juvenile killer whale (*Orcinus orca*) and a 3.6-meter (12-foot) adult great white shark (*Carcharodon carcharias*). The video clearly shows the killer whale biting and completely severing the shark's head! If this is representative of the way these two animals interact in the wild, then the killer whale is the top carnivore in the ocean. It is believed that the killer whale's superior maneuverability and use of echolocation helped it defeat the shark.

---
[13]Baleen (also called whalebone) was used for such items as buggy whips and corset stays before synthetic materials—mostly plastics—were substituted.

Figure 14.25 Baleen.
Diagrammatic views of **(a)** a cross-section through the head of a typical baleen whale and **(b)** how a baleen whale feeds. Photos of **(c)** a rack of gray whale baleen and **(d)** an individual slat of right whale baleen.

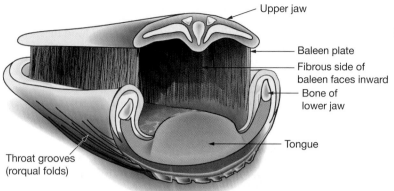

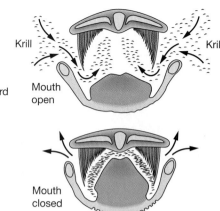

(a) Cross-sectional view through the head of a baleen whale, showing the baleen plates hanging from the upper jaw, which form a sieve that allows these whales to concentrate and eat large quantities of smaller organisms.

(b) A baleen whale takes krill and other small prey into its mouth (*above*) and then closes its mouth to force the water through its baleen (*below*), capturing its prey.

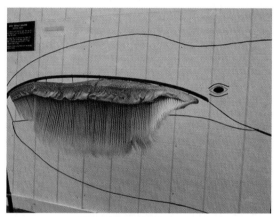

(c) A rack of baleen from a bottom-feeding gray whale (*Eschrichtius robustus*).

(d) Individual slat of baleen from a surface-skimming North Atlantic right whale (*Eubalaena glacialis*).

Web Animation
Feeding in Baleen Whales

Figure 14.26 Feeding by rorqual whales. Rorqual whales such as this Bryde's whale (*Balaenoptera edeni*) off Baja California, Mexico, feed by engulfing an entire patch of prey with their huge mouths. They fill their mouths with a body's weight of water and prey that expands their pleated throat and then sift their prey from seawater using their baleen plates.

Figure 14.27 Humpback whale bubblenet feeding in Alaskan waters. Humpback whales (*Megaptera novaeangliae*) often feed by swimming in a circle underwater and emitting a curtain of bubbles to corral prey. Then each whale swims through the concentrated prey with their mouth open and surfaces in a vertical lunge. Baleen plates are used to filter prey items from the water and occur as parallel rows on the upper jaw that are separated by the roof of the mouth (*pink*). Also visible on the head are round bumps, which are called nodules or tubercules, each holding a single long hair. Whalers called these bumps *stove bolts* because they looked like the stove bolts that hold a ship together.

ESSENTIAL CONCEPT 14.4

Marine mammals include orders Carnivora (sea otters, polar bears, and pinnipeds—walrus, seals, sea lions, and fur seals), Sirenia (manatees and dugongs), and Cetacea (whales, dolphins, and porpoises).

Web Animation

Humpback Whales
Bubblenet Feeding

2. The **rorqual**[14] **whales**, which have short baleen, many ventral grooves, and feed by gulping large mouthfuls of water containing small food items. The rorqual whales are divided into two subfamilies:

   a. The *balaenopterids*, which have long, slender bodies, small sickle-shaped dorsal fins, and flukes with smooth edges (minke, Bryde's, sei, fin, and blue whales).

   b. The *megapterids*, or humpback whales, which have a more robust body, long flippers, flukes with uneven trailing edges, tiny dorsal fins, and nodules or tubercles (bumps that each hold a single long hair) on the head.

3. The **right whales**,[15] which have long, fine baleen, broad triangular flukes, no dorsal fin, and no ventral grooves. Of the four species of right whales, the North Atlantic right whale and the North Pacific right whale are among the most critically endangered whales in the world. A third species of right whale, the Southern right whale, inhabits the Southern Ocean. The fourth species of right whale is the bowhead whale, which lives near the edge of the Arctic pack ice. Right whales feed by moving through surface waters with their mouths open, skimming for small surface-dwelling prey.

Baleen Whales and Sounds Baleen whales produce sound, too, but at much lower frequencies than toothed whales. For example, gray whales produce pulses (possibly for echolocation) and moans that may help them maintain contact with other gray whales. Rorqual whales produce moans that last from one to many seconds. These sounds are extremely low in frequency and are probably used to communicate over distances of up to 50 kilometers (31 miles). Blue whales have been documented to produce sounds that may travel along the SOFAR channel across entire ocean basins.[16] Songs of humpback whales are thought to be a form of sexual display, but it is unclear whether their main purpose is to repel other males or to attract females.

CONCEPT CHECK 14.4

1 What common characteristics do all organisms in class Mammalia share?

2 Describe marine mammals within the order Carnivora, including their adaptations for living in the marine environment.

3 How can true seals be differentiated from the eared seals (sea lions and fur seals)?

4 Describe the marine mammals within the order Sirenia, including their distinguishing characteristics.

5 How can dolphins be differentiated from porpoises?

6 Describe differences between cetaceans of the suborder Odontoceti (toothed whales) with those of the suborder Mysticeti (baleen whales). Be sure to include examples from each suborder.

7 Describe the process by which the sperm whale produces echolocation clicks.

8 Discuss how sound reaches the inner ear of toothed whales.

9 Describe the mechanism by which baleen whales feed.

---

[14]The term *rorqual* refers to the longitudinal grooves on the lower jaw called rorqual folds; *rorqual* means "furrow whale" in Norwegian.

[15]The right whales are so named because they swim slowly, live close to shore, float when dead, and yield plenty of oil, meat, and whalebone, so, from a 19th century whaler's point of view, they were the "right" whale to take.

[16]For more details about the SOFAR channel, see Diving Deeper 16.1.

14.5 An Example Of Migration: Why Do Gray Whales Migrate?

Many marine organisms—such as fish, squid, sea turtles, and marine mammals—undertake seasonal migrations. Some of the longest migrations known in the open ocean are those of baleen whales, and one of the best-studied is that of the gray whale (*Eschrichtius robustus*, often called the Pacific gray whale or California gray whale). Gray whales, which are a medium-sized, slow-moving coastal whale with a mottled gray appearance, can live 60 years, grow to 15 meters (50 feet) in length, and weigh up to 36 metric tons (40 short tons).

Migration Route

The migratory routes of commercially important baleen whales (including gray whales) have been well known since the mid-1800s. Marine mammals are relatively easy to track because they must surface periodically for air. Gray whales are particularly easy to track because they spend their entire lives in nearshore areas. Radio tracking of individual gray whales as well as visual tracking of gray whales from shore has enabled scientists to delineate the route and timing of gray whale migration.

Most gray whales undertake a 22,000-kilometer (13,700-mile) round-trip journey every year, the longest-known migration of any mammal. These gray whales feed primarily on sea floor organisms in cold high-latitude waters in the coastal Arctic Ocean and the far northern Pacific Ocean, near Alaska; they breed and give birth in warm tropical lagoons along the west coast of Baja California and mainland Mexico (**Figure 14.28**). Feeding occurs during the highly productive summer, when long hours of sunlight produce an abundant feast of crustaceans, clams, and other bottom-dwelling organisms in the shallow portions of the Bering, Chukchi, and Beaufort Seas. Without this bountiful food, the whales could not sustain themselves during the long migrations and the mating and calving season, when feeding is minimal.

Although most gray whales migrate, a resident population of about 200 gray whales stays along the eastern Pacific coast from Canada to northern California throughout the summer, not making the farther trip to Alaskan waters. This summer resident group takes advantage of the abundance of food along the coast caused by coastal upwelling. These whales are known as the *Pacific coast Feeding Group* (Figure 14.28).

Reasons for Migration

Initially, gray whales were thought to migrate so far because the physical environment of their cold-water feeding grounds did not meet the needs of young gray whales. Recent research on the physiology of newborn gray whales indicates, however, that gray whale calves can survive in much colder water. So why do they migrate? One hypothesis is that the migration is a relic from the Ice Age, when sea level was lower. During that time, the feeding grounds that are so productive today were above sea level. Hence, gray whales could not feast on the abundant food and probably gave birth to smaller calves that could not survive in the cold water. This necessitated the migration to warmer water regions, which continues to this day despite the abundant food supply.

Alternatively, gray whales may have left the colder waters to avoid killer whales, which are more numerous there and are a major threat to young whales. This may also explain why only those lagoons in Mexico with shallow entrances are used for calving. Killer whales have been seen near the lagoons, however, and have also been observed to feed in extremely shallow water (see Web Figure 14.1). In fact, killer whales take an estimated 30% of gray whale calves born each year.

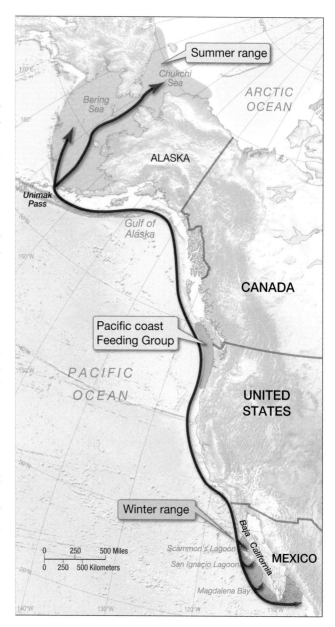

Figure 14.28 Gray whale migration route. Gray whales (*Eschrichtius robustus*) undertake as much as a 22,000-kilometer (13,700-mile) annual migration, the longest of any mammal. Most gray whales migrate from Arctic summer feeding areas in the Bering and Chukchi Seas to warmer winter breeding and calving lagoons offshore Mexico.

Timing of Migration

The timing of the gray whale migration is closely linked to physical oceanographic conditions. The migration usually begins in September, after the high-latitude summer bloom in productivity has peaked. By this time, the whales have stored enough fat to last until they return to high-latitude waters. Gray whales have been observed to feed during their migration, however, when the opportunity presents itself. When Arctic pack ice begins to form over their feeding grounds along the continental shelf, the whales begin to move south.

Pregnant females are the first to leave on the 2-month journey south. They are followed in steady succession by nonpregnant mature females, immature females, mature males, and then immature males. After navigating through passages between the Aleutian Islands, they follow the coast throughout their southbound journey. Traveling about 200 kilometers (125 miles) per day, most reach the lagoons of Baja California by December and early January; the peak birthing season is the middle of January.

In these warm-water lagoons, the pregnant females give birth to calves that are about 4.6 meters (15 feet) long and weigh about 1 metric ton (1.1 short tons). The calves nurse on milk that is almost half butterfat and has the consistency of cheese, allowing the calves to put on weight quickly during the next two months. While the calves are nursing, the mature males breed with the mature females that did not bear calves. Producing large offspring (the gestation period is up to one year) and providing them with fat-rich milk for several months requires enormous amounts of energy, so it is not uncommon for females to give birth only once every two or three years.

Beginning in late February, gray whales embark on their northbound journey beginning with males and females without new calves. Pregnant females and nursing mothers with their newborns are the last to depart, leaving only when their calves are ready for the journey, which is usually from late March to mid-April. Often, a few mothers linger in the protective lagoons with their young calves well into May. Depending on the extent of Arctic pack ice, early-arriving whales may have a difficult time getting to the most productive feeding areas. Most of the whales are back in their high-latitude feeding grounds by the end of June, when shallow continental shelf waters are ice-free and the high-latitude summer productivity bloom has begun. The whales are able to feed once again on the prodigious quantities of bottom-dwelling organisms found here to replenish their depleted stores of fat and blubber before their next yearly trip south.

Are Gray Whales an Endangered Species?

Although many other whales exist in smaller populations than before whaling (Figure 14.29) and are endangered species, the North Pacific gray whale was removed from the endangered species list in 1993, when their numbers exceeded 20,000, which surpassed the estimated size of their population before whaling. Out of nearly 1400 species listed as endangered since 1973, gray whales have become one of only 13 species to be removed from the endangered species list. Other populations of gray whales were not so fortunate. For instance, the gray whales that used to inhabit the North Atlantic Ocean were hunted to extinction several centuries ago, and the gray whales that live in the Western Pacific near Japan are critically endangered (less than 130 whales including only about 30 reproductive females). Recent research indicates that there is some intermixing between the populations of Western Pacific and Eastern Pacific gray whales. And, a confirmed sighting in 2010 of a gray whale in the Mediterranean Sea off Israel suggests that gray whales may be expanding their range into old breeding grounds that have not been used for centuries.

These slow-moving whales of suborder Mysticeti spend most of their lives in coastal waters, which made them easy targets for whalers. In the mid-1800s, they were traced to their birthing and breeding lagoons and were hunted to the brink of extinction. A common strategy was to harpoon a calf, which then lured its vigilant

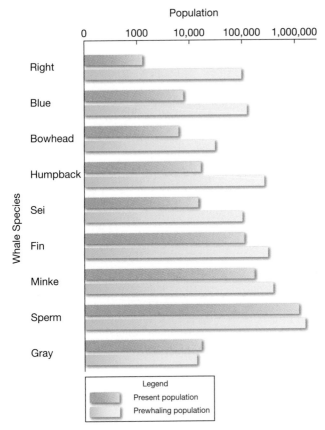

Figure 14.29 World whale populations (present and prewhaling). Bar graph showing present world whale populations (*red*) compared to prewhaling populations (*green*) for various species. Note that population size is shown on a logarithmic scale. Full-scale commercial whaling was suspended in 1986, when the plunging numbers of many whale species moved the International Whaling Commission to declare a moratorium. Today, however, certain countries are permitted to catch designated species for research or cultural reasons.

mother to the whalers, too. During this time, gray whales were known as "devilfish" because the adults would often capsize small whaling boats when they came to the aid of their young. By the late 1800s, the number of gray whales had diminished to the point that they were difficult to find during their annual migration. Fortunately, these low numbers also made it difficult for whalers to hunt them successfully.

In 1938, the International Whaling Treaty banned the taking of gray whales, which were thought to be nearly extinct. This protection has allowed them to steadily increase in number to this day and to become the first marine creature to be removed from endangered status. Their repopulation is truly one of the most impressive success stories of how protecting animals can ensure their continued survival. What is perhaps most surprising is how friendly they are now toward people in boats (Figure 14.30) in the same lagoons where they were hunted to near extinction over 150 years ago. Devilfish, indeed!

Whaling and the International Whaling Commission

From the mid-1800s to the mid-1900s, many stocks of large whales in the world were overhunted and their numbers severely depleted (see Figure 14.29). The **International Whaling Commission (IWC)** was established in 1946 to manage the subsistence and commercial hunting of large whales. In 1986, the 72 nations that are members of the IWC passed a ban on commercial whaling in order to allow whales to recover from overhunting and to give researchers time to develop methods for assessing whale populations. Today, the ban on whaling is still in effect, although some countries such as Japan, Norway, and Iceland have caught more than 33,000 whales since the 1986 whaling moratorium. Recently, there has been a push by some nations to end the ban on whaling and establish annual whale quotas. The ban can be reversed only by a three-quarters majority vote of member nations.

According to the IWC, there are currently three ways to engage in legal hunting of whales:

1. **Whaling by objection.** Countries can continue to hunt whales legally under an objection to the IWC ban on commercial whaling. (Notable countries doing this: Norway, Iceland.)
2. **Scientific whaling.** Countries may decide to kill whales for scientific research. Under the IWC convention, meat from whales hunted for this research may be sold commercially. (Notable countries doing this: Japan.[17])
3. **Aboriginal subsistence whaling.** Subsistence whaling by native cultures is allowed by the IWC in special cases where there is a long-standing cultural tradition of whaling and where whale meat satisfies the nutritional needs of the community. (Notable countries doing this: Greenland, Russia, the United States, and the Caribbean nation of Saint Vincent and the Grenadines.)

CONCEPT CHECK 14.5
1. Discuss reasons why gray whales leave their cold-water feeding grounds during the winter season.
2. Is the gray whale an endangered species? Explain.
3. Why did the International Whaling Commission (IWC) invoke a ban on commercial whaling?
4. What three ways exist to legally hunt whales? Which countries are doing each?

Figure 14.30 **Gray whale friendly behavior.** Gray whales (*Eschrichtius robustus*) exhibit friendly behavior by approaching a boat and initiating contact in Scammon's Lagoon, Baja California, Mexico.

STUDENTS SOMETIMES ASK . . .

How do whales, you know, DO IT?!

Whale sex is truly an incredible achievement. Imagine trying to copulate as whales do when you have nothing with which to grasp a mate and, in the ocean, you have nothing to push against for support! Copulation in large whales has been infrequently studied and witnessed in only a remarkably few cases. In fact, the world's largest whale, the blue whale, has never been observed copulating.

From coastal-dwelling whale species such as the gray whale, here's what is known: Whales are one of the few animals besides humans and bonobo chimpanzees that copulate belly-to-belly. Mating begins when the female is in a forward position horizontally near the water surface. She turns toward the male, who emerges from below and enters her as they spiral on their long axis through the water, which helps push them together. As with many other mammals, mating takes only a few seconds. Both a whale's penis and its testes are internal to enhance streamlining, but the tapered fibro-elastic tip of the penis is especially suited for the tricky act of copulation with no limbs. Once it is extruded outside the body for mating, it is dexterous enough to maneuver itself into a female's genital slit to deposit sperm.

ESSENTIAL CONCEPT 14.5
Gray whales undertake the longest migration of any mammal, traveling from high-latitude Arctic summer feeding areas to low-latitude winter birthing and breeding lagoons in Mexico.

[17]In 2007, Japan had planned to kill nearly 1000 humpback, fin, and minke whales in South Pacific waters for scientific purposes, but international pressure reduced the take to about 550 minkes.

Essential Concepts Review

14.1 How are marine organisms able to stay above the ocean floor?

▶ *Pelagic animals that comprise the majority of the ocean's biomass remain mostly within the upper surface waters of the ocean*, where their primary food source exists. Those animals that are not planktonic (floating forms such as microscopic zooplankton) depend on *buoyancy* or their *ability to swim* to help them remain in food-rich surface waters.

▶ *The rigid gas containers in some cephalopods and the expandable swim bladders in some fish help increase buoyancy.* Other organisms maintain their positions near the surface with *gas-filled floats* (such as those of the Portuguese man-of-war) and *soft bodies that lack high-density hard parts* (such as the jellies).

▶ *Nekton—squid, fish, and marine mammals—are strong swimmers* that depend on their swimming ability to avoid predators and obtain food. Squid swim by trapping water in their body cavities and forcing it out through a siphon. Most fish swim by creating a wave of body curvature that passes from the front of the fish to the back and provides a forward thrust.

▶ *The caudal (rear) fin provides the most thrust, while the paired pelvic and pectoral (chest) fins are used for maneuvering. The dorsal (back) and anal fins serve primarily as stabilizers.* A rounded caudal fin is flexible and can be used for maneuvering at slow speeds. The lunate fin is rigid and is of little use in maneuvering but produces thrust efficiently for fast swimmers such as tuna.

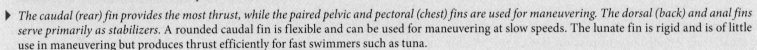

Study Resources
Online Study Guide Quizzes, Web Video

Critical Thinking Question

Explain how a swim bladder works.

14.2 What adaptations do pelagic organisms possess for seeking prey?

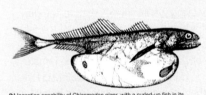

(b) Ingestion capability of *Chiasmodon niger*, with a curled-up fish in its stomach that is longer than it is.

▶ *Fish can be categorized as lungers* (such as groupers) or *cruisers* (such as tuna). Lungers sit motionless and lunge at passing prey. They have mostly white muscle tissue, which fatigues more quickly than red muscle tissue. Cruisers swim constantly in search of prey and possess mostly red, myoglobin-rich muscle tissue.

▶ *Fish swim slowly when cruising, fast when hunting for prey, and fastest when trying to escape from predators.* Although *most fish are cold-blooded*, the fast-swimming tuna *Thunnus* is homeothermic, meaning that it maintains its body temperature well above water temperature.

▶ *Deep-water nekton have special adaptations—such as good sensory devices* and *bioluminescence—*that allow them to survive in this still and completely dark environment. *Bioluminescence—*the ability to organically produce light—*has many uses in the deep ocean.*

Study Resources
Online Study Guide Quizzes, Web Video

Critical Thinking Question

What are the major structural and physiological differences between the fast-swimming cruisers and lungers that patiently lie in wait for their prey?

14.3 What adaptations do pelagic organisms possess to avoid being prey?

(b) Mutualism occurs when both participants benefit, such as this clown anemonefish and sea anemone.

▶ *Many marine organisms such as fish, squid, and crustaceans exhibit schooling*, probably because it increases their chances of avoiding predation compared to swimming alone and serves to preserve the species. Some organisms live closely together in *symbiotic relationships*.

Study Resources
Online Study Guide Quizzes

Critical Thinking Question

List several reasons so many marine species use schooling.

14.4 What characteristics do marine mammals possess?

▶ Good fossil evidence shows that *marine mammals evolved from land-dwelling animals* about 50 million years ago. *Marine mammals are warm-blooded, breathe air, have hair or fur, bear live young, and the females have mammary glands.* Marine mammals belong to orders *Carnivora, Sirenia,* and *Cetacea.*

▶ *Marine mammals in order Carnivora* have prominent canine teeth and include *sea otters, polar bears,* and the *pinnipeds (walruses, seals, sea lions,* and *fur seals). Marine mammals of order Sirenia,* which include *manatees* and *dugongs* ("sea cows"), have toenails (manatees only) and sparse hairs covering their bodies, and they are vegetarians.

▶ The mammals best adapted to life in the open ocean are those of the *order Cetacea,* which includes *whales, dolphins,* and *porpoises.* Cetaceans have *highly streamlined bodies* so that they are fast swimmers. Other adaptations—such as being able to absorb 90% of the oxygen they inhale, storing large quantities of oxygen, reducing the use of oxygen by noncritical organs, and having collapsible ribs and lungs—*allow them to dive deeply* and minimize the effects of nitrogen narcosis and decompression sickness, although recent research indicates that they are not immune to it.

▶ *Cetaceans are divided into suborder Odontoceti (the toothed whales) and suborder Mysticeti (the baleen whales). Odontocetes use echolocation* to find their way through the ocean and locate prey. They emit clicking sounds and can determine the size, shape, internal structure, and distance of the objects from the nature of the returning signals and the time elapsed.

▶ *Mysticetes,* which include the largest whales in the world, *separate their small prey from seawater using their baleen plates as a strainer.* Baleen whales include the *gray whale,* the *rorqual whales,* and the *right whales.*

Study Resources
Online Study Guide Quizzes, Web Diving Deeper 14.1, Web Animations, Web Videos

Critical Thinking Question

List the modifications that are thought to give some cetaceans the ability to (a) increase their swimming speed, (b) dive to great depths without suffering the bends, and (c) stay submerged for long periods.

14.5 An example of migration: Why do gray whales migrate?

▶ *Gray whales migrate from their cold-water summer feeding grounds in the Arctic to warm, low-latitude lagoons in Mexico during winter for breeding and birthing purposes.* This behavior may have evolved to allow their young to be born into warm water during the last ice age, when lower sea level eliminated today's highly productive Arctic feeding areas.

▶ *The International Whaling Commission (IWC) was established in 1946 to manage the subsistence and commercial hunting of large whales,* which had experienced severe population decreases. *Commercial whaling was banned* in 1986, although *whaling can still be done under objection to the ban, for scientific research, or by native cultures.*

Study Resources
Online Study Guide Quizzes

Critical Thinking Question

Using a calendar and tying your answer to productivity considerations, explain the seasonal cycle of migration for gray whales.

MasteringOceanography™

www.masteringoceanography.com

Looking for additional review and test prep materials? Visit the Study Area in MasteringOceanography to enhance your understanding of this chapter's content by accessing a variety of resources, including Self-Study Quizzes, Geoscience Animations, RSS feeds, flashcards, Web links, and an optional Pearson eText.

A sea kayaker explores tide pools teeming with life. A sea kayaker explores shallow tide pools in southeastern Alaska. Many types of bottom-dwelling marine life can often be found in great abundance in tide pools.

Before you begin reading this chapter, use the glossary at the end of this book to discover the meanings of any of these words you don't already know:

15

ANIMALS OF THE BENTHIC ENVIRONMENT

ESSENTIAL CONCEPTS

At the end of this chapter, you should be able to:

15.1 Describe characteristics of the communities that exist along rocky shores

15.2 Describe characteristics of the communities that exist along sediment-covered shores

15.3 Describe characteristics of the communities that exist on the shallow offshore ocean floor

15.4 Describe characteristics of the communities that exist on the deep-ocean floor

O f the 250,000 known species that inhabit the marine environment, more than 98% (about 245,000) live in or on the ocean floor. Ranging from the rocky, sandy, and muddy intertidal zone to the muddy deposits of the deepest ocean trenches, the ocean floor provides a tremendously varied environment that is home to a diverse group of specially adapted organisms.

The distribution of benthic **biomass**[1] shown in **Figure 15.1** closely matches the distribution of chlorophyll in surface waters (which is an approximation of primary productivity; compare Figures 15.1 and 13.5). Mostly, life on the ocean floor depends on the productivity of the ocean's surface waters; as a result, great abundances of benthic life are found beneath areas of high primary productivity.

The vast majority of known benthic species live on the continental shelf, where the water is often shallow enough to allow sunlight to penetrate to the ocean bottom and support photosynthesis. Recent expeditions to the deep-ocean floor, however, suggest that a huge number of undescribed species live there.

The number of benthic species found at similar latitudes on opposite sides of an ocean basin depends on how ocean surface currents affect coastal water temperature—one of the most important variables affecting species diversity. The Gulf Stream, for example, warms the European coast from Spain to the northern tip of Norway, giving rise to more than three times the number of benthic species than are found in similar latitudes along the Atlantic coast of North America, where the Labrador Current cools the water as far south as Cape Cod, Massachusetts.

Living at or near the interface of the ocean floor and seawater, an organism's success is closely related to its ability to cope with the physical conditions of the water, the ocean floor, and the other organisms that inhabit its environment. In this chapter, we'll examine a variety of benthic communities and the animals that inhabit them.

15.1 What Communities Exist along Rocky Shores?

If you visit any rocky shoreline around the world, you're likely to find an abundance of marine organisms that live on the surface of the ocean floor. These organisms, called **epifauna** (*epi* = upon, *fauna* = animal) are either permanently attached to the

---

[1]Remember that *biomass* is the mass of living organisms.

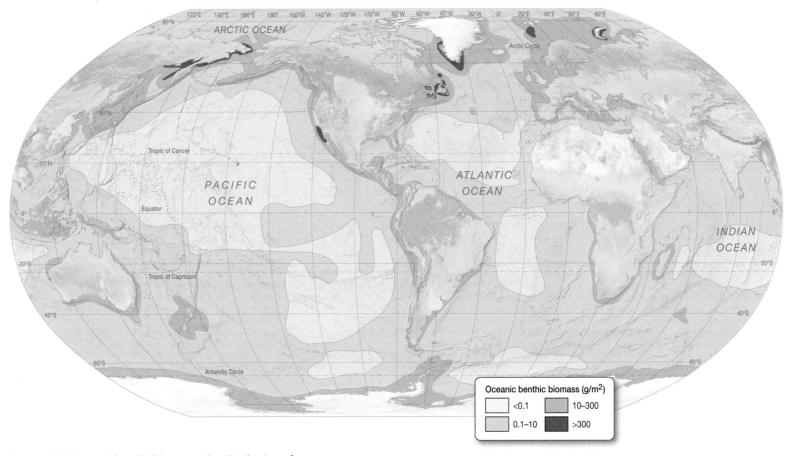

Figure 15.1 Oceanic benthic biomass. The distribution of oceanic benthic biomass (in grams per square meter) shows that the ocean's lowest biomass is beneath the centers of subtropical gyres and the highest values are in high-latitude continental shelf areas. Note the similarity in distribution to that of surface chlorophyll (see Figure 13.5), which suggests that most of the benthic community receives its food from surface waters.

bottom (for example, marine algae) or move over it (for example, crabs). **Table 15.1** lists some of the special adaptations these organisms have to withstand the rigors of life on rocky shores.

Diversity of species that inhabit rocky shores varies widely. Overall, rocky intertidal ecosystems have a moderate diversity of species compared to other benthic environments. The greatest animal diversity is in the lower (tropical) latitudes, while the diversity of algae is greater in the middle latitudes, probably because of better availability of nutrients.[2]

Intertidal Zonation

A typical rocky shore (**Figure 15.2a**) can be divided into a **spray zone**, which is above the spring high tide line and is covered by water only during storms, and an **intertidal zone**, which lies between the high and low tidal extremes. Along most shores, the intertidal zone can be clearly separated into the following subzones (Figure 15.2a):

- The **high tide zone**, which is relatively dry and is covered only by the highest high tides
- The **middle tide zone**, which is alternately covered by all high tides and exposed during all low tides
- The **low tide zone**, which is usually wet but is exposed during the lowest low tides

---

[2]As discussed in Chapter 13, this increased nutrient supply is a result of the lack of a permanent thermocline in the middle latitudes.

TABLE 15.1   **ADVERSE CONDITIONS OF ROCKY INTERTIDAL ZONES AND ORGANISM ADAPTATIONS**

Adverse conditions of rocky intertidal zones	Organism adaptations	Examples of organisms
Drying out during low tide	• Ability to seek shelter or withdraw into shells • Thick exterior or exoskeleton to prevent water loss • External surfaces covered with rock or shell fragments to prevent water loss • Adapted to periodic drying out without dying	Sea slugs, snails, crabs, kelp
Strong wave activity	• In algae: Strong holdfasts to prevent being washed away • In animals: Seeking shelter or employing strong attachment threads, biological adhesives, a muscular foot, multiple legs, or hundreds of tube feet to allow them to attach firmly to the bottom • In both: Hard structures adapted to withstand wave energy; clustering closely together	Kelp, snails, sea stars, mussels, sea urchins
Predators occupy area during low tide/high tide	• Firm attachment of body parts, including a hard shell • Stinging cells • Camouflage • Inking response • Ability to break off body parts and regrow them later (regenerative capability)	Mussels, sea anemones, sea slugs, octopi, sea stars
Difficulty finding mates for attached species	• Release of large numbers of eggs/sperm into the water column during reproduction • Long organs to reach others for sexual reproduction	Abalones, sea urchins, barnacles
Rapid changes in temperature, salinity, pH, and oxygen content	• Ability to withdraw into shells to minimize exposure to rapid changes in environment • Ability to exist in varied temperature, salinity, pH, and low-oxygen environments for extended periods	Snails, limpets, mussels, barnacles
Lack of space or attachment sites	• Overtake another organism's space • Attach to other organisms • Planktonic larval forms that inhabit new areas, which limits parental and offspring competition for the same space	Bryozoans, coral, barnacles, limpets

The intertidal zone has remarkably different characteristics from place to place. For example, some of the physical characteristics that change in a matter of just a few vertical centimeters are the amount of wave energy, exposure to the atmosphere, and changes in temperature and salinity. As a result, intertidal organisms have evolved specific adaptations to cope with the environmental conditions they face. Not surprisingly, then, the subzones of the intertidal zone can be delineated based on characteristic populations of benthic organisms found within each zone. In addition, this zonation of marine organisms along rocky shores produces the most finely delineated biozones in the marine environment.

The Spray (Supratidal) Zone: Organisms and Their Adaptations

Organisms that live within the spray zone, which is also known as the **supratidal** (*supra* = above, *tidal* = the tides) **zone**, must avoid drying out during the long periods when they are above water level. Consequently, many animals, such as the periwinkle snail (genus *Littorina*; **Figure 15.2b**), have shells, and few species of marine algae are found.

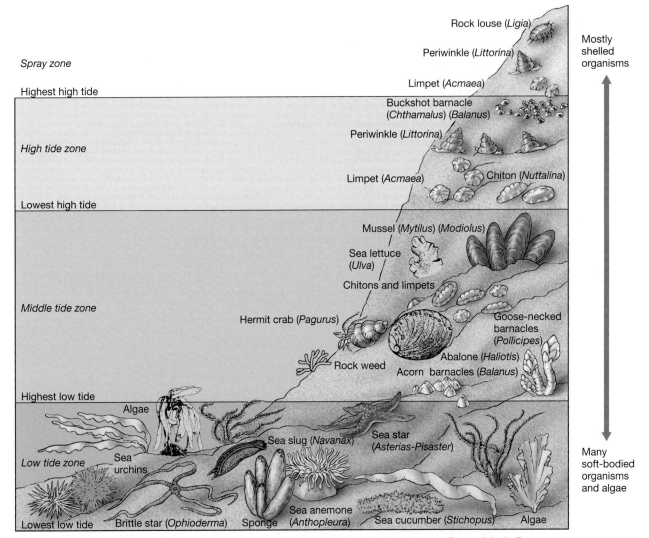

Spray zone

Highest high tide

High tide zone

Lowest high tide

Middle tide zone

Highest low tide

Low tide zone

Lowest low tide

Rock louse (*Ligia*)

Periwinkle (*Littorina*)

Limpet (*Acmaea*)

Buckshot barnacle (*Chthamalus*) (*Balanus*)

Periwinkle (*Littorina*)

Limpet (*Acmaea*)   Chiton (*Nuttalina*)

Mussel (*Mytilus*) (*Modiolus*)

Sea lettuce (*Ulva*)

Chitons and limpets

Hermit crab (*Pagurus*)

Goose-necked barnacles (*Pollicipes*)

Abalone (*Haliotis*)

Rock weed   Acorn barnacles (*Balanus*)

Algae

Sea urchins

Sea slug (*Navanax*)   Sea star (*Asterias-Pisaster*)

Brittle star (*Ophioderma*)   Sponge   Sea anemone (*Anthopleura*)   Sea cucumber (*Stichopus*)   Algae

Mostly shelled organisms

Many soft-bodied organisms and algae

(a) Diagrammatic view of rocky shore tide zones from the spray zone (*top*) to the low tide zone (*bottom*), including representative organisms (not to scale).

(b) Periwinkles (*Littorina*), spray zone to upper high tide zone.

(c) Rock louse (*Ligia*), spray zone.

Figure 15.2 Rocky shore zonation and common organisms. (a) Diagrammatic view of rocky shore tide zones, including representative organisms (not to scale). **(b–i)** Photos of some common rocky shore organisms.

(d) Limpets (*Acmaea*), high tide zone.

(e) Buckshot barnacles (*Chthamalus*), high tide zone.

(f) Rock weed (*Fucus*), middle tide zone.

(g) Acorn barnacles (*Balanus*), middle tide zone.

(h) Blue mussel bed (*Mytilus*), including gooseneck barnacles (*Pollicipes*) and acorn barnacles (*Balanus*), middle tide zone.

(i) Sea anemone (*Anthopleura*), low tide zone.

Figure 15.2 (Continued).

Found among the cobbles and boulders that typically cover the floors of sea caves well above the high tide line are rock lice or sea roaches (isopods of the genus *Ligia*) (**Figure 15.2c**). These scavengers reach lengths of 3 centimeters (1.2 inches) and scurry about at night, feeding on organic debris. During the day, they mostly hide in crevices.

A distant relative of the periwinkle snail, the limpet is also found in the spray zone (genus *Acmaea*; **Figure 15.2d**). Both limpets and periwinkle snails feed on marine algae. The limpet has a flattened conical shell and a muscular foot with which it clings tightly to rocks.

The High Tide Zone: Organisms and Their Adaptations

Like animals within the spray zone, most animals that inhabit the high tide zone have a protective covering to prevent them from drying out. Periwinkles, for example, have a protective shell and can move between the spray zone and the high tide zone. Buckshot barnacles (**Figure 15.2e**) have a protective shell, too, but they cannot live above the high tide shoreline because they filter-feed from seawater, and their larval form is planktonic.

The most conspicuous algae in the high tide zone are rock weeds, members of the genus *Fucus* that live in colder latitudes (**Figure 15.2f**) and *Pelvetia* that live in warmer latitudes. Both have thick cell walls to reduce water loss during periods of low tide.

Rock weeds are among the first organisms to colonize a rocky shore. Later, **sessile** (*sessilis* = sitting on) animal forms—those attached to the bottom, such as barnacles and mussels—begin to establish themselves, competing for attachment sites with the rock weeds.

The Middle Tide Zone: Organisms and Their Adaptations

Seawater constantly bathes the middle tide zone, so more types of marine algae and soft-bodied animals can live there. The total biomass is much greater than in the high tide zone, so there is much greater competition for rock space among sessile forms.

Shelled organisms inhabiting the middle tide zone include acorn barnacles (*Balanus*) (**Figure 15.2g**); gooseneck barnacles (*Pollicipes*) (**Figure 15.2h**), which attach themselves to rocks with a long, muscular neck; and various mussels (genera *Mytilus* and *Modiolus*) (**Figure 15.2h**). Mussels attach to bare rock, algae, or barnacles during their planktonic stage and remain in place by means of strong *byssus* threads.

Mussels are often grouped together into a distinctive *mussel bed* that appears as a pronounced band or layer (**Figure 15.3a**) and can often be one of the most

Figure 15.3 Middle tide zone mussel bed and sea star.

(a) The distincitve black band covering the rock surface is a bed of mussels (*Mytilus*), which is a common feature of middle tide zones along rocky shores such as this one in Glacier Bay, Alaska. Note how the diffuse band can also be seen on the beach to the left.

(b) An ochre sea star (*Pisaster*), feeds on a mussel by first enveloping a mussel with its tube feet and prying apart the mussel's shell. The star can then digest the mussel's soft tissues with its own everted stomach.

recognizable features of middle tidal zones along rocky coasts. The mussel bed thickens toward the bottom until it reaches an abrupt bottom limit, where physical conditions restrict mussel growth. Often protruding from the mussel bed are numerous gooseneck barnacles; crowded in with the mussels are less-conspicuous species, such as acorn barnacles, crustaceans, marine worms, rock-boring clams, sea stars, and algae.

Carnivorous snails and sea stars (such as genera *Pisaster* and *Asterias*) feed upon mussels in the mussel beds. To pry open the mussel shell, sea stars pull on either side with hundreds of tubelike feet. The mussel eventually becomes fatigued and can no longer hold its shell halves closed. When the shell opens ever so slightly, the sea star turns its stomach inside out, slips it through the crack in the mussel shell, and digests the edible tissue inside (Figure 15.3b).

Where the rock surface flattens out within the middle tidal zone, tide pools trap water as the tide ebbs. These pools support microecosystems containing a wide variety of organisms. The most conspicuous member of this community is often the sea anemone (see Figure 15.2i), which is a relative of jellies.

Shaped like a sack, an anemone has a flat foot disk that provides a suction attachment to the rock surface. Directed upward, the open end of the sack is the mouth, which leads directly to the gut cavity and is surrounded by rows of tentacles (Figure 15.4). The tentacles are covered with stinging needlelike cells called **nematocysts** (*nemato* = thread, *cystis* = bladder) (Figure 15.4, *inset*), which inject the victim with a potent neurotoxin. When an organism brushes against a sea anemone's tentacles, the nematocysts are automatically released (except for certain organisms, such as clownfish that live in a symbiotic relationship with a sea anemone).

Hermit crabs (*Pagurus*) inhabit tide pools, too. They have a well-armored pair of claws and upper body but a soft, unprotected abdomen, which they protect by inhabiting an abandoned snail shell (Figure 15.5a). They can often be seen scurrying around the tide pool area or fighting with other hermit crabs for new shells. Their abdomen has even evolved a curl to the right to make it fit properly into snail shells. Once in the snail shell, the crab can protect itself by closing off the shell's opening with its large claws.

In tide pools near the lower limit of the middle tide zone, sea urchins may be found feeding on algae (Figure 15.5b). A sea urchin has a five-toothed mouth centered on the bottom side of its hard spherical shell, consisting of fused calcium carbonate plates perforated to allow tube feet and water to pass through. Resembling a pincushion, the shell of a sea urchin has numerous spines for protection and to scrape out protective holes in rocks.

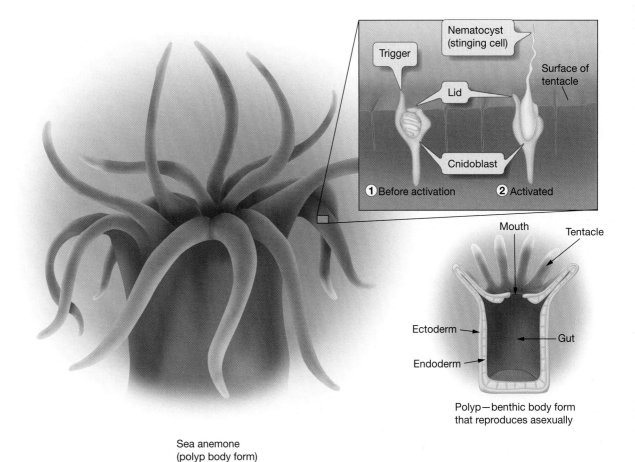

Figure 15.4 Sea anemone structure and operation of its stinging cells. Sea anemone morphology (*left*), with detail of its stinging nematocysts (*inset*), which are used to sting and incapacitate prey. An internal view of the basic sea anemone body form is shown at right.

(a) Hermit crab (*Pagurus*) that has taken up residence in a *Maxwellia gemma* shell.

(b) Purple sea urchins (*Echinus*) that have burrowed into the bottom of a rocky tide pool in the middle tide zone.

Figure 15.5 Hermit crab and sea urchins. (a) Hermit crab (*Pagurus*) and **(b)** purple sea urchins (*Echinus*); both inhabit the middle tide zone.

Figure 15.6 Surf grass. Green surf grass (*Phyllospadix*) and various species of brown algae are exposed during an extremely low tide in this California tide pool. When the tide rises, the anchored surf grass floats and provides a protective hiding place for many low tide zone organisms.

The Low Tide Zone: Organisms and Their Adaptations

The low tide zone is almost always submerged, and an abundance of algae is typically present. A diverse community of animals live here, too, but they are hidden by the great variety of marine algae and surf grass (*Phyllospadix*) (**Figure 15.6**). Various types of encrusting red algae (such as *Lithophyllum* and *Lithothamnium*), which are also seen in middle-zone tide pools, become very abundant in lower tide pools. In temperate latitudes, moderate-sized red and brown algae provide a drooping canopy beneath which much of the animal life can hide during low tide.

Scampering from crevice to crevice and in and out of tide pools across the full range of the intertidal zone are various species of shore crabs (**Figure 15.7**). These scavengers help keep the shore clean. Shore crabs spend most of the day hiding in cracks or beneath overhangs. At night, they eat algae as rapidly as they can tear them from the rock surface with their large front claws, called *chelae* (*khele* = claw). Their hard exoskeleton prevents them from drying out too quickly, so they can spend long periods of time out of water.

(a) Sally lightfoot crab (*Grapsus grapsus*), which is a brightly colored and speedy crab found in tropical regions.

(b) Shore crab (*Pachygrapsus crassipes*). This female is carrying eggs (orange mass under her abdomen).

Figure 15.7 Shore crabs. (a) Sally lightfoot crab (*Grapsus grapsus*) and **(b)** shore crab (*Pachygrapsus crassipes*).

15.2 What Communities Exist along Sediment-Covered Shores?

Even though most sediment-covered shores have intertidal zones similar to rocky shores, life on and in sediment-covered shores requires very different adaptations than on rocky shores. Sediment-covered shores, for example, are composed of unconsolidated materials that often change shape and so require specific adaptations for organisms. In addition, there is much less species diversity in sediment-covered shores, but the organisms are usually found in great numbers. In the low tide zone of some beaches and on mud flats, for example, as many as 5000 to 8000 burrowing clams have been counted in only 1 square meter (10.8 square feet).

Nearly all large organisms that inhabit sediment-covered shores are called **infauna** (*in* = inside, *fauna* = animal) because they can burrow into the sediment. Sediment-covered shores also contain large numbers of microbial communities, particularly in quiet environments such as salt marshes and mud flats that tend to accumulate organic matter.

Physical Environment Of the Sediment

Sediment-covered shores include *coarse boulder beaches, sand beaches, salt marshes,* and *mud flats,* which represent progressively lower-energy environments and are consequently composed of progressively finer sediment. The energy level that a shore experiences is related to the strength of waves and longshore currents. Along shores that experience low energy levels, particle size becomes smaller, the sediment slope decreases, and overall sediment stability increases. Thus, the sediment in a fine-grained mud flat is more stable than that of a high-energy sandy beach.

A large quantity of water from breaking waves rapidly sinks into the sand and brings a continual supply of nutrients and oxygen-rich water for the animals that live there. This supply of oxygen also enhances bacterial decomposition of dead tissue. The sediment in salt marshes and mud flats is not nearly as rich in oxygen, however, so decomposition occurs more slowly.

Intertidal Zonation

The intertidal zone of the sediment-covered shore consists of supratidal, high tide, middle tide, and low tide zones, as shown in **Figure 15.8**. These zones are best developed on steeply sloping, coarse-sand beaches and are less distinct on the more gentle sloping, fine-sand beaches. On mud flats, the tiny, clay-sized particles form a deposit with essentially no slope, so zonation is not possible in this protected, low-energy environment.

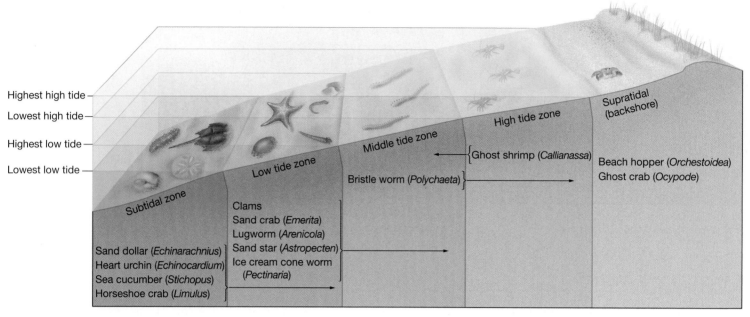

Highest high tide
Lowest high tide
Highest low tide
Lowest low tide

Subtidal zone

Low tide zone

Middle tide zone

High tide zone

Supratidal (backshore)

Sand dollar (*Echinarachnius*)
Heart urchin (*Echinocardium*)
Sea cucumber (*Stichopus*)
Horseshoe crab (*Limulus*)

Clams
Sand crab (*Emerita*)
Lugworm (*Arenicola*)
Sand star (*Astropecten*)
Ice cream cone worm (*Pectinaria*)

Bristle worm (*Polychaeta*)

Ghost shrimp (*Callianassa*)

Beach hopper (*Orchestoidea*)
Ghost crab (*Ocypode*)

Figure 15.8 Intertidal zonation and common organisms of sediment-covered shores. Profile view of a coarse sand-covered shore, showing intertidal zones, which are related to the amount of exposure during low tide, and common organisms, which usually occur only in specific tide zones. Beaches composed of finer sediment have more gentle slopes and show less distinct intertidal zonation.

The species of animals differ from zone to zone. As in intertidal rocky shores, however, the maximum *number* of species and the greatest *biomass* in intertidal sediment-covered shores are found near the low tide shoreline, and both diversity and biomass decrease toward the high tide shoreline.

Sandy Beaches: Organisms and Their Adaptations

Most animals at the beach burrow into the sand because there is no stable, fixed surface (as on rocky shores) to which they can attach. As a result, organisms are much less obvious at beaches than in other environments. By burrowing only a few centimeters beneath the surface, they encounter a much more stable environment where they are not bothered by fluctuations of temperature and salinity or the threat of drying out.

BIVALVE MOLLUSKS A **bivalve** (*bi* = two, *valva* = a valve) is an animal that has two hinged shells, such as a clam or a mussel. A **mollusk** is a member of the phylum Mollusca (*molluscus* = soft), characterized by a soft body and either an internal or external hard calcium carbonate shell.

Bivalve mollusks are well adapted to living within sediment. A single foot digs into the sediment to pull the creature down into the sand. The method by which clams bury themselves is shown in **Figure 15.9**. How deeply a bivalve can bury itself depends on the length of its siphons, which must reach above the sediment surface to pull in water for food (plankton) and oxygen. Indigestible matter is forced back out the siphon periodically by quick muscular contractions.

The greatest biomass of clams is burrowed into the low tide region of sandy beaches and decreases where the sediment becomes muddier.

Figure 15.9 How a clam burrows. Diagrammatic view of a clam burrowing into sediment.

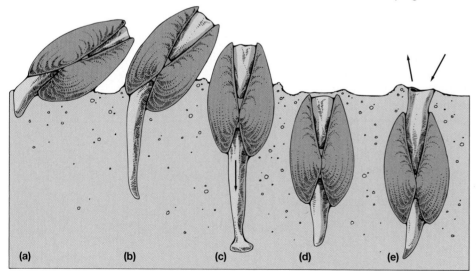

(a)   (b)   (c)   (d)   (e)

A clam burrows into sediment by **(a)** extending its pointed foot into the sediment and **(b)** forcing its foot deeper into the sediment and using this increasing leverage to bring the exposed, shell-clad body toward vertical. When the foot has penetrated deeply enough, a bulbous anchor forms at the bottom **(c)**, and a quick muscular contraction pulls the entire animal into the sediment **(d)**. The siphons are then pushed up above the sediment to pump in water **(e)**, from which the clam extracts food and oxygen.

ANNELID WORMS A variety of **annelids** (*annelus* = ring)—worms—are well adapted to life in the sediment.

The lugworm (*Arenicola*), for example, constructs a vertical, U-shaped burrow, the walls of which are strengthened with mucus. The worm moves forward to feed and extends its proboscis (snout) up into the head shaft of the burrow to loosen sand with quick, pulsing movements. A cone-shaped depression forms at the surface over the head end of the burrow as sand continually slides into the burrow and is ingested by the worm. As the sand passes through the worm's digestive tract, the sand's *biofilm* (coating of organic matter) is digested, and the processed sand is deposited back at the surface.

CRUSTACEANS Crustaceans (*crusta* = shell)—such as crabs, lobsters, shrimps, and barnacles—include predominately aquatic animals that are characterized by a segmented body, a hard exoskeleton, and paired, jointed limbs. On most sandy beaches, numerous crustaceans called *beach hoppers* feed on kelp cast up by storm waves or high tides. A common genus is *Orchestoidea*, which is only 2 to 3 centimeters (0.8 to 1.2 inches) long but can jump more than 2 meters (6.6 feet) high. Laterally flattened, beach hoppers usually spend the day buried in the sand or hidden in kelp. They become particularly active at night, when large groups may hop at the same time and form clouds above the piles of kelp on which they feed.

Sand crabs (*Emerita*) (**Figure 15.10**) are a type of crustacean common to many sandy beaches. Ranging in length from 2.5 to 8 centimeters (1 to 3 inches), they move up and down the beach near the shoreline. They bury their bodies in the sand and leave their long, curved, V-shaped antennae pointing up the beach slope. These little crabs filter food particles from the water and can be located by looking in the lower intertidal zone for a V-shaped pattern in the swash as it runs down the beach face.

ECHINODERMS Echinoderms (*echino* = spiny, *derma* = skin) that live in beach deposits include the sand star (*Astropecten*) and heart urchin (*Echinocardium*). Sand stars prey on invertebrates that burrow into the low tide region of sandy beaches. The sand star is well designed for moving through sediment: It has a smooth back and five tapered legs with spines.

More flattened and elongated than the sea urchins of the rocky shore, heart urchins live buried in the sand near the low tide line (**Figure 15.11**). They gather sand grains into their mouths, where the biofilm of organic matter that coats sand grains is scraped off and ingested.

MEIOFAUNA Meiofauna (*meio* = lesser, *fauna* = animal) are small marine organisms that live in the spaces between sediment particles. These organisms, generally only 0.1 to 2 millimeters (0.004 to 0.08 inch) long, feed primarily on bacteria removed from the surface of sediment particles. Meiofauna include polychaetes,

Figure 15.10 Sand crab. A sand crab (*Emerita*) emerges from the sand. Sand crabs can often be found just beneath the surface of sandy beaches within the lower intertidal zone.

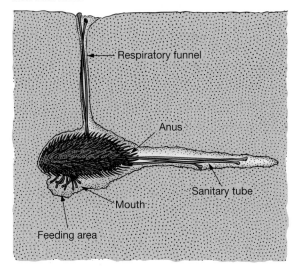

Figure 15.11 Heart urchin. Feeding and respiratory structures of a heart urchin (*Echinocardium*), which feeds on the biofilm of organic matter that covers sand grains.

(a) Head of a nematode, magnified 800 times. The pit on its left side and numerous projections are sensory structures.

(b) Amphipod, which has a length of 3 millimeters (0.1 inch) and is related to crustaceans such as lobsters, crabs, and krill.

(c) Polychaete worm, which has a length of 1 millimeter (0.04 inch) and is shown with its proboscis (mouth) extended (*left*).

Figure 15.12 Scanning electron micrographs of meiofauna. Various examples of meiofauna, which are small marine organisms that live in the spaces between sediment particles.

Figure 15.13 Fiddler crab. A male fiddler crab (*Uca*), which uses its large claw for protection and for attracting mates.

mollusks, arthropods, and nematodes (Figure 15.12) and live in sediment from the intertidal zone to deep-ocean trenches.

Mud Flats: Organisms and Their Adaptations

Eelgrass (*Zostera*) and turtle grass (*Thalassia*) are widely distributed in the low tide zone of mud flats and the adjacent shallow coastal regions. Numerous openings at the surface of mud flats attest to a large population of bivalve mollusks and other invertebrates.

Fiddler crabs (*Uca*) live in burrows that may exceed 1 meter (3 feet) deep in mud flats. Relatives of the shore crabs, they usually measure no more than 2 centimeters (0.8 inch) in width. A male fiddler crab has one small claw and one oversized claw, which is up to 4 centimeters (1.6 inches) long (Figure 15.13). Fiddler crabs get their

name because they wave around their large claws as if they were playing imaginary fiddles. The females have two normal-sized claws. The large claw of the male is used to court females and to fight competing males.

15.3 What Communities Exist on the Shallow Offshore Ocean Floor?

The shallow offshore ocean floor extends from the spring low tide shoreline to the seaward edge of the continental shelf. It is mainly sediment covered, but rocky exposures may occur locally near shore. Rocky exposures feature many types of marine algae, which have adaptations (such as gas-filled floats) for reaching from the shallow sea floor to near the sunlit surface waters.

The sediment-covered shelf has moderate to low species diversity. Surprisingly, the diversity of benthic organisms is *lowest* beneath upwelling regions. This is because upwelling waters that are rich in nutrients produce high pelagic production, so large amounts of dead organic matter are produced. When this matter rains down on the bottom and decomposes, it consumes oxygen, so the oxygen supply can be locally depleted, thereby limiting benthic populations. However, kelp beds associated with rocky bottoms are a specialized shallow-water community with higher diversity.

Rocky Bottoms (Subtidal): Organisms and Their Adaptations

A rocky bottom within the shallow inner **subtidal** (*sub* = under, *tidal* = the tides) **zone** is usually covered with various types of marine macro algae.

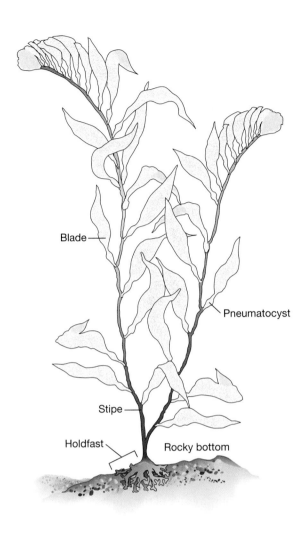

(a) Structure of the giant brown bladder kelp (*Macrocystis*), a common species in kelp forests.

(b) Underwater photo of a kelp forest, which provides habitat and protection for many marine organisms.

Figure 15.14 Kelp and kelp forests.

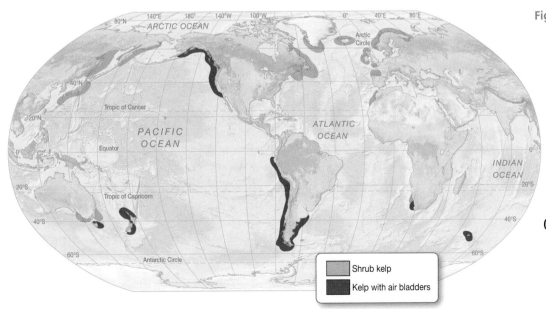

(c) Map showing distribution of kelp forests. Shrub kelp (*orange*) includes smaller species of kelp such as *Sargassum* and rock weed (*Fucus, Pelvetia*). Kelp with air bladders (*red*) includes larger species such as *Macrocystis* and bull kelp (*Nereocystis*).

Shrub kelp

Kelp with air bladders

STUDENTS SOMETIMES ASK . . .

What is an urchin barren?

An urchin barren is created when the population of sea urchins goes unchecked and the urchins devour entire areas of giant brown bladder kelp (*Macrocystis*), one of the main types of algae in kelp forests. The urchins chew through the holdfast structure that holds the kelp in place and set it adrift. In California, the severe reduction of animals that prey on urchins (such as the wolf eel and sea otter) has upset the natural balance of ocean food webs. Consequently, sea urchins have proliferated, and urchin barrens now exist where there were once lush kelp beds.

STUDENTS SOMETIMES ASK . . .

How big was the largest lobster ever caught?

The largest specimen on record of the larger American lobster, *Homarus americanus*, measured 1.1 meters (3.5 feet) from the end of its tail fan to the tip of its large claw and weighed 20.1 kilograms (44.4 pounds). It was caught off Nova Scotia, Canada, in 1977 and was later sold to a New York restaurateur.

KELP AND KELP FORESTS Along the North American Pacific coast, the giant brown bladder **kelp** (*Macrocystis*) uses a rootlike anchor called a *holdfast* (**Figure 15.14a**) to attach to rocky bottoms as deep as 30 meters (100 feet). The holdfast is so strong that only large storm waves can break the algae free. The *stipes* and *blades* of the algae are supported by gas-filled floats called *pneumatocysts* (*pneumato* = breath, *cystis* = bladder), which allow the algae to grow upward and extend for another 30 meters (100 feet) along the surface to allow for good exposure to sunlight. Under ideal conditions, *Macrocystis* can grow up to 0.6 meter (2 feet) per day, making it the fastest-growing algae in the world.

The giant brown bladder kelp and bull kelp (*Nereocystis*), another fast-growing kelp, often form beds called **kelp forests** along the Pacific coast (**Figure 15.14b**). Smaller tufts of red and brown algae are found on the bottom and also live on the kelp blades.

Kelp forests are highly productive ecosystems that provide shelter for a wide variety of organisms living within or directly upon the kelp as epifauna. These organisms are an important food source for many of the animals living in and near the kelp forest, including mollusks, sea stars, fishes, octopi, lobsters, and marine mammals. Surprisingly, very few animals feed directly on the living kelp. Among those that do are sea urchins and a large sea slug called the sea hare (*Aplysia*). The distribution of kelp forests is shown in **Figure 15.14c**.

LOBSTERS Large crustaceans—including lobsters and crabs—are common along rocky bottoms. The spiny lobsters are named for their spiny covering and have two very large, spiny antennae (**Figure 15.15a**). These antennae serve as feelers and are equipped with noise-making devices near their base that are used for protection. The genus *Panulirus*, which reaches lengths to 50 centimeters (20 inches), is considered a delicacy and lives in water deeper than 20 meters (65 feet) along the European coast. The Caribbean lobster (*Panulirus argus*) sometimes exhibits a remarkable behavior of migrating single file across the sea floor in lines that are several kilometers long.

Panulirus interruptus is the spiny lobster of the American West Coast. All spiny lobsters are taken for food, but none are as highly regarded as the so-called true lobsters (genus *Homarus*), which include the American (Maine) lobster, *Homarus americanus* (**Figure 15.15b**). Although they are scavengers like their spiny relatives, the true lobsters also feed on live animals, including mollusks, crustaceans, and other lobsters.

Figure 15.15 Spiny and American lobsters. Photographs showing morphological differences between **(a)** a spiny lobster (*Panulirus interruptus*) and **(b)** an American, or Maine, lobster (*Homarus americanus*).

(a) Spiny lobster (*Panulirus interruptus*), which lacks large claws but has long spiny antennae. It is found along rocky bottoms in the Caribbean and along the West Coast of North America.

(b) American, or Maine, lobster (*Homarus americanus*), which has large claws that are used for feeding and for defense. It is found along the East Coast of North America from Labrador, Canada, to North Carolina in the United States.

OYSTERS   Oysters are thick-shelled grow best where there is a steady flow of clean water to provide plankton and oxygen.

Oysters are food for sea stars, fishes, crabs, and snails that bore through the shell and rasp away the soft tissue inside (**Figure 15.16**). In fact, this may be one of the main reasons that oysters have such a thick shell.[3] Oysters also have great commercial importance to humans throughout the world as a food source.

Oyster beds are composed of empty shells of many previous generations that are cemented to a hard substrate or to one another, with the living generation on top. Each female produces many millions of eggs each year, which become planktonic larvae when fertilized. After a few weeks as plankton, the larvae attach themselves to the bottom. As a material upon which to anchor, oyster larvae prefer (in order) live oyster shells, dead oyster shells, and rock.

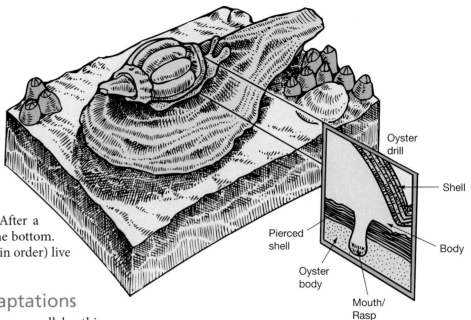

Figure 15.16 **An oyster-drill snail feeding on an oyster.** An oyster-drill snail uses its rasplike mouthpiece (*inset*) to drill through an oyster's shell and feed on the oyster's soft body.

Coral Reefs: Organisms and Their Adaptations

Individual **polyps** (*poly* = many, *pous* = foot), which are small benthic marine animals that feed with stinging tentacles and are related to jellies. Most species of corals are about the size of an ant, live in large colonies, and construct hard calcium carbonate structures for protection. Coral species are found throughout the ocean (and even in cold, deep water), but coral accumulations that are classified as **coral reefs**, which are wave-resistant structures produced by corals and other organisms, are restricted to shallow warmer water regions. See Chapter 2 for a description of the stages of development in coral reefs.

CONDITIONS NECESSARY FOR CORAL REEF DEVELOPMENT   Corals are very temperature sensitive and require warm water to survive. In fact, corals need water where the average monthly temperature exceeds 18°C (64°F) throughout the year (**Figure 15.17**). Furthermore, water that is too warm can kill corals: They cannot survive for long when the water temperature exceeds 30°C (86°F). That's why warmer-than-normal sea surface temperatures—such as those experienced during severe El Niños or other warming episodes—tend to stress corals and are linked to outbreaks of coral bleaching and other diseases, as discussed later in this section.

Water warm enough to support coral growth is found primarily in the tropics. Reefs also grow as far north and south as 35 degrees latitude on the western margins of ocean basins, however, where warm-water currents raise average sea surface temperatures (Figure 15.17).

The map in Figure 15.17 also shows the greater diversity of reef-building corals on the western side of ocean basins. More than 50 genera of corals thrive in a broad area of the western Pacific Ocean and a narrow belt of the western Indian Ocean. Fewer than 30 genera, however, occur in the Atlantic Ocean, with the greatest diversity occurring in the Caribbean Sea. This pattern is related to the positions of the continents prior to about 30 million years ago, when the warm equatorial Tethys Sea connected the world's tropical oceans and provided a highway for the worldwide distribution of coral species and reef-associated organisms. With time, tectonic changes in landmass position closed the Tethys Sea and were accompanied by changes in ocean currents and climate that reduced coral reef biodiversity in areas such as the Atlantic. Further, the presence of numerous tropical islands in the western Pacific provides a variety of habitats that favors coral speciation.

STUDENTS SOMETIMES ASK . . .

I've heard about the recent discovery of deep-water corals. How are they different from shallow-water corals?

Although coral reefs are generally associated with shallow tropical seas, recent deep-ocean exploration using advanced acoustics and submersibles has revealed unexpectedly widespread and diverse coral ecosystems in deep, cold waters. These corals are found below sunlit surface waters—at a record depth of 6,328 meters (20,800 feet)—on continental shelves, continental slopes, seamounts, and mid-ocean ridge systems around the world. Since the corals don't live in particularly deep water, the term *cold-water coral* is more appropriate. They lack the symbiotic zooxanthellae algae that their shallow-water cousins have but can be brightly colored and use their stinging tentacles to capture tiny plankton or detritus concentrated by ocean currents. They build their skeletons of calcium carbonate as shallow corals do and create large reef structures or coral mounds that provide a habitat for many other species. Some species of deep-water coral may be thousands of years old. What is remarkable about them is that they have remained unnoticed for so long. New cold-water coral species, in fact, have recently been discovered in waters off highly populated Southern California!

---

[3]This is an example of *co-evolution*, where a weapon possessed by one species creates evolutionary pressure in another species for a defense to defeat it. This, in turn, causes other evolutionary traits for new weapons, creating a co-evolutionary "arms race."

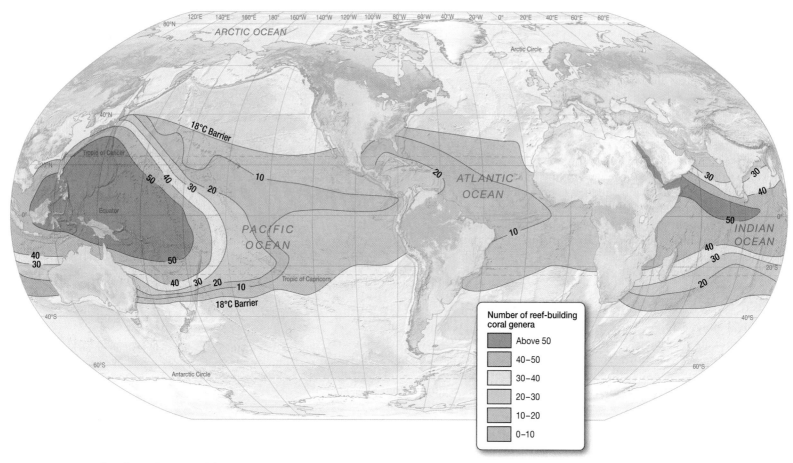

Figure 15.17 Coral reef distribution and diversity. Coral reefs are restricted to warm tropical waters between the two 18°C (64°F) temperature barriers. On the western side of each ocean basin, the coral reef belt is wider and the diversity of coral genera is greater, which is a result of ocean surface circulation patterns and the presence of numerous tropical islands that favor speciation.

ESSENTIAL CONCEPT 15.3A

Corals are small colonial animals with stinging cells that are found primarily in shallow tropical waters and need strong sunlight, wave or current action, lack of turbidity, normal salinity seawater, and a hard substrate for attachment.

Besides warm water, other environmental conditions necessary for coral growth include the following:

- *Strong sunlight.* The sunlight isn't needed by the corals themselves, which are animals and can exist in deeper water, but by the symbiotic photosynthetic microscopic dinoflagellate algae called **zooxanthellae** that live within the coral's tissues.[4]

- *Strong wave or current action* to bring nutrients and oxygen.

- *Lack of turbidity.* Suspended particles in the water tend to absorb radiant energy, interfere with the coral's filter-feeding capability, and can even bury corals, so corals are not usually found close to areas where major rivers drain into the sea.

- *Salt water.* Corals die if the water is too fresh, which is another reason coral reefs do not form near the mouths of freshwater rivers.

- *A hard substrate* for attachment. Corals cannot attach to a muddy bottom, so they often build upon the hard skeletons of their ancestors, creating coral reefs that are several kilometers thick.

SYMBIOSIS OF CORAL AND ALGAE   "Coral reefs" are more than just coral. Algae, mollusks, and foraminifers contribute to the reef structure, too. Individual reef-building corals are **hermatypic** (*herma* = pillar, *typi* = type) because they have a *mutualistic relationship*[5] with microscopic algae (zooxanthellae) that live within the tissue of the coral polyp. The algae provide their coral hosts with a continual supply of food, and the corals provide the zooxanthellae with nutrients. Although coral

---

[4]It is zooxanthellae (*zoo* = animal, *xanthos* = yellow, *ella* = small) algae that give corals their distinctive bright coloration (which can be many colors besides yellow).

[5]See Chapter 14 for a discussion of the types of symbiosis, including mutualism.

(a) Coral polyps, which are nourished by internal zooxanthellae algae and also by extending their tentacles to capture tiny planktonic organisms from the surrounding water.

(b) The blue-gray sponge *Niphates digitalis* (*left*) and the brown sponge *Agelas* (*right*), which contains symbiotic algae or bacteria.

(c) A giant clam (*Tridacna gigas*), which depends on symbiotic algae living within its mantle tissue.

Figure 15.18 Coral reef inhabitants that rely on symbiotic algae. Photographs of **(a)** coral polyps, **(b)** two species of sponges, and **(c)** a giant clam (*Tridacna gigas*).

polyps capture tiny planktonic food with their stinging tentacles, most reef-building corals receive up to 90% of their nutrition from symbiotic zooxanthellae algae. In this way, corals are able to survive in the nutrient-poor waters that are characteristic of tropical oceans. As we'll see, the mutualistic relationship between coral and algae is very sensitive to subtle changes in the environment, such as increased ocean temperature, salinity, and light.

Other reef animals also have a symbiotic relationship with various types of marine algae. Those that derive part of their nutrition from their algae partners are called **mixotrophs** (*mixo* = mix, *tropho* = nourishment) and include coral, foraminifers, sponges, and mollusks (**Figure 15.18**). The algae not only nourish the coral but may also contribute to its calcification by extracting carbon dioxide from the coral's body fluids.

Coral reefs actually contain up to three times as much algal biomass as animal biomass. Zooxanthellae, for example, account for up to 75% of the biomass of reef-building coral. Nevertheless, zooxanthellae account for less than 5% of the reef's overall algal mass; most of the rest is filamentous green algae.

CORAL REEF ZONATION On many large coral reefs, there is a well-developed vertical and horizontal zonation of the reef slope (**Figure 15.19**), which is caused by changes in sunlight, wave energy, salinity, water

Figure 15.19 Coral reef zonation. Coral reefs exhibit zonation because of the decrease in both wave energy and sunlight intensity with increasing ocean depth. As a result, massive branching corals occur at depths less than 20 meters (66 feet) below the surface, where wave energy is high; corals become more delicate with increasing depth, until around 150 meters (500 feet) below the surface, where too little solar radiation is available to allow the survival of their symbiotic zooxanthellae algae.

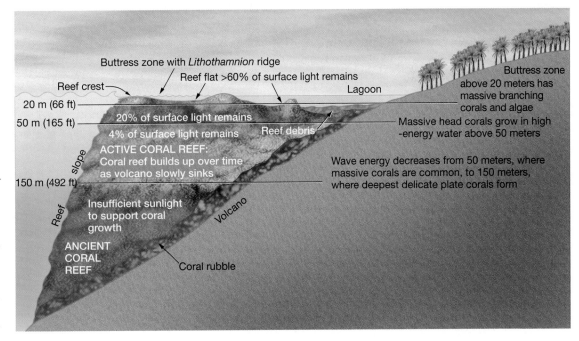

Buttress zone with *Lithothamnion* ridge

Reef flat >60% of surface light remains

Reef crest

Lagoon

20 m (66 ft)

50 m (165 ft)

20% of surface light remains

4% of surface light remains

Reef debris

ACTIVE CORAL REEF: Coral reef builds up over time as volcano slowly sinks

150 m (492 ft)

Reef slope

Insufficient sunlight to support coral growth

Volcano

ANCIENT CORAL REEF

Coral rubble

Buttress zone above 20 meters has massive branching corals and algae

Massive head corals grow in high-energy water above 50 meters

Wave energy decreases from 50 meters, where massive corals are common, to 150 meters, where deepest delicate plate corals form

depth, temperature, and other factors. These zones can readily be identified by the types of coral present and the assemblages of other organisms found in and near the reef.

Because algae need sunlight for photosynthesis, the greatest depth to which active coral growth can occur is about 150 meters (500 feet). Water motion is less at these depths, so relatively delicate plate corals can live on the outer slope of the reef from 150 meters (500 feet) up to about 50 meters (165 feet) below the surface, where light intensity is as low as 4% of the surface intensity (Figure 15.19).

From 50 meters (164 feet) to about 20 meters (66 feet) below the surface, water motion from breaking waves increases on the side of the reef facing into the prevailing current flow. Correspondingly, the mass of coral growth and the strength of the coral structure supporting it increase toward the top of this zone, where light intensity is as low as 20% of the surface value.

The reef flat may have a water depth of a few centimeters to a few meters at low tide, so it has at least 60% of the surface light intensity. Many species of colorful reef fish inhabit this shallow water, as well as sea cucumbers, worms, and mollusks. In the protected water of the reef lagoon live gorgonian coral, anemones, crustaceans, mollusks, and echinoderms (**Figure 15.20**).

THE IMPORTANCE OF CORAL REEFS Coral reefs are some of the largest structures created by living creatures on Earth. (Australia's Great Barrier Reef, for example, is more than 2000 kilometers [1250 miles] long.) Although coral reefs cover less than half a percent of the ocean's surface area, they are home to 25% of all marine species, including almost one-third of the world's estimated 20,000 species of marine fish. Coral reefs provide shelter, food, and breeding grounds that attract numerous other species, including sea anemones, sea stars, crabs, sea slugs, clams, sponges, sea turtles, marine mammals, and sharks. In fact, coral reefs foster a diversity of species that surpasses even that of tropical rainforests, making coral reefs the most diverse communities in the marine environment.

Coral reefs provide many benefits to people. Some 100 million people worldwide depend on healthy reef systems, which provide a multi-billion-dollar tourist industry. In fact, many tropical countries that have coral reefs receive more than 50% of their gross national product as tourism related to reefs. Fisheries associated with reefs supply more than one-sixth of all fish from the sea. Recently, pharmacologists and marine chemists have discovered a storehouse of new medical compounds from coral reef inhabitants that fight maladies such as cancer and infections. In addition, reefs help prevent shoreline erosion and protect coastal communities from storm waves and tsunami. The hard calcium carbonate skeletons of coral have even been used in some human bone grafts.

Figure 15.20 Non-reef-building inhabitants. Photographs of **(a)** a Guineafowl (spotted) puffer (*Arothron meleagris*) and **(b)** a non-reef-building soft gorgonian coral (*Nicella schmitti*).

(a) Coral reefs supply habitat and protection for many fishes, including this Guineafowl (spotted) puffer (*Arothron meleagris*).

(b) Unlike reef-building corals, some corals do not secrete a hard calcium carbonate structure, such as this soft gorgonian coral (*Nicella schmitti*), which has feeding polyps (*light purple*) extending from its branches. Each polyp is about 1 centimeter (0.4 inch) across.

CORAL REEFS AND NUTRIENT LEVELS   When human populations increase on land adjacent to coral reefs, the reefs deteriorate. Fishing, trampling, boat collisions with the reef, sediment increase due to development, and removal of reef inhabitants by visitors all damage a reef. One of the more subtle effects is the inevitable increase in the nutrient levels of the reef waters from sewage discharge and farm fertilizers.

As nutrient levels increase in reef waters, the dominant benthic community changes in the following ways:

- At low nutrient levels, hermatypic corals and other reef animals that contain algal symbiotic partners thrive.
- Moderate nutrient levels favor fleshy benthic plants and algae, and high nutrient levels favor suspension feeders such as clams.
- At high nutrient levels, the phytoplankton mass exceeds the benthic algal mass, so benthic populations tied to the phytoplankton food web dominate.

Increased phytoplankton biomass reduces the clarity of the water, which interferes with the coral's filter-feeding capability. The fast-growing members of the phytoplankton-based ecosystem destroy the reef structure by overgrowing the slow-growing coral and through *bioerosion*, which is erosion of the reef by organisms. Bioerosion by sea urchins and sponges is particularly damaging to many coral reefs.

THE CROWN-OF-THORNS PHENOMENON   The crown-of-thorns (*Acanthaster planci*) is a sea star (**Figure 15.21**) that has greatly proliferated and destroyed living coral on many reefs throughout the western Pacific Ocean since 1962 and more recently in the Indian Ocean and Red Sea. The sea star moves across reefs and eats the coral polyps, devouring as much as 13 square meters (140 square feet) of coral in a single year. Normally, however, the coral can grow back if it has enough time to do so.

Though a natural part of reef ecosystems, the sea stars can cause extensive damage because, for reasons not fully understood, a small group can suddenly multiply to millions. For example, vast numbers of crown-of-thorns sea stars have recently decimated many coral communities, particularly on the Great Barrier Reef. Initially, divers were employed to smash the crown-of-thorns, but sea stars (which have tremendous regenerating capabilities) can easily produce new individuals from various body parts, so it only made the problem worse.

Some investigators suggest that the proliferation of the crown-of-thorns sea star is a modern phenomenon brought about by the activities of humans, although there is little supporting evidence. There is, however, speculation that the crown-of-thorns sea star has multiplied because large reef fish that prey on them have been removed by overfishing. Other studies show that during the past 80,000 years, the crown-of-thorns sea star has been even more abundant than it is today. Thus, the sea star may be an integral part of the reef ecology in this region, and its increase may be part of a long-term natural cycle rather than a destructive event triggered by human actions.

CORAL BLEACHING AND OTHER DISEASES   **Coral bleaching** describes the loss of color in corals that causes them to turn white, as if bleach has been poured on the reef (**Figure 15.22**). Coral bleaching often occurs in response to elevated water temperatures and causes the coral's colorful symbiotic partner—zooxanthellae algae—to be killed outright, leave, or become toxic to their hosts, in which case they are expelled. Once bleached, the coral no longer receives nourishment from the algae, and if the coral does not regain its symbiotic algae within a matter of weeks, it will die. Bleaching often occurs in surface waters—the top 2 to 3 meters (7 to 10 feet)—but has also been observed at depths of 30 meters (100 feet) and can occur as quickly as overnight.

Florida's coral reefs have experienced at least eight widespread bleachings since the early 1900s, and at least 70% of the corals along the Pacific Central American

STUDENTS SOMETIMES ASK . . .

How do corals synchronize their spawning?

Although some corals spawn internally, most coral species employ a technique called *broadcast spawning*, in which sacs containing both eggs and sperm are released directly into seawater synchronously with neighboring colonies. The corals appear to use three cues to begin their mass spawning: the full moon; sunset, which they sense through photoreceptors; and a chemical that allows them to "smell" each other spawning. For night divers lucky enough to witness coral spawning, it's a phenomenal sight! Researchers are trying to determine what will happen to fertilization rates as depleted coral colonies become fewer and farther between.

Figure 15.21   Crown-of-thorns sea star. Crown-of-thorns sea stars (*Acanthaster planci*) have plagued Australia's Great Barrier Reef.

Easter Island

March 1999   **March 2000**

(a) Normal coral   **(b)** Bleached coral

Figure 15.22 Normal and bleached coral. Paired photographs of the same area offshore of Easter Island showing **(a)** normal coral (*Pocillopora verrucosa*) and **(b)** bleached coral one year later, after exceptionally high sea surface temperatures.

coast died due to bleaching associated with the severe El Niño event of 1982–1983. For example, coral reefs around the Galápagos Islands thrive in ocean water at or below 27°C (81°F). If the water is even 1°C or 2°C (2°F or 4°F) warmer for an extended period, however, the coral may expel the algae, in effect "bleaching" itself. The warming during the 1982–1983 El Niño was so severe and long-lived that two species of Panamanian coral became extinct. The bleaching episode of 1987 affected coral reefs worldwide, especially those in Florida and throughout the Caribbean. Since then, widespread bleachings have occurred with increasing frequency and intensity. For instance, the El Niño event of 1997–1998 raised water temperatures several degrees higher than normal and has been blamed for the most geographically widespread bleaching ever recorded, including in the equatorial eastern Pacific Ocean, the Yucatán coast, the Florida Keys, and the Netherlands Antilles. Since then, bleaching has been observed in every ocean basin with increased frequency and severity. For example, in 2010, reefs in southeastern Asia bleached at record levels, part of the Hawaiian Islands experienced mild bleaching, and significant to severe bleaching occurred throughout the Caribbean.

Not only will algae leave their coral hosts or be expelled because of abnormally high surface water temperatures (such as El Niño events), but also due to elevated ultraviolet radiation levels, a decrease in sunlight-blocking particles high in the atmosphere, pollution, salinity changes, invasion of disease, or a combination of factors. Some researchers suggest that reactive oxygen builds up in the coral's tissues and becomes toxic when temperatures are excessively warm. Then the algae are expelled or, perhaps, the algae leave with dead tissue. The strong correlation between coral bleaching and elevated water temperature concerns marine scientists because sea surface temperatures are becoming warmer as a result of human-induced climate change (see Chapter 16, "The Oceans and Climate Change"). Whatever the cause, coral bleaching indicates that corals are experiencing severe environmental stress.

James Porter, a coral reef ecologist at the University of Georgia, and his colleagues study diseases that affect corals. They have been monitoring the health of corals in the Florida Keys since 1995 and have discovered the reappearance of *white plague disease* as well as a dozen new diseases, such as *white band disease, white pox, black band disease, yellow band (yellow blotch) disease, patchy necrosis,* and *rapid wasting disease.*

The cause of most of these diseases is still being investigated, and it is not known whether the new diseases are from the invasion of microorganisms—bacteria, viruses, or fungi—or related to environmental stress, as coral bleaching is. As human population has increased along the Florida Keys, the coral reefs of the Keys have begun to show signs of stress, thus making them more susceptible to a host of diseases. The increased nutrient levels and water turbidity resulting from soil runoff and improper sewage disposal in the Keys may contribute to the problem, too.

CORAL REEFS IN DECLINE   Studies concerning the health of coral reefs worldwide show that they are in rapid decline due to various human and environmental factors. A recent survey of coral reef ecosystems, for example, shows that only 30% are healthy now, down from 41% in 2000. Another study estimated that more than one-third of the major reef-building coral species are currently at high risk of extinction, with severe consequences for entire reef ecosystems. In the Caribbean, the area of sea floor covered by live hard coral has decreased by a staggering 80% in the past 30 years. Even the Great Barrier Reef, widely regarded as one of the world's most pristine coral reefs, has lost over half of its coral cover during the past 40 years.

The most serious threat to coral reefs—overshadowing natural cataclysms such as hurricanes, floods, and tsunami—is human activity. Overfishing, for example, has depleted the populations of many of the fishes that graze on algae and keep it from smothering the reefs. Runoff laden with sediment and pollutants further fuels the growth of algae and spreads harmful bacteria. Even more threatening to coral reefs are increased levels of human-caused atmospheric carbon dioxide, which is absorbed into the ocean and increases ocean acidity, making it more difficult for corals to create their calcium carbonate skeletons.[6] In addition, global warming as a result of human activities has been shown to have increased ocean surface temperatures, affecting the growth rate of temperature-sensitive corals worldwide and making them more prone to disease and bleaching episodes. In addition, the predicted rise in sea level as a result of global warming may wreak havoc on corals by submerging them more deeply, in effect reducing the amount of sunlight they receive. The future for healthy coral reefs looks bleak unless immediate and dramatic conservation measures are enacted to preserve them.

CONCEPT CHECK 15.3

1 Discuss the dominant species of kelp, their epifauna, and animals that feed on kelp in Pacific coast kelp forests.

2 Describe the environmental conditions required for development of coral reefs.

3 Describe the zones of the reef slope, the characteristic coral types, and the physical factors related to its zonation.

4 What is coral bleaching? How does it occur? What other diseases affect corals?

15.4 What Communities Exist on the Deep-Ocean Floor?

The vast majority of the ocean floor lies submerged below several kilometers of water. Less is known about life in the deep ocean than about life in any of the shallower nearshore environments because investigating the deep sea is difficult and expensive. For example, just obtaining samples from the deep-ocean floor requires a specially designed submersible or a properly equipped research vessel that has a spool of high-strength cable at least 12 kilometers (7.5 miles) long. In the past, the inaccessibility of ocean depths fueled much debate about the existence of life in the deep ocean (see Web Diving Deeper 15.1).

Even today, collecting samples by submersible or with a biological dredge is a time-consuming process. Because the supply of oxygen is limited, submersibles with humans can stay down for only about 12 hours, and it may take 8 of those hours to descend and ascend. To send a dredge from a ship to the deep-ocean floor and retrieve it from the depths takes about 24 hours.

Robotics and remotely operated vehicles (ROVs) are making it possible to observe and sample even the deepest reaches of the ocean more easily. Because they are unmanned, ROVs are cheaper to operate and can stay beneath the surface for months, if necessary. These developments should lead to further discoveries in one of Earth's least-known habitats.

---

[6]For more information on the recent increase in ocean acidity and other climate change issues, see Chapter 16, "The Oceans and Climate Change."

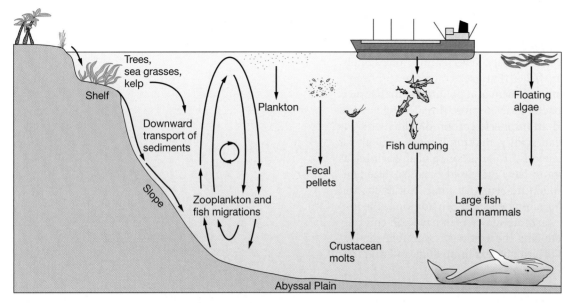

Figure 15.23 Food sources for deep-sea organisms. Most deep-sea organisms obtain their food from surface water after it slowly settles through the water column to the sea floor. The food supply is usually limited, except when large fish or mammals (such as whales) sink to the bottom.

The Physical Environment

The deep-ocean floor includes bathyal, abyssal, and hadal zones.[7] Here, the physical environment is much different than at the surface; it is quite stable and uniform. Light is present in only the lowest concentrations down to a maximum of 1000 meters (3300 feet) and absent below this depth. The temperature rarely exceeds 3°C (37°F) at the deep-ocean floor, and it falls as low as −1.8°C (28.8°F) in the high latitudes. Salinity remains at slightly less than 35‰.[8] Oxygen content is constant and relatively high. Pressure exceeds 200 kilograms per square centimeter (200 atmospheres [2940 pounds per square inch]) on the oceanic ridges, exceeds 300 to 500 kilograms per square centimeter (300 to 500 atmospheres [4410 to 7350 pounds per square inch]) on the deep-ocean abyssal plains, and exceeds 1000 kilograms per square centimeter (1000 atmospheres [14,700 pounds per square inch]) in the deepest trenches.[9] Bottom currents are generally slow but more variable than once believed. For example, **abyssal storms** created by warm- and cold-core eddies of surface currents affect certain areas, lasting several weeks and causing bottom currents to reverse and/or increase in speed.

A thin layer of sediment covers much of the deep-ocean floor. On abyssal plains and in deep trenches, sediment is composed of mudlike abyssal clay deposits. The accumulation of oozes—composed of dead planktonic organisms that have sunk through the water column—occurs on the flanks of oceanic ridges and rises. On the continental rise, there may be some coarse sediment from nearby land sources. Sediment may be absent on steep areas of the continental slope. It may also be absent near the crest of the mid-ocean ridge and along the slopes of seamounts and oceanic islands, where it has not had enough time to accumulate on newly formed ocean floor.

Food Sources and Species Diversity

Because of the lack of light at the deep-ocean floor, photosynthetic primary production cannot occur. Except for the chemosynthetic productivity that occurs around hydrothermal vents, all benthic organisms receive their food from the surface waters above. Only about 1% to 3% of the food produced in the euphotic (sunlit) zone reaches the deep-ocean floor, so the scarcity of food that drifts down from the sunlit surface waters—not low temperature or high pressure—limits deep-sea benthic biomass. However, some variability in the supply of food is caused by seasonal phytoplankton blooms at the surface. **Figure 15.23** shows the food sources for deep-sea organisms.

Many of the organisms that inhabit the deep sea have special adaptations to help them detect food using chemical clues. Once food is found, these organisms are efficient at consuming it (**Diving Deeper 15.1**).

---

[7]The bathyal, abyssal, and hadal zones are described in Chapter 12.

[8]Recall that average surface seawater salinity is 35 parts per thousand (‰).

[9]The pressure is 1 atmosphere (1 kilogram per square centimeter [14.7 pounds per square inch]) at the ocean surface and increases by 1 atmosphere for each 10 meters (33 feet) of depth. Thus, a pressure of 1000 atmospheres is 1000 times that at the ocean's surface.

Diving Deeper 15.1 Research Methods in Oceanography

HOW LONG WOULD YOUR REMAINS REMAIN ON THE SEA FLOOR?

What happens to people who are buried at sea? How long do their remains remain on the sea floor? How long do the remains of a large organism such as a whale remain on the sea bottom? Oceanographers who study deep-sea biocommunities have conducted experiments in the deep sea to help answer these questions.

One experiment was conducted along the sea floor in the Philippine Trench, at a depth of 9600 meters (31,500 feet), in 1975. Several whole fish were placed on the sea floor, and an underwater camera positioned above them took a picture every few minutes to observe how long the bait remained there (**Figure 15A**). *Hirondellea gigas*, a scavenging benthic shrimplike amphipod, discovered the bait after only a few hours. The bait was swarming with amphipods after 9 hours, and it was stripped of flesh in 16 hours! Other studies obtained similar results, suggesting that organisms the size of humans would have their soft tissue devoured within a day on the deep-ocean floor.

For deep-sea organisms, large food falls represent an unpredictable but intense nutrient supply, instantaneously depositing

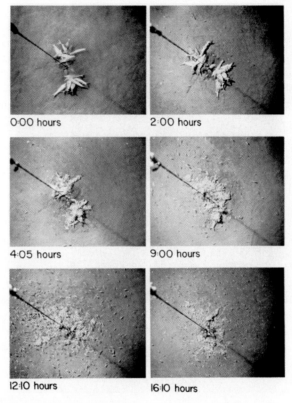

0:00 hours 2:00 hours

4:05 hours 9:00 hours

12:10 hours 16:10 hours

Figure 15A Time-sequence photography of fish remains on the deep-ocean floor.

as much food as would be delivered to the deep sea by the normal rain of detritus in about 2000 years. Deep-sea scavengers such as amphipods, hagfish, and sleeper sharks use special chemoreceptive sensory devices to identify and quickly locate food on the sea floor. In addition, whale carcasses can support a thriving ecosystem of benthic organisms—including some species normally associated with hydrothermal vents—that arrive on deep-ocean currents as drifting larvae.

To test how long a whale's remains lasted on the sea floor, researchers used two dead juvenile gray whales that washed up on the beach in Southern California in 1996 and 1997. With permission from the National Marine Fisheries Service, the dead whales—each of which weighed about 5 metric tons (5.5 short tons)—were intentionally weighted and sunk in the San Diego Trough. Researchers in a deep-diving submersible visited the carcasses at regular intervals and found that the gray whales were completely stripped of flesh in four months! Other studies indicate that even blue whales, which can weigh 25 times more than a gray whale, are stripped of their flesh in as little as six months.

For many years, it was believed that the species diversity of the deep-ocean floor was quite low compared with shallow-water communities. Researchers studying sediment-dwelling animals in the North Atlantic, however, discovered an unexpectedly large diversity of species. An area of 21 square meters (225 square feet) contained 898 species, of which 460 were new to science. After analyzing 200 samples, new species were being discovered at a rate that suggested millions of deep-sea species!

It turns out that deep-sea species diversity—especially for small infaunal deposit feeders—rivals that of tropical rainforests. It also appears, however, that the distribution of deep-sea life is patchy and depends to a large degree on the presence of certain microenvironments.

ESSENTIAL CONCEPT 15.4A

The deep-ocean floor is a stable environment of darkness, cold water, and high pressure, but it still supports life. The food for most deep-sea organisms comes from sunlit surface waters.

Deep-Sea Hydrothermal Vent Biocommunities: Organisms and Their Adaptations

The surprising discovery of deep-sea **hydrothermal** (*hydro* = water, *thermo* = heat) **vents** and their biocommunities has been one of the most important finds in the modern history of ocean discovery. Life at hydrothermal vents may provide insights into both the origin of life on Earth and its possible existence elsewhere in the solar system.

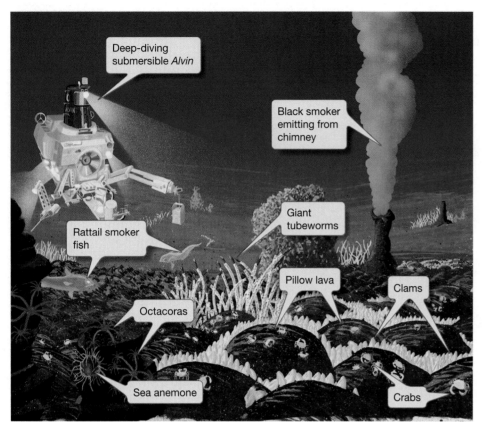

DISCOVERY OF HYDROTHERMAL VENT BIOCOMMU-NITIES   An active hydrothermal vent field on the ocean floor was visited for the first time in 1977, during a dive of the submersible *Alvin*. The field exists in complete darkness, in water below 2500 meters (8200 feet) in the Galápagos Rift, near the equator in the eastern Pacific Ocean (**Figures 15.24** and **15.25**). Hot water was observed spewing from cracks in the sea floor and from tall chimneys. Water temperature near the vents was 8 to 12°C (46 to 54°F), whereas normal water temperature at these depths is about 2°C (36°F).

These vents supported the first known **hydrothermal vent biocommunities**, consisting of organisms that had been unknown to science and

Figure 15.24 *Alvin* **approaches a hydrothermal vent biocommunity.** Schematic drawing of the deep-diving submersible *Alvin* approaching a hydrothermal vent area. Note the active black smoker spewing hot (350°C [662°F]), sulfide-rich water from a chimney, pillow lava from underwater volcanic eruptions, and abundant vent organisms.

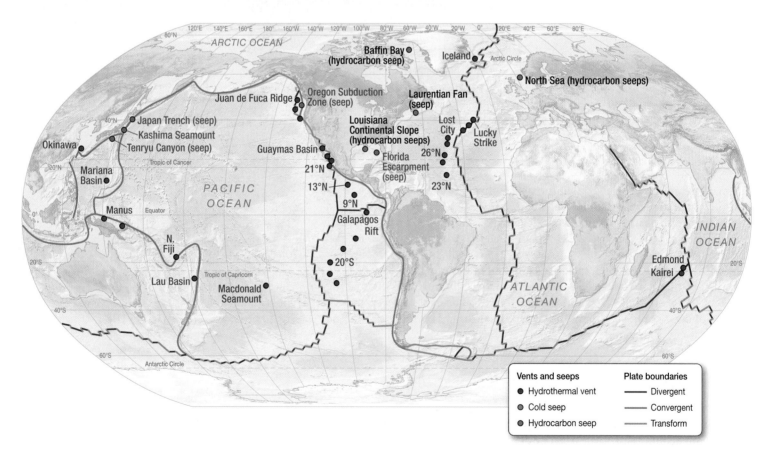

Figure 15.25 Vents and seeps known to support deep-sea biocommunities. Location map of major hydrothermal vents (*red dots*), cold seeps (*blue dots*), and hydrocarbon seeps (*brown dots*). Colored lines indicate plate boundaries.

Figure 15.26
Chemosynthetic life. Photos of various hydrothermal vent organisms that rely on chemosynthesis, including **(a)** tubeworms, **(b)** sulfur-oxidizing archaea, and **(c)** a hydrothermal vent biocommunity.

(a) Tubeworms (*Riftia*) up to 1 meter (3.3 feet) long are found at the Galápagos Rift and other deep-sea hydrothermal vents.

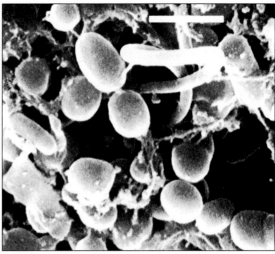

(b) Sulfur-oxidizing archaea (enlarged 20,000 times) that live symbiotically within the tissue of tubeworms, clams, and mussels found at hydrothermal vents.

(c) A hydrothermal vent biocommunity from the Mariana Back-Arc Basin, which included a new genus and species of sea anemone (*Marianactis bythios*); the gastropod *Alviniconcha hessleri*, which is the first known conch to contain chemosynthetic bacteria; and the galatheid crab (*Munidopsis marianica*).

were unusually large for those depths. The most prominent species include giant tubeworms up to 1.8 meters (6 feet) long (**Figure 15.26a**), clams as large as 25 centimeters (10 inches) across, large mussels, two varieties of white crabs, and extensive microbial mats. These biocommunities had up to 1000 times more biomass than the rest of the deep-ocean floor. In a region of scarce nutrients and small populations of organisms, these hydrothermal vents are truly the oases of the deep ocean.

In 1979, south of the tip of Baja California at 21 degrees north latitude on the East Pacific Rise, tall underwater chimneys were discovered belching hot vent water (350°C [662°F]) so rich in metal sulfides that it colored the water black. These chimney vents, which are composed primarily of sulfides of copper, zinc, and silver, came to be called **black smokers** because of their resemblance to factory smokestacks.

A typical hydrothermal vent may emit fluids that are not only hot (upward of 350°C [662°F]) but also acidic (pH = 3–4) and extremely toxic, with high concentrations of dissolved hydrogen sulfide and heavy metals, such as cadmium, arsenic, and lead. Yet the dark water venting from black smokers has been found to be rich in microbes.

CHEMOSYNTHESIS   The most important members of hydrothermal vent biocommunities are microscopic **archaea** (*archaeo* = ancient) (**Figure 15.26b**), which are primitive single-celled organisms that resemble bacteria but have chemical similarities to multicelled organisms. Archaea thrive on sea floor chemicals—most notably hydrogen sulfide—and perform **chemosynthesis** (*chemo* = chemistry, *syn* = with, *thesis* = an arranging), manufacturing carbohydrates (sugar) from water, carbon dioxide, and dissolved oxygen; sulfuric acid is produced as a byproduct

Figure 15.27 Chemosynthesis (*top*) and representative reaction viewed chemically (*bottom*). The process of chemosynthesis, which is accomplished by archaea, is represented in the top panel and is shown chemically in the bottom panel. Compare this figure with photosynthesis (Figure 13.1).

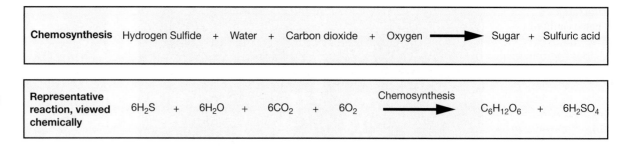

Chemosynthesis	Hydrogen Sulfide	+	Water	+	Carbon dioxide	+	Oxygen	⟶	Sugar	+	Sulfuric acid

Representative reaction, viewed chemically	$6H_2S$	+	$6H_2O$	+	$6CO_2$	+	$6O_2$	$\xrightarrow{\text{Chemosynthesis}}$	$C_6H_{12}O_6$	+	$6H_2SO_4$

(Figure 15.27). By chemosynthesizing, archaea form the base of the food web for vent ecosystems. Although some animals feed directly on archaea and larger prey, many of them depend primarily on a symbiotic relationship with archaea. Tubeworms and giant clams, for instance, depend entirely on sulfur-oxidizing archaea that live symbiotically within their tissues. The microbes inside the tubeworms are provided habitat and bathed in hydrogen sulfide, which they chemosynthesize into sugars. The tubeworms, in return, get a steady supply of food and grow prolifically, sometimes gaining 80 centimeters (31 inches) in a year.

Recent analysis of the genome of the symbiotic archaea that lives inside tubeworms reveals that the microbe is remarkably versatile. For example, researchers found that it can use two different methods to metabolize carbon dioxide and can switch back and forth to accommodate fast-changing environmental conditions. Such metabolic flexibility is a valuable asset in deep-sea vent habitats characterized by fluctuating flows of hot fluids.

DISCOVERY OF OTHER HYDROTHERMAL VENT FIELDS In 1981, humans in a submersible first visited the Juan de Fuca Ridge biocommunity offshore of Oregon. Although vent fauna at this site are less abundant than at the Galápagos Rift and on the East Pacific Rise, the metallic sulfide deposits from the vents aroused much interest because they are the only active hydrothermal vent deposits in U.S. waters.

In 1982, the first hydrothermal vents beneath a thick layer of sediment were discovered during a submersible dive in the Guaymas Basin of the Gulf of California. In this region, a spreading center is actively working to rift apart the sea floor as it is being covered with sediment. Sediment samples recovered in this region were high in sulfide and saturated with hydrocarbons, which may have entered the food chain through bacteria. The abundance and diversity of life discovered here may exceed those of the Galápagos Rift and along the East Pacific Rise.

Like the Guaymas Basin, the Mariana Basin of the western Pacific has a small spreading center beneath a sediment-filled basin. A research dive via submersible in 1987 revealed many new species of hydrothermal vent organisms (Figure 15.26c). Subsequent exploration has revealed numerous hydrothermal vent biocommunities in other parts of the Pacific Ocean (see Figure 15.25) and additional new species. In fact, more than 400 new species have been found so far at vent sites worldwide.

In 1985, the first active hydrothermal vents with associated biocommunities in the Atlantic Ocean were discovered at depths below 3600 meters (11,800 feet), near the axis of the Mid-Atlantic Ridge between 23 and 26 degrees north latitude. The predominant fauna of these vents consists of shrimp that have no eye lens but can detect levels of light emitted by the black smoker chimneys that are invisible to the human eye (Figure 15.28).

In 1993, a hydrothermal vent community was discovered on a flat-topped volcano rising to 1525 meters (5000 feet)—well above the walls of the rift valley along the Mid-Atlantic Ridge. Called the "Lucky Strike" vent field, it is about 1000 meters (3300 feet) shallower than most other sites. It is the only Mid-Atlantic Ridge site known to contain the mussels that are common at many other vent sites and is the only location where a new species of pink sea urchin has been found.

In 2000, Japanese investigators discovered the Indian Ocean's first hydrothermal vent biocommunity, which is associated with black smokers spewing water up to 365°C (689°F). The vents are covered by shrimp similar to those found in Atlantic

Figure 15.28 Atlantic Ocean hydrothermal vent organisms. Swarm of particulate-feeding shrimp, the predominant animals observed at hydrothermal vents near 26 degrees north latitude on the Mid-Atlantic Ridge. Most of these shrimp are about 5 centimeters (2 inches) long.

fields, while sea anemones mark the ambient temperature boundaries beyond. In between are clusters of animals similar to those found at other hydrothermal vents.

Also in 2000, researchers discovered the "Lost City" hydrothermal vent field about 15 kilometers (9 miles) west of the Mid-Atlantic Ridge. Remarkably, it has much lower temperatures (about 90°C [194°F]), a pH between 9 and 11 (much more alkaline than most black smokers), and tall chimneys made of calcium carbonate (instead of metal sulfides), and it releases methane and hydrogen into the surrounding water (not hydrogen sulfide or dissolved metals, which are the major outputs of volcanic black smokers). Unlike the activity at black smokers, which is driven by water superheated by shallow magma, Lost City vent activity is driven by serpentinization reactions of underlying mantle rock. Scientists revisited the site in 2003 to collect samples, confirming that Lost City supports vastly different life-forms than black smokers—including a variety of microorganisms that live in, on, and around the vents. Scientists have suggested that life on Earth may have originated in warm, alkaline settings similar to those found today in Lost City.

Vents often differ dramatically from each other in chemical and geological characteristics. However, even vents that are physically similar can host distinctly different communities of organisms. For example, giant tubeworms are found only in Pacific vents. In the North Atlantic, meanwhile, shrimp and mussels dominate. And, on the deep-sea floor near Antarctica, the yeti crab (see Figure 12.8) is found in large numbers. Researchers are currently studying dispersal patterns and determining biogeographic relationships between vent sites to help explain this phenomenon.

Today, researchers continue to use submersibles to study deep-sea hydrothermal vents. Although there is direct evidence of about 300 hydrothermal vents worldwide, researchers estimate that another 700 or so await discovery. Every visit to a vent site—even a repeat visit—reveals new information about how vents work and the uniquely adapted microbes and other organisms that live there. In fact, the research effort is so intense at some sites that researchers have recently adopted a self-enforced code of conduct to ensure that humans do not alter vent ecosystems.

LIFE SPAN OF HYDROTHERMAL VENTS   Because the hot-water plumbing of the sea floor is controlled by sporadic volcanic activity associated with mid-ocean ridge spreading centers, a vent may remain active for only limited periods—years or sometimes decades. For instance, a hydrothermal vent field called the Coaxial Site along the Juan de Fuca Ridge offshore of Washington that had been active was revisited a few years later and found to be inactive. Inactive sites such as this one are identified by an accumulation of large numbers of dead hydrothermal vent organisms. When the vent becomes inactive and the hydrogen sulfide that serves as the source of energy for the community is no longer available, organisms of the community die if they cannot move elsewhere.

Other sites indicate an increase in volcanic activity. For example, at a site along the East Pacific Rise known as Nine-Degrees North, a large number of tubeworms were cooked by lava flowing into their midst, in what has been described as a "tubeworm barbecue." The discovery of newly formed and ancient vent areas along spreading centers indicates that hydrothermal vents can suddenly appear or cease to operate. Moreover, areas of active venting may lie tens or even hundreds of kilometers apart.

Hydrothermal vent organisms are well adapted to the temporary nature of hydrothermal vents. Most have high metabolic rates, for example, which cause them to mature rapidly so they can reproduce while the vent is still active.

Studies of several hydrothermal vent sites suggest that species diversity is low. In fact, only about 300 animal species have been identified to date. Many species, however, are common to widely separated hydrothermal vent fields. Although hydrothermal vent animals typically release drifting larvae into the water, it is not clear how the larvae—which can survive a few months at most—are able to survive the journey to hydrothermal vents that lie at such great distances from one another.

One idea, called the *dead whale hypothesis*, suggests that when large animals die, they may sink to the deep-ocean floor, decompose, and provide an energy source in stepping-stone fashion for the larvae of hydrothermal vent organisms. The organisms

STUDENTS SOMETIMES ASK . . .

Are the mussels found at hydrothermal vents edible?

Not by humans. The microbes that form the base of the food web use hydrogen sulfide gas (which has a characteristic "rotten egg" odor) as a source of energy. To most organisms, sulfide is a deadly poison, even at low levels, and tends to concentrate in tissues. Although organisms within the hydrothermal biocommunities can ingest sulfide and have mechanisms of getting rid of it, hydrothermal vent organisms would be toxic to humans. Even if they were edible, though, mussels found at hydrothermal vents would be expensive to harvest because they live at such great depths.

settle and grow where a large dead animal rests on the deep-ocean floor; then they breed and release their own larvae, some of which make it to the next hydrothermal vent field. Other studies suggest that the rift valleys of mid-ocean ridges act as passageways that drifting larvae traverse to inhabit new vent fields. Still other recent research that involves dispersing and then tracking chemical tracers indicates that deep-ocean currents are strong enough to transport drifting larvae to new sites. By whatever means they travel, they colonize new hydrothermal vents soon after the vents are created. In 2006, for example, a well-studied hydrothermal vent community along the East Pacific Rise had been wiped out by seafloor eruptions. In 2007 and 2008, researchers returned to the area and discovered that tubeworms and other life-forms had already established themselves.

HYDROTHERMAL VENTS AND THE ORIGIN OF LIFE Life is thought to have begun in the oceans, and environments similar to those of the hydrothermal vents must have been present in the early history of the planet. The uniformity of conditions and abundant energy of the vents, therefore, have led some scientists to propose that hydrothermal vents would have provided an ideal habitat for the origin of life. In fact, hydrothermal vents may represent one of the oldest life-sustaining environments because hydrothermal activity occurs wherever there are both volcanoes and water. The presence of bacteria-like archaea, which have ancient genetic makeup, helps support this idea.

The ancient status of deep-sea life was strengthened by the recent discovery of deep-sea microbes with identical genes to those of microbes found in the human body. Researchers isolated two previously unknown bacterial species from hydrothermal vents near Japan and compared the new species' genomes to the genomes of two common gut pathogens, one that causes ulcers and another that causes diarrhea. The comparison shows that despite eons of evolutionary change, the deep-sea species and the pathogens share genes that enable them to colonize animal hosts. According to the researchers, genes that likely help deep-sea bacteria maintain symbiotic relationships with other vent-dwelling organisms assist their gut-dwelling relatives in evading immune systems. The researchers suggest that the human-harming microbes evolved from deep-sea ancestors and later acquired more virulence factors while living in symbiosis with animals.

Figure 15.29 Hypersaline seep biocommunity at the base of the Florida Escarpment. (a) Location map, **(b)** seismic reflection profile, and **(c)** underwater photograph of the Florida Escarpment hypersaline seep.

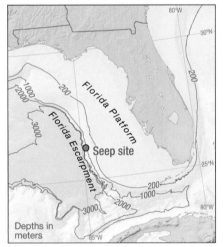

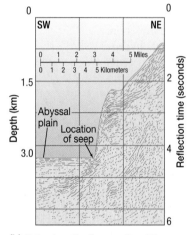

(a) Location of the Florida Escarpment hypersaline seep.

(b) Seismic reflection profile of the Florida Escarpment, showing the location of the hypersaline seep at its base.

Low-Temperature Seep Biocommunities: Organisms and Their Adaptations

Three additional submarine seep environments—locations where water trickles out of the sea floor—have been found that chemosynthetically support biocommunities similar to hydrothermal vent communities. Because they exist at much lower temperatures, these environments are known as **cold seeps**.

HYPERSALINE SEEPS In 1984, a hypersaline seep was studied in water depths below 3000 meters (9800 feet) at the base of the Florida Escarpment in the Gulf of Mexico (**Figure 15.29a**). The water from this seep had a salinity of 46.2‰, but its temperature was not warmer than normal. Researchers had discovered a **hypersaline seep biocommunity** similar in many respects to a hydrothermal vent biocommunity. The seeping water appears to flow from fractures at the base of a limestone escarpment (**Figure 15.29b**) and move out across the clay deposits of the abyssal plain at a depth of about 3200 meters (10,500 feet).

The hydrogen sulfide–rich waters support a number of white microbial growths called *mats*, which conduct chemosynthesis

(c) Florida Escarpment seep biocommunity of dense mussel beds. White dots are small gastropods on mussel shells. Tubeworms (*lower right*) are covered with hydrozoans and galatheid crabs.

in a fashion similar to archaea at hydrothermal vents. These and other chemosynthetic microbes may provide most of the sustenance for a diverse community of animals, including sea stars, shrimp, snails, limpets, brittle stars, anemones, tubeworms, crabs, clams, mussels, and a few species of fish (Figure 15.29c).

HYDROCARBON SEEPS   Also observed in 1984 were dense biological communities associated with oil and gas seeps on the Gulf of Mexico continental slope (Figure 15.30). Trawls at depths of between 600 and 700 meters (2000 and 2300 feet) recovered fauna similar to those observed at hydrothermal vents and at the hypersaline seep in the Gulf of Mexico. Subsequent investigations have identified nearly 100 seeps on the continental slope that have the potential of hosting chemosynthetic biocommunities; 10 of these have been visited with submersibles to depths of 2775 meters (9100 feet), where chemosynthetic bacteria and a host of other organisms were discovered.

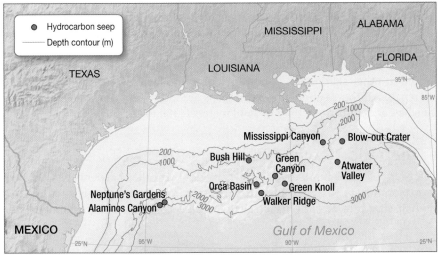

(a) Map showing locations of known hydrocarbon seeps that contain biocommunities in the Gulf of Mexico.

Figure 15.30 Hydrocarbon seeps on the continental slope in the Gulf of Mexico. (a) Location map and **(b** and **c)** underwater photographs of the Gulf of Mexico hydrocarbon seeps.

(b) Photo showing abundant mussels and tubeworms at Neptune's Gardens (Alaminos Canyon site).

(c) Close-up photo of chemosynthetic mussels and tubeworms from the Bush Hill seep.

Carbon-isotope analysis indicates that these **hydrocarbon seep biocommunities** are based on chemosynthesis that derives its energy from hydrogen sulfide and/or methane. Microbial oxidation of methane produces calcium carbonate slabs found here and at other hydrocarbon seeps (see Figure 15.25).

SUBDUCTION ZONE SEEPS In 1984, a **subduction zone seep biocommunity** was discovered during one of *Alvin*'s dives to study folding of the sea floor in a subduction zone. The seep is located near the Cascadia subduction zone of the Juan de Fuca Plate, at the base of the continental slope off the coast of Oregon (**Figure 15.31a**). The trench is filled with sediments, which are folded into a ridge at the seaward edge of the slope. At the crest of this ridge, water slowly flows from the 2-million-year-old folded sedimentary rocks into a thin overlying layer of soft sediment on the sea floor. Eventually, the water is released from the sediment through seeps on the ocean floor.

At a depth of 2036 meters (6678 feet), the seeps produce water that is only about 0.3°C (0.5°F) warmer than seawater at that depth. The vent water contains methane that is probably produced by decomposition of organic material in the sedimentary rocks. Microbes oxidize the methane, chemosynthetically producing food for themselves and the rest of the community, which contains many of the same genera found at other vent and seep sites (**Figure 15.31b**).

Since the detection of subduction zone seeps, similar communities have been discovered in other subduction zones, including the Japan Trench and the Peru–Chile Trench. All these subduction zone seeps are located on the landward side of the trenches, at depths from 1300 to 5640 meters (4300 to 18,500 feet).

The Deep Biosphere: A New Frontier

The discovery of the rich microbe communities at hydrothermal vents has resulted in the exploration of the **deep biosphere**, an environment that exists within the sea floor itself. Only recently have scientists even considered that microbial life

---

ESSENTIAL CONCEPT 15.4B

Hydrothermal vent biocommunities occur near black smokers and rely on chemosynthetic archaea for food. Other deep-sea cold seep biocommunities that depend on chemosynthesis exist around hypersaline, hydrocarbon, and subduction zone seeps.

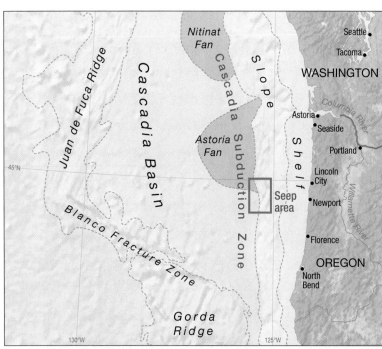

(a) Map showing sea floor features and the location of vent biocommunities off the coast of Oregon. These communities are associated with the Cascadia subduction zone, where sediment filling the trench is folded into a ridge with vents at its crest.

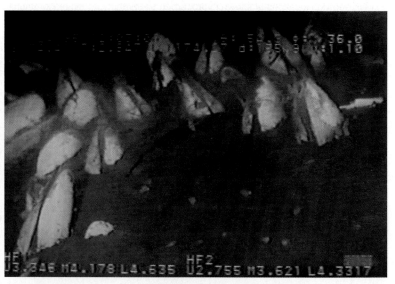

(b) Photo of giant white clams (*Calyptogena soyoae*) half buried in methane-rich mud at a depth of 1100 meters (3600 feet) near the Japan Trench. The clams host sulfide-oxidizing microorganisms, which supply the clams with food.

Figure 15.31 Subduction zone seep biocommunities. (a) Location map and **(b)** underwater photograph of a subduction zone seep biocommunity.

might exist deep within Earth, and in 2002, researchers made the first expedition to study life in this environment. Cores were drilled up to 420 meters (1380 feet) deep into the sea floor off Peru, in water depths between 150 and 5300 meters (490 and 17,400 feet). At these depths, researchers discovered a host of diverse and active microbial communities living within circulating fluids that pass through the porous sea floor. Other subsequent research confirms the abundance and diversity of microbes in deep-sea floor rocks and associated sediments that rival the rich microbial ecosystems found in soil.

These studies suggest that as much as two-thirds of Earth's entire bacterial biomass might exist in the deep biosphere. Moreover, deep biosphere microbes fuel their metabolisms by taking advantage of the chemical energy stored in various minerals. In addition, the findings raise new questions about the role of the deep-sea floor in the evolution of life on Earth. Intriguingly, other bodies in the solar system have similar subsurface conditions, so they might harbor microbes, too. Earth's deep biosphere will therefore continue to be an area of active research.

CONCEPT CHECK 15.4

1 Where does the food come from to supply organisms living on the deep-ocean floor? How does this affect benthic biomass?

2 Describe the characteristics of hydrothermal vents. What evidence suggests that hydrothermal vents have short life spans?

3 What is the "dead whale hypothesis"? What other ideas have been suggested to help explain how organisms from hydrothermal vent biocommunities populate new vent sites?

4 What are the major differences between the conditions and biocommunities of hydrothermal vents and cold seeps? How are they similar?

Essential Concepts Review

15.1 What communities exist along rocky shores?

▶ *More than 98% of the 250,000 known marine species live in diverse environments within or on the ocean floor. Species diversity* of these benthic organisms *depends on their ability to adapt to the conditions of their environment,* particularly temperature. With few exceptions, the *biomass of benthic organisms closely matches that of photosynthetic productivity in surface waters above.*

▶ *Many adverse conditions exist in the intertidal zone of rocky shores, but organisms have adapted* so they can densely populate these environments. Influenced by the tides, rocky shores can be divided into a *high tide zone* (mostly dry), a *middle tide zone* (equally wet and dry), and a *low tide zone* (mostly wet). The *intertidal zone* is bounded by the *supratidal zone,* which is covered only by storm waves, and the *subtidal zone,* which extends below the low tide shoreline.

▶ *Each of these zones contains characteristic types of marine life.* Periwinkle snails, rock lice, and limpets can be found in the subtidal zone. Sessile organisms, such as buckshot barnacles, can be found in the high tide zone. *Algae become more abundant in the middle tide zone, and the diversity and abundance of the flora and fauna increase toward the lower intertidal zone.* Acorn barnacles, gooseneck barnacles, mussels, and sea stars are commonly found in the middle tide zone, as are sea anemones, fishes, hermit crabs, and sea urchins. The low tide zone in temperate latitudes has a variety of moderately sized red and brown algae that provide a drooping canopy for animal life.

Study Resources
Online Study Guide Quizzes

Critical Thinking Question

From memory, construct a diagram that shows a rocky-shore intertidal region. Name each zone and include characteristic organisms that are typically found there.

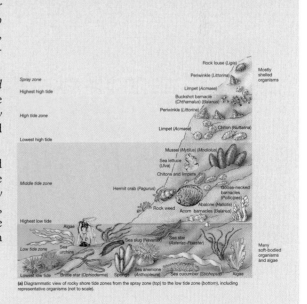

(a) Diagrammatic view of rocky shore tide zones from the spray zone (top) to the low tide zone (bottom), including representative organisms (not to scale).

Essential Concepts Review (cont.)

15.2 What communities exist along sediment-covered shores?

▶ *Many varieties of burrowing infauna are common along sediment-covered shores* (that is, beaches, salt marshes, and mud flats). Compared with rocky shores, however, *sediment-covered shores have less diversity of species.* As with the rocky shore, *the diversity of species and abundance of life on the sediment-covered shore increase toward the low tide shoreline.*

▶ In more protected segments of the shore, wave energy is lower, so sand and mud are deposited. Sand deposits are usually well oxygenated compared to mud deposits. *The intertidal region of sediment-covered shores has high, middle, and low tide zones, similar to rocky shores.* Organisms characteristic of *sandy beaches* include bivalve (two-shelled) mollusks, lugworms, beach hoppers, sand crabs, sand stars, and heart urchins. Organisms characteristic of *mud flats* include eelgrass, turtle grass, bivalve mollusks, and fiddler crabs.

Study Resources
Online Study Guide Quizzes

Critical Thinking Question

From memory, construct a diagram that shows a sand-covered intertidal region. Name each zone and include characteristic organisms that are typically found there.

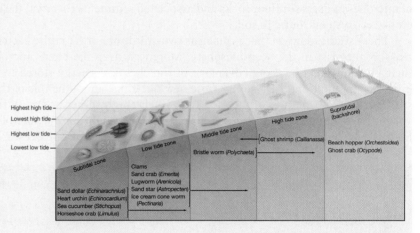

15.3 What communities exist on the shallow offshore ocean floor?

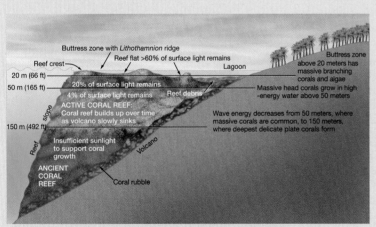

▶ *Attached to the rocky subtidal bottom just beyond the shoreline is a band of algae that often creates kelp forests.* Kelp forests are home to many organisms, including other varieties of algae, mollusks, sea stars, fishes, octopus, lobsters, marine mammals, sea hares, and sea urchins.

▶ *Spiny lobsters* are common to rocky bottoms in the Caribbean and along the West Coast, and the *American (Maine) lobster* is found from Labrador, Canada, to Cape Hatteras, North Carolina. Oyster beds found in estuarine environments consist of individuals that attach themselves to the bottom or to the empty shells of previous generations.

▶ *Coral reefs consist of large colonies of coral polyps and many other species that need warm water and strong sunlight to live. Coral reefs are usually found in nutrient-poor tropical waters.* Reef-building corals and other mixotrophs are *hermatypic,* containing *symbiotic algae (zooxanthellae)* in their tissues. Delicate varieties are found at 150 meters (500 feet), and they become more massive near the surface, where wave energy is higher. *The potentially lethal "bleaching" of coral reefs is caused by the removal or expulsion of symbiotic algae,* probably under the stress of elevated sea surface temperatures.

Study Resources
Online Study Guide Quizzes, Web Video

Critical Thinking Question

Describe the relationship between corals and algae and explain how corals benefit from having internal algae.

15.4 What communities exist on the deep-ocean floor?

▶ *The physical conditions of the deep-ocean floor* are much different from those of shallow water. There is *no light*, and the *water is
uniformly cold. The primary food source is from the surface waters
above*, which limits biomass. Species diversity in the deep ocean,
however, is much higher than was previously thought.

▶ *Primary production in hydrothermal vent communities near black
smokers is due to chemosynthesis.* Some evidence suggests that
*hydrothermal vents may have been some of the first regions where life
became established on Earth,* despite the short life span of individual
vents. Chemosynthesis has also been identified in low-temperature
seep biocommunities near *hypersaline, hydrocarbon,* and *subduction zone seeps.* Studies of the *deep biosphere* below the ocean floor
reveal a host of microbes.

Study Resources
Online Study Guide Quizzes, Web Diving Deeper 15.1

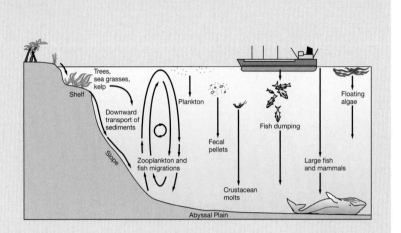

Critical Thinking Question

Describe changes in the physical environment that occur as one
moves from the shoreline to the deep-ocean floor.

MasteringOceanography™

www.masteringoceanography.com

Looking for additional review and test prep materials? Visit the Study
Area in MasteringOceanography to enhance your understanding of
this chapter's content by accessing a variety of resources, including
Self-Study Quizzes, Geoscience Animations, RSS feeds, flashcards,
Web links, and an optional Pearson eText.

Smokestacks spew human-caused emissions into the atmosphere. Emissions from the combustion of fossil fuels by power plants, cars, and factories are released into the atmosphere. These human-caused emissions are affecting all parts of the environment and especially the oceans.

THE OCEANS AND CLIMATE CHANGE

Before you begin reading this chapter, use the glossary at the end of this book to discover the meanings of any of these words you don't already know:

Before you begin reading this chapter, use the glossary at the end of this book to discover the meanings of any of these words you don't already know:

Proxy Feedback loop
Positive-feedback loop pH Greenhouse gases Paleoclimatology
Carbon dioxide Sunspots Sequester Heat budget Greenhouse effect
Faculae Ocean acidification
Intergovernmental Panel on Climate Change (IPCC)
Iron hypothesis fossil fuel Biological pump Kyoto Protocol Methane Thermocline
Phytoplankton Climate system Global warming
Global engineering
Negative-feedback loop

ESSENTIAL CONCEPTS

At the end of this chapter, you should be able to:

16.1 Describe the components of Earth's climate system

16.2 Discuss whether Earth's recent climate change is natural or caused by human influence

16.3 Demonstrate an understanding of how the atmosphere's greenhouse effect works

16.4 Describe the changes that are occurring in the oceans as a result of global warming

16.5 Discuss what should be done to reduce greenhouse gases

"Human-induced climate change is a reality, not only in remote polar regions and in small tropical islands, but everyplace around the country, in our own backyards. It's happening. It's happening now. It's not just a problem for the future. We are beginning to see its impacts in our daily lives. More than that, humans are responsible for the changes that we are seeing, and our actions now will determine the extent of future change and the severity of the impacts."

—Jane Lubchenco, marine ecologist and NOAA chief administrator (2009)

Climate change and global warming are topics that have received much media attention recently. These topics are often in public opinion polls and in newspaper headlines; as such, they have spurred intense debate on whether climate change is natural or human caused and what climate changes are likely to occur in the future. These topics have also become the subject of numerous international conferences and have spurred complicated discussions among journalists and scientists. Human-caused climate change continues to be one of the most studied aspects of climate science.

Throughout its long history, Earth has experienced both warmer and cooler global climates as compared to today's climate. In fact, evidence from fossils, sea floor sediments, and rocks on land suggest that many places on Earth have experienced dramatic swings in climate over geologic time. For example, some regions are known to have remained at cooler high latitudes (after taking into account the movement of tectonic plates over time), yet they exhibit fossils such as corals and coal deposits that are indicative of warmer temperatures. Alternatively, there are other regions that are known to have been located at low latitudes, yet they exhibit sediments likely to have been produced by glaciation.

Numerous climate science studies suggest that human activities—not natural variability—are the cause of the recent documented changes in Earth's climate. Unlike past climate changes, modern climate change is dominated by human influences so large and occurring so rapidly that they exceed the bounds of any natural factors that influence Earth's climate. Moreover, these changes are likely to continue for at least the next 1000 years. Climate changes can be very disruptive not only to humans but to many other life-forms as well, especially if they occur as rapidly as some scientists predict.

In this chapter, we will examine Earth's climate system, the science that indicates Earth's recent and dramatic climate change, how the greenhouse effect works, what effects are being felt in the oceans today, and what can be done about this urgent problem.

16.1 What Comprises Earth's Climate System?

Climate is defined as the conditions of Earth's atmosphere—including temperature, precipitation, and wind—that characteristically prevail in a particular region over extended time spans.

Figure 16.1 Major components of Earth's climate system. Schematic view of the major components of Earth's climate system. Yellow boxes show both natural and human-induced climate changes; orange boxes show interactions between various components of Earth's climate system.

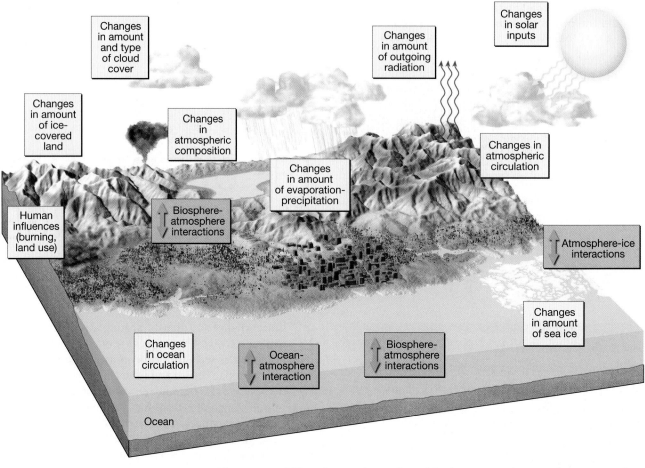

Obtaining a full understanding of Earth's climate involves studying more than just the atmosphere. Earth's climate is a complex and interacting system that includes five spheres: the atmosphere, hydrosphere, geosphere, biosphere, and cryosphere.[1] Earth's **climate system** involves the exchanges of energy and moisture that occur among the five spheres. These exchanges link the atmosphere to the other spheres so that the entire system functions as an interactive unit. Changes to the climate system do not occur in isolation. Rather, when one part of it changes, the other components also react. The major components of Earth's climate system are shown in **Figure 16.1**. Notice that the oceans are the most massive part of Earth's climate system.

Changes in Earth's climate system are large scale, complex, and involve many **feedback loops**, which are processes that modify initial changes. For example, **Figure 16.2** (*left*) shows that warmer surface temperatures increase evaporation rates. This in turn increases water vapor in the atmosphere, which absorbs heat emitted from Earth's surface before it gets released into space. Therefore, the more water vapor in the air, the less heat escapes and the warmer the planet becomes. This type of feedback loop is called a **positive-feedback loop**, or *reinforcing loop*, because it has an amplifying effect that fortifies or adds to an initial change. That is, A produces more of B, which in turn produces more of A.

Alternatively, a **negative-feedback loop**, or *balancing loop*, has a diminishing effect that tends to counteract or subtract from an initial change. One such example is the formation of clouds (Figure 16.2, *right*). A probable result of a global temperature rise is an accompanying increase in cloud cover due to the higher moisture content of the atmosphere. Most clouds are good reflectors of incoming solar energy, thus diminishing the amount of solar energy available to heat Earth's surface and warm the atmosphere. In this way, clouds can cause a decrease in overall air temperature. In this instance, A produces more of B, which in turn produces *less* of A.

Web Video
The Science Behind a
Climate Headline

STUDENTS SOMETIMES ASK . . .

What's the difference between weather and climate?

*W*eather describes the conditions of the atmosphere at a given place and time, whereas *climate* is the long-term average of weather. For example, the expected weather conditions on a particular day will help you determine if you'll wear shorts or thermal underwear that day; the ratio of shorts to thermal underwear in your drawer reflects the climate of the region. Or, as Mark Twain once said about the difference between the two, "Climate is what we expect, weather is what we get."

[1]The cryosphere (*kruos* = icy cold, *sphere* = globe) refers to the ice and snow that exists at Earth's surface.

These two examples of increased water vapor in the atmosphere show that it can be both a positive-feedback loop and a negative-feedback loop (Figure 16.2). Which effect, if either, is stronger? Studies show that the negative effect of higher reflectivity is dominant, so clouds have an overall cooling effect. Therefore, the net result of an increase in atmospheric moisture should be a decrease in air temperature. The magnitude of this negative-feedback loop, however, is not likely to be as great as the feedback caused by other positive-feedback loops between other parts of Earth's climate system. Thus, although increases in atmospheric moisture and cloud cover may partly offset a global temperature increase, climate models show that the overall effect will still be a temperature increase. In fact, climate models backed by measurements of Earth's climate indicate that increasing levels of human-caused[2] emissions will lead to a warmer planet with a different distribution of climate patterns than what currently exist on Earth today.

Other feedback loops may have even stronger effects on future climate. For example, if rising temperatures cause Arctic sea ice to melt (a trend that is already evident), the lack of highly reflective ice cover will mean that the Arctic Ocean waters will absorb far more of the Sun's incoming radiation. This would create a positive-feedback loop whereby warming the ocean leads to further melting of sea ice.

The global climate system contains many feedback loops, such as the role of clouds at different altitudes, the presence of fine atmospheric particles called *aerosols*,[3] the shading effect from air pollution, the addition of water vapor in the atmosphere, the reflectivity of ice, and heat uptake by the oceans. In addition, many of these feedback loops influence other feedbacks. As such, the successful modeling of Earth's climate and its feedback loops is one of the biggest scientific challenges today—even using some of the world's most powerful computers.

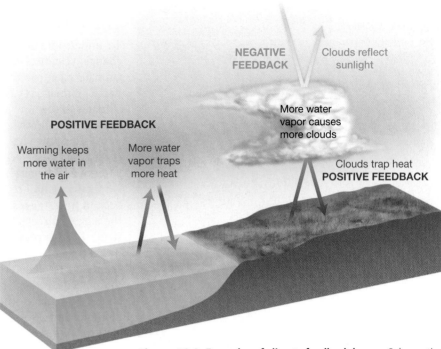

Figure 16.2 Examples of climate feedback loops. Schematic drawing of a positive-feedback loop (*left*) that amplifies an initial change and a negative-feedback loop (*right*) that reduces an initial change. Note that the negative-feedback loop can also produce a positive feedback (*lower right*).

ESSENTIAL CONCEPT 16.1

Earth's climate system consists of exchanges of energy and moisture between the atmosphere, hydrosphere, geosphere, biosphere, and cryosphere. The global climate system contains many complex feedback loops.

CONCEPT CHECK 16.1

1. List the five parts of Earth's climate system.

2. What are the two types of climate feedback loops? Give an example of each type.

3. An example of a feedback loop occurs when melting snow exposes more dark ground, which in turn absorbs additional heat and causes more snow to melt. Is this a positive-feedback loop or a negative-feedback loop? Explain your reasoning.

16.2 Earth's Recent Climate Change: Is It Natural or Caused by Human Influence?

Records of past climate change reveal that natural events have influenced climate throughout Earth's history. Skeptics of global climate change point out that because Earth's climate has fluctuated in the past, the recent climate change observed on Earth could be a natural event. How can scientists tell whether this is true?

Web Video
Inside an Antarctic Time Machine

---

[2]Human-caused influence is also known by the term *anthropogenic* (*anthro* = human, *generare* = to produce).

[3]*Aerosols* come from both human-made sources, such as coal-burning power plants and biomass burning, and natural sources, such as sea spray, dust, and volcanoes.

Rings are wider when conditions favor growth

Rings are narrower when conditions inhibit growth

Figure 16.3 Tree growth rings as climate proxies. Tree rings are wider when conditions favor growth and narrower when times are difficult. By combining multiple tree-ring studies with other climate proxy records, scientists have been able to estimate regional and global climates for hundreds to thousands of years in the past.

Determining Earth's Past Climate: Proxy Data and Paleoclimatology

To understand the behavior of the atmosphere and to anticipate future climate change, scientists must be able to discern how climate has changed in the past. Climate scientists use three closely connected methods to understand changes in Earth's climate: (1) They look at geologic evidence and other records of Earth's past climates to see how and why climate changed in the past, (2) they build computer models that allow them to see how the climate works, and (3) they closely monitor Earth's current vital signs with an array of instruments, including historical records, ship data, deep-sea thermometers, and data from space-based satellites (the latter of which have been available for only a few decades).

Instrumental records go back only a couple centuries at most, and the further back in time, the less complete and more unreliable the data become. To overcome the lack of direct measurements in the past, scientists must decipher and reconstruct Earth's previous climates using indirect evidence. Such **proxy** (*proxum* = nearest) data come from natural recorders of climate variability such as sea floor sediments (see Chapter 4), tree-growth rings (**Figure 16.3**), trapped air bubbles in the annual layers of glacial ice (**Figure 16.4**), fossil pollen, coral reefs, cave deposits, and even historical documents. The data are cross-checked between the various methods and also matched with recent instrumental measurements (where overlap exists) to ensure accuracy. Scientists who analyze proxy data and reconstruct past climates are engaged in the study of **paleoclimatology** (*paleo* = ancient, *climate* = climate, *ology* = the study of). The main goal of such work is to understand Earth's past climate in order to gain insight into Earth's current climate and future climate. For example, climatologists have identified both warmer and cooler periods in Earth's recent past, such as the Medieval Warm Period (approximately A.D. 950–1250) and the Little Ice Age (approximately A.D. 1400–1700). As we will see, climatologists have used proxy data to construct a detailed history of Earth's climate that extends back in time over the past several hundred thousand years.

Natural Causes Of Climate Change

Natural factors that affect Earth's climate include changes in solar energy, variations in Earth's orbit, volcanic eruptions, and even the movement of Earth's tectonic plates. Let's examine how each of these factors affects global climate.

CHANGES IN SOLAR ENERGY Among the most persistent hypotheses of climate change have been those based on the idea that the Sun's output of energy varies through time such that increases in solar output cause global warming, while reductions in solar energy result in global cooling. This notion is appealing because solar activity is known to influence Earth's temperature, and solar variation can be used to explain climate change of any length or intensity. However, an increase in solar output falls short of explaining recent warming. Earth-orbiting satellites have been making precise measurements of the Sun's output since the 1980s, and while the Sun's luminosity has increased by a small amount (0.04%), the observed changes were not large enough to account for the warming recorded during the same period. Even proxy data of solar brightness over the past 1000 years do not show a correlation with changes in climate.

Several proposals for climate change are based on solar variability related to **sunspots**, which are cooler dark areas that occur periodically on the Sun's surface (**Figure 16.5a**). **Faculae** (*facula* = bright torch), which are bright spots on the Sun, are also more abundant when sunspots are most active. Sunspots and faculae are associated with huge magnetic storms that extend from the Sun's interior to its surface and cause the Sun's ejection of particles (**Figure 16.5b**). These particles can disrupt satellite communications and also produce the *aurora* (*Aurora* = Roman goddess of dawn),

Figure 16.4 Researchers extract an ice core from its drilling tube. To reconstruct past temperatures and atmospheric conditions, scientists use natural records of climate change such as this ice core that was collected in Antarctica. The annual layers of snow packed in glacial ice preserve a record of climate that stretches back hundreds of thousands of years.

(a) Visual wavelength image of the Sun from the Helioseismic and Magnetic Imager (HMI), showing sunspots.

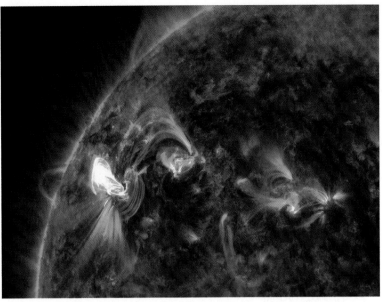

(b) Matching extreme ultraviolet wavelength image of the Sun from the Atmospheric Imaging Assembly (AIA), showing the ejection of solar particles.

Figure 16.5 Sunspots. Matching images of the Sun, showing sunspots and accompanying ejection of particles taken on March 5, 2012.

which is a phenomenon caused by charged solar particles that interact with Earth's magnetic field and produce lights in the sky. In the Northern Hemisphere, these lights are known as the *aurora borealis*, or *northern lights*, and they have a matching component in the Southern Hemisphere, called the *aurora australis*, or *southern lights*.

When solar activity increases, as it does every 11 years or so, both sunspots and faculae become more numerous. But during the peak of a cycle, the faculae brighten the Sun more than sunspots dim it. The number of sunspots and faculae affects the amount of *total solar irradiance*, which is a measure of the amount of sunlight striking Earth (**Figure 16.6**, *blue curve*). The next *solar maximum* of solar

STUDENTS SOMETIMES ASK . . .

In lay terms, what is a proxy? A *proxy* is a substitution or an approximation for something else. A good example of a proxy is an avatar, which is a cyber-representation of you that competes in video games. There are even proxies at school. For example, if you send someone to attend class in your absence, that person is acting as a proxy for you. And, it works both ways: A substitute teacher is a proxy who conducts your class when your regular teacher can't be there.

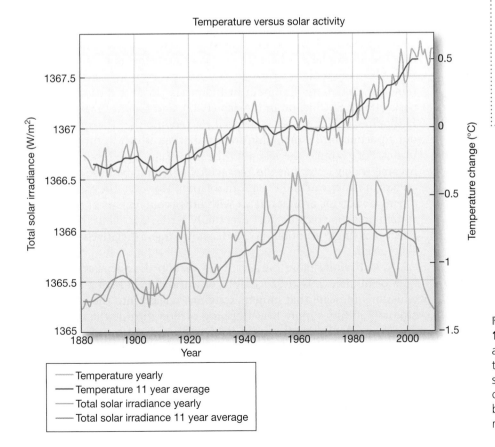

Figure 16.6 Earth temperature and solar activity since 1880. Graph showing average Earth temperature (*red curve*) and total solar irradiance (*blue curve*), which is a measure of the amount of sunlight striking Earth. The red and blue curves show smoothed data on an 11-year average. Note the 11-year cycle of total solar irradiance (*green curve*), which is affected by sunspots. Total solar irradiance is in watts per square meter, and temperature change is in degrees centigrade.

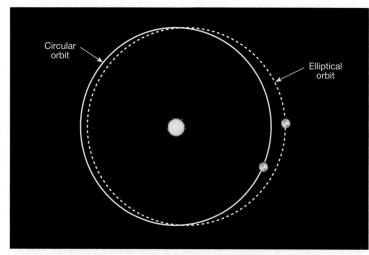

(a) The shape of the Earth's orbit (*eccentricity*) gradually changes from nearly circular to one that is more elliptical and then back again during a cycle of about 100,000 years. Note that Earth's elliptical orbit is greatly exaggerated in this figure.

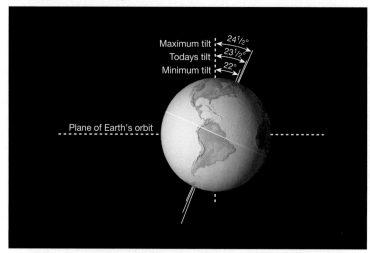

(b) The tilt of Earth's axis of rotation with respect to the plane of Earth's orbit (*obliquity*) varies from 22° to 24.5° during a cycle of about 41,000 years. Currently, Earth's tilt is 23.5°.

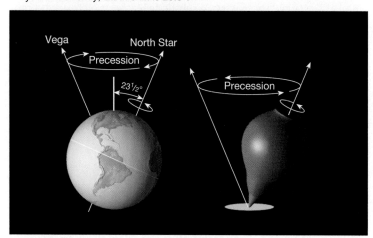

(c) Earth's axis wobbles like that of a spinning top (*precession*); as a result, the axis points to different spots in the sky during a cycle of about 26,000 years.

Figure 16.7 Variations in Earth's orbit. Diagram showing the three factors that cause variations in Earth's orbit.

activity is predicted to occur in May 2013, but it is expected to have a below-average number of sunspots. During the last solar maximum in 2001, sunspots produced powerful solar flares that caused large geomagnetic disturbances and disrupted some space-based technology on Earth.

The graph in Figure 16.6 also shows the lack of correlation between solar activity and average Earth temperate since 1880 (*red curve*). Since 1980, in fact, average solar activity has been decreasing. In addition, many studies have shown that there is no significant correlation between solar activity and climate on such short timescales.

VARIATIONS IN EARTH'S ORBIT Another natural mechanism of climate change involves changes in Earth's orbit. Changes in the shape of the orbit (*eccentricity*), variations in the angle that Earth's axis makes with the plane of its orbit (*obliquity*), and the wobbling of the axis (*precession*) cause fluctuations in the seasonal and latitudinal distribution of solar radiation reaching Earth (**Figure 16.7**). These variations have cycles of about 100,000 years, 41,000 years, and 26,000 years, respectively; when they coincide with one another, they tend to amplify each other and cause climate variations on Earth. This idea, first developed by Serbian astrophysicist Milutin Milankovitch, is called a *Milankovitch cycle*. It is now well established that these variations have contributed to the alternating glacial and interglacial episodes that characterize the most recent ice age, which occurred during the past few million years.

Although it is well established that Milankovitch cycles are responsible for long-term climate change associated with the most recent ice age on Earth, these changes take many thousands of years to manifest themselves. In contrast, the dramatic and rapid climate change that is occurring on our planet cannot be explained by these long-term variations in Earth's orbit.

VOLCANIC ERUPTIONS Explosive volcanic eruptions emit huge quantities of gases and fine-grained debris into the atmosphere (**Figure 16.8**). The largest eruptions are sufficiently powerful to inject material high into the atmosphere, where it spreads around the globe and remains aloft for many months or even years. As was seen with historic eruptions such as Mount Tambora in Indonesia (1815), Krakatoa in Indonesia (1883), El Chichón in Mexico (1982), and Mount Pinatubo in the Philippines (1991), volcanic material ejected into the atmosphere filters out a portion of the incoming solar radiation, which in turn cools the planet. For example, the year after the 1815 eruption of Mount Tambora became widely known as the *Year without Summer* because of its effect on North American and European weather. However, the gases emitted during a volcanic eruption react with other components of the climate system and the volcanic dust eventually settles out. Thus, the cooling effect of a single eruption, no matter how large, is relatively small and short-lived.

Which emits more heat-trapping carbon dioxide: Earth's volcanoes or human activities? Analysis of worldwide human and volcanic emissions indicate that human activities release into the atmosphere at least 130 times more heat-trapping carbon dioxide than volcanoes do (**Figure 16.9**). As discussed later in this chapter, carbon dioxide emissions make the greatest human contribution to warming Earth's atmosphere.

If volcanism is to have a pronounced impact over an extended period, many great eruptions closely spaced in time would need to occur.

If this happened, the upper atmosphere would be loaded with enough gases to alter the composition of the atmosphere and enough volcanic dust to seriously diminish the amount of solar radiation reaching the surface. Because no such period of explosive volcanism is known to have occurred in historic times, it is unlikely to be responsible for the recent observable climate changes. In the distant past, however, large and long-lasting volcanic eruptions may have been influential in contributing to Earth's climate shifts.

MOVEMENT OF EARTH'S TECTONIC PLATES   As described in Chapter 2, Earth's tectonic plates have moved great distances. During the geologic past, plate movements have accounted for many dramatic climate changes as landmasses shifted in relation to one another and moved to different latitudinal positions. As landmasses have moved, they have changed ocean circulation, altering the transport of heat and moisture and consequently the climate. For example, the opening of Drake Passage between South America and Antarctica about 41 million years ago caused a fundamental reorganization of ocean currents in the Southern Hemisphere, leading to the isolation of Antarctica, which caused it to become much cooler and develop a permanent ice cap. However, the rate of plate movement is very slow—only a few centimeters per year—and so appreciable changes in the positions of continents occur only over great spans of geologic time. Thus, climate changes triggered by shifting plates are extremely gradual and happen on a scale of millions of years.

CAN NATURAL CLIMATE CHANGE FACTORS EXPLAIN RECENTLY OBSERVED CLIMATE CHANGES?   It is clear that natural factors have changed Earth's climate in the past and that they will undoubtedly change it in the future. For example, natural climate change has been definitively linked to global climate shifts such as the Pleistocene Ice Age, the Medieval Warm Period, and the Little Ice Age.

During the past million years, the biggest temperature swings on Earth have been the ice ages. When ice ages have ended, it has taken about 5000 years for the planet to warm between 4°C and 7°C (7°F and 13°F). In the past century alone, the temperature has climbed about 0.7°C (1.3°F), which is roughly eight times faster than previous warming. In fact, proxy studies show that humanity is altering Earth's climate 5000 times faster than the pace of the most rapid natural warming episode in our planet's past.

Over the past century, scientists from all over the world have been collecting data on natural factors that influence climate—such as changes in the Sun's brightness, variations in Earth's orbit, major volcanic eruptions, and other factors. These observations have failed to show any long-term changes that could fully account for the recent rapid warming of Earth's temperature. Scientists agree that the observed warming in recent decades is occurring more quickly and with greater magnitude than can be explained by any natural factors. In essence, the release of human-caused emissions is the only viable explanation for the documented and observable recent climate changes—including the increased average temperature of Earth's surface.

IS THERE SCIENTIFIC CONSENSUS ABOUT HUMAN-CAUSED CLIMATE CHANGE?   Today, there is a clear scientific consensus that human-induced emissions are responsible for the observed warming on Earth. In fact, the warming trends observed since 1950 cannot be explained without accounting for human-caused emissions.

Human-caused climate change has also received broad support by major scientific organizations, including the U.S. National Academy of Sciences (NAS), the American Meteorological Society (AMS), the American Geophysical Union (AGU), and the American Association for the Advancement of Science (AAAS). Scientific consensus about human-caused climate change was clearly established more than 20 years ago and has not changed since then. However,

Figure 16.8 Volcanic eruptions spew volcanic debris and gases into the atmosphere. This 1991 eruption of Mount Pinatubo in the Philippines shows that volcanoes have the ability to inject into the atmosphere large quantities of volcanic dust and gases, which can circle the globe and block incoming solar radiation, thereby cooling the planet.

Web Animation
Orbital Variations and Climate Change

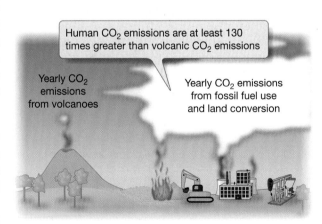

Figure 16.9 A comparison of human and volcanic carbon dioxide emissions. Worldwide, human carbon dioxide emissions are at least 130 times greater than volcanic emissions. Note that human-caused carbon dioxide emissions make the greatest human contribution to warming Earth's atmosphere.

Figure 16.10 Scientific consensus on human-caused climate change. Bar graph showing various groups' responses to the question "*Do you think human activity is a significant contributing factor in changing mean global temperatures?*" Note that climatologists who are active publishers on climate change had the highest percentage of affirmative response.

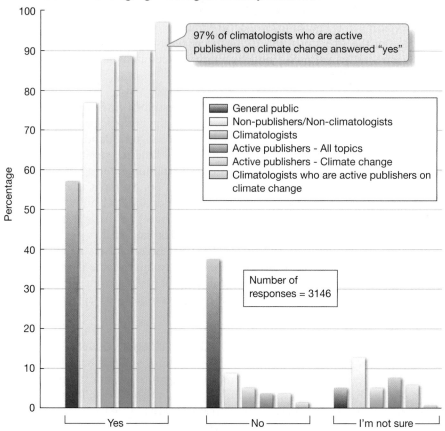

STUDENTS SOMETIMES ASK . . .

Did the stolen e-mails from researchers at East Anglia prove that climate change is just an elaborate hoax invented by climate scientists who tampered with scientific data?

No. The Climatic Research Unit (CRU) e-mail controversy, which became known as *Climategate*, began in November 2009 with the external hacking of a server at the University of East Anglia's CRU in Norwich, England. Hacked e-mails and documents from climate researchers were posted to various locations on the Internet, and climate change deniers alleged that the e-mails showed that global warming was a grand scientific conspiracy and that scientists manipulated climate data in an attempt to suppress critics. The CRU rejected these accusations and stated that the e-mails had been taken out of context and merely reflected an honest exchange of ideas.

In response to the controversy, the American Geophysical Union (AGU) issued a statement that it found "it offensive that these e-mails were obtained by illegal cyber-attacks and they are being exploited to distort the scientific debate about the urgent issue of climate change." Other scientific organizations, including the American Association for the Advancement of Science, American Meteorological Society, and Union of Concerned Scientists (UCS), released statements supporting the scientific consensus that the Earth's average surface temperature has been rising for decades. In the aftermath, eight independent committees—including panels from the U.S. and British governments—investigated the allegations and found no evidence of fraud, scientific misconduct, or other behavior that undermined the conclusions of the scientific consensus that Earth's climate is warming as a result of human activities.

a recent poll (**Figure 16.10**) shows that even though the vast majority of scientists—especially climatologists—accept human-caused climate change, social acceptance of the idea by the U.S. general public lags far behind. Figure 16.10 also shows that the most highly trained experts in climate science (those who are active publishers on climate change) have the highest level of confidence that human activities are causing the recent changes in Earth's climate.

Although it makes interesting news stories to suggest that there is disagreement among scientists about the reality of human-caused climate change, there is overwhelming scientific consensus that human emissions are causing changes in Earth's climate. The existing scientific disagreements are, in fact, about specific future impacts and consequences.

The IPCC: Documenting Human-Caused Climate Change

In 1988, the United Nations Environment Programme and the World Meteorological Organization sponsored the **Intergovernmental Panel on Climate Change (IPCC)**, a worldwide group of atmospheric and climate scientists that began studying the human effects on climate change and global warming. The IPCC utilizes peer-reviewed literature to analyze all aspects of climate change—including science, impacts, adaptation, and mitigation—to provide independent scientific advice about climate change. Since 1990, the group has published a series of assessment reports (**Figure 16.11**) that are highly regarded by both scientists and policymakers and have sparked international movement on climate change.

THE IPCC ASSESSMENT REPORTS   The IPCC's first assessment report, released in 1990, became the basis for the United Nations Framework Convention on Climate Change, an international treaty in which signatories agreed to the idea of reducing concentrations of greenhouse gases in the atmosphere. The IPCC's second assessment report, published in 1995, states that "the balance of evidence suggests a discernible human influence on global climate" and that global warming "is unlikely to be entirely due to natural causes."

In 2001, a third IPCC assessment report was published under the guidance of 426 scientists and was unanimously accepted by more than 160 delegates from 100 countries. The report states: "There is new and stronger evidence that most of the warming observed over the past 50 years is attributable to human activities." The report notes that recent regional climate changes already have affected many physical and biological systems on Earth and that projected climate change—as well as changes in climate extremes—could have major consequences. The report also revised the estimate of the world's expected temperature increase for the period between 1990 and 2100. Previously, the amount of predicted warming had been 1.0°C to 3.5°C (1.8°F to 6.3°F); the new report revised it upward, to 1.4°C to 5.8°C (2.5°F to 10.4°F), based on new climate models.

In 2005, an international consortium of science academies, including the U.S. National Academy of Sciences, issued this statement: "The scientific understanding of climate change is now sufficiently clear to justify nations taking prompt action.... As the United Nations Framework Convention on Climate Change (UNFCCC) recognizes, a lack of full scientific certainty about some aspects of climate change is not a reason for delaying an immediate response that will, at a reasonable cost, prevent dangerous [human-induced] interference with the climate system."

In 2007, a fourth IPCC assessment report—created by more than 600 authors from 40 countries and reviewed by more than 600 individuals—was published. The report was accepted and approved by representatives from 113 countries. The fourth IPCC report confirmed what many scientists had long suspected: Human-caused climate change is already altering Earth. In fact, *climate change models can mimic present-day conditions only if human emissions are taken into account*. Some of the documented changes specifically mentioned in the report include the warming of oceans and land, temperature extremes, melting of snow and ice, changing wind patterns, changing water patterns, and a variety of changes to a large assortment of organisms. The report also states that the temperature increases observed since the mid-20th century are very likely due to human-caused emissions, with the probability of human influence upgraded to greater than 90% certainty. This IPCC report clearly documents the fact that by adding emissions to the atmosphere, humans are altering global climate and are producing significant impacts on physical and biological systems worldwide. A fifth IPCC assessment report is expected to be published in 2013.

The IPCC assessment reports provide strong documentation of the planet's human-induced climate changes, such as global warming. In recognition of that fact, the IPCC was named a co-recipient of the 2007 Nobel Peace Prize, along with former U.S. Vice President Al Gore, Jr., for his work on the documentary film *An Inconvenient Truth*. When it bestowed the award, the Nobel Committee noted, "Through the scientific reports it has issued over the past two decades, the IPCC has created an ever-broader informed consensus about the connection between human activities and global warming."

In summary, the IPCC's four reports spanning 20 years have documented the broad scientific consensus of leading climate experts from around the globe, affirming the reality of human-caused climate change—now no longer as a prediction but

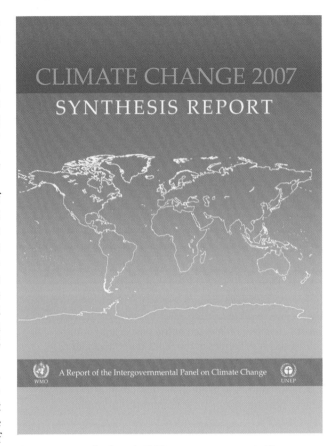

Figure 16.11 The fourth IPCC assessment report. The Intergovernmental Panel on Climate Change (IPCC), an international consortium of climate scientists, has published four assessment reports since 1990, confirming that human-induced emissions are altering Earth's climate. The fifth assessment report is scheduled to be published in 2013.

as an observational reality. In addition, the IPCC reports state that global warming poses substantial risks to societies and ecosystems on all continents.

OTHER SCIENTIFIC REPORTS CONFIRM IPCC FINDINGS   A host of subsequent reports have confirmed the findings of the IPCC. In 2009, for example, the U.S. Global Change Research Program issued a 190-page interagency report entitled *Global Climate Change Impacts in the United States*. The report states that "global warming is unequivocal and primarily human-induced." The report also notes that "global average temperature has risen by about 1.5°F [0.8°C] since 1900. By 2100, it is projected to rise another 2 to 11.5°F [1.1 to 6.4°C]. Increases at the lower end of this range are more likely if global heat-trapping gas emissions are cut substantially. If emissions continue to rise at or near current rates, temperature increases are more likely to be near the upper end of the range." The report warns that climate change will have numerous impacts on water resources, ecosystems, agriculture, coastal areas, human health, and other sectors.

In 2011, the U.S. National Research Council issued the final report of a five-volume series about climate change, titled *America's Climate Choices*. The reports, which were requested by the U.S. Congress, reaffirm that the preponderance of scientific evidence points to human activities—especially the release of carbon dioxide and other greenhouse gases into the atmosphere—as the most likely cause of the vast majority of global warming that has occurred over the past several decades. The reports also reiterate the pressing need for substantial action to limit the magnitude of climate change and to prepare to adapt to its impacts. The reports indicate that actions taken now "can reduce the risk of major disruptions to human and natural systems; inaction could serve to increase these risks, especially if the rate or magnitude of climate change is particularly large."

CONCEPT CHECK 16.2

1. What are proxy data? List several examples. Why are such data necessary for paleoclimatology studies?
2. List several examples of natural climate change. Do natural climate change mechanisms account for the recent climate changes that Earth is experiencing? Explain.
3. Is there scientific consensus about human-caused climate change? Explain.
4. What is the IPCC? What role does it have in documenting human-caused climate change?

16.3 What Causes the Atmosphere's Greenhouse Effect?

Numerous scientific studies indicate that human-caused emissions are responsible for the recent and dramatic climate changes experienced on Earth, including the increase in average worldwide temperature, which is called **global warming**. Although the **greenhouse effect** is a natural process that influences the temperature of Earth's surface and atmosphere, it is now being altered by human emissions, a phenomenon that is often referred to as the *anthropogenic greenhouse load* or the *enhanced greenhouse effect*.

The greenhouse effect gets its name because it keeps Earth's surface and lower atmosphere warm in a way similar to a greenhouse that keeps plants warm enough to grow, regardless of outside conditions (**Figure 16.12**). Energy radiated by the Sun covers the full electromagnetic spectrum, but most of the energy that reaches Earth's surface is short wavelengths, in and near the visible portion of the spectrum. In a greenhouse, shortwave sunlight passes through the glass or plastic covering, where it strikes the plants, the floor, and other objects inside and is converted

Figure 16.12 How a greenhouse works. The glass of a greenhouse allows incoming sunlight to pass through but traps heat. Similarly, gases such as water, carbon dioxide, and methane in Earth's atmosphere act just like the glass of a greenhouse by allowing sunlight to pass through but trapping heat.

into longer-wavelength infrared radiation (heat). Some of this heat energy escapes from the greenhouse and some is trapped for a while by the glass or plastic covering, which keeps the greenhouse nice and snug—much like what happens in Earth's atmosphere.[4]

Earth's Heat Budget and Changes in Wavelength

Figure 16.13 shows the various components of Earth's **heat budget**, which describes all the ways in which heat is added to and subtracted from Earth. In the upper atmosphere, most solar radiation within the visible spectrum penetrates the atmosphere to Earth's surface, like sunlight coming through greenhouse glass. After scattering by atmospheric molecules and reflection off clouds, about 47% of the solar radiation that is directed toward Earth is absorbed by the oceans and continents. About 23% is absorbed by the atmosphere and clouds, and about 30% is reflected into space by atmospheric backscatter, clouds, and Earth's surface.

Figure 16.14 shows that most of the energy coming to Earth from the Sun is within the visible spectrum and peaks at a wavelength of 0.48 micrometer (0.0002 inch).[5] The atmosphere is transparent to much of this radiation, but it is absorbed by materials such as water and rocks at Earth's surface. These surface materials absorb and then emit radiation away from Earth's surface and toward space as longer-wavelength infrared (heat) radiation, with a peak at a wavelength of 10 micrometers (0.004 inch). Molecules of atmospheric gases such as water vapor, carbon dioxide, and other gases intercept the heat radiation that attempts to leave the planet, thus heating the atmosphere. This trapping of heat radiation and heating of the atmosphere is known as the greenhouse effect.

In summary, most of the solar radiation that is not reflected back to space passes through the atmosphere and is absorbed at Earth's surface. Earth's surface, in turn, emits longer-wavelength infrared radiation (heat). A portion of this energy is absorbed by certain heat-trapping gases in the atmosphere, thus producing the greenhouse effect. Thus, *the change of wavelengths from visible to infrared is the key to understanding how the greenhouse effect works.*

Which Gases Contribute to the Greenhouse Effect?

Earth's greenhouse effect is caused by an array of atmospheric gases, many of which have both natural and human-caused sources. Take, for example, water vapor, which contributes more to the greenhouse effect than any other gas. In fact, water vapor is the single most important absorber of heat—its contribution to the greenhouse effect is between 36% and 66% of the greenhouse effect, and together with clouds, it comprises about 75% of the greenhouse effect.

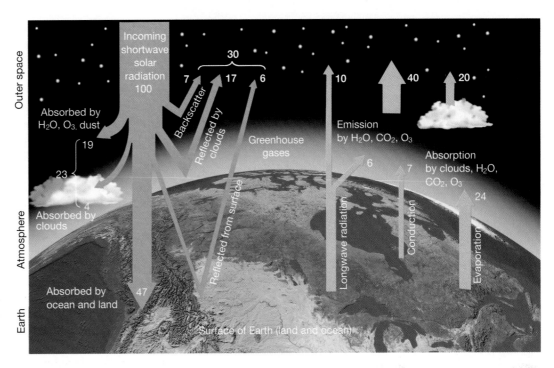

Figure 16.13 Earth's heat budget. One hundred units of shortwave solar radiation from the Sun (mostly visible light) are reflected, scattered, and absorbed by various components of the Earth–atmosphere system. The absorbed energy is radiated back into space from Earth as longwave infrared radiation (heat). If this infrared radiation does not leave Earth, global warming will occur.

Web Animation
Global Warming

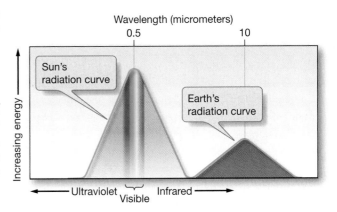

Figure 16.14 Energy radiated by the Sun and Earth. The intensity of energy radiated by the Sun peaks at a wavelength of 0.48 micrometer (0.0002 inch), which is in the visible part of the spectrum. Some of this energy is absorbed or reflected while some reradiates from Earth in the infrared (heat) range at a wavelength of 10 micrometers (0.004 inch).

---

[4]Recent studies have indicated that an additional factor in keeping a greenhouse warm is that the greenhouse covering prevents mixing of air inside with cooler air outside. Although this is different from how the atmosphere works, the term *greenhouse effect* is still commonly used to describe the atmosphere's warming process.

[5]A micrometer (μm), or micron, is one-millionth of a meter.

STUDENTS SOMETIMES ASK . . .

What would Earth's temperature be like if there were no natural greenhouse effect?

In a word: *freezing*! The worldwide average temperature of Earth and the lowermost atmosphere (troposphere) is about 15°C (59°F). If the atmosphere contained no heat-trapping greenhouse gases or warming cloud cover, the average worldwide temperature would be about –5°C (23°F). At this temperature, most of our planet would very likely have a frozen surface. Instead, the atmosphere's naturally occurring greenhouse gases help create and sustain the moderate temperatures that make Earth habitable.

Web Animation
Atmospheric Energy Balance

Water vapor enters the atmosphere mostly through evaporation and other natural processes. Although atmospheric water vapor concentrations fluctuate regionally, studies suggest that human activity does not significantly affect water vapor concentrations except at local scales, such as near irrigated fields. And even then, the water vapor does not stay in the atmosphere very long. In essence, human activities do not directly affect the amount of water vapor in the atmosphere on a global scale. However, humans may be indirectly affecting the amount of water vapor in the atmosphere: Satellite measurements of atmospheric water vapor show a 4% increase in global specific humidity at sea level since 1970, likely caused by a warming climate.

Still, atmospheric water vapor plays an important role in warming. For example, a recent study suggests that an increase in stratospheric water vapor due to natural processes between 1980 and 2000 may have amplified the rapid warming of that period by as much as 30%. Conversely, a 10% decrease in the amount of water vapor in the stratosphere since 2000 may have slowed global warming by causing global temperatures to level off despite the continued rise in human emissions.

Table 16.1 shows the concentration of **greenhouse gases**—so called because of their heat-trapping capacity—that have been increasing as a result of human activities. Remarkably, these gases exist in very small amounts in the atmosphere, yet they have a profound effect on heating. And unlike water vapor that does not stay in the atmosphere very long, many of these gases stay in the atmosphere for long periods and continue to trap heat. Some of these greenhouse gases are released by both human and natural sources (such as carbon dioxide, which has been a component of the atmosphere since long before human activities). Others, however, have no natural source and thus are clearly human induced (such as the chemicals known as chlorofluorocarbons).

CARBON DIOXIDE   Of all the human-caused gases, carbon dioxide makes the greatest relative contribution to increasing the greenhouse effect (Table 16.1). Carbon dioxide enters the atmosphere as a result of combustion of carbon compounds with oxygen. It is a colorless and odorless gas that is the same one we exhale from our lungs. The conversion of **fossil fuels** (oil and natural gas) into energy by cars, factories, and power plants accounts for the majority of the annual human

TABLE 16.1   HUMAN-CAUSED GREENHOUSE GASES AND THEIR CONTRIBUTION TO INCREASING THE GREENHOUSE EFFECT

Atmospheric gas	Human-caused sources of gas	Pre-industrial (circa 1750) concentration (ppbv[a])	Present concentration (ppbv[a])	Current rate of increase or decrease (% per year)	Relative contribution to increasing the greenhouse effect (%)	Infrared radiation absorption per molecule (number of times greater than CO_2)
Carbon dioxide (CO_2)	Combustion of fossil fuels	280,000	395,000	+0.5	60	1
Methane (CH_4)	Leakage, domestic cattle, rice agriculture	700	1775	+1.0	15	25
Nitrous oxide (N_2O)	Combustion of fossil fuels, industrial processes	270	315	+0.2	5	200
Tropospheric ozone (O_3)	Byproduct of combustion	0	10–80	+0.5	8	2000
Chlorofluorocarbon (CFC-11)	Refrigerants, industrial uses	0	0.26	−1.0	4	12,000
Chlorofluorocarbon (CFC-12)	Refrigerants, industrial uses	0	0.54	0.0	8	15,000
Total					100	

[a]ppbv = parts per billion by volume (not by weight).

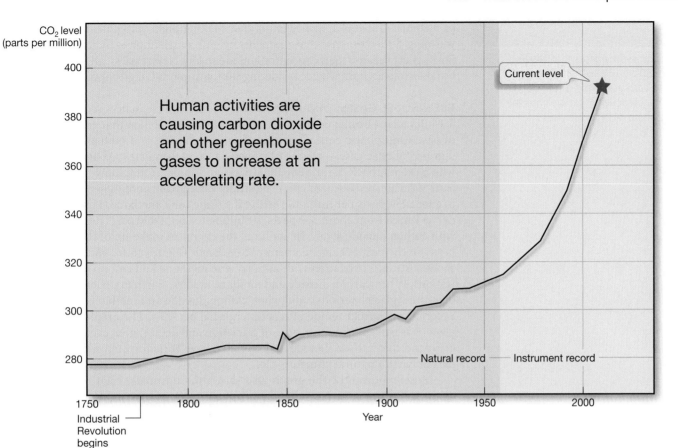

Human activities are causing carbon dioxide and other greenhouse gases to increase at an accelerating rate.

Figure 16.15
Amount of carbon dioxide in the atmosphere since 1750. There has been a dramatic increase of average worldwide atmospheric carbon dioxide since the Industrial Revolution began in the late 1700s. Values for 1958 to the present are from instrumental measurement of carbon dioxide at Mauna Loa Observatory in Hawaii; natural record values prior to 1958 are from air bubbles in polar ice cores.

contribution to carbon dioxide emissions, with industrialized nations contributing the most. As a result of human activities, therefore, atmospheric concentration of carbon dioxide has increased more than 40% over the past 250 years (**Figure 16.15**). What concerns scientists is that over the past 250 years—and especially in the past 50 years—human activities have been responsible for raising the concentration of greenhouse gases in the atmosphere at an ever-increasing rate. As of 2013, the average yearly concentration of atmospheric carbon dioxide is 395 parts per million and is increasing by about 2 parts per million each year; this rate of increase is double what occurred only 50 years ago. In terms of sheer numbers, humans are now pumping into the atmosphere more than 33 billion metric tons (36.3 billion short tons) of carbon dioxide each year.[6] And the last time that carbon dioxide levels were as high as they are today was about 15 million years ago.

METHANE   Methane is the second-most-abundant human-caused greenhouse gas (Table 16.1). It is produced by leakage from decomposing trash in landfills, by methane-belching domestic cattle, and by agriculture (particularly the cultivation of rice). Even though methane has a lower concentration in the atmosphere than carbon dioxide, it has a greater ability to produce warming on a per-molecule basis. Since the Industrial Revolution began in about 1750, the concentration of methane in the atmosphere has increased by 250%.

OTHER GREENHOUSE GASES   The other trace gases shown in Table 16.1—nitrous oxide, tropospheric ozone, and chlorofluorocarbons—are present in far lower concentrations than carbon dioxide and methane. Yet they are still very important because they absorb many times more infrared radiation per molecule than carbon dioxide

[6]On average, each person on Earth emits more than 4.7 metric tons (5.2 short tons) of carbon dioxide per year. Of course, this number is several times greater for those living in industrial nations than for those living in developing countries.

STUDENTS SOMETIMES ASK . . .

Is carbon dioxide causing the hole in the ozone layer?

Here's the short answer: Definitely not! The ozone layer occurs within the atmosphere's stratosphere and is composed of ozone molecules (O_3) that absorb most of the Sun's ultraviolet radiation. Without it, unhealthy levels of ultraviolet radiation would reach Earth's surface, making the planet largely uninhabitable. The main ozone hole (actually, a seasonal thinning of the ozone layer) occurs above the South Pole, with a smaller one above the North Pole. Both are caused by chemical reactions with natural and human-generated compounds, particularly the now-banned chemicals CFC-11 and CFC-12, but not carbon dioxide. CFCs are also strong greenhouse gases (see Table 16.1), so CFCs are thus double threats to the environment: Their buildup in the atmosphere leads to the destruction of the ozone layer and, at the same time, contributes to global warming. Notice also from Table 16.1 that tropospheric (lower atmosphere) ozone, which is a byproduct of combustion, is a potent greenhouse gas. With the ban of CFCs firmly in place, scientists predict that the ozone layer will achieve its normal thickness by the middle of the 21st century. However, new research reports that human-generated nitrous oxide, which destroys ozone, is now the single greatest ozone-depleting substance.

or methane (Table 16.1, *last column*), thus making them very potent contributors to warming. Still, these gases have a smaller overall contribution to increasing the greenhouse effect because their concentrations are so low. Nevertheless, all these gases must be taken into account when considering the total amount of greenhouse warming.

GREENHOUSE GASES: PAST AND FUTURE   In 2005, researchers recovered a nearly 3.2-kilometer (2-mile) continuous ice core from Antarctica that contains a record of past atmospheric concentrations of carbon dioxide and methane—two important greenhouse gases—that get trapped as ice accumulates. Analysis of the core, which extends back in time 800,000 years ago (**Figure 16.16**), shows that the average level of carbon dioxide (*red curve*) naturally varied from about 180 parts per million to about 280 parts per million. During the same time, methane (Figure 16.16, *green curve*) varied from about 350 parts per billion to about 800 parts per billion, in sync with carbon dioxide levels. In addition, the chemical make-up of the ice provides a proxy of the past average temperature on Earth (Figure 16.16, *black curve*), which shows a strong correlation with atmospheric methane and carbon dioxide concentrations: When carbon dioxide and methane are low, Earth experiences cooler temperatures (glacial periods), and when carbon dioxide and methane are high, Earth experiences warmer temperatures (interglacial periods). The graph shows that our planet has passed through a cycle of glaciation and deglaciation every 100,000 years or so and that carbon dioxide and methane have varied in step with it. This is all part of Earth's natural climate cycle.

Most importantly, the graph also shows the unusually high levels of carbon dioxide and methane in today's atmosphere (*colored stars*). In fact, the current atmospheric concentrations of carbon dioxide (395 parts per million) and methane (1775 parts per billion) are at their highest levels by far in at least the past 800,000 years.

Analysis of proxy data of Earth's geologic past shows that Earth has experienced times of much higher average surface temperatures than it is experiencing today. One of the largest-known events of this type is called the *Paleocene–Eocene Thermal Maximum*, or PETM, which occurred about 56 million years ago, when Earth's climate was much warmer than it is today. Chemical indicators point to a huge release of carbon dioxide that caused Earth to warm by at least 5°C (9°F) over a few thousand years. The PETM is marked by massive marine ecosystem changes that were caused by an increase in ocean temperatures, which caused the depletion

Figure 16.16 Ice core data of atmospheric composition and global temperature. (a) Antarctic ice core data for the past 800,000 years, showing atmospheric carbon dioxide (*red curve*), methane (*green curve*), and average global temperature (*black curve*). Today's carbon dioxide and methane levels are shown by the stars above the graph. Atmospheric composition is from analysis of trapped air bubbles in ice; temperature reconstruction is derived from the chemical make-up of the ice. **(b)** Enlargement showing data for the most recent 2000 years. Medieval Warm Period and Little Ice Age are also shown.

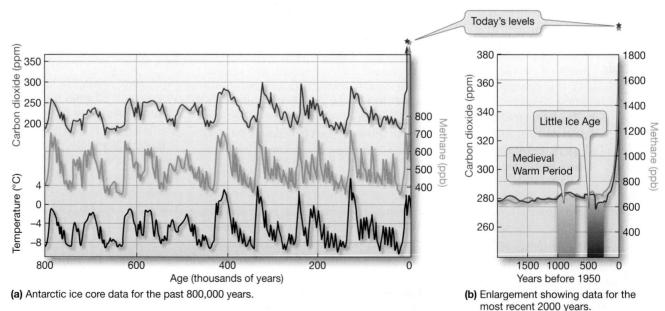

(a) Antarctic ice core data for the past 800,000 years.

(b) Enlargement showing data for the most recent 2000 years.

of dissolved oxygen in surface waters because warmer water can't hold as much oxygen. In addition, higher amounts of carbon dioxide from the atmosphere dissolved in the oceans, making them more acidic (which is also currently happening; this is discussed later in this chapter). As a result, these ancient warmer periods such as the PETM are typically marked by the extinction of many marine organisms.

What will the level of carbon dioxide—with its known ability to trap heat in Earth's atmosphere—be like in the future? Based on sophisticated computer models, the IPCC has made predictions to the year 2100 based on various scenarios (**Figure 16.17**). The predictions are based on greenhouse gas emission scenarios, which are affected by factors such as population growth, economic development, technological changes, and cultural and social interactions. For example, in scenario A2 (Figure 16.17, *red curve*), population increases at the current growth rate accompanied by an economic theme of self-reliance and preservation of local identities. In this "business-as-usual" scenario, carbon dioxide reaches its highest levels and global surface temperature is expected to increase by 4°C (7.2°F). Not since the end of the last ice age 10,000 years ago has the average temperature changed as dramatically as the change that this scenario predicts in less than 100 years.

In scenario A1B, (Figure 16.17, *blue curve*), there is very rapid economic growth, global population peaks in midcentury and declines thereafter, and there is rapid introduction of new and more efficient technologies (such as fuel cells, solar panels, and wind energy) that replace much of today's fossil fuel combustion. This "middle-of-the-road" scenario results in a moderate amount of atmospheric carbon dioxide that equates to a warming of about 2.5°C (4.5°F).

In scenario B1 (Figure 16.17, *green curve*), there is the same global population as in the A1B scenario but with rapid changes in economic structures toward a service and information economy, accompanied by reduction in material intensity and the widespread adoption of clean and resource-efficient technologies. In this scenario, countries come together to use both technology and general environmental controls to decrease emissions. This "best case" scenario results in the lowest amounts of atmospheric carbon dioxide and a warming of only about 1.0°C (1.8°F).

Even if greenhouse gas concentrations stabilized today with no additional increase, the planet would continue to warm by about 0.6°C (1.1°F) over the next century because it takes years for Earth's climate system to fully react to increases in greenhouse gases (including the stabilization of various feedback loops). This future warming is often referred to as Earth's *commitment to warming*. Most importantly, it is clear that the choices made now will determine the amount of atmospheric carbon dioxide and associated warming for the remainder of the 21st century and beyond.

What Documented Changes Are Occurring Because Of Global Warming?

Melting glaciers and ice caps, shorter winters, shifts in species distribution, and a steady rise in average global and sea surface temperatures are just some of the indications that additional human-induced greenhouse warming is occurring. Take,

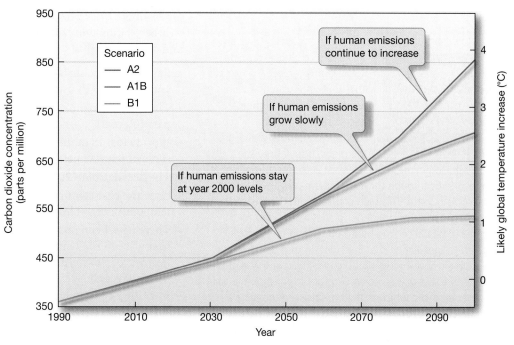

Figure 16.17 Scenarios for future atmospheric carbon dioxide levels and corresponding global temperature increase. Graph showing the predicted atmospheric carbon dioxide levels and likely global temperature increase for various scenarios. Higher carbon dioxide and temperature values are associated with the A2 scenario (*red curve*), which assumes that human emissions will increase (people will burn more and more fossil fuel). Intermediate carbon dioxide and temperature values are associated with the A1B scenario (*blue curve*), which assumes greenhouse gas emissions will grow slowly. The lowest carbon dioxide and temperature values are associated with the B1 scenario (*green curve*), which assumes that greenhouse gases will stay at year 2000 levels.

ESSENTIAL CONCEPT 16.3B

Trapped air bubbles in ancient ice reveal that today's levels of heat-trapping carbon dioxide and methane are greater than at any time within the past 800,000 years. Scenarios show that human emissions will cause continued warming on Earth, but the choices we make now will determine the amount of future warming.

STUDENTS SOMETIMES ASK . . .

I've heard that scientists were predicting an oncoming ice age only a few decades ago. What's changed?

Skeptics of global warming have pointed out that in the 1970s, climate scientists were warning of an imminent ice age. In reality, the idea lacked scientific consensus, although the media popularized the notion with sometimes quite alarmist accounts. It is true that from the 1950s through the 1970s, there was an observed global cooling trend. This fact, coupled with the idea that ice ages are cyclical and that our planet was poised for millennia of global cooling based on variations in Earth's orbit, led to speculation that Earth may return to an ice age. However, it is now known that the recent cooling was probably due to a substantial increase in soot from smokestacks and other aerosols, which at the time masked the telltale signature of global warming. Looking at a longer view of global temperature since that time (see Figure 16.18), Earth's recent warming trend can be clearly discerned.

for example, these observations about Earth's temperature, which are based on weather stations on land, satellite data, and, for earlier measurements, proxy data and data from ships:

- Earth's average surface temperature has risen 0.6°C (1.1°F) over the past 30 years and 0.8°C (1.4°F) over the past 140 years (**Figure 16.18**).
- The rate of warming in the past 50 years is double the rate observed over the past 100 years.
- Worldwide, 18 of the 20 warmest years on record have occurred since 1990, with some of the warmest years occurring most recently.
- The decade 2000–2010 was the warmest decade on record.
- Over the past century, the planet has experienced the largest increase in surface temperature in at least 1300 years.
- There have been increasing occurrences of severe heat waves, such as the 2012 heat wave in the United States (with the hottest July on record) and the 2010 heat wave in eastern Europe and Russia (the region's greatest heat wave in the past 500 years). Such events have recently been linked to human-caused climate warming, and models project that severe heat waves will become more likely in the future.

As global temperatures continue to increase, researchers use sophisticated climate models to forecast what changes will occur on Earth. Because of the complexity of the climate system and its feedback loops, not all models agree about what or how severe those changes will be. However, there are points on which the models do agree: a strong warming of high northern latitudes, a moderate warming of middle latitudes, and relatively little warming in low latitudes. Other predicted changes—many of which are already happening—include the following:

- Earlier summertime seasons, with higher summer temperatures, including longer and more intense heat waves.
- More extreme precipitation events, such as severe droughts in certain areas and increased chances of flooding in other areas.
- The worldwide retreat of ice fields and mountain glaciers, which is already being observed.

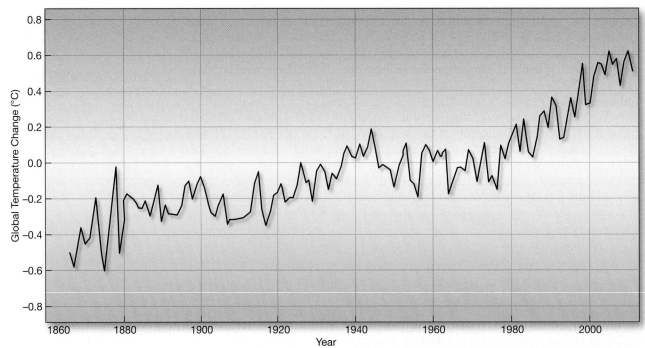

Figure 16.18 Instrumental temperature record since 1865. The record of global average surface air temperature from thermometer readings indicates a global warming of at least 0.8°C (1.4°F) over the past 140 years. The peaks and troughs indicate the natural year-to-year variability of climate.

- Water contamination issues that lead to larger outbreaks of waterborne infectious diseases such as malaria, yellow fever, and dengue fever.
- Shifts in the distribution of plant and animal communities that affect entire ecosystems and may drive certain species to extinction.

Yet not all predicted changes have negative consequences. For instance, increased warming is expected to provide a longer growing season for some crops, increased atmospheric carbon dioxide should help promote **productivity** in plants, and ice-free Arctic waters have higher biological productivity. However, most research suggests that the negative impacts of a changing climate far outweigh the positive impacts. In addition, a number of uncertainties remain in understanding regional effects of climate change, and various components of the climate system may respond to these changes in unanticipated and surprising ways.

CONCEPT CHECK 16.3

1. Describe the fundamental difference between solar radiation absorbed at Earth's surface and the radiation that is primarily responsible for heating Earth's atmosphere.

2. Discuss the greenhouse gases in terms of their relative concentrations in the atmosphere and their relative contributions to global warming.

3. Why has the carbon dioxide level of the atmosphere been rising for more than 150 years?

4. Explain what a commitment to warming means. What commitment to warming will Earth experience in the future?

5. Describe how atmospheric temperatures and other factors are likely to change as carbon dioxide levels continue to increase.

STUDENTS SOMETIMES ASK . . .

Why was this winter so cold when there's supposed to be global warming?

One of the documented changes of a warmer world is increased variation in both temperature and precipitation extremes. This means that while the climate is warming, not only will there be a wider range of temperatures—including both warmer and colder temperature extremes—but also wetter or drier conditions. In essence, global warming increases the chances of extreme events such as heat waves, record cold snaps, storms, droughts, intense rains, unusual amounts of snow, and lack of snow altogether. It's no wonder that many scientists prefer the term *global weirding*.

Also, remember that climate is the long-term average of weather, so although it might be colder during one season, the climate can still be warming. What matters is not what happens on any given day or season but what the trend is over a period of years. On this, the data are clear: Earth is experiencing long-term global warming.

Web Video
Global Temperatures
1880–2010

16.4 What Changes Are Occurring in the Oceans as a Result Of Global Warming?

The oceans are a key component in the global climate system that is currently experiencing dramatic changes. Let's examine some of the observed and predicted effects of global warming in the oceans.

Increasing Ocean Temperatures

Studies have revealed that the oceans have absorbed the majority of the increased heat in the atmosphere. Indeed, millions of ocean temperature observations at various depths reveal that there has been an overall increase in surface temperature (**Figure 16.19**). These measurements indicate that global sea surface temperatures have risen by about 0.6°C (1.1°F) because of global warming, mainly since about 1970. However, this warming has not been uniform throughout the ocean. The greatest temperature increases have been experienced in the Arctic Ocean, near the Antarctic Peninsula, and in tropical waters. Even deep waters are showing signs of warming: In some places, warming has been documented to a depth of 0.8 kilometer (0.5 mile) or more. And deeper waters are warming faster than expected. To determine the extent of warming experienced in the oceans, scientists have initiated global monitoring of ocean temperature (**Diving Deeper 16.1**).

Figure 16.19 Global temperature change. Map showing the change in land and ocean surface temperature in 2008 compared to the 1950–1980 baseline period. Most of the globe is anomalously warm (*red areas*) as a result of greenhouse warming, with the greatest temperature increases in the Arctic Ocean, Siberia, the Antarctic Peninsula, and tropical regions. A persistent cool-water La Niña event (*blue area*) is affecting the values in the tropical Pacific Ocean. Gray areas indicate no data.

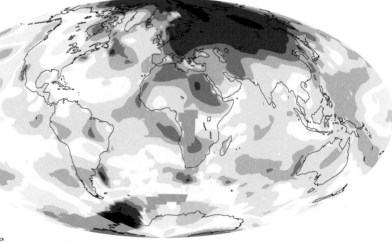

2008 Surface Temperature Anomaly (degrees Celsius)

-3.5 -2.5 -1.5 -1.0 -0.6 -0.2 0.2 0.6 1.0 1.5 2.5 3.5

Diving Deeper 16.1 Focus on the Environment

THE ATOC EXPERIMENT: SOFAR SO GOOD?

Worldwide, a layer exists in the ocean at a depth of about 1000 meters (3300 feet) formed by temperature and pressure conditions that cause sound originating above and below it to become refracted, or bent, into the layer (**Figure 16A**, *lower inset*). Once in this layer, called the *SOFAR channel* (an acronym for *sound fixing and ranging*), or *deep sound channel*, sound is efficiently trapped and transmitted long distances. For instance, certain whales may use the SOFAR channel to send sounds across entire ocean basins.

The idea of sending sounds through the SOFAR channel in an effort to detect the amount of ocean warming as a result of the

greenhouse effect was initiated by Walter Munk of the Scripps Institute of Oceanography (Figure 16A, *upper inset*). His experiment, named **Acoustic Thermometry of Ocean Climate (ATOC)**, is designed to accurately measure the travel time of similar low-frequency sound signals through the SOFAR channel now and in the future. The speed of sound in seawater increases as temperature increases, so sound should take less time to travel the same distance in the future if, in fact, the oceans are warming.

In 1991, Munk's group successfully tested ATOC at Heard Island in the southern Indian Ocean, from which sound can reach many

different receiving sites along straight-line paths (Figure 16A, map). The researchers deployed an underwater array similar to a series of loud-speakers that sent *acoustical* (*akouein* = to hear) signals for six days. These signals refracted into the ocean's SOFAR channel and were transmitted throughout the oceans. After traveling for many hours and as far as 19,000 kilometers (11,800 miles), the signals were received at shipboard recording stations.

The success of ship-based testing at Heard Island led to the establishment of a fixed sound source near Pioneer Seamount off California and an array of fixed receivers throughout the Pacific Ocean. In 1995, ATOC sound signals were sent again. Even though many precautions were taken to avoid any unwanted effects on marine mammals, three humpback whales were found dead in the area a few days after the sound transmissions had begun. The sounds were halted while the U.S. National Marine Fisheries Service (NMFS) conducted research to determine whether the sounds affected the hearing of nearby whales and contributed to their deaths. After extensive scientific study, their results indicated that marine mammals are unaffected by the transmissions and that the dead whales were an unfortunate coincidence. The long-term effect of ATOC signals on marine mammals—including marine mammal communication through the SOFAR channel—is poorly understood but is also likely to be minimal. The NMFS approved the project, and transmissions from offshore California and Hawaii have been conducted without incident, producing a wealth of data that indicate a warming trend in the Pacific and establishing an important baseline for comparison with future measurements.

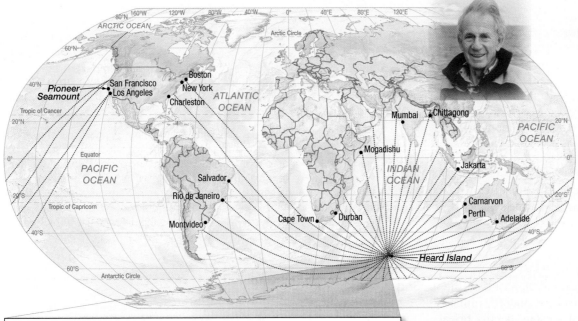

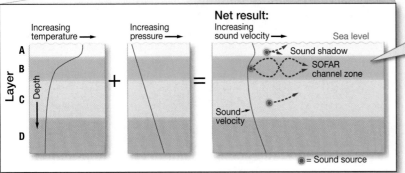

Sound waves created in layers A and C are bent into the SOFAR channel and remain within it, allowing the sound to be transmitted across entire ocean basins

Figure 16A The ATOC experiment. Map showing Heard Island sound travel paths, photograph of oceanographer Walter Munk (*upper inset*), and diagram of the SOFAR channel (*lower inset*). Temperature and pressure conditions in the ocean combine to produce the SOFAR channel (*layer B*), which traps and transmits sound energy.

The impacts of a warmer ocean are far-reaching and will persist for several centuries. For example, increased seawater temperatures are likely to affect temperature-sensitive organisms such as corals. Coral reefs are already in decline, and as discussed in Chapter 15, warmer sea surface temperatures have been implicated in widespread coral bleaching events. Warmer waters could also shift or disrupt the spawning cycles of corals. In addition, studies of coral distribution have found that some corals are already migrating with the spread of warmer waters to areas where they haven't been seen before.

Other studies have documented how warmer ocean waters are changing the physical characteristics of the ocean. For example, research using the Argo program of free-drifting floats (see Chapter 7) has shown that the world's hydrologic cycle is accelerating as warming speeds up evaporation of ocean surface waters, thereby affecting ocean salinity. As mentioned earlier in this chapter, the melting of sea ice forms a positive-feedback loop that accelerates ocean warming in high latitudes because ice-free water absorbs much more solar radiation than reflective sea ice. In addition, increased seawater temperatures are likely to affect the ocean's deep-water circulation pattern, El Niño/La Niña events, and the development of hurricanes.

Increasing Hurricane Activity

Many scientists have suggested that warmer oceans would most certainly cause an increase in the general level of storminess because additional heat accelerates evaporation, which fuels hurricanes. In fact, the frequency and severity of recent hurricanes—especially those in the Atlantic Ocean (see the section on hurricanes in Chapter 6)—have led some to speculate that global warming has already enhanced the formation of hurricanes. Although the recent landfall of several large Atlantic hurricanes leads to a general impression that hurricanes have increased, there have been conflicting reports on the topic in the scientific literature (**Figure 16.20**). For example, some research articles have attributed increases in hurricane intensity, numbers, and wind speeds to warmer sea surface temperatures, while others have claimed that changes in data-gathering methods and instrumentation are responsible for the trends. Other studies suggest that the apparent increase in hurricanes falls within the statistical limits of normal.

Although the number of tropical storms has not increased worldwide, the scientific consensus is that global warming has led to more intense hurricanes. In the most comprehensive study of recent hurricane activity to date, researchers demonstrate that there have been significant increases in tropical storm intensity and duration around the world since 1970 and that these trends are strongly related to rising sea surface temperatures. Another study of historical Atlantic hurricanes over the past 1500 years suggests that times of peak hurricane activity are related to increased sea surface temperatures and the reinforcing effects of La Niña–like climate conditions. Other research explicitly shows that the most energetic storm levels—those with Category 4 and 5 designations—have already increased significantly, particularly in the North Atlantic and northern Indian Oceans. In addition, sophisticated climate models suggest that the number of Category-4 and Category-5 storms in the western tropical Atlantic could double by the end of the century, despite a drop in the overall number of storms.

Changes In Deep-Water Circulation

Evidence from deep-sea sediments and computer models indicates that changes in the global deep-water circulation pattern can dramatically and abruptly affect climate. Circulation in the North Atlantic Ocean, which provides an important

Figure 16.20 Storm surge from Hurricane Sandy hits the East Coast. Waves crash over homes along the shoreline in Milford, Connecticut, as a result of the storm surge caused by Hurricane Sandy on October 29, 2012.

North Atlantic Ocean Circulation

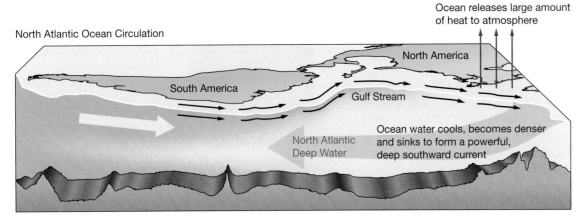

Ocean releases large amount of heat to atmosphere

North America

Gulf Stream

North Atlantic Deep Water

Ocean water cools, becomes denser and sinks to form a powerful, deep southward current

South America

Figure 16.21 North Atlantic Ocean circulation. Perspective view of circulation in the North Atlantic Ocean, showing that the Gulf Stream carries a tremendous amount of heat northward that warms the North Atlantic region. As this water cools, it generates a huge volume of cold, salty, dense water called North Atlantic Deep Water that sinks into the deep-ocean basin and flows southward. Disruption of this circulation pattern could have severe effects on global climate.

Web Animation
Disruption of North Atlantic
Ocean Circulation

source of deep water (**Figure 16.21**), is particularly sensitive to these changes. What drives deep-water circulation is the sinking of cold, high-salinity, dense surface waters at high latitudes, particularly in the North Atlantic. If surface waters stopped sinking because they were too warm and/or too diluted by melting ice (and thus low in density), then the oceans would absorb and redistribute heat from solar radiation much less efficiently. This would likely cause even warmer surface water temperatures and much higher land temperatures than are experienced now.

Many scientific studies suggest that the buildup of greenhouse gases in the atmosphere will change ocean circulation. One way in which this could happen is that warmer air temperatures will increase the rate at which glaciers in Greenland melt, forming a pool of fresh, low-density surface water in the North Atlantic Ocean. This freshwater could inhibit the downwelling that generates North Atlantic Deep Water, reorganizing global circulation patterns and causing a corresponding change in climate. Many climate experts warn that a large outflow of freshwater from Greenland could take the North Atlantic system of currents to a tipping point, causing rapid reorganization of deep-water currents and related changes in climate. Indeed, evidence suggests that an outburst from a North American ice-dammed lake flooded the North Atlantic with freshwater about 8000 years ago, causing rapid global climate change. This event may be the type of scenario that the North Atlantic could experience again because of increased precipitation and melting of ice.

Melting Of Polar Ice

Computer models have predicted that global warming will affect Earth's polar regions in a very dramatic way. One fundamental difference between the two polar regions is that in the Northern Hemisphere, the polar region is dominated by the Arctic Ocean and its cover of drifting sea ice (an ocean surrounded by land), whereas the South Pole is dominated by the continent of Antarctica and its thick ice cap, including shelf ice that extends into the ocean (land surrounded by an ocean).

The Arctic is one of the locations where the effects of global warming are being most keenly felt (see Figure 16.19) and likely will experience quite dramatic changes in the future; this phenomenon is called *Arctic amplification*. Since 1978, satellite analysis of the extent of Arctic Ocean sea ice indicates that it is dramatically shrinking and thinning at an accelerating rate (**Figure 16.22**). In the past decade alone, there has been a loss of over 2 million square kilometers (800,000 square miles) of Arctic sea ice. In fact, measurements of the ice cap in 2012 revealed that it had shrunk to its smallest size since researchers began collecting satellite measurements; summer Arctic sea ice is now about half the size it was only 30 years ago. In addition, thick multiyear ice in the Arctic has been disappearing and is being replaced by thinner first-year ice that is less likely to survive the summer melt season. As a result, Arctic ice is now unusually thin and spread out, causing wide patches of ice-free ocean during the summer—even at the North Pole. A recent study of proxy data showed that the current decline of Arctic sea ice is unprecedented for at least the past 1450 years.

Climate models are in general agreement that one of the strongest signals of greenhouse warming will be a loss of Arctic sea ice. Indeed, during the past 15 years, the decline in Arctic sea ice has occurred much faster than models had predicted. In fact, the Arctic has been warming twice as quickly as the Northern

STUDENTS SOMETIMES ASK . . .

The movie The Day After Tomorrow *depicts rapid climate change as a result of alterations in deep-water circulation. Could this really happen?*

Although Hollywood movies are well known for dramatization, what is interesting about this blockbuster 2004 film is that its story line is based on recent scientific findings: The ocean's deep-water circulation helps drive ocean currents around the globe and is important to world climate. In fact, strong evidence suggests that the North Atlantic Deep Water circulation has already weakened and may further weaken during this century, resulting in unpleasant effects on our climate—but certainly not as rapid or as cataclysmic as the situation portrayed in the film. Computer models suggest that the continued weakening of the North Atlantic Deep Water circulation will result in some long-term cooling, particularly over parts of northern Europe.

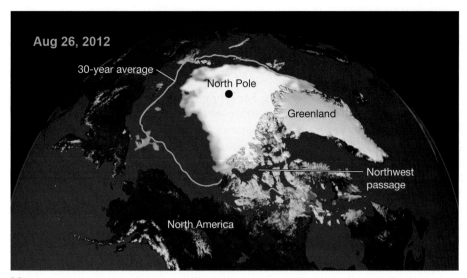

(a) Perspective view of the Arctic based on satellite data, showing the extent of Arctic sea ice in August 2012 compared with the 30-year average extent of sea ice (*yellow line*). Note the opening of a new ice-free Northwest Passage.

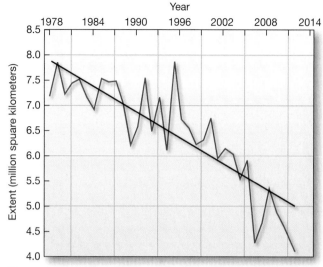

(b) Graph showing the decrease in Arctic sea ice extent based on satellite data.

Figure 16.22 Arctic sea ice decline. The substantial decrease in extent and thickness of Arctic Ocean sea ice is due to human-induced warming of the Arctic.

Hemisphere average. Cycles of natural variability are known to play a role in Arctic sea ice extent, but the sharp decline observed in the past two decades cannot be explained by natural variability alone. The accelerated Arctic sea ice melting appears to be linked to shifts in Northern Hemisphere atmospheric circulation patterns that have caused the region to experience unusually rapid warming. As a result, ocean temperatures in the Arctic Ocean have also increased, causing sea ice to melt from below. Disappearance of sea ice is likely to enhance future warming in the region because lower amounts of sea ice will reflect less of the Sun's radiation back into space, creating a positive-feedback loop and exacerbating the problem as heat is absorbed by the newly uncovered ocean. Researchers fear that the Arctic may be on the verge of a fundamental transition, or "tipping point," that will lead to the Arctic having only seasonal ice cover. Some models, for example, suggest that the Arctic could experience the complete disappearance of summer sea ice as early as 2030.

The decrease of sea ice in the Arctic has already had profound effects on Arctic ecosystems. Polar bears (*Ursus maritimus*; **Figure 16.23**), for example, are excellent swimmers but do not hunt in the water. Instead, they require a platform of floating sea ice to capture their prey, which are mainly ringed and bearded seals. As the Arctic Ocean becomes more ice-free and the ice habitat shrinks, polar bears will have more difficulty finding adequate food and making dens. As a result, polar bear breeding and survival rates may decline below the point needed to maintain the population. Their habitat destruction has been so severe that polar bears were listed as a threatened species in 2008, according to the U.S. Endangered Species Act. Studies reveal that polar bears are likely to lose nearly half of their summer sea ice habitat by the middle of the 21st century, which would in turn reduce the world's polar bear population—currently estimated at 25,000—by two-thirds.

In addition, human inhabitants of the Arctic who depend on subsistence living are being affected by the lack of sea ice. This is because some of the inhabitants' food sources are more difficult to obtain now that marine species live further from shore to be near the ice edge. Arctic residents have also claimed that their weather is changing, and a recent study supports their observations. The study determined that years with the least amount of Arctic sea ice have had significantly stronger Arctic storms. One new opportunity that now exists, however, involves the creation of a new "Northwest Passage" shipping lane linking the North Pacific and North Atlantic Oceans through the largely ice-free portions of the Arctic Ocean (Figure 16.22a).

Web Animation
Arctic Sea Ice Decline

Web Video
Loss of Arctic Sea Ice

Figure 16.23 Habitat destruction threatens polar bears. Polar bears (*Ursus maritimus*) are reliant on floating sea ice as a feeding platform and for building dens. As a result of habitat destruction caused by shrinking Arctic sea ice, polar bears were designated a threatened species in 2008.

While all these changes are taking place in the Arctic, different but equally dramatic ones are taking place in Antarctica, particularly in the western part of Antarctica that includes the Antarctic Peninsula. As discussed in Chapter 6, Antarctica produces many icebergs from glaciers on land. The rate at which Antarctica is producing icebergs—especially large icebergs the size of small U.S. states—has increased in recent years. For example, the Antarctic Peninsula's Larsen Ice Shelf has decreased by more than 40% over the past decade, including a huge release of 3250 square kilometers (1250 square miles) of ice during two months in 2002 (see Figure 6.26d). In 2006, Antarctica lost nearly 200 billion metric tons (220 billion short tons) of ice; during 10 days in 2008, the Wilkins Ice Shelf lost over 400 square kilometers (160 square miles) of ice. The rate of thinning of the Pine Island Glacier—the largest stream of fast-moving ice on the West Antarctic Ice Sheet—quadrupled from 1995 to 2006. And, in the past 30 years, there have been 10 major Antarctic ice shelf collapses—including the disappearance of the ice shelves knows as Jones, Larsen A, Muller, and Wordie—after some 400 years of relative stability. Scientists attribute this catastrophic retreat to warming in Antarctica; the Antarctic Peninsula has experienced some of the greatest warming worldwide (see Figure 16.19). In fact, Antarctica has warmed at a rate of about 0.12°C (0.22°F) per decade since 1957, for a total average temperature rise of 0.5°C (1.0°F) (**Figure 16.24**).

Recent Increase in Ocean Acidity

The human-induced increase in the amount of carbon dioxide in the atmosphere has some severe implications for ocean chemistry and for marine life. Recent studies show that a little less than half of the carbon dioxide released by the burning of

Figure 16.24 Antarctic warming trends. This satellite image shows the amount of warming that Antarctica has experienced since 1957. The data are from satellites and weather station measurements. Notice that western Antarctica has experienced the greatest temperature change.

Temperature change per decade (degrees Celsius)

| 0 | 0.05 | 0.10 | 0.15 | 0.20 | 0.25 |

(a) Coccolithophores, which are a type of phytoplankton (diameter of each sphere = 20 microns, or 0.0008 in.)

(b) Pteropod, which is a zooplankton that is a small swimming snail with a shell (diameter = 2 mm, or 0.08 in.)

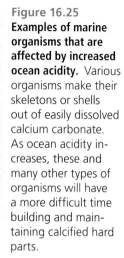

Figure 16.25
Examples of marine organisms that are affected by increased ocean acidity. Various organisms make their skeletons or shells out of easily dissolved calcium carbonate. As ocean acidity increases, these and many other types of organisms will have a more difficult time building and maintaining calcified hard parts.

(c) Sea urchins

(d) Corals

fossil fuels stays in the atmosphere and about one-third currently ends up in the oceans, readily dissolving into seawater at the ocean surface. This "sink" reduces global warming—but at the expense of acidifying the sea.

When large amounts of carbon dioxide enter the ocean, it overwhelms the ocean's natural ability to buffer itself.[7] This absorbed carbon dioxide forms carbonic acid in seawater, lowering the ocean's pH (that is, increases its acidity in a process called **ocean acidification**) and changing the balance of carbonate and bicarbonate ions. In fact, the oceans have already absorbed enough carbon dioxide for surface waters to have experienced a pH decrease of 0.1 pH unit since pre-industrial times. Although 0.1 pH unit sounds like a small amount, pH is a logarithmic scale, so a difference of 0.1 pH unit represents about a 30% increase in hydrogen ion concentration; every decrease of 1 pH unit is equivalent to a *tenfold* increase in hydrogen ion concentration. Another study has confirmed a 0.04 pH unit decrease in the North Pacific Ocean during just the past two decades.

Moreover, this shift toward increased acidity and the ensuing changes in ocean chemistry makes it more difficult for certain marine creatures to build and maintain hard parts out of easily dissolved calcium carbonate.[8] The decline in pH thus threatens a diverse assortment of calcifying organisms—creatures that grow calcium carbonate skeletons or shells—such as coccolithophores, foraminifers, pteropods, calcareous algae, sea urchins, mollusks, and corals (**Figure 16.25**). These organisms provide essential food and habitat to many other species, so

Web Video
Increasing Ocean Acidity

---

[7]For a discussion of the ocean's buffering system and the pH scale, see Chapter 5.

[8]Chemically, more acidic waters create more hydrogen ions that bind with carbonates. That leaves fewer carbonates available for carbonate-forming marine organism to make their shells.

Figure 16.26 Pteropods dissolved by acidic waters. Six-week time-lapse sequence of shell dissolution of the Antarctic pteropod *Limacima helicina* in water simulating surface ocean acidity projected for the Southern Ocean by 2100 if carbon dioxide emissions continue to increase.

Figure 16.27 Historical and projected dissolved carbon dioxide and ocean pH. As the concentration of carbon dioxide in the ocean increases, so does ocean acidity (causing pH to decline).

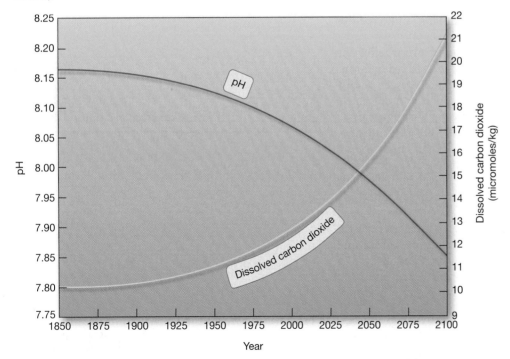

their demise could affect entire ocean ecosystems. For example, research shows that the projected future level of ocean acidity prevents Antarctic krill eggs from hatching, which would affect entire Antarctic food webs. Other studies have shown that in the past 20 years, ocean acidification has already caused a 15% decrease in the growth rate of corals in Australia's Great Barrier Reef. And, laboratory experiments that expose organisms to conditions that mimic the ocean's future acidity clearly show the negative impact on the shells of calcite-secreting organisms (**Figure 16.26**).

Ocean acidification affects more than just organisms that make hard parts out of calcium carbonate. Studies show that increasing ocean acidity can interfere with basic bodily functions for all marine animals, shelled or not. For example, research has shown that the larvae of Atlantic cod (*Gadus morhua*) exposed to acidic waters experience severe tissue damage in many internal organs, thus increasing their mortality rate. Other studies of marine organisms exposed to acidic waters show a decrease in reproductive rates. By disrupting processes as fundamental as growth and reproduction, ocean acidification threatens marine animals' health and even the survival of species.

The graph in **Figure 16.27** shows the projected increase of carbon dioxide in the ocean and the resulting decrease in ocean pH (rise in acidity). The graph shows that if the current trend of human-induced carbon dioxide emissions continues, by 2100, the ocean will experience a pH decrease of at least 0.3 pH unit; some studies indicate that pH could decrease by as much as 0.6 pH unit. Even at the lower value of 0.3 pH unit, this reduction of ocean pH represents an increase in hydrogen ion concentration of 100% above what it was in pre-industrial times. An additional concern is that deep currents will eventually transmit this increase in acidity to the deep-ocean floor, where it will likely affect deep-sea species, whose stable environment leaves them ill-equipped to adapt to change.

Geologic records show that comparable ocean pH changes wiped out many species of sea life, especially bottom-dwelling organisms. Remarkably, today's alteration of ocean chemistry is unprecedented on Earth: An analysis of the past 300 million years of Earth history fails to find a time when ocean chemistry is changing as fast as it is today.

Several processes influence the amount of carbon dioxide that is absorbed by the ocean and how much remains in the atmosphere. The ocean and atmosphere are two reservoirs for carbon dioxide storage, but carbon dioxide is not equally divided between the two. This is because carbon dioxide gas is readily absorbed by the ocean and, as a result, the ocean contains a much larger amount of carbon dioxide than the atmosphere.[9] The amount of carbon dioxide from the atmosphere that is dissolved in the ocean varies with the chemistry of seawater but is also controlled by positive-feedback loops. For example, one positive-feedback loop is that as the ocean approaches saturation of carbon dioxide, it will absorb less of the gas, which means that more will remain in the atmosphere. Another positive-feedback loop is that as the ocean warms, it will further reduce the amount of carbon dioxide that goes into the oceans because warmer water can't hold as much dissolved gas. Yet another positive-feedback loop

---

[9]Of the three places where carbon dioxide resides—atmosphere, ocean, and land biosphere—approximately 93% is found in the ocean. The atmosphere, in contrast, contains the smallest amount.

involves the rate at which deep waters mix with surface waters: The more rapid the mixing, the more it facilitates the uptake of carbon dioxide from the atmosphere. If deep-water circulation slows as predicted, this will also slow the uptake of carbon dioxide from the atmosphere. In addition, another positive-feedback loop is that ocean acidification inhibits the ability of marine organisms to make calcium carbonate hard parts, so less carbon dioxide in the form of calcium carbonate can be stored by organisms and effectively removed from the environment. All of these positive-feedback loops cause more of the carbon dioxide being released now to stay in the atmosphere, where it will likely create additional human-caused warming.

Rising Sea Level

Analysis of worldwide tide records indicates that there has been a rise in global sea level of between 10 and 25 centimeters (4 and 10 inches) over the past 100 years. At certain tide-recording stations where data go back well into the 19th century, there has been an increase in relative sea level of as much as 40 centimeters (16 inches) over the past 150 years (**Figure 16.28**). More recently, satellite altimeter data since 1993 indicate a global increase in sea level of about 3 millimeters (0.1 inch) per year (**Figure 16.29**). Studies have shown that the current rate of sea level rise is occurring faster than at any other time over the past 4000 years, and the rate is expected to increase with additional warming.

Two main factors contribute to the global rise in sea level: (1) thermal expansion of ocean water as it warms and (2) an increase in the amount of water in the ocean from the melting of ice on land. Note that the melting of floating sea ice (such as in the Arctic Ocean) or floating ice shelves (such as those that fringe Antarctica) does not contribute to sea level rise because that ice/water is already in the ocean. More specifically, in order of their overall contribution to the observed global rise in sea level (**Figure 16.30**), the main contributors are:

1. The melting of the Antarctic and Greenland Ice Sheets
2. The thermal expansion of ocean surface waters
3. The melting of land glaciers and small ice caps
4. The thermal expansion of deep-ocean waters

An additional factor that affects global sea level is the storage of water on land in reservoirs. A recent study indicates that the amount of this storage has varied but has generally increased since 1900; the study suggests that without reservoirs, sea level would have risen even more.

Although the current rate of sea level rise might seem inconsequential, even a small amount of sea level rise can severely affect regions that have a gently sloping shoreline, such as the U.S. Atlantic and Gulf coasts. Surprisingly, sea level rise does not occur equally everywhere. For example, a recent study identified a zone along the U.S. East Coast from Cape Cod to Cape Hatteras that has experienced rates of sea level rise three to four times higher than the global average since 1950. Hazards associated with sea level rise include drowning of beaches, accelerated coastal erosion, permanent inland flooding, alteration of coastal ecosystems, loss of protective coastal wetlands, and increased damage from destructive storms. In addition, if global warming increases the intensity of hurricanes (as discussed above), damage to coastal regions will be even greater.

The current rate of global sea level rise, though small, is likely to increase as the Greenland and Antarctic Ice Sheets experience additional warming and make a greater contribution to sea level rise. In fact, detailed studies of Antarctic ice thickness reveal that the rate at which the ice is thinning near the coast has doubled since the 1990s. Neither the Greenland nor Antarctic Ice Sheets are likely to disappear before 2100, but there is the danger that global warming could trigger massive and

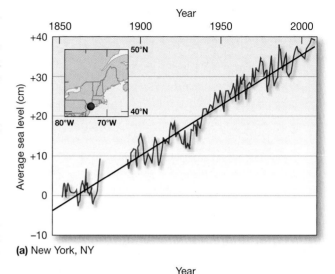

(a) New York, NY

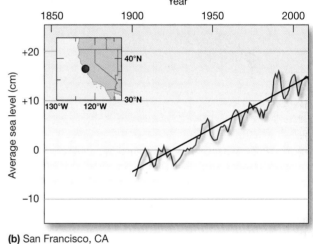

(b) San Francisco, CA

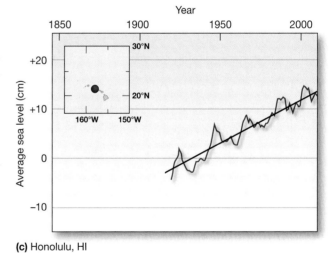

(c) Honolulu, HI

Figure 16.28 Measured relative sea level rise from tide gauges. Sea level data from **(a)** New York, New York, **(b)** San Francisco, California, and **(c)** Honolulu, Hawaii, all of which show an increase in sea level. While some of the documented rise is due to local effects (such as changes in the elevation of Earth's crust caused by tectonic movement or isostatic adjustment), the majority is caused by the addition of water from the melting of continental ice caps and glaciers as well as from thermal expansion of warmer ocean water.

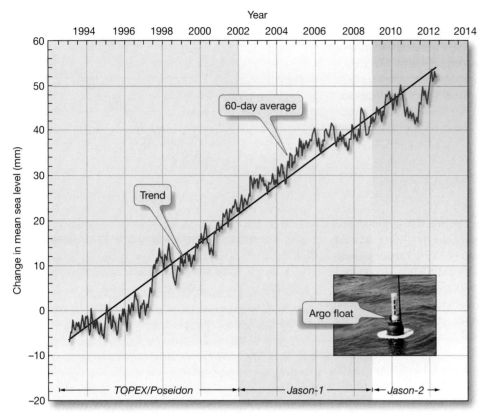

Figure 16.29 Sea level rise determined by satellites. Radar altimeter data from *TOPEX/Poseidon*, *Jason-1*, and *Jason-2* satellites combined with Argo drifting floats (inset; see also Chapter 7) reveal that sea level has risen 3 millimeters (0.1 inch) per year since 1993. Scientists attribute about half of this increase to melting ice and the other half to thermal expansion as the ocean absorbs excess heat from the atmosphere.

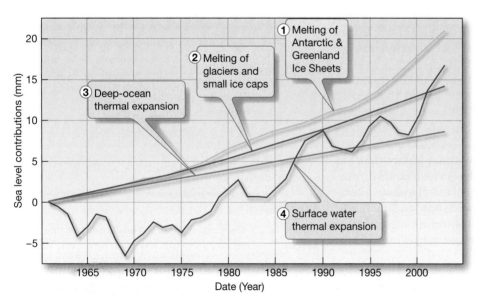

Figure 16.30 Main components that contribute to global sea level rise. Graph showing how much each of the four main components contributes to overall sea level rise. The four main components are (1) melting of the Antarctic and Greenland Ice Sheets (*light blue curve*), (2) melting of glaciers and small ice caps (*dark blue curve*), (3) deep-ocean thermal expansion (*brown curve*), and (4) surface water thermal expansion (*red curve*).

catastrophic discharges of ice from these ice sheets. A recent study determined that if the West Antarctic Ice Sheet were to collapse, it would raise global sea level by about 3.2 meters (10.5 feet).

According to recent models, the rate of sea level rise will increase with increased global warming. Studies of sea level rise that include both thermal expansion and ice-sheet contributions project a rise in sea level by 2100 of between 0.6 and 1.6 meters (2.0 and 5.2 feet), which is particularly worrisome for low-lying coastal regions. In addition, increased amounts of human settlement and development in coastal regions compound the problem. And projections suggest that sea level may well rise by several meters in the next few hundred years.

Other Predicted and Observed Changes

Several other ocean changes are predicted to result from global warming, and some changes are being observed now. One predicted change is lower dissolved oxygen in seawater. As discussed in Chapter 12, dissolved oxygen in seawater is vital for most marine animals, which extract dissolved oxygen directly from seawater. As the ocean warms, its ability to hold and carry dissolved oxygen is diminished, while at the same time the metabolic rate of marine organisms increases, which means they need higher levels of dissolved oxygen. In addition, the warming of surface waters will limit the critical overturning process that brings oxygen to deep waters. At current rates of carbon dioxide emissions, studies predict severe oxygen-depleted zones both at the surface and into deeper waters for thousands of years. Reduced oxygen levels will likely have dramatic consequences for marine ecosystems and coastal regions that already experience oxygen-depleted "dead" zones (see the section on dead zones in Chapter 13). Reduced dissolved oxygen in seawater has already been documented in both coastal waters and open-ocean environments.

Wind speeds and wave heights have been affected by climate change, too. A recent analysis of 23 years of satellite altimeter measurements suggests that windiness has been increasing on a global scale, with severe winds increasing the most. Although the study found that wave heights have not increased globally, the heights of larger waves had a slight increase in higher latitudes.

Still another effect of ocean warming is the alteration of oceanic productivity, which has implications for the distribution of essentially all marine organisms. As ocean surface waters warm, ocean stratification will increase and a stronger **thermocline** (*thermo* = heat, *cline* = slope) will develop.[10] This stratification would make it harder for nutrients to reach the surface and also limit nutrient recycling with deeper waters. As a result, productivity is expected to decrease because of diminished **upwelling**, and the water that is brought to the surface by upwelling would be more nutrient depleted. Recall from Chapter 13 that **phytoplankton** (*phyto* = plant, *planktos* = wandering)—which include marine algae such as *diatoms* and *coccolithophores*—comprise the base of most marine food webs and thus support other larger organisms in the oceans, including commercial fish species. In fact, studies already show a decline of global phytoplankton biomass related to ocean warming. Researchers estimate that, by 2100, productivity could decrease by as much as 20% relative to pre-industrial times.

Warmer waters can also affect marine organisms directly. Many types of phytoplankton and other organisms are very sensitive to changes in water temperature. For example, a large study of phytoplankton in the North Atlantic shows that ocean warming has increased the abundance of phytoplankton in cooler regions and decreased them in warmer waters. Another example is a recent study in waters off California that shows a decrease in cool-water organisms caused by a deep, penetrative warming not observed in the past 1400 years. In these organisms' place, there has been a boost in the population of 25 fish groups that prefer warmer waters as surface waters have shifted from cold to warm temperatures over recent decades.

In response to rising ocean temperatures, marine organisms have also begun to migrate into deeper waters and toward the poles. As verification of this trend, a recent study of fish species in the North Sea suggests that many commercially important fish, such as cod, whiting, and anglerfish, have shifted northward as much as 800 kilometers (500 miles). The report notes that if these climate trends continue, some species of fish may withdraw completely from the North Sea by 2050. Another report suggests that high-latitude regions are expected to benefit from anticipated changes in ocean fisheries, but the warm-water tropics are likely to suffer declines in fish stocks. In other cases, warm-water species are moving into frigid waters that have been previously inaccessible for millions of years. For example, king crabs (*Neolithodes yaldwyni*) have invaded Antarctic waters, where they are feeding on fragile creatures such as sea cucumbers, sea lilies, and brittle stars that have little resistance to clawed predators. Changes such as these have severe implications for the health of marine ecosystems and the sustainability of marine fisheries.

In summary, climate change is clearly and fundamentally altering the physical characteristics of the ocean, including entire ocean ecosystems. There is also great concern among the scientific community that the human-induced greenhouse effect may bring unpleasant surprises, such as abrupt and unpredicted changes in marine systems.

ESSENTIAL CONCEPT 16.4

Changes occurring in the ocean due to increased global warming include increased ocean temperatures, more intense hurricane activity, changes in deep-water circulation, melting of polar ice, increased ocean acidity, and rising sea level.

CONCEPT CHECK 16.4

1 Describe several changes that are already occurring in the oceans because of global warming.

2 What physical conditions produce a SOFAR channel, or sound channel, below the ocean's surface? How is the SOFAR channel being used to detect warming in the oceans?

3 Why is the Arctic one of the locations where the effects of global warming are being most keenly felt? Why are polar bears threatened by global warming?

4 Explain how excess carbon dioxide in the atmosphere affects the pH of seawater. Why is increasing ocean acidity such a problem in the ocean?

---

[10]For more details on the thermocline (the layer of rapidly changing temperature), see Chapter 5.

16.5 What Should Be Done to Reduce Greenhouse Gases?

Numerous scientific studies suggest that the planet has warmed due to human-caused greenhouse gas emissions. Recently, there have been serious discussions about deliberately manipulating Earth's climate system to counteract human-caused global warming and its unwanted effects. These manipulations—called **global engineering**, or *geoengineering*—are very controversial. Some of the concerns about any type of global engineering include the justification for intentionally altering any of Earth's systems on a global scale, the potential of triggering harmful and unpredicted side effects, whether it would need to be constantly maintained, and who should pay for its implementation. Most importantly, skeptics of global engineering point out that it distracts from perhaps the most immediate way of limiting human-caused climate change: reducing the amount of greenhouse gases that humans release into the atmosphere.

Most global engineering proposals fall into two general types: (1) reducing the amount of sunlight reaching Earth or (2) removing human-caused greenhouse gases from the atmosphere and disposing of them somewhere else. Examples of the first type include spraying sulfate aerosols into the atmosphere to mimic the cooling effect of a major volcanic eruption or installing thousands of reflective sunshades in orbit to block incoming sunlight. Examples of the second type include removing atmospheric carbon dioxide and placing it into subsurface layers within Earth or into the deep ocean. Even if reasonably effective and beneficial overall, global engineering is unlikely to alleviate all of the serious impacts from increasing greenhouse gas emissions. For example, efforts to dim the amount of sunlight reaching Earth would not diminish the effects of ocean acidification.

Still, the threat of climate change is serious. Agreements to reduce human emissions have been debated and global engineering proposals have been advanced, although often viewed as a last resort. In this section, let's examine the ocean's role in reducing global warming, some global engineering ideas that involve the ocean, and international agreements that seek to reduce human emissions.

The Ocean's Role in Reducing Global Warming

Because the oceans naturally absorb vast amounts of carbon dioxide from the atmosphere, the oceans play a vital role in reducing the greenhouse effect. In fact, the vast majority of carbon dioxide in the ocean–atmosphere system is found in the ocean because carbon dioxide is approximately 30 times more soluble in water than are other common gases. Currently, a little less than half of the carbon dioxide humankind releases into the atmosphere stays there; about one-third enters the oceans, and the rest is absorbed by land plants.

THE OCEAN'S BIOLOGICAL PUMP What happens to the carbon dioxide that enters the ocean? Most of it is incorporated into organisms through photosynthesis and through their secretion of carbonate shells. Moreover, carbon dioxide is cycled very effectively from the atmosphere to the ocean. In fact, more than 99% of the carbon dioxide added to the atmosphere in the geologic past by volcanic activity has been removed by the ocean and deposited in marine sediments as biogenous calcium carbonate and fossil fuels. Thus, the ocean acts as a *repository* (or *sink*) for carbon dioxide, soaking it up and removing it from the environment as sea floor deposits. The way in which material is removed from the euphotic zone to the sea floor is called a **biological pump** because it "pumps" carbon dioxide and nutrients from the upper ocean and concentrates them in deep-sea waters and sea floor sediments (**Figure 16.31**). As noted earlier in this chapter, however, the capacity of this biological pump to remove carbon dioxide will be reduced as the oceans become warmer and more acidic.

THE OCEAN AS A THERMAL SPONGE   Due to its unique thermal properties,[11] water can absorb large quantities of heat without much change in temperature. Thus, the thermal properties of the ocean make it ideal for minimizing any increase in global temperature. Further, the oceans serve as the planet's biggest single reservoir for surplus energy. If not for the oceans, the planet would be experiencing much greater global warming. In essence, the oceans act as a "thermal sponge," absorbing heat but not increasing much in temperature, and so the oceans minimize the amount of warming experienced. Even so, recent studies have documented that the heat content of the oceans has been increasing (**Figure 16.32**).

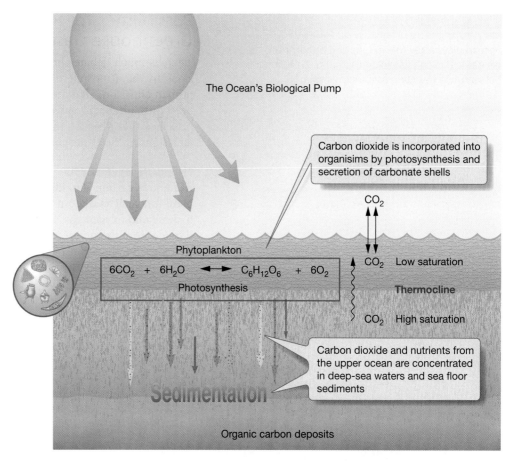

Figure 16.31 The ocean's biological pump. Schematic view of the ocean's biological pump, which removes carbon dioxide from the atmosphere.

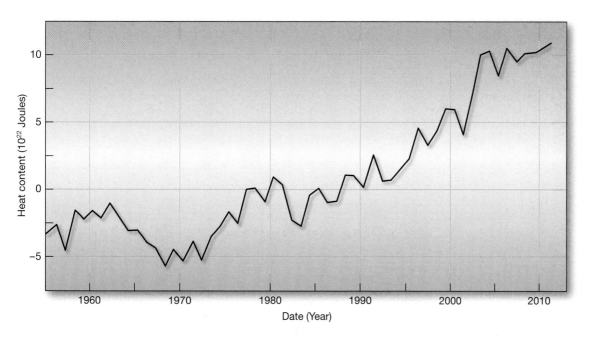

Figure 16.32 Heat content of the oceans. Graph showing how the heat content of the oceans has increased since the 1950s. The short-term variability of the curve shows the natural variation of the ocean's heat content and is influenced by factors such as El Niño/La Niña events.

---

[11]For a discussion of water's unique thermal properties, such as latent heat, see Chapter 5.

Possibilities for Reducing Greenhouse Gases

There is much debate about what to do about the increasing human-caused emissions in the atmosphere. Of course, one suggestion is to reduce the amount of human emissions that are spewed into the atmosphere. However, this scenario has been difficult for humans to implement. Let's examine two global engineering approaches that propose to use the ocean to reduce the amount of human-caused emissions in the atmosphere.

THE IRON HYPOTHESIS Stimulating productivity in the ocean has been proven to remove carbon dioxide from the atmosphere. Through photosynthesis, phytoplankton such as diatoms convert carbon dioxide dissolved in the ocean to carbohydrate and oxygen gas. By capturing and removing additional amounts of carbon dioxide from the ocean, the ocean can, in turn, absorb more heat-trapping carbon dioxide from the atmosphere, thus cooling the planet.

Areas of the ocean that have relatively low productivity, such as the tropics, are a good place to stimulate productivity and thus increase the amount of carbon dioxide removed from the atmosphere. In 1987, oceanographer John Martin determined that the absence of the essential nutrient iron limited productivity in tropical oceans. Iron is required for numerous processes within cells, including photosynthesis, respiration, and nutrient uptake, yet because of its chemical properties, iron is present in only vanishingly small quantities in seawater. Martin proposed fertilizing the ocean with iron to increase its productivity, thereby removing carbon dioxide from the atmosphere and reducing global warming. Subsequently, this idea became known as the **iron hypothesis** (Figure 16.33). In reference to its presumed effectiveness, Martin once famously quipped, "give me a half a tanker of iron and I'll give you the next ice age." Unfortunately, Martin died in 1993, before his idea could be tested in the open ocean.

In late 1993, Martin's associates tested his idea by adding a suspension of finely ground iron to a small patch of ocean near the Galápagos Islands in the Pacific

Figure 16.33 The iron hypothesis. Seeding the ocean with the essential nutrient iron stimulates productivity and draws in heat-trapping carbon dioxide from the atmosphere that is used by phytoplankton. Some of this carbon dioxide gets tied up as tissue or in fecal pellets and sinks toward the sea floor, thereby removing it from the environment. Photo (*inset*) shows a view from a research vessel involved in applying finely ground iron to the ocean near Antarctica.

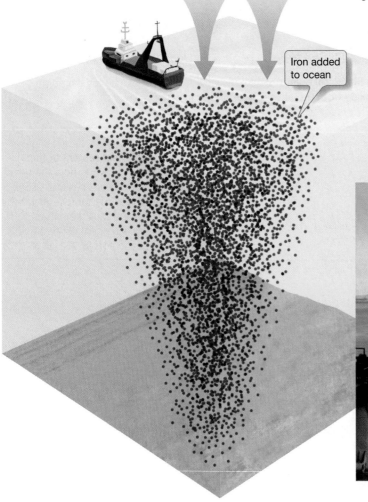

Carbon dioxide in atmosphere

Iron added to ocean

Ocean. Their results, combined with the results of about a dozen other open-ocean experiments worldwide since 1993, confirmed that adding iron to the ocean increased phytoplankton productivity up to 30 times. Remarkably, some of the patches of iron-enriched high productivity were visible to Earth-orbiting satellites. More importantly, the open-ocean experiments showed that the phytoplankton bloom removed vast amounts of carbon dioxide from the atmosphere. As the plankton die, they remove carbon from surface waters and deposit it into ocean bottom waters and sea floor sediments, essentially locking it up for centuries or more.

Although the results from these small-scale open-ocean experiments proved that fertilizing the ocean with iron did indeed draw carbon dioxide out of the atmosphere, the scientists had to overcome problems such as grinding the iron fine enough, dispersing the iron, and keeping it in suspension within surface waters long enough for organisms to use it. More serious problems, however, are the long-term global environmental effects of increasing the amount of carbon dioxide in the ocean and stimulating biological productivity. For example, there is concern that severe oxygen depletion of seawater could occur in areas where iron fertilization is conducted because the large population of algae that is produced eventually dies and decomposes, thus consuming dissolved oxygen. This decomposition also releases byproducts such as carbon dioxide and nitrous oxide—both important greenhouse gases. As previously discussed, higher amounts of carbon dioxide in the ocean lead to increased ocean acidity, which is a problem for marine life and marine ecosystems. Another concern is that there will likely be interference with the natural ecology of large oceanic regions, including associated effects such as changes in microorganism species distribution, removal of other dissolved nutrients, and reduction of marine fish stocks. In addition, when altering a complex marine ecosystem that includes a multitude of feedback loops, there are likely to be unforeseeable consequences. Nevertheless, even larger open-ocean experiments have been proposed, and several private companies have filed patents for ocean-fertilization technologies to enable ocean fertilization to occur on a commercial scale.

SEQUESTERING EXCESS CARBON DIOXIDE IN THE OCEANS Successful experiments have involved capturing emissions either before they are released into the atmosphere or directly from the atmosphere and then pumping the gas into the deep ocean or underground reservoirs. This process—called **sequestering** (*sequester* = depository) carbon dioxide—removes carbon dioxide from the atmosphere and, as a result, reduces global warming. If the deep ocean is used as a disposal site for carbon dioxide, however, there are concerns about how it will impact deep-sea chemistry and, as a result, marine ecosystems. In addition, it is unclear how long the sequestered carbon dioxide will stay in the deep ocean because of deep-water circulation patterns. However, the capture of carbon dioxide emissions and their subsequent disposal into deep reservoirs beneath the sea floor is already occurring at a handful of test sites worldwide.

The Kyoto Protocol: Limiting Greenhouse Gas Emissions

The mounting scientific evidence that global warming and some of its side effects are occurring has led to international efforts to address the human contribution to the greenhouse effect. A number of international conferences have resulted in an agreement to voluntarily limit greenhouse gas emissions. This agreement, which is called the **Kyoto Protocol** (because it was created at an international conference held in 1997 in Kyoto, Japan), sets target reductions for each country. For example, industrialized countries (such as the United States, which constitutes only about 5% of world population but is responsible for about 20% of worldwide carbon

STUDENTS SOMETIMES ASK . . .

The iron hypothesis seems like it could cause dangerous changes in the oceans. Why would scientists support the idea?

The iron hypothesis is a controversial idea and, like many other ideas that have far-reaching and unknown effects, some scientists support it, while others strongly oppose it. In 2007, for example, the parties to the London Convention—an international treaty governing ocean dumping—considered iron fertilization and agreed that it is not yet justified, given gaps in scientific knowledge. Further, in 2008, the United Nations Convention on Biological Diversity (CBD) called for a ban on open-ocean fertilization activities, which caused a German research vessel in 2009 to suspend its ocean fertilization experiment. (The project eventually went forward, but its supporting institution will not conduct any further ocean fertilization studies.) In 2010, participants at the CBD international conference supported a moratorium on global engineering approaches "until there is an adequate scientific basis on which to justify such activities and appropriate consideration for the associated risks." The CBD does, however, grant exceptions to the moratorium for small studies conducted in controlled settings.

In spite of the objections, many scientists think it is advisable to explore how global engineering proposals might feasibly be undertaken as a last resort if global warming accelerates and causes disastrous consequences. The fifth IPCC assessment report, due out in 2013, will include an analysis of peer-reviewed scientific literature on the full range of global engineering proposals.

STUDENTS SOMETIMES ASK . . .

What can I do to help with the problem of global warming?

Even though global warming is a worldwide problem, there are many things concerned individuals can do to help curb greenhouse gas emissions:

1. *Drive smart*: Tune up your car and drive with properly inflated tires.
2. *Write your politicians*: Urge them to raise fuel economy standards for vehicles.
3. *Go green*: Support clean, renewable, non-fossil-fuel energy sources.
4. *Switch lights*: Replace incandescent light bulbs with energy-efficient compact fluorescent bulbs.
5. *Check your house*: Insulate and weather-strip; ask your utility company for a free energy audit.
6. *Become a smart water consumer*: Turn your water heater down and use low-flow faucets.
7. *Buy energy-efficient electronics and appliances*: Replace older, inefficient refrigerators and air conditioners.
8. *Plant a tree, protect a forest*: Preserve plant communities, which absorb carbon dioxide.
9. *Reduce, reuse, and recycle*: Support recycling efforts by choosing recycled products.
10. *Educate and vote*: Educate others and support measures that encourage energy conservation.

Note that these measures are all sound practices for preserving the environment that make sense no matter what the future levels of global warming might be.

ESSENTIAL CONCEPT 16.5B

Possibilities for reducing greenhouse gases include adding iron to the ocean to stimulate productivity and absorbing carbon dioxide from the atmosphere, sequestering excess carbon dioxide in the deep ocean, and limiting greenhouse gas emissions.

dioxide emissions[12]) are to reduce their collective emission of six greenhouse gases by at least 5% by 2012 compared to 1990 levels. Although 122 nations (representing 44% of the total worldwide carbon dioxide emissions) ratified the treaty, the U.S. government never formally agreed to the protocol, citing potential harm to the world economy. Further, the U.S. has lagged behind other countries in establishing emission reductions of greenhouse gases. But the United States is not alone: In 2011, Canada became the first country to withdraw from the Kyoto Protocol.

In an effort to forge a successor agreement to the Kyoto Protocol, an international meeting of delegates from both industrialized nations and major emerging economies met in 2009 at the United Nations Climate Change Conference in Copenhagen, Denmark, and formally recognized a mandate to curb human-induced greenhouse gas emissions. The Copenhagen Accord cites a goal of holding the global rise in average global temperatures to 2°C (3.6°F). However, it does not specify long-term methods for reducing human emissions, set emissions-reduction targets, or require nations to adopt it. As a result, the accord is not legally binding, and so it is widely acknowledged as lacking the ability to produce meaningful results.

In 2011, international delegates met in Durban, South Africa, and agreed to the Durban Platform, which commits the world to negotiating a new climate treaty by 2015. Crucially, that treaty would legally require all nations—including the two biggest emitters, China and the United States—to meet as-yet-unspecified emissions targets. It also extends the Kyoto Protocol by a period of between five and eight years, with the final term to be decided at the next annual conference of the United Nations Framework Convention on Climate Change in Doha, Qatar, at the end of 2012.

At this point, it is very clear that human activities—such as burning fossil fuels, releasing emissions, and land use changes—are altering the environment on a global scale. Although the changes cannot be completely eliminated, many human impacts can be reduced. For instance, activities that affect the environment most severely—such as the high rate of consumption of fossil fuels and their subsequent harmful emissions—need to be decreased. Plant communities and ocean ecosystems, which provide a vital role in absorbing human emissions from the atmosphere, must also be preserved. In addition, a high priority must be placed on research that improves our understanding of how Earth's climate system works.

Ultimately, global warming will impact life on Earth in many ways, but the extent of the change is largely up to us. Scientists have shown that human emissions of greenhouse gases are pushing global temperatures up, and many aspects of climate are responding to the warming in the way that scientists predicted they would. This offers hope. Since people are causing global warming, people can also diminish it, if we act in time. Greenhouse gases are long-lived, so the planet will continue to warm and changes will continue to happen far into the future, but the extent to which global warming changes life on earth depends on our decisions now.

CONCEPT CHECK 16.5

1 How does the ocean's biological pump work to help reduce global warming?

2 How does the ocean act as a thermal sponge, helping minimize the amount of global warming experienced?

3 What strategies exist to reduce future warming?

4 What is the Kyoto Protocol? Why has the United States not agreed to support it?

5 What other international agreements have been established to reduce human emissions since the Kyoto Protocol?

---

[12]China, which is a highly populated and rapidly industrializing country, recently surpassed the United States as the leading carbon dioxide emitter worldwide. The United States is currently second, with India and Russia next.

Essential Concepts Review

16.1 What comprises Earth's climate system?

▶ *Earth's climate system includes the atmosphere, hydrosphere, geosphere, biosphere, and cryosphere* (the ice and snow that exists at Earth's surface). This system involves *exchanges of energy and moisture* that occur among the five spheres. Earth's climate system is modified by both *positive-feedback loops* that reinforce changes and *negative-feedback loops* that counteract changes.

Study Resources
Online Study Guide Quizzes

Critical Thinking Question

To help reinforce your knowledge of Earth's climate system, construct and label your own diagram similar to Figure 16.1 from memory.

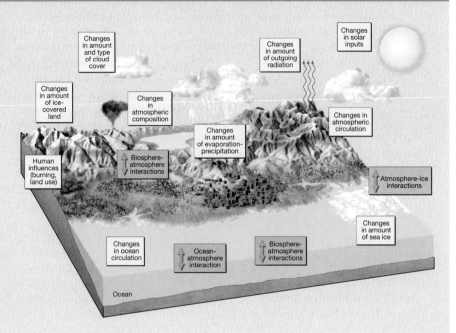

16.2 Earth's recent climate change: Is it natural or caused by human influence?

▶ *Paleoclimatology*, the study of Earth's past climate, uses *proxy data*, which consist of *natural recorders of past climate*, including *sea floor sediments, tree-growth rings, trapped air bubbles in glacial ice, fossil pollen, coral fossils, cave deposits*, and information contained in *historical documents*.

▶ *Several explanations have been formulated to explain Earth's climate change.* Current hypotheses for *natural factors* (causes unrelated to human activities) of climate change include *changes in the Sun's output associated with sunspots, variations in Earth's orbit, volcanic activity, and movement of Earth's tectonic plates*. Although natural mechanisms have altered climate in the past, *recent climate change is greater than can be explained by any natural factors*.

▶ Assessment reports published by the *Intergovernmental Panel on Climate Change (IPCC)* clearly document that the *current climate changes observed on Earth are due primarily to human activities that release heat-trapping greenhouse gas emissions to the atmosphere*.

Study Resources
Online Study Guide Quizzes, Web Animation

Critical Thinking Question

Some friends tell you that they've heard that the recent climate change Earth is experiencing is just part of a natural cycle. What would you explain to them—including using any figures from this chapter—to help convince them otherwise?

Essential Concepts Review (cont.)

16.3 What causes the atmosphere's greenhouse effect?

▶ *The increase in average worldwide temperature is called global warming.* Although the warming of Earth's surface and atmosphere is a natural process controlled by the *greenhouse effect*, it is also being *altered by human greenhouse gas emissions*, a phenomenon that is often referred to as the *enhanced greenhouse effect*.

▶ *The greenhouse effect is produced by incoming sunlight that changes in wavelength, thus heating Earth. Energy reaching Earth from the Sun is mostly in the ultraviolet and visible regions* of the electromagnetic spectrum, whereas *energy radiated back to space from Earth is primarily in the infrared (heat) region. Water vapor, carbon dioxide, methane,* and other trace gases absorb infrared radiation and thus heat the atmosphere.

▶ Heat-trapping gases in the atmosphere are responsible for a *commitment to warming,* and various scenarios can be used to *predict the amount of future warming. The choices made now will determine the amount of atmospheric carbon dioxide and associated warming for the remainder of the 21st century and beyond.*

▶ *Earth's average surface temperature has warmed over the past century,* and there is strong evidence that it is *due primarily to human-caused increases of certain heat-trapping gases such as carbon dioxide and methane.* Models predict that there will be a *strong warming of high northern latitudes,* a moderate warming of middle latitudes, and relatively little warming in low latitudes. Other predicted changes caused by global warming include *changes in the length of seasons, more intense heat waves, changes in both average and extreme temperature and precipitation, retreat of mountain glaciers, alteration of the range of infectious diseases, and shifts in the distribution of plants and animals.*

Study Resources
Online Study Guide Quizzes, Web Animations

Critical Thinking Question
Sketch a labeled diagram that shows how Earth's greenhouse effect works.

16.4 What changes are occurring in the oceans as a result of global warming?

▶ *Changes that have been observed in the oceans due to global warming include increased ocean temperatures, more intense hurricane activity, changes in deep-water circulation, melting of polar ice, increase in ocean acidity, and rising sea level.* These changes in the ocean *will continue for centuries.*

▶ *The ocean removes carbon dioxide from the atmosphere by its biological pump,* which pumps carbon dioxide into sea floor deposits. The ocean also acts as a *thermal sponge by soaking up excess heat from the atmosphere,* thereby minimizing the amount of warming experienced.

Study Resources
Online Study Guide Quizzes

Critical Thinking Question
Of the changes that are occurring in the oceans as a result of global climate change, which one(s) is/are the most serious threat(s) to the marine ecosystem?

16.5 What should be done to reduce greenhouse gases?

▶ *The ocean removes carbon dioxide from the atmosphere via its biological pump*, which pumps carbon dioxide into sea floor deposits. The ocean also acts as a *thermal sponge by soaking up excess heat from the atmosphere*, thereby minimizing the amount of warming experienced.

▶ *The ocean can be used as a repository for some of society's carbon dioxide emissions* through either the *iron hypothesis, which involves stimulating phytoplankton productivity with the addition of finely ground iron*, or through *sequestering excess carbon dioxide by pumping it directly into the deep ocean*. However, the *effect* that these proposals might cause in the ocean is largely *unknown*.

▶ *The Kyoto Protocol establishes limits on greenhouse gas emissions*, but the *United States has not agreed to support it* because of potential harm to the economy.

Carbon dioxide in atmosphere

Iron added to ocean

Study Resources
Online Study Guide Quizzes

Critical Thinking Question

Describe the iron hypothesis and discuss the relative merits and dangers of undertaking such a global engineering project that could cause dramatic changes to the environment worldwide.

MasteringOceanography™

www.masteringoceanography.com

Looking for additional review and test prep materials? Visit the Study Area in MasteringOceanography to enhance your understanding of this chapter's content by accessing a variety of resources, including Self-Study Quizzes, Geoscience Animations, RSS feeds, flashcards, Web links, and an optional Pearson eText.

AFTERWORD

*"In the end, we will conserve only what we love.
We love only what we understand. We will
understand only what we are taught."*
— Baba Dioum, Senegalese conservationist (1968)

At the end of our journey through this book together, it seems fitting to examine people's perceptions of the ocean. Many people describe the ocean as "powerful," "awe inspiring," "moving," "serene," "abundant," and "majestic." Others call it "vast," "infinite," or "boundless." These are all appropriate descriptions because even though humans have been exploring and studying the ocean for centuries, the ocean still holds many secrets. The oceans are constantly surprising researchers with unusual features, newly discovered species, and geologic wonders that exist within its watery world. In fact, less than 5% of the ocean has been seen by humans and even less has been scientifically explored.

Despite the ocean's impressive size, it is beginning to feel the effects of human activities. For instance, every ocean contains large areas of floating plastic, and even remote beaches are littered with trash. Organisms living in the ocean also feel the effect of humankind's use of the ocean. Whaling in the 19th and 20th centuries, for example, pushed many great whale populations to near extinction. Restrictions on whaling and the development of substitutes for whale products helped whales survive this threat, but now they face destruction of their feeding and breeding grounds. Overfishing has degraded entire marine ecosystems, and human-caused climate change is altering marine chemistry on a global scale. The oceans are approaching irreversible, potentially catastrophic change, suggesting that the ocean is not quite as "vast," "infinite," or "boundless" as most people believe.

Humans have become one of the most important agents of change on the planet. This is largely due to our rapidly expanding human population (**Figure Aft.1**). Today, more than 7 billion people occupy the planet, and births now exceed deaths by about two-and-a-half times. At the present rate of population increase—which sounds small enough, at about 1.1%, but produces truly staggering numbers—there are more than 2 new people on the planet each second, or about 9000 more people each hour. Each year there are about 77 million more of us, a total equal to the combined populations of the six largest cities in the world (Mumbai, Shanghai, Karachi, Delhi, Istanbul, and São Paulo). The number of people has quadrupled since 1900 and, if the present trend continues, world population is expected to peak at about 9.2 billion around 2050. Some scientists, in fact, have proposed calling today's time period the *Anthropocene* (*anthro* = human, *cene* = new), or a "sixth mass extinction," in recognition of human activities that are degrading vital Earth systems and the evidence of those impacts that will be preserved in the geologic record. Clearly, our burgeoning human population is the greatest environmental threat of all.

Human-induced changes in the marine environment are broad and far-reaching. Examples include pollution, shoreline development, overfishing, introduction of non-native species, biodiversity decline, ecosystem degradation, and perhaps most serious of all,

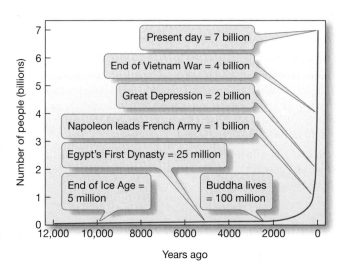

Figure Aft.1 Human population growth. Graph showing world population, which has experienced rapid growth in recent decades. It took 4 million years for humanity to reach the 2 billion mark but only 50 years to double that total. The current world population recently surpassed 7 billion people and is increasing at a rate of over 1% per year.

climate change. A 2008 study of the cumulative impact of 17 of the most urgent land- and ocean-based threats to worldwide marine ecosystems shows that no areas of the ocean are untouched by human activities. The study revealed that fully one-third of the oceans are strongly affected by multiple factors and that the most heavily impacted ecosystems are continental shelves, rocky reefs, coral reefs, seagrass beds, and deep-sea seamounts. More recently, a 2012 study reported that one-fifth of all invertebrate species worldwide are at risk of extinction, including almost one-third of reef-building corals.

Several commissions have suggested ways to protect the marine environment. In 2003, for example, the Pew Oceans Commission recommended significant changes aimed at guiding the way in which the federal government should manage American's marine environment. In 2004, the U.S. Commission on the Oceans submitted 212 recommendations for a coordinated and comprehensive national ocean policy. In 2008, the Joint Ocean Commission issued a report card assessing the nations' collective progress in ocean policy during 2007 (its overall grade: a "C," with some categories receiving a "D"). A 2012 study of the health and benefits of the global ocean produced an index of 10 diverse public goals for a healthy coupled human–ocean system, including an integrative assessment of factors such as food provision, carbon storage, tourism value, and biodiversity. The study gave the world ocean a score of 60 out of 100, with developed countries generally performing better than developing countries. (The United States achieved a score of 63.) These reports urge an integrated research and education ecosystem-based management approach to support ocean environmental health and resource sustainability. Another recommendation is the establishment of marine protected areas.

What Are Marine Protected Areas?

Concerns about the health and longevity of the oceans have led to the establishment of *marine protected areas (MPAs)*, which cover a wide range of marine areas with some level of restriction to protect living, nonliving, cultural, and/or historic resources (**Figure Aft.2**). MPAs can be established for a multitude of reasons: to protect a certain species, to benefit fisheries management, or to protect full ecosystems, rare habitat, or nursing grounds for fish. MPAs are also established to protect historical sites such as shipwrecks and important cultural sites such as aboriginal fishing grounds. MPAs can be very large (for example, Australia's 2000-kilometer [1200-mile] Great Barrier Reef) or very small (for example, Italy's 13,500-hectare [33,360-acre] coastal Marina Protetta Capo Rizzuto). Because the term *MPA* has been used widely around the globe, its meaning in any one region may be quite different than in another. MPAs offer various levels of protection that include areas designated as marine sanctuaries, marine reserves, special protected areas, marine parks, no-take refuges, and areas of special conservation, all of which have specific types of restrictions associated with them, as defined by the laws of the governing body.

Studies have shown that MPAs can help increase biodiversity, protect habitats, and aid in the recovery of fish populations when placed and managed effectively. While MPAs may not solve all the issues that threaten various parts of the ocean, they have been shown to help restore marine ecosystems on a local scale and act as buffers to larger-scale impacts.

In 1972, the U.S. Congress began establishing a type of MPA called a *national marine sanctuary* to protect vital pockets of the ocean from further degradation. Today, 14 national marine sanctuaries cover more than 47,000 square kilometers (18,000 square miles) in areas such as the Florida Keys, the Stellwagen Bank off Massachusetts, the Channel Islands off Southern California, the Flower Garden Banks in the Gulf of Mexico, and the Hawaiian Islands (**Figure Aft.3**). Many activities that degrade the marine environment, however, are still allowed in marine sanctuaries, such as fishing, recreational boating, and even mining of some resources.

Many scientists who recognize the importance of preserving vital marine habitats are calling for governments to more fully protect marine areas by establishing large no-take refuges called *marine reserves*, in which fishing and other activities are prohibited. Such action would allow the recovery of heavily overfished stocks and protect sea floor communities decimated by trawling the sea floor with nets. In 2006, the United States established the 360,000-square-kilometer (139,000-square-mile) Northwest Hawaiian Islands Marine National Monument, creating the world's largest marine reserve at the time. The reserve, which in 2006 was renamed the Papahānaumokuākea Marine National Monument, encompasses 362,073 square kilometers (139,797 square miles) of the Pacific Ocean—an area larger than all U.S. National Parks combined.

Other countries are beginning to realize the economic benefit of fully protecting ocean resources, too. A healthy coral reef, for example, can be worth more as a tourist draw than it might be worth

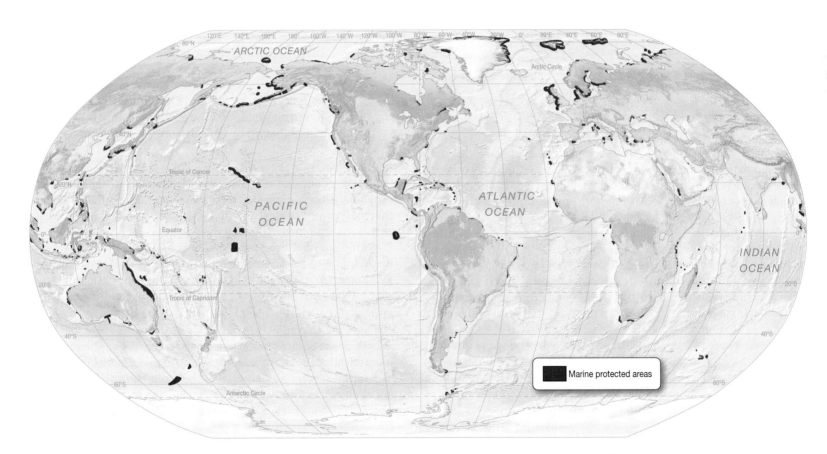

Figure Aft.2 Marine protected areas. Worldwide, more than 4000 marine protected areas (MPAs) have been established, offering various levels of protection to living, nonliving, cultural, and/or historic resources. MPAs cover about 1% of the world ocean.

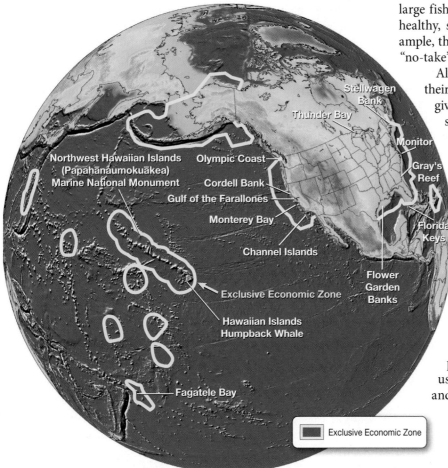

Figure Aft.3 **U.S. national marine sanctuaries.** The United States has established 14 U.S. national marine sanctuaries, which protect natural and/or cultural resources. Also shown is the boundary of the 200-nautical-mile (370-kilometer) exclusive economic zone (*yellow lines*).

large fish that help restore the natural predator–prey relationships of healthy, self-sustaining ecosystems. And yet, issues remain. For example, the temptation exists for poachers to illegally fish within these "no-take" reserves, and proper enforcement is a concern.

Although the oceans are feeling the effects of human activities, their huge size and physical properties make them very resilient and give them a tremendous ability to withstand change. For example, studies reveal that when human impact is reduced, ocean areas often return to a nearly pristine state because natural ocean processes tend to disperse and eventually remove many types of pollutants. However, the ocean's natural resistance is beginning to be seriously compromised. Pollution, habitat loss, and overfishing are dangerous threats on their own, but when these factors converge, they can destroy marine ecosystems.

What Can I Do?

Because the ocean is vast and capable of absorbing many substances, it has been used as a dumping ground for many of society's wastes. Even today, humans are adding pollutants to the ocean at staggering rates. What can each of us do to help? Some ways to help the environment in general and the ocean in particular include the following:

- **MINIMIZE YOUR IMPACT ON THE ENVIRONMENT.** Reduce the amount of waste you generate by making wise consumer choices. Avoid products with excessive packaging and support companies that have good environmental records. Use nontoxic or less hazardous products around the house. Conserve resources. Reuse and recycle items and also help close the recycling loop by buying goods made of recycled materials. Do simple things that make a positive impact on the environment (**Diving Deeper Aft.1**).

- **BECOME POLITICALLY AWARE.** Many ocean-related issues come before the public and require a majority of voters to approve a proposal before it is enacted into law. A common political saying is that the "majority rules" and this situation is true for local as well as national and international issues.

  Within our lifetimes, for instance, we may very well decide whether to spend large amounts of money to add finely ground iron to the ocean to reduce the amount of carbon dioxide in the atmosphere. Many political issues in the future will involve the ocean.

- **EDUCATE YOURSELF ABOUT HOW THE OCEAN WORKS.** A recent poll indicates that more than 90% of the American public consider themselves scientifically illiterate. With our society becoming more scientifically advanced, people need to understand how science operates. Science is not meant to be comprehended by an elite few. Rather, science is for everyone. By studying oceanography, you have begun to understand how the ocean works. It is our hope that you will be a lifelong student of the ocean. And if you are considering participating in the scientific study of the oceans, look for more details on working in marine science in Appendix V, "Careers in Oceanography."

as a source of seafood. In 2008, for example, the Republic of Kiribati, an ocean nation in the central Pacific approximately midway between Australia and Hawaii, established the Phoenix Islands Protected Area, which is about the same size as the state of California, making it the largest MPA in the Pacific Ocean.

Today, more than 4000 marine reserves exist worldwide, but they still cover only about 1% of the world ocean. (In 2010, the United Nations Convention on Biological Diversity set an ambitious target of 10% ocean protection by 2020.) Although the commercial and recreational fishing industries have historically opposed creating marine reserves, research indicates that marine reserves "spill over" and provide benefits to surrounding regions by acting as nurseries for fish larvae that are later spread by ocean currents, thereby boosting certain populations of commercially important fish outside their borders. For example, studies show that within two decades of the prohibition of fishing at Cabo Pulmo, a small Mexican marine reserve off the southern end of the Baja California Peninsula, the total mass of marine organisms quintupled, demonstrating the tremendous potential of habitat protection. Other research has used DNA analysis to show that marine reserves supply about half of all young fish that live outside marine reserve boundaries. Marine reserves are also important in protecting

Diving Deeper AFT.1 Focus on the Environment

TEN SIMPLE THINGS YOU CAN DO TO HELP PREVENT MARINE POLLUTION

"Nobody made a greater mistake than he who did nothing because he could only do a little."

—*Edmund Burke (circa 1790)*

There are many simple things you can do every day to help prevent marine pollution, such as the following:

1. **Snip plastic six-pack rings.** Plastic six-pack rings entangle many marine organisms, so snip each circle with scissors before you toss them into the garbage—or, better yet, recycle them along with the cans or bottles to which they are attached. If you find any at a beach, pick them up, snip them, and recycle them, too.

2. **Use lawn fertilizers, yard chemicals, and laundry detergent sparingly.** Lawn fertilizers contain nitrates and phosphates, which cause harmful algal blooms when runoff from land enters the ocean through storm drains. Detergents also contain phosphates, so read detergent labels to find one that is phosphate free and use a smaller amount than recommended. Use yard chemicals such as pesticides and weed killers only when absolutely necessary.

3. **Clean up after your pet.** Dog and cat feces have high bacteria levels. When they wash into a stream or storm drain and eventually to the ocean, they also provide nitrates and phosphates, which create harmful algal blooms.

4. **Make sure your car doesn't leak oil.** Oil that leaks from automobiles is responsible for a large percentage of the oil that gets into the ocean. Annually, the amount of oil that enters the ocean as runoff from road sources (nonpoint source pollution) through storm drains is greater than a major oil spill. If you change your car's oil yourself, be sure to recycle the used oil at an appropriate recycling center.

5. **Drive less and carpool more.** Reducing the amount of gasoline you use reduces the amount of oil that must be transported across the ocean, which minimizes the potential for oil spills.

6. **Take your own bags to the grocery store.** Paper bags are biodegradable, but plastic bags are not, and plastics are an increasing problem in the ocean, especially for animals like sea turtles that eat plastic bags when they mistake them for jellyfish. That's why plastic bags—as well as Styrofoam containers—have been banned in certain countries and coastal communities.

7. **Don't release balloons.** Balloons that are released far from the ocean can still wind up there, where they deflate, quickly lose their color, and resemble drifting jellyfish that marine animals can ingest.

8. **Don't litter.** Any material that is carelessly discarded on land can become nonpoint source pollution when it washes down a storm drain, into a stream, and eventually into the ocean.

9. **Pick up trash at the beach or volunteer for an organized beach cleanup.** It is important to remove trash that washes up at the beach so it can't endanger marine organisms. Trash arrives on beaches from nonpoint source pollution, ships, recreational boaters, beachgoers, and other sources. It is truly surprising (and also somewhat horrifying) to discover what ends up on beaches as the result of human activities.

10. **Inform and educate others.** Many people are unaware that their actions have a negative influence on the environment—especially the marine environment.

Web Animation
Sylvia Earle on Protecting Our Oceans

Figure Aft.4 **Sunset at the ocean in the Gulf of California, Mexico.**

APPENDIX I Metric and English Units Compared

How many inches are there in a mile? How many cups in a gallon? How many pounds in a ton? In our daily lives, we often need to convert between units. Worldwide, the metric system of measurement is the most widely used system. Besides the United States, only *two other countries in the world*—Liberia and Myanmar (formerly Burma)—still use English units as their primary system of measurement. The metric system has many advantages over the English system. It is simple, logical, and makes conversion between units easy. For those of us in the United States, it is only a matter of time before the change to the metric system occurs.

Benjamin Franklin was one of the first to propose that the United States adopt the sensible metric system. Interestingly, the United States was officially declared a metric nation by the Secretary of the Treasury in 1893, but Americans have never embraced the idea.

On December 23, 1975, U.S. President Gerald R. Ford signed the Metric Conversion Act of 1975. It defined the metric system as the International System of Units (officially called the *Système International d'Unités* [SI]), as interpreted in the United States by the secretary of commerce. The Trade Act of 1988 and other legislation declared the metric system the preferred system of weights and measures for U.S. trade and commerce, called for the federal government to adopt metric specifications, and mandated the Commerce Department to oversee the program.

Although the metric system has not become the system of choice in most Americans' daily use, and there is great resistance to using it, the United States must change over to remain competitive in world markets. Many of the sciences are leading the way in this changeover. In fact, oceanographers all over the world have been using the metric system for years.

The English System

The English system actually consists of two related systems—the U.S. Customary System[1] (used in the United States and its dependencies) and the British Imperial System (used throughout Great Britain). Ironically, Great Britain has now largely converted to the metric system. The basic unit of length in the English system is the *yard*; the basic unit of weight (not mass) is the *pound*.[2]

In the English system, the units of length were initially based arbitrarily on dimensions of the body. The yard as a measure of length, for example, can be traced to the early Saxon kings. They wore a sash around their waists that could be used as a convenient measuring device. Thus, the word *yard* comes from the Saxon word *gird* (like a girdle), in reference to the circumference of a person's waist. The circumference of a person's waist varies from person to person, however, and even varies from time to time on the *same* person, so it is of limited use. It had to be standardized to be useful, so King Henry I of England decreed that the yard should be the distance from the tip of his nose to the end of his thumb!

Romans initially defined the mile (*mille passuum* = 1000 paces) as an even 5000 feet. The early Tudor rulers in England, however, defined a *furlong* ("furrow-long") as 220 yards, based on the length of agricultural fields. To facilitate the conversion between miles and furlongs, Queen Elizabeth I declared in the 16th century that the traditional Roman mile of 5000 feet would be forever replaced by one of 5280 feet, making the mile exactly 8 furlongs. Today, a furlong is an archaic unit of measurement (although still used in horse races), but a mile is still 5280 feet.

The Metric System

The metric system of weights and measures was devised in France and adopted there in 1799. It is based on a unit of length called the *meter* and a unit of mass called the *kilogram*. Originally defined as one ten-millionth the distance from the North Pole to the equator, the meter is now defined as the distance light travels through a vacuum in $1/299,792,458$ second. The kilogram was originally related to the volume of 1 cubic meter of water. It is now defined as the mass of the International Prototype Kilogram, a precision-fabricated cylinder of platinum–iridium alloy about the size of a plum that is kept at Sèvres (near Paris), France, but work is currently under way to base the kilogram on a fundamental property of nature. Other metric units can be defined in terms of the meter and the kilogram.

Fractions and multiples of the metric units are related to each other by powers of 10, allowing conversion from one unit to a multiple of it simply by shifting the decimal point. The lengthy arithmetical operations required with English units can be avoided. Even the names of the metric units indicate how many are in a larger unit. For instance, how many cents are there in a dollar? It is the same as the number of *centi*grams in a gram. The prefixes listed in **Table A1.1** have been accepted for designating multiples and fractions of the meter, the gram (which equals $1/1000$ of a kilogram), and other units.

Despite many fears that changing to the metric system will be an economic burden on businesses in the United States, the conversion has already begun. Examples include 35-millimeter film, 2-liter soft drink bottles, 750-milliliter wine bottles, computers

---

[1] The names of the units and the relationships between them are generally the same in the U.S. Customary System and the British Imperial System, but the sizes of the units differ, sometimes considerably.

[2] "Mass" is commonly equated with "weight," but technically the "mass" of an object refers to the amount of matter in it, whereas its "weight" is caused by the gravitational attraction between the object and Earth. Furthermore, the basic unit of mass in the English system is the *slug*, although it is rarely used.

TABLE A1.1 SOME COMMON PREFIXES FOR BASIC METRIC UNITS

Factor	Name	Prefix name
$10^{12} = 1{,}000{,}000{,}000{,}000$	trillion	tera
$10^9 = 1{,}000{,}000{,}000$	billion	giga
$10^6 = 1{,}000{,}000$	million	mega
$10^3 = 1000$	thousand	kilo
$10^2 = 100$	hundred	hecto
$10^1 = 10$	ten	deka
$10^{-1} = 0.1$	tenth	deci
$10^{-2} = 0.01$	hundredth	centi
$10^{-3} = 0.001$	thousandth	milli
$10^{-6} = 0.000001$	millionth	micro
$10^{-9} = 0.000000001$	billionth	nano
$10^{-12} = 0.000000000001$	trillionth	pico

with gigabytes of memory, and 10-kilometer runs. American cars are now manufactured using metric measurements in order to compete with other countries. Most people in the United States will experience the increasing use of the metric system in their daily lives. Some ways to ease yourself into this unfamiliar (yet sensible) system of units include noticing distances on road signs in kilometers, using a meterstick instead of a yardstick, and charting your weight in kilograms.

Temperature Scales

The Celsius (centigrade) temperature scale was invented in 1742 by a Swedish astronomer named Anders Celsius. The scale was designed so that the freezing point of pure water—the temperature at which it changes state from a liquid to a solid—is set as 0 degrees. Similarly, the boiling point of pure water—the temperature at which it changes state from a liquid to a gas—is set at 100 degrees. Doing so made the liquid range of water—0 degrees to 100 degrees—an even 100 units, giving the scale its name: centigrade (*centi* = 100, *grade* = graduations).

The Fahrenheit temperature scale, on the other hand, sets water's freezing and boiling points at 32 degrees and 212 degrees, respectively, for a spread of 180 units. This scale was devised by a German-born physicist named Gabriel Daniel Fahrenheit, who invented the mercury thermometer in 1714. He initially designed the scale so that normal body temperature was set at 100°F. However, because of inaccuracies in early thermometers or how the scale was initially devised, normal human body temperature is now known to be 98.6°F (37°C).

Conversion Tables

The following tables show conversions between various units:

Length	
1 micrometer	0.001 millimeter 0.0000394 inch
1 millimeter (mm)	1000 micrometers 0.1 centimeter 0.001 meter 0.0394 inch
1 centimeter (cm)	10 millimeters 0.01 meter 0.394 inch
1 meter (m)	100 centimeters 39.4 inches 3.28 feet 1.09 yards 0.547 fathom
1 kilometer (km)	1000 meters 1093 yards 3280 feet 0.62 statute mile 0.54 nautical mile
1 inch (in)	25.4 millimeters 2.54 centimeters
1 foot (ft)	12 inches 30.5 centimeters 0.305 meter
1 yard (yd)	3 feet 0.91 meter
1 fathom (fm)	6 feet 2 yards 1.83 meters
1 statute mile (mi)	5280 feet 1760 yards 1609 meters 1.609 kilometers 0.87 nautical mile
1 nautical mile (nm)	1 minute of latitude 6076 feet 2025 yards 1852 meters 1.15 statute miles
1 league (lea)	5280 yards 15,840 feet 4805 meters 3 statute miles 2.61 nautical miles

Area

1 square centimeter (cm^2)	0.155 square inch 100 square millimeters
1 square meter (m^2)	10,000 square centimeters 10.8 square feet
1 square kilometer (km^2)	100 hectares 247.1 acres 0.386 square mile 0.292 square nautical mile
1 square inch (in^2)	6.45 square centimeters
1 square foot (ft^2)	144 square inches 929 square centimeters

Volume

1 cubic centimeter (cc; cm^3)	1 milliliter 0.061 cubic inch
1 liter (l)	1000 cubic centimeters 61 cubic inches 1.06 quarts 0.264 gallon
1 cubic meter (m^3)	1,000,000 cubic centimeters 1000 liters 264.2 gallons 35.3 cubic feet
1 cubic kilometer (km^3)	0.24 cubic mile 0.157 cubic nautical mile
1 cubic inch (in^3)	16.4 cubic centimeters
1 cubic foot (ft^3)	1728 cubic inches 28.32 liters 7.48 gallons

Mass

1 gram (g)	0.035 ounce
1 kilogram (kg)	2.2 pounds[†] 1000 grams
1 metric ton (mt)	2205 pounds 1000 kilograms 1.1 U.S. short tons
1 pound[†] (lb)	16 ounces 454 grams 0.454 kilogram
1 U.S. short ton (ton; t)	2000 pounds 907.2 kilograms 0.91 metric ton

[†] The pound is a weight unit, not a mass unit, but it is often used as such.

Pressure

1 kilogram per square centimeter (at sea level)	1 atmosphere (atm) 760 millimeters of mercury 101,300 Pascal 1013 millibars 14.7 pounds per square inch 29.9 inches of mercury 33.9 feet of freshwater 33 feet of seawater

Speed

1 centimeter per second (cm/s)	0.0328 foot per second
1 meter per second (m/s)	2.24 statute miles per hour 1.94 knots 3.28 feet per second 3.60 kilometers per hour
1 kilometer per hour (kph)	27.8 centimeters per second 0.62 mile per hour 0.909 foot per second 0.55 knot
1 statute mile per hour (mph)	1.61 kilometers per hour 0.87 knot
1 knot (kt)	1 nautical mile per hour 51.5 centimeters per second 1.15 miles per hour 1.85 kilometers per hour

Oceanographic Data

Velocity of sound in 34.85‰ seawater	4945 feet per second 1507 meters per second 824 fathoms per second
Seawater with 35 grams of dissolved substances per kilogram of seawater	3.5 percent (%) 35 parts per thousand (‰) 35,000 parts per million (ppm) 35,000,000 parts per billion (ppb)

Temperature

Exact formula	Approximation (easy way)
$°C = \dfrac{(°F - 32)}{1.8}$	$°C = \dfrac{(°F - 30)}{2}$
$°F = (1.8 \times °C) + 32$	$°F = (2 \times °C) + 30$

Some useful equivalent temperatures

100°C = 212°F	(boiling point of pure water)
40°C = 104°F	(heat wave conditions)
37°C = 98.6°F	(normal body temperature)
30°C = 86°F	(very warm—almost hot)
20°C = 68°F	(room temperature)
10°C = 50°F	(a warm winter day)
3°C = 37°F	(average temperature of deep water)
0°C = 32°F	(freezing point of pure water)

In degrees centigrade

Thirty is hot, twenty is pleasing;
Ten is not, and zero is freezing.

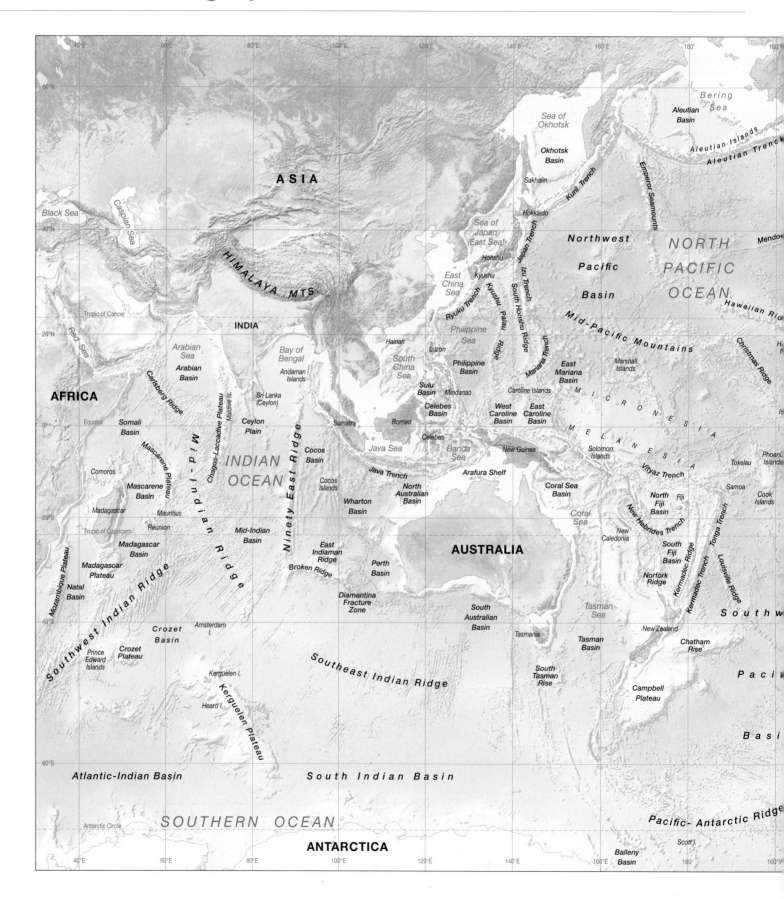

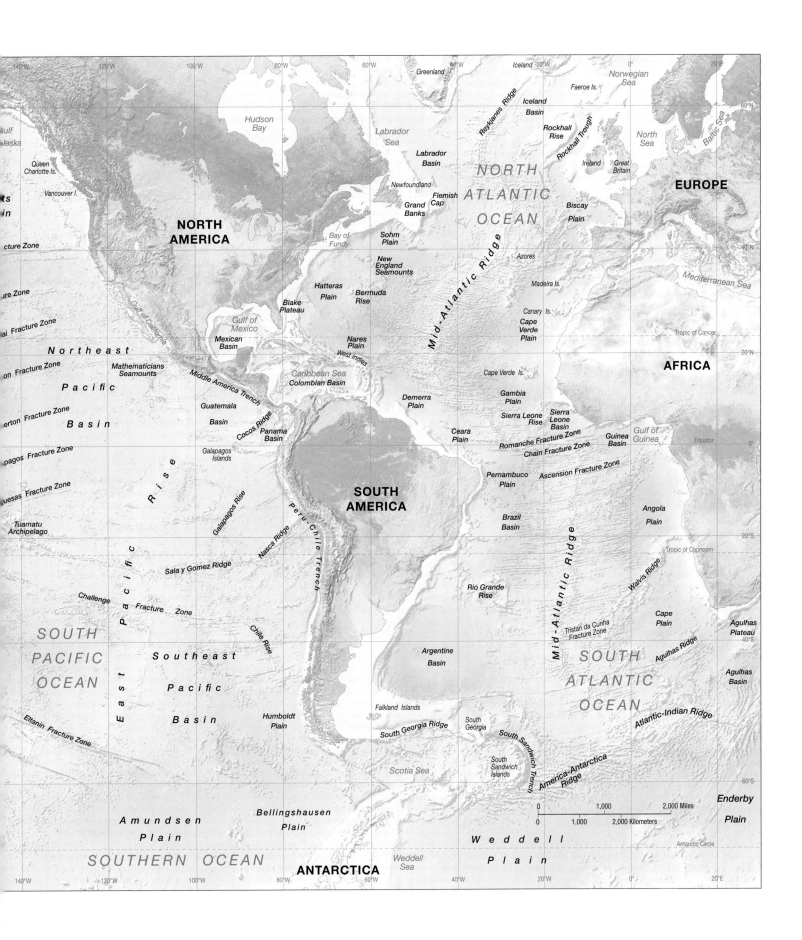

APPENDIX III Latitude and Longitude on Earth

Suppose that you find a great fishing spot or a shipwreck on the sea floor. How would you remember where it is—far from any sight of land—so that you might return? How do sailors navigate a ship when they are at sea? Even using sophisticated equipment such as the Global Positioning System (GPS), which uses satellites to determine location, how is this information reported?

To solve problems like these, a navigational grid—a series of intersecting lines across a globe—is used. Once starting places are fixed for a grid, a location can be identified based on its position within the grid. Some cities, for example, use the regular numbering of streets as a grid. Even if you have never been to the intersection of 5th Avenue and 42nd Street in New York City, you would be able to locate it on a map, or know which way to travel if you were at 10th Avenue and 42nd Street. Although many grid (or coordinate) systems are used today, the one most universally accepted uses *latitude* and *longitude*.

A series of north–south and east–west lines that together comprise a grid system can be used to locate points on Earth's surface whether at sea or on land. The north–south lines of the grid are called *meridians* and extend from pole to pole (**Figure A3.1**). The meridian lines converge at the poles and are spaced farthest apart at the equator. The east–west lines of the grid are called *parallels* because they are parallel to one another. The longest parallel is the *equator* (so called because it divides the globe into two equal hemispheres), and the parallels at the poles are a single point (Figure A3.1).

Latitude and Longitude

Latitude is the angular distance (in degrees of arc) measured north or south of the starting line (the equator) from the center of Earth. All points that lie along the same parallel are an identical distance from the equator, so they all have the same latitude. The latitude of the equator is 0 degrees, and the North and South Poles lie at 90 degrees north and 90 degrees south, respectively. **Figure A3.2** shows that the latitude of New Orleans is an angular distance of 30 degrees north from the equator.

Longitude is the angular distance (in degrees of arc) measured east or west of the starting line from the center of Earth. All meridians are identical (none is longer or shorter than the others, and they all pass through both poles), so the starting line of longitude has been chosen arbitrarily. Before it was agreed upon by all nations, different countries used different starting points for 0 degrees longitude.

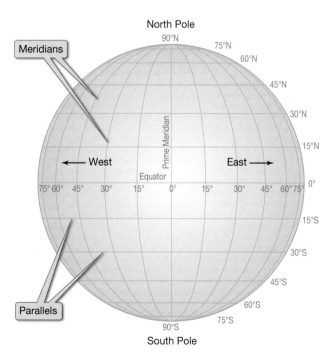

Figure A3.1 Earth's grid system. Earth's grid system is composed of parallels (which extend east–west and are measured north or south of the equator) and meridians (which extend north–south and are measured west or east of the Prime Meridian). Parallels are lines of latitude while meridians are lines of longitude. Note that while all meridians are the same length, the parallels are not (the longest parallel is the equator).

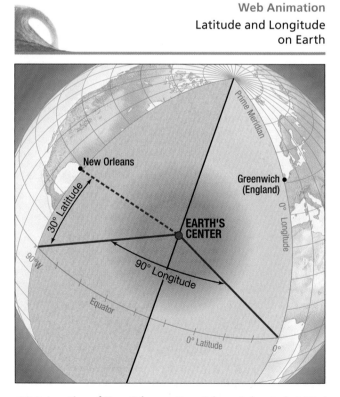

Figure A3.2 Location of New Orleans. New Orleans is located at 30 degrees north latitude and 90 degrees west longitude. Note the angles of each relative to the center of Earth.

Examples include the Canary Islands, the Azores, Rome, Copenhagen, Jerusalem, St. Petersburg, Pisa, Paris, and Philadelphia. At the 1884 International Meridian Conference, it was agreed that the meridian that passes through the Royal Observatory at Greenwich, England, would be used as the zero or starting point of longitude. This zero degree line of longitude is also called the *Prime Meridian*, and the longitude for any place on Earth is measured east or west from this line. Longitude can vary from 0 degrees along the Prime Meridian to 180 degrees (either east or west), which is halfway around the globe and is known as the *International Date Line*. As shown in Figure A3.2, New Orleans is 90 degrees west of the Prime Meridian.

A particular latitude and longitude defines a unique location on Earth, provided that the direction from the starting point is given, too. For instance, 42 degrees north and 120 degrees west defines a unique spot on Earth because it is different from 42 degrees south and 120 degrees west and from 42 degrees north and 120 degrees east.

A degree of latitude or longitude can be divided into smaller units. One degree (°) of arc (angular distance) is equal to 60 minutes (') of arc. One minute of arc is equal to 60 seconds (") of arc. When using a small-scale map or a globe, it may be difficult to estimate latitude and longitude to the nearest degree or two. When using a large-scale map, however, it is often possible to determine the latitude and longitude to the nearest fraction of a minute.

Determination Of Latitude and Longitude

Today, latitude and longitude can be determined very precisely by using satellites that remain in orbit around Earth in fixed positions. How did navigators determine their position before this technology was available?

Latitude was determined from the positions of particular stars. Initially, navigators in the Northern Hemisphere measured the angle between the horizon and the North Star (Polaris), which is directly above the North Pole. The latitude north of the equator is the angle between the two sightings (**Figure A3.3**). In the Southern Hemisphere, the angle between the horizon and the Southern Cross was used because the Southern Cross is directly overhead at the South Pole. Later, the angle of the Sun above the horizon, corrected for the date, was also used.

There was no method of determining longitude until one based on time was developed at the end of the 18th century. As Earth turns on its rotational axis, it moves through 360 degrees of arc every 24 hours (one complete rotation on its axis) (**Figure A3.4a**). Earth, therefore, rotates through 15 degrees of longitude per hour (360 degrees ÷ 24 hours = 15 degrees per hour). As a result, a navigator needed only to know the time at the prime or Greenwich meridian (Figure A3.4b) at exactly the same time the Sun was at its highest point (local noon) at the ship's location. In this way, a navigator aboard a ship could calculate the ship's longitude each day at noon. That's why the development of John Harrison's chronometer (see Box 1.2 in Chapter 1) was so crucial for navigation.

Suppose a ship sets sail west across the Atlantic Ocean from Europe, checking its longitude each day at noon local time. One day when the Sun is at the noon position (noon local time), the chronometer reads 16:18 hours (0:00 is midnight and 12:00 is noon). What is the ship's longitude (**Figure A3.4c**)?

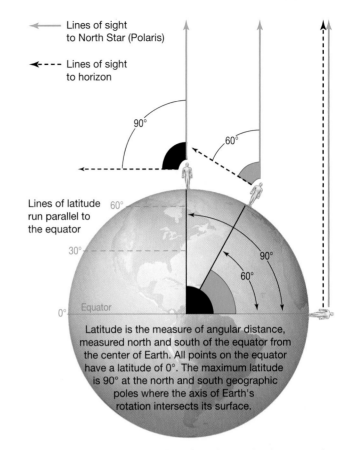

Figure A3.3 **Determining latitude based on the North Star.** Latitude can be determined by noting the angular difference between the horizon and the North Star, which is directly over the North Pole of Earth.

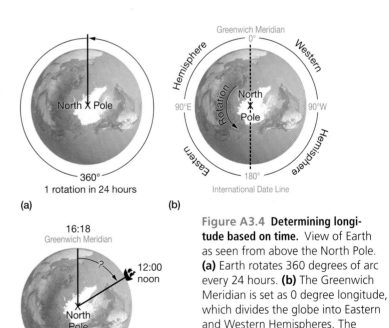

Figure A3.4 **Determining longitude based on time.** View of Earth as seen from above the North Pole. **(a)** Earth rotates 360 degrees of arc every 24 hours. **(b)** The Greenwich Meridian is set as 0 degree longitude, which divides the globe into Eastern and Western Hemispheres. The International Date Line is 180 degrees from the Greenwich Meridian. **(c)** An example of how a ship at sea can determine its longitude using time.

If the clock is keeping good time (which Harrison's chronometer did), then we know that the ship is 4 hours and 18 minutes behind (west of) Greenwich time. To determine the longitude, we must convert time into longitude. Each hour represents 15 degrees of longitude based on the rotation of Earth. Thus, the four hours represents 60 degrees of longitude (4 hours × 15 degrees of longitude = 60 degrees). One degree is divided into 60 minutes of arc, so Earth rotates through 1/4 degree (15′) of arc per minute of time. Thus, 18 minutes of time multiplied by 15 minutes of arc per minute of time equals 270 minutes of arc. To convert 270 minutes of arc into degrees, it must be divided by 60 minutes per degree, which gives 4.5 degrees of longitude. Therefore, the answer is 60 degrees plus 4.5 degrees, or 64.5 degrees west longitude.

Web Video
Latitude and Longitude
on Earth

APPENDIX IV A Chemical Background: Why Water Has 2 H's and 1 O

An element (*elementum* = a first principle) is a substance comprised entirely of like particles that cannot be broken into smaller particles by chemical means. The atom (*a* = not, *tomos* = cut) is the smallest particle of an element that can combine with similar particles of other elements to produce compounds. The periodic table of elements, shown in **Figure A4.1**, lists the elements and describes their atoms. A compound (*compondre* = to put together) is a substance containing two or more elements combined in fixed proportions. A molecule (*molecula* = a mass) is the smallest particle of an element or compound that, in the free state, retains the characteristics of the substance.

As an illustration of these terms, consider Sir Humphrey Davey's use of electrical dissociation to break the compound water into its component elements, hydrogen and oxygen. Atoms of the elements hydrogen (H) and oxygen (O) combine in the proportion 2 to 1, respectively, to produce molecules of water (H_2O). As an electric current is passed through the water, the molecules dissociate into hydrogen atoms that collect near the cathode (negatively charged electrode) and oxygen atoms that collect near the anode (positively charged electrode). Here they combine to form the diatomic gaseous molecules of the elements hydrogen (H_2) and oxygen (O_2). Because there are twice as many hydrogen atoms as oxygen atoms in a given volume of water, twice as many molecules of hydrogen gas (H_2) as oxygen gas (O_2) are formed. Further, the volume of gas under identical conditions of temperature and pressure is proportional to the number of gas particles (molecules) present, so two volumes of hydrogen gas are produced for each volume of oxygen gas.

A Look at the Atom

Building on earlier discoveries, the Danish physicist Niels Bohr (1884–1962) developed his model of the atom as a small solar system in which a positively charged nucleus takes the place of the Sun and the planets that orbit around it are represented by negatively charged electrons. Although this model has since been diagrammatically altered, it is still commonly used to demonstrate the arrangement of electrons and nuclear particles in the atom.

Bohr's earliest concern was with the atom of hydrogen, which he considered to consist of a single positively charged proton (*protos* = first) in its nucleus orbited by a single negatively charged electron (*electro* = electricity). Since the mass of the electron is only about $1/1840$ the mass of the proton, it will be considered negligible in our discussion of atomic masses.

According to the Bohr model, the number of protons—or units of positive charge in the nucleus—coincides with the atomic number of the element. Thus hydrogen, with an atomic number of 1, has a nucleus (*nucleos* = a little nut) containing a single proton. Helium,

the next heavier element, having an atomic number of 2, contains two nuclear protons, and so forth. The atomic number also indicates the number of electrons in a normal atom of any element, because this number is equal to the number of protons in the nucleus (**Figure A4.2**).

An isotope (*isos* = equal, *topos* = place) is an atom of an element that has a different atomic mass than other atoms of the same element. As we will see, chemical properties of atoms are determined by the electron arrangement that surrounds the nucleus. This arrangement, in turn, is determined by the number of protons in the nucleus. Because some isotopes have different atomic masses but identical chemical characteristics, there must be a nuclear particle in the nucleus other than the proton that can influence the atomic mass of an atom but does not affect the electron structure surrounding the nucleus (**Figure A4.3**).

Nuclear research has discovered the additional particle, the neutron (*neutr* = neutral) which was postulated by Ernest Rutherford in 1920 and was first detected by his associate James Chadwick in 1932. It has a mass very similar to that of the proton but no electrical charge. This characteristic of being electrically neutral has made it one of the particles most utilized by nuclear physicists, who continue to explore the atomic nucleus and to make new discoveries of nuclear particles. We will not consider these particles, as knowledge of them is not necessary for our understanding of the chemical nature of atoms.

An atom of a given element can be changed to an ion of that element by adding or taking away one or more electrons, and it is changed to an isotope of the element by adding or taking away one or more neutrons. Adding or taking away one or more protons will change the atom to an atom of a different element.

The atom can be divided into two parts—the nucleus, which contains neutrons and protons, and the electron cloud surrounding the nucleus that is involved in chemical reactions. The random motion of electrons makes it impossible to determine the precise location of electrons in this cloud at any instant, but it is possible to estimate the most probable position of an electron in this cloud. The regions in which there would more likely be particular electrons can be viewed as concentric spheres—or shells—that surround the nucleus (see Figure A4.2).

Chemical Bonds

In considering the chemical reactions in which atoms are involved, we will be concerned primarily with the distribution of electrons in the outer shell. When atoms combine to form compounds, they usually do so by forming one of two bonds:

1. *Ionic* (*ienai* = to go) *bonds*, where electrons are either gained or lost
2. *Covalent* (*co* = with, *valere* = to be strong) *bonds*, where electrons are shared between atoms

Periodic Table of the Elements

Light Metals

Atomic number
Symbol of element
Atomic weight
Name of element

1
H
1.0080
Hydrogen

Transitional Elements

Heavy Metals

- Inert Gas
- Gas
- Liquid

Nonmetals

	IA	IIA	IIIB	IVB	VB	VIB	VIIB		VIIIB		IB	IIB	IIIA	IVA	VA	VIA	VIIA	VIIIA
1	1 **H** 1.0080 Hydrogen																	2 **He** 4.003 Helium
2	3 **Li** 6.939 Lithium	4 **Be** 9.012 Beryllium											5 **B** 10.81 Boron	6 **C** 12.011 Carbon	7 **N** 14.007 Nitrogen	8 **O** 15.994 Oxygen	9 **F** 18.998 Fluorine	10 **Ne** 20.1863 Neon
3	11 **Na** 22.990 Sodium	12 **Mg** 24.31 Magnesium											13 **Al** 26.98 Aluminum	14 **Si** 28.09 Silicon	15 **P** 30.974 Phosphorus	16 **S** 32.064 Sulphur	17 **Cl** 35.453 Chlorine	18 **Ar** 39.948 Argon
4	19 **K** 39.102 Potassium	20 **Ca** 40.08 Calcium	21 **Sc** 44.96 Scandium	22 **Ti** 47.90 Titanium	23 **V** 50.94 Vanadium	24 **Cr** 52.00 Chromium	25 **Mn** 54.94 Manganese		26 **Fe** 55.85 Iron	27 **Co** 58.93 Cobalt	28 **Ni** 58.71 Nickel	29 **Cu** 63.54 Copper	30 **Zn** 65.37 Zinc	31 **Ga** 69.72 Gallium	32 **Ge** 72.59 Germanium	33 **As** 74.92 Arsenic	34 **Se** 78.96 Selenium	35 **Br** 79.909 Bromine	36 **Kr** 83.80 Krypton
5	37 **Rb** 85.47 Rubidium	38 **Sr** 87.62 Strontium	39 **Y** 88.91 Yttrium	40 **Zr** 91.22 Zirconium	41 **Nb** 92.91 Niobium	42 **Mo** 95.94 Molybdenum	43 **Tc** (99) Technetium		44 **Ru** 101.1 Ruthenium	45 **Rh** 102.90 Rhodium	46 **Pd** 106.4 Palladium	47 **Ag** 107.870 Silver	48 **Cd** 112.40 Cadmium	49 **In** 114.82 Indium	50 **Sn** 118.69 Tin	51 **Sb** 121.75 Antimony	52 **Te** 127.60 Tellurium	53 **I** 126.90 Iodine	54 **Xe** 131.30 Xenon
6	55 **Cs** 132.91 Cesium	56 **Ba** 137.34 Barium	57 TO 71	72 **Hf** 178.49 Hafnium	73 **Ta** 180.95 Tantalum	74 **W** 183.85 Tungsten	75 **Re** 186.2 Rhenium		76 **Os** 190.2 Osmium	77 **Ir** 192.2 Iridium	78 **Pt** 195.09 Platinum	79 **Au** 197.0 Gold	80 **Hg** 200.59 Mercury	81 **Tl** 204.37 Thallium	82 **Pb** 207.19 Lead	83 **Bi** 208.98 Bismuth	84 **Po** (210) Polonium	85 **At** (210) Astatine	86 **Ra** (222) Radon
7	87 **Fr** (223) Francium	88 **Ra** 226.05 Radium	89 TO 103															

Lanthanide series

57 **LA** 138.91 Lanthanum	58 **Ce** 140.12 Cerium	59 **Pr** 140.91 Praseodymium	60 **Nd** 144.24 Neodymium	61 **Pm** (147) Promethium	62 **Sm** 150.35 Samarium	63 **Eu** 157.25 Europium	64 **Gd** 158.92 Gadolinium	65 **Tb** 158.92 Terbium	66 **Dy** 162.50 Dysprosium	67 **Ho** 164.93 Holmium	68 **Er** 167.26 Erbium	69 **Tm** 168.93 Thulium	70 **Yb** 173.04 Ytterbium	71 **Lu** 174.97 Lutetium

Actinide series

89 **Ac** (227) Actinium	90 **Th** 232.04 Thorium	91 **Pa** (231) Protactinium	92 **U** 238.03 Uranium	93 **Np** (237) Neptunium	94 **Pu** (242) Plutonium	95 **Am** (243) Americium	96 **Cm** (247) Curium	97 **Bk** (249) Berkelium	98 **Cf** (251) Californium	99 **Es** (254) Einsteinium	100 **Fm** (253) Fermium	101 **Md** (256) Mendelevium	102 **No** (256) Nobelium	103 **Lw** (257) Lawrencium

Figure A4.1 The periodic table of the elements.

540

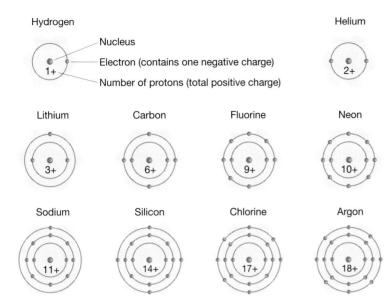

Hydrogen

Nucleus

Electron (contains one negative charge)

1+

Number of protons (total positive charge)

Helium

2+

Lithium 3+   Carbon 6+   Fluorine 9+   Neon 10+

Sodium 11+   Silicon 14+   Chlorine 17+   Argon 18+

Figure A4.2 Bohr orbital models for atoms. Each atom is composed of a positively charged nucleus with negatively charged electrons around it. The nucleus, which occupies very little space, contains most of the mass of the atom. The atomic number is equal to the number of protons (positively charged nuclear particles) in the nucleus. Note that the first shell holds only two electrons and other shells can hold eight (or more) electrons.

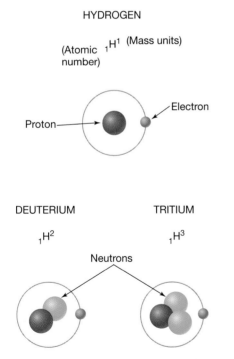

HYDROGEN

(Atomic number) $_1H^1$ (Mass units)

Proton — Electron

DEUTERIUM $_1H^2$ TRITIUM $_1H^3$

Neutrons

Figure A4.3 Hydrogen isotopes. Isotopes are atoms of an element that have different atomic masses. The hydrogen atom ($_1H^1$) accounts for 99.98% of the hydrogen atoms on Earth and has a nucleus that contains only one proton. Deuterium ($_1H^2$) contains one neutron and one proton in its nucleus and combines with oxygen to form "heavy water" with a molecular mass of 20. Tritium ($_1H^3$) is a very rare radioactive isotope of hydrogen that has a nucleus containing one proton and two neutrons.

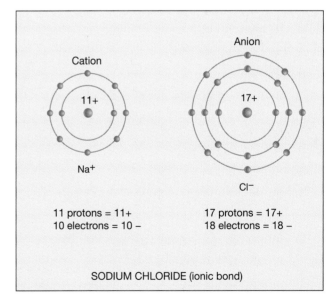

Figure A4.4 Ionic bonds in sodium chloride (table salt). Ions of sodium (left) and chlorine (right) combine to form sodium chloride, producing common table salt. Notice the contrast in the outer shells of the ions of sodium (Na^+) and chlorine (Cl^-) with the atoms of these elements shown in Figure A4.2.

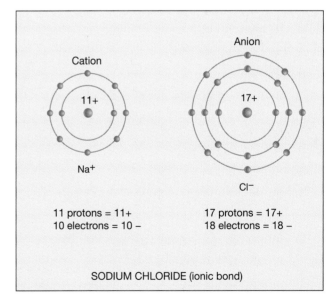

Cation

11+

Na^+

Anion

17+

Cl^-

11 protons = 11+
10 electrons = 10 –

17 protons = 17+
18 electrons = 18 –

SODIUM CHLORIDE (ionic bond)

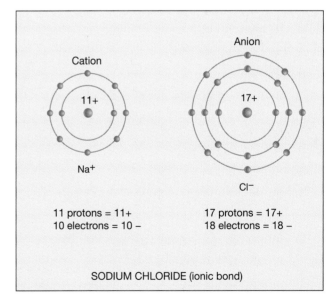

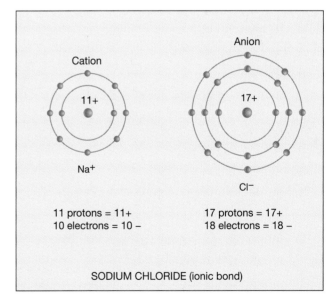

Ionic bonding produces an ion, an electrically charged atom that no longer has the properties of a neutral atom of the element it represents. A positively charged ion, a cation (*kation* = something going down) is produced by the loss of electrons from the outer shell, the positive charge being equal to the number of electrons lost. A negatively charged ion, an anion (*anienai* = to go up), is produced by the gain of electrons in the outer shell of an atom, and its charge is equal to the number of electrons gained.

An example of a compound formed by ionic bonding is common table salt, which is sodium chloride (**Figure A4.4**). In the formation of this compound, the sodium atom, which has one electron in its outer shell, loses this electron and forms a sodium ion with a positive electrical charge of one. The chlorine atom, which contains seven electrons in its outer shell, completes its shell by gaining an electron and becoming a chloride ion with a negative charge of one. These two ions are held in close proximity by an electrostatic attraction between the two ions of equal and opposite charge.

Moreover, it is the tendency of an individual atom to assume the outer-shell electron content of the inert gases such as helium (2), neon (8), and argon (8) that make it chemically reactive (see Figure A4.2). Normally, if an atom can assume this configuration by either sharing one or two electrons with an atom of another element or by losing or gaining one or two electrons, the elements are highly reactive. By contrast, the elements are less reactive if three or four electrons must be shared, gained, or lost to achieve the desired configuration.

Valence (*valentia* = capacity) is the number of hydrogen atoms with which an atom of a given element can combine with either ionic or covalent bonds. Elements with lower valences of one or two, for

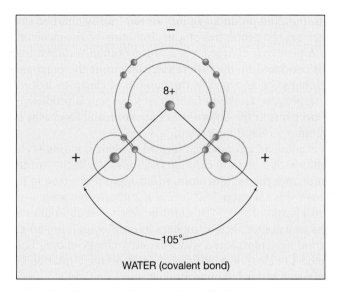

Figure A4.5 Covalent bonds in water. Water (H_2O) is composed of 2 atoms of hydrogen and 1 atom of oxygen, held together by covalent bonds.

example, combine chemically in a more highly reactive manner than do those with higher valences of three or four or, in certain instances, more than four. Although those elements with higher valences do not react as violently as low-valence elements, they have a greater combining power and can gather about them larger numbers of atoms of other elements than can the atoms with lower valence values.

An example of a covalent bond is the sharing of electrons by hydrogen and oxygen atoms in the water molecule (**Figure A4.5**). In the formation of this molecule, both hydrogen atoms and the oxygen atom assume the inert gas configuration they seek by sharing electrons.

Many people think a career in oceanography consists of swimming with marine animals at a marine life park or snorkeling in crystal-clear tropical waters, studying coral reefs. In reality, these kinds of jobs are extremely rare, and there is intense competition for them. Most oceanographers use science to answer questions about the ocean, such as the following:

- What is the role of the ocean in limiting human-caused climate change?
- What kinds of pharmaceuticals can be found naturally in marine organisms?
- How does sea floor spreading relate to the movement of tectonic plates?
- What economic deposits are there on the sea floor?
- Can rogue waves be predicted?
- What is the role of longshore transport in the distribution of sand on the beach?
- How does a particular pollutant affect organisms in the marine environment?

Preparation for a Career in Oceanography

Preparing yourself for a career in oceanography is an interesting and rewarding, yet difficult, path to travel. The study of oceanography is typically divided into four different academic disciplines (or subfields) of study:

- *Geological oceanography* is the study of the structure of the sea floor and how the sea floor has changed through time, the creation of sea floor features, and the history of sediments deposited on it.
- *Chemical oceanography* is the study of the chemical composition and properties of seawater, how to extract certain chemicals from seawater, and the effects of pollutants.
- *Physical oceanography* is the study of waves, tides, and currents; the ocean–atmosphere relationship that influences weather and climate; and the transmission of light and sound in the oceans.
- *Biological oceanography* is the study of oceanic life forms and their relationships to one another, adaptations to the marine environment, and development of ecologically sound methods of harvesting seafood.

Other disciplines include ocean engineering, marine archaeology, and marine policy. Oceanography is an *interdisciplinary* science because it utilizes all the disciplines of science as they apply to the oceans. Some of the most exciting work and best employment opportunities combine two or more of these disciplines.

Individuals working in oceanography and marine-related fields need a strong background (typically an undergraduate degree) in at least one area of basic science (for example, geology, physics, chemistry, or biology) or engineering. In almost all cases, mathematics is required as well. Marine archaeology requires a background in archaeology or anthropology; marine policy studies require a background in at least one of the social sciences (such as law, economics, or political science).

The ability to speak and write clearly and critical thinking skills are prerequisites for any career. Fluency in computers—preferably PCs, not Apple— and the ability to debug computer systems is now a necessity. Because many job opportunities in oceanography require trips on research vessels (**Figure A5.1**), any shipboard experience is also desirable. Mechanical ability (such as the ability to fix equipment while on board a vessel without having to return to port) is a plus. Depending on the type of work that is required, other traits that may

Figure A5.1 Oceanographers at work on a research vessel. Oceanographers deploy a sonar device on a cable using the A-frame on Scripps Institute of Oceanography's R/V *Melville.*

be desirable include the ability to speak one or more foreign languages, certification as a scuba diver, the ability to work for long periods of time in cramped conditions, physical stamina, physical strength, and a high tolerance to motion sickness.

Oceanography is a relatively new science with much room left for discovery, so most people enter the field with an advanced degree (master's or doctorate). Work as a marine technician, however, usually requires only a bachelor's degree or applicable experience. Although it does take a large commitment to achieve an advanced degree, don't feel overwhelmed. Many others like you have faced a similar journey. In the end, what makes all the hard work worthwhile is the incredible satisfaction of discovering new things about the ocean as well as all the interesting people you'll meet along the way.

Job Duties Of Oceanographers

There has been and will continue to be enormous expansion in the number of ocean-related jobs. Job opportunities for oceanographers exist with scientific research institutions (universities), various government agencies, and private companies that are engaged in searching for economic sea floor deposits, investigating areas for sea farming, and evaluating natural energy production from waves, currents, and tides. The duties of oceanographers vary from job to job but generally can be described as follows:

- *Geological oceanographers and geophysicists* explore the ocean floor and map submarine geologic structures. Studies of the physical and chemical properties of rocks and sediments give us valuable information about Earth's history. The results of their work help us understand the processes that created the ocean basins and the interactions between the ocean and the sea floor.

- *Chemical oceanographers and marine geochemists* investigate the chemical composition of seawater and its interaction with the atmosphere and the sea floor. Their work may include analysis of seawater components, desalination of seawater, and study of the effects of pollutants. They also examine chemical processes operating within the marine environment and work with biological oceanographers to study living systems. Their study of trace chemicals in seawater helps us understand how ocean currents move seawater around the globe and how the ocean affects climate.

- *Physical oceanographers* investigate ocean properties such as temperature, density, wave motions, tides, and currents. They study ocean–atmosphere relationships that influence weather and climate, the transmission of light and sound through water, and the ocean's interactions with its boundaries at the sea floor and the coast.

- *Biological oceanographers, marine biologists, and fisheries scientists* study marine plants and animals. They want to understand how marine organisms develop, relate to one another, adapt to their environment, and interact with their environment. They develop ecologically sound methods of harvesting seafood and study biologic responses to pollution. New fields associated with biological oceanography include marine biotechnology (the use of natural marine resources in the development of new industrial and biomedical products) and molecular biology (the study of the structure

and function of bioinformational molecules—such as DNA, RNA, and proteins—and the regulation of cellular processes at the molecular level). Because marine biology is the best-known field of oceanography (and because the larger marine animals have such wide appeal), it is currently the most competitive, too.

- *Marine and ocean engineers* apply scientific and technical knowledge to practical uses. Their work ranges from designing sensitive instruments for measuring ocean processes to building marine structures that can withstand ocean currents, waves, tides, and severe storms. Subfields include acoustics; robotics; electrical, mechanical, civil, and chemical engineering; and naval architecture. These engineers often use highly specialized computer techniques.

- *Marine archaeologists* systematically recover and study material evidence, such as shipwrecks, graves, buildings, tools, and pottery remaining from past human life and culture that is now covered by the sea. Marine archaeologists use state-of-the-art technology to locate underwater archaeological sites.

- *Marine policy experts* combine their knowledge of oceanography and social sciences, law, or business to develop guidelines and policies for the wise use of the ocean and coastal resources.

Other job opportunities for oceanographers include work as science journalists specializing in marine science, teachers at various grade levels, and aquarium and museum curators.

Sources Of Career Information

For sources of career information, consult the catalog of any college or university that offers a curriculum in oceanography or marine science. Online pamphlets that feature oceanography career information include:

- *Marine Science Careers: A Sea Grant Guide to Ocean Opportunities* by NOAA's Sea Grant College; available at http://marinecareers.net.

- *Preparing for a Career in Oceanography* by the Scripps Institute of Oceanography at the University of California, San Diego; available at www.siommg.ucsd.edu/slices/.

- *The SeaWorld/Bush Gardens Guide to Zoological Park Careers* by SeaWorld's Education Department; available at www.seaworld.org/infobooks/zoocareers/home.html.

- *Careers in Science and Engineering: A Student Planning Guide to Graduate School and Beyond* by the National Academies Press; available at www.nap.edu/openbook.php?record_id=5129.

Other online sources of oceanography career information include the following:

- The Oceanography Society (TOS), which has a listing of oceanography career profiles at www.tos.org/resources/career_profiles.html.

- The U.S. Navy's Website, which has information on careers in oceanography at www.oc.nps.edu/careers.html.

REPORT FROM A STUDENT/OCEANOGRAPHER

One of the true pleasures of teaching is that some of your students become so interested in the subject matter you teach that they pursue a career in it. One of Al Trujillo's former students, Joseph Fell-McDonald (**Figure A5A**), works as a ranger and interpretive specialist on Maui for the State of Hawaii's Department of Land and Natural Resources, Division of Forestry and Wildlife, Natural Area Reserve System. Joe writes,

Aloha:

My career in oceanography-related work has followed a long and sinuous path. After high school I attended a trade school for mechanical engineering, where I learned how to fix cars and build houses. With this background, I worked in industry for many years yet I had a yearning for something satisfying, enjoyable, and well-paid. I bounced from place to place and job to job not really settling on a career but beginning to realize that I enjoyed a lifestyle near the world's oceans where I could surf, swim, lounge, and play. Living in various coastal areas nearly my whole life, I was always amazed at the interaction of the oceans. I decided fixing and building things was not for me. So I looked into colleges and attended nearby Palomar Community College in California, where I took many courses in geography, geology, and biology. I also enrolled in an oceanography course and realized how important the oceans are to life in general. It was during this time that it hit me: I longed for a career in ocean science!

Because I was formally trained in mechanical engineering and construction, I knew I could get a job just about anywhere. I decided to give the island of Maui a try just after finishing my courses in oceanography. There was no formal work in oceanography here so I applied for a job as a vessel mechanic at a local eco-tour company. The company offered volunteer opportunities, including photo identification and public interpretation of the seasonal humpback whales that migrate to Maui each year. The one thing I realized on Maui was that not one organization—be it a government branch, non-profit organization, or company—could control, regulate, or even understand the broad basis of the world's oceans. I became fascinated with my surroundings and volunteered at many places, realizing my best attribute was an understanding of Earth's terrain and oceanic interactions. I volunteered and studied native coastal restoration projects and got hired by a local botanical garden to promote native Hawaiian plants and Hawaiian plant restoration.

Figure A5A Former oceanography student Joseph Fell-McDonald (*at far right on vessel and inset*) at work in the Hawaiian Islands Humpback Whale National Marine Sanctuary, Maui, Hawaii.

I was then recruited to help restore a high erosion area at a National Marine Sanctuary. The National Oceanic and Atmospheric Administration (NOAA) Hawaiian Islands Humpback Whale National Marine Sanctuary (sister sanctuary to Monterey Bay) hired me to do building maintenance, coastal dune restoration, vessel repair for scientific exploration, and general public liaison. I was asked to assist with the National Marine Fisheries Service in helping stranded marine life—particularly the green sea turtle (*Chelonia mydas*), which often strands itself when it succumbs to disease—and habitat restoration and monitoring of the critically endangered hawksbill turtle (*Eretmochelys imbricata*). I also became involved with rescuing stranded Hawaiian monk seals (*Monachus schauinslandi*) and entangled humpback whales (*Megaptera novaeangliae*). I credit my success to a well-rounded education, being open to new opportunities, volunteering, and lots of hard work.

I currently work as a ranger for the State of Hawaii's Department of Land and Natural Resources, Division of Forestry and Wildlife, Natural Area Reserve System (NARS). The NARS program is the highest level of protection that Hawaii can offer its lands. I work on Maui at the Ahihi-Kinau Natural Area Reserve and Surrounding State Protected Lands Kenioio and Makena. We follow recommendations by residents and local native Hawaiian kapuna (elders) to bring conservation back to Ahihi-Kinau. Much of my work is enforcement of protection

laws, although I do a fair amount of public education, restoration, cultural inventory, biologic surveying, and other protection of natural resources. The reserve is approximately 2000 land acres and about 800 marine protected acres out to a water depth of 100 fathoms. I patrol the area via four-wheel drive vehicle, jet ski, ATV, and even on horseback.

The most common question I get asked is, "How does someone get a job like yours?" Truthfully, I wake up with a smile on my face every day and enjoy the location, pay, and benefits. Sometimes it can be a tough and demanding job but I actually enjoy it. For many people, it is hard to believe that studying a broad range of sciences can lead to an enjoyable and fulfilling career. I also met my wife and companion here, who was working as a marine biologist with the U.S. Fish and Wildlife Service. Due to my compassion and deep understanding of a broad range of sciences, I was able to apply my education, experience, and curiosity into the career of my dreams. I encourage anyone who wants to learn more about natural resources or careers on Maui to contact me.

Joseph Fell-McDonald, Ahihi-Kinau Ranger
State of Hawaii Department of Land
and Natural Resources
Division of Forestry and Wildlife
54 High Street, Room 101
Wailuku, Maui, HI 96793
Joseph.C.Fell-McDonald@hawaii.gov

- The Website of Peter Brueggeman of the Scripps Institute of Oceanography, which offers a comprehensive list of information about careers in oceanography, marine science, and marine biology, including many links, at http://ocean.peterbrueggeman.com/career.html.

- The Palomar College Oceanography Program Careers in Oceanography Website, which includes this appendix, at www.palomar.edu/oceanography/links/Careers.html.

- The SUNY Stony Brook Biological Sciences Department, which offers a list of worldwide marine laboratories and institutions at http://life.bio.sunysb.edu/marinebio/mblabs.html.

- Woods Hole Oceanographic Institution's Website on women in oceanography, which is devoted to the achievements of women oceanographers and includes unique perspectives of women scientists, at www.womenoceanographers.org.

GLOSSARY

Abiotic environment The nonliving components of an ecosystem.

Abyssal clay Deep-ocean (oceanic) deposits containing less than 30% biogenous sediment. Often oxidized and red in color, thus commonly termed red clay.

Abyssal hill A volcanic peak rising less than 1 kilometer (0.6 mile) above the ocean floor.

Abyssal hill province A deep-ocean region, particularly in the Pacific Ocean, where oceanic sedimentation rates are so low that abyssal plains do not form and the ocean floor is covered with abyssal hills.

Abyssal plain A flat depositional surface extending seaward from the continental rise or oceanic trenches.

Abyssal storm Stormlike occurrences of rapid current movement affecting the deep ocean floor. They are believed to be caused by warm- and cold-core eddies of surface currents.

Abyssal zone The benthic environment between 4000 and 6000 meters (13,000 and 20,000 feet).

Abyssopelagic The open-ocean (oceanic) environment below 4000 meters (13,000 feet) in depth.

Acid A substance that releases hydrogen ions (H^+) in solution.

Acoustic Thermometry of Ocean Climate (ATOC) The measurement of oceanwide changes in water properties such as temperature by transmitting and receiving low-frequency sound signals.

Active margin A continental margin marked by a high degree of tectonic activity, such as those typical of the Pacific Rim. Types of active margins include convergent active margins (marked by plate convergence) and transform active margins (marked by transform faulting).

Adiabatic Pertaining to a change in the temperature of a mass resulting from compression or expansion. It requires no addition of heat to or loss of heat from the substance.

Age of Discovery The 30-year period from 1492 to 1522 when Europeans explored the continents of North and South America and the globe was circumnavigated for the first time.

Agulhas Current A warm current that carries Indian Ocean water around the southern tip of Africa and into the Atlantic Ocean.

Air mass A large area of air that has a definite area of origin and distinctive characteristics.

Albedo The fraction of incident electromagnetic radiation reflected by a surface.

Algae Primarily aquatic, eukaryotic, photosynthetic organisms that have no root, stem, or leaf systems. Can be microscopic or macroscopic.

Alkaline A substance that releases hydroxide ions (OH^-) in solution. Also called basic.

Alveoli A tiny, thin-walled, capillary-rich sac in the lungs where the exchange of oxygen and carbon dioxide takes place.

Amino acid One of more than 20 naturally occurring compounds that contain NH_2 and COOH groups. They combine to form proteins.

Amnesic shellfish poisoning Partial or total loss of memory resulting from poisoning caused by eating shellfish with high levels of domoic acid, a toxin produced by a diatom.

Amphidromic point A nodal, or "no-tide," point in the ocean or sea around which the crest of the tide wave rotates during one tidal period.

Amphipoda A crustacean order containing laterally compressed members such as the "beach hoppers."

Anadromous Pertaining to a species of fish that spawns in freshwater and then migrates into the ocean to grow to maturity.

Anaerobic respiration Respiration carried on in the absence of free oxygen (O_2). Some bacteria and protozoans carry on respiration this way.

Anchovy (anchoveta) A small silvery fish (*Engraulis ringens*) that swims through the water with its mouth open to catch its food.

Andesite A gray, fine-grained volcanic rock composed chiefly of plagioclase feldspar.

Anhydrite A colorless, white, gray, blue, or light purple evaporite mineral (anhydrous calcium sulfate, $CaSO_4$) that usually occurs as layers associated with gypsum deposits.

Animalia A kingdom of many-celled animals.

Anion An atom that has gained one or more electrons and has an electrical negative charge.

Annelida A phylum of elongated segmented worms.

Anomalistic month The time required for the Moon to go from perigee to perigee, 27.5 days.

Anoxic Without oxygen.

Antarctic Bottom Water A water mass that forms in the Weddell Sea, sinks to the ocean floor, and spreads across the bottom of all oceans.

Antarctic Circle The latitude 66.5° south.

Antarctic Circumpolar Current The eastward-flowing current that encircles Antarctica and extends from the surface to the deep-ocean floor. The largest volume current in the oceans. Also called the West Wind Drift.

Antarctic Convergence The zone of convergence along the northern boundary of the Antarctic Circumpolar Current where the southward-flowing boundary currents of the subtropical gyres converge on the cold Antarctic waters.

Antarctic Divergence The zone of divergence separating the westward-flowing East Wind Drift and the eastward-flowing Antarctic Circumpolar Current.

Antarctic Intermediate Water Antarctic zone surface water that sinks at the Antarctic convergence and flows north at a depth of about 900 meters (2950 feet) beneath the warmer upper-water mass of the South Atlantic Subtropical Gyre.

Anthophyta Seed-bearing plants.

Anticyclone An atmospheric system characterized by the rapid, outward circulation of air masses about a high-pressure center that is associated with sinking air. Anticyclones circulate clockwise in the Northern Hemisphere and counterclockwise in the Southern Hemisphere and are usually accompanied by dry, clear, fair weather.

Anticyclonic flow The flow of air around a region of high pressure clockwise in the Northern Hemisphere.

Antilles Current A warm current that flows north seaward of the Lesser Antilles from the North Equatorial Current of the Atlantic Ocean to join the Florida Current.

Antinode A zone of maximum vertical particle movement in standing waves where crest and trough formation alternate.

Aphelion The point in the orbit of a planet or comet where it is farthest from the Sun.

Aphotic zone A zone without light. The ocean is generally in this state below 1000 meters (3280 feet).

Apogee The point in the orbit of the Moon or an artificial satellite that is farthest from Earth.

Aragonite A form of $CaCO_3$ that is less common and less stable than calcite. Pteropod shells are usually composed of aragonite.

Archaea One of the three major domains of life. The domain consists of simple microscopic bacterialike creatures (including methane producers and sulfur oxidizers that inhabit deep-sea vents and seeps) and other microscopic life-forms that prefer environments of extreme conditions of temperature and/or pressure.

Archipelago A large group of islands.

Arctic Circle The latitude 66.5° north.

Arctic Convergence A zone of converging currents similar to the Antarctic Convergence but located in the Arctic.

Argo A global array of free-drifting profiling floats that move vertically and measure the temperature, salinity, and other water characteristics of the upper 2000 meters (6600 feet) of the ocean.

Aspect ratio The index of propulsive efficiency obtained by dividing the square of fin height by fin area.

Asthenosphere A plastic layer in the upper mantle 80 to 200 kilometers (50 to 124 miles) deep that may allow lateral movement of lithospheric plates and isostatic adjustments.

Atlantic-type margin See *passive margin*.

Atoll A ring-shaped coral reef growing upward from a submerged volcanic peak. It may have low-lying islands composed of coral debris.

Atom A unit of matter, the smallest unit of an element, having all the characteristics of that element and consisting of a dense, central, positively charged nucleus surrounded by a system of electrons.

Atomic mass The mass of an atom, usually expressed in atomic mass units.

Atomic number The number of protons in an atomic nucleus.

Autotroph Algae, plants, and bacteria that can synthesize organic compounds from inorganic nutrients.

Autumnal equinox The passage of the Sun across the equator as it moves from the Northern Hemisphere into the Southern Hemisphere, approximately September 23. During this time, all places in the world experience equal lengths of night and day. Also called fall equinox.

B

Backshore The inner portion of the shore, lying landward of the mean spring tide high water line. Acted upon by the ocean only during exceptionally high tides and storms.

Backwash The flow of water down the beach face toward the ocean from a previously broken wave.

Bacteria One of the three major domains of life. The domain includes unicellular, prokaryotic microorganisms that vary in terms of morphology, oxygen and nutritional requirements, and motility.

Bacterioplankton Bacteria that live as plankton.

Bacteriovore An organism that feeds on bacteria.

Bar-built estuary A shallow estuary (lagoon) separated from the open ocean by a bar deposit such as a barrier island. The water in these estuaries usually exhibits vertical mixing.

Barrier flat An area that lies between the salt marsh and dunes of a barrier island and that is usually covered with grasses or forests if protected from overwash for a sufficient length of time.

Barrier island A long, narrow, wave-built island separated from the mainland by a lagoon.

Barrier reef A coral reef separated from the nearby landmass by open water.

Barycenter The center of mass of a system.

Basalt A dark-colored volcanic rock characteristic of the ocean crust. Contains minerals with relatively high iron and magnesium content.

Base See *alkaline*.

Bathyal zone The benthic environment between the depths of 200 and 4000 meters (660 and 13,000 feet). It includes mainly the continental slope and the oceanic ridges and rises.

Bathymetry The measurement of ocean depth.

Bathypelagic zone The pelagic environment between the depths of 1000 and 4000 meters (3300 and 13,000 feet).

Bathyscaphe A specially designed deep-diving submersible.

Bathysphere A specially designed deep-diving submersible that resembles a sphere.

Bay barrier A marine deposit attached to the mainland at both ends and extending entirely across the mouth of a bay, separating the bay from the open water. Also known as a bay-mouth bar.

Beach Sediment seaward of the coastline through the surf zone that is in transport along the shore and within the surf zone.

Beach compartment A series of rivers, beaches, and submarine canyons involved in the movement of sediment to the coast, along the coast, and down one or more submarine canyons.

Beach face The wet, sloping surface that extends from the berm to the shoreline. Also known as the low tide terrace.

Beach replenishment The addition of beach sediment to replace lost or missing material. Also called beach nourishment.

Beach starvation The interruption of sediment supply and resulting narrowing of beaches.

Beaufort Wind Scale A standardized wind scale that describes the appearance of the sea surface from dead calm conditions to hurricane-force winds.

Benguela Current The cold eastern boundary current of the South Atlantic Subtropical Gyre.

Benthic Pertaining to the ocean bottom.

Benthos The forms of marine life that live on the ocean bottom.

Berm The dry, gently sloping region on the backshore of a beach at the foot of the coastal cliffs or dunes.

Berm crest The area of a beach that separates the berm from the beach face. The berm crest is often the highest portion of a berm.

Bicarbonate ion (HCO_3^-) An ion that contains the radical group HCO_3^-.

Bioaccumulation The accumulation of a substance, such as a toxic chemical, in various tissues of a living organism.

Bioerosion Erosion of reef or other solid bottom material by the activities of organisms.

Biofilm Coating of organic matter such as that found on sand grains.

Biogenous sediment Sediment containing material produced by plants or animals, such as coral reefs, shell fragments, and housings of diatoms, radiolarians, foraminifers, and coccolithophores.

Biogeochemical cycle The natural cycling of compounds among the living and nonliving components of an ecosystem.

Biological pump The movement of CO_2 that enters the ocean from the atmosphere through the water column to the sediment on the ocean floor by biological processes—photosynthesis, secretion of shells, feeding, and dying.

Bioluminescence Light organically produced by a chemical reaction. Found in bacteria, phytoplankton, and various fishes (especially deep-sea fish).

Biomagnification Concentration of impurities as animals are eaten and the impurity is passed through food chains.

Biomass The total mass of a defined organism or group of organisms in a particular community or in the ocean as a whole.

Biomass pyramid A representation of trophic levels that illustrates the progressive decrease in total biomass at successive higher levels of the pyramid.

Bioremediation The technique of using microbes to assist in cleaning toxic spills.

Biotic community The living organisms that inhabit an ecosystem.

Biozone A region of the environment that has distinctive biological characteristics.

Bivalve A mollusk, such as an oyster or a clam, that has a shell consisting of two hinged valves.

Black smoker A hydrothermal vent on the ocean floor that emits a black cloud of hot water filled with dissolved metal particles.

Body wave A longitudinal or transverse wave that transmits energy through a body of matter.

Boiling point The temperature at which a substance changes state from a liquid to a gas at a given pressure.

Bore A steep-fronted tide crest that moves up a river in association with an incoming high tide.

Boundary current The northward- or southward-flowing currents that form the western and eastern boundaries of the subtropical current gyres.

Brackish Low-salinity water caused by the mixing of freshwater and saltwater.

Brazil Current The warm western boundary current of the South Atlantic Subtropical Gyre.

Breaker zone A region where waves break at the seaward margin of the surf zone.

Breakwater An artificial structure constructed roughly parallel to shore and designed to protect a coastal region from the force of ocean waves.

Brittle Descriptive term for a substance that is likely to fracture when force is applied to it.

Bryozoa A phylum of colonial animals that often share one coelomic cavity. Encrusting and branching forms secrete a protective housing (zooecium) of calcium carbonate or chitinous material.

Buffering The process by which a substance minimizes a change in the acidity of a solution when an acid or base is added to the solution.

Buoyancy The ability or tendency to float or rise in a liquid.

Bycatch Marine organisms that are caught incidentally by fishers seeking commercial species.

C

Calcareous Containing calcium carbonate.

Calcite A mineral with the chemical formula $CaCO_3$.

Calcite compensation depth (CCD) The depth at which the amount of calcite ($CaCO_3$) produced by the organisms in the overlying water column is equal to the amount of calcite the water column can dissolve. No calcite deposition occurs below this depth, which, in most parts of the ocean, is at a depth of 4500 meters (15,000 feet).

Calcium carbonate ($CaCO_3$) A chalklike substance secreted by many organisms in the form of coverings or skeletal structures.

California Current The cold eastern boundary current of the North Pacific Subtropical Gyre.

Calorie A unit of heat, defined as the amount of heat required to raise the temperature of 1 gram of water 1°C.

Calving The process by which a glacier breaks at an edge, so that a portion of the ice separates and falls from the glacier.

Canary Current The cold eastern boundary current of the North Atlantic Subtropical Gyre.

Capillarity The action by which a fluid, such as water, is drawn up in small tubes as a result of surface tension.

Capillary wave An ocean wave whose wavelength is less than 1.74 centimeters (0.7 inch). The dominant restoring force for such waves is surface tension.

Carapace (1) A chitinous or calcareous shield that covers the cephalothorax of some crustaceans. (2) The dorsal portion of a turtle shell.

Carbohydrate An organic compound containing the elements carbon, hydrogen, and oxygen with the general formula $(CH_2O)_n$.

Carbonate ion (CO_3^{2-}) An ion that contains the radical group CO_3^{2-}.

Caribbean Current The warm current that carries equatorial water across the Caribbean Sea into the Gulf of Mexico.

Carnivora The order of marine mammals that includes the sea otter, polar bear, and pinnipeds.

Carnivore An animal that depends on other animals solely or chiefly for its food supply.

Carotin An orange-yellow pigment found in plants.

Cation An atom that has lost one or more electrons and has an electrical positive charge.

Caulerpa taxifolia A tropical seaweed; a cold-water clone was introduced into the aquarium industry and found its way into coastal waters of the Mediterranean and Southern California. It continues to spread in the Mediterranean, but has been eradicated in Southern California.

Centigrade temperature scale A temperature scale based on the freezing point (0°C = 32°F) and boiling point (100°C = 212°F) of pure water. Also known as the Celsius scale after its founder.

Centripetal force A center-seeking force that tends to make rotating bodies move toward the center of rotation.

Cephalopoda A class of the phylum Mollusca whose members have a well-developed pair of eyes and a ring of tentacles surrounding the mouth. The shell is absent or internal on most members. The class includes the squid, octopus, and nautilus.

Cetacea An order of marine mammals that includes the whales, dolphins, and porpoises.

Chalk A soft, compact form of calcite, generally gray-white or yellow-white in color and derived chiefly from microscopic fossils.

Chemical energy A form of potential energy stored in the chemical bonds of compounds.

Chemosynthesis A process by which bacteria or archaea synthesize organic molecules from inorganic nutrients using chemical energy released from the bonds of a chemical compound (such as hydrogen sulfide) by oxidation.

Chloride ion (Cl^-) A chlorine atom that has become negatively charged by gaining one electron.

Chlorinity The amount of chloride ion and ions of other halogens in ocean water expressed in parts per thousand (‰) by weight.

Chlorophyll A group of green pigments that make it possible for plants to carry on photosynthesis.

Chlorophyta Green algae. Characterized by the presence of chlorophyll and other pigments.

Chondrite A stony meteorite composed primarily of silicate rock material and containing chondrules (spheroidal granules). They are the most commonly found meteorites.

Chronometer An exceptionally precise timepiece.

Chrysophyta An important phylum of planktonic algae that includes the diatoms. The presence of chlorophyll is masked by the pigment carotene that gives the plants a golden color.

Ciguatera A type of seafood poisoning caused by ingestion of certain tropical reef fish (most notably barracuda, red snapper, and grouper) that have high levels of naturally occurring dinoflagellate toxins.

Cilium A short, hairlike structure common on lower animals. Beating in unison, cilia may create water currents that carry food toward the mouth of an animal or may be used for locomotion.

Circadian rhythm Behavioral and physiological rhythms of organisms related to the 24-hour day. Sleeping and waking patterns are an example.

Circular orbital motion The motion of water particles caused by a wave as the wave is transmitted through water.

Circumpolar Current An eastward-flowing current that extends from the surface to the ocean floor and encircles Antarctica.

Clay (1) A particle size between silt and colloid. (2) Any of various hydrous aluminum silicate minerals that are plastic, expansive, and have ion exchange capabilities.

Climate The meteorological conditions, including temperature, precipitation, and wind that characteristically prevail in a particular region; the long-term average of weather.

Cnidaria A phylum that contains some 10,000 species of predominantly marine animals with a sacklike body and stinging cells on tentacles that surround the single opening to the gut cavity. There are two basic body forms. The medusa is a pelagic form represented by the jellyfish. The polyp is a predominantly benthic form found in sea anemones and corals. Previously named Coelenterata.

Cnidoblast A stinging cell of phylum Cnidaria that contains the stinging mechanism (nematocyst) used in defense and capturing prey.

Coast A strip of land that extends inland from the coastline as far as marine influence is evidenced in the landforms.

Coastal geostrophic current See *geostrophic current.*

Coastal plain estuary An estuary formed by rising sea level flooding a coastal river valley.

Coastal upwelling The movement of deeper nutrient-rich water into the surface water mass as a result of windblown surface water moving offshore.

Coastal waters The relatively shallow water areas that adjoin continents or islands.

Coastal wetland A biologically productive region bordering estuaries and other protected coastal areas; typically as salt marshes in latitudes greater than 30° and as mangrove swamps in lower latitudes.

Coastline The landward limit of the effect of the highest storm waves on the shore.

Coccolith Tiny calcareous discs averaging about 3 micrometers (0.00012 inch) in diameter that form the cell wall of coccolithophores.

Coccolithophore A microscopic planktonic form of algae encased by a covering composed of calcareous discs (coccoliths).

Coelenterata A phylum of radially symmetrical animals that includes two basic body forms, the medusa and the polyp. Includes jellyfish (medusoid) and sea anemones (polypoid). Preferred name is now Cnidaria.

Cohesion The intermolecular attraction by which the elements of a body are held together.

Cold core ring A circular eddy of a surface current that contains cold water in its center and rotates counterclockwise in the Northern Hemisphere.

Cold front A weather front in which a cold air mass moves into and under a warm air mass. It creates a narrow band of intense precipitation.

Cold-blooded See *poikilothermic* and *ectothermic.*

Colonial animal An animal that lives in groups of attached or separate individuals. Groups of individuals may serve special functions.

Columbia River estuary An estuary at the border between the states of Washington and Oregon that has been most adversely affected by the construction of hydroelectric dams.

Comb jelly Common name for members of the phylum Ctenophora. See also *ctenophore.*

Commensalism A symbiotic relationship in which one party benefits and the other is unaffected.

Compensation depth for photosynthesis The depth at which net photosynthesis becomes zero; below this depth, photosynthetic organisms can no longer survive. This depth is greater in the open ocean (up to 100 meters or 330 feet) than near the shore due to increased turbidity that limits light penetration in coastal regions.

Compound A substance containing two or more elements combined in fixed proportions.

Condensation The conversion of water from the vapor to the liquid state. When it occurs, the energy required to vaporize the water is released into the atmosphere. This is about 585 calories per gram of water at 20°C.

Conduction The transmission of heat by the passage of energy from particle to particle.

Conjunction The apparent closeness of two heavenly bodies. During the new moon phase, Earth and the Moon are in conjunction on the same side of the Sun.

Constant proportions, principle of A principle which states that the major constituents of ocean-water salinity are found in the same relative proportions throughout the ocean, independent of salinity.

Constructive interference A form of wave interference in which two waves come together in phase, for example, crest to crest, to produce a greater displacement from the still-water line than that produced by either of the waves alone.

Consumer An animal within an ecosystem that consumes the organic mass produced by the producers.

Continent About one-third of Earth's surface that rises above the deep-ocean floor to be exposed above sea level. Continents are composed primarily of granite, an igneous rock of lower density than the basaltic oceanic crust.

Continental accretion Growth or increase in size of a continent by gradual external addition of crustal material.

Continental arc An arc-shaped row of active volcanoes produced by subduction that occurs along convergent active continental margins.

Continental borderland A highly irregular portion of the continental margin that is submerged beneath the ocean and is characterized by depths greater than those characteristic of the continental shelf.

Continental drift A term applied to early theories supporting the possibility the continents are in motion over Earth's surface.

Continental effect Describes areas that are less affected by the sea and therefore having a greater range of temperature differences (both daily and yearly).

Continental margin The submerged area next to a continent comprising the continental shelf, continental slope, and continental rise.

Continental rise A gently sloping depositional surface at the base of the continental slope.

Continental shelf A gently sloping depositional surface extending from the low water line to the depth of a marked increase in slope around the margin of a continent or island.

Continental slope A relatively steeply sloping surface lying seaward of the continental shelf.

Convection Heat transfer in a gas or liquid by the circulation of currents from one region to another.

Convection cell A circular-moving loop of matter involved in convective movement.

Convergence The act of coming together from different directions. There are polar, tropical, and subtropical regions of the oceans where water masses with different characteristics come together. Along these lines of convergence, the denser masses will sink beneath the others.

Convergent plate boundary A lithospheric plate boundary where adjacent plates converge, producing ocean trench–island arc systems, ocean trench–continental volcanic arcs, or folded mountain ranges.

Copepoda An order of microscopic to nearly microscopic crustaceans that are important members of zooplankton in temperate and subpolar waters.

Coral A group of benthic anthozoans that exist as individuals or in colonies and secrete $CaCO_3$ external skeletons. Under the proper conditions, corals may produce reefs composed of their external skeletons and the $CaCO_3$ material secreted by varieties of algae associated with the reefs.

Coral bleaching The loss of color in coral reef organisms that causes them to turn white. Coral bleaching is caused by the expulsion of the coral's symbiotic zooxanthellae algae in response to high water temperatures or other adverse conditions.

Coral reef A calcareous organic reef composed significantly of solid coral and coral sand. Algae may be responsible for more than half of the $CaCO_3$ reef material. Found in waters where the minimum average monthly temperature is 18°C or higher.

Core (1) The deep, central layer of Earth, composed primarily of iron and nickel. It has a liquid outer portion 2270 kilometers (1410 miles) thick and a solid inner core with a radius of 1216 kilometers (756 miles). (2) A cylinder of sediment and/or rock material usually obtained by drilling.

Coriolis effect An apparent force resulting from Earth's rotation causes particles in motion to be deflected to the right in the Northern Hemisphere and to the left in the Southern Hemisphere.

Cosmogenous sediment Sediment derived from outer space.

Cotidal line A line connecting points where high tide occurs simultaneously.

Counterillumination Camouflaging by using bioluminescence to match the color and intensity of dim filtered sunlight from above and obliterate a telltale shadow.

Countershading Protective coloration in an animal or insect, characterized by darker coloring of areas exposed to light and lighter coloring of areas that are normally shaded.

Covalent bond A chemical bond formed by the sharing of one or more electrons, especially pairs of electrons, between atoms.

Crepuscular A term that describes animals that are active primarily during twilight (dawn and dusk).

Crest (wave) The portion of an ocean wave that is displaced above the still-water level.

Cruiser Fish (such as the bluefin tuna) that constantly cruise pelagic waters in search of food.

Crust (1) The uppermost outer layer of Earth's structure that is composed of basaltic oceanic crust and granitic continental crust. The average thickness of the crust ranges from 8 kilometers (5 miles) beneath the ocean to 35 kilometers (22 miles) beneath the continents. (2) A hard covering or surface layer of hydrogenous sediment.

Crustacea A class of subphylum Arthropoda that includes barnacles, copepods, lobsters, crabs, and shrimp.

Crystalline rock Igneous or metamorphic rocks. These rocks are made up of crystalline particles with orderly molecular structures.

Ctenophore A member of the phylum of gelatinous organisms that are more or less spheroidal with biradial symmetry. These exclusively marine animals have eight rows of ciliated combs for locomotion, and most have two tentacles for capturing prey.

Current A physical movement of water.

Cyclone An atmospheric system characterized by the rapid, inward circulation of air masses about a low-pressure center that is associated with rising air. Cyclones circulate counterclockwise in the Northern Hemisphere and clockwise in the Southern Hemisphere and are usually accompanied by stormy, often destructive, weather.

Cyclonic flow The flow of air around a region of low pressure counterclockwise in the Northern Hemisphere.

D

Davidson Current A northward-flowing current along the Washington–Oregon coast that is driven by geostrophic effects on a large freshwater runoff.

DDT An insecticide (dichlorodiphenyltrichloroethane) that caused damage to marine bird populations in the 1950s and 1960s. Its use is now banned throughout most of the world.

Dead zone A region of hypoxic conditions that kills off most marine organisms that cannot escape. It is usually the result of eutrophication caused by runoff from land-based fertilizer applications.

Decay distance The distance over which waves change from a choppy "sea" to uniform swell.

Declination The angular distance of the Sun or Moon above or below the plane of Earth's equator.

Decomposer Primarily bacteria that break down non-living organic material, extract some of the products of decomposition for their own needs, and make available the compounds needed for primary production.

Decompression sickness A serious condition that occurs in divers when they ascend too rapidly, causing nitrogen bubbles to form in the blood and tissue, resulting in great pain and sometimes death. Also known as the bends.

Deep biosphere The microbe-rich region beneath the sea floor.

Deep boundary current A relatively strong deep current flowing across the continental rise along the western margin of ocean basins.

Deep current A density-driven circulation that is initiated at the ocean surface by temperature and salinity conditions that produce a high-density water mass, which sinks and spreads slowly beneath surface waters.

Deep-ocean Assessment and Reporting of Tsunamis (DART) A system that utilizes sea floor sensors capable of picking up the small yet distinctive pressure pulse from a tsunami at the surface.

Deep-ocean basin Areas of the ocean floor that have deep water, are far from land, and are underlain by basaltic crust.

Deep scattering layer (DSL) A layer of marine organisms in the open ocean that scatter signals from an echo sounder. It migrates daily from depths of slightly over 100 meters (330 feet) at night to more than 800 meters (2600 feet) during the day.

Deep-sea fan A large fan-shaped deposit commonly found on the continental rise seaward of such sediment-laden rivers as the Amazon, Indus, or Ganges-Brahmaputra. Also known as a submarine fan.

Deep-sea fish Any of a large group of fishes that lives within the aphotic zone and has special adaptations for finding food and avoiding predators in darkness.

Deep-sea system The bathyal, abyssal, and hadal benthic environments.

Deep water The water beneath the permanent thermocline (and resulting pycnocline) that has a uniformly low temperature.

Deep-water wave An ocean wave traveling in water that has a depth greater than one-half the average wavelength. Its speed is independent of water depth.

Delta A low-lying deposit at the mouth of a river, usually having a triangular shape as viewed from above.

Density The mass per unit volume of a substance. Usually expressed as grams per cubic centimeter (g/cm^3). For ocean water with a salinity of 35‰ at 0°C, the density is $1.028 \, g/cm^3$.

Density stratification A layering based on density, where the highest density material occupies the lowest space.

Deposit feeder An organism that feeds on food items that occur as deposits, including detritus and various detritus-coated sediment.

Depositional-type shore A shoreline dominated by processes that form deposits (such as sand bars and barrier islands) along the shore.

Desalination The removal of salt ions from ocean water to produce pure water.

Destructive interference A form of wave interference in which two waves come together out of phase, for example, crest to trough, and produce a wave with less displacement than the larger of the two waves would have produced alone.

Detritus (1) Any loose material produced directly from rock disintegration. (2) Material resulting from the disintegration of dead organic remains.

Diatom A member of the class Bacillariophyceae of algae that possesses a wall of overlapping silica valves.

Diatomaceous earth A deposit composed primarily of the tests of diatoms mixed with clay. Also called diatomite.

Diffraction A change in the direction or intensity of a wave after passing an obstacle that cannot be interpreted as refraction or reflection.

Diffusion A process by which materials move through fluids by random molecular movement from areas of high concentration to areas in which they are in lower concentrations, thus becoming evenly distributed.

Dinoflagellate A single-celled microscopic planktonic organism that may possess chlorophyll and belong to the phylum Pyrrophyta (autotrophic) or may ingest food and belong to the class Mastigophora of the phylum Protozoa (heterotrophic).

Dipolar Having two poles. The water molecule possesses a polarity of electrical charge with one pole being more positive and the other more negative in electrical charge.

Discontinuity An abrupt change in a property such as temperature or salinity at a line or surface.

Disphotic zone The dimly lit zone, corresponding approximately to the mesopelagic, in which there is not enough light to support photosynthetic organisms; sometimes called the twilight zone.

Disruptive coloration A marking or color pattern that confuses prey.

Dissolved oxygen Oxygen that is dissolved in ocean water.

Distillation A method of purifying liquids by heating them to their boiling point and condensing the vapor.

Distributary A small stream flowing away from a main stream. Such streams are characteristic of deltas.

Disturbing force The energy that causes waves to form.

Diurnal inequality The difference in the heights of two successive high tides or two successive low tides during a lunar (tidal) day.

Diurnal tidal pattern A tidal pattern exhibiting one high tide and one low tide during a tidal day; a daily tide.

Divergence A horizontal flow of water from a central region, as occurs in upwelling.

Divergent plate boundary A lithospheric plate boundary where adjacent plates diverge, producing an oceanic ridge or rise (spreading center).

Doldrums A global belt of light, variable winds near the equator, resulting from the vertical flow of low-density air masses upward within this equatorial belt. Associated with much precipitation.

Dolphin (1) A brilliantly colored fish of the genus *Coryphaena*. (2) The name applied to the small, beaked members of the cetacean family Delphinidae.

Dorsal Pertaining to the back or upper surface of most animals.

Downwelling In the open or coastal ocean, where Ekman transport causes surface waters to converge or impinge on the coast, surface water that moves down beneath the surface.

Drift bottle Equipment used to study ocean current movement by drifting with currents.

Driftnet A fishing net made of monofilament fishing line that catches organisms by entanglement.

Drifts Thick sediment deposits on the continental rise produced where the deep boundary current slows and loses sediment when it changes direction to follow the base of the continental slope.

Drowned beach An ancient beach now beneath the coastal ocean because of rising sea level or subsidence of the coast.

Drowned river valley The lower part of a river valley that has been submerged by rising sea level or subsidence of the coast.

Dune (coastal) A coastal deposit of sand lying landward of the beach and deriving its sand from onshore winds that transport beach sand inland.

Dynamic topography A surface configuration resulting from the geopotential difference between a given surface and a reference surface of no motion. A contour map of this surface is useful in estimating the nature of geostrophic currents.

E

Earthquake A sudden release of energy inside Earth that creates seismic waves and is usually caused by either faulting (movement along a fracture in Earth's crust) or volcanic activity.

East Australian Current The warm western boundary current of the South Pacific Subtropical Gyre.

East Pacific Rise A fast-spreading divergent plate boundary extending southward from the Gulf of California through the eastern South Pacific Ocean.

East Wind Drift The coastal current driven in a westerly direction by the polar easterly winds blowing off Antarctica.

Eastern boundary current Equatorward-flowing cold drifts of water on the eastern side of all subtropical gyres.

Eastern Pacific Garbage Patch An area in the eastern part of the North Pacific Gyre that serves as an accumulation area for floating plastic and other trash; it is about twice the size of Texas.

Ebb current The flow of water seaward during a decrease in the height of the tide.

Echinodermata A phylum of animals that have bilateral symmetry in larval forms and usually a five-sided radial symmetry as adults. Benthic and possessing rigid or articulating exoskeletons of calcium carbonate with spines, this phylum includes sea stars, brittle stars, sea urchins, sand dollars, sea cucumbers, and sea lilies.

Echo sounder A device that transmits sound from a ship's hull to the ocean floor where it is reflected back to receivers. The speed of sound in the water is known, so the depth can be determined from the travel time of the sound signal.

Echolocation A sensory system in odontocete cetaceans in which usually high-pitched sounds are emitted and their echoes interpreted to determine the direction and distance of objects.

Ecliptic The plane of the center of the Earth–Moon system as it orbits around the Sun.

Ecosystem All the organisms in a biotic community and the abiotic environmental factors with which they interact.

Ectothermic Of or relating to an organism that regulates its body temperature largely by exchanging heat with its surroundings; cold-blooded.

Eddy A current of any fluid forming on the side of a main current. It usually moves in a circular path and develops where currents encounter obstacles or flow past one another.

Ekman spiral A theoretical consideration of the effect of a steady wind blowing over an ocean of unlimited depth and breadth and of uniform viscosity. The result is a surface flow at 45° to the right of the wind in the Northern Hemisphere. Water at increasing depth below the surface will drift in directions increasingly more slowly and to the right until at about 100 meters (330 feet) depth it may move in a direction opposite to that of the wind.

Ekman transport The net transport of surface water set in motion by wind. Due to the Ekman spiral phenomenon, it is theoretically in a direction 90° to the right and 90° to the left of the wind direction in the Northern Hemisphere and Southern Hemisphere, respectively.

El Niño A southerly flowing warm current that generally develops off the coast of Ecuador around Christmastime. Occasionally it will move farther south into Peruvian coastal waters and cause the widespread death of plankton, fish, and other organisms such as marine mammals that depend on fish for food.

El Niño–Southern Oscillation (ENSO) The correlation of El Niño events with an oscillatory pattern of pressure change in a persistent high-pressure cell in the southeastern Pacific Ocean and a persistent low-pressure cell over the East Indies.

Electrical conductivity The ability or power to conduct or transmit electricity.

Electrolysis A separation process by which salt ions are removed from saltwater through water-impermeable membranes toward oppositely charged electrodes.

Electromagnetic energy Energy that travels as waves or particles with the speed of light. Different kinds possess different properties based on wavelength. The longest wavelengths belong to radio waves, up to 100 kilometers (60 miles) in length. At the other end of the spectrum are cosmic rays with greater penetrating power and wavelengths of less than 0.000001 micrometer.

Electromagnetic spectrum The spectrum of radiant energy emitted from stars and ranging between cosmic rays with wavelengths of less than 10 to 11 centimeters (4 to 4.3 inches) and very long waves with wavelengths in excess of 100 kilometers (60 miles).

Electron A subatomic particle that orbits the nucleus of an atom and has a negative electric charge.

Electron cloud The diffuse area surrounding the nucleus of an atom where electrons are found.

Electrostatic force A force caused by electric charges at rest.

Element One of a number of substances, each of which is composed entirely of like particles—atoms—that cannot be broken into smaller particles by chemical means.

Emerging shoreline A shoreline resulting from the emergence of the ocean floor relative to the ocean surface. It is usually rather straight and characterized by marine features usually found at some depth.

ENSO See *El Niño–Southern Oscillation (ENSO)*.

ENSO index An index showing the relative strength of El Niño and La Niña conditions.

Environment The sum of all physical, chemical, and biological factors to which an organism or community is subjected.

Environmental bioassay An environmental assessment technique that determines the concentration of a pollutant that causes 50% mortality among a specific group of test organisms.

Epicenter The point on Earth's surface that is directly above the focus of an earthquake.

Epifauna Animals that live on the ocean bottom, either attached or moving freely over it.

Epipelagic zone A subdivision of the oceanic province that extends from the surface to a depth of 200 meters (660 feet).

Equator The imaginary great circle around Earth's surface, equidistant from the poles and perpendicular to Earth's axis of rotation. It divides Earth into the Northern Hemisphere and the Southern Hemisphere.

Equatorial Pertaining to the equatorial region.

Equatorial countercurrent Eastward-flowing currents found between the North and South Equatorial Currents in all oceans but particularly well developed in the Pacific Ocean.

Equatorial current Westward-flowing currents that travel along the equator in all ocean basins, caused by the trade winds. They are called North or South Equatorial Currents, depending on their position north or south of the equator.

Equatorial low A band of low atmospheric pressure that encircles the globe along the equator.

Equatorial upwelling The movement of deeper nutrient-rich water into the surface water mass as a result of divergence of currents along the equator.

Erosion The group of natural processes, including weathering, dissolution, abrasion, corrosion, and transportation, by which material is worn away from Earth's surface.

Erosional-type shore A shoreline dominated by processes that form erosional features (such as cliffs and sea stacks) along the shore.

Estuarine circulation pattern A flow pattern in an estuary characterized by a net surface flow of low-salinity water toward the ocean and an opposite net subsurface flow of seawater toward the head of the estuary.

Estuary A partially enclosed coastal body of water in which salty ocean water is significantly diluted by freshwater from land runoff. Examples of estuaries include river mouths, bays, inlets, gulfs, and sounds.

Eukarya One of the three major domains of life. The domain includes single-celled or multicellular organisms whose cells usually contain a distinct membrane-bound nucleus.

Euphotic zone A layer that extends from the surface of the ocean to a depth where enough light exists to support photosynthesis, rarely deeper than 100 meters (330 feet).

Euryhaline A descriptive term for organisms with a high tolerance for a wide range of salinity conditions.

Eurythermal A descriptive term for organisms with a high tolerance for a wide range of temperature conditions.

Eustatic sea level change A worldwide raising or lowering of sea level.

Eutrophic A region of high productivity.

Eutrophication The enrichment of waters by a previously scarce nutrient.

Evaporation The process of changing from the liquid to the vapor state at a temperature below the boiling point of a substance.

Evaporite A sedimentary deposit that is left behind when water evaporates. Evaporite minerals include gypsum, calcite, and halite.

Evolution The change of groups of organisms with the passage of time, mainly as a result of natural selection, so that descendants differ morphologically and physiologically from their ancestors.

Exclusive economic zone (EEZ) A coastal zone that generally extends 200 nautical miles (370 kilometers) from shore and establishes coastal nation jurisdiction including mineral resources, fish stocks, and pollution. If the continental shelf extends beyond the 200-mile EEZ, the EEZ is extended to 350 nautical miles (648 kilometers) from shore.

Extrusive rock Igneous rock that flows out onto Earth's surface before cooling and solidifying (lava).

Eye (1) An organ of vision or of light sensitivity. (2) The circular low-pressure area of relative calm at the center of a hurricane.

F

Fact Something having real, demonstrable existence. A scientific fact is an occurrence that has been repeatedly confirmed.

Faculae A bright spot on the Sun that is associated with magnetic storms.

Fahrenheit temperature scale (°F) A temperature scale whereby the freezing point of water is 32° and the boiling point of water is 212°.

Falcate Curved and tapering to a point; sickle-shaped.

Falkland Current A northward-flowing cold current found off the southeastern coast of South America.

Fall bloom A middle-latitude bloom of phytoplankton that occurs during the fall and is limited by the availability of sunlight.

Fall equinox See *autumnal equinox*.

Fan A gently sloping, fan-shaped feature normally located near the lower end of a canyon. Also known as a submarine fan.

Fat An organic compound formed from alcohol, glycerol, and one or more fatty acids; a lipid, it is a solid at atmospheric temperatures.

Fathom (fm) A unit of depth in the ocean, commonly used in countries using the English system of units. It is equal to 1.83 meters (6 feet).

Fault A fracture or fracture zone in Earth's crust along which displacement has occurred.

Fault block A crustal block bounded on at least two sides by faults. Usually elongate; if it is down-dropped, it produces a graben; if uplifted, it is a horst.

Fauna The animal life of any particular area or of any particular time.

Fecal pellet The excrement of planktonic crustaceans that assist in speeding up the descent rate of sedimentary particles by combining them into larger packages.

Feedback loop A self-repeating processes in which an initial change is modified. A positive-feedback loop amplifies an initial change, and a negative-feedback loop counteracts an initial change.

Ferrel cell The large atmospheric circulation cell that occurs between 30° and 60° latitude in each hemisphere.

Ferromagnesium Minerals rich in iron and magnesium.

Fetch (1) Pertaining to the area of the open ocean over which the wind blows with constant speed and direction, thereby creating a wave system. (2) The distance across the fetch (wave-generating area) measured in a direction parallel to the direction of the wind.

Filter feeder An organism that obtains its food by filtering seawater for other organisms. Also known as suspension feeder.

Fishery Fish caught from the ocean by commercial fishers.

Fishery management The organized effort directed at regulating fishing activity with the goal of maintaining a long-term fishery.

Fissure A long, narrow opening; a crack or cleft.

Fjord A long, narrow, deep, U-shaped inlet that usually represents the seaward end of a glacial valley that has become partially submerged after the melting of the glacier.

Flagellum A whiplike living structure used by some cells for locomotion.

Floe A piece of floating ice other than fast ice or icebergs. May range in maximum horizontal dimension from about 20 centimeters (8 inches) to more than 1 kilometer (0.6 mile).

Flood current A tidal current associated with increasing height of the tide, generally moving toward the shore.

Flood tide A rising tide.

Flora The plant life of any particular area or of any particular time.

Florida Current A warm current flowing north along the coast of Florida that merges into the Gulf Stream.

Folded mountain range A mountain range formed as a result of the convergence of lithospheric plates. Folded mountain ranges are characterized by masses of folded sedimentary rocks that formed from sediments deposited in the ocean basin that was destroyed by the convergence.

Food chain The passage of energy materials from producers through a sequence of a herbivore and a number of carnivores.

Food web A group of interrelated food chains.

Foraminifer An order of planktonic and benthic protozoans that possess protective coverings, usually composed of calcium carbonate.

Forced wave A wave that is generated and maintained by a continuous force such as the gravitational attraction of the Moon.

Foreshore The portion of the shore lying between the normal high and low water marks; the intertidal zone.

Fossil Any remains, print, or trace of an organism that has been preserved in Earth's crust.

Fracture zone An extensive linear zone of unusually irregular topography of the ocean floor, characterized by large seamounts, steep-sided or asymmetrical ridges, troughs, or long, steep slopes. Usually represents ancient, inactive transform fault zones.

Free wave A wave created by a sudden rather than a continuous impulse that continues to exist after the generating force is gone.

Freeze separation The desalination of seawater by multiple episodes of freezing, rinsing, and thawing.

Freezing The process by which a liquid is converted to a solid at its freezing point.

Freezing point The temperature at which a liquid becomes a solid under any given set of conditions. The freezing point of water is 0°C at one atmosphere pressure.

Fringing reef A reef that is directly attached to the shore of an island or continent. It may extend more than 1 kilometer (0.6 mile) from shore. The outer margin is submerged and often consists of algal limestone, coral rock, and living coral.

Fucoxanthin The reddish-brown pigment that gives brown algae its characteristic color.

Full moon The phase of the Moon that occurs when the Sun and Moon are in opposition; that is, they are on opposite sides of Earth. During this time, the lit side of the Moon faces Earth.

Fully developed sea The maximum average size of waves that can be developed for a given wind speed when it has blown in the same direction for a minimum duration over a minimum fetch.

Fungi Any of numerous eukaryotic organisms of the kingdom Fungi, which lack chlorophyll and vascular tissue and range in form from a single cell to a body mass of branched filamentous structures that often produce specialized fruiting bodies. The kingdom includes the yeasts, molds, lichens, and mushrooms.

Fur seal Any of several eared seals of the genera *Callorhinus* or *Arctocephalus*, having thick, soft underfur.

Fusion reaction A type of nuclear reaction where hydrogen atoms are converted to helium atoms, thereby releasing large amounts of energy.

G

Galápagos Rift A divergent plate boundary extending eastward from the Galápagos Islands toward South America. The first deep-sea hydrothermal vent biocommunity was discovered here in 1977.

Galaxy One of the billions of large systems of stars that make up the universe.

Gas hydrate A latticelike compound composed of water and natural gas (usually methane) formed in high-pressure and low-temperature environments such as those found in deep ocean sediments. Also known as clathrates because of their cagelike chemical structure.

Gaseous state A state of matter in which molecules move by translation and only interact through chance collisions.

Gastropoda A class of mollusks, most of which possess an asymmetrical, spiral one-piece shell and a well-developed flattened foot. A well-developed head will usually have two eyes and one or two pairs of tentacles. Includes snails, limpets, abalone, cowries, sea hares, and sea slugs.

Geostrophic current A current that grows out of Earth's rotation and is the result of a near balance between gravitational force and the Coriolis effect.

Gill A thin-walled projection from some part of the external body or the digestive tract used for respiration in a water environment.

Gill net See *driftnet*.

Glacial deposit A sedimentary deposit formed by a glacier and characterized by poor sorting.

Glacier A large mass of ice formed on land by the recrystallization of old, compacted snow. It flows from an area of accumulation to an area of wasting where ice is removed from the glacier by melting.

Glauconite A group of green hydrogenous minerals consisting of hydrous silicates of potassium and iron.

Global engineering Deliberately manipulating one or more of Earth's systems for the benefit of humankind. Also called geoengineering.

Global Positioning System (GPS) A system of satellites that transmit microwave signals to Earth, allowing people at the surface to accurately determine their locations.

Globigerina **ooze** An ooze that contains a large percentage of the calcareous tests of the foraminifer *Globigerina*.

Goiter An enlargement of the human thyroid gland, visible as a swelling at the front of the neck, often associated with iodine deficiency.

Gondwanaland A hypothetical protocontinent of the Southern Hemisphere named for the Gondwana region of India. It included the present continental masses Africa, Antarctica, Australia, India, and South America.

Graded bedding Stratification in which each layer displays a decrease in grain size from bottom to top.

Gradient The rate of increase or decrease of one quantity or characteristic relative to a unit change in another. For example, the slope of the ocean floor is a change in elevation (a vertical linear measurement) per unit of horizontal distance covered. Commonly measured in meters per kilometer.

Grain size The average size of the grains of material in a sample. Also known as fragment or particle size.

Granite A light-colored igneous rock characteristic of the continental crust. Rich in nonferromagnesian minerals such as feldspar and quartz.

Gravitational force The force of attraction that exists between any two bodies in the universe that is proportional to the product of their masses and inversely proportional to the square of the distance between the centers of their masses.

Gravity wave A wave for which the dominant restoring force is gravity. Such waves have a wavelength of more than 1.74 centimeters (0.7 inch), and their speed of propagation is controlled mainly by gravity.

Gray whale A slow-moving, coastal-dwelling baleen whale (*Eschrichtius robustus*) of northern Pacific waters, having grayish-black coloring with white blotches. It undertakes the longest yearly migration of any mammal.

Greenhouse effect The heating of Earth's atmosphere that results from the absorption by components of the atmosphere such as water vapor and carbon dioxide of infrared radiation from Earth's surface.

Groin A low artificial structure built perpendicular to the shore and designed to interfere with longshore transportation of sediment so that it traps sand and widens the beach on its upstream side.

Groin field A series of closely spaced groins.

Gross ecological efficiency The amount of energy passed on from a trophic level to the one above it divided by the amount it received from the one below it.

Gross primary production The total carbon fixed into organic molecules through photosynthesis or chemosynthesis by a discrete autotrophic community.

Grunion A small fish (*Leuresthes tenuis*) of coastal waters of California and Mexico that spawns at night along beaches during the high tides of spring and summer. Grunion time their reproductive activities to coincide with tidal phenomena and are the only fish that comes completely out of water to spawn.

Gulf Stream The high-intensity western boundary current of the North Atlantic Subtropical Gyre that flows north off the East Coast of the United States.

Guyot See *tablemount*.

Gypsum A colorless, white, or yellowish evaporite mineral, $CaSO_4 \cdot 2H_2O$.

Gyre A large, horizontal, circular-moving loop of water. Used mainly in reference to the circular motion of water in each of the major ocean basins centered in subtropical high-pressure regions.

H

Habitat A place where a particular plant or animal lives. Generally refers to a smaller area than environment.

Hadal Pertaining to the deepest ocean environment, specifically that of ocean trenches deeper than 6 kilometers (3.7 miles).

Hadal zone Pertaining to the deepest ocean benthic environment, specifically that of ocean trenches deeper than 6 kilometers (3.7 miles).

Hadley cell The large atmospheric circulation cell that occurs between the equator and 30° latitude in each hemisphere.

Half-life The time required for half the atoms of a sample of a radioactive isotope to decay to an atom of another element.

Halite A colorless or white evaporite mineral, NaCl, which occurs as cubic crystals and is used as table salt.

Halocline A layer of water in which a high rate of change in salinity in the vertical dimension is present.

Hard stabilization Any form of artificial structure built to protect a coast or to prevent the movement of sand along a beach. Examples include groins, breakwaters, and seawalls.

Harmful algal bloom (HAB) See *red tide*.

Headland A steep-faced irregularity of the coast that extends out into the ocean.

Heat Energy moving from a high temperature system to a lower temperature system. The heat gained by the one system may be used to raise its temperature or to do work.

Heat budget (global) The equilibrium that exists on the average between the amounts of heat absorbed by Earth and that returned to space.

Heat capacity The amount of heat energy required to raise the temperature of a substance by 1°C. See also *specific heat*.

Heat energy Energy of molecular motion. The conversion of higher forms of energy such as radiant or mechanical energy to heat energy within a system increases the heat energy within the system and the temperature of the system.

Heat flow (flux) The quantity of heat flow to Earth's surface per unit of time.

Hemoglobin A red pigment found in red blood corpuscles that carries oxygen from the lungs to tissue and carbon dioxide from tissue to lungs.

Herbivore An animal that relies chiefly or solely on plants for its food.

Hermatypic coral Reef-building corals that have symbiotic algae in their ectodermal tissue. They cannot produce a reef structure below the euphotic zone.

Heterotroph Animals and bacteria that depend on the organic compounds produced by other organisms as food. Organisms not capable of producing their own food by photosynthesis.

High slack water The period of time associated with the peak of high tide when there is no visible flow of water into or out of bays and rivers.

High tide zone The portion of the intertidal zone that lies between the lowest high tides and highest high tides that occur in an area. It is, on average, exposed to desiccation for longer periods each day than it is covered by water.

High water (HW) The highest level reached by the rising tide before it begins to recede.

Higher high water (HHW) The higher of two high waters occurring during a tidal day where tides exhibit a mixed tidal pattern.

Higher low water (HLW) The higher of two low waters occurring during a tidal day where tides exhibit a mixed tidal pattern.

Highly stratified estuary A relatively deep estuary in which a significant volume of marine water enters as a subsurface flow. A large volume of freshwater stream input produces a widespread low surface–salinity condition that produces a well-developed halocline throughout most of the estuary.

Holoplankton Organisms that spend their entire life as members of the plankton.

Homeothermic Of or relating to an animal that maintains a precisely controlled internal body temperature using its own internal heating and cooling mechanisms; warm-blooded.

Horse latitudes The global belts between about 30 and 35 degrees north and south latitude, where winds are light and variable as a result of the vertical movement of air masses downward at these latitudes, causing a hot, dry climate. These belts are associated with the major continental and maritime deserts of the world.

Hotspot The relatively stationary surface expression of a persistent column of molten mantle material rising to the surface.

Hurricane A tropical cyclone in which winds reach speeds in excess of 120 kilometers (74 miles) per hour. Generally applied to such storms in the North Atlantic Ocean, eastern North Pacific Ocean, Caribbean Sea, and Gulf of Mexico. Such storms in the western Pacific Ocean are called typhoons and those in the Indian Ocean are known as cyclones.

Hydration The condition of being surrounded by water molecules, as when sodium chloride dissolves in water.

Hydrocarbon Any organic compound consisting only of hydrogen and carbon. Crude oil is a mixture of hydrocarbons.

Hydrocarbon seep biocommunity A deep bottom-dwelling community of organisms associated with a hydrocarbon seep from the ocean floor. The community depends on methane and sulfur-oxidizing bacteria as producers. The bacteria may live free in the water, on the bottom, or symbiotically in the tissues of some of the animals.

Hydrogen bond An intermolecular bond that forms within water because of the dipolar nature of water molecules.

Hydrogenous sediment Sediment that forms from precipitation from ocean water or ion exchange between existing sediment and ocean water. Examples are manganese nodules, metal sulfides, and evaporites.

Hydrologic cycle The cycle of water exchange among the atmosphere, land, and ocean through the processes of evaporation, precipitation, runoff, and subsurface percolation. Also called the water cycle.

Hydrothermal spring Vents of hot water found primarily along the spreading axes of oceanic ridges and rises.

Hydrothermal vent Ocean water that percolates down through fractures in recently formed ocean floor is heated by underlying magma and surfaces again through these vents. They are usually located near the axis of spreading along the mid-ocean ridge.

Hydrothermal vent biocommunity A deep bottom-dwelling community of organisms associated with a hydrothermal vent. The hot water vent is usually associated with the axis of a spreading center, and the community is dependent on sulfur-oxidizing bacteria that may live free in the water, on the bottom, or symbiotically in the tissue of some of the animals of the community.

Hydrozoa A class of cnidarians that characteristically exhibit alternation of generations, with a sessile polypoid colony giving rise to a pelagic medusoid form by asexual budding.

Hypersaline Waters that are highly or excessively saline.

Hypersaline lagoon A shallow lagoon that may reach high salinity levels due to little tidal flushing, high evaporation rates, and low freshwater input.

Hypersaline seep A seep of high-salinity water that trickles out of the sea floor.

Hypersaline seep biocommunity A bottom-dwelling community of organisms associated with a hypersaline seep that depend on methane and sulfur-oxidizing bacteria as producers. The bacteria may live free in the water, on the bottom, or symbiotically in the tissue of some of the animals within the biocommunity.

Hypertonic Pertaining to the property of an aqueous solution having a higher osmotic pressure (salinity) than another aqueous solution from which it is separated by a semipermeable membrane that will allow osmosis to occur. The hypertonic fluid will gain water molecules through the membrane from the other fluid.

Hypothesis A tentative, testable statement about the general nature of the phenomenon observed.

Hypotonic Pertaining to the property of an aqueous solution having a lower osmotic pressure (salinity) than another aqueous solution from which it is separated by a semipermeable membrane that will allow osmosis to occur. The hypotonic fluid will lose water molecules through the membrane to the other fluid.

Hypsographic curve A curve that displays the relative elevations of the land surface and depths of the ocean.

I

Ice Age The most recent glacial period, which occurred during the Pleistocene Epoch.

Iceberg A massive piece of glacier ice that has broken from the front of the glacier (calved) into a body of water. It floats with its tip at least 5 meters (16 feet) above the water's surface and at least four-fifths of its mass submerged.

Ice floe See *floe*.

Ice rafting (1) The movement of trapped sediment within or on top of ice by flotation. (2) The forcing of some ice flows on top of others as a result of expanding floes running out of open water on which to float as they grow.

Ice sheet An extensive, relatively flat accumulation of ice.

Igneous rock One of the three main classes into which all rocks are divided (igneous, metamorphic, and sedimentary). Rock that forms from the solidification of molten or partly molten material (magma).

In situ In place; *in situ* density of a sample of water is its density at its original depth.

Incidental catch See *bycatch*.

Indian Ocean Subtropical Gyre The large, counterclockwise-flowing subtropical gyre that exists in the Indian Ocean.

Inertia Newton's first law of motion. It states that a body at rest will stay at rest and a body in motion will remain in uniform motion in a straight line unless acted on by some external force.

Infauna Animals that live buried in the soft substrate (sand or mud).

Infrared radiation Electromagnetic radiation lying between the wavelengths of 0.8 micrometers (0.000003 inch) and about 1000 micrometers (0.04 inch). It is bounded on the shorter wavelength side by the visible spectrum and on the long side by microwave radiation.

Inner sublittoral zone The zone on the inner continental shelf, above the intersection with the euphotic zone, where attached plants grow.

Insolation The rate at which solar radiation is received per unit of surface area at any point at or above Earth's surface.

Integrated Ocean Drilling Program (IODP) A drilling program that replaced the Ocean Drilling Program in 2003 with a new drill ship that has riser technology, enabling cores to be collected from deep within Earth's interior.

Interface A surface separating two substances of different properties, such as density, salinity, or temperature. In oceanography, it usually refers to a separation of two layers of water with different densities caused by significant differences in temperature and/or salinity.

Interface wave An orbital wave that moves along an interface between fluids of different density. An example is ocean surface waves moving along the interface between the atmosphere and the ocean.

Interference (wave) The overlapping of different wave groups, either in phase (constructive interference, which results in larger waves), out of phase (destructive interference, which results in smaller waves), or some combination of the two (mixed interference).

Intermolecular bond A relatively weak bond that forms between molecules of a given substance. The hydrogen bond and the van der Waals bonds are intermolecular bonds.

Internal wave A wave that develops below the surface of a fluid, the density of which changes with increased depth. This change may be gradual or occur abruptly at an interface.

International Whaling Commission (IWC) An international body that was established in 1946 to manage the subsistence and commercial hunting of large whales. In 1986, the IWC passed a ban on commercial whaling in order to allow whales to recover from overhunting and to give researchers time to develop methods for assessing whale populations.

Intertidal zone The ocean floor within the foreshore region that is covered by the highest normal tides and exposed by the lowest normal tides, including the water environment of tide pools within this region.

Intertropical Convergence Zone (ITCZ) The global zone where northeast trade winds and southeast trade winds converge. It occurs close to the equator but averages about 5 degrees north latitude in the Pacific and Atlantic Oceans and 7 degrees south latitude in the Indian Ocean.

Intraplate feature Any feature that occurs within a tectonic plate and not along a plate boundary.

Intrusive rock An igneous rock such as granite that cools slowly beneath Earth's surface.

Invertebrate An animal without a backbone.

Ion An atom that becomes electrically charged by gaining or losing one or more electrons. The loss of electrons produces a positively charged cation, and the gain of electrons produces a negatively charged anion.

Ionic bond A chemical bond formed as a result of the electrical attraction.

Irminger Current A warm current that branches off from the Gulf Stream and moves up along the west coast of Iceland.

Iron hypothesis A hypothesis that states that an effective way of increasing productivity in the ocean is to fertilize the ocean by adding the only nutrient that appears to be lacking—iron. Adding iron to the ocean also increases the amount of carbon dioxide removed from the atmosphere.

Irons A meteorite consisting essentially of iron, but it may also contain up to 30% nickel.

Island arc A linear arrangement of islands, many of which are volcanic, usually curved so that the concave side faces a sea separating the islands from a continent. The convex side faces the open ocean and is bounded by a deep-ocean trench.

Island mass effect An effect that occurs as surface current flows past an island and causes surface water to be carried away from the island on the downcurrent side. This water is replaced in part by upwelling of water on the downcurrent side of the island.

Isohaline Of the same salinity.

Isopoda An order of dorsoventrally flattened crustaceans that are mostly scavengers or parasites on other crustaceans or fish.

Isopycnal Of the same density.

Isostasy A condition of equilibrium, comparable to buoyancy, by which Earth's brittle crust floats on the plastic mantle.

Isostatic adjustment The adjustment of crustal material due to isostasy.

Isostatic rebound The upward movement of crustal material due to isostasy.

Isotherm A line connecting points of equal temperature.

Isothermal Of the same temperature.

Isotonic Pertaining to the property of having equal osmotic pressure. If two such fluids were separated by a semipermeable membrane that will allow osmosis to occur, there would be no net transfer of water molecules across the membrane.

Isotope One of several atoms of an element that has a different number of neutrons, and therefore a different atomic mass, than the other atoms, or isotopes, of the element.

J

Jellyfish (1) Free-swimming, umbrella-shaped medusoid members of the cnidarian class Scyphozoa. (2) Also frequently applied to the medusoid forms of other cnidarians.

Jet stream An easterly moving air mass at an elevation of about 10 kilometers (6 miles). Moving at speeds that can exceed 300 kilometers (185 miles) per hour, the jet stream follows a wavy path in the middle latitudes and influences how far polar air masses may extend into the lower latitudes.

Jetty A structure built from the shore into a body of water to protect a harbor or a navigable passage from being closed off by the deposition of longshore drift material.

Juan de Fuca Ridge A divergent plate boundary off the Oregon–Washington coast.

K

K–T event An extinction event marked by the disappearance of the dinosaurs that occurred 65 million years ago at the boundary between the Cretaceous (K) and Tertiary (T) Periods of geologic time.

Kelp Large varieties of Phaeophyta (brown algae).

Kelp forest An extensive bed of various species of macroscopic brown algae that provides a habitat for many other types of marine organisms.

Key A low, flat island composed of sand or coral debris that accumulates on a reef flat.

Kinetic energy Energy of motion, which increases as the mass or speed of the object in motion increases.

Knot (kt) A unit of speed equal to 1 nautical mile per hour, approximately 1.15 statute (land) miles per hour.

Krill A common name frequently applied to members of crustacean order Euphausiacea (euphausids).

Kuroshio Current The warm western boundary current of the North Pacific Subtropical Gyre.

Kyoto Protocol An agreement among 60 nations to voluntarily limit greenhouse gas emissions, signed in 1997 in Kyoto, Japan.

L

La Niña An event where the surface temperature in the waters of the eastern South Pacific falls below average values. It often follows an El Niño event.

Labrador Current A cold current flowing south along the coast of Labrador in the northwest Atlantic Ocean.

Lagoon A shallow stretch of seawater partly or completely separated from the open ocean by an elongate narrow strip of land such as a reef or barrier island.

Laguna Madre A hypersaline lagoon located landward of Padre Island along the south Texas coast.

Laminar flow The manner in which a fluid flows in parallel layers or sheets such that the direction of flow at any point does not change with time. Also known as nonturbulent flow.

Land breeze The seaward flow of air from the land caused by differential cooling of Earth's surface.

Langmuir circulation A cellular circulation set up by winds that blow consistently in one direction with speeds in excess of 12 kilometers (7.5 miles) per hour. Helical spirals running parallel to the wind direction are alternately clockwise and counterclockwise.

Larva An embryo that has a different form before it assumes the characteristics of the adult of the species.

Latent heat The quantity of heat gained or lost per unit of mass as a substance undergoes a change of state (such as liquid to solid) at a given temperature and pressure.

Latent heat of condensation The heat energy that must be removed from one gram of a substance to convert it from a vapor at a given temperature below its boiling point. For water, it is 585 calories at 20°C.

Latent heat of evaporation The heat energy that must be added to one gram of a liquid substance to convert it to a vapor at a given temperature below its boiling point. For water, it is 585 calories at 20°C.

Latent heat of freezing The heat energy that must be removed from one gram of a substance at its melting point to convert it to a solid. For water, it is 80 calories.

Latent heat of melting The heat energy that must be added to one gram of a substance at its melting point to convert it to a liquid. For water, it is 80 calories.

Latent heat of vaporization The heat energy that must be added to one gram of a substance at its boiling point to convert it to a vapor. For water, it is 540 calories.

Lateral line system A sensory system running down both sides of fishes to sense subsonic pressure waves transmitted through ocean water.

Latitude Location on Earth's surface based on angular distance north or south of the equator. Equator = 0 degrees; North Pole = 90 degrees north; South Pole = 90 degrees south.

Laurasia An ancient landmass of the Northern Hemisphere. The name is derived from Laurentia, pertaining to the Canadian Shield of North America, and Eurasia, of which it was composed.

Lava Fluid magma coming from an opening in Earth's surface, or the same material after it solidifies.

Law of gravitation See *gravitational force.*

Leeuwin Current A warm current that flows south out of the East Indies along the western coast of Australia.

Leeward The direction toward which the wind is blowing or waves are moving.

Levee (1) Natural low ridges on either side of river channels that result from deposition during flooding. (2) Artificial ridges built by humans to control water flow.

Light Electromagnetic radiation that has a wavelength in the range from about 4000 (violet) to about 7700 (red) angstroms and may be perceived by the normal unaided human eye.

Light-year The distance traveled by light during one year at a speed of 300,000 kilometers (186,000 miles) per second. It equals 9.8 trillion kilometers (6.2 trillion miles).

Limestone A class of sedimentary rocks composed of at least 80% carbonates of calcium or magnesium. Limestones may be either biogenous or hydrogenous.

Limpet A mollusk of the class Gastropoda that possesses a low conical shell that exhibits no spiraling in the adult form.

LIMPET 500 The world's first commercial wave power plant that can generate up to 500 kilowatts of power. It is located on Islay, a small island off the west coast of Scotland, and began generating electricity in November 2000.

Linnaeus, Carolus Latinized name of Swedish botanist Carl von Linné, the father of taxonomic classification and binomial nomenclature.

Liquid state A state of matter in which a substance has a fixed volume but no fixed shape.

Lithify A process by which sediment becomes hardened into sedimentary rock.

Lithogenous sediment Sediment composed of mineral grains derived from the weathering of rock material and transported to the ocean by various mechanisms of transport, including running water, gravity, the movement of ice, and wind.

Lithosphere The outer layer of Earth's structure, including the crust and the upper mantle to a depth of about 200 kilometers (124 miles). Lithospheric plates are the major components involved in plate tectonic movement.

***Lithothamnion* ridge** A feature common to the windward edge of a reef structure, characterized by the presence of the red algae *Lithothamnion.*

Littoral zone The benthic zone between the highest and lowest spring tide shorelines; also known as the intertidal zone.

Lobster A large marine crustacean considered a delicacy. *Homarus americanus* (American or Maine lobster) possesses two large chelae (pincers) and is found offshore from Labrador to North Carolina. Various species of *Panulirus* (spiny lobsters or rock lobsters) have no chelae but possess long, spiny antennae effective in warding off predators. *P. argus* is found off the coast of Florida and in the West Indies, whereas *P. interruptus* is common along the coast of Southern California.

Longitude Location on Earth's surface based on angular distance east or west of the Prime (Greenwich) Meridian (0 degrees longitude). 180 degrees longitude is the International Date Line.

Longitudinal wave A wave phenomenon when particle vibration is parallel to the direction of energy propagation.

Longshore bar A deposit of sediment that forms parallel to the coast within or just beyond the surf zone.

Longshore current A current located in the surf zone and running parallel to the shore as a result of waves breaking at an angle to the shore.

Longshore drift The load of sediment transported along the beach from the breaker zone to the top of the swash line in association with the longshore current. Also called longshore transport or littoral drift.

Longshore trough A low area of the beach that separates the beach face from the longshore bar.

Lophophore A horseshoe-shaped feeding structure bearing ciliated tentacles characteristic of the phyla Bryozoa, Brachiopoda, and Phoronidea.

Low slack water The period of time associated with the peak of low tide when there is no visible flow of water into or out of bays and rivers.

Low tide terrace See *beach face.*

Low tide zone The portion of the intertidal zone that lies between the lowest low tide shoreline and the highest low tide shoreline.

Low water (LW) The lowest level reached by the water surface at low tide before the rise toward high tide begins.

Lower high water (LHW) The lower of two high waters occurring during a tidal day where tides exhibit a mixed tidal pattern.

Lower low water (LLW) The lower of two low waters occurring during a tidal day where tides exhibit a mixed tidal pattern.

Lunar day The time interval between two successive transits of the Moon over a meridian, approximately 24 hours and 50 minutes of solar time. Also called a tidal day.

Lunar month The time it takes for one orbit of the Moon around Earth; 291/2 days Also known as the lunar cycle or synodic month.

Lunar tide The part of the tide caused solely by the tide-producing force of the Moon.

Lunger Fish (such as grouper) that sit motionless on the ocean floor waiting for prey to appear. A lunger uses quick bursts of speed over short distances to capture prey.

Lysocline The level in the ocean at which calcium carbonate begins to dissolve, typically at a depth of about 4000 meters (13,100 feet). Below the lysocline, calcium carbonate dissolves at an increasing rate with increasing depth until the calcite compensation depth (CCD) is reached.

M

Macroplankton Plankton larger than 2 centimeters (0.8 inch) in their smallest dimension.

Magma Fluid rock material from which igneous rock is derived through solidification.

Magnetic anomaly A distortion of the regular pattern of Earth's magnetic field resulting from the various magnetic properties of local concentrations of ferromagnetic minerals in Earth's crust.

Magnetic dip The dip of magnetite particles in rock units of Earth's crust relative to sea level. It is approximately equivalent to latitude. Also called magnetic inclination.

Magnetic field A condition found in the region around a magnet or an electric current, characterized by the existence of a detectable magnetic force at every point in the region and by the existence of magnetic poles.

Magnetic inclination See *magnetic dip.*

Magnetite The mineral form of black iron oxide, Fe_3O_4, that often occurs with magnesium, zinc, and manganese and is an important ore of iron.

Magnetometer A device used for measuring the magnetic field of Earth.

Manganese nodule A concretionary lump containing oxides of manganese, iron, copper, cobalt, and nickel found scattered over the ocean floor.

Mangrove swamp A marshlike environment dominated by mangrove trees. They are restricted to latitudes below 30 degrees.

Mantle (1) The zone between the core and crust of Earth; rich in ferromagnesian minerals. (2) In pelecypods, the portion of the body that secretes shell material.

Mantle plume A rising column of molten magma from Earth's mantle.

Marginal sea A semi-enclosed body of water adjacent to a continent and floored by submerged continental crust.

Mariculture The application of the principles of agriculture to the production of marine organisms.

Marine effect Describes locations that experience the moderating influences of the ocean, usually along coastlines or islands.

Marine Mammals Protection Act An act by U.S. Congress in 1972 that specifies rules to protect marine mammals in U.S. waters.

Marine Protected Area (MPA) An area of the ocean in which there is some level of restriction to protect living, nonliving, cultural, and/or historic resources.

Marine reserve A type of MPA in which all species and their habitat are fully protected.

Marine sanctuary A type of MPA in which biologic or cultural resources are protected. In some marine sanctuaries, fishing, recreational boating, and mining are allowed.

Marine terrace A wave-cut bench that has been exposed above sea level.

MARPOL (Marine Pollution) An international treaty that banned the disposal of all plastics and regulated the dumping of most other garbage at sea.

Marsh An area of soft, wet, flat land that is periodically flooded by salt water and common in portions of lagoons.

Maximum sustainable yield (MSY) The maximum fishery biomass that can be removed yearly and still be sustained by the fishery ecosystem.

Mean high water (MHW) The average height of all the high waters occurring over a 19-year period.

Mean higher high water (MHHW) The average height of the daily higher of the high waters occurring over a 19-year period where tides exhibit a mixed tidal pattern.

Mean low water (MLW) The average height of all the low waters occurring over a 19-year period.

Mean lower low water (MLLW) The average height of the daily lower of the low waters occurring over a 19-year period where tides exhibit a mixed tidal pattern.

Mean sea level (MSL) The mean surface water level determined by averaging all stages of the tide over a 19-year period, usually determined from hourly height observations along an open coast.

Mean tidal range The difference between mean high water and mean low water.

Meander A sinuous curve, bend, or turn in the course of a current.

Mechanical energy Energy manifested as work being done; the movement of a mass some distance.

Mediterranean circulation Circulation characteristic of bodies of water with restricted circulation with the ocean that results from an excess of evaporation as compared to precipitation and runoff similar to the Mediterranean Sea. Surface flow is into the restricted body of water with a subsurface counterflow as exists between the Mediterranean Sea and the Atlantic Ocean.

Medusa A free-swimming, bell-shaped cnidaria body form with a mouth at the end of a central projection and tentacles around the periphery.

Meiofauna Small species of animals that live in the spaces among particles in a marine sediment.

Melon A fatty organ located forward of the blowhole on certain odontocete cetaceans that is used to focus echolocation sounds.

Melting point The temperature at which a solid substance changes to the liquid state.

Mercury A silvery white poisonous metallic element, liquid at room temperature and used in thermometers, barometers, vapor lamps, and batteries and in the preparation of chemical pesticides.

Meridian of longitude Great circles running through the North and South Poles.

Meroplankton Planktonic larval forms of organisms that are members of the benthos or nekton as adults.

Mesopelagic zone That portion of the oceanic province 200 to 1000 meters (660 to 3300 feet) deep. Corresponds approximately with the disphotic (twilight) zone.

Mesosaurus An extinct, presumably aquatic, reptile that lived about 250 million years ago. The distribution of its fossil remains helps support plate tectonic theory.

Mesosphere The middle region of Earth below the asthenosphere and above the core.

Metal sulfide A compound containing one or more metals and sulfur.

Metamorphic rock Rock that has undergone recrystallization while in the solid state in response to changes of temperature, pressure, and chemical environment.

Meteor A bright trail or streak that appears in the sky when a meteoroid is heated to incandescence by friction with Earth's atmosphere. Also called a falling or shooting star.

Meteorite A stony or metallic mass of matter that has fallen to Earth's surface from outer space.

Methane hydrate A white compact icy solid made of water and methane. The most common type of gas hydrate.

Microplankton Net plankton. Plankton not easily seen by the unaided eye, but easily recovered from the ocean with the aid of a silk-mesh plankton net.

Mid-Atlantic Ridge A slow-spreading divergent plate boundary running north–south and bisecting the Atlantic Ocean.

Mid-ocean ridge A linear, volcanic mountain range that extends through all the major oceans, rising 1 to 3 kilometers (0.6 to 2 miles) above the deep-ocean basins. Averaging 1500 kilometers (930 miles) in width, rift valleys are common along the central axis. Source of new oceanic crustal material.

Middle tide zone The portion of the intertidal zone that lies between the highest low tide shoreline and the lowest high tide shoreline.

Migration Long journeys undertaken by many marine species for the purpose of successful feeding and reproduction.

Minamata Bay, Japan The site of the occurrence of human poisoning in the 1950s by mercury contained in marine organisms that were consumed by victims.

Minamata disease A degenerative neurological disorder caused by poisoning with a mercury compound found in seafood obtained from waters contaminated with mercury-containing industrial waste.

Mineral An inorganic substance occurring naturally on Earth and having distinctive physical properties and a chemical composition that can be expressed by a chemical formula. The term is also sometimes applied to organic substances such as coal and petroleum.

Mixed interference A pattern of wave interference in which there is a combination of constructive and destructive interference.

Mixed surface layer The surface layer of the ocean water mixed by wave and tide motions to produce relatively isothermal and isohaline conditions.

Mixed tidal pattern A tidal pattern exhibiting two high tides and two low tides per tidal day with a marked diurnal inequality. Coastal locations that experience such a tidal pattern may also show alternating periods of diurnal and semidiurnal patterns. Also called mixed semidiurnal.

Mixotroph An organism that depends on a combination of autotrophic and heterotrophic behavior to meet its energy requirements. Many coral reef species exhibit such behavior.

Mohorovicíc discontinuity (Moho) A sharp compositional discontinuity between the crust and mantle of Earth. It may be as shallow as 5 kilometers (3 miles) below the ocean floor or as deep as 60 kilometers (37 miles) beneath some continental mountain ranges.

Molecular motion Molecules move in three ways: vibration, rotation, and translation.

Molecule A group of two or more atoms bound together by ionic or covalent bonds.

Mollusca A phylum of soft, unsegmented animals usually protected by a calcareous shell and having a muscular foot for locomotion. Includes snails, clams, chitons, and octopi.

Monera A kingdom of organisms that do not have nuclear material confined within a sheath but spread throughout the cell. Includes bacteria, blue-green algae, and Archaea.

Mononodal Pertaining to a standing wave with only one nodal point or nodal line.

Monsoon A name for seasonal winds derived from the Arabic word for season, mausim. The term was originally applied to winds over the Arabian Sea that blow from the southwest during summer and the northeast during winter.

Moraine A deposit of unsorted material deposited at the margins of glaciers. Many such deposits have become important economically as fishing banks after being submerged by the rising level of the ocean.

Mud Sediment consisting primarily of silt and clay-sized particles smaller than 0.06 millimeters (0.002 inch).

Mutualism A symbiotic relationship in which both participants benefit.

Myoglobin A red, oxygen-storing pigment found in muscle tissue.

Mysticeti The baleen whales.

N

Nadir The point on the celestial sphere directly opposite the zenith and directly beneath the observer.

Nannoplankton Plankton less than 50 micrometers (0.002 inch) in length that cannot be captured in a plankton net and must be removed from the water by centrifuge or special microfilters.

Nansen bottle A device used by oceanographers to obtain samples of ocean water from beneath the surface.

National Flood Insurance Program (NFIP) A program financed by the U.S. government that was intended to prevent costly federal aid after a natural disaster but instead encourages building in risk-prone areas.

Natural selection The process in nature by which only the organisms best adapted to their environment tend to survive and transmit their genetic characters in increasing numbers to succeeding generations while those less adapted tend to be eliminated.

Nauplius A microscopic, free-swimming larval stage of crustaceans such as copepods, ostracodes, and decapods. Typically has three pairs of appendages.

Neap tide Tides of minimal range occurring about every two weeks when the Moon is in either first- or third-quarter moon phase.

Nearshore The zone of a beach that extends from the low tide shoreline seaward to where breakers begin forming.

Nebula A diffuse mass of interstellar dust and/or gas.

Nebular hypothesis A model that describes the formation of the solar system by contraction of a nebula.

Negative-feedback loop See *feedback loop*.

Nektobenthos Those members of the benthos that can actively swim and spend much time off the bottom.

Nekton Pelagic animals such as adult squids, fish, and mammals that are active swimmers to the extent that they can determine their position in the ocean by swimming.

Nematath A linear chain of islands and/or seamounts that are progressively older in one direction. It is created by the passage of a lithospheric plate over a hotspot.

Nematocyst The stinging mechanism found within the cnidoblast of members of the phylum Cnidaria.

Neritic province That portion of the pelagic environment from the shoreline to where the depth reaches 200 meters (660 feet).

Neritic sediment That sediment composed primarily of lithogenous particles and deposited relatively rapidly on the continental shelf, continental slope, and continental rise.

Net primary production The primary production of producers after they have removed what is needed for their metabolism.

Neutral A state in which there is no excess of either the hydrogen or the hydroxide ion.

Neutron An electrically neutral subatomic particle found in the nucleus of atoms that has a mass approximately equivalent to that of a proton.

New moon The phase of the Moon that occurs when the Sun and the Moon are in conjunction; that is, they are both on the same side of Earth. During this time, the dark side of the Moon faces Earth.

New production Primary production supported by nutrients supplied from outside the immediate ecosystem by upwelling or other physical transport.

Newton's law of universal gravitation An equation that quantifies gravitational force between two bodies; it states that the gravitational force is directly proportional to the product of the masses of the two bodies and is inversely proportional to the square of the distance between the two masses.

Niche The ecological role of an organism and its position in the ecosystem.

Niigata, Japan The site of mercury poisoning of humans in the 1960s by ingestion of contaminated seafood.

Nitrogen narcosis A sickness that affects divers. It results from too much nitrogen gas being dissolved in the blood and reducing the flow of oxygen to tissues. The threat of this problem increases with increasing pressure (depth).

Node The point on a standing wave where vertical motion is lacking or minimal. If this condition extends across the surface of an oscillating body of water, the line of no vertical motion is a nodal line.

Non-native species Species that are introduced into waters in which they are alien and often cause severe problems by displacing native species. Also called exotic, alien, or invasive species.

Nonpoint source pollution Any type of pollution entering the ocean from multiple sources rather than from a single discrete source, point, or location. Examples include urban runoff, trash, pet waste, lawn fertilizer, and other types of pollution generated by a multitude of sources. Also called poison runoff.

North Atlantic Current The northernmost current of the North Atlantic Subtropical Gyre.

North Atlantic Deep Water A deep-water mass that forms primarily at the surface of the Norwegian Sea and moves south along the floor of the North Atlantic Ocean.

North Atlantic Subtropical Gyre The large, clockwise-flowing subtropical gyre that exists in the North Atlantic Ocean.

North-East Pacific Time-series Undersea Networked Experiments (NEPTUNE) A cutting-edge sea floor observatory system designed to monitor tectonic activity along the Juan de Fuca tectonic plate in the northeast Pacific Ocean.

North Pacific Current The northernmost current of the North Pacific Subtropical Gyre.

North Pacific Subtropical Gyre The large, clockwise-flowing subtropical gyre that exists in the North Pacific Ocean.

Northern boundary current The northern boundary current of Northern Hemisphere subtropical gyres.

Northeast Monsoon A northeast wind that blows off the Asian mainland onto the Indian Ocean during the winter season.

Norwegian Current A warm current that branches off from the Gulf Stream and flows into the Norwegian Sea between Iceland and the British Isles.

Nucleus The positively charged central region of an atom, composed of protons and neutrons and containing almost all of the mass of the atom.

Nudibranch A sea slug. A member of the mollusk class Gastropoda that has no protective covering as an adult. Respiration is carried on by gills or other projections on the dorsal surface.

Nurdle A small, rounded piece of pre-production plastic.

Nutrient Any of a number of organic or inorganic compounds used by primary producers. Nitrogen and phosphorus compounds are important examples.

O

Observation An occurrence that can be measured with one's senses.

Ocean The entire body of salt water that covers more than 71% of Earth's surface.

Ocean acidification The process by which the ocean's pH is lowered, which increases its acidity.

Ocean acoustical tomography A method by which changes in water temperature may be determined by changes in the speed of transmission of sound. It has the potential to help map ocean circulation patterns over large ocean areas.

Ocean beach The beach on the open-ocean side of a barrier island.

Ocean Drilling Program (ODP) A program that replaced the Deep Sea Drilling Project in 1983, focusing on drilling the continental margins using the drill ship *JOIDES Resolution*.

Ocean thermal energy conversion (OTEC) A technique that involves generating energy by using the difference in temperature between surface waters and deep waters in low latitude regions.

Oceanic Common Water Deep water found in Pacific and Indian Oceans as a result of mixing of Antarctic Bottom Water and North Atlantic Deep Water.

Oceanic crust A mass of rock with a basaltic composition that is about 5 kilometers (3 miles) thick.

Oceanic province The division of the pelagic environment where the water depth is greater than 200 meters (660 feet).

Oceanic ridge A portion of the global mid-ocean ridge system that is characterized by slow spreading and steep slopes.

Oceanic rise A portion of the global mid-ocean ridge system that is characterized by fast spreading and gentle slopes.

Oceanic sediment The inorganic abyssal clays and the organic oozes that accumulate slowly on the deep-ocean floor.

Ocelli A light-sensitive organ around the base of many medusoid bells.

Odontoceti The toothed whales.

Offset Separation that occurs due to movement along a fault.

Offshore The comparatively flat submerged zone of variable width extending from the breaker line to the edge of the continental shelf.

Oligotrophic Exhibiting low levels of biological production, such as the centers of subtropical gyres.

Omnivore An animal that feeds on both plants and animals.

Oolite A deposit formed of small spheres from 0.25 to 2 millimeters (0.01 to 0.08 inch) in diameter. They are usually composed of concentric layers of calcite.

Ooze A pelagic sediment containing at least 30% skeletal remains of pelagic organisms, the balance being clay minerals. Oozes are further defined by the chemical composition of the organic remains (siliceous or calcareous) and by their characteristic organisms (e.g., diatomaceous ooze, foraminifer ooze).

Opal An amorphous form of silica ($SiO_2 \cdot nH_2O$) that usually contains from 3 to 9% water. It forms the shells of radiolarians and diatoms.

Ophiolite An assemblage of mafic and ultramafic igneous rocks from Earth's oceanic crust and the underlying upper mantle that has been uplifted and exposed above sea level and often emplaced onto continental crustal rocks by plate tectonic processes. Ophiolites are generally composed of green-colored metamorphic rocks such as peridotite and serpentinite.

Opposition The separation of two heavenly bodies by 180 degrees relative to Earth. The Sun and Moon are in opposition during the full moon phase.

Orbital wave A wave phenomenon in which energy is moved along the interface between fluids of different densities. The wave form is propagated by the movement of fluid particles in orbital paths.

Orthogonal line A line constructed perpendicular to a wave front and spaced so that the energy between lines is equal at all times. Orthogonals are used to help determine how energy is distributed along the shoreline by breaking waves.

Osmosis The process by which water molecules move through a semipermeable membrane from higher water molecule concentration (lower salinity) to lower water molecule concentration (higher salinity).

Osmotic pressure A measure of the tendency for osmosis to occur. It is the pressure that must be applied to the more concentrated solution to prevent the passage of water molecules into it from the less concentrated solution.

Osmotic regulation Physical and biological processes used by organisms to counteract the osmotic effects of differences in osmotic pressures of their body fluids and the water in which they live.

Ostracoda An order of crustaceans that are minute and compressed within a bivalve shell.

Otocyst Gravity-sensitive organs around the bell of a medusa.

Outer sublittoral zone The continental shelf below the intersection with the euphotic zone where no plants grow attached to the bottom.

Outgassing The process by which gases are removed from within Earth's interior.

Overfishing A situation that occurs when adult fish in a population are harvested faster than their natural rate of reproduction.

Oxygen compensation depth The depth in the ocean at which marine plants receive just enough solar radiation to meet their basic metabolic needs. It marks the base of the euphotic zone.

Oxygen minimum layer (OML) A zone of low dissolved oxygen concentration that occurs at a depth of about 700 to 1000 meters (2300 to 3280 feet).

P

Pacific Decadal Oscillation (PDO) A natural oscillation in the Pacific Ocean that lasts 20 to 30 years and appears to influence sea surface temperatures.

Pacific Ring of Fire An extensive zone of volcanic and seismic activity that coincides roughly with the borders of the Pacific Ocean.

Pacific Warm Pool A large region of warm surface water on the western side of the Pacific Ocean.

Pacific-type margin See *active margin*.

Paleoceanography The study of how the ocean, atmosphere, and land have interacted to produce changes in ocean chemistry, circulation, biology, and climate.

Paleogeography The study of the historical changes of shapes and positions of the continents and oceans.

Paleomagnetism The study of Earth's ancient magnetic field.

Pancake ice Circular pieces of newly formed sea ice from 0.3 to 3 meters (1 to 10 feet) in diameter that form in the early fall in polar regions.

Pangaea An ancient supercontinent of the geologic past that contained all Earth's continents.

Panthalassa A large, ancient ocean that surrounded Pangaea.

Paralytic shellfish poisoning (PSP) Paralysis resulting from poisoning caused by eating shellfish contaminated with the toxic dinoflagellate *Gonyaulax*.

Parasitism A symbiotic relationship between two organisms in which one benefits at the expense of the other.

Parts per thousand (‰) A unit of measurement used in reporting salinity of water equal to the number of grams of dissolved substances in 1000 grams of water. For example, 1‰ is equivalent to 0.1%, or 1000 ppm.

Passive margin A continental margin that lacks a plate boundary and is marked by a low degree of tectonic activity, such as those typical of the Atlantic Ocean.

PCBs A group of industrial chemicals (polychlorinated biphenyls) used in a variety of products; responsible for several episodes of ecological damage in coastal waters.

Peat deposit Partially carbonized organic matter found in bogs and marshes that can be used as fertilizer and fuel.

Pelagic environment The open ocean environment, which is divided into the neritic province (water depth 0 to 200 meters or 656 feet) and the oceanic province (water depth greater than 200 meters or 656 feet).

Pelecypoda A class of mollusks characterized by two more or less symmetrical lateral valves with a dorsal hinge. These filter feeders pump water through the filter system and over gills through posterior siphons. Many possess a hatchet-shaped foot used for locomotion and burrowing. Includes clams, oysters, mussels, and scallops.

Perigee The point on the orbit of an Earth satellite (Moon) that is nearest Earth.

Perihelion That point on the orbit of a planet or comet around the Sun that is closest to the Sun.

Permeability Capacity of a porous rock or sediment for transmitting fluid.

Peru Current The cold eastern boundary current of the South Pacific Subtropical Gyre.

Petroleum A naturally occurring liquid hydrocarbon.

Pfiesteria piscicida A species of toxic dinoflagellate that has been known to cause fish kills.

pH scale A measure of the acidity or alkalinity of a solution, numerically equal to 7 for neutral solutions, increasing with increasing alkalinity and decreasing with increasing acidity. The pH scale commonly in use ranges from 0 to 14.

Phaeophyta Brown algae characterized by the carotinoid pigment fucoxanthin. Contains the largest members of the marine algal community.

Phosphate Any of a number of phosphorus-bearing compounds.

Phosphorite A sedimentary rock composed primarily of phosphate minerals.

Photic zone The upper ocean in which the presence of solar radiation is detectable. It includes the euphotic and disphotic zones.

Photophore One of several types of light-producing organs found primarily on fishes and squids inhabiting the mesopelagic and upper bathypelagic zones.

Photosynthesis The process by which plants and algae produce carbohydrates from carbon dioxide and water in the presence of chlorophyll, using light energy and releasing oxygen.

Phycoerythrin A red pigment characteristic of the Rhodophyta (red algae).

Phytoplankton Algal plankton. One of the most important communities of primary producers in the ocean.

Picoplankton Small plankton within the size range of 0.2 to 2.0 micrometers (0.000008 to 0.00008 inch) in size. Composed primarily of bacteria.

Pillow basalt A basalt exhibiting pillow structure. See *pillow lava*.

Pillow lava A general term for those lavas displaying discontinuous pillow-shaped masses (pillow structure) caused by the rapid cooling of lava as a result of underwater eruption of lava or lava flowing into water.

Ping A sharp, high-pitched sound made by the transmitting device of many sonar systems.

Pinniped A group of marine mammals that have prominent flippers; includes the sea lions/fur seals, seals, and walruses.

Plane of the ecliptic The surface connecting all points in Earth's orbit.

Plankter Informal term for plankton.

Plankton Passively drifting or weakly swimming organisms that are not independent of currents. Includes mostly microscopic algae, protozoa, and larval forms of higher animals.

Plankton bloom A very high concentration of phytoplankton, resulting from a rapid rate of reproduction as conditions become optimal during the spring in high latitude areas. Less obvious causes produce blooms that may be destructive in other areas.

Plankton net A plankton-extracting device that is cone shaped and typically of a silk material. It is towed through the water or lifted vertically to extract plankton down to a size of 50 micrometers (0.0002 inch).

Plantae A kingdom of many-celled plants.

Plastic (1) Capable of being shaped or formed. (2) Composed of plastic or plastics.

Plate tectonics Global dynamics having to do with the movement of a small number of semirigid sections of Earth's crust, with seismic activity and volcanism occurring primarily at the margins of these sections. This movement has resulted in changes in the geographic positions of continents and the shape and size of ocean basins.

Plume A rising column of molten mantle material that is associated with a hotspot when it penetrates Earth's crust.

Plunging breaker Impressive curling breakers that form on moderately sloping beaches.

Pneumatic duct An opening into the swim bladder of some fishes that allows rapid release of air into the esophagus.

Poikilothermic An organism whose body temperature varies with and is largely controlled by its environment; cold-blooded.

Polar Pertaining to the polar regions.

Polar cell The large atmospheric circulation cell that occurs between 60 and 90 degrees latitude in each hemisphere.

Polar easterly wind belt A global wind belt that moves away from the polar regions toward the polar front at about 60 degrees north or south latitude in each hemisphere. These winds move from a northeasterly direction in the Northern Hemisphere and from a southeasterly direction in the Southern Hemisphere.

Polar front The boundary between the global wind belts' prevailing westerlies and polar easterlies that is centered at about 60 degrees latitude in each hemisphere and is characterized by rising air and much precipitation.

Polar high The region of high atmospheric pressure that occurs at the poles in both hemispheres.

Polar wandering curve A curve that shows the change in the position of a pole through time.

Polarity Intrinsic polar separation, alignment, or orientation, especially of a physical property (such as magnetic or electrical polarity).

Pollution (marine) The introduction of substances that result in harm to the living resources of the ocean or humans who use these resources.

Polychaeta A class of annelid worms that includes most of the marine segmented worms.

Polyp A single individual of a colony or a solitary attached cnidarian.

Population A group of individuals of one species living in an area.

Porifera A phylum of sponges. Supporting structure composed of $CaCO_3$ or SiO_2 spicules or fibrous spongin. Water currents created by flagella-waving choanocytes enter tiny pores, pass through canals, and exit through a larger osculum.

Porosity The ratio of the volume of all the empty spaces in a material to the volume of the whole.

Positive-feedback loop See *feedback loop*.

Potential energy The energy of a particle or system of particles derived from position or condition rather than motion.

Precession Describes the change in the attitude of the Moon's orbit around Earth as it slowly changes its direction. The cycle is completed every 18.6 years and is accompanied by a clockwise rotation of the plane of the Moon's orbit that is completed in the same time interval.

Precipitate To cause a solid substance to be separated from a solution, usually due to a change in physical or chemical conditions.

Precipitation In a meteorological sense, the discharge of water in the form of rain, snow, hail, or sleet from the atmosphere onto Earth's surface.

Prevailing westerly wind belt A global wind belt that moves from a subtropical high-pressure belt at about 30 degrees north or south latitude toward the polar front at about 60 degrees north or south latitude. These winds move from a southwesterly direction in the Northern Hemisphere and from a northwesterly direction in the Southern Hemisphere.

Primary productivity The rate at which energy is stored by organisms through the formation of organic matter (carbon-based compounds) using energy derived from solar radiation (photosynthesis) or chemical reactions (chemosynthesis). Also known simply as productivity.

Prime meridian The meridian of longitude 0 degrees used as a reference for measuring longitude that passes through the Royal Observatory at Greenwich, England. Also known as the Greenwich Meridian.

Producer The autotrophic component of an ecosystem that produces the food that supports the biocommunity.

Productivity See *primary productivity*.

Progressive wave A wave in which the waveform progressively moves.

Propagation The transmission of energy through a medium.

Protein A very complex organic compound made up of large numbers of amino acids. Proteins make up a large percentage of the dry weight of all living organisms.

Protoctista A kingdom of organisms that includes any of the unicellular eukaryotic organisms and their descendant multicellular organisms. Includes single-celled and multicelled marine algae as well as single-celled animals called protozoa.

Protoearth The young, early developing Earth.

Proton A positively charged subatomic particle found in the nucleus of atoms that has a mass approximately equivalent to that of a neutron.

Protoplanet Any planet that is in its early stages of development.

Protoplasm The self-perpetuating living material making up all organisms, mostly consisting of the elements carbon, hydrogen, and oxygen combined into various chemical forms.

Protozoa A phylum of one-celled animals with nuclear material confined within a nuclear sheath.

Proxigean A tidal condition of extremely large tidal range that occurs when spring tides coincide with perigee. Also called "closest of the close moon" tides.

Pseudopodia An extension of protoplasm in a broad, flat, or long needlelike projection used for locomotion or feeding. Typical of amoeboid forms such as foraminifers and radiolarians.

Pteropoda An order of pelagic gastropods in which the foot is modified for swimming and the shell may be present or absent.

Purse seine net A style of large fishing net that resembles a purse. It is set around a grouping of organisms (such as tuna) and the bottom is drawn tight to capture the organisms.

Pycnocline A layer of water in which a high rate of change in density in the vertical dimension is present.

Pyrrophyta A phylum of dinoflagellates that possess flagella for locomotion.

Q

Quadrature The state of the Moon during the first- and third-quarter moon phases when the Sun and the Moon are at right angles relative to Earth.

Quarter moon First- and third-quarter moon phases, which occur when the Moon is in quadrature about one week after the new moon and full moon phases, respectively. The third-quarter moon phase is also known as the last quarter moon phase.

Quartz A very hard mineral composed of silica, SiO_2.

R

Radiata A grouping of phyla with primary radial symmetry—phyla Cnidarian and Ctenophora.

Radioactivity The spontaneous breakdown of the nucleus of an atom resulting in the emission of radiant energy in the form of particles or waves.

Radiolaria An order of planktonic and benthic protozoans that possess protective coverings usually made of silica.

Radiometric age dating A technique that involves the use of radioactive half-lives to determine the age of a rock.

Ray A cartilaginous fish in which the body is dorsoventrally flattened, eyes and spiracles are on the upper surface, and gill slits are on the bottom. The tail is reduced to a whiplike appendage. Includes electric rays, manta rays, and stingrays.

Recreational beach The area of a beach above shoreline, including the berm, berm crest, and the exposed part of the beach face.

Red clay See *abyssal clay*.

Red muscle fiber Fine muscle fibers rich in myoglobin that are abundant in cruiser-type fishes.

Red tide A reddish-brown discoloration of surface water, usually in coastal areas, caused by high concentrations of microscopic organisms, usually dinoflagellates. It probably results from increased availability of certain nutrients. Toxins produced by the dinoflagellates may kill fish directly; decaying plant and animal remains or large populations of animals that migrate to the area of abundant plants may also deplete the surface waters of oxygen and cause asphyxiation of many animals.

Reef A strip or ridge of rocks, sand, coral, or human-made objects that rises to or near the surface of the ocean and creates a navigational hazard.

Reef flat A platform of coral fragments and sand on the lagoon side of a reef that is relatively exposed at low tide.

Reef front The upper seaward face of a reef from the reef edge (seaward margin of reef flat) to the depth at which living coral and coralline algae become rare, 16 to 30 meters (50 to 100 feet).

Reflection (wave) The process in which a wave has part of its energy returned seaward by a reflecting surface.

Refraction (wave) The process by which the part of a wave in shallow water is slowed down, causing the wave to bend and align itself nearly parallel to the shore.

Regenerated production The portion of gross primary production that is supported by nutrients recycled within an ecosystem.

Relict beach A beach deposit laid down and submerged by a rise in sea level. It is still identifiable on the continental shelf, indicating that no deposition is presently taking place at that location on the shelf.

Relict sediment A sediment deposited under a set of environmental conditions that still remains unchanged although the environment has changed, and it remains unburied by later sediment. An example is a beach deposited near the edge of the continental shelf when sea level was lower.

Relocation The strategy of moving a structure that is threatened by being claimed by the sea.

Residence time The average length of time a particle of any substance spends in the ocean. It is calculated by dividing the total amount of the substance in the ocean by the rate of its introduction into the ocean or the rate at which it leaves the ocean.

Respiration The process by which organisms use organic materials (food) as a source of energy. As the energy is released, oxygen is used and carbon dioxide and water are produced.

Restoring force A force such as surface tension or gravity that tends to restore the ocean surface displaced by a wave to that of a still water level.

Resultant force The difference between the provided gravitational force of various bodies and the required centripetal force on Earth. The horizontal component of the resultant force is the tide-producing force.

Reverse osmosis A method of desalinating ocean water that involves forcing water molecules through a water-permeable membrane under pressure.

Reversing current The tide current as it occurs at the margins of landmasses. The water flows in and out for approximately equal periods of time separated by slack water where the water is still at high and low tidal extremes.

Rhodophyta A phylum of algae composed primarily of small encrusting, branching, or filamentous plants that receive their characteristic red color from the presence of the pigment phycoerythrin. With a worldwide distribution, they are found at greater depths than other algae.

Rift valley A deep fracture or break, about 25 to 50 kilometers (15 to 30 miles) wide, extending along the crest of a mid-ocean ridge.

Rifting The movement of two plates in opposite directions such as along a divergent boundary.

Right whale A surface-feeding baleen whale of the family Balaenidae that were a favorite target of early whalers.

Rip current A strong narrow surface or near-surface current of short duration and high speed flowing seaward through the breaker zone at nearly right angles to the shore. It represents the return to the ocean of water that has been piled up on the shore by incoming waves.

Rip-rap Large blocky material used to armor coastal structures.

Rogue wave An unusually large wave that occurs unexpectedly amid other waves of smaller size. Also known as a superwave, monster wave, sleeper wave, or freak wave.

Rorqual whale A large baleen whale with prominent ventral groves (rorqual folds) of the family Balaenopteridae: the minke, Baird's, Bryde's, sei, fin, blue, and humpback whales.

Rotary current A tidal current that is observed in the open ocean and makes one complete rotation during a tidal period.

Rotary drilling Drilling involving the use of a long, hollow pipe with a drill bit on its end that is rotated to crush the rock around the outside and retain a cylinder of rock (a core sample) on the inside of the pipe.

S

Saffir-Simpson scale A scale of hurricane intensity that divides tropical cyclones into categories based on wind speed and damage.

Salinity A measure of the quantity of dissolved solids in ocean water. Formally, it is the total amount of dissolved solids in ocean water in parts per thousand (‰) by weight after all carbonate has been converted to oxide, the bromide and iodide to chloride, and all the organic matter oxidized. It is normally computed from conductivity, refractive index, or chlorinity.

Salinometer An instrument that is used to determine the salinity of seawater by measuring its electrical conductivity.

Salpa A genus of pelagic tunicates that are cylindrical, transparent, and found in all oceans.

Salt A substance that yields ions other than hydrogen or hydroxyl. Salts are produced from acids by replacing the hydrogen with a metal.

Salt marsh A relatively flat area of the shore where fine sediment is deposited and salt-tolerant grasses grow. One of the most biologically productive regions of Earth.

Salt wedge estuary A very deep river mouth with a very large volume of freshwater flow beneath which a wedge of saltwater from the ocean invades. The Mississippi River is an example.

San Andreas Fault A transform fault that cuts across the state of California from the northern end of the Gulf of California to Point Arena north of San Francisco.

Sand Particle size of 0.0625 to 2 millimeters (0.002 to 0.08 inch). It pertains to particles that lie between silt and granules on the Wentworth scale of grain size.

Sargasso Sea A region of convergence in the North Atlantic lying south and east of Bermuda where the water is a very clear, deep blue color, and contains large quantities of floating *Sargassum*.

Sargassum A brown alga characterized by a bushy form, substantial holdfast when attached, and a yellow-brown, green-yellow, or orange color. The two dominant species of macroscopic algae in the Sargasso Sea are *S. fluitans* and *S. natans*.

Scarp A linear steep slope on the ocean floor separating gently sloping or flat surfaces.

Scavenger An animal that feeds on dead organisms.

Schooling A well-defined large groups of fish, squid, and crustaceans that apparently aid them in survival.

Scientific method The principles and empirical processes of discovery and demonstration considered characteristic of or necessary for scientific investigation, generally involving the observation of phenomena, the formulation of a hypothesis concerning the phenomena, experimentation to demonstrate the truth or falseness of the hypothesis, and a conclusion that validates or modifies the hypothesis.

Scuba An acronym for *s*elf-*c*ontained *u*nderwater *b*reathing *a*pparatus, a portable device containing compressed air that is used for breathing underwater.

Scyphozoa A class of cnidarians that includes the true jellyfish in which the medusoid body form predominates and the polyp is reduced or absent.

Sea (1) A subdivision of an ocean, generally enclosed by land and usually composed of salt water. Two types of seas are identifiable and defined. They are the Mediterranean seas, where a number of seas are grouped together collectively as one sea, and adjacent seas, which are connected individually to the ocean. (2) A portion of the ocean where waves are being generated by wind.

Sea anemone A member of the class Anthozoa whose bright color, tentacles, and general appearance make it resemble flowers.

Sea arch An opening through a headland caused by wave erosion. Usually develops as sea caves are extended from one or both sides of the headland.

Glossary **561**

Sea breeze The landward flow of air from the sea caused by differential heating of Earth's surface.

Sea cave A cavity at the base of a sea cliff formed by wave erosion.

Sea cow See *Sirenia*.

Sea cucumber A common name given to members of the echinoderm class Holotheuroidea.

Sea floor spreading A process producing the lithosphere when convective upwelling of magma along the oceanic ridges moves the ocean floor away from the ridge axes at rates between 2 to 12 centimeters (0.8 to 5 inches) per year.

Sea ice A form of ice originating from the freezing of ocean water.

Sea lion Any of several eared seals with a relatively long neck and large front flippers, especially the California sea lion *Zalophus californianus* of the northern Pacific. Along with the fur seals, these marine mammals are known as eared seals.

Sea otter A seagoing otter that has recovered from near extinction along the North Pacific coasts. It feeds primarily on abalone, sea urchins, and crustaceans.

Sea snake A reptile belonging to the family Hydrophiidae with venom similar to that of cobras. They are found primarily in the coastal waters of the Indian Ocean and the western Pacific Ocean.

Sea stack An isolated, pillarlike rocky island that is detached from a headland by wave erosion.

Sea turtle Any of the reptilian order Testudinata found widely in warm water.

Sea urchin An echinoderm belonging to the class Echinoidea possessing a fused test (external covering) and well-developed spines.

Seaknoll See *abyssal hill*.

Seal (1) Any of several earless seals with a relatively short neck and small front flippers. Also known as true seals. (2) A general term that describes any of the various aquatic, carnivorous marine mammals of the families Phocidae and Otariidae (true seals and eared seals), found chiefly in the Northern Hemisphere and having a sleek, torpedo-shaped body and limbs that are modified into paddlelike flippers.

Seamount An individual volcanic peak extending over 1000 meters (3300 feet) above the surrounding ocean floor.

Seasonal thermocline A thermocline that develops due to surface heating of the oceans in middle to high latitudes. The base of the seasonal thermocline is usually above 200 meters (656 feet).

Seawall A wall built parallel to the shore to protect coastal property from the waves.

SeaWiFS An instrument aboard the SeaStar satellite launched in 1997 that measures the color of the ocean with a radiometer and provides global coverage of ocean chlorophyll levels as well as land productivities every two days.

Secchi disk A light-colored disk-shaped device that is lowered into water in order to measure the water's ability to transmit light.

Sediment Particles of organic or inorganic origin that accumulate in loose form.

Sediment maturity A condition in which the roundness and degree of sorting increase and clay content decreases within a sedimentary deposit.

Sedimentary rock Rock resulting from the consolidation of loose sediment, or rock resulting from chemical precipitation, such as sandstone and limestone.

Seep An area where water of various temperature trickles out of the sea floor, such as warm seeps and cold seeps.

Seiche A standing wave of an enclosed or semienclosed body of water that may have a period ranging from a few minutes to a few hours, depending on the dimensions of the basin. The wave motion continues after the initiating force has ceased.

Seismic Pertaining to an earthquake or Earth vibration, including those that are artificially induced.

Seismic moment magnitude (M_w) A scale used for measuring earthquake intensity based on energy released in creating very long-period seismic waves.

Seismic sea wave See *tsunami*.

Seismic surveying The use of sound-generating techniques to identify features on or beneath the ocean floor.

Semidiurnal tidal pattern A tidal pattern exhibiting two high tides and two low tides per tidal day with small inequalities between successive highs and successive lows; a semidaily tide.

Sessile Permanently attached to the substrate and not free to move about.

Sewage sludge The semisolid material precipitated by sewage treatment.

Shallow-water wave A wave on the surface having a wavelength of at least 20 times water depth. The bottom affects the orbit of water particles and speed is determined by water depth.

Shelf break The depth at which the gentle slope of the continental shelf steepens appreciably. It marks the boundary between the continental shelf and continental rise.

Shelf ice Thick shelves of glacial ice that push out into Antarctic seas from Antarctica. Large tabular icebergs calve at the edge of these vast shelves.

Shoal To become shallow.

Shore The area seaward of the coast, which extends from the highest level of wave action during storms to the low water line.

Shoreline The line marking the intersection of water surface with the shore. Migrates up and down as the tide rises and falls.

Side-scan sonar A method of mapping the topography of the ocean floor along a strip up to 60 kilometers (37 miles) wide using computers and sonar signals that are directed away from both sides of the survey ship.

Silica Silicon dioxide (SiO_2).

Silicate A mineral whose crystal structure contains SiO_4 tetrahedra.

Siliceous A condition of containing abundant silica (SiO_2).

Sill A submarine ridge partially separating bodies of water such as fjords and seas from one another or from the open ocean.

Silt A particle size of 0.008 to 0.0625 millimeter (0.0003 to 0.002 inch). It is intermediate in size between sand and clay.

Siphonophora An order of hydrozoan cnidarians that form pelagic colonies containing both polyps and medusae. An example is *Physalia*.

Sirenia An order of large, vegetarian marine mammals that includes dugongs and manatees, which are also known as sea cows.

Slack water A situation that occurs when a reversing tidal current changes direction at high tide (high slack water) or low tide (low slack water). During slack water, the speed of the current is zero.

Slick A smooth patch on an otherwise rippled surface caused by a monomolecular film of organic material that reduces surface tension.

Slightly stratified estuary An estuary of moderate depth in which marine water invades beneath the freshwater runoff. The two water masses mix so the bottom water is slightly saltier than the surface water at most places in the estuary.

SOFAR channel An acronym for the *sound fixing and ranging channel*, which is a low-velocity sound travel zone that coincides with the permanent thermocline in low and middle latitudes.

Solar day The 24-hour period during which Earth completes one rotation on its axis.

Solar distillation A process by which ocean water can be desalinated by evaporation and the condensation of the vapor on the cover of a container. The condensate then runs into a separate container and is collected as freshwater. Also called solar humidification.

Solar humidification See *solar distillation*.

Solar system The Sun and the celestial bodies, asteroids, planets, and comets that orbit around it.

Solar tide The partial tide caused by the tide-producing forces of the Sun.

Solid state A state of matter in which the substance has a fixed volume and shape. A crystalline state of matter.

Solstice The time during which the Sun is directly over one of the tropics. In the Northern Hemisphere the summer solstice occurs on June 21 or 22 as the Sun is over the Tropic of Cancer, and the winter solstice occurs on December 21 or 22 when the Sun is over the Tropic of Capricorn.

Solute A substance dissolved in a solution. Salts are the solute in saltwater.

Solution A state in which a solute is homogeneously mixed with a liquid solvent. Water is the solvent for the solution that is ocean water.

Solvent A liquid that has one or more solutes dissolved in it.

Somali Current A current that flows north along the Somali coast of Africa during the southwest monsoon season.

Sonar An acronym for *sound navigation and ranging*, a method that uses sound to determine the distance of objects in the ocean.

Sorting A texture of sediments, where a well-sorted sediment is characterized by having great uniformity of grain sizes.

Sounding A measured depth of water beneath a ship.

South Atlantic Subtropical Gyre The large, counterclockwise-flowing subtropical gyre that exists in the South Atlantic Ocean.

South Pacific Subtropical Gyre The large, counterclockwise-flowing subtropical gyre that exists in the South Pacific Ocean.

Southern boundary current The southern boundary current of Southern Hemisphere subtropical gyres.

Southern Oscillation The periodic change in the pressure differential between the Southeastern Pacific high pressure and the Western Pacific equatorial low pressure that occurs in concert with El Niño–Southern Oscillation events.

Southwest Monsoon A southwest wind that develops during the summer season. It blows off the Indian Ocean onto the Asian mainland.

Southwest Monsoon Current During the southwest monsoon season, an eastward-flowing current that replaces the west-flowing North Equatorial Current in the Indian Ocean.

Space dust Micrometeoroid space debris.

Species A fundamental category of taxonomic classification, ranking below a genus or subgenus and consisting of related organisms capable of interbreeding.

Species diversity The number or variety of species found in a subdivision of the marine environment.

Specific gravity The ratio of density of a given substance to that of pure water at 4°C and at 1 atmosphere pressure.

Specific heat The amount of heat energy required to raise the temperature of 1 gram of a substance by 1°C. Also known as specific heat capacity.

Spermaceti organ A large fatty organ located within the head region of sperm whales (*Physeter macrocephalus*) that is used to focus echolocation sounds.

Spermatophyta See *Anthophyta*.

Spherule A cosmogenous microscopic globular mass composed of silicate rock material (tektites) or of iron and nickel.

Spicule A minute, needlelike calcareous or siliceous projection found in sponges, radiolarians, chitons, and echinoderms that acts to support the tissue or provide a protective covering.

Spilling breaker A type of breaking wave that forms on a gently sloping beach, which gradually extracts the energy from the wave to produce a turbulent mass of air and water that runs down the front slope of the wave.

Spit A small point, low tongue, or narrow embankment of land commonly consisting of sand deposited by longshore currents and having one end attached to the mainland and the other terminating in open water.

Splash wave A long-wavelength wave created by a massive object or series of objects falling into water; a type of tsunami.

Sponge See *Porifera*.

Spray zone The shore zone lying between the high tide shoreline and the coastline. It is covered by water only during storms.

Spreading center A divergent plate boundary.

Spreading rate The rate of divergence of plates at a spreading center.

Spring bloom A middle-latitude bloom of phytoplankton that occurs during the spring and is limited by the availability of nutrients.

Spring equinox See *vernal equinox*.

Spring tide A tide of maximum range occurring about every two weeks when the Moon is in either new or full moon phase.

Stack An isolated mass of rock projecting from the ocean off the end of a headland from which it has been separated by wave erosion.

Standard seawater Ampules of ocean water for which the chlorinity has been determined by the Institute of Oceanographic Services in Wormly, England. The ampules are sent to laboratories all over the world so that equipment and reagents used to determine the salinity of ocean water samples can be calibrated by adjustment until they give the same chlorinity as is shown on the ampule label.

Standing stock (crop) The mass of fishery organisms present in an ecosystem at a given time.

Standing wave A wave, the form of which oscillates vertically without progressive movement. The region of maximum vertical motion is an antinode. On either side are nodes, where there is no vertical motion but maximum horizontal motion.

Stenohaline Pertaining to organisms that can withstand only a small range of salinity change.

Stenothermal Pertaining to organisms that can withstand only a small range of temperature change.

Stick chart A device made of sticks or pieces of bamboo that was used by early navigators at sea.

Still water level The horizontal surface halfway between crest and trough of a wave. If there were no waves, the water surface would exist at this level. Also known as zero energy level.

Storm An atmospheric disturbance characterized by strong winds accompanied by precipitation and often by thunder and lightning.

Storm surge A rise above normal water level resulting from wind stress and reduced atmospheric pressure during storms. Consequences can be more severe if it occurs in association with high tide.

Strait of Gibraltar The narrow opening between Europe and Africa through which the waters of the Atlantic Ocean and Mediterranean Sea mix.

Stranded beach An ancient beach deposit found above present sea level because of a lowering of sea level.

Streamlining The shaping of an object so it produces the minimum of turbulence while moving through a fluid medium. The teardrop shape displays a high degree of streamlining.

Stromatolite A calcium carbonate sedimentary structure in which algal assemblages trap sediment and bind it into forms that are often dome shaped. They are known to form only in shallow-water environments.

Subduction The process by which one lithospheric plate descends beneath another as they converge.

Subduction zone A long, narrow region beneath Earth's surface in which subduction takes place.

Subduction zone seep biocommunity Animals that live in association with seeps of pore water squeezed out of deeper sediments. They depend on sulfur-oxidizing bacteria that act as producers for the ecosystem.

Submarine canyon A steep, V-shaped canyon cut into the continental shelf or slope.

Submarine fan See *deep-sea fan*.

Submerged dune topography Ancient coastal dune deposits found submerged beneath the present shoreline because of a rise in sea level or submergence of the coast.

Submerging shoreline A shoreline formed by the relative submergence of a landmass in which the shoreline is on landforms developed under subaerial processes. It is characterized by bays and promontories and is more irregular than a shoreline of emergence.

Subneritic province The benthic environment extending from the shoreline across the continental shelf to the shelf break. It underlies the neritic province of the pelagic environment.

Suboceanic province The benthic environments seaward of the continental shelf.

Subpolar Pertaining to the oceanic region that is covered by sea ice in winter. The ice melts away in summer.

Subpolar gyre A small, circular-moving loop of water that is centered at about 60 degrees latitude in both hemispheres. Subpolar gyres rotate in a counterclockwise orientation in the Northern Hemisphere and clockwise in the Southern Hemisphere.

Subpolar low A global belt of low atmospheric pressure located at about 60 degrees north or south latitude that is associated with upward vertical flow of low-density air and abundant precipitation.

Substrate The base on which an organism lives and grows.

Subsurface current A current usually flowing below the pycnocline, generally at slower speed and in a different direction from the surface current.

Subtidal zone That portion of the benthic environment extending from low tide to a depth of 200 meters (660 feet); considered by some to be the surface of the continental shelf.

Subtropical Pertaining to the oceanic region poleward of the tropics (about 30 degrees latitude).

Subtropical Convergence The zone of convergence that occurs within all subtropical gyres as a result of Ekman transport driving water toward the interior of the gyres.

Subtropical gyre A large, circular-moving loop of water that is centered at about 30 degrees latitude and is initiated by the trade winds and the prevailing westerlies. A total of five subtropical gyres exist, with rotation clockwise in the Northern Hemisphere and counterclockwise in the Southern Hemisphere.

Subtropical high A region of high atmospheric pressure located at about 30 degrees latitude.

Sulfur A yellow mineral composed of the element sulfur. It is commonly found in association with hydrocarbons and salt deposits.

Sulfur-oxidizing bacteria Bacteria that support many deep-sea hydrothermal vents and cold-water seep biocommunities by using energy released by oxidation to synthesize organic matter chemosynthetically.

Summer solstice In the Northern Hemisphere, it is the instant when the Sun moves north to the Tropic of Cancer before changing direction and moving southward toward the equator approximately June 21.

Summertime beach A beach that is characteristic during summer months. It typically has a wide sandy berm and a steep beach face.

Sunspot A dark spot on the Sun that is associated with magnetic storms.

Superwave See *rogue wave*.

Supratidal zone The splash or spray zone above the spring high tide shoreline.

Surf beat An irregular wave pattern caused by mixed interference that results in a varied sequence of larger and smaller waves.

Surf zone The nearshore zone of breaking waves.

Surface tension The tendency for the surface of a liquid to contract owing to intermolecular bond attraction.

Surfing The sport of riding on the crest or along the tunnel of a wave, especially while standing or lying on a surfboard.

Surging breaker A compressed breaking wave that builds up over a short distance and surges forward as it breaks. It is characteristic of abrupt beach slopes.

Suspension feeder See *filter feeder*.

Suspension settling The process by which fine-grained material that is being suspended in the water column slowly accumulates on the sea floor.

Sverdrup (Sv) A unit of flow rate equal to 1 million cubic meters per second. Named after Norwegian oceanographer Harald Sverdrup.

Swash A thin layer of water that washes up over exposed beach as waves break at the shore.

Swell A free ocean wave by which energy put into ocean waves by wind in the sea is transported with little energy loss across great stretches of ocean to the margins of continents where the energy is released in the surf zone.

Swim bladder A gas-containing, flexible, cigar-shaped organ that aids many fishes in attaining neutral buoyancy.

Symbiosis A relationship between two species in which one or both benefit or neither or one is harmed. Examples are commensalism, mutualism, and parasitism.

Syzygy Either of two points in the orbit of the Moon (full or new moon phase) when the Moon lies in a straight line with the Sun and Earth.

T

Tablemount A conical volcanic feature on the ocean floor resembling a seamount except that it has had its top truncated to a relatively flat surface.

Taxonomy The classification of organisms in an ordered system that indicates natural relationships.

Tectonic estuary An estuary, the origin of which is related to tectonic deformation of the coastal region.

Tectonics Deformation of Earth's surface by forces generated by heat flow from Earth's interior.

Tektite See *spherule*.

Temperate Pertaining to the oceanic region where pronounced seasonal change occurs (about 40 to 60 degrees latitude). Also known as the middle latitudes.

Temperature A direct measure of the average kinetic energy of the molecules of a substance.

Temperature of maximum density The temperature at which a substance reaches its highest density. For water, it is 4°C.

Temperature–salinity (T–S) diagram A diagram with axes representing water temperature and salinity, whereby the density of the water can be determined.

Terrigenous sediment Sediment produced from or of the Earth. Also called lithogenous sediment.

Territorial sea A strip of ocean, 12 nautical miles wide, adjacent to land over which the coastal nation has control over the passage of ships.

Test The supporting skeleton or shell (usually microscopic) of many invertebrates.

Tethys Sea An ancient body of water that separated Laurasia to the north and Gondwanaland to the south. Its location was approximately that of the present Alpine–Himalayan mountain system.

Texture The general physical appearance of an object.

Theory A well-substantiated explanation of some aspect of the natural world that can incorporate facts, laws (descriptive generalizations about the behavior of an aspect of the natural world), logical inferences, and tested hypotheses.

Thermal contraction The reduction in size as a result of lowering of temperature.

Thermocline A layer of water beneath the mixed layer in which a rapid change in temperature can be measured in the vertical dimension.

Thermohaline circulation The vertical movement of ocean water driven by density differences resulting from the combined effects of variations in temperature and salinity; produces deep currents.

Tidal bore A steep-fronted wave that moves up some rivers when the tide rises in the coastal ocean.

Tidal bulge The theoretical mound of water found on both sides of Earth caused by the relative positions of the Moon (lunar tidal bulges) and the Sun (solar tidal bulges).

Tidal day See *lunar day*.

Tidal period The time that elapses between successive high tides. In most parts of the world, it is 12 hours and 25 minutes.

Tidal range The difference between high tide and low tide water levels over any designated time interval, usually one lunar day.

Tide The periodic rise and fall of the surface of the ocean and connected bodies of water resulting from the gravitational attraction of the Moon and Sun acting unequally on different parts of Earth.

Tide-generating force The magnitude of the centripetal force required to keep all particles of Earth having identical mass moving in identical circular paths required by the movements of the Earth–Moon system is identical. This required force is provided by the gravitational attraction between the particles and the Moon. This gravitational force is identical to the required centripetal force only at the center of Earth. For ocean tides, the horizontal component of the small force that results from the difference between the required and provided forces is the tide-generating force on that individual particle. These forces are such that they tend to push the ocean water into bulges toward the tide-generating body on one side of Earth and away from the tide-generating body on the opposite side of Earth.

Tide wave A long-period gravity wave generated by tide-generating forces and manifested as rising and falling tides.

Tissue An aggregate of cells and their products developed by organisms for the performance of a particular function.

Tombolo A sand or gravel bar that connects an island with another island or the mainland.

Topography The configuration of a surface. In oceanography it refers to the ocean bottom or the surface of a mass of water with given characteristics.

Toxic compound A poisonous substance capable of causing injury or death, especially by chemical means.

Trade winds A global wind belt that moves from a subtropical high-pressure belt at about 30 degrees north or south latitude toward the equatorial region. These winds move from a northeasterly direction in the Northern Hemisphere and from a southeasterly direction in the Southern Hemisphere.

Transform fault A fault with side-to-side motion that offsets segments of a mid-ocean ridge. Oceanic transform faults occur wholly on the ocean floor, while continental transform faults occur on land.

Transform plate boundary The boundary between two lithospheric plates formed by a transform fault.

Transitional wave A wave moving from deep water to shallow water that has a wavelength more than twice the water depth but less than 20 times the water depth. Particle orbits are beginning to be influenced by the bottom.

Transverse wave A wave in which particle motion is at right angles to energy propagation.

Trench A long, narrow, and deep depression on the ocean floor with relatively steep sides that is caused by plate convergence.

Trophic level A nourishment level in a food chain. Plant producers constitute the lowest level, followed by herbivores and a series of carnivores at the higher levels.

Tropic of Cancer The latitude 23.5 degrees north, which is the furthest location north that receives vertical rays of the Sun.

Tropic of Capricorn The latitude 23.5 degrees south, which is the furthest location south that receives vertical rays of the Sun.

Tropical Pertaining to, characteristic of, occurring in, or inhabiting the tropics.

Tropical cyclone See *hurricane*.

Tropical tide A tide that occurs twice monthly when the Moon is at its maximum declination near the Tropic of Cancer or the Tropic of Capricorn.

Tropics The region of Earth's surface lying between the Tropic of Cancer and the Tropic of Capricorn. Also known as the Torrid Zone.

Troposphere The lowermost portion of the atmosphere, which extends from Earth's surface to 12 kilometer (7 miles). It is where all weather is produced.

Trough (wave) The part of an ocean wave that is displaced below the still-water level.

Tsunami A seismic sea wave. A long-period gravity wave generated by a submarine earthquake or volcanic event. Not noticeable on the open ocean but builds up to great heights in shallow water.

Turbidite deposit A sediment or rock formed from sediment deposited by turbidity currents characterized by both horizontally and vertically graded bedding.

Turbidity A state of reduced clarity in a fluid caused by the presence of suspended matter.

Turbidity current A gravity current resulting from a density increase brought about by increased water turbidity. Possibly initiated by some sudden force such as an earthquake, the turbid mass continues under the force of gravity down a submarine slope.

Turbulent flow Flow in which the flow lines are confused heterogeneously due to random velocity fluctuations.

Typhoon See *hurricane*.

U

Ultraplankton Plankton for which the greatest dimension is less than 5 micrometers (0.0002 inch). They are very difficult to separate from the water.

Ultrasonic Sound frequencies above those that can be heard by humans (above 20,000 cycles per second).

Ultraviolet (UV) radiation Electromagnetic radiation shorter than visible radiation and longer than X-rays. The approximate range is 1 to 400 nanometers.

Upper water An area of the ocean near the surface that includes the mixed layer and the permanent thermocline. It is approximately the top 1000 meters (3300 feet) of the ocean.

Upwelling The process by which deep, cold, nutrient-laden water is brought to the surface, usually by diverging equatorial currents or coastal currents that pull water away from a coast.

V

Valence The combining capacity of an element measured by the number of hydrogen atoms with which it will combine.

van der Waals force A weak attractive force between molecules resulting from the interaction of one molecule and the electrons of another.

Vapor The gaseous state of a substance that is liquid or solid under ordinary conditions.

Vent An opening on the ocean floor that emits hot water and dissolved minerals.

Ventral Pertaining to the lower or under surface.

Vernal equinox The passage of the Sun across the equator as it moves from the Southern Hemisphere into the Northern Hemisphere, approximately March 21. During this time, all places in the world experience equal lengths of night and day. Also known as the spring equinox.

Vertebrata The subphylum of chordates that includes those animals with a well-developed brain and a skeleton of bone or cartilage; includes fish, amphibians, reptiles, birds, and land animals.

Vertically mixed estuary Very shallow estuaries such as lagoons in which freshwater and marine water are totally mixed from top to bottom so that the salinity at the surface and the bottom is the same at most places within the estuary.

Virioplankton Viruses that live as plankton.

Viscosity A property of a substance to offer resistance to flow caused by internal friction.

Volcanic arc An arc-shaped row of active volcanoes directly above a subduction zone. Can occur as a row of islands (island arc) or mountains on land (continental arc).

W

Walker Circulation Cell The pattern of atmospheric circulation that involves the rising of warm air over the East Indies low-pressure cell and its descent over the high-pressure cell in the southeastern Pacific Ocean off the coast of Chile. It is the weakening of this circulation that accompanies an El Niño event, which has led to the development of the term El Niño–Southern Oscillation event.

Walrus A large marine mammal (*Odobenus rosmarus*) of Arctic regions belonging to the order Pinnipedia and having two long tusks, tough wrinkled skin, and four flippers.

Waning crescent The Moon when it is between third-quarter and new moon phases.

Waning gibbous The Moon when it is between full moon and third-quarter phases.

Warm-blooded See *homeothermic*.

Warm front A weather front in which a warm air mass moves into and over a cold air mass producing a broad band of gentle precipitation.

Water mass A body of water identifiable from its temperature, salinity, or chemical content.

Wave A disturbance that moves over the surface or through a medium with a speed determined by the properties of the medium.

Wave base The depth at which circular orbital motion becomes negligible. It exists at a depth of one-half wavelength, measured vertically from still water level.

Wave-cut beach A gently sloping surface produced by wave erosion and extending from the base of the wave-cut cliff out under the offshore region.

Wave-cut cliff A cliff produced by landward cutting by wave erosion.

Wave dispersion The separation of waves as they leave the sea area by wave size. Larger waves travel faster than smaller waves and thus leave the sea area first, to be followed by progressively smaller waves.

Wave frequency (f) The number of waves that pass a fixed point in a unit of time (usually one second). A wave's frequency is the inverse of its period.

Wave height (H) The vertical distance between a crest and the adjoining trough.

Wave period (T) The elapsed time between the passage of two successive wave crests (or troughs) past a fixed point. A wave's period is the inverse of its frequency.

Wave speed (S) The rate at which a wave travels. It can be calculated by dividing a wave's wavelength (L) by its period (T).

Wave steepness Ratio of wave height (H) to wavelength (L). If a 1:7 ratio is ever exceeded by the wave, then the wave breaks.

Wave train A series of waves from the same direction. Informally known as a wave set.

Wavelength (L) The horizontal distance between two corresponding points on successive waves, such as from crest to crest.

Waxing crescent The Moon when it is between new moon and first-quarter phases.

Waxing gibbous The Moon when it is between first-quarter and full moon phases.

Weather The state of the atmosphere at a given time and place, with respect to variables such as temperature, moisture, wind velocity, and barometric pressure.

Weathering A process by which rocks are broken down by chemical and mechanical means.

Wentworth scale of grain size A logarithmic scale for size classification of sediment particles.

West Australian Current A cold current that forms the eastern boundary current of the Indian Ocean Subtropical Gyre. It is separated from the coast by the warm Leeuwin Current except during El Niño–Southern Oscillation events when the Leeuwin Current weakens.

West Wind Drift See *Antarctic Circumpolar Current*.

Western boundary current Poleward-flowing warm currents on the western side of all subtropical gyres.

Western boundary undercurrent (WBUC) A bottom current that flows along the base of the continental slope eroding sediment from it and redepositing the sediment on the continental rise. It is confined to the western boundary of deep-ocean basins.

Western intensification Pertaining to the intensification of warm western boundary currents of each subtropical gyre that are faster, narrower, and deeper than their corresponding eastern boundary currents.

Wetland See *coastal wetland*.

Whirlpool A rapidly rotating current of water; a vortex.

White muscle fiber Thick muscle fibers with relatively low concentrations of myoglobin that make up a large percentage of the muscle fiber in lunger-type fishes.

White smoker A smoker similar to a black smoker but emits water of a lower temperature that is white in color.

Wilson cycle A model that uses plate tectonic processes to show the distinctive life cycle of ocean basins during their formation, growth, and destruction.

Wind-driven circulation A movement of ocean water that is driven by winds. This includes most horizontal movements in the surface waters of the world's oceans.

Windward The direction from which the wind is blowing.

Winter solstice The instant the southward-moving Sun reaches the Tropic of Cancer before changing direction and moving north back toward the equator, approximately December 21.

Wintertime beach A beach that is characteristic during winter months. It typically has a narrow rocky berm and a flat beach face.

Z

Zebra mussel A non-native species released into U.S. and Canadian waters of the Great Lakes region.

Zenith The point on the celestial sphere directly over the observer.

Zooplankton Animal plankton.

Zooxanthellae A form of algae that lives as a symbiont in the tissue of corals and other coral reef animals and provides varying amounts of their required food supply.

CREDITS AND ACKNOWLEDGEMENTS

Cover Photo
Splash News/Corbis.

Table of Contents
vi Alan P. Trujillo. **vii** Planetary Visions Ltd / Photo Researchers, Inc. **viii** Image courtesy of Quade Paul. **ix** Rebecca Jackrel / DanitaDelimont.com Danita Delimont Photography/ Newscom. **x** Yusuke Okada/amanaimages/Alamy. **xi** Image Source/Alamy.

About the Authors
Trujillo Alan P. Trujillo. **Thurman** Hal Thurman.

Introduction
Opener Ian Cartwright/Getty. **Figure I-4** NASA Images. **Figure I-5** NASA.

Chapter 1
Opener NASA Earth Observing System. **Figure 1A** Alan P. Trujillo. **Figure 1B** Alan P. Trujillo. **Figure 1C** Science and Society/SuperStock. **Figure 1D** DAVID PARRY/PA Photos /Landov. **Figure 1D (Inset)** HIP / Art Resource, NY. **Figure 1.1** From Challenger Report, Great Britain, 1895. **Figure 1.3** Data from M.J. Kennish (1994) and National Geographic Society. **Figure 1.4** Office of Naval Research, U.S. Navy. **Figure 1.5** MARK THIESSEN/AFP/Getty Images/Newscom. **Figure 1.7** AP Images. **Figure 1.10** U.S. Naval Historical Center Photography. **Figure 1.11** Reprinted by permission from Tarbuck, E. J. and Lutgens, F. K., Earth Science, 6th ed. (Figs. 19.1 and 19.2), Macmillan Publishing Company, 1991. **Figure 1.13** NASA, ESA & Mohammad Heydari-Malayeri (Observatoire de Paris, France). **Figure 1.15** Steve Munsinger / Photo Researchers, Inc. **Figure 1.21** REUTERS/Stringer. **Figure 1.24** After Zimmer, C., 2001, How old is it? National Geographic 200:3, 92. **Figure 1.25** After Tarbuck, E. J. and Lutgens, F. K., The Earth: An Introduction to Physical Geology, 7th ed. (Fig. 8.15), Prentice Hall, 2002. Data from the Geological Society of America, the U.S. Geological Survey, and the International Commission on Stratigraphy. **Excerpt 1.1** Courtesy of Loren Shriver. **Excerpt 1.44** Dobzhansky, Theodosius. "Nothing in Biology Makes Sense Except in the Light of Evolution", American Biology Teacher vol. 35 (March 1973).

Chapter 2
Opener Alan P. Trujillo. **Figure 2A** Lawrence Cruciana/Shutterstock. **Figure 2B, 2C** Randolph A. Koski, U.S. Geological Survey. **Figure 2.1** bpk, Berlin/Art Resource, NY. **Figure 2.2** After Dietz, R. S. and Holden, J. C., 1970, Reconstruction of Pangaea: Breakup and dispersion of continents, Permian to present, Journal of Geophysical Research 75:26, 4939–4956. **Figure 2.3** From Continental Drift by Don and Maureen Tarling, © 1971 by G. Bell&Sons, Ltd. Reprinted by permission of Doubleday & Co., Inc. **Figure 2.4** After Tarbuck, E. J. and Lutgens, F. K., Earth Science, 8th ed. (Fig. 7.6), Prentice Hall, 1997. **Figure 2.5** After Tarbuck, E. J. and Lutgens, F. K., Earth Science, 8th ed. (Fig. 7.4), Prentice Hall, 1997. **Figure 2.6** After Tarbuck, E. J. and Lutgens, F. K., Earth Science, 8th ed. (Fig. 7.4), Prentice Hall, 1997. **Figure 2.7** Reprinted by permission from Tarbuck, E. J. and Lutgens, F. K., The Earth: An Introduction to Physical Geology, 3d ed. (Fig. 18.8 and Fig. 18.9), Merrill Publishing Company, 1990. **Figure 2.8** After Tarbuck, E. J. and Lutgens, F. K., Earth Science, 8th ed. (Fig. 7.17), Prentice Hall, **Figure 2.9** After Tarbuck, E. J. and Lutgens, F. K., Earth Science, 8th ed. (Fig. 7.18), Prentice Hall, 1997. **Figure 2.10** Reprinted by permission from Tarbuck, E. J. and Lutgens, F. K., The Earth: An Introduction to Physical Geology, 3d ed. (Fig. 18.11), Merrill Publishing Company, 1990. **Figure 2.11** After Tarbuck, E. J. and Lutgens, F. K., Earth Science, 8th ed. (Fig. 7.20), Prentice Hall, 1997. **Figure 2.12** After Tarbuck, E. J. and Lutgens, F. K., The Earth: An Introduction to Physical Geology, 3d ed. (Fig. 19.16), Merrill Publishing Company, 1990. After The Bedrock Geology of the World, by R. L. Larson et al., Copyright © 1985 by W. H. Freeman. **Figure 2.13a** After Tarbuck, E. J. and Lutgens, F. K., The Earth: An Introduction to Physical Geology, 3d ed. (Fig. 5.11), Merrill Publishing Company, 1990. Data from National Geophysical Data Center/ NOAA. **Figure 2.13b** After Tarbuck, E. J. and Lutgens, F. K., The Earth: An Introduction to Physical Geology, 3d ed. (Fig. 5.11), Merrill Publishing Company, 1990; modified with data from W. B. Hamilton, U. S. Geological Survey. **Figure 2.14** Reprinted by permission from Tarbuck, E. J. and Lutgens, F. K., The Earth: An Introduction to Physical Geology, 3d ed. (Fig. 6.7), Merrill Publishing Company, 1990. **Figure 2.15** After Tarbuck, E. J. and Lutgens, F. K., Earth Science, 8th ed. (Fig. 7.10), Prentice Hall, 1997. **Figure 2.16** Patricia Deen, Palomar College. **Figure 2.17** After Tarbuck, E. J. and Lutgens, F. K., Earth Science, 8th ed. (Fig. 7.12), Prentice Hall, 1997. **Figure 2.18a** After Tarbuck, E. J. and Lutgens, F. K., Earth Science, 8th ed. (Fig. 7.11), Prentice Hall, 1997. **Figure 2.18c** Copyright © 2013 Cornelis Klein and Tony Philpotts. Reprinted with the permission of Cambridge University Press. **Figure 2.19** Scripps Institution of Oceanography Geological Collections, University of California at San Diego. **Figure 2.20** Reprinted by permission from Tarbuck, E. J. and Lutgens, F. K., Earth Science, 5th ed. (Fig. 6.11), Merrill Publishing Company, 1988. **Figure 2.21c** Austin Post/U.S. Geological Survey, Denver. **Figure 2.22** After Tarbuck, E. J. and Lutgens, F. K., Earth Science, 8th ed. (Fig. 7.15), Prentice Hall, 1997. **Figure 2.22c** Ric Ergenbright/ CORBIS. **Figure 2.23** After Tarbuck, E. J. and Lutgens, F. K., The Earth: An Introduction to Physical Geology, 3d ed. (Figs. 18.22, 18.23), Merrill Publishing Company, 1990. **Figure 2.25** Data from Robert A. Duncan. **Figure 2.26** After Tarbuck, E. J. and Lutgens, F. K., The Earth: An Introduction to Physical Geology, 6th ed. (Fig. 19.30), Prentice Hall, 1999. **Figure 2.30** Courtesy of National Aeronautic Space Administration. **Figure 2.31** Courtesy Paul C. Lowman Jr. and others, NASA/ Goddard Space Flight Center, Greenbelt, Maryland. **Figure 2.33** After Dietz, R. S. and Holden, J. C., 1970, The breakup of Pangaea, Scientific American 223: 4, 30–41. **Figure 2.34** Adapted from Wilson, J. T., American Philosophical Society Proceedings 112, 309–320, 1968; Jacobs, J. A., Russell, R. D., and Wilson, J. T., Physics and Geology, McGraw-Hill, New York, 1971. **Excerpt 2.1** Wegener, Alfred. The Origins of Continents and Oceans. Braunschweig. 1929.

Chapter 3
Opener Planetary Visions Ltd / Photo Researchers, Inc. **Figure 3A** After Tarbuck, E. J. and Lutgens, F. K., Earth Science, 6th ed. (Fig. 10.2), Macmillan Publishing Company, 1991. **Figure 3B** Photo© Woods Hole Oceanographic Institution. **Figure 3.1** Courtesy Peter A. Rona, Hudson Laboratories of Columbia University. **Figure 3.2** Erika Mackay/ National Institute of Water and Atmospheric Research. **Figure 3.2b** Photo Ifremer; from Manaute Cruise; Courtesy of Jean-Marie Auzende. Photo: Fred N. Spiess. **Figure 3.3** Copyright by the American Geophysical Union. Photo courtesy of Daniel J. Fornari, Lamont-Doherty Earth Observatory, Columbia University. Reprinted with permission of the American Geophysical Union. **Figure 3.4** After Gross, M. G., Oceanography, 6th ed. (Fig. 16.10), Prentice Hall, 1993. **Figure 3.5** Courtesy David Sandwell, Scripps Institution of Oceanography, University of California at San Diego. **Figure 3.6** Scripps Institution of Oceanography Geological Collections, University of California at San Diego. **Figure 3.7** Courtesy of the Deep Sea Drilling Project, Scripps Institution of Oceanography, University of California at San Diego; with thanks to Jerry Bode. **Figure 3.8** After Tarbuck, E. J. and Lutgens, F. K., The Earth: An Introduction to Physical Geology, 5th ed. (Fig. 19.3), Prentice Hall, 1996. **Figure 3.9** After Plummer, C. C. and McGeary, D., Physical Geology, 5th ed. (Fig. 18.6), Wm. C. Brown, 1991. **Figure 3.10** After Tarbuck, E. J. and Lutgens, F. K., The Earth: An Introduction to Physical Geology, 5th ed. (Fig. 19.4), Prentice Hall, 1996. **Figure 3.12a** After Tarbuck, E. J. and Lutgens, F. K., The Earth: An Introduction to Physical Geology, 5th ed. (Fig. 19.6), Prentice Hall, 1996. **Figure 3.12b** Gerald Kuhn Archives. **Figure 3.12c** Alan P. Trujillo. **Figure 3.14b** (c) 2005 The University of New HanlDshire and its Center for Coastal & Ocean Mapping. **Figure 3.15** Used with permission from the papers of Charles Davis Hollister, © Woods Hole Oceanographic Institution Data Library and Archives. **Figure 3.18** Alcoa Technical Center. **Figure 3.19a** Ken C. Macdonald, University of California, Santa Barbara/Photo Researchers. **Figure 3.19b** A.E.J. Engel and the Geological Collections/Scripps Institution of Oceanography, UC San Diego. **Figure 3.19c** Jeffrey Grover/Cuesta College. **Figure 3.20a** After Tarbuck, E. J. and Lutgens, F. K., The Earth: An Introduction to Physical Geology, 5th ed. (Fig. 21.23), Prentice Hall, 1996. Dudley Foster/Fred N. Spiess/Scripps Institution of Oceanography, UC San Diego. **Excerpt 3-1** Maury, Matthew Fontaine. The Physical Geography of the Sea. New York: Harper & Brothers. 1855. **QR code 3.4** Courtesy of the Oceanographic and Atmospheric Administration.

Chapter 4
Opener Alfred Wegener Institute for Polar and Marine Research. **Figure 4A** World Minerals Inc. **Figure 4.1** RISE Project Group, Fred N. Spiess et al/Scripps Institution of Oceanography, UC San Diego. **Figure 4.2** Courtesy of the Ocean Drilling Program, Texas A& M University. Modified after Sverdrup, H. U., et al., 1942. Divins, D. L., NGDC Total Sediment Thickness of the World's Oceans & Marginal Seas, Retrieved date 4/10/2006,

http://www.ngdc.noaa. gov/ mgg/ http://www.ngdc.noaa.gov/mgg/sedthick/sedthick. html. **Figure 4.4** Alan P. Trujillo. **Figure 4.6a** NASA. **Figure 4.6b** U.S. Marine Corps photo by Cpl. Alicia M. Garcia. **Figure 4.6c** Alan P. Trujillo. **Figure 4.6d** Alan P. Trujillo. **Figure 4.7** Walter N. Mack, Michigan State University. **Figure 4.8** Provided by the Sea-WiFS Project, NASA/Goddard Space Flight Center, and ORBIMAGE. **Figure 4.9a** Reprinted by permission of VSP International Science Publishers, Zeist, Netherlands, from Gustaaf M. Hallegraeff, Plankton: A Microscopic World, 1988, p. 46. **Figure 4.9b** Scripps Institution of Oceanography, UC San Diego. **Figure 4.9c** Taken from Celite Corporation's Diatomite Mine in Lompoc, California. Photo courtesy of World Minerals Inc. **Figure 4.10a** Reprinted by permission of Brill Academic Publishers, Leiden, Netherlands, from Gustaaf M. Hallegraeff, Plankton: A Microscopic World, 1988, p. 8. **Figure 4.10b** Brill Academic Publishers Reprinted by permission of Brill Academic Publishers, Leiden, Netherlands, from Gustaaf M. Hallegraeff, Plankton: A Microscopic World, 1988, p. 16. **Figure 4.10c** Memorie Yasuda/Scripps Institution of Oceanography, UC San Diego. **Figure 4.10d** Scripps Institution of Oceanography Deep Sea Drilling Project. **Figure 4.11** David Hughes/Robert Harding. **Figure 4.11 (Inset)** Steve Gschmeissner / Photo Researchers, Inc. **Figure 4.12** Jarrod Boord/Shutterstock. **Figure 4.16** After Biscaye, P. E., et al., 1976; Berger, W. H., et al., 1976; and Kolla V. and Biscaye, P. E., 1976. **Figure 4.17a** Charles D. Winters / Photo Researchers, Inc. **Figure 4.17b** Jimmy Greenslate/Scripps Institution of Oceanography, UC San Diego. Charles D Winters/Photo Researchers/Getty Images. **Figure 4.18** Alan P. Trujillo. **Figure 4.19** Dr. Carlos A. Alvarez Zarikian/IODP—Integrated Ocean Drilling Program. **Figure 4.21** After Sverdrup, H. U., et al., 1942. **Figure 4.23** Photo© Woods Hole Oceanographic Institution. **Figure 4.24** National Geophysical Data Center. **Figure 4.25** Scott Gibson/Corbis. **Figure 4.26a** Alan P. Trujillo. **Figure 4.26b** Alan P. Trujillo. **Figure 4.28** Patricia Deen, Palomar College. **Figure 4.29** Sunoco, Inc. **Figure 4.30** With thanks to Joseph Holliday, El Camino College. After Cronan, D. S., 1977, Deep sea nodules: Distribution and geochemistry, in Glasby, G. P., ed., Marine Manganese Deposits, Elsevier Scientific Publishing Co. **Excerpt 4.1** Berger, Wolf. Oceans: Reflections on a Century of Exploration. University of California Press. 2009. Used by permission. **Excerpt 4.19** Darwin, Charles. Origin of the Species. London: John Murray. 1859. **QR code 4.1** Courtesy of the Integrated Ocean Drilling Program.

Chapter 5

Opener Image courtesy of Quade Paul. **Figure 5A** Dr. Ivan Polunin/Bruce Coleman Inc./Alamy. **Excerpt 5.1** Caglioti, Luciano. The Two Faces of Chemistry. MIT Press. 1985. **Figure 5.1** After Tarbuck, E. J. and Lutgens, F. K., The Earth: An Introduction to Physical Geology, 5th ed. (Fig. 2.4), Prentice Hall, 1996. **Figure 5.11** After data from Jet Propulsion Laboratory, NASA. **Figure 5.13** Electron Microscopy Laboratory, Agricultural Research Service, USDA. **Figure 5.14** Martyn F. Chillmaid/Photo Researchers, Inc. **Figure 5.17** Rafael Ben-Ari/Chameleons Eye Chameleons Eye Witness/Newscom. **Figure 5.18** After Tarbuck, E. J. and Lutgens, F. K., The Earth: An Introduction to Physical Geology, 4th ed. (Fig. 10.2), Macmillan Publishing Company, 1993. **Figure 5.23** After Pickard, G. L., Descriptive Physical Oceanography, Pergamon Press Ltd., 1963. **Figure 5.24** NASA EarthObservatory. **Figure 5.26** After Pickard, G. L., Descriptive Physical Oceanography, Pergamon Press Ltd., 1963.

Chapter 6

Opener Dale Wilson/Getty Images. **Figure 6A** Morgan P. Sanger/Columbus Foundation in the British Virgin Islands. **Figure 6.6** After Gross, M. G., Oceanography, 6th ed. (Fig. 5.2), Prentice Hall, 1993. **Figure 6.15** After Gross, M. G., Oceanography, 6th ed. (Fig. 5.17), Prentice Hall, 1993. **Figure 6.16** Reprinted by permission from Lutgens, F. K. and Tarbuck, E. J., The Atmosphere, 6th ed. (Fig. 8.2), Prentice Hall, 1995. **Figure 6.19** NASA. **Figure 6.20** Photoshot. **Figure 6.21** NOAA. **Figure 6.22** Chris Seward/Raleigh News & Observer/MCT/Landov. **Figure 6.23a** National Oceanic and Atmospheric Administration NOAA. **Figure 6.23b** Vincent Laforet/POOL/EPA/Newscom. **Figure 6.25a** Doug Allan/Photodisc/Getty Images. **Figure 6.25b** NASA. **Figure 6.26a** William Bacon / Photo Researchers, Inc. **Photo 6.26c** Josh Landis/National Science Foundation. **Figure 6.26d** NASA's Earth Observatory. **Figure 6.27** NASA's Earth Observatory. **Figure 6.28** Photoshot/Photoshot Holdings Ltd/Alamy. **Excerpt 6.1** Coleridge, Samuel. Rime of the Ancient Mariner. 1798.

Chapter 7

Opener NASA Earth Observing System. **Figure 7A** Courtesy of Eos Transactions AGU 73: 34, 361 (1992). Copyright by the American Geophysical Union. Copyright by the American Geophysical Union. AP Photo/Maritime New Zealand. Dave Ingraham. **Figure 7.1a** Douglas Alden/Scripps Institution of Oceanography, UC San Diego. **Figure 7.1b** Dann S Blackwood/USGS. **Figure 7.2** Jet Propulsion Laboratory NASA Headquarters. **Figure 7.3b** Alan P. Trujillo. Courtesy of National Aeronautic Space Administration. **Figure 7.3** With thanks to Holly Dodson, Sierra College. **Figure 7.3** Data courtesy Argo Information Center at www.argo.ucsd.edu. **Figure 7.9** After data from NASA and NOAA. **Figure 7.17** O. Brown, R. Evans, and M. Carle, University of Miami Rosenstiel School of Marine and Atmospheric Science, Miami, Florida/NOAA. **P7.20a** NASA Earth Observing System. **Photo 7.20b** NASA Earth Observing System. **Figure 7.20** Courtesy of National Aeronautic Space Administration. **Figure 7.23** Maps courtesy of the International Research Institute for Climate and Society (IRI), Lamont-Doherty Earth Observatory, Columbia University. Data from Reynolds, R. W., N. A. Rayner, T. M. Smith, D. C. Stokes, and W. Wang, 2002: An Improved In Situ and Satellite SST Analysis for Climate. J. Climate, 15, 1609–1625. **Figure 7.24** NOAA/ESRL/Physical Sciences Division—University of Colorado Boulder/CIRES/CDC.

Chapter 8

Opener Rebecca Jackrel / DanitaDelimont.com Danita Delimont Photography/Newscom. **Figure 8.2** Kinsman, Blair. 1965. Wind Waves: Their Generation and Propagation on the Ocean Surface. Englewood Cliffs, NJ: Prentice-Hall. **Figure 8.4a** Modified from The Tasa Collection: Shorelines. Published by Macmillan Publishing Co., New York. Copyright © 1986 by Tasa Graphic Arts, Inc. **Figure 8.6** Kyodo News/AP Figure. **Photo 8.8b** Handout/Reuters. **Figure 8.13** National Archives and Records Administration. **Figure 8.11** Courtesy of National Aeronautic Space Administration. **Table 8.1** After Bowditch, N., 1958. Tables from American Practical Navigator. U.S. Government Printing Office. **Figure 8.16** After Gross, M. G., Oceanography, 6th ed. (Fig. 8.4), Prentice Hall, 1993. **Figure 8.17a** Courtesy of 1st Officer Claudio Suttora/www.michelangelo-raffaello.com. **Figure 8.17b** Courtesy of 1st Officer Claudio Suttora/www.michelangelo-raffaello.com. **Figure 8.19** Modified from The Tasa Collection: Shorelines. Published by Macmillan Publishing, New York. Copyright © 1986 by Tasa Graphic Arts, Inc. All rights reserved. **Figure 8.20b** irabel8/Shutterstock. **Figure 8.20c** SweetParadise / Alamy. **P8.21c** Rich Reid/National Geographic/Getty Images. **Figure 8.21b** Adapted from Tarbuck, E. J. and Lutgens, F. K., Earth Science, 6th ed. (Fig. 6.14) Macmillan Publishing Company, 1991. **Figure 8.22** Mark Rightmire/ZUMA Press/Newscom. **Figure 8.24b** JOANNE DAVIS/AFP/GETTY IMAGES/Newscom. **Figure 8.25** CORBIS. **Figure 8.26** After González, F. I., 1999, Tsunami! Scientific American 280: 5, 59. Modified 2009. **Figure 8.27a** USGS. **Figure 8.27b** USGS. **Figure 8.30** Alan P. Trujillo. **Figure 8.31** Image courtesy of Voith Hydro Wavegen Ltd. **Figure 8.32** Courtesy of Pelamis Wave Power. **Figure 8.33** Map constructed from data in U. S. Navy Summary of Synoptic Meteorological Observations (SSMO). Adapted from Sea Secrets, Sea Frontiers 33: 4, 260–261, International Oceanographic Foundation, 1987.

Chapter 9

Opener Laszlo Podor/Alamy. **Figure 9A** New Brunswick Department of Tourism. **Figure 9B** Reuters/Landov. **Figure 9.10** Modified from The Tasa Collection: Shorelines. Published by Macmillan Publishing Co., New York. Copyright © 1986 by Tasa Graphic Arts, Inc. **Figure 9.15** After von Arx,W. S.,1962; original by H. Poincaré 1910, Leçons de Mécanique Céleste, a Gauther-Crofts,Vol. 3. **Figure 9.16** After Hauge, C., Tides, currents, and waves, California Geology, July 1972. **Figure 9.18a** Nova Scotia Department of Tourism and Culture. **Figure 9.18b** Nova Scotia Department of Tourism and Culture. **Figure 9.20** Peter McBride/Aurora/Getty Images. **Figure 9.21** Mark Conlin /VWPICS/Visual&Written SL/Alamy. **Figure 9.22** Michel Brigaud/La Mediatheque EDF, Boulogne, France. **Excerpt 9.1** Newton, Sir Isaac. Philosophiae Naturalis Principia Mathematica (Philosophy of Natural Mathematical Principles). 1686.

Chapter 10

Opener Alan P. Trujillo. **Figure 10A** Alan P. Trujillo. **Figure 10B** Alan P. Trujillo. **Figure 10.2a** Alan P. Trujillo. **Figure 10.2b** Alan P. Trujillo. **Figure 10.03** University of Washington Libraries, Special Collections, John Shelton Collection, KC14461. **Figure 10.5** Travel Pictures/Alamy. **Figure 10.6** University of Washington Libraries, Special Collections, John Shelton Collection, 5902. **Figure 10.8a** USDA/FSA/Farm Service Agency. **Figure 10.8b** Martin Beebee / Alamy. **Figure 10.9** After Tarbuck, E. J. and Lutgens, F. K., The Earth: An Introduction to Physical Geology, 4th ed. (Fig. 14.12), Macmillan Publishing Company, 1993. **Figure 10.9c** USDA/FSA/Farm Service Agency. **Figure 10.12a** NASA Earth Observing System. **Figure 10.12b** NASA's Earth Observatory. **Figure 10.18** NASA Headquarters. **Figure 10.19** University of Washington Libraries, Special Collections, John Shelton Collection, 8974. **Figure 10.21** University of Washington Libraries, Special Collections, John Shelton Collection, S2-31. **Figure 10.22** After Tarbuck, E. J. and Lutgens, F. K., Earth Science, 5th ed., Merrill Publishing Company, 1988. **Figure 10.23** U.S. Army Corps of Engineers, Washington. **Figure 10.24** Peter Titmuss/Alamy. **Figure 10.26a** The Fairchild Aerial Photography Collection at Whittier College, Flight C-1670, Frame 4. **Figure 10.26b** The Fairchild Aerial Photography Collection at Whittier

College, Flight C-14180, Frame 3:61. **Figure 10.28a** Alan P. Trujillo. **Figure 10.28b** Alan P. Trujillo. **Figure 10.29** Alan P. Trujillo. **Excerpt 10.1** Byron, Lord. The Works of Lord Byron: Letters and Journals. John Murray. 1922.

Chapter 11

Opener mZUMA Press/Newscom. **Figure 11A** Chuck Cook/AP Images. **Figure 11B** SAUL LOEB/AFP/Getty Images/Newscom. **Figure 11.1a** Tony Souter/Dorling Kindersley. **Figure 11.5a** NASA Headquarters. **Figure 11.5b** Eye Ubiquitous/Glow Images. **Figure 11.5c** USDA/FSA/Farm Service Agency. **Figure 11.5d** M-Sat Ltd / Science Source. **Figure 11.8** After Officer, C. B., et al., 1984. Chesapeake Bay Anoxia. Science 23: 22-27. **Figure 11.11** Adapted from Encyclopedia of Oceanography, edited by Rhodes Fairbridge, © 1966. **Figure 11.12b** tbkmedia.de / Alamy. **Figure 11.12c** Alan P. Trujillo. **Figure 11.14b** NATALIE B. FOBES/National Geographic Stock. **Figure 11.14c** NOAA Central Library Photo Collection. **Figure 11.15** Data from The Oil Spill Intelligence Report at http://www.iosc.org/papers/01480.pdf. **Figure 11.16** After Mitchell, J. G., 1999, In the wake of the spill: Ten years after the Exxon Valdez. National Geographic 195:3, 96–117. **Figure 11.16b** Bob Jordan/AP Photo. **Figure 11.17** National Oceanic and Atmospheric Administration. **Figure 11.19** Charlie Riedel, File/AP images. **Figure 11.20** © 1994 John Trever, Albuquerque Journal. Reprinted by permission. **Figure 11.22** (c) 1994 John Trever, Albuquerque Journal. Reprinted by permission. **Figure 11.23** Jack Smith/AP Images. **Figure 11.26** age fotostock / SuperStock. **Figure 11.27** AP Images. **Figure 11.30** Alan P. Trujillo. **Figure 11.33** David W. Leindecker/Shutterstock. **Figure 11.34a** Wayne Perryman/National Marine Fisheries Service. **Figure 11.34b** Photoshot. **Figure 11.34c** David Liittschwager/National Geographic/Getty Images. Susan Middleton. **Figure 11.35** Tony Freeman / PhotoEdit. **Figure 11.36** Alan P. Trujillo **Figure 11.38** Alan P. Trujillo. **Excerpt 11.29** Courtesy of the World Health Organization.

Chapter 12

Opener Jeff Rotman/Alamy. **Figure 12.2** The Art Archive / Alamy. **Figure 12.8** IFREMER/AFP/Getty Images/Newscom. **Figure 12.11** After Sverdrup H.U., M.W. Johnson, and R.H. Flemming, 1942: The Oceans: Their Physics, Chemistry, and General Biology. Printice-Hall, Englewood Cliffs, New Jersey, 1087pp. **Figure 12.13** Reprinted by permission of Brill Academic Publishers, Leiden, Netherlands, from Gustaaf M. Hallegraeff, Plankton: A Microscopic World, 1988, p. 21. **Figure 12.19** cbimages/Alamy. **Figure 12.20** Mircea Bezergheanu/Shutterstock. **Figure 12.21a** Eugene Sim/Fotolia. **Figure 12.21b** Alan P. Trujillo. **Figure 12.23** Alan P. Trujillo. **Figure 12.27** Peter David / Taxi / Getty Images. **Figure 12.28** Photo© Woods Hole Oceanographic Institution. **P12.A** Scripps Institution of Oceanography, UC San Diego. **Excerpt 12.1** Wilson, E.O. The Diversity of Life. Belknap. **2010.**

Chapter 13

Opener Yusuke Okada/amanaimages/Alamy. **Figure 13A** Photo by WilArt Studios, courtesy of Monroe County Public Library. **Figure 13C** Loren McClenachan, Scripps Institution of Oceanography, UC San Diego. **Figure 13.2** Scripps Institution of Oceanography Explorations, Scripps Institution of Oceanography, University of California at San Diego. Photo by Elizabeth L. Venrick. **Figure 13.4** Patricia Deen, Palomar College. **Figure 13.5** SeaWiFS Project, NASA/Goddard Space Flight Center and ORBIMAGE. **Figure 13.7** Alan P. Trujillo. **Figure 13.8a** Alan P. Trujillo. **Figure 13.8b** Alan P. Trujillo. **Figure 13.8c** Harold V. Thurman. **Figure 13.8d** Alan P. Trujillo. **Figure 13.9** Ng Han Guan/AP Images. **Figure 13.10** Miller, S. Scientific and Regulatory Aspects of Quality Control for Chiral Drugs. Presented at 18th International Symposium on Chirality, Busan, Korea, June 2006. http://www.fda.gov/downloads/AboutFDA/CentersOffices/CDER/ucm103532.pdf (accessed online August 2011). Reprinted by permission of Brill Academic Publishers, Leiden, Netherlands, from Gustaaf M. Hallegraeff, Plankton: A Microscopic World, 1988, p. 43. **Figure 13.10b** Wuchang Wei/Scripps Institution of Oceanography, UC San Diego. **Figure 13.10c** Reprinted by permission of Brill Academic Publishers, Leiden, Netherlands, from Gustaaf M. Hallegraeff, Plankton: A Microscopic World, 1988, p. 91. **Figure 13.10d** Reprinted by permission of Brill Academic Publishers, Leiden, Netherlands, from Gustaaf M. Hallegraeff, Plankton: A Microscopic World, 1988, p. 81. **Table 13.10** Data after Strahler, A. H. and Strahler, A. N., Modern Physical Geography, 4th ed. (Table 25.1),Wiley, 1992. **Figure 13.11** Carleton Ray/Photo Researchers, Inc. **Figure 13.13** Everett Collection Inc. **Figure 13.15** Courtesy of National Aeronautic Space Administration. **Figure 13.16** American Geophysical Union. Reproduced/modified by permission of American Geophysical Union. **Figure 13.17** Claire Ting/SPL Photo Researchers, Inc. **Figure 13.18d** Scripps Institution of Oceanography, UC San Diego. **Figure 13.25** Walter E Harvey / Photo Researchers / Getty Images. **Figure 13.33** © 2007 BRIAN J. SKERRY/National Geographic Stock. **Figure 13.31** Data from the United Nations Food and Agriculture Organization. **Figure 13.32** Data from

the United Nations Food and Agriculture Organization. **Figure 13.34** W. High/National Marine Fisheries Service. **Figure 13.35** With kind thanks to Pete Pederson. **Figure 13.36** Data from the United Nations Food and Agriculture Organization. **Figure 13.37** Data from Mora, C., et al., 2009, Management Effectiveness of the World's Marine Fisheries. PLoS Biol 7(6): e1000131. **Figure 13.38** Courtesy Monterey Bay Aquarium. **Excerpt 13.1** Ward, Barbara. Who Speaks for Earth? W.W. Norton & Co Inc. 1973.

Chapter 14

Opener Image Source/Alamy. **Figure 14A** Michael Patrick O'Neill/Alamy. **Figure 14.3** Kozo Takahashi, Kyushu University, Fukuoka, Japan. **Figure 14.4** Siim Sepp (sandatlas.org <http://sandatlas.org>). **Figure 14.5** From Giesbrecht, W., Fauna und Flora des Golfes von Neapel und der Angrenzenden Meeres-Abschnitte,1892, Berlin, Verlag Von R. Friedländer and Sohn. Courtesy of Scripps Institution of Oceanography Explorations, Scripps Institution of Oceanography, University of California at San Diego. **Figure 14.6** Alan P. Trujillo. **Figure 14.6** Image Quest Marine. **Figure 14.7** WaterFrame/Alamy. **Figure 14.10a** tania zbrodko/shutterstock. **Figure 14.10b** lioneldivepix/fotolia. **Figure 14.10c** stephan kerkhofs / Fotolia. **Figure 14.10d** NaturePL/Superstock. **Figure 14.10e** cbpix/Fotolia. **Figure 14.11a** Amar and Isabelle Guillen/Guillen Photography/Alamy. **Figure 14.11b** National Marine Fisheries Service. **Figure 14.12** Lalli, C. M. and Parsons,T. R., Biological Oceanography: An Introduction (Fig. 6.5), Pergamon Press, 1993. **Figure 14.12h** Fred McConnaghey/Photo Researchers. **Figure 14.13** After Lalli, C. M. and Parsons,T. R., Biological Oceanography: An Introduction (Figs. 6.6b and 6.7a), Pergamon Press, 1993. **Figure 14.14** FredMcConnaughey/Photo Researchers. **Figure 14.15a** Jonathan Bird/Peter Arnold/Getty Images. **Figure 14.15b** cbpix/Fotolia. **Figure 14.15c** Image Quest Marine. **Table 14A** Adapted from the International Shark Attack File at http://www.flmnh.ufl.edu/fish/Sharks/isaf/isaf.htm; fatality data from U.S. Center for Disease Control and Prevention. **Figure 14.17a** Photoshot Holdings Ltd./Bruce Coleman. **Figure 14.17b** Fritz Poelking/Alamy. **Figure 14.17c** Photoshot/Alamy. **Figure 14.17d** Arco Images/Delpho M/Alamy. **14.17e** Irina Moskalev/Shutterstock. **Figure 14.19a** Cornforth Images / Alamy. **Figure 14.19b** helmut corneli/imagebroker/Alamy. **Figure 14.22** Alan P. Trujillo. **Figure 14.25c** Alan P. Trujillo **Figure 14.25d** Alan P. Trujillo. **Figure 14.26** naturepl/SuperStock. **Figure 14.27** Justin Hofman. **Figure 14.29** Data from National Marine Fisheries Service. **Figure 14.30** Kevin Schafer/NHPA/Photoshot. **Excerpt 14.14** Benchley, Peter. "Great white sharks". National Geographic: 12. (April 2000). **Excerpt 14.14** Benchley, Peter. "Oceans in Peril." in Ocean Planet: Writings and Images of the Sea. Smithsonian Institution. 2005.

Chapter 15

Opener Joel W. Rogers/CORBIS. **Figure 5A** Reprinted from R.R. Hessler, C.L. Ingram, A.A. Yayanos, and B.R. Burnett, Scavenging amphipods from the floor of the Philippine Trench, Deep-Sea Research 25:1029-1047, Fig. 1. (c) 1978, with permission from Elsevier Science. Photos supplied by Robert R. Hessler. **Figure 15.1** After Alexandrovitch Zenkevitch. Marine Biology. December 1971, Volume 11, Issue 4, pp 299-305. **Figure 15.2b** Harold V. Thurman. **Figure 15.2c** Harold V. Thurman. **Figure 15.2d** Alan P. Trujillo. **Figure 15.02e** Harold V. Thurman. **Figure 15.2f** Alan P. Trujillo. **Figure 15.2g** Harold V. Thurman. **Figure 15.2h** Alan P. Trujillo. **Figure 15.2i** Alan P. Trujillo. **Figure 15.3a** Alan P. Trujillo. **Figure 15.3b** Photoshot Holdings Ltd./Bruce Coleman. **Figure 15.5a** James R. McCullagh. **Figure 15.5b** Harold V. Thurman. **Figure 15.6** Alan P. Trujillo. **Figure 15.7a** Alan P. Trujillo. **Figure 15.7b** Premaphotos/Alamy. **Figure 15.10** Walter Dawn/National Audubon Society/Photo Researchers. **Figure 15.12a** Howard J. Spero, University of California, Davis. **Figure 15.12b** Howard J. Spero, University of California, Davis. **Figure 15.12c** Howard J. Spero, University of California, Davis. **Figure 15.13** blickwinkel/woike/alamy. **Figure 15.14b** David Hall/Photo Researchers. **Figure 15.15a** Alan P. Trujillo. **Figure 15.15b** Andrew J. Martinez/Photo Researchers. **Figure 15.17** After Stehli, F. G. and Wells, J. H., Diversity and Age Patterns in Hermatypic Corals, Systematic Zoology 20, 115–126, 1971. **Figure 15.18a** flpa/alamy. **Figure 15.18b** Photoshot. **Figure 15.20a** Kerry L Werry/Shutterstock. **Figure 15.20b** Charles Stirling/Diving/Alamy. **Figure 14.21** Franco Banfi/Steve Bloom Images/Alamy. **Figure 14.22a** Alan P. Trujillo. **Figure 14.22b** Alan P. Trujillo. **Figure 15.23** After Lalli, C. M. and Parsons,T. R., Biological Oceanography: An Introduction (Fig. 8.16), Pergamon Press, 1993. **Figure 14.26a** J. Frederick Grassle, Institute of Marine and Coastal Sciences, Rutgers University. **Figure 14.26b** Photo© Woods Hole Oceanographic Institution. **Figure 14.26c** Robert Hessler/Scripps Institution of Oceanography, UC San Diego. **Figure 15.27** With thanks to Ron Johnson, Old Dominion University. **Figure 15.28** National Oceanic and Atmospheric Administration (NOAA). **Figure 15.29c** C.K. Paull/Scripps Institution of Oceanography, UC San Diego. **Figure 15.30b** Charles R. Fisher, Pennsylvania State University. **Figure 15.30c** Charles R. Fisher, Pennsylvania State University. **Figure 15.31b** Y. Fujiwara JAMSTEC/Japan Marine Science and Technonology Center.

Chapter 16

Opener Corbis Premium RF / Alamy. **Figure 16A** Jeff Cordia/Scripps Institution of Oceanography. **Figure 16.1** Reprinted with the kind permission of Dennis Tasa; modified from Tarbuck, E. J. ed. and Lutgens, F. K., Earth: An Introduction to Physical Geology, 9th (Fig. 21.2), Pearson Prentice Hall, 2008. **Figure 16.3** Alan P. Trujillo. **Figure 16.4** Reto Stockli/NASA Goddard Space Flight Center. **Figure 16.5a** Images by NASA/SDO/HMI. **Figure 16.5b** NASA/SDO/AIA. **Figure 16.8** InterNetwork Media/Photodisc/Getty Images. **Figure 16.11** Modified from Climate Change 2007: Synthesis Report. Contribution of Working Groups I, II and III to the Fourth Assessment Report of the Intergovernmental Panel on Climate Change, cover page. IPCC, Geneva, Switzerland. **Figure 16.7** Reprinted with the kind permission of Dennis Tasa; from Tarbuck, E. J. and Lutgens, F. K., Earth: An Introduction to Physical Geology, 9th ed. (Fig. 18.33), Pearson Prentice Hall, 2008. **Figure 16.16** From Brock, E., 2008, Palaeoclimate: Windows on the greenhouse, Nature 453, 291-292 (15 May 2008). **Figure 16.18** Adapted from Our Changing Climate, Reports to the Nation on Our Changing Planet, Fall 1997, No. 4, University Corporation for Atmospheric Research. Updated with data to 2009 from NASA Earth Observatory. **16.19a** REUTERS/Michelle McLoughlin. **Figure 16.20** Adapted from an illustration in Oceanus (11/16/2006) by E. Paul Oberlander, Woods Hole Oceanographic Institution. **Figure 16.21** NASA. **Figure 16.22** Fritz Poelking/Elvele Images Ltd/Alamy. **Figure 16.23** NASA. **Figure 16.24a** CSIRO Plant Industry. **Figure 16.24b** Sinclair Stammers/Photo Researchers, Inc. **Figure 16.24c** Jeff Rotman/Alamy. **Figure 16.24d** Dobermaraner/Shutterstock. **Figure 16.25a** David Liittschwager/National Geographic/Getty Images. **Figure 16.25b** David Liittschwager/National Geographic/Getty Images. **Figure 16.25c** David Liittschwager/National Geographic/Getty Images. **Figure 16.26** After Doney, S. C., 2006,The dangers of ocean acidification. Scientific American 294:3, 62. **Figure 16.27** After Neumann, J. E., et al., Sea-level rise and global climate change: A review of impacts to U.S. coasts, PEW Center on Global Climate Change, 2000. **Figure 16.27b and c** After Miller, L. and Douglas, B. C., 2004, Mass and volume contributions to twentieth-century global sea level rise. Nature 428:5656,407. **Figure 16.28** NASA Earth Observatory; modified to include recent data. Alan P. Trujillo. **Figure 16.29** NASA Earth Observatory **Figure 16.31** NASA Earth Observatory. **Figure 16.32** Ken Buesseler/Woods Hole Oceanographic Institution. **Excerpt 16.1** Courtesy of Jane Lubchenco. **Excerpt 16.15** The National Academies Press. **Excerpt 16.17** Global Climate Change Impacts in the United States. U.S. Global Change Research Program. 2009.

Afterword

Figure Aft.4 Alan P. Trujillo

Appendices

Figure A5.1 Alan P. Trujillo. **Figure A5.A** Joseph Fell-McDonald. **Figure A5.A (Inset)** Joseph Fell-Mcdonald.

INDEX

Note: page numbers followed by f or t refer to Figures or Tables

Simplified Phylogenetic Tree of Marine Life

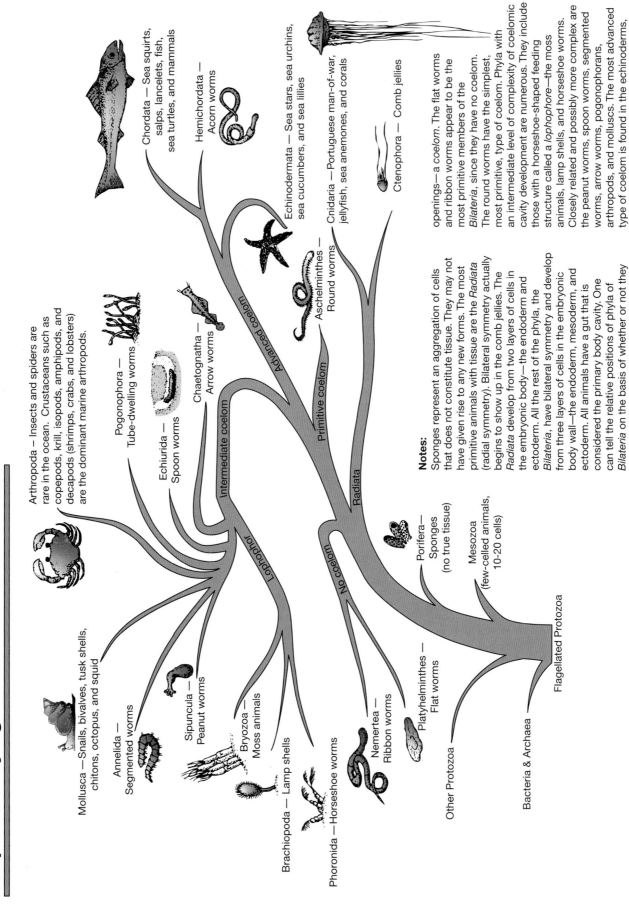

Chordata — Sea squirts, salps, lancelets, fish, sea turtles, and mammals

Hemichordata — Acorn worms

Echinodermata — Sea stars, sea urchins, sea cucumbers, and sea lillies

Cnidaria — Portuguese man-of-war, jellyfish, sea anemones, and corals

Ctenophora — Comb jellies

Arthropoda – Insects and spiders are rare in the ocean. Crustaceans such as copepods, krill, isopods, amphipods, and decapods (shrimps, crabs, and lobsters) are the dominant marine arthropods.

Pogonophora — Tube-dwelling worms

Echiurida — Spoon worms

Chaetognatha — Arrow worms

Aschelminthes — Round worms

Advanced coelom

Intermediate coelom

Primitive coelom

Lophophore

Radiata

No coelom

Mollusca — Snails, bivalves, tusk shells, chitons, octopus, and squid

Annelida — Segmented worms

Sipuncula — Peanut worms

Bryozoa — Moss animals

Brachiopoda — Lamp shells

Phoronida — Horseshoe worms

Nemertea — Ribbon worms

Platyhelminthes — Flat worms

Porifera— Sponges (no true tissue)

Mesozoa (few-celled animals, 10-20 cells)

Other Protozoa

Flagellated Protozoa

Bacteria & Archaea

Notes:
Sponges represent an aggregation of cells that does not constitute tissue. They may not have given rise to any new forms. The most primitive animals with tissue are the *Radiata* (radial symmetry). Bilateral symmetry actually begins to show up in the comb jellies. The *Radiata* develop from two layers of cells in the embryonic body—the endoderm and ectoderm. All the rest of the phyla, the *Bilateria*, have bilateral symmetry and develop from three layers of cells in the embryonic body wall—the endoderm, mesoderm, and ectoderm. All animals have a gut that is considered the primary body cavity. One can tell the relative positions of phyla of *Bilateria* on the basis of whether or not they have a secondary body cavity with no external openings—a *coelom*. The flat worms and ribbon worms appear to be the most primitive members of the *Bilateria*, since they have no coelom. The round worms have the simplest, most primitive, type of coelom. Phyla with an intermediate level of complexity of coelomic cavity development are numerous. They include those with a horseshoe-shaped feeding structure called a *lophophore*—the moss animals, lamp shells, and horseshoe worms. Closely related and possibly more complex are the peanut worms, spoon worms, segmented worms, arrow worms, pogonophorans, arthropods, and molluscs. The most advanced type of coelom is found in the echinoderms, acorn worms, and chordates.

Commonly Used Conversion Factors

Length

1 millimeter (mm)	1,000 micrometers 0.1 centimeter 0.001 meter 0.0394 inch
1 centimeter (cm)	10 millimeters 0.01 meter 0.394 inch
1 meter (m)	100 centimeters 39.4 inches 3.28 feet 1.09 yards 0.547 fathom
1 kilometer (km)	1000 meters 1093 yards 3280 feet 0.62 statute mile 0.54 nautical mile
1 inch (in)	25.4 millimeters 2.54 centimeters
1 foot (ft)	12 inches 30.5 centimeters 0.305 meter
1 fathom (fm)	6 feet 2 yards 1.83 meters
1 statute mile (mi)	5280 feet 1760 yards 1609 meters 1.609 kilometers 0.87 nautical mile

Area

1 square centimeter (cm^2)	0.155 square inch 100 square millimeters
1 square meter (m^2)	10,000 square centimeters 10.8 square feet
1 square kilometer (km^2)	100 hectares 247.1 acres 0.386 square mile 0.292 square nautical mile
1 square inch (in^2)	6.45 square centimeters
1 square foot (ft^2)	144 square inches 929 square centimeters

Temperature

Exact formula	Approximation (easy way)
$°C = \dfrac{(°F - 32)}{1.8}$	$°C = \dfrac{(°F - 30)}{2}$
$°F = (1.8 \times °C) + 32$	$°F = (2 \times °C) + 30$

Volume

1 cubic centimeter (cc; cm^3)	1 milliliter 0.061 cubic inch
1 liter (l)	1000 cubic centimeters 61 cubic inches 1.06 quarts 0.264 gallon
1 cubic meter (m^3)	1,000,000 cubic centimeters 1000 liters 264.2 gallons 35.3 cubic feet
1 cubic kilometer (km^3)	0.24 cubic mile 0.157 cubic nautical mile
1 cubic inch (in^3)	16.4 cubic centimeters
1 cubic foot (ft^3)	1728 cubic inches 28.32 liters 7.48 gallons

Mass

1 gram (g)	0.035 ounce
1 kilogram (kg)	2.2 pounds[†] 1000 grams
1 metric ton (mt)	2205 pounds 1000 kilograms 1.1 U.S. short tons
1 pound[†] (lb) (mass)	16 ounces 454 grams 0.454 kilogram
1 U.S. short ton (ton; t)	2000 pounds 907.2 kilograms 0.91 metric ton

[†]The pound is a weight unit, not a mass unit, but it is often used as such.

Speed

1 centimeter per second (cm/s)	0.0328 foot per second
1 meter per second (m/s)	2.24 statute miles per hour 1.94 knots 3.28 feet per second 3.60 kilometers per hour
1 kilometer per hour (kph)	27.8 centimeters per second 0.62 mile per hour 0.909 foot per second 0.55 knot
1 statute mile per hour (mph)	1.61 kilometers per hour 0.87 knot
1 knot (kt)	1 nautical mile per hour 51.5 centimeters per second 1.15 miles per hour 1.85 kilometers per hour